**4**th edition

# Plant
# Propagation

## PRINCIPLES AND PRACTICES

Hudson T. Hartmann
University of California, Davis

Dale E. Kester
University of California, Davis

PRENTICE-HALL, INC., Englewood Cliffs, New Jersey 07632

*Library of Congress Cataloging in Publication Data*

HARTMANN, HUDSON THOMAS
   Plant propagation.

   Bibliography: p.
   Includes index.
   1. Plant propagation.  I. Kester, Dale E.  II. Title.
SB119.H3  1983     631.5′3     82-16551
ISBN  0-13-681007-1

© 1983, 1975, 1968, 1959 by Prentice-Hall, Inc.
Englewood Cliffs, New Jersey 07632

Printed in the United States of America

10  9  8  7  6  5  4  3  2  1

*Editorial/production supervision*
   *and interior design by Joan L. Stone*
*Manufacturing buyer: John Hall*
*Cover design by Judith Winthrop*

ISBN 0-13-681007-1

PRENTICE-HALL INTERNATIONAL, INC., *London*
PRENTICE-HALL OF AUSTRALIA PTY. LIMITED, *Sydney*
EDITORA PRENTICE-HALL DO BRASIL LTDA., *Rio de Janeiro*
PRENTICE-HALL CANADA, INC., *Toronto*
PRENTICE-HALL OF INDIA PRIVATE LIMITED, *New Delhi*
PRENTICE-HALL OF JAPAN, INC., *Tokyo*
PRENTICE-HALL OF SOUTHEAST ASIA PTE. LTD., *Singapore*
WHITEHALL BOOKS LIMITED, *Wellington, New Zealand*

# Contents

# Preface

This book was written primarily as a text for university-level plant propagation courses. It provides information concerning the fundamental principles involved in plant propagation and, in addition, serves as a manual that describes the many useful techniques for propagating plants. It is presumed that the students using this book will have had some background training, either in high school or college biology or botany courses, although the chapters covering the techniques of plant propagation can usually be handled successfully without such background.

This book covers all aspects of the propagation of higher plants, both sexual and asexual, especially in reference to *human* efforts to increase plant numbers, as contrasted to plant reproduction *in nature*.

The text is organized into five major units: The first, consisting of two chapters, covers general information pertaining to the various types of propagation, such as the facilities, equipment, and supplies needed, nomenclature, and organizations supporting propagation activities. The second unit, consisting of five chapters, deals with the production of seeds and the methods of using them in sexually producing new plants. The third unit, consisting of seven chapters, considers all the types of asexual or vegetative propagation—the use of cuttings; grafting and budding; layering, suckers, and runners; as well as the use of specialized structures, such as bulbs, corms, rhizomes, and tubers, to produce new plants. The fourth unit is comprised of two chapters that deal with micropropagation, a relatively new and exciting field using aseptic culture methodology to grow tiny plant pieces for the sexual or asexual regeneration, under confined, controlled conditions, of vast numbers of new plants. The fifth unit consists of three chapters giving, in encyclopedic format, the accepted propagation methods for the important fruit and nut crops, the principal ornamental trees, shrubs, and vines, as well as the common annual and herbaceous perennial plants used as ornamentals.

In considering the various ways by which plants can be propagated—as seeds, cuttings, grafting, etc.—this book has separated the theoretical implications from the techniques involved into different, consecutive chapters. For example, the basic botanical principles underlying grafting and budding are stressed in one chapter, while the techniques of the various grafting and budding methods are described in detail in the two following chapters. Although some readers are very much concerned with the fundamental principles of plant propagation, others may be interested primarily in the actual techniques involved.

An extensive bibliography of the important research literature is included for each subject. In addition, suggested supplementary readings are listed for most of the subjects. These specialized works deal with the topic more extensively than is possible in this book. These references should be valuable for those wishing to study the subject in greater detail.

The term *cultivar* (**culti**vated **var**iety)—now widely used throughout the world in the horticultural literature—is used in this book rather than the earlier and less precise term *variety,* which does not distinguish between botanical varieties and the commonly cultivated varieties of horticultural interest.

The Latin plant names used generally conform to those given in *Hortus Third* (Macmillan, 1976).

In preparing the fourth edition of this book, we have depended upon the asssitance of authorities in the various fields of propagation and related subjects. They gave their time most generously in reading sections of the manuscript and offering suggestions. The responsibility, however, for the final version of the text is the authors'.

We also thank our wives, Hazel M. Hartmann and Daphne Kester, for their great assistance in typing and proofreading throughout the various stages of preparing the fourth edition of this book. We wish to thank, too, Marilyn Hartmann, who made some of the drawings used in this book.

—*Hudson T. Hartmann*
—*Dale E. Kester*

Plant propagation is the multiplication of plants by both sexual and asexual means. A study of plant propagation has three different aspects. First of all, successful propagation of plants requires a knowledge of mechanical manipulations and technical skills that take a certain amount of practice and experience to master, such as how to bud or graft or how to make cuttings. This aspect may be considered as the art of propagation.

Second, successful propagation requires a knowledge of plant growth and structure. This may be said to be the science of propagation. The propagator can obtain some of this information empirically by working with the plants themselves, but it should be supplemented, if possible, with formal courses in botany, horticulture, genetics, and plant physiology. Such knowledge aids the propagator in understanding why he does the things he does, in doing them better, and in coping with unexpected problems.

A third aspect of successful plant propagation is a knowledge of the different kinds of plants and the various possible methods by which certain plants can be propagated. To a large extent the method selected must be related to the responses of the kind of plant being propagated and to the situation at hand.

The propagation of plants has been a fundamental occupation of mankind since civilization began. Agriculture started some 10,000 years ago when ancient peoples learned to plant and grow kinds of plants that fulfilled nutritional needs for themselves and their animals (*9*). As civilization advanced, people added to the variety of plants, cultivating not only additional food crops but also those that provided fibers, medicine, recreational opportunities, and beauty (*8*). From the great diversity and variation in plant life it has been possible to select kinds of plants particularly useful to the welfare of humans and their animals (*3, 12*).

Progress in agricultural development has involved an interplay of two separate activities. One is the selection of specific kinds of plants; the second is the reproduction of these kinds in a manner to retain their valuable characteristics in culture.

Much progress in plant improvement was made long before the modern period of plant breeding (*1, 6*). Our cultivated plants originated mainly by three general methods: First, some kinds of plants were selected directly from wild species but, under the selective hand of man, evolved into types that differ radically from their wild relatives. Examples of this group are lima bean, tomato, barley, and rice. Second, other kinds of plants arose as hybrids between species, accompanied by changes in chromosome number. These plants are completely unique to cultivation and have no single wild relative. Examples of

# 1

# Introduction

this group include maize (corn), wheat, tobacco, and strawberry. Third, another group of plants occurs naturally as rare monstrosities. Although unadapted to a native environment they are useful to man. Among these are heading cabbage, broccoli, and Brussels sprouts.

This progress in plant improvement would have been of little significance, however, without methods whereby improved forms could be maintained in cultivation. Consequently, there has been a process of invention and development of techniques for plant propagation. Most cultivated plants either would be lost or would revert to less desirable forms if they were not propagated under controlled conditions that preserve the unique characteristics that make them useful. Throughout history, as new kinds of plants became available, the knowledge and techniques to preserve them had to be learned; conversely, as advances in propagation techniques developed, the number of plants that became available for cultivation increased.

Listed below are the general methods of propagating plants. Many, if not most, of them antedate recorded history. It is probably no mere coincidence that some of the oldest fruit crops—grape, olive, mulberry, quince, pomegranate, and fig—are also the easiest to propagate by means of simple techniques using hardwood cuttings. To grow most other tree fruits, budding and grafting procedures had to be developed. Invention of glasshouses in the nineteenth century made the rooting of leafy cuttings possible. More recently, the discovery of root-inducing chemicals and the development of mist propagation has greatly enhanced many nursery procedures. Likewise, production of seed crops has been revolutionized by the discovery of genetic principles leading to, among other advances, the production of hybrid seed. A significant step forward occurred with the development of the aseptic micropropagation and in vitro tissue culture techniques described in chapters 16 and 17.

### Methods of propagating plants with typical examples

**I.** Sexual
  **A.** Propagation by seed—annuals, biennials, and many perennial plants
  **B.** In vitro culture systems
    **1.** Ovule culture—carnation, tobacco, petunia
    **2.** Embryo culture—iris, olive
    **3.** Seed culture—orchid
    **4.** Pollen—tobacco, *Datura*
    **5.** Spores—ferns
**II.** Apomictic (asexual)
  **A.** Seed
    **1.** Nucellar embryos—citrus, mango
    **2.** Adventitious embryony—Kentucky bluegrass
  **B.** In vitro culture systems
    **1.** Nucellar embryos—citrus
    **2.** Embryogenesis from cells and callus—tobacco, carrot
**III.** Vegetative (asexual)
  **A.** Propagation by cuttings
    **1.** Stem cuttings
      **a.** Hardwood—fig, grape, crape myrtle, rose, willow, poplar

        **b.** Semihardwood—lemon, camellia, holly, rhododendron

        **c.** Softwood—lilac, forsythia, weigela, crape myrtle

        **d.** Herbaceous—geranium, coleus, chrysanthemum, sugar cane

    **2.** Leaf cuttings—*Begonia rex, Bryophyllum, Sansevieria,* African violet

    **3.** Leaf-bud cuttings—blackberry, hydrangea

    **4.** Root cuttings—red raspberry, horseradish, phlox

**B.** Propagation by grafting

    **1.** Root grafting

        **a.** Whip graft—apple and pear

    **2.** Crown grafting

        **a.** Whip graft—Persian walnut

        **b.** Cleft graft—camellia

        **c.** Side graft—narrow-leaved evergreens

    **3.** Top grafting

        **a.** Cleft graft—various fruit trees

        **b.** Saw-kerf or notch graft—various fruit trees

        **c.** Bark graft—various fruit trees

        **d.** Side graft—various fruit trees

        **e.** Whip graft—various fruit trees

    **4.** Approach grafting—mango

**C.** Propagation by budding

    **1.** T-budding—stone and pome fruit trees, rose

    **2.** Patch budding—walnut and pecan

    **3.** Ring budding—walnut and pecan

    **4.** I-budding—walnut and pecan

    **5.** Chip budding—grape, mango, fruit trees

**D.** Propagation by layering

    **1.** Tip—trailing blackberry, black raspberry

    **2.** Simple—honeysuckle, spirea, filbert

    **3.** Trench—apple, pear, cherry

    **4.** Mound or stool—gooseberry, apple

    **5.** Air (pot or Chinese)—India rubber plant, lychee

    **6.** Compound or serpentine—grape, honeysuckle

**E.** Propagation by runners—strawberry, *Chlorophytum comosum*

**F.** Propagation by suckers—red raspberry, blackberry

**G.** Separation

    **1.** Bulbs—hyacinth, lily, narcissus, tulip

    **2.** Corms—gladiolus, crocus

**H.** Division

    **1.** Rhizomes—canna, iris

    **2.** Offsets—houseleek, pineapple, date palm

    **3.** Tubers—Irish potato

    **4.** Tuberous roots—sweet potato, dahlia

    **5.** Crowns—everbearing strawberry, phlox

I. In vitro culture systems
    1. Shoot-tip culture—orchid, carnation, asparagus, chrysanthemum, fern, strawberry
    2. Adventitious shoot formation—rhododendron, African violet, lily
    3. Micrografting—citrus, apple, plum
    4. Tissue and cell culture—tobacco, carrot

## CELLULAR BASIS FOR PROPAGATION

Plant propagation involves the control of two basically different types of development life cycle—**sexual** (p. 59) and **asexual** (p. 199). Preservation of the unique characteristics of a plant or group of plants depends upon the transmission from one generation to the next of a particular combination of genes present on the chromosomes in the cells. The sum total of these genes make up the **genotype** of the plant. The genotype, in combination with the environment, produces a plant of a given outward appearance (the **phenotype**). The function of any plant propagation technique is *to preserve a particular genotype or population of genotypes* that will reproduce the particular kind of plant being propagated.

### Meiosis and Sexual Reproduction

Sexual reproduction involves the union of male and female sex cells, the formation of seeds, and the creation of a population of seedling individuals with new and differing genotypes (Figures 1–1 and 1–2). The cell division **(meiosis)** which produces the sex cells involves reduction division of the chromosomes, in which their number is reduced by half. The original chromosome number is restored during fertilization, resulting in new individuals containing chromosomes from both the male and the female parents. Offspring may resemble either, neither, or both of the parents, depending upon their genetic similarities. Among the progeny from a particular combination of parents, considerable variation may occur.

The outward appearance (phenotype) of a plant and the way characteristics are inherited from generation to generation are controlled through the action of genes present on the chromosomes. Some traits are controlled by a single gene as shown for pea height in Figure 1–3. Figure 1–4 shows the more complex case where two independent genes affect the appearance of peach fruit. The analysis of inheritance for traits controlled by many genes, which is the usual case, is much more complex and requires statistical analysis to show how closely the offspring resembles the parents. The kind of genes present and how much the environment influences their effect also must be taken into account (*1, 5*).

The two terms **homozygous** and **heterozygous** are useful to describe the genotype of a particular plant. If a high proportion of the genes present on one chromosome are the same as those on the opposite member of the chromosome pair **(homologous chromosomes),** the plant is homozygous and will "breed true" if self-pollinated or if the other parent is genetically similar. This means that the particular traits or characteristics that the plant possesses will be transmitted to its offspring, and the offspring will resemble the parent. On the other hand, if a sufficient number of genes on one chromosome differ from those on the other member of the chromosome pair, then the plant is said to be

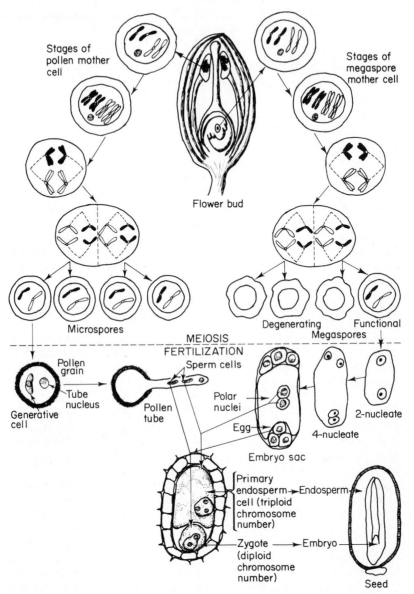

**Figure 1-1**     Diagrammatical representation of the sexual cycle in angiosperms. Meiosis occurs in the flower bud in the anther (male) and the pistil (female) during the bud stage. During this process the pollen mother cells and the megaspore mother cells, both diploid, undergo a reduction division in which homologous chromosomes segregate to different cells. This is followed immediately by a mitotic division which produces four daughter cells, each with half the chromosomes of the mother cells. In fertilization a male gamete unites with the egg to produce a zygote, in which the diploid chromosome number is restored. A second male gamete unites with the polar nuclei to produce the endosperm.

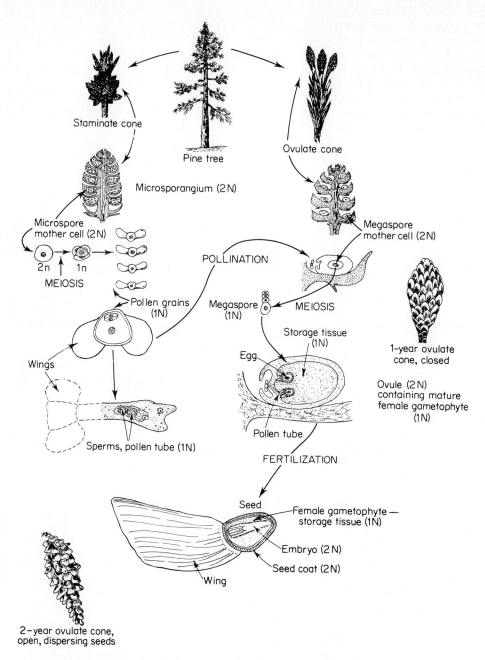

Staminate cone

Pine tree

Ovulate cone

Microsporangium (2N)

Microspore mother cell (2N)

2n   1n

MEIOSIS

Megaspore mother cell (2N)

POLLINATION

Pollen grains (1N)

Megaspore (1N)

MEIOSIS

1-year ovulate cone, closed

Storage tissue (1N)

Egg

Wings

Ovule (2N) containing mature female gametophyte (1N)

Sperms, pollen tube (1N)

Pollen tube

FERTILIZATION

Seed

Female gametophyte — storage tissue (1N)

Embryo (2N)

Seed coat (2N)

Wing

2-year ovulate cone, open, dispersing seeds

**Figure 1–2**   Diagrammatical representation of the sexual cycle in a gymnosperm (pine) showing meiosis and fertilization. It differs from that in the angiosperms by the formation of a "naked" seed, and the development of a haploid female gametophyte that comprises the storage tissue of the seed.

**Figure 1-3** Inheritance involving a single gene pair in a monohybrid cross. In garden peas tallness (D) is dominant over dwarfness (d). A tall pea plant is either homozygous (DD) or heterozygous (Dd). Segregation occurs in the $F_2$ generation to produce three genotypes (DD, Dd, dd) and two phenotypes (tall and dwarf).

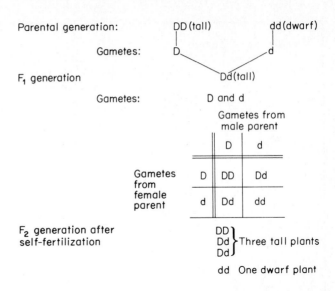

heterozygous. In such a case, important phenotypic traits of the parent may not be transmitted to its offspring, and the seedlings may differ in appearance not only from the parent but also from each other. The amount of seedling variation may be very great in some kinds of plants.

**Figure 1-4** Inheritance in a dihybrid cross involving peach (*Prunus persica*). Fuzzy skin (G) of a peach is dominant over the glabrous (i.e., smooth) skin of the nectarine (g). White flesh color (Y) is dominant over yellow flesh color (y). In the example shown, the phenotype of the $F_1$ generation is different from either parent. Segregation in the $F_2$ generation produces nine genotypes and four phenotypes.

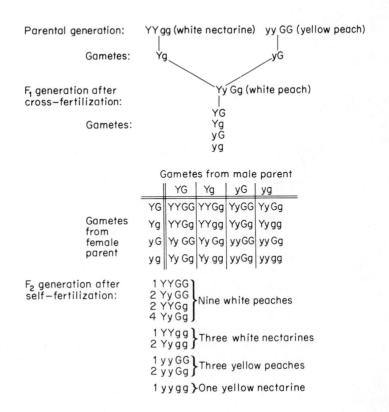

To minimize variation and to be sure that seedling offspring will possess the particular characteristics for which they are to be grown, certain procedures must be followed during seed production. These are discussed in chapter 4.

## Mitosis and Asexual Reproduction

Asexual propagation is possible because each cell of the plant contains all the genes necessary for growth and development and, during the cell division (**mitosis**) that occurs during growth and regeneration, the genes are replicated in the daughter cells. Regeneration of a new organism by asexual methods occurs readily in higher plants but not in higher animals. In some forms of lower animal life, however, such as the flatworm, *Planaria,* in the phylum Platyhelminthes, asexual multiplication can take place. A flatworm cut in half will develop into two worms, each half regenerating the missing part.

The details of mitosis are shown in Figure 1–5, its principal feature being that individual chromosomes split longitudinally, the two identical parts going to two daughter cells. As a result, the complete chromosome system of an individual cell is duplicated in each of its two daughter cells (with certain exceptions). The chromosomes produced will be the same as in the cell from which they came. Consequently the characteristics of the new plant that grows will be the same as that from which it originated.

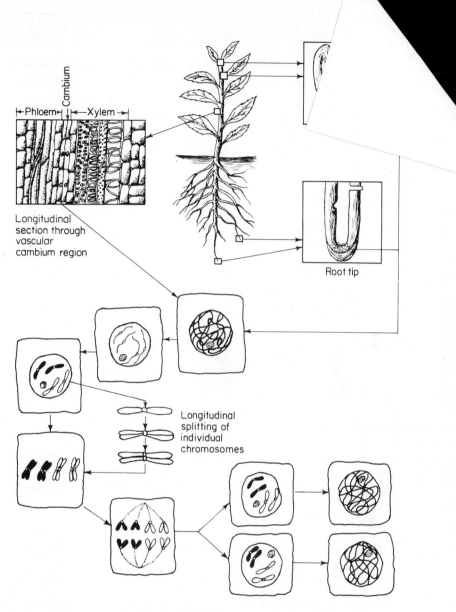

**Figure 1–5** Diagrammatical representation of the process by which growth and asexual reproduction take place in a dicotyledonous plant. Mitosis occurs in three principal growing regions of the plant: the stem tip, the root tip of primary and secondary roots, and the cambium. A meristematic cell is shown dividing to produce two daughter cells whose chromosomes will (usually) be identical with those of the original cell.

Mitosis occurs in specific growing points or areas of the plant to produce growth. (See Figure 1–5.) These are the **shoot apex,** the **root apex,** the **cambium,** and the **intercalary zones** (internode bases of monocotyledonous

**Figure 1–6**    Types of regeneration occurring in asexual propagation. *Left:* adventitious shoots growing from a root cutting. *Center:* adventitious roots developing from the base of a stem cutting. *Right:* callus tissue produced to give healing of a graft union.

plants). Mitosis also occurs when **callus** forms on a wounded plant part and when new growing points are initiated on root and stem pieces. Callus parenchyma consists of new cells proliferating from cut tissues in response to wounding. When new growing points are initiated on a vegetative structure, such as root, stem, or leaf, they are referred to as **adventitious roots** or **adventitious shoots.** (See Figure 1–6 and chapter 9.)

Adventitious roots are those that arise from aerial plant parts, from underground stems, or from relatively old roots. All roots other than those arising from the embryo axis and all their branches formed in normal sequence can be considered adventitious roots. Adventitious shoots are those appearing on roots or internodally on stems after the terminal and lateral growing points are produced. New shoots **(watersprouts)** sometimes arise from latent growing points or buds that are not adventitious but originated along with the branch on which they occur. These are common on old branches of woody plants and can be stimulated into active growth if the part terminal to it is removed.

Mitosis is the basic process of normal vegetative growth, regeneration, and wound healing which makes possible such vegetative propagation techniques as cuttage, graftage, layerage, separation, and division. These methods of propagation are important because they permit large-scale multiplication of an **individual plant** into as many separate plants as the amount of parent material will permit. Each separate plant produced by such means is (in most cases) genetically identical with the plant from which it came. The primary reason for using these vegetative propagation techniques is to reproduce exactly the genetic characteristics of any individual plant, although there may be additional advantages from the standpoint of culture.

# PLANT NOMENCLATURE

Since plant propagation involves the preservation of genotypes that are important to man, it is essential to have some means of labeling them. A system of nomenclature that has gradually evolved through the efforts of botanists and horticulturists provides the basis for uniform worldwide plant identification. The system is embodied in the *International Code of Botanical Nomenclature* (*11*).

## Botanical Classification

Classification of plants in nature is the function of taxonomists. The system of classification is based upon increasing specialization and complexity in structure and organization resulting from evolutionary processes. For example, the plant **kingdom** is divided into **divisions**—first, with one-celled plants—i.e., Schizophyta (bacteria and blue-green algae)—through somewhat more complex plants, such as the Bryophyta (mosses, liverworts), to specialized higher plants, Pterophyta.

This text is principally concerned with the Pterophyta, which includes three **classes:** Filicinae (ferns), Gymnospermae (gymnosperms, as *Ginkgo* and conifers), and Angiospermae (flowering plants). Classification within these classes is based principally on flower structure. The gymnosperms produce seeds that are not enclosed. The angiosperms include plants in which the seed is produced within an enclosed structure—the ovary. The angiosperms are divided into two **subclasses**—the monocotyledons (e.g., grasses, palms, orchids) and the dicotyledons (e.g., beans, roses, peaches). These subclasses are divided into **orders,** the orders into **families,** families into **genera,** and genera into **species.**

The **species** is the fundamental unit customarily used by taxonomists to designate groups of plants that can be recognized as distinct kinds. In nature, individuals within one species normally interbreed freely but do not interbreed with a separate species because they are separated either by distance or by some other physiological, morphological, or genetic barrier that prevents the interchange of genes between the two species (*14*). Consequently it is possible to reproduce a species by seed and to maintain it through propagation. However, if one reproduces a species from individual plants or from different parts of the natural range of that species, there may be a great deal of natural variability in appearance and adaptability among individuals of that species (*7, 10, 13, 15*). To get a total picture of species variability, one should examine individuals from all parts of the range rather than relying on individuals from only a narrow selection.

A distinct morphological subgrouping within a species (usually resulting from geographical separation) may be recognized taxonomically as a **botanical variety.** The botanical code also makes provision for certain other naturally occurring subdivisions within the species such as **subspecies, subvariety, group, form,** and **individual.**

Natural variation among native plants of a species can also be described with the two terms, **cline** and **ecotype** (*10, 13*). A cline refers to the continuous differences in genetically controlled physiological and morphological characteristics that occur within a species in different parts of its range. The differences are related to continuous variations in the environment and result from the evolution of populations of plants adapted to these variations. Where the differences are distinct and discontinuous, the term ecotype is used.

## Classification of Cultivated Plants

The classification and naming of those special kinds of cultivated plants has a different basis from that for plants growing wild in nature. The basis for separating one kind from another in cultivation is not because of naturally occurring variation but because each kind has some practical significance in agriculture and horticulture. The group of plants representing each kind has invariably arisen as a variant, either within a species or as a hybrid between one or more species. Very often it is derived from a single plant that is reproduced asexually. Such a group of plants representing a single type whose unique characteristics are reproduced during propagation is known as a **cultivar,** which is synonymous with such older terms as *variety* (English), *variete* (French), *variedad* (Spanish), *sorte* (German), *sort* (Scandinavian), *ras* (Russian), or *razza* (Italian). Such groups of plants are labelled by a non-Latin name, usually bestowed by its originator.

Naming of cultivated plants can be handled best if it conforms to a widely agreed upon system of nomenclature. The International Code of Nomenclature for Cultivated Plants (*4*) covers the special categories and names of cultivated plants. First, it recognizes the **cultivar** as a special taxonomic category for cultivated plants which is defined as ''an assemblage of cultivated plants which is clearly distinguished by any characters (morphological, physiological, cytological, chemical, or others) and which when reproduced, sexually or asexually, retains its distinguishing characters.'' Second, it establishes guidelines by which cultivar names can be applied, and third, it establishes registration procedures by which cultivar names can be recorded.

The word **cultivar** is a contraction of the words ''**cultiv**ated **var**iety,'' and should be distinguished from the analogous, naturally occurring category—botanical variety—which is a population of plants within a species that has unique characteristics and occurs naturally. A cultivar is also a unique population of plants, but it is artificially maintained by human efforts and is named. Many cultivars would not continue to exist without human efforts.

The complete scientific name of any plant developed or maintained under cultivation includes the: (a) genus, (b) species, and (c) cultivar name, the first two in the usual Latin form and the latter in words of a common language. The cultivar name can be set off by the abbreviation **cv.,** or by single quotation marks, but not both. It can also just be attached to the common name.

*Syringa vulgaris* cv. Mont Blanc
*Syringa vulgaris* 'Mont Blanc'
Lilac 'Mont Blanc'

A cultivar should have a proper name that can be recognized by horticulturists everywhere, one that will not confuse its identification with other cultivars. Multiple names for the same cultivar, or one name for several similar cultivars, arising either by accident or from deliberate changes in name, can only lead to confusion and misrepresentation. The most important principle in naming cultivars is that normally the earliest name applied should have priority. A second principle is that once a name is correctly applied, it should be changed only for exceptional reasons. There are additional rules which can be of assistance in choosing a plant name; for these the Code of Nomenclature for Cultivated Plants should be consulted (4).

## Categories of Cultivars

**Sexually reproduced cultivars** are propagated by seed. The basic category is the **line,** which is a population of seed-propagated plants in which genetic variability is controlled and the uniformity is maintained to a standard appropriate for that cultivar. An example is *Triticum aestivum* 'Marquis' wheat, in which the uniformity is readily maintained in propagation because it is homozygous and self-pollinated.

As knowledge of genetic principles developed, plant breeders have formulated various other types of seed-propagated lines that require special methods of parent selection and use of seed production procedures to maintain. These categories of cultivars include **inbred lines, composite** or **synthetic lines,** and **hybrid lines.** In addition, certain mixtures, as well as natural variants, known as **provenances** (see p. 96), can be classed as cultivars. Definitions of these special categories and their underlying genetic basis are discussed in chapter 4.

**Asexually reproduced cultivars** are propagated by various vegetative methods, as by cuttings, division, or grafting. The basic category of this type of cultivar is the **clone.** It is defined in the Code (4) as "a genetically uniform assemblage of individuals (which may be chimeral in nature), derived originally from a single individual by asexual propagation, for example, by cuttings, division, grafts, or obligate apomixis. Individuals propagated from a distinguishable bud mutation form a cultivar distinct from the parent plant." (Chimeras are discussed on p. 210 and apomixis on p. 74.) Examples of clones are 'Redhaven' peach, 'King Alfred' daffodil, and 'Burbank Russet' potato. Concepts of the clone and maintenance of its uniformity during propagation are discussed in chapter 8.

Another type of asexual cultivar is comprised of those groups of seedling plant populations reproduced by **obligate apomixis.** This concept is discussed in chapter 3 (p. 74).

Plants showing special or unique growth phases (see p. 202) within clones and which can be reproduced by vegetative means may also be classed as separate cultivars. Examples of these are *Sequoia sempervirens* 'Prostrata', a prostrate form of the redwood tree, and *Picea abies* 'Pygmaea', a witches' broom type of spruce.

## PLANT PROPAGATION ORGANIZATIONS

There are several kinds of organizations that promote sexual or asexual propagation of plants. Some of the principal ones are described below:

*The International Plant Propagators' Society* (IPPS)[1] was organized in 1951 and now has regional sections in the Eastern, Western, and Southern United States, Great Britain and Ireland, Australia, and New Zealand. Each region holds an-

[1] P.O. Box 3131, Boulder, Colo. 80307 U.S.A.

nual meetings and the papers presented at these meetings are published in an annual hardbound *Combined Proceedings.* This society has about 2000 members, and membership is confined to individuals active in some phase of plant propagation. A requirement for continued membership is participation in the society's activities, for example, by attending a regional meeting at least once in four years or writing an article for publication in the society's newsletter, *The Plant Propagator.* Publications are available only to members and libraries.

*The International Dwarf Fruit Tree Association*[2] has about 1600 members worldwide, and holds an annual meeting, usually in Michigan. Papers are presented on various aspects of fruit tree rootstocks and propagation as well as general orchard culture. These papers are published in the annual proceedings, *Compact Fruit Tree.* The association also sponsors an annual orchard tour in different fruit-growing areas.

*International Association for Plant Tissue Culture*[3] has about 2500 members representing 63 countries, including 300 members in the United States. Any person interested in in vitro plant cell, tissue, or organ culture is eligible for membership. The association promotes the interests of plant tissue culture workers by publishing a newsletter and by sponsoring an International Congress of Plant Tissue Culture every four years. Distributed three times a year, the newsletter contains feature articles on plant tissue culture and work underway in various laboratories, book reviews, bibliography listings of articles on plant tissue culture, and names of new members. Members represent industry as well as academic institutions and their interest includes commercial propagation; freeing stock plants from virus, bacterial, and fungal infestation; plant improvement by new genetic technology; and secondary product production in vitro.

*Bedding Plants, Inc.*[4] is an association of more than 3000 members growing or interested in bedding plants. Its annual meetings are held in various parts of the United States, papers are published in an annual proceedings. A *Buyers' Guide* is published every two years to help bedding plant growers obtain their supplies. A newsletter is published several times a year.

*The American Association of Nurserymen* (AAN),[5] organized in 1875, is a national trade organization of the U.S. nursery and landscape industry. It serves about 3200 member firms involved in the nursery business—wholesale growers, garden center retailers, landscape firms, mail-order nurserymen, and allied suppliers to the horticultural community.

Through information, educational meetings, and other services, the AAN helps its members to manage their businesses more effectively. Its twice-a-month newsletter, *Up-date,* keeps members informed of legislative and regulatory decisions with which they must comply. Through active committees,

[2] Department of Horticulture, Mich. State University, East Lansing, Mich. 48824 U.S.A.

[3] Department of Plant Sciences, Texas A & M University, College Station, Tex. 77843 U.S.A.

[4] P.O. Box 286, Okemos, Mich. 48864 U.S.A.

[5] 230 Southern Building, 15th & H Streets, N.W., Washington D.C. 20005 U.S.A.

the association maintains an up-to-date system for sizing and describing plants for ease in trading, and information on plant nomenclature, quarantines, pesticides, and the like.

The association publishes the *American Standard for Nursery Stock* ( *2* ) prepared by a committee of nurserymen, landscape architects, landscape contracters, and others trading in or specifying nursery plants. These standards are reviewed and endorsed by interested national and regional societies, associations, and governmental agencies, and approved by the American National Standards Institute. Categories of nursery stock for which standards are established are: shade and flowering trees; deciduous shrubs; coniferous evergreens; broadleaf evergreens; roses; vines and ground covers; fruit trees; small fruits; lining out stock; seedling trees and shrubs; bulbs, corms, and tubers; and Christmas trees. The American Association of Nurserymen also holds annual meetings in various U.S. cities.

Several organizations deal with plant reproduction and growing by the use of seeds. Some of the principal ones are:

*The International Seed Testing Association (ISTA)*[6] is an intergovernmental association with worldwide membership accredited by the governments of 59 countries. There are 137 official seed-testing stations that belong to the association. The chief purposes of the association are to develop, adopt, and publish standard procedures for sampling and testing seeds, and to promote uniform application of these procedures for evaluation of seeds moving in international trade.

Secondary purposes of ISTA are to promote research in all areas of seed science and technology, to encourage cultivar certification, and to participate in conferences and training courses aimed at promoting these objectives.

At a general meeting held every three years, there is a symposium for scientific and technical papers presented by specialists working in seed technology.

ISTA publishes the journal, *Seed Science and Technology;* a quarterly newsletter, *ISTA News Bulletin;* and a number of technical handbooks such as the *Handbook for Seed Evaluation.*

*Association of Official Seed Analysts*[7] is an organization of member laboratories, serving both private seed analysts and workers on seed research problems in recognized government seed laboratories. Most members are in continental North America. The objectives of the organization are to improve all branches of seed testing and to make it more useful to agriculture and society.

The association holds an annual meeting and publishes the proceedings in the *Journal of Seed Technology.* It also publishes a quarterly newsletter along with several other publications, such as *Rules for Testing Seeds.*

*American Seed Trade Association, Inc. (ASTA),*[8] an organization of seed companies, has been serving the industry since 1883. ASTA holds a general meeting each year, as well as conferences on farm seed, lawn seed, soybean seed, the garden seed industry, and hybrid corn and sorghum industry research. The

---

[6] P.O. Box 412, CH–8046, Zurich, Switzerland
[7] P.O. Box 5425, Mississippi State, Miss. 39762 U.S.A.
[8] 1030 15th Street, N.W., Washington, D.C. 20005 U.S.A.

association publishes a semimonthly newsletter, an annual yearbook, and proceedings of the farm seed, soybean seed, and hybrid corn and sorghum industry conferences. ASTA becomes involved in regulatory activities, for example, by participating in the drafting of amendments to the Recommended Uniform State Seed Law, by evaluating proposed changes in tariffs or amendments to the Federal Seed Act, studying and evaluating benefits of international seed certification, and formulating trade rules for domestic and foreign commerce in seeds. ASTA also maintains close liaison with seed-testing authorities and evaluates the benefits of international seed certification. It serves as the industry's center of information on current business facts and problems and it actively promotes industry sales in both U.S. and overseas markets.

*The Canadian Seed Trade Association,* similar to ASTA, promotes the seed industry of Canada.[9]

*The Association of Official Seed Certifying Agencies (AOSCA)*[10] is an organization whose members are U.S. and Canadian agencies responsible for seed certification in their respective areas. AOSCA was organized in 1919 as the International Crop Improvement Association. These agencies maintain a close working relationship with the seed industry, seed regulatory agencies, governmental agencies involved in international seed market development and movement, agricultural research and extension services.

## REFERENCES

1. Allard, R. W. 1960. *Principles of plant breeding.* New York: John Wiley.

2. American Association of Nurserymen. 1980. *American standard for nursery stock.* Washington, D.C.: Amer. Assn. Nurs.

3. Baker, H. G. 1978. *Plants and civilization* (3rd ed.). Belmont, Calif.: Wadsworth.

4. Brickell, C. D., ed. 1980. International code of nomenclature for cultivated plants—1980. *Regnum Vegetabile* 104:7–32 (Obtainable from *Amer. Hort. Soc.,* Mt. Vernon, VA 22121 U.S.A.).

5. Briggs, F. N., and P. F. Knowles. 1967. *Introduction to plant breeding.* New York: Reinhold.

6. Frey, K. J., ed. 1981. *Plant breeding II.* Ames, Iowa: Iowa State Univ. Press.

7. Harlan, J. R. 1956. Distribution and utilization of natural variability in cultivated plants. In *Genetics in plant breeding.* Brookhaven Symposia in Biol. 9:191–206.

8. ———. 1976. The plants and animals that nourish man. *Scient. Amer.* 235(3): 88–97.

9. Hartmann, H. T., W. J. Flocker, and A. M. Kofranek. 1981. *Plant science: Growth, development, and utilization of cultivated plants.* Englewood Cliffs, N.J.: Prentice-Hall.

10. Langlet, O. 1962. Ecological variability and taxonomy of forest trees. In *Tree growth,* T. T. Kozlowski, ed. New York: Ronald Press, pp. 357–69.

[9] 408 Gertrude Avenue, Winnepeg, Manitoba, Canada
[10] 3709 Hillsborough Street, Raleigh, N.C. 27607 U.S.A.

11. Lanjouw, J., ed. 1966. International code of botanical nomenclature. *Regnum Vegetabile,* 46:402.

12. Sauer, C. O. 1969. *Agricultural origins and dispersals.* Cambridge, Mass.: Massachusetts Institute of Technology Press.

13. Stebbins, G. L. 1950. *Variation and evolution in plants.* New York: Columbia Univ. Press.

14. ———. 1971. *Processes of organic evolution* (2nd ed.). Englewood Cliffs, N.J.: Prentice-Hall.

15. Vavilov, N. I. 1930. Wild progenitors of the fruit trees of Turkestan and the Caucasus and the problem of the origin of fruit trees. *Rpt. and Proc. IX Inter. Hort. Cong.,* London, pp. 271–86.

## SUPPLEMENTARY READING

BAILEY, L. H., E. Z. BAILEY, and STAFF OF L. H. BAILEY HORTORIUM. 1976. *Hortus third.* New York: Macmillan.

FRANKEL, O. H., and E. BENNETT. 1970. *Genetic resources in plants: Their exploration and conservation.* Oxford: Blackwell Scientific Company.

HARTMANN, H. T., W. J. FLOCKER, and A. M. KOFRANEK. 1981. *Plant science: Growth, development, and utilization of cultivated plants.* Englewood Cliffs, N.J.: Prentice-Hall.

HEISER, C. B., JR. 1973. *Seed to civilization.* San Francisco: W. H. Freeman & Company Publishers.

HODGSON, R. E. 1961. Germ plasm resources. *American Association for Advancement of Science Publ. No. 66.* Washington, D.C.

SCHERY, R. W. 1972. *Plants for man* (2nd ed.). Englewood Cliffs, N.J.: Prentice-Hall.

SCHWANITZ, F. 1966. *The origin of cultivated plants* (English translation from German edition of 1957). Cambridge, Mass.: Harvard Univ. Press.

WEIER, T. E., C. R. STOCKING, M. G. BARBOUR, and T. L. ROST. 1982. *Botany: An introduction to plant biology* (6th ed.) New York: John Wiley.

In propagating and growing young nursery plants, the facilities and procedures are best arranged so as to optimize the response of the plants to the five fundamental environmental factors influencing their growth and development: **light, water, temperature, gases,** and **mineral nutrients.** In addition, young nursery plants require protection from pathogens and other pests, as well as control of salinity levels in the media.

The propagation structures, equipment, and procedures described in this chapter, if handled properly, maximize the plants' growth and development by controlling their environment.

Facilities required for propagating plants by seed, cuttings, and grafting include two basic units. One is a structure with temperature control and ample light, such as a greenhouse or hotbed, where seeds can be germinated or cuttings rooted. The second unit is a structure into which the young, tender plants can be moved for hardening, preparatory to transplanting out-of-doors. Cold frames or lathhouses are useful for this purpose. Any of these structures may, at certain times of the year and for certain species, serve both purposes.

## PROPAGATING STRUCTURES

### Greenhouses

There are several types of greenhouses. A simple one is a shed-roof, lean-to structure utilizing one side, preferably the south or east, of another building as one wall. In fact, as a solar heat utilization measure, greenhouses may be constructed as part or all of the south wall of a dwelling, the heat accumulated in the greenhouse being used to heat the house (*63, 88, 89*).

Small, inexpensive greenhouses can also be constructed from a number of standard 3-ft by 6-ft hotbed sashes fastened to a 2-by-4 wood framework. Such a

# 2

# Propagation Structures, Media, Fertilizers, Sanitation, and Containers

framework can also be covered with fiberglass panels or polyethylene sheeting to construct an inexpensive greenhouse (*114*).

Commercial greenhouses are usually independent structures of even-span, gable-roof construction, proportioned so that the space is well utilized for convenient walkways and propagating benches (*40*). In large operations, several single greenhouse units are often attached side by side, eliminating the cost of glassing-in the adjoining walls (see Figures 2–1 and 2–2).

**Figure 2–1**    *Above:* Air-supported plastic greenhouse (*right*) in comparison with a range of conventional glass-covered greenhouses. *Below:* Interior of air-supported greenhouse where a crop of tomatoes is being grown. Courtesy W. L. Bauerle and T. H. Short, Ohio Agricultural Research and Development Center.

**Figure 2-2** Interior of a large, well-kept greenhouse used to grow foliage plants for sale. Efficient use of expensive space is accomplished by hanging ferns above lower benches. Note sanitary practice of keeping hose nozzle off the floor where it might pick up disease pathogens.

Arrangements of benches in greenhouses vary considerably. Some propagation installations do not have permanently attached benches, their placement varying according to the type of equipment, such as lift trucks or electric carts, used to move flats and plants (*114*). An innovation that can reduce aisle space and increase the useable space in a greenhouse is the use of rolling benches, which are pushed together until one needs to get between them, and then rolled apart (Figure 2-3).

Greenhouse construction begins with a wood or metal framework, to which are fastened wood or metal sash bars to support panes of glass embedded in putty. All-metal prefabricated greenhouses also are widely used and are available from several manufacturers.[1] In Europe, and to some extent in the United States, a translucent type of glass, which tends to give a uniform, diffuse light, is used for greenhouse construction.

Ventilation to provide air movement and air exchange with the outside is necessary in all greenhouses to aid in controlling temperature and humidity. A mechanism for manual opening of panels at the ridge or at the sides is used in

[1] Some manufacturers of commercial greenhouses in the United States are: Double A Truss Mfg. Co., 320 Wetmore, Manteca, Calif. 95336; Ickes-Braun Glasshouses Div., Roper Corporation, P.O. Box 147, Deerfield, Ill. 60015; Lord and Burnham Div., Burnham Corporation, Irvington, N.Y. 10533; National Greenhouse Co., P.O. Box 100, Pana, Ill. 62557; Nexus Corporation, 2250 19th Street, Denver, Colo. 80202; Rough Brothers, Inc., P.O. Box 16010, Cincinnati, Ohio 45217.

**Figure 2–3**    For more efficient use of costly greenhouse space, movable benches on rollers have been installed to reduce aisle space.

smaller greenhouses, but most larger installations use forced-air fan ventilation controlled by thermostats.

Traditionally greenhouses have been heated by steam or hot water from a central boiler through banks of pipes (some finned to increase radiation surface) suitably located in the greenhouse. Sometimes unit heaters for each house, with fans for improved air circulation, also are used. If oil or gas heaters are used, they must be vented to the outside because the combustion products are toxic to plants. In large greenhouses heated air is often blown into large—30 to 60 cm (12 to 24 in.)—4-mil polyethylene tubes hung overhead and running the length of the greenhouse. Small—5 to 7.5 cm (2 to 3 in.)—holes spaced throughout the length of these tubes allow the hot air to escape, thus giving uniform heating throughout the house (see Figure 2–4). These same tubes can be used for forced air ventilation in summer, eliminating the need for manual side and top vents.

Solar heating of greenhouses occurs naturally. The increasing cost of fossil fuels has evoked considerable interest in methods of conserving day-time solar heat for night heating (*6, 52, 73, 89, 105*). Otherwise, high heating costs may eventually make winter use of greenhouses in the colder regions economically unfeasible, relegating greenhouse operations to areas with relatively warm winters (*28, 109*). Consequently, better conservation of heat in the greenhouse is essential.

Most heat loss in greenhouses takes place through the roof. One method of reducing heat loss in winter is to install double-layered, sealed polyethylene sheeting outside over the glass.[2] This form of insulation is very effective. The

[2] Some sources of double polyethylene panels are: Monsanto Chemical Co., St. Louis, Mo.; I.C.C. Industries, Primex Plastic, C Division, Oakland, N.J. 07436.

**Figure 2–4**    Completely automated heating and cooling systems installed in fiberglass-covered greenhouse. *Upper left:* hot air from hot-water heaters (at top) is blown into polyethylene distribution tubes. *Upper right:* distribution tubes, which have outlet holes spaced to disperse heated air uniformly, extend the length of the house. *Lower left and right:* the opposite end wall of greenhouse has an insert of a wettable pad (*right*) through which air is pulled by exhaust fans for cooling. Automatic closure panels (*left*) shut off outside air movement into the house through this pad when heating is required. All components of both heating and cooling systems are thermostatically controlled.

two layers are kept separate by an air cushion from a low-pressure blower. Energy savings from the use of this system are substantial—more than 50 percent reduction in fuel compared to conventional greenhouses—but the lower light intensity with the double-layer plastic cover may lower yields of the greenhouse crops (*13, 14, 38*).

Another device that reduces heat loss dramatically, is a movable thermal curtain[3] (Figure 2–5), which at night comes between the crop and the

[3] Some suppliers of greenhouse thermal blankets and related equipment are: Ball Seed Co., West Chicago, Ill. 60185; Shade Corporation of America, 8232 N. W. 56th, Miami, Fla. 33166; and X. S. Smith, Inc., Drawer X, Red Bank, N.J. 07701.

**Figure 2–5**    Movable insulating blankets above the plants in a greenhouse (see arrow). On cold winter nights they automatically close to conserve heat.

greenhouse roof and walls (*50, 98, 104, 110*). Insulating the north wall reduces heat loss without appreciably lowering the available light.

Greenhouses can be cooled mechanically in the summer by the use of large evaporative cooling units, as shown in Figure 2–4. The "pad and fan" system, in which a wet pad of some material, such as aspen-wood excelsior or Kool-Cel pads, is installed at one side (or end) of a greenhouse and large exhaust fans at the other, has proved to be the best method of cooling large greenhouses (*5*).

Greenhouses are often sprayed on the outside at the onset of warm spring weather with a thin layer of whitewash or a white, cold-water paint (see p. 439). This coating reflects much of the heat from the sun thus preventing excessively high temperatures in the greenhouse. Too heavy a coating of whitewash, however, can reduce the light intensity to undesirably low levels.

Thermostat control is recommended for greenhouse heating and evaporative cooling systems. Although varying with the species, a minimum night temperature of 13° to 15.5°C (55° to 60°F) is suggested. The evaporator cooling thermostats should be set to start the blowers at about 24°C (75°F).

Glass-covered greenhouses are expensive, but for a permanent long-term installation, are likely to be more satisfactory than the lower-cost plastic-covered houses, which require renewal of the plastic every few years.

PLASTIC-COVERED GREENHOUSES

Lightweight frames covered with plastic materials are popular for small home-garden structures as well as for large commercial installations (*77, 96*). Worldwide, the area covered by plastic greenhouses is about three times that under glass-covered greenhouses (*111*). Two types of plastic are available and in general use—polyethylene (polythene) and fiberglass. Both are lightweight and relatively inexpensive compared to glass. Their light weight permits a less expensive supporting framework than is required for glass. A third material, polyvinyl chloride (PVC), has been used but tends to darken prematurely in sunlight.

Plastic-covered greenhouses tend to be much tighter than glass-covered ones, with a consequent increase in humidity and, especially in winter, an undesirable water drip on the plants. This problem can be overcome, however, by maintaining adequate ventilation (*24*).

### POLYETHYLENE (POLYTHENE)

Polyethylene is widely available and is the least expensive covering material, but it has the shortest life. It breaks down in summer and must be replaced after one or two years, generally in the fall in preparation for winter. Ultraviolet-ray-resisting polyethylene lasts longer (14 to 30 months) but costs somewhat more. A thickness of 4 to 6 mils (1 mil = 0.001 in.) is recommended. For better insulation and lowered winter heating costs, a double layer of UV-inhibited co-polymer material is available with a 2.5 cm (1 in.) air gap between layers, kept separated by air pressure from a small blower. Laid over conventional glass greenhouses, this material substantially reduces winter heat loss (*6*).

A single-layer polyethylene-covered greenhouse loses more heat at night or in winter than a glass-covered house since polyethylene allows passage of heat energy from the soil and plants inside the greenhouse much more readily than glass. Glass stops most infrared radiation, whereas polyethylene is transparent to it. Polyethylene is available in widths up to 12 meters (40 ft). Only materials especially prepared for greenhouse covering should be used. Many installations, especially in windy areas, use a supporting material, usually welded wire mesh, for the polyethylene film. Occasionally other supporting materials, such as Saran cloth, are used.

Polyethylene transmits about 85 percent of the sun's light and passes all wavelengths of light required for the growth of plants (*101*). A tough, white, opaque film consisting of a mixture of polyethylene and vinyl plastic is available. This film stays more flexible under low winter temperatures than does polyethylene but is more expensive. Because temperature fluctuates less under such film than under clear plastic, it is suitable for winter storage of container-grown plants.

Polyethylene permits the passage of oxygen and carbon dioxide, necessary for the growth processes of plants, while reducing the passage of water vapor.

### FIBERGLASS

Rigid panels, corrugated or flat, of polyester resin reinforced with fiberglass are widely used for greenhouse construction. This material is strong, long-lasting, lightweight, and easily applied, and comes in a variety of widths, lengths, and thicknesses. Only the clear material—especially made for greenhouses and in a thickness of 0.096 cm (0.038 in.) or more and weighing 4 to 5 oz per sq ft—should be used. When new this material transmits about 80 to 90 percent of the available light, but light transmission decreases over the years, which can be a serious problem. Since the polyester in the panels burns rapidly, an entire greenhouse may be consumed by fire in 10 minutes. Fiberglass is more expensive than polyethylene.

## Hotbeds

The hotbed is a small, low structure, used for the same purposes as a greenhouse. Seedlings can be started and leafy cuttings rooted in hotbeds early in the season. Heat is provided below the propagating medium by electric heating cables, hot water, steam pipes, hot air flues, or fermenting manure. As in the greenhouse, close attention must be paid in hotbeds to shading and ventilation, as well as to temperature and humidity control.

The hotbed is a large wood box or frame with a sloping, tight-fitting lid made of window sash or, preferably, regular hotbed sash. It should be placed in a sunny but protected and well-drained location. The size of the frame conforms to the size of the glass sash available. A standard size is 0.9 by 1.8 meters (3 ft by 6 ft). If polyethylene or fiberglass is used as the covering, any convenient dimensions can be used. The frame can be built easily of 1-in. or 2-in. lumber nailed to 4-by-4 corner posts set in the ground. Decay-resistant wood such as redwood, cypress, or cedar should be used, and preferably treated with a wood preservative, such as copper naphthenate. This compound retards decay for many years and does not give off fumes toxic to plants. Creosote should not be used on wood structures in which plants will be grown, since the fumes released, particularly on hot days, are toxic to plants. Publications are available giving in detail the construction of such equipment ( *97* ). Hotbeds can be used throughout the year, except in areas with severe winters, where their use may be restricted to spring, summer, and fall.

Lead- or plastic-covered electric soil-heating cables are quite satisfactory for providing bottom heat in hotbeds. Automatic temperature control can be obtained with inexpensive thermostats. For a hotbed 1.8 by 1.8 meters (6 by 6 ft) about 18 meters (60 ft) of heating cable is required. The details of a typical installation are shown in Figure 2–6. Low-voltage soil-heating systems are

**Figure 2–6** Construction of a hotbed showing the installation of an electric heating cable and thermostat. Courtesy General Electric Co.

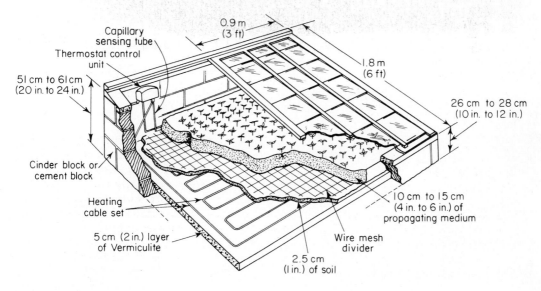

metimes used, especially in Europe and Australia. A transformer reduces the
gular line voltage to about 30 v, lessening the danger of electrical shock. The
eating element consists of low-cost No. 8 or No. 10 bare, galvanized wire (*27*,
*79*).

The hotbed is filled with 10 to 15 cm (4 to 6 in.) of a rooting or seed-
germinating medium over the heating cables. Alternatively, flats containing the
medium can be used; these are placed directly on a thin layer of sand covering
wire netting, placed over the heating cables to protect them from tools.

### Cold Frames (*97*)

Cold-frame construction (Figures 2–7 and 2–8) is the same as for hotbeds,
except that no provision is made for supplying bottom heat. The standard glass
0.9 by 1.8 meter (3 by 6 ft) hotbed sash is often used as a covering for the frame,
although lightweight, less expensive frames can be constructed with polyethylene or fiberglass covering. The covered frames should fit tightly in order to retain heat and obtain high humidity. Cold frames should be placed in locations
protected from winds, with the sash cover sloping down from north to south.

A primary use of cold frames is in conditioning or hardening rooted cuttings
or young seedlings preceding field, nursery-row, or container planting. They
may be used also for starting new plants in late spring, summer, or fall when no
external supply of heat is necessary. In cold frames only the heat of the sun, retained by the transparent covering, is utilized.

Close attention to ventilation, shading, watering, and winter protection is
necessary for success with cold frames. When young, tender plants are first
placed in a cold frame, the coverings are generally kept closed tightly to maintain a high humidity, but as the plants become adjusted, the sash frames are
gradually raised to permit more ventilation and dryer conditions. Frequent
sprinkling of the plants in a cold frame is essential to maintain humid conditions. During sunny days temperatures can build up to excessively high levels in
closed frames unless ventilation and shading are provided. Spaced lath, cloth-

**Figure 2–7**  Banks of cold frames that
were used for starting tender plants.
Frames are opened after protection is no
longer required. Kew Gardens, Rich-
mond, England.

**Figure 2–8**  Commercial use of glass-
covered cold frames in propagating
ground cover plants by cuttings.

covered frames, or reed mats are useful to lay over the sash to provide protection from the sun.

In areas where extremely low temperatures occur, plants being over-wintered in cold frames may require additional protective coverings.

## Lathhouses

Lathhouses (Figure 2–9) provide outdoor shade and protect container-grown plants from high summer temperatures and high light intensities. They reduce moisture stress and decrease the water requirements of plants. Lathhouses have many uses in propagation, particularly in conjunction with transplanting, and with maintenance of shade-requiring or tender plants. At times a lathhouse, in which watering needs are relatively low, is used by nurseries simply to hold plants for sale.

Lathhouse construction varies widely. Aluminum prefabricated lathhouses are available but may be more costly than wood structures. More commonly, wood or pipe supports are used, set in concrete with the necessary supporting cross-members. Sometimes shade is provided by thin wood strips about 5 cm (2 in.) wide, placed to give one-third to two-thirds cover, depending on the need. Both sides and the top are usually covered. Rolls of snow fencing attached to a supporting framework can be utilized for inexpensive construction.

Woven plastic material—Saran or polypropylene fabric—is widely used in covering structures to provide shade. These materials are available in different densities, thus allowing various intensities of light on the plants. They are lightweight and can be attached to heavy wire fastened to supporting posts.

As an alternative to shading, applying water from overhead sprinklers on hot days may be more economical in regions having naturally cool summers.

**Figure 2–9** Lathhouses are often covered by Saran or polypropylene shade cloth supported by wires stretched between metal upright poles. These may extend to cover many acres.

Propagation Structures, Media, Fertilizers, Sanitation, and Containers

# Miscellaneous Propagating Structures

### FLUORESCENT LIGHT BOXES

Young plants of many species grow satisfactorily under the artificial light from fluorescent lamps or other light sources. These units may be used to start young seedlings and to root cuttings (*62*). By enclosing the lamps in boxes, it is possible to maintain high humidity. The "cool-white" fluorescent tubes are generally preferable to other types (*60*). With such equipment it is often helpful to provide bottom heat, from either thermostatically controlled soil-heating cables or a heater or a lamp in the air space below the rooting medium. Although adequate growth of many plant species may be obtained under fluorescent lamps, the results rarely equal or surpass those obtained under good greenhouse conditions.

### PROPAGATING FRAMES

Even in a greenhouse, humidity is not always high enough to permit satisfactory rooting of certain kinds of leafy cuttings. Enclosed frames covered with glass or one of the plastic materials may be necessary for successful rooting (see Figure 2–10). There are many variations of such devices, called Wardian cases in earlier days. Such enclosed frames are useful also for completed grafts of small potted nursery stock, since they retain high humidity during the healing process.

Bell jars (large inverted glass jars) can be set over a container of cuttings to be rooted. Humidity can be kept high in such devices, but shading and ventilation are necessary as soon as rooting starts. As shown in Figure 2–11, polyethylene plastic bags can be placed over a simple wire framework set in the rooting container to provide an inexpensive protective cover that maintains high humidity when rooting a few cuttings.

In using all such structures, care is necessary to avoid the buildup of pathogenic organisms. The warm, humid conditions, combined with lack of air movement and relatively low light intensity, provide excellent conditions for the growth of various fungi and bacteria. Cleanliness of all materials placed in such

**Figure 2–10** Polyethylene-covered beds used in a greenhouse to maintain high humidity surrounding the cuttings during rooting.

**Figure 2-11**    A small unit for rooting a few cuttings. A polyethylene bag has been placed over a framework made from two wire coat hangers. The bag is folded under the pot to give a tight seal. This unit should be set in a fairly light place but never in the direct sunlight, which would cause overheating within the bag.

units is important but, in addition, use of fungicides is sometimes necessary (see section on sanitation).

# MEDIA FOR PROPAGATING AND GROWING NURSERY PLANTS

Various materials and mixtures of materials are used for germinating seeds and rooting cuttings. For good results the following characteristics of the medium are required (*82*):

**1.** The medium must be sufficiently firm and dense to hold the cuttings or seeds in place during rooting or germination. Its volume must be fairly constant when either wet or dry; that is, excessive shrinkage after drying is undesirable.

**2.** It must retain enough moisture so that watering does not have to be too frequent.

**3.** It must be sufficiently porous so that excess water drains away, permitting adequate aeration.

**4.** It must be free from weed seeds, nematodes, and various pathogens.

**5.** It must not have a high salinity level.

**6.** It should be capable of being pasteurized with steam or chemicals without harmful effects.

**7.** It must provide adequate nutrients in situations where plants remain for a long period.

SOIL

A soil is composed of materials in the solid, liquid, and gaseous states. For satisfactory plant growth these materials must exist in the proper proportions. The solid portion of a soil is comprised of both inorganic and organic forms.

The inorganic part consists of the residue from parent rock after decomposition resulting from the chemical and physical process of weathering. Such inorganic components vary in size from gravel down to extremely minute colloidal particles of clay, the texture of the soil being determined by the relative proportions of particles of different size. The coarser particles serve mainly as a supporting framework for the remainder of the soil, whereas the colloidal clay fractions of the soil serve as storehouses for nutrients that are absorbed by plants. The organic portion of the soil consists of both living and dead organisms. Insects, worms, fungi, bacteria, and plant roots generally constitute the living organic matter, whereas the remains of such animal and plant life in various stages of decay make up the dead organic material. The residue from such decay (termed **humus**) is largely colloidal and assists in holding the water and plant nutrients.

The liquid part of the soil, the soil solution, is made up of water containing dissolved minerals in various quantities, as well as dissolved oxygen and carbon dioxide. Mineral elements, water, and possibly some carbon dioxide enter the plant from the soil solution.

The gaseous portion of the soil is important to good plant growth. In poorly drained, waterlogged soils, water replaces the air, thus depriving plant roots, as well as certain desirable aerobic microorganisms, of the oxygen necessary for their existence.

The **texture** of a soil depends upon the relative proportions of *sand* (0.05 to 2 mm particle diameter), *silt* (0.05 to 0.002 mm particle diameter), and *clay* (less than 0.002 mm particle diameter). The principal texture classes are sand, loamy sand, sandy loam, loam, silt loam, clay loam, and clay. A typical sandy loam might consist of 70 percent sand, 20 percent silt, and 10 percent clay; whereas a clay loam might have 35 percent sand, 35 percent silt, and 30 percent clay.

In contrast to soil *texture,* which refers to the proportions of individual soil particles, soil **structure** refers to the arrangement of those particles in the entire soil mass. These individual soil grains are held together in aggregates of various sizes and shapes. Maintenance of a favorable granular and crumb soil structure is very important. For example, working heavy clay soils when they are too wet can so change the soil structure that the heavy clods formed may remain for years.

### SAND

Sand consists of small rock grains, 0.05 to 2.0 mm in diameter, formed as the result of the weathering of various rocks, its mineral composition depending upon the type of rock. Quartz sand, consisting chiefly of a silica complex, is generally used for propagation purposes. The type used in plastering is the grade ordinarily the most satisfactory for rooting cuttings. Sand is the heaviest of all rooting media used, a cubic foot of dry sand weighing about 100 lb. It should preferably be fumigated or heat-treated before use, as it may contain weed seeds and various harmful pathogens. Sand contains virtually no mineral nutrients and has no buffering capacity. It is used mostly in combination with organic materials.

### PEAT

Peat consists of the remains of aquatic, marsh, bog, or swamp vegetation, which has been preserved under water in a partially decomposed state. The lack of oxygen in the bog slows bacterial and chemical decomposition of the plant material. Composition of different peat deposits varies widely, depending upon the vegetation from which it originated, state of decomposition, mineral content, and degree of acidity (*21, 64, 74, 78*).

There are three types of peat as classified by the U.S. Bureau of Mines: moss peat, reed sedge, and peat humus.

*Moss peat* (usually referred to in the market as "peat moss") is the least decomposed of the three types and is derived from sphagnum, hypnum, or other mosses. It varies in color from light tan to dark brown. It has a high moisture-holding capacity (15 times its dry weight), has a high acidity (pH of 3.2 to 4.5), and contains a small amount of nitrogen (about 1 percent) but little or no phosphorus or potassium. This type of peat generally comes from Canada or Europe, although some is produced in the northern United States.

*Reed sedge peat* consists of the remains of grasses, reeds, sedges, and other swamp plants. This type of peat varies considerably in composition and in color, ranging from reddish-brown to almost black. The pH ranges from about 4.0 to 7.5 and its water-holding capacity is about 10 times its own dry weight.

*Peat humus* is in such an advanced state of decomposition that the original plant remains cannot be identified; it can originate from either hypnum moss or reed sedge peat. It is dark brown to black in color with a low moisture-holding capacity but with 2.0 to 3.5 percent nitrogen.

When peat moss is to be used in mixtures, it should be broken apart and moistened before adding to the mixture. Continued addition of coarse organic materials such as peat moss or sphagnum moss to greenhouse soil mixtures can cause a decrease in wettability. Water will not penetrate easily, and many of the soil particles will remain dry even after watering. No good method for preventing this nonwettability is known, although the repeated use of commercial wetting agents may improve water penetration (*65*).

Peat, as used in propagation, is not a uniform product and can be a source of weed seed, insects, and disease inoculum. Peat should be pasteurized along with the other media components (*16, 25*).

SPHAGNUM MOSS

Commercial sphagnum moss is the dehydrated young residue or living portions of acid-bog plants in the genus *Sphagnum,* such as *S. papillosum, S. capillaceum,* and *S. palustre.* It is relatively sterile, light in weight, and has a very high water-holding capacity, being able to absorb 10 to 20 times its weight of water. The stem and leaf tissues of sphagnum moss consist largely of groups of water-holding cells. This material is generally shredded, either by hand or mechanically, before it is used in a propagating or growing medium. It contains such small amounts of minerals that plants grown in it for any length of time require added nutrients. Sphagnum moss has a pH of about 3.5 to 4.0. It contains a specific fungistatic substance, or substances, which accounts for its ability to inhibit damping-off of seedlings germinated in it (*26, 31*).

VERMICULITE

Vermiculite is a micaceous mineral that expands markedly when heated. Extensive deposits are found in Montana and North Carolina. Chemically, it is a hydrated magnesium-aluminum-iron silicate. When expanded, vermiculite is very light in weight, 90 to 150 kg per cubic meter (6 to 10 lb per cu ft), neutral in reaction with good buffering properties, and insoluble in water; it is able to ab-

·b large quantities of water—3 to 4 gal per cu ft. Vermiculite has a relatively h cation exchange capacity and thus can hold nutrients in reserve and later ase them. It contains enough magnesium and potassium to supply most ts.

In the crude vermiculite ore, the particles consist of a great many very thin, separate layers with microscopic quantities of water trapped between them. When run through furnaces at temperatures near 1090°C (2000°F), the water turns to steam, popping the layers apart, forming small, porous, spongelike kernels. Heating to this temperature provides complete sterilization. Horticultural vermiculite is graded to four sizes: No. 1 has particles from 5 to 8 mm in diameter; No. 2, the regular horticultural grade, from 2 to 3 mm; No. 3, from 1 to 2 mm; No. 4, which is most useful as a seed-germinating medium, from 0.75 to 1 mm. Expanded vermiculite should not be compacted when wet, as pressing destroys its desirable porous structure. Do not use nonhorticultural (construction grade) vermiculite as it is treated with chemicals toxic to plant tissues.

### PUMICE

Chemically pumice is mostly silicon dioxide and aluminum oxide, with small amounts of iron, calcium, magnesium, and sodium in the oxide form. It is mined in several regions in the western states of the United States, one source being in the Sierra Nevada mountains near Bishop, California. Pumice is screened to different size grades but is not heat-treated. It increases aeration and drainage in a rooting mix and can be used alone or mixed with peat moss (51).

### PERLITE

Perlite, a gray-white silicaceous material, is of volcanic origin, mined from lava flows. The crude ore is crushed and screened, then heated in furnaces to about 760°C (1400°F), at which temperature the small amount of moisture in the particles changes to steam, expanding the particles to small, sponge-like kernels that are very light, weighing only 5 to 8 lb per cu ft. The high processing temperature provides a sterile product. Usually a particle size of 1.6 to 3.0 mm (1/16 to 1/8 in.) in diameter is used in horticultural applications. Perlite holds three to four times its weight of water. It is essentially neutral with a pH of 6.0 to 8.0 but with no buffering capacity; unlike vermiculite, it has no cation exchange capacity and contains no mineral nutrients. It is most useful in increasing aeration in a mixture. Perlite, in combination with peat moss, is a very popular rooting medium for cuttings (23).

### SYNTHETIC PLASTIC AGGREGATES

These materials are used, especially in Europe, as substitutes for sand or perlite.

*Expanded polystyrene flakes* improve drainage and aeration and decrease bulk density. They are chemically neutral, do not absorb water, and do not decay.

*Urea-formaldehyde foam* consists of sponge-like particles that have a high water-holding capacity and a 30 percent nitrogen content, which is slowly released over a period of several years. This material should not be used in plant growing media until the odor of formaldehyde has disappeared.

### SHREDDED BARK, SAWDUST, AND WOOD SHAVINGS

Shredded bark, sawdust, and wood shavings from redwood, cedar, fir, pine, or various hardwood species can be used as a component in growing and propagating mixes, serving much the same purposes as peat moss (*21, 33, 69, 71, 80, 117*). Additional nitrogen may be needed in an amount sufficient for the decomposition requirements of the material, plus an additional amount for use by the plants (*118*). Rate of decomposition varies with the wood species. Because of their relatively low cost, light weight, and availability, these materials are widely used in soil mixes for container-grown plants, but supplementary nutrients must be added. Some types, when fresh, may contain materials toxic to plants, such as phenols, resins, terpenes, and tannins, so they require composting for 10 to 14 weeks before using (*19*).

#### COMPOST

Composting can be defined as the biological decomposition of bulk organic wastes under controlled conditions, which takes place in piles or bins. The process occurs in three steps: (1) an initial stage lasting a few days in which decomposition of easily degradable soluble materials occurs; (2) a second stage, lasting several months, during which high temperatures occur and cellulose compounds are broken down; (3) a final stabilization stage when decomposition decreases, temperatures lower, and microorganisms recolonize the material. Microorganisms include bacteria, fungi, and nematodes; larger organisms, such as millipedes, soil mites, beetles, springtails, earthworms, earwigs, slugs, sowbugs, and fruit flies, can often be found in compost piles in great numbers. Compost prepared largely from leaves may have a high soluble salt content, which will inhibit plant growth, but can be lowered by leaching with water before use (*84*).

In the home garden a compost mixture may be useful as a moisture-holding humus material. Mixed with soil, compost adds organic matter. To start a compost, leaves and garden refuse are accumulated and allowed to decompose, preferably in a 1.2 by 2.4 meter (4 by 6 ft) bin with slatted sides to give good aeration. Moisture should be added from time to time during the dry seasons; decomposition is hastened if some nitrogenous fertilizer is sprinkled through each batch of newly added material. The mass of material should be stirred once a week to ensure even decomposition. Several bins are preferable—one for newly started material, one for material undergoing decomposition, and one for completely decomposed compost, ready to use. Twelve to 24 months may be necessary for complete decomposition to humus. Since compost may contain weed seeds and nematodes, as well as noxious insects and pathogens, preferably it should be pasteurized (*2, 39, 49, 54, 84*).

## MIXTURES FOR CONTAINER GROWING

In propagation procedures, young seedlings or rooted cuttings (liners) are sometimes planted directly in the field, but frequently they are started in a blended mix in some type of container. Container growing of young seedlings and rooted cuttings is an important alternative for field growers. For this purpose special growing mixes are needed (*21, 33, 69, 71, 115*).

To provide uniform potting mixtures of better textures, sand and some organic matter, such as peat moss or sawdust or shredded bark, are usually added to a loam soil. In preparing these mixtures, the soil should be screened to make it uniform and to eliminate large particles. If the materials are very dry, they should be moistened slightly; this applies particularly to peat, which if mixed when dry, absorbs moisture very slowly. The soil should not be wet and sticky, however. In mixing, the various ingredients may be arranged in layers in a pile and turned with a shovel. A power-driven cement mixer, soil shredder, or skip-loader is used in large-scale operations.

Preparation of the soil mixture should preferably take place at least a day prior to use. During the ensuing 24 hours the moisture tends to become equalized throughout the mixture. The soil mixture should be just slightly moist at the time of use so that it does not crumble; on the other hand, it should not be sufficiently wet to form a ball when squeezed in the hand. Recommended potting mixtures using soil and mixed by volume are:

1. Heavy soils, such as clay loams or clay
   2 parts perlite or sand
   1 part soil
   2 parts peat moss (or composted shredded bark, sawdust, or leaf mold)
2. Medium soils, such as silt loams
   1 part perlite or sand
   1 part soil
   1 part peat moss (or composted shredded bark, sawdust, or leaf mold)
3. Light soils, such as sandy loams
   1 part peat moss (or composted shredded bark, sawdust, or leaf mold)
   1 part soil

For each bushel (35 liters) of the above mixes add:
224 g (8 oz) dolomitic limestone
280 g (10 oz) 20 percent superphosphate
(For plants requiring acid soils, substitute calcium sulfate for the limestone.)

After starting new plants by rooting cuttings or germinating seeds, commercial producers of nursery stock grow many of these plants to a salable size in containers, using growing media that may or may not contain soil. These mixes vary widely throughout the nursery industry, but generally include a fine sand mixed in varying proportions with such materials as peat moss, sawdust, or shredded fir, pine, or hardwood bark. Such mixes require fertilizer supplements and continued feeding of the plants until they become established in their permanent locations. For example, one successful mix for small seedlings, rooted cuttings, and bedding plants consists of one part each of shredded fir bark, peat moss, perlite, and sand. To this mixture is added gypsum, superphosphate, dolomite lime, and potash. Nitrogen is added subsequently in the irrigation water.

In summary, nurseries have generally been changing from loam-based growing media, as exemplified by the John Innes composts (3) developed in England in the 1930s, to mixes incorporating such materials as sand, peat, perlite, vermiculite, sawdust, and shredded bark in varying proportions. The trend away from loam-based mixes is due to a lack of suitable uniform soils, the added costs of having to pasteurize soil mixes, and the costs of handling and shipping the heavier soils compared to the lighter materials.

The U.C. potting mixtures were developed by plant pathologists and others at the University of California, Los Angeles, to provide growing media that could be readily prepared in large quantities for the commercial nurseryman as an integral part of a pathogen-free propagating and cultural program (*68*). Since the U.C. mixes are based upon materials that are uniform, generally available, and require no previous preparation, they can easily be duplicated. The basic components are:

1. An inert type of fine sand, and
2. Finely shredded peat moss—mixed with each other in varying proportions—plus
3. Fertilizer mixtures as described below.

The sand consists of round, wind-blown particles, uniform in size and relatively small (0.05 to 0.5 mm in diameter), thus having a rather high moisture-holding capacity. Such sand does not tend to compact even though the particles are small, because of their round shape and uniformity. The absence of colloidal clay particles in sand tends to prevent compaction or shrinkage. Disadvantages of the U.C. mixes are their heavy weight and poor aeration and drainage.

The chief purpose of the peat moss in this soil mixture is to increase its moisture and nutrient-holding capacities. In a mixture of equal parts of sand and peat moss, the maximum moisture-holding capacity is about 48 percent.

Basic fertilizer additives recommended (*68*) for a U.C. mix of 50 percent fine sand and 50 percent peat moss are as follows:

1. *If the mix is to be stored for an indefinite period before using.* This supplement furnishes a moderate supply of available nitrogen, but the plants will soon require further feeding. To each cubic yard (0.76 m$^3$) of the mix, add

| | |
|---|---|
| 112 g (4 oz) potassium nitrate | 3.38 kg (7½ lb) dolomite lime |
| 112 g (4 oz) potassium sulfate | 1.13 kg (2½ lb) calcium |
| 1.13 kg (2½ lb) single | carbonate lime |
| superphosphate | |

2. *If the mix is to be planted within one week of preparation.* This supplement furnishes available nitrogen as well as a moderate nitrogen reserve. To each cubic yard (0.76 m$^3$) of the mix, add

| | |
|---|---|
| 1.13 kg (2½ lb) hoof and horn or blood meal (13 percent nitrogen) | 1.13 kg (2½ lb) single superphosphate |
| 112 g (4 oz) potassium nitrate | 3.38 kg (7½ lb) dolomite lime |
| 112 g (4 oz) potassium sulfate | 1.13 kg (2½ lb) calcium carbonate lime |

The organic nitrogen is omitted if the mix is to be stored before using, since such organic forms break down during storage, releasing a high content of water soluble nitrogen, which may cause plant injury. Other forms of organic nitrogen, such as cottonseed meal (7 percent nitrogen) or fish tankage (6 to 10 percent nitrogen), may be substituted for the hoof and horn or blood meal, provided that a comparable amount of nitrogen is supplied.

In preparing the U.C. mix, the fine sand, shredded peat moss, and fertilizer must be mixed together thoroughly. The peat moss should be moistened before mixing. If the mixing is done well, the peat moss will not separate and float to the top when the mixture is saturated with water. This mixture, including the fertilizer, can be safely sterilized by steam or chemicals without the subsequent harmful effects to the plants that often occur when other soils or mixes are sterilized.

If large amounts of superphosphate are added to a growing mix, it may contain toxic levels of fluoride as an impurity. The fluoride can be "fixed" in an insoluble form, however, by the addition of lime (*87*).

### THE CORNELL PEAT-LITE MIXES

The Cornell Peat-Lite mixes are artificial "soils" used primarily for seed germination and for container-growing of bedding plants and annuals (*17, 87*). The components are lightweight, uniform, and readily available, and have chemical and physical characteristics suitable for the growth of plants. Excellent results have been obtained with these mixes, and many commercial ready-mixed preparations[4] are available (*94*). Mixes of this type have an important advantage of usually not requiring any decontamination before use. It may be desirable, however, to pasteurize the peat moss before use to eliminate any disease inoculum or other plant pests.

**Peat-Lite mix A:** to make 0.9 m³ (1 cubic yard)

0.39 m³ (11 bu) shredded German or Canadian sphagnum peat moss
0.39 m³ (11 bu) horticultural grade vermiculite (No. 2 or 4)
2.25 kg (5 lb) ground limestone (preferably dolomitic), finely ground
0.45 to 0.9 kg (1 to 2 lb) single superphosphate (20 percent), preferably powdered
0.45 kg (1 lb) calcium nitrate
84 g (3 oz) fritted trace elements (FTE 555)
56 g (2 oz) iron sequestrene (330)
84 g (3 oz) wetting agent

**Peat-Lite mix B** (same as A, except that horticultural perlite is substituted for the vermiculite)

**Peat-Lite mix C** (for germinating seeds)

0.35 m³ (1 bu) shredded German or Canadian sphagnum peat moss
0.35 m³ (1 bu) horticultural grade vermiculite No. 4 (fine)
42 g (1½ oz) (4 level tbsp) ammonium nitrate
42 g (1½ oz) (2 level tbsp) superphosphate (20 percent), powdered
210 g (7½ oz) (10 level tbsp) finely ground dolomitic limestone

The materials should be mixed thoroughly, with special attention to wetting the peat moss during mixing. Adding a nonionic wetting agent, such as Aqua-Gro (28 g [1 oz] per 6 gal of water) to the initial wetting usually aids in wetting the peat moss.

[4] Burpee Planting Formula, Burpee Seed Co., Burpee Bldg., Warminster, Pa. 18974; Pro-Mix, Premier Brands, Inc., 350 Madison Ave., New York, N.Y. 10017; Terra-Lite, W. R. Grace & Co., Cambridge, Mass. 02140.

Proprietary micronutrient materials, such as Esmigran, FTE 555, or Micromax, consisting of combinations of minor elements, are available for adding to growing media. Adding a slow release fertilizer such as Osmocote, Mag Amp, or Nutriform, to the basic Peat-Lite mix is useful if the plants are to be grown in it for an extended period of time.

## PREPLANTING TREATMENTS OF SOIL AND SOIL MIXES

Soils may contain weed seeds, nematodes, and various fungi and bacteria harmful to plant tissue. The so-called damping-off commonly encountered in seed-beds is caused by soil fungi, such as species of *Pythium, Phytophthora, Rhizoctonia,* and *Fusarium.* To avoid loss from these pathogens, it is desirable to treat the soil, or mixtures with soil or leaf mold in them, before using.

Soil can be heated or fumigated with chemicals to eliminate weeds, insects, nematodes, and disease organisms. Heating soil mixes high in leaf mold or compost hastens decomposition of the organic matter, especially if it is already partially rotted, and leads to the formation of toxic compounds necessitating leaching with water or a three- to six-week delay in planting. Undecomposed materials, like the brown types of peat moss, are relatively unaffected.

Certain of the complex chemical compounds in the soil are broken down by excessive heat—above 85°C (185°F)—increasing the amounts of soluble salts of nitrogen, manganese, phosphorus, potassium, and others. Some, particularly nitrogen in the form of ammonia, may be present during the first few weeks after steaming in such quantities as to be toxic to the plants, particularly in mixes high in organic nitrogen forms. Later the ammonia is converted to nitrate nitrogen, which reaches a peak in about six weeks. Added superphosphate in the soil mix is useful for tying up excessive manganese released during heating, particularly of acid soils—thus preventing injury from manganese toxicity.

### Heat Treatment

Although the term soil ''sterilization'' has been commonly used, a more accurate word is ''**pasteurization**,'' since the recommended heating processes do not kill all organisms ( *7, 8, 9, 10, 11* ).

Pasteurization of growing media with steam is generally preferable to fumigation with chemicals. After treatment with steam, the medium can be used much sooner. Steam is nonselective for pests whereas chemicals are highly selective. Steam is much less dangerous to use than fumigant chemicals, both to plants and to the operator. Chemicals do not vaporize well at low temperatures, but steam pasteurization can be used for cold, wet media.

The moist heat is advantageous; it can be injected directly into the soil in covered bins or benches from perforated pipes placed 15 to 20 cm (6 to 8 in.) below the surface. In heating the soil, which should be moist but not wet, a temperature of 82°C (180°F) for 30 minutes has been a standard recommendation, since this procedure kills most harmful bacteria and fungi as well as nematodes, insects, and most weed seeds, as indicated in Figure 2–12. However, a lower temperature, such as 60°C (140°F) for 30 minutes, is more desirable since it kills pathogens but leaves many beneficial organisms that prevent explosive growth of harmful organisms if recontamination occurs. The

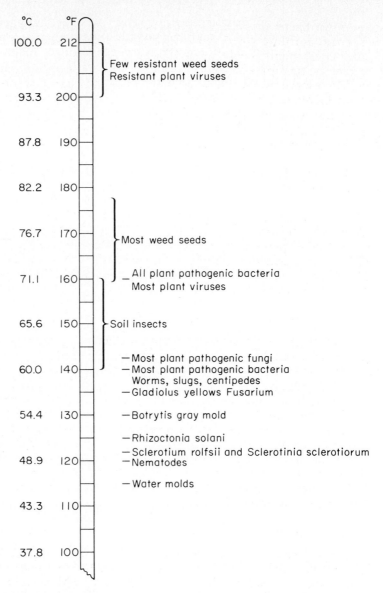

**Figure 2–12** Soil temperatures required to kill weed seeds, insects, and various plant pathogens. Temperatures given are for 30 min under moist conditions. From University of California Division of Agricultural Sciences, *Manual 23 (7)*.

lower temperature also tends to avoid toxicity problems, such as the release of excess ammonia and nitrite, as well as manganese injury (*15, 108, 112*), which is often encountered at higher steaming temperatures.

Air mixed with steam (**aerated steam**) in the ratio of 4.1 to 1 by volume gives a temperature of 60°C (140°F) (*11*). This temperature will not kill many weed seeds but, if held for 30 minutes, and if the soil is moist, it will kill most pathogenic bacteria and fungi. Designing equipment to mix air and steam in the proper proportions to give this temperature presents some difficulties, but successful equipment has been developed (1, 20, 43).

Electric heat pasteurizers are in use for amounts of soil up to 0.4 m³ (½ cu yd). Microwave ovens can be used effectively for small quantities of soil. They do not have the drying effect of oven heating and will kill insects, disease organisms, weed seed, and nematodes.

## Fumigation with Chemicals

Chemical fumigation kills organisms in the propagating mixes without disrupting their physical and chemical characteristics to the extent occurring with heat treatments (75, 99).[5] Ammonia production may increase following chemical fumigation, however, because of the removal of organisms antagonistic to the ammonifying bacteria. The mixes should be moist (between 40 percent and 80 percent of field capacity) and at temperatures of 18° to 24°C (65° to 75°F) for satisfactory results. Before using the mixture after chemical fumigation, allow a waiting period of two days to two weeks, depending upon the material, for dissipation of the fumes.

### FORMALDEHYDE

Formaldehyde is a good fungicide with strong penetrating powers. It kills some weed seeds, but is not reliable for killing nematodes or insects. A mixture of 3.8 liters (1 gal) of commercial formalin (37 percent strength) with 190 liters (50 gal) of water is applied to the soil at the rate of 21 to 42 liters per sq meter (2 to 4 qt per sq ft or 1 gal per bu of soil). The treated area should be covered immediately with an airtight material and left for 24 hours or more. Following this treatment, about two weeks should be allowed for drying and airing, but the soil should not be planted until all odor of formaldehyde has disappeared.

For small-scale treatments, commercial formalin can be applied at a rate of 2½ tbsp per bu of a light soil mixture or 1 tbsp per standard-size flat. Dilute with five to six parts of water, apply to soil, and mix thoroughly. Let stand 24 hours, plant seeds, and water thoroughly.

### CHLOROPICRIN (TEAR GAS)

Chloropicrin is a liquid ordinarily applied with an injector, which should put 2 to 4 ml into holes 3 to 6 in. deep, spaced 9 to 12 in. apart. It may also be applied at the rate of 5 ml per cu ft of soil. Chloropicrin changes to a gas that penetrates through the soil. The gas should be confined by sprinkling the soil surface with water and then covering it with an airtight material, which is then

[5] Recommendations on the pesticide labels must be followed to conform to the permitted usages.

Chloropicrin and methyl bromide are hazardous materials to use, especially in confined areas. They should be applied only by persons trained in their use, who must take the necessary precautions as stated in the instructions on the containers or in the accompanying literature.

left for three days. Seven to ten days are required for thorough aeration of the soil before it can be planted. Chloropicrin is effective against nematodes, insects, some weed seeds, verticillium, and most other resistant fungi. Chloropicrin fumes are very toxic to living plant tissue.

### METHYL BROMIDE

Methyl bromide is an odorless material very volatile and very toxic to humans. It should be mixed with other materials and applied only by those trained in its use. Most nematodes, insects, weed seeds, and some fungi are killed by methyl bromide, but it will not kill verticillium. It is often used by injecting the material from pressurized containers into an open vessel placed under a plastic sheet which covers the soil to be treated. The cover is sealed around the edges with soil, and should be kept in place for 48 hours. Penetration is very good, and its effect extends to a depth of about 30 cm (12 in.). For treating bulk soil, methyl bromide at 333 ml per cu meter (10 ml per cu ft) or 0.6 kg per cu meter (4 lb per 100 cu ft) can be used.

### METHYL BROMIDE AND CHLOROPICRIN MIXTURES

Proprietary materials are available containing both methyl bromide and chloropicrin. Such combinations are more effective than either material alone in controlling weeds, insects, nematodes, and soil-borne pathogens. Aeration for 10 to 14 days is required following applications of methyl bromide-chloropicrin mixtures.

### VAPAM®

Vapam® (sodium N-methyl dithiocarbamate dihydrate) is a water-soluble soil fumigant that kills weeds, germinating weed seeds, most soil fungi, and under the proper conditions, nematodes. It undergoes rapid decomposition to produce a very penetrating gas. Vapam® is applied by sprinkling on the soil surface, through irrigation systems, or with standard injection equipment. For seed-bed fumigation, 0.95 liter (1 qt) of the liquid formulation of Vapam® in 7.6 to 11.4 liters (2 to 3 gal) of water is used, sprinkled uniformly over 9 sq meters (100 sq ft) of area. After application, the Vapam® is sealed with additional water or with a roller. Three weeks after application the soil can be planted. Although Vapam® has a relatively low toxicity to humans, care should be taken to avoid inhaling fumes or splashing the solution on the skin.

### D-D MIXTURE

*D-D mixture (1,3-dichloropropene + 1,2-dichloropropane)* is a soil fumigant widely used for nematode control in nursery fields; 380 to 570 gal per hectare (40 to 60 gal per acre) is the usual dosage, and a one- to two-week aeration period is required before planting can begin.

### FUNGICIDAL SOIL DRENCHES

Fungicidal soil drenches can be applied to soil in which young plants are growing or are to be grown to inhibit growth of many soil-borne fungi. These materials may be applied either to the soil or to the plants. Preferably, a wetting agent should be added to the chemicals before application. It is very important in using such chemicals to read and follow the manufacturer's directions carefully, and to try the chemicals on a limited number of plants first before going to large-scale applications. Examples of these materials are:

*Diazoben* (*p*-dimethylaminobenzene diazo sodium sulfonate) will control the water molds, *Phytophthora* and *Pythium*.

*Benomyl* is a systemic fungicide that inhibits growth of such soil pathogens as *Rhizoctonia, Cylindrocladium, Fusarium,* and *Verticillium*. It is ineffective against the water molds, *Pythium* and *Phytophthora*. A 50 percent wettable powder is available for use on ornamentals.

*Captan* added to the rooting or potting medium at about 500 ppm is effective against *Pythium* and *Fusarium* but gives only slight inhibition of *Rhizoctonia*. Drenching cuttings being rooted under mist every two weeks with Captan at 236 g per 100 liters (2 lb per 100 gal) water helps control pathogens. Wash excess material from leaves after drenching with the fungicide.

*Truban*, incorporated into the root medium at about 50 ppm, gives good control of *Pythium* and *Phytophthora* and some control of *Fusarium* and *Rhizoctonia*.

## SANITATION IN PROPAGATION

During propagation procedures losses of young seedlings, rooted cuttings, and grafted nursery plants to various pathogens and insect pests can sometimes be devastating, especially under the warm, humid conditions found in greenhouses (*36, 61, 70, 76, 83*). It is preferable, by far, to operate in a pathogen-free and insect-free environment, rather than continually attempting to suppress such pests on infected plants and propagating and growing plants by the continual use of pesticides.

In recent years the importance of sanitation during propagation and growing procedures has become widely accepted and recognized as an essential part of nursery operations. Some large nurseries have detailed pest management programs with crews supervised by trained plant pathologists and entomologists (*22*). Such programs involve primarily the prevention of plant diseases and the avoidance of insects, mites, and weed problems. (Virus control is considered in chapter 8.)

Harmful pathogens and other pests are best eliminated by dealing with the three situations where they can enter and become a problem during propagation procedures:

**1.** The physical propagation facilities—propagating room, pots, flats, knives, shears, working surfaces, hoses, greenhouse benches, and the like.

**2.** The propagation media—rooting and growing mixes for cuttings and seedlings.

**3.** The plant material used for propagation—seeds, cutting material, scion and stock material for grafting.

If pathogens and other pests are eliminated in each of these areas it is likely that the young plants can be propagated and grown to a salable size with no disease, insect, or mite infestations. The pathogens most likely to cause disease development during propagation are species of *Pythium, Phytophthora, Fusarium, Cylindrocladium, Thielariopsis, Sclerotinia,* and *Rhizoctonia solani*. These are all soil-borne organisms that attack plant roots. They can best be controlled by pasteurization of the propagating and growing mixes, general hygiene of the plants and facilities, avoidance of overwatering, good drainage of excess water, and the use, if necessary, of the proper fungicides (*72, 102*).

## Physical Propagation Facilities

The space where the actual propagation (making cuttings, planting seeds, grafting, and so on) takes place should be a light, very clean, cool room, completely separated from areas where the soil mixing, pot and flat storage, growing, and other operations take place. Traffic and visitors in this room should be kept to a minimum. At the end of each working day all plant debris and soil should be cleaned out, the floors hosed down, and working surfaces washed with a formaldehyde solution (1 part formalin, 37 percent, to 50 parts water) or a sodium hypochlorite solution (Clorox, diluted 1 part to 9 parts water).

Flats and pots coming into this room should have been washed thoroughly and, if used previously, sterilized with steam or chemicals, for example, a 30-minute soak in sodium hypochlorite (Clorox) diluted 1 to 9. No dirty flats or pots should be allowed in the propagating area. Knives, shears, and other equipment used in propagation should be sterilized periodically during the propagation day by dipping in a disinfectant.

Mist propagating and growing areas in greenhouses, cold frames, and lathhouses should be kept clean and dead plant debris should be removed. The bench tops and frames should be painted annually with 2 percent copper naphthenate, thus providing self-disinfecting surfaces that help control algae, fungi, and bacteria. Water to be used for misting should be free of pathogens.

## Propagation Media

Since certain components of propagation mixes, particularly leaf mold, soil, sand, and peat moss (*16, 25*) can contain harmful pathogens, they should be pasteurized, preferably by aerated steam (*11, 43*), otherwise, by chemicals before being brought into the ''clean'' propagation area. The containers (bins, flats, pots) for such pasteurized mixes should, of course, have been treated to eliminate pathogens. Never put pasteurized mixes into dirty containers.

New materials such as vermiculite and perlite, which have been heat-treated during their manufacture, need not be pasteurized until reuse.

## Plant Material

In selecting propagating material, use only those source plants that are disease and insect-free. Some nurseries maintain stock plant blocks, which they keep meticulously ''clean.'' However, stock plants of particularly disease-prone plants, such as euonymus, might well be sprayed with a suitable fungicide several days before cuttings are taken.

It is best to select cutting material from the upper portion of stock plants rather than from near the ground where the material could possibly be covered with soil pathogens.

As cutting material is being collected it should be placed in new plastic bags (or in used ones that have been washed and rinsed in a weak [30 mg/1] chlorine solution).

After the cuttings have been made and before they are stuck in the flats they should be dipped in a weak (5 mg/1) chlorine solution, followed by a dip in a fungicide solution (such as Captan at about 3½ g/1).

# SUPPLEMENTARY FERTILIZERS

Even with a good soil mixture, complete with added mineral nutrients, continued growth of plants in containers necessitates the addition at intervals of supplementary minerals, especially nitrogen. Artificial, nonsoil mixes especially must have added fertilizers.

A satisfactory feeding program for growing container plants is to combine a slowly available dry fertilizer in the original soil mix with a liquid fertilizer applied at frequent intervals during the growing season or with controlled-release fertilizers added as top dressings at intervals, as needed (37).

Of the three major elements—nitrogen, phosphorus, and potassium—nitrogen has the most control on the amount of vegetative growth. Some organic fertilizers which supply nitrogen are blood meal or hoof and horn meal applied at about 1 heaping tsp for a plant growing in a 3.8-liter (1-gal) container. Twice this amount of cottonseed meal should be used.

To supply nitrogen, phosphorus, and potassium in dry form, 2 heaping tsp of the following mixture is recommended (68) for plants growing in 3.8-liter (1-gal) containers:

- 1.8  kg (4 lb) hoof and horn or blood meal
- 1.8  kg (4 lb) single superphosphate
- 0.45 kg (1 lb) potassium sulfate

As a supplement to this, a dilute inorganic nutrient feeding at weekly intervals throughout the growing season is desirable. A simple solution may be prepared by dissolving 1 tsp of potassium nitrate and 1 tsp of ammonium nitrate in 3.8 liters (1 gal) of water. Adding 2 tsp of a mixed fertilizer, such as 10-6-4, to 3.8 liters (1 gal) of water will also make a satisfactory solution. A complete nutrient solution for fertilizing container-grown plants is:

| | |
|---|---|
| water | 380 liters (100 gal) |
| ammonium nitrate (or urea) | 168 grams (6 oz) |
| monoammonium phosphate | 168 grams (6 oz) |
| potassium nitrate | 168 grams (6 oz) |

For large-scale operations, it is more practical to prepare a liquid concentrate and inject it into the regular watering or irrigating system by the use of a proportioner (18)—"fertigation." For this, a nutrient concentrate formula such as the following can be used but, since soluble forms of phosphorus are expensive, this chemical may be added as superphosphate in the soil mix and only potassium nitrate and ammonium nitrate used in the liquid concentrate. Indicator dyes are sometimes added to serve as a visual guide to the fertilizer flow. The following nutrient concentrate formula is used in large-scale operations:

| | |
|---|---|
| water | 38   liters (10 gal) |
| ammonium nitrate | 6.75 kg (15 lb) |
| monoammonium phosphate[6] | 1.8  kg (4 lb) |
| potassium chloride | 2.7  kg (6 lb) |

[6] Phosphoric acid is sometimes used instead since it aids in controlling the pH of the irrigation water. A final pH of about 6.7 is satisfactory.

The above chemicals should be thoroughly dissolved. *This concentrate should then be diluted—1 part to 200 parts water—before being applied to the plants.* Some nurseries use a combination of top-dressing with dry fertilizers, plus liquid-feeding through the irrigation system. Large nurseries usually have periodic analyses made of their soil mixes and irrigation water to insure that the proper nutrient levels are being maintained. Satisfactory nutrient levels in the irrigation water are:

| | |
|---|---|
| nitrogen | 100–200 ppm |
| phosphorus | 20 ppm |
| potassium | 100 ppm |

### Controlled-Release Fertilizers

Controlled-release fertilizers provide nutrients to the plants gradually over a long period and reduce the possibility of injury from excessive applications (*55, 66, 67, 107, 113*). They are expensive, however, and are used chiefly on high-value container-grown plants.

There are three types of slow-release fertilizers: (1) coated water-soluble pellets or granules; (2) inorganic materials that are slowly soluble; (3) organic materials of low solubility that gradually decompose by biological breakdown or by chemical hydrolysis.

Examples of the coated type are Osmocote and sulfur-coated urea. The Osmocote pellets contain soluble fertilizers enclosed in membranes consisting of resin and various polymers. After a period of time—three or four months—the fertilizer has completely diffused out of the pellets. Sulfur-coated urea granules consist of urea coated with a sulfur-wax mixture, consisting of about 82 percent urea, 13 percent sulfur, and 2 percent wax.

An example of the slowly soluble, inorganic type is MagAmp ( magnesium ammonium phosphate), an inorganic material of low water solubility. Added to the soil mix, it supplies nutrients slowly for up to two years. Similarly, potassium glass frit is a relatively soft glass that slowly supplies potassium at adequate levels for up to 18 months.

An example of the organic, slow-release type is urea-formaldhyde (UF), which will supply nitrogen slowly over a long period of time (*56*). Another organic slow-release fertilizer is isobutylidene diurea (IBDU), which is a condensation product between urea and isobutylaldehyde, having 31 percent nitrogen.

The diffusion and/or breakdown mechanisms for all these products are not well understood.

## SALINITY IN SOIL MIXES

Excessive salts in the propagating or growing mixes or in irrigation water (over 2 milli-mho per cm)[7] can reduce plant growth, burn the foliage, or even kill

[7] Salinity levels in water extracts of the soil (saturation-extract method) can be measured by electrical conductivity using a "Solubridge." Such readings are expressed as milli-mhos per centimeter. Readings of less than 2 (1400 ppm) indicate no salinity problem; readings much over 4 indicate a level at which most plants are likely to be affected. At readings over 8, only salt-tolerant plants will grow.

the plants (*29, 30, 32, 55, 90*). The required fertilization programs also contribute to salt accumulations. Overfertilization causes rapid and severe salinity symptoms, starting with foliage wilting and tip and marginal leaf burning. These symptoms may be accompanied by a white surface accumulation of salts on the medium. To prevent salt buildup, the containers or flats should be leached with water periodically. If the irrigation water contains 250 ppm of salts, the leaching should be done every 12 weeks; for 500 ppm, every six weeks; and for 1000 ppm every three weeks. In addition, fertilizers should not be used that tend to contribute to excess salinity; for example, use potassium nitrate rather than potassium chloride. There is considerable variability among the various plant species in their tolerance to salts. Some such as *Araucaria, Bougainvillea,* and *Callistemon,* have very high tolerance, while others such as *Rosa, Mahonia,* and *Photinia* show extreme sensitivity (*32, 93*).

## WATER QUALITY

Water quality is an important factor in rooting cuttings, germinating seeds, and growing-on the young plants (*29, 53*).

For good results the available water supply should not contain total soluble salts in excess of 1400 ppm (approximately 2 milli-mhos per cm)—ocean water averages about 35,000 ppm. The salts are combinations of such cations as sodium, calcium, and magnesium with such anions as sulfate, chloride, and bicarbonate. Water containing a high proportion of sodium to calcium and magnesium can adversely affect the physical properties and water-absorption rates of soils and should not be used for irrigation purposes.

Although not itself detrimental to plant tissue, so-called "hard" water contains relatively high amounts of calcium and magnesium (as bicarbonates and sulfates) and can be a problem in mist-propagating units or in evaporative water cooling systems as deposits build up wherever evaporation occurs. When water is much over 6 grains per gal (100 ppm) in hardness it is often run through a water softener for household use. Some types of equipment are based upon the replacement of the calcium and magnesium in the water by sodium ions. Such "soft," high-sodium water is toxic to plant tissue and should never be used for watering plants.

A better, but more costly, method of improving water quality is the **deionization** process. Here calcium, magnesium, and sodium are removed by substituting hydrogen ions for them. Water passes over an absorptive medium charged with hydrogen ions, which absorbs calcium and other ions in exchange for hydrogen. For further deionization the water is passed through a second filter charged with hydroxyl (OH) ions, which replace carbonates, sulfates, and chlorides. The proper nutritive ions may then be added back in suitable amounts (*12*).

Boron salts are not removed by deionization units and if present in water in excess of 1 ppm can cause plant injury. There is no satisfactory method for removing excess boron from water. The best solution is to acquire another water source.

Another method for improving water quality is **reverse osmosis,** a process in which pressure is applied to a solvent to force it through a semipermeable membrane from a more concentrated solution to a less concentrated solution (*12*).

Municipal treatment of water supplies with chlorine (0.1 to 0.6 ppm) is not sufficiently high to cause plant injury. However, the addition of fluoride to water supplies at 1 ppm can cause leaf damage to a few tropical foliage plants (*47*).

When the water source is a pond, well, lake, or river, contamination by weed seeds, mosses, or algae can be a problem. Chemical contamination from drainage into the water source from herbicides applied to adjoining fields or from excess fertilizers on crop fields can also be damaging to nursery plants (*100*).

Nurseries using such water for their plants should treat the water before use. A good procedure is to:

**1.** Use strainers to remove large debris.

**2.** Run the water through sand filters with automatic back flushing. This removes coarse particles and weed seed.

**3.** Add chlorine to suppress algae and disease pathogens at about ½ ppm (0.5 mg/1). This water can then be used for field watering of container nursery plants after soluble fertilizers have been injected into the system.

**4.** If the water has a high salt content, it can be improved by running it through a deionization or reverse osmosis unit.

## SOIL pH

Soil reaction (or pH) is a measure of the concentration of hydrogen ions in the soil. Although not directly influencing plant growth, it has a number of indirect effects, such as the availability of various nutrients and the activity of beneficial microbial activity. A pH range of 5.5 to 7.0 is best for growth of most plants (7.0 is neutral—below this level is acid and above is alkaline). To lower pH of an alkaline soil, use ammonium sulfate fertilizer; to raise the pH of acid soils, use calcium nitrate.

## CARBON DIOXIDE ($CO_2$) ENRICHMENT IN THE GREENHOUSE

Carbon dioxide is one of the required ingredients for the basic photosynthetic process that accounts for the dry-weight materials produced by the plant (*35, 57, 58, 59, 116*).

$$6CO_2 + 12H_2O \xrightarrow[\text{green plant cell}]{\text{light energy}} C_6H_{12}O_6 + 6O_2 + 6H_2O$$

Carbon dioxide exists normally in the atmosphere at about 300 ppm (0.03 percent). Sometimes the concentration in winter in closed greenhouses may drop to 200 ppm, or lower, during the sunlight hours, owing to its use by the plants. Under conditions where $CO_2$ is limiting the photosynthetic rate (at adequate light intensities and relatively high temperatures—about 29.5°C [85°F] an increase in the $CO_2$ concentration, to between 1000 and 2400 ppm, can be expected to result in an increase in photosynthesis, as much as 200 percent over the rate found at 300 ppm. To take full advantage of this potential increase in dry-weight production, plant spacing must be adequate to prevent shading of over-crowded leaves. When supplementary $CO_2$ is used during periods of sunny weather, the temperature in the greenhouse should be kept relatively high. Added $CO_2$ would be of little benefit whenever the light intensity drops to very low levels. Adding $CO_2$ at night is of no value. Good circulation of air in the greenhouse will prevent undesirable lowering of the $CO_2$ level just at the leaf surface. A tightly closed greenhouse is necessary to be able to increase the ambient $CO_2$.

# CONTAINERS
# FOR PROPAGATING
# AND GROWING
# YOUNG PLANTS

## Flats

Flats are shallow wood, plastic, or metal trays, with drainage holes in the bottom. They are useful for germinating seeds or rooting cuttings, since they permit young plants to be moved around easily when necessary. Durable kinds of wood, such as cypress, cedar, or redwood, are preferable for flats. Galvanized-iron and plastic flats are available in various sizes, but zinc released from the galvanized flats can cause toxicity to plants. Both of the latter types will nest, and thus require relatively little storage space.

## Clay Pots

The familiar red clay flower pots, long used for growing young plants, are heavy and porous and lose moisture readily. They are easily broken, and their round shape is not economical of space. After continued use, toxic salt accumulations build up, requiring soaking in water before reuse. Clay pots are rarely used today in large-scale, commercial propagation.

## Plastic Pots

Plastic pots, round and square, have numerous advantages; they are non-porous, reusable, lightweight, and use little storage space since they will "nest." Some types are fragile, however, and require careful handling, although other types, made from polyethylene, are flexible and quite sturdy. Square pots are also made up into "packs" of eight or 12 for easier handling. Plastic pots (and flats) cannot be steam sterilized, but some of the more common plant pathogens can be eliminated by a hot water dip, 70°C (158°F), for three minutes without damage to the plastic.

Figure 2–13 shows a folding type of plastic container widely used in reforestation procedures.

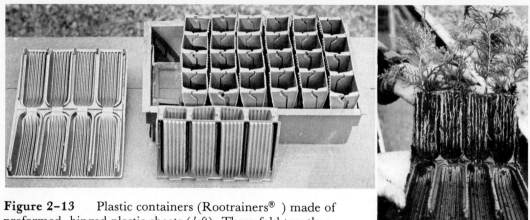

**Figure 2-13**　Plastic containers (Rootrainers® ) made of preformed, hinged plastic sheets (*left*). These fold together and lock to form a set of four containers that fits into a special plastic tray. The vertical grooves on the sides of the containers reduce the likelihood of undesirable root spiralling. The containers can be opened to permit inspection of the roots or removal of the plants. This type of container is widely used in reforestation for propagating and growing seedlings.

## Fiber Pots

Containers of various sizes, round or square, are pressed into shape from peat plus wood fiber, with fertilizer added. They are dry and will keep indefinitely. Since these pots are biodegradable, they are set in the soil along with the plants. Peat pots find their best use where plants are to be held for a relatively short time and then put in a larger container or in the field. Small peat pots with plants growing in them eventually deteriorate because of constant moisture, and may fall apart when moved. On the other hand, unless the pots are kept moist, roots will fail to penetrate the walls of the pot and will grow into an undesirable spiral pattern. Units of six or twelve square peat pots fastened together are available. When large numbers of plants are involved, time and labor are saved in handling by the use of these units.

Larger pots made of molded asphalt-impregnated fiber are available for growing nursery stock; these can be used for direct planting as they also deteriorate in the soil.

## Peat or Fiber Blocks

Blocks of solid material, sometimes with a prepunched hole (Figure 2-14) have become popular as a germinating medium for seeds or as a rooting medium for cuttings, especially for such plants as chrysanthemums and poinsettias. Fertilizers are usually incorporated into the material. One type (Figure 2-15) is made of highly compressed peat and, when water is added, swells to its usable size and is soft enough for the cutting or seed to be inserted. Such blocks become a part of the plant unit and are set in the soil along with the plant. These blocks replace not only the pot but the propagating mix also.

Synthetic rooting blocks are becoming more widely used in the nursery in-

**Figure 2-14**    Solid blocks of compressed fibers with a prepunched hole are sometimes used for cuttings of easily rooted plants. After rooting, the block plus the rooted cutting is planted into a growing container or the field nursery.

**Figure 2-15**    Use of solid block rooting medium. *Left:* compressed sphagnum peat discs, encased in a plastic netting and containing some added mineral nutrients. *Right:* adding water causes peat to swell to size shown. Chrysanthemum cuttings inserted into full size pellets rooted rapidly and are ready to be planted in soil medium.

dustry, being well adapted to automation. Other advantages are their light weight, reproducibility, and sterile condition. Watering must be carefully controlled to provide constant moisture, while maintaining adequate aeration.

## Metal Containers

Hundreds of thousands of nursery plants are grown and marketed each year in 3.8-liter (1-gal) and—to a lesser extent—in 11-liter (3-gal) and 19-liter (5-gal) cans. Such cans are tapered for nesting and the drainage holes are punched. Some containers are salvaged from canneries, restaurants, and bakeries by suppliers that make a business of this. These are enameled to prevent rusting. Machine planters have been developed utilizing such containers, in which rooted cuttings or seedlings can be planted as rapidly as 10,000 or more a day.

Plants are easily removed from tapered containers by inverting and tapping. Untapered metal containers must be cut down each side with can shears or tin snips to permit removal of the plant.

In areas having high summer temperatures, use of light-colored (white or aluminum) containers may improve root growth by avoiding heat damage to the roots, often encountered with dark-colored containers that absorb considerable heat if exposed to the sun. In addition, soil temperatures in metal cans tend to be higher than in plastic containers.

### Polyethylene Bags

Polyethylene bags are widely used in Europe, Australia, and the tropics, but to a much lesser extent in the United States, for growing-on rooted cuttings or seedlings to a salable size. They are considerably less expensive than rigid metal or plastic containers and seem to be satisfactory (see Figure 2-16) but some types deteriorate rapidly. They are usually black, but one kind is black on the inside and light-colored on the outside. The lighter color reflects heat and lowers the root temperature (*106*). After planting, however, they cannot be stacked as easily as the rigid containers for truck transportation.

## HANDLING CONTAINER-GROWN PLANTS

Watering of container-grown nursery stock is a major expense. Hand watering of individual cans with a large-volume, low-pressure applicator on a hose several times a week is expensive, and is used only for small-scale operations. In large operations, overhead sprinklers are usually used, although much runoff waste occurs. Watering of container plants by trickle or drip irrigation (Figure 2-17), also widely used, results in less waste (*103*). Setting the containers on damp mats or beds of fine sand over plastic with water moving upward into the soil by capillarity is another method of supplying water to plants (*4, 81*). This system, however, may spread pathogens such as *Phytophthora* from container to container.

**Figure 2-17** Automatic watering system for container-grown plants. A small ''spaghetti'' tube carries water from the plastic feeder tubes to each plant. Automatic control of the main lines can water hundreds of thousands of plants on a given schedule.

Fertilizer solutions are usually injected into the irrigation system in large commercial nurseries. After the container stock leaves the wholesale nursery the retailer should maintain the stock by adequate irrigation and fertilization until the plants have been purchased by the ultimate grower (*34*).

In areas with severe winters attention must be given to the problems of winter injury (*45, 85, 91*). The amount of injury varies with the species. The chances of cold injury are lessened if the plants are well established in the containers before the onset of winter. In addition, setting the plants close together in large groups (jamming or bunching) tends to prevent damage from rapid temperature fluctuation. Wrapping heavy brown paper around the outer rows of cans is helpful. In cold winter regions, however, some additional form of winter protection must be used, for example, placing mulch covering, such as straw or hay, over the tops of the containers or placing the containers inside a protective structure, such as a cold frame or plastic-covered greenhouse. The most reliable type of winter protection, as shown in Figure 2–18, is a temporary frame constructed over the plants and covered with opaque polyethylene sheeting, 4 to 6 mils in thickness (*48*). Often two layers of polyethylene, separated by a 2.5 to 5 cm (1 to 2 in.) air space, are used. Soaking the soil in the structure with water before the plants are covered enhances the protection. Condensation of moisture inside the cover as the temperature drops reduces radiant heat loss.

**Figure 2–18** Winter protection of broad- and narrow-leaved evergreen nursery stock in severe winter areas. Plants are placed close together in beds and covered in late fall with 6 mil white polyethylene on pipe framing. *Below, right:* a basin made from polyethylene is filled with water, which releases heat upon freezing and absorbs heat upon thawing, acting as a buffer against sudden temperature changes. Courtesy The Conard-Pyle Co., West Grove, Pa.

**Figure 2-19** One disadvantage of growing trees in containers is the possibility of producing poorly shaped root systems. Here a defective, twisted root system resulted from holding the young nursery tree too long in a container before transplanting. Such spiraling roots retain this shape after planting and are unable to anchor the tree firmly in the ground.

Overhead sprinkling is also an effective method of protecting tender plants from subfreezing temperatures. Water must be present continually, changing to ice as long as the subfreezing temperatures occur. When liquid water changes to ice approximately 144 Btu of heat is released per pound of water (*46, 92*).

In most woody plant species the roots do not develop as much winter hardiness as the tops, hence low-temperature damage of container stock occurs primarily to the root system. Winterhardiness of roots varies with the species (*44, 95*).

As shown in Figure 2-19, plants kept in containers too long will form an undesirable constricted root system from which they may never recover when planted in their permanent location (*41*). The plants should be shifted to larger containers before such "root spiralling" occurs. Judicious root pruning, early transplanting, and careful potting during the early transplanting stages can do much to develop a good root system by the time the young plant is ready for setting in its permanent location (*42*). Some types of plastic containers have vertical grooves along the sides which tend to prevent horizontal spiralling of the roots (Figure 2-13).

Often when a container plant, grown in one of the synthetic, lightweight mixes in the nursery is transplanted into the home garden—into a much heavier loam or clay soil—a problem is encountered in maintaining sufficient moisture to the plant. If the root ball is covered with the heavier native soil, and water is then applied to the plant, possibly in a shallow basin at ground level, the water tends to stay in the heavier soil with its smaller, more absorptive, pores, while the coarser-textured mix containing the roots remains completely dry. To avoid this, the soil mix in the root ball should remain exposed at the top so that applied water must pass through it and so wet the roots.

## REFERENCES

*1.* Aldrich, R. A., and P. E. Nelson. 1969. Equipment for aerated steam treatment of small quantities of soil and soil mixes. *Plant Dis. Rpt.* 53(10):784–88.

*2.* Ammon, G. L. 1978. Composting and use of hardwood bark media for container growing. *Proc. Inter. Plant Prop. Soc.* 28:368–70.

3. Alvey, N. G. 1961. Soil for John Innes composts. *Jour. Hort. Sci.* 36:228–40.

4. Auger, E., C. Zafonte, and J. J. McGuire. 1977. Capillary irrigation of container plants. *Proc. Inter. Plant Prop. Soc.* 27:467–73.

5. Augsburger, N. D., H. R. Bohanon, and J. L. Calhoun. 1978. *The greenhouse climate control handbook.* Muskogee, Okla.: Acme Eng. & Mfg. Co.

6. Baird, C. D., and W. E. Waters. 1979. Solar energy and greenhouse heating. *HortScience* 14(2):147–51.

7. Baker, K. F., and C. N. Roistacher. 1957. Heat treatment of soil. Section 8 in *Calif. Agr. Exp. Sta. Man. 23.*

8. ———. 1957. Principles of heat treatment of soil. Section 9 in *Calif. Agr. Exp. Sta. Man. 23.*

9. ———. 1957. Equipment for heat treatment of soil. Section 10 in *Calif. Agr. Exp. Sta. Man. 23.*

10. ———. 1962. Principles of heat treatment of soil and planting material. *Jour. Australian Inst. Agr. Sci.* 28(2):118–26.

11. ———. 1976. Aerated steam treatment of nursery soils. *Proc. Inter. Plant Prop. Soc.* 26:52–59.

12. Barnstead Co. 1971. *The Barnstead basic book on water.* Boston: Barnstead Co.

13. Bauerle, W. L., and T. H. Short. 1977. Conserving heat in glass greenhouses with surface-mounted air-inflated plastic. *Ohio Agr. Res. and Develop. Center Spec. Cir. 101.*

14. ———. 1978. Greenhouse energy conservation and effects on plant response. *Ohio Rpt. on Res. and Develop.* 63(5):74–76.

15. Birch, P. D. W., and D. J. Eagle. 1969. Toxicity to seedlings of nitrite in sterilized composts. *Jour. Hort. Sci.* 44:321–30.

16. Bluhm, W. L. 1978. Peat, pests, and propagation. *Proc. Inter. Plant Prop. Soc.* 28:66–70.

17. Boodley, J. W., and R. Sheldrake, Jr. 1964. Cornell "Peat-Lite" mixes for container growing. Dept. Flor. and Orn. Hort., Cornell Univ. Mimeo. Rpt.

18. Boodley, J. W., C. F. Gortzig, R. W. Langhans, and J. W. Layer. 1966. Fertilizer proportioners for floriculture and nursery crop production management. *Cornell Ext. Bul. 1175.*

19. Branson, R. L., J. P. Martin, and W. A. Dost. 1977. Decomposition rate of various organic materials in soil. *Proc. Inter. Plant Prop. Soc.* 27:94–96.

20. Brazelton, R. W. 1968. Sterilizing soil mixes with aerated steam. *Agric. Eng.* 49(7):400–401.

21. Bunt, A. C. 1976. *Modern potting composts.* London: G. Allen & Unwin.

22. Conner, D. 1977. Propagation at Monrovia Nursery Company: Sanitation. *Proc. Inter. Plant Prop. Soc.* 27:102–6.

23. Cooke, C. D., and B. L. Dunsby. 1978. Perlite for propagation. *Proc. Inter. Plant Prop. Soc.* 28:224–28.

24. Cotter, D. J., and J. N. Walker. 1966. Climate-humidity relationships in plastic greenhouses. *Proc. Amer. Soc. Hort. Sci.* 89:584–93.

25. Coyier, D. L. 1978. Pathogens associated with peat moss used for propagation. *Proc. Inter. Plant Prop. Soc.* 28:70–72.

26. Creech, J. L., R. F. Dowdle, and W. O. Hawley. 1955. Sphagnum moss for plant propagation. *USDA Farmers' Bul. 2058.*

27. De Lance, P. 1976. Electric soil and hot house heating. *Proc. Inter. Plant Prop. Soc.* 26:369–72.

28. Duncan, G. A., and J. N. Walker. 1980. How to save energy in the greenhouse. *Amer. Nurs. CLII* (10):13, 90–112.

29. Fireman, M., and H. E. Hayward. 1955. Irrigation water and saline and alkali soils. *USDA yearbook of agriculture—water,* pp. 321–27.

30. Fireman, M., and R. L. Branson. 1963. Salinity in greenhouse soils. *Calif. Agr. Ext. Ser. OSA 68* (rev.).

31. Fleming, G., and C. E. Hess. 1965. The isolation of a damping-off inhibitor from sphagnum moss. *Proc. Inter. Plant Prop. Soc.* 14:153–54.

32. Francois, L. E. 1980. Salt injury to ornamental shrubs and ground covers. *USDA SEA Home and Garden Bul. No. 231.*

33. Furuta, T. 1970. Soil mixtures. Section 6 in *Nursery management handbook,* Univ. Calif. Agr. Ext. Ser.

34. ———. 1971. Fertilizer and irrigation for plants on retail display. *Univ. Calif. Agr. Ext. Ser. AXT—361.*

35. Gaastra, P. 1963. Climatic control of photosynthesis and respiration. In *Environmental control of plant growth,* L. T. Evans, ed. New York: Academic Press.

36. Geard, I. D. 1979. Fungal diseases in plant propagation. *Proc. Inter. Plant Prop. Soc.* 29:589–94.

37. Gilliam, C. H., and E. M. Smith. 1980. How and when to fertilize container nursery stock. *Amer. Nurs. CLI* (2)7, 117–27.

38. Goldsberry, K. L. 1979. Greenhouse heat conservation and the effect of wind on heat loss. *HortScience* 14(2):152–55.

39. Goluek, C. G. 1972. *Composting: A study of the process and its principles.* Emmaus, Pa.: Rodale Press.

40. Hanan, J., W. D. Holley, and K. L. Goldsberry. 1978. Greenhouse construction. Chapter 3 in *Greenhouse management.* New York: Springer-Verlag.

41. Harris, R. W., D. Long, and W. B. Davis. 1967. Root problems in nursery liner production. *Calif. Agr. Ext. Ser. AXT-244.*

42. Harris, R. W., W. B. Davis, N. W. Stice, and D. Long. 1971. Effects of root pruning and time of transplanting in nursery liner production. *Calif. Agr.* 25(12):8–10.

43. Hartmann, H. T., and J. E. Whisler. 1979. Mobile aerated steam soil pasteurizer unit. *Proc. Inter. Plant Prop. Soc.* 29:36–41.

44. Havis, J. T. 1974. Tolerance of plant roots in winter storage. *Amer. Nurs.* 139(1):10.

45. Havis, J. T., and R. D. Fitzgerald. 1976. Winter storage of nursery plants. *Mass. Coop. Ext. Ser. Publ. No. 125.*

46. Hendershott, C. H. 1979. Cold protection of low growing plants. *Proc. Inter. Plant Prop. Soc.* 29:533–36.

47. Henley, R. W., R. T. Poole, and C. A. Conover. 1976. Injury to selected plants due to fluoride toxicity. *Proc. Inter. Plant Prop. Soc.* 26:185–89.

48. Hawley, M. A., and D. E. Hamilton. 1980. How to build a poly house for overwintering nursery stock. *Amer. Nurs. CLII* (8):7–9, 111–15.

49. Hoitink, H. A. J., and H. A. Poole. 1980. Factors affecting quality of composts for utilization in container media. *HortScience* 15(2):171–73.

50. Huang, K. T., and J. J. Hanan. 1976. Theoretical analysis of internal and external covers for greenhouse heat conservation. *HortScience* 11(6):582–83.

51. Inose, K. 1971. Pumice as a rooting medium. *Proc. Inter. Plant Prop. Soc.* 21:82–83.

52. Jensen, M. H. 1977. Energy alternative and conservation for greenhouses. *HortScience* 12(1):14–24.

53. Johnson, C. R. 1977. Some water quality problems faced by horticulturists. *Proc. Inter. Plant Prop. Soc.* 27:202–6.

54. Johnson, C. E. 1980. The wild world of composts. *National Geographic* 158(2):273–84.

55. Kelley, J. D. 1960. Effects of over-fertilization on container-grown plants. *Proc. Plant Prop. Soc.* 10:58–63.

56. ———. 1962. Response of container-grown woody ornamentals to fertilization with urea-formaldehyde and potassium frit. *Proc. Amer. Soc. Hort. Sci.* 81:544–51.

57. Kohl, H. C. 1966. Carbon dioxide fertilization. *Proc. Inter. Plant Prop. Soc.* 15:300–306.

58. Krizek, D. T. 1970. Controlled atmospheres for plant growth. *Trans. Amer. Soc. Agr. Eng.* 13(3):237–68.

59. Krizek, D. T., W. A. Bailey, H. H. Klueter, and H. M. Cathey. 1968. Controlled environments for seedling production. *Proc. Inter. Plant Prop. Soc.* 18:273–80.

60. LaCroix, L. J., D. T. Canvin, and J. Walker. 1966. An evaluation of three fluorescent lamps as sources of light for plant growth. *Proc. Amer. Soc. Hort. Sci.* 89:714–22.

61. Lambe, R. C., and W. H. Wills. 1979. The major diseases of holly in the nursery. *Proc. Inter. Plant Prop. Soc.* 29:536–44.

62. Lechner, A., D. Sprague, E. Schaufler, B. Shalucha, R. Langhans, and P. Hammer. 1972. The Cornell automated plant grower. *Cornell Agr. Ext. Ser. Bul. 40.*

63. Likums, A. 1980. Greenhouse-residence: A new place in the sun. *Agr. Res.* 29(1):4–7.

64. Lucas, R. E., P. E. Riecke, and R. S. Farnham. 1971. Peats for soil improvement and soil mixes. *Mich. Coop. Ext. Ser. Bul. No. E–516.*

65. Lunt, O. R., R. H. Sciaroni, and W. Enomoto. 1963. Organic matter and wettability for greenhouse soils. *Calif. Agr.* 17(4):6.

66. Lunt, O. R., A. M. Kofranek, and S. B. Clark. 1964. Nutrient availability in soil: Availability of minerals from magnesium-ammonium-phosphates. *Agr. and Food Chem.* 12:497–504.

67. Lunt, O. R. 1965. Controlled availability fertilizers. *Farm Tech.* 21(4):11, 26.

68. Matkin, O. A., and P. A. Chandler. 1957. The U.C. type soil mixes. Section 5 in *Calif. Agr. Exp. Sta. Man. 23.*

69. Mastalerz, J. W. 1977. Growing media. Chapter 6 in *The greenhouse environment.* New York: John Wiley.

70. McCain, A. H. 1977. Sanitation in plant propagation. *Proc. Inter. Plant Prop. Soc.* 27:91–93.

71. Matkin, O. A. 1971. Soil mixes today. *Proc. Inter. Plant Prop. Soc.* 21:162–63.

72. McCully, A. J., and M. B. Thomas. 1977. Soil-borne diseases and their role in propagation. *Proc. Inter. Plant Prop. Soc.* 27:339–50.

73. Mears, D. R., ed. 1979. *Proc. 4th Ann. Conf. on Solar Energy for Heating of Greenhouses.* New Brunswick, N.J.: Dept. Biol. and Agr. Eng., Rutgers Univ.

74. Miller, N. 1981. Bogs, bales, and BTU's: A primer on peat. *Horticulture* 49(4):38–45.

75. Munnecke, D. E. 1957. Chemical treatment of nursery soils. Section 11 in *Calif. Agr. Exp. Sta. Man. 23.*

76. Ormrod, D. J. 1975. Fungicides and their spectra. *Proc. Inter. Plant Prop. Soc.* 25:112–15.

77. Parsons, R. A. 1971. Small plastic greenhouses. *Univ. Calif. Agr. Ext. Ser. AXT-328.*

78. Patek, J. M. 1965. Peat moss. *Amer. Hort. Mag.* 44:132–41.

79. Peterson, H. 1955. Low voltage heating. *N. Y. State Flower Growers' Bul. 115.*

80. Pokorny, F. A. 1979. Pine bark container media—an overview. *Proc. Inter. Plant Prop. Soc.* 29:484–95.

81. Richards, M. 1978. Capillary watering of container plants. *Proc. Inter. Plant Prop. Soc.* 28:411–13.

82. Richards, S. J., J. E. Warneke, and F. K. Aljibury. 1964. Physical properties of soil mixes used by nurseries. *Calif. Agr.* 18(5):12–13.

83. Rumbal, J. M. 1977. Aspects of propagation hygiene. *Proc. Inter. Plant Prop. Soc.* 27:323–24.

84. Sawhney, B. L. 1976. Leaf compost for container-grown plants. *HortScience* 11(1):34–35.

85. Self, R. 1977. Winter protection of nursery plants. *Proc. Inter. Plant Prop. Soc.* 77:303–7.

86. ———. 1978. Pine bark in potting mixes: Grades and age, disease and fertility problems. *Proc. Inter. Plant Prop. Soc.* 28:363–68.

87. Sheldrake, R., G. E. Doss, L. E. St. John, Jr., and D. J. Lisk. 1978. Lime and charcoal amendments reduce fluoride absorption by plants cultured in a peat-perlite medium. *Jour. Amer. Soc. Hort. Sci.* 103(2):268–70.

88. Sherwood, G. 1981. The many faces of the modern greenhouse. *Horticulture* 59(11):43–48.

89. Short, T. H., and Bauerle, W. L. 1980. Greenhouse production with lower fuel costs. In *USDA 1980 yearbook of agriculture,* J. Hayes, ed. Washington, D.C.: U.S. Govt. Printing Office.

90. Schoonover, W. R., and R. H. Sciaroni. 1957. The salinity problem in nurseries. Section 4 in *Calif. Agr. Exp. Sta. Man. 23.*

91. Smith, E. M., ed. 1977. *Proc. Woody Ornamentals Winter Storage Symposium.* Coop. Ext., Ohio State Univ.

92. Steavenson, H. 1976. Using water to provide cold weather protection and conserve energy. *Amer. Nurs. CXLIII* (8):10, 65–67.

93. Skimina, C. A. 1980. Salt tolerance of ornamentals. *Proc. Inter. Plant Prop. Soc.* 30:113–18.

94. Stone, G. 1981. Hold the dirt. *Horticulture* 59(11):17–22.

95. Studer, E. J., P. L. Steponkus, G. L. Good, and S. C. Wiest. 1978. Root hardiness of container-grown ornamentals. *HortScience* 11(1):34–35.

96. Trickett, E. S., and J. D. S. Goulden. 1958. The radiation transmission and heat conserving properties of some plastic films. *Jour. Agr. Eng. Res.* 3(4):281–87.

97. U.S. Dept. of Agriculture. 1965. Hotbed and propagating frames. *USDA Misc. Publ. 986.*

98.  U.S. Dept. of Energy. 1978. Energy audit for growers: A self-inspection guide to reduce energy costs. Alexandria, Va.: Soc. Amer. Florists.

99.  Vaartaja, O. 1964. Chemical treatment of seedbeds to control nursery diseases. *Bot. Rev.* 30:1–91.

100.  Vance, B. F. 1975. Water quality and plant growth. *Proc. Inter. Plant Prop. Soc.* 25:136–41.

101.  Walker, J. N., and D. C. Stack. 1970. Properties of greenhouse covering materials. *Trans. Agr. Soc. Amer. Eng.* 13(5):682–84.

102.  Ward, J. 1980. An approach to the control of *Phytophthora cinnamomi*. *Proc. Inter. Plant Prop. Soc.* 30:230–37.

103.  Weatherspoon, D. M., and C. C. Harrell. 1980. Evaluation of drip irrigation for container production of woody landscape plants. *HortScience* 15(4):488–89.

104.  Weiler, T. C. 1977. Survival strategies of Northern Europe's greenhouse industry. *HortScience* 12(1):30–32.

105.  Whitcomb, C. E. 1978. A self-contained solar-heated greenhouse. *HortScience* 13(1):30–32.

106.  ———. 1979. Growing plants in poly bags. *Amer. Nurs., CXLIX* (12):10–11, 97–98.

107.  White, D. P., and B. G. Ellis. 1965. Nature and action of slow release fertilizers as nutrient sources for forest tree seedlings. *Mich. Quart. Bul.* 47(4):606–14.

108.  White, J. W. 1971. Interaction of nitrogenous fertilizers and steam on soil chemicals and carnation growth. *Jour. Amer. Soc. Hort. Sci.* 96(2):134–37.

109.  ———. 1979. Energy efficient growing structures for controlled environment agriculture. In *Horticultural reviews,* Vol. 1, J. Janick, ed. Westport, Conn: AVI Publ. Co.

110.  White, J. B., and R. A. Aldrich. 1980. *Greenhouse energy conservation.* University Park, Pa.: Pennsylvania State Univ.

111.  White, R. A. J. 1978. Some comparisons between plastic and glass-covered greenhouses. *Proc. Inter. Plant Prop. Soc.* 28:273–79.

112.  Wiebe, J. 1958. Phytotoxicity as a result of heat treatment of soil. *Proc. Amer. Soc. Hort. Sci.* 72:331–38.

113.  Williams, D. J. 1980. How slow-release fertilizers work. *Amer. Nurs., CLI* (6):90–97.

114.  Williams, T. J. 1978. *How to build and use greenhouses.* San Francisco: Ortho Books, Chevron Chemical Co.

115.  Wilson, G. C. S. ed. 1980. Symposium on substrates in horticulture other than soils in situ. *Acta Hort.* 99:1–248.

116.  Wittwer, S. H., and W. Robb. 1964. Carbon dioxide enrichment of greenhouse atmospheres for food crop production. *Econ. Bot.* 18:34–56.

117.  Worrall, R. 1976. The use of sawdust in potting mixes. *Proc. Inter. Plant Prop. Soc.* 26:379–81.

118.  Yates, N. L., and M. N. Rogers. 1981. Effects of time, temperature, and nitrogen source on the composting of hardwood bark for use as a plant growing medium. *Jour. Amer. Soc. Hort. Sci.* 106(5):589–93.

# SUPPLEMENTARY READING

ALDHOUS, J. R. 1972. *Nursery practice.* Forestry Commission Bul. 43. London: Her Majesty's Stationery Office.

AUGSBURGER, N. D., H. R. BOHANON, and J. L. CALHOUN. 1978. *The greenhouse climate control handbook.* Muskogee, Okla.: Acme Eng. and Mfg. Co.

BAKER, K. F., ed. 1957. The U.C. system for producing healthy container-grown plants. *Calif. Agr. Exp. Sta. Man. 23.*

BALL, V., ed. 1975. *The Ball red book* (13th ed.). West Chicago, Ill.: Geo. J. Ball, Inc.

BOODLEY, J. W. 1981. *The commercial greenhouse handbook.* New York: Van Nostrand Reinhold.

BUNT, A. C. 1976. *Modern potting composts.* London: G. Allen & Unwin.

CATHEY, H. M., and L. E. CAMPBELL. 1980. Light and lighting systems for horticultural plants. *Hort. Rev.* 2:491–537.

CORRELL, P. G., and J. G. PEPPER, eds. 1977. *Energy conservation in greenhouses.* Newark, Del.: Univ. of Delaware.

DAVIDSON, H., and R. MECKLENBURG. 1981. *Nursery management: Administration and culture.* Englewood Cliffs, N.J.: Prentice-Hall.

EDMONDS, J. 1980. *Container plant manual.* London: Grower Books.

FURUTA, T. 1970. *Nursery management handbook.* Berkeley: Univ. Calif. Agr. Ext. Ser.

HANAN, J. J., W. D. HOLLEY, and K. L. GOLDSBERRY. 1978. *Greenhouse management.* New York: Springer-Verlag.

LAMB, J. G. D., J. C. KELLY, and P. BOWBRICK. 1975. *Nursery stock manual.* London: Grower Books.

LANGHANS, R. W. 1980. *Greenhouse management.* Ithaca, N.Y.: Halcyon Press.

LAURIE, A., D. C. KIPLINGER, and K. S. NELSON. 1979. *Commercial flower forcing* (8th ed.). New York: McGraw-Hill.

MASTALERZ, J. W., ed. 1976. *Bedding plants* (2nd ed.). Univ. Park, Pa.: Pennsylvania Flower Growers.

———. 1977. *The greenhouse environment.* New York: John Wiley.

McGUIRE, J. J. 1972. Growing ornamental plants in containers: A handbook for the nurseryman. *Univ. of Rhode Island Coop. Ext. Ser. Bul. 197.*

NELSON, P. V. 1978. *Greenhouse operation and management.* Reston, Va.: Reston.

TINUS, P. W., and S. E. McDONALD. 1979. *How to grow tree seedlings in containers in greenhouses.* Gen. Tech. Rpt. RM-60. Ft. Collins, Colo.: Rocky Mt. For. and Range Exp. Sta., For. Ser., USDA.

Propagation by seeds is the major method by which plants reproduce in nature and one of the most efficient and widely used propagation methods for cultivated crops. The plants produced are referred to as **seedlings,** a term that in horticulture may be applied throughout the life of that plant although in botany the term is applied more specifically to the immediate period following seed germination.

The planting of the seed is the physical beginning of seedling propagation. The seed itself, however, is the end product of a process of growth and development within the parent plant. This developmental process begins with the fusion of male and female gametes to form a single cell (the **zygote**) within the ovule of the flower. The zygote has the property of **totipotency,** that is, it has all of the genetic information needed to reproduce the full-sized plant and to initiate the seedling cycle of the next generation. However, embryos may also develop in some plants by **asexual reproduction** in the process called **apomixis,** described later in this chapter. Furthermore, **somatic** or **vegetative embryos** can be produced in vitro in tissue and suspension cultures. In each of these cases, the seedling plants, whether originating sexually or asexually, have a characteristic life cycle that contrasts to the life cycle of a vegetatively propagated plant.

### SEEDLING LIFE CYCLE

The life cycle of a seedling plant is shown in chapter 1 as involving, first, the vegetative growth (Figure 1–5) of the plant from various growing points or regions (apical growing points and cambium) and, second, the formation of flowers and production of a seed (Figures 1–1 and 1–2).

The pattern of growth and development by which these consecutive processes occur, and the ability to grow from new vegetative shoots annually, determine the length of the life cycle and the use of seedling plants in propagation. Variations in these cycles are adaptations to natural seasonal or climatic cycles.

# 3

# The Development
# of Seeds
# and Spores

**Annual** plants go through the entire sequence, from germination of seeds to flowering and production and dissemination of seeds, in one growing season and then they die.

**Biennial** plants have a two-year life cycle. These plants are vegetative and grow as a low clump or rosette of leaves the first season. During the second season the plants are said to "bolt," producing flower stalks, flowers, and seeds; then the plants die. The transition from vegetative to reproductive stage is often a response to some environmental trigger such as chilling **(vernalization)** or a certain length of day **(photoperiod).**

**Perennial** plants live for more than two years and repeat the vegetative-reproductive cycle annually. Consecutive growth and dormancy cycles are related either to warm-cold or to wet-dry weather changes.

**Herbaceous perennials** are those in which the shoots die during the winter or in dry periods. The plants survive such dormant periods as specialized underground structures, such as bulbs, rhizomes, or crowns (see chapters 14 and 15). **Woody perennial** plants continue to increase in size each year by growth of the shoot and root tips or by lateral cambium growth, or both. Only a portion of the shoot growing points become reproductive while the others remain vegetative to continue shoot growth of the plant.

PHASES OF THE SEEDLING CYCLE

Figure 3–1 shows further aspects of the life cycle of seedling plants and introduces the concept of **phase change** (7).

**Figure 3–1**   Model illustrating phase changes that occur in a seedling life cycle. The upper part of the cycle illustrates the continuous development of a meristem (growing point) in an annual or biennial plant leading to the formation of flowers and seeds and completion of the life cycle. The lower part illustrates the production of vegetative shoots in a perennial plant in which a regular pattern of growth and development occurs to maintain consecutive vegetative→reproductive cycles continuously (30).

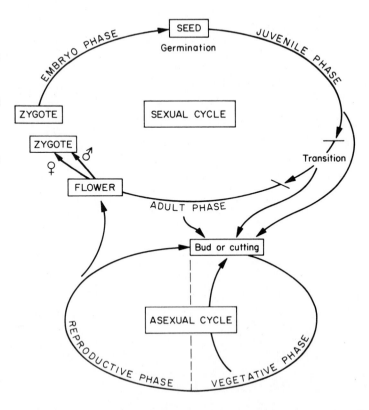

The cycle begins with the formation of a **zygote** in a flower. This single-celled structure undergoes growth and development within the ovule and ovary to produce an **embryo** enclosed within the **seed** and fruit. During this period the embryo is parasitic on the mother plant and undergoes characteristic stages of development.

With the germination of the seed, the embryo develops into a seedling plant to begin the **juvenile** phase. Vegetative growth predominates as the seedling plant enlarges in size through elongation in stems and roots and increases in cross-sectional area.

Juvenile plants continue to be vegetative for a period of time and are unable to respond to flower-inducing stimuli. Eventually the **transitional** phase begins in which the plant loses its characteristic juvenility (Figure 3–2). In the succeeding **adult** or **mature** phase reproduction by seed predominates. In herbaceous annuals and biennials the plant reaches its ultimate size, flowers, and then dies. Once the perennial (herbaceous or woody) plant reaches the adult phase, it regularly develops flowers from then on and continues to regenerate itself by production of vegetative shoots.

The shift from juvenile to adult (or mature) has been called **phase change** (*7*) or **maturation** (*27*). Such a transition is present in all plants but in some (annuals, for example) it is morphologically indistinct and may, therefore, not be readily recognized; in others, particularly woody and herbaceous perennials, the juvenile phase may last for many years and in some cases may be morphologically distinct.

**Figure 3–2**    Phase changes are illustrated in English ivy (*Hedera helix*) in which the **juvenile** phase is a vine that, as it grows into a vertical form, can undergo a **transition** into the **mature** form in which flowers and fruits are produced.

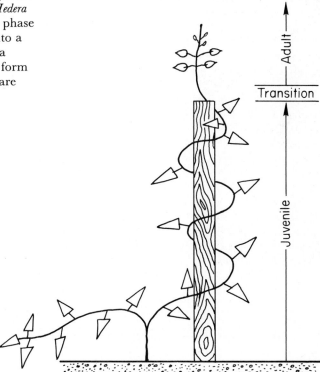

# PRODUCTION OF
# THE FLOWER

The production of a flower involves two distinct, contrasting morphological stages, **vegetative** and **reproductive.** When the plant, or a plant part such as a branch, is vegetative, a shoot elongates by apical and lateral buds, producing a series of nodes and internodes. Apical and lateral vegetative buds develop in patterns characteristic of the individual species and cultivar. In the second stage various vegetative growing points differentiate to become reproductive and develop into flowers. The initiation of flowering marks the end of the vegetative stage of a particular growing point. The flowering stage begins with **flower induction,** which is an internal physiological change in the individual growing point that precedes any morphological change. Only some of the growing points of the stem are induced and subsequently develop into flowers; others may remain vegetative.

In biennials and in some woody perennials the development of flowers takes place in the elongating stem in the growth period following an exposure to chilling (**vernalization**). In other kinds of plants, particularly many deciduous woody perennials, differentiation begins during the warm summer growing period, but the flower buds must be subjected to a subsequent regime of chilling temperatures which alters the natural hormone levels, thus allowing normal progression of flower development and blooming. Many spring flowering woody perennial plants of the temperate zone behave in this manner.

Control of flowering in herbaceous and woody perennials is complicated by differences between the responses given by the seedling plant (a) in the juvenile phase of its life cycle and (b) in the mature phase after it attains reproductive maturity. This difference is illustrated in the upper cycle in Figure 3–1 contrasted to the lower cycle.

The juvenile phase is extensive in some woody and herbaceous perennials and the length of time before seed crops are produced is very long. Plants of some bamboo species remain vegetative for decades, then suddenly produce flowers and seeds and die. The so-called century plant (*Agave americana*) follows a similar pattern. Trees of many forest species do not produce seed until they are 15 to 20 years old (*46*). Some fruit tree species, such as apple or pear, often require five to seven years before they flower. Seedlings of many orchid and bulb species do not start blooming until they are five to seven years old.

In order to begin the transition phase, plants of most species must grow to a certain size or age to attain the ability to flower. In some species, maturation is more precisely described in terms of *numbers of nodes* produced rather than chronological age. The physiological and genetic basis of the particular maturation phase resides in the cells of the individual growing point. The control is complex, however, and hormonal influences from leaves and roots also may be involved (*27*).

Early induction of maturation and flowering requires that the plant grow rapidly through the required juvenile phase. Therefore, exposure of the plant to optimum growing conditions to maximize vegetative growth will bring on the mature stage more quickly, particularly among woody perennials. As an example, in controlled experiments apple seedlings have been caused to flower after only 16 months by keeping the plants growing continuously in a greenhouse, compared to three to eight years for similar seedlings grown out-of-doors. Similar results have been reported for pear, crabapple, birch, and spruce seedlings (57).

Once past the juvenile phase, the mature plant begins the flowering process by responding to various flower induction stimuli. These may be environmental signals, such as particular photoperiods (either long or short days), or particular temperature regimes (cool or warm). Conditions that slow or reduce vegetative growth, such as girdling or decreasing nitrogen, may cause flower initiation, whereas excessively vigorous growth depresses it. These effects are opposite to those which favor transition to mature phase from the juvenile phase. Because conditions required for flower induction vary with individual species, one must be familiar with their growth cycles to be able to induce early flowering (27).

Many woody perennial plants have a biennial cycle in which shoots are vegetative one season and reproductive the following. Annual cropping is produced because other shoots on the same plant alternate in their cycle. Erratic seed production often seen in forest trees may result from a variety of adverse conditions that inhibit flower induction (46). These conditions include: (a) competition from an excessively large seed crop the preceding season, (b) stress produced by inadequate nutrition or water, or excessively low or high temperatures, or (c) defoliation from insects or disease. However, if the plant is growing satisfactorily, a check in vegetative growth at an appropriate stage by girdling or root pruning often induces flowering.

# FORMATION OF THE FRUIT, SEED, AND EMBRYO

The miniature flower parts subsequently develop into complete flowers. The sexual cycle includes development of the male (pollen) and female (embryo sac) structures of the flower, as shown for angiospermous flowers in Figure 1-1. In this part of the cycle reduction division of chromosomes occurs to produce the haploid (N) chromosome number. Figure 3-3 shows the basic structure of the flower and Figure 3-4 shows more details of the pollen, ovary, and ovule structure. Pollen grains and an eight-celled embryo sac contain haploid (1N) male and female gametes, respectively. During flowering, pollen is transferred from the anther to the stigma **(pollination),** where it germinates. A pollen tube grows down the style into the ovary until it reaches the embryo sac within the ovule (Figure 3-4). Two male gametes from the pollen tube are discharged into the embryo sac—one to unite with a female gamete **(fertilization)** to produce the **zygote** and another to unite with the two polar nuclei to produce the **endosperm.**

The zygote is diploid (2N) and divides to become the embryo; the endosperm is triploid (3N) and develops into a nutritive tissue for the developing embryo. Both structures are enclosed within the **nucellus** (inside the ovule) which functions initially as a nurse tissue for the developing embryo and endosperm. In some species the nucellus develops into a food storage tissue in the seed.

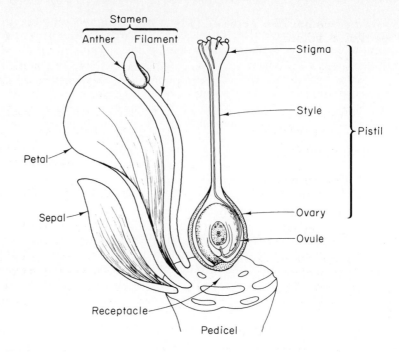

Figure 3-3    Basic flower structure of an angiospermous plant.

Figure 3-4    *Left:* pollen grain germinating on stigma; notice the two male haploid generative nuclei. *Right:* an ovary showing a single ovule containing a mature embryo sac and its eight haploid nuclei. It is surrounded by the diploid nucellus that is enclosed within the two integuments, all of which are genetically the same as the mother plant.

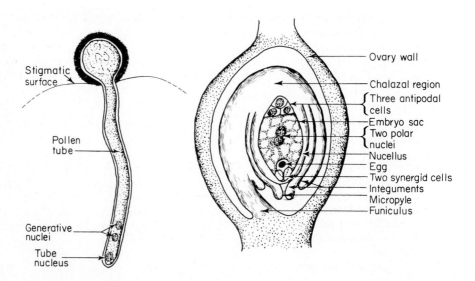

The relationships between flower structure and the parts of the fruit and seed are as follows:

Ovary································· > fruit (sometimes composed of more than one ovary plus additional tissues)

Ovule································· > seed (sometimes coalesces with fruit)
integuments····················· > testa (seed coats)
nucellus ························· > perisperm (usually absent or reduced; sometimes storage tissue)

2 polar nuclei +
sperm nucleus··················· > endosperm (triploid–3N)
egg nucleus + sperm
nucleus······ > zygote······ > embryo (diploid–2N)

Embryo development in the gymnosperms is somewhat different (Figure 1–2). The ovule is not produced in an enclosed ovary but is exposed within the **ovulate cone.** The developing embryo is nourished by the haploid (1N) female gametophyte, which develops into the storage organ of the mature seed ( *47* ).

# FRUIT, SEED, AND EMBRYO DEVELOPMENT

Fruit and seed development involves five separate processes:

1. Morphological development of fruit, seed, and embryo
2. Acquisition of the ability of the embryo to germinate
3. Accumulation of stored foods
4. Development of internal germination controls ( **primary dormancy** )
5. Methods of fruit and seed dispersal.

To produce a viable seed, both pollination and fertilization must take place. In some cases, however, the fruit may mature and contain only shriveled and empty seed coats with no embryo or with one that is thin and shrunken. Such "seedlessness" may result from several causes: (a) **parthenocarpy** (the development of the fruit without pollination or fertilization); (b) **embryo abortion** (the death of the embryo during its development); or (c) the inability of the embryo to accumulate the required food reserves. If embryo abortion occurs early, it is most likely that the fruit will soon drop or will not grow to its normal size ( *38* ).

## Morphological Development

Figures 3–5 and 3–6 show growth patterns of different parts of the fruit and seed. The basic sequence of development applies to all species but variations establish unique patterns that are characteristic of particular plant families. These morphological variations have major effects on the physiology of seed germination.

Three developmental stages are illustrated by the lettuce seed in Figure 3–5:

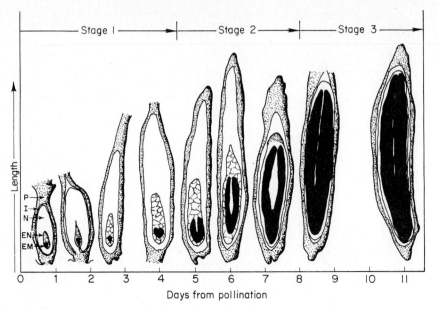

**Figure 3–5**    Growth of the fruit (an indehiscent achene) and seed of lettuce. P—pericarp, I—integuments, N—nucellus, EN—endosperm, EM—embryo. Redrawn from Jones (*29*).

Stage 1 involves the initial increase in size of the ovary (fruit) and the ovule (seed) during the first four days. The endosperm develops a cellular consistency but remains small and encloses the very small, more or less globular embryo (**proembryo** stage). Toward the end of the four-day period, the endosperm has enlarged somewhat and the embryo becomes heart-shaped as its cotyledons begin to enlarge.

At Stage 2 growth of the ovary and ovule ceases. Rapid growth and enlargement in size of the endosperm is followed by the growth of the embryo within the ovule (five to eight days in Figure 3–5).

The endosperm functions in a nutritional capacity for the embryo, although no vascular connections exist between the two structures (*8,55*). Growth of the embryo is preceded by growth of the endosperm, which digests the inner nucellar tissue as it grows. The endosperm is, in turn, digested by the developing embryo. Both nucellus and endosperm are reduced to a remnant in the mature seed of lettuce and in many other species. In plants of other families these structures are not consumed in embryo development but remain to function as storage tissue.

Failure of the endosperm to develop properly results in retardation or arrest of embryo development, and embryo abortion can result. This phenomenon commonly occurs when two genetically different individuals are hybridized, either from different species (*13, 45, 55*) or from two individuals of different polyploid constitution. It therefore can be a barrier to hybridization in angiosperms but not in gymnosperms (*47*), since the "endosperm" in these plants is the haploid female gametophyte.

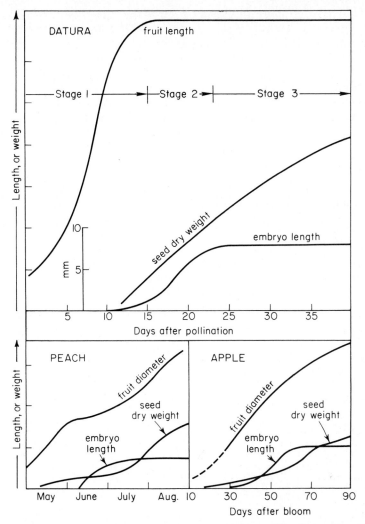

**Figure 3-6**  Comparative development of three different types of fruits: *Datura* (dry capsule), peach (fleshy drupe), and apple (fleshy pome). Note that changes in seed dry weight and embryo length are essentially the same for all species shown. *Datura* data from Rietsema et al. (*42*); apple data in part from Luckwill (*33*).

The stages of the lettuce seed illustrate the basic processes of embryo and seed development. However, variations from this pattern occur and are characteristic of specific plant families. For example, variations in pericarp development patterns (Figure 3-6) lead to different fruit structures. *Datura* is a dry capsule and has a pattern like lettuce (*42*). In the apple (a pome fruit) the fruit tissue is fleshy and continues to enlarge through Stages 2 and 3. In the peach, the outer portion of the pericarp, the fleshy **mesocarp,** enlarges in Stage 3 starting sometimes in Stage 2. The inner part of the pericarp, the **endocarp,** begins to harden in Stage 2 and becomes the "**pit**" or "**stone**" surrounding the seed.

Embryo sizes and structures in the mature seed differ among various plant families. These characteristics are not only important for seed identification (*35*) but are also significant for understanding physiological behavior during germination. Differences result from the extent to which the embryo develops in the seed. In some seeds embryo development is arrested at the end of Stage 1 and produces **rudimentary** embryos. In other families the embryo develops further in size, does not digest the endosperm or perisperm tissues, and is relatively small in the mature seed. These differences are the basis of morphological seed classification. The differences are due to genetic factors controlling seed development.

Various environmental agents can prevent normal embryo development, even though the fruit itself continues to develop. Numerous kinds of insects (*6*) attack the developing seed and fruit, particularly in forest trees. The developing seed of carrot and other Umbelliferous plants is attacked by *Lygus* bugs, which can penetrate the fruit and feed on the embryo. Adverse weather, such as frosts during early fruit development, sometimes kills the embryo, but the fruit itself continues to develop (*23, 48*). Growth tensions, causing the hardening endocarp to split or crack, can result in abortion of stone fruit seeds (*16*). This condition is associated with excessive vigor, reduced crop density, root restriction, or tree injuries resulting in girdling.

## Acquisition of the Ability to Germinate

Morphological and physiological development of the embryo is called **embryogenesis** (*4, 38*). This process either takes place in Stages 1 and 2 or is completed in species with rudimentary or linear embryos after the seed is separated from the plant. During seed development, cell division and expansion occur in various parts of the embryo structure. In seed germination, cell division occurs at the ends of the polarized embryo axis, that is, the root and shoot tips.

If the embryo is excised at various times during its development and placed under proper aseptic culture conditions, embryogenesis ceases and the immature embryo shows **precocious germination.** Any seedling growth that is produced tends to be abnormal, however (*28, 38*). The more the embryo has developed before excision the greater the embryo's capacity to germinate normally.

Control of embryogenesis is complex. Both hormonal controls and innate potentiality of the embryo cells to divide and differentiate in a particular way seem to be involved. There is evidence that control of both embryogenesis and germination potential is mediated through specific kinds of **mRNA** (messenger ribonucleic acid) that direct the hormonal balance within the seed (*19, 20, 21*). Such specific mRNA molecules appear to be carried over into the mature seed and may function to initiate the first stages of germination.

Embryo culture as a propagation technique is described in chapter 16.

## Accumulation of Food Reserves

Some increase in dry weight in the seed can be measured in the early part of the fruit development period because of the increase in size. After the seed has attained its full size, further increase in dry weight is a measure of the accumulation of storage materials into the seed. This takes place largely in Stage 2 near the end of the fruit development process. These reserve materials originate as carbohydrates produced by photosynthesis in the leaves and are translocated to the fruits and seeds, where they are converted to complex storage products—carbohydrates, fats, and proteins (53).

This accumulative process must take place properly and reach a minimum level of completeness if high-quality seeds are to be produced (15). Such seeds should be plump and heavy for their size. Since the initial growth of the seedling depends upon these reserve materials, heavier seeds should result in better germination and produce more vigorous seedlings. If conditions interfere with this storing process so that few reserve materials accumulate, the seeds will be thin, shrunken, and light in weight. The more severe this condition, the less the seeds can survive storage periods, the poorer the germination, and the weaker the seedlings that are produced (17, 40).

Interference with proper food storage accumulation is mainly due to adverse growing conditions, such as poor nutrition, moisture stress, disease or insect damage, and excessively low or high temperatures. The chief cause of the lack of reserve foods is immaturity of seed at harvest (2, 17, 22).

### Development of Internal Germination Controls

Once the embryo has attained the ability to germinate, it is essential for the survival of the species that seed germination take place at a time and place favorable for the growth and survival of the seedling plant. Mechanisms must be present to prevent germination while the seed is on the plant. The two principal means by which this is prevented are *control of moisture content* and *imposition of dormancy.*

*Vivipary* The phenomenon in which seeds germinate in the fruit while still attached to the plant is known as **vivipary** and the germinating embryos are called **viviparous embryos.** This characteristic is controlled both genetically and environmentally. For example, germinating seeds are occasionally found in mature citrus fruits. In mangrove (*Rhizophora mangle*), a swamp-growing tree species, vivipary is an adaptation to its environment. Embryos germinate directly on the tree to produce seedlings with a long javelin-shaped root. These eventually fall and become embedded in the mud below (49).

For most plant species, however, vivipary is undesirable. Premature seed sprouting may occur in some grain crops in periods of wet weather during or just before harvest (24, 41).

The tendency toward vivipary is a varietal characteristic in grain crops (56); such a tendency is automatically selected against as a defective character (18).

*Moisture content* Control of moisture content of the seed is an important aspect of seed handling. There are three general patterns of development in relation to moisture levels.

1. In most species both seeds and fruit become dehydrated naturally during ripening and dissemination. Moisture content drops to 30 percent or less on the plant and the seed is then dried further during harvest and prior to storage. Germination cannot occur at this level.

2. In other species, seeds that dry below a given level (about 30 to 50 percent) rapidly lose their ability to germinate. These plants include species whose fruits ripen early in summer with seeds germinating immediately (some maples, poplar,

elm), species whose seeds mature in autumn and remain in moist soil over winter (oak), and species from warm, humid tropics (citrus).

**3.** In other species, seeds become dry but are enclosed inside a fleshy fruit. This fleshy tissue should be removed promptly to prevent damage from spontaneous heating or inhibiting substance.

*Primary dormancy*    In most kinds of seeds, internal regulatory controls that prevent germination develop during ripening and persist in the seed for a period of time after harvest (see chapters 6 and 7). It is important to note that the origin of these germination controls is part of the seed development process. Two principal mechanisms interact to impose dormancy on the seed: (a) the accumulation of **chemical growth inhibitors** into different tissues of the fruits and seeds, and (b) the development of **seed coverings** that control the uptake of water, the permeability to gases, and the leaching of inhibitors.

Many naturally occurring chemicals that inhibit germination have been extracted from different parts of the seed and fruit. Demonstrating a role in natural dormancy control does not always follow, however. Inhibitors are present in fleshy fruits, such as tomato, lemon, and strawberry, but the inhibiting effects have been attributed also to high sugar concentration. In some desert plants high salt concentration has been suggested as the inhibitor. Probably the most important naturally occurring inhibitor is **abscisic acid,** which accumulates in seeds and fruits during development and is associated with ripening ( *20, 41*). The control of germination is more likely to involve interactions of inhibiting and promoting substances.

Seed coverings play a crucial role in germination control. The morphological basis for these effects is established during later ripening stages. Coverings arise primarily from the outer integument layer, which becomes hard, fibrous, or mucilaginous during ripening and dehydration (Figure 3–7). In addition, layers of the fleshy fruit may dry and become part of the seed covering as in *Cotoneaster* or hawthorn ( *Crataegus*). In some drupe fruits, as the olive or *Prunus* species, these layers become the hardened endocarp (pit or stone). In other seeds, such as walnut, they become the surrounding shell. In others, such as caryopsis or achenes in grains or grasses, the fruit covering becomes fibrous and coalesces with the seed.

In various plant families, such as Leguminoseae, the outer seed coats harden and become suberized and thus impervious to water (Figure 3–7A). Cells of the outer integument become rearranged, coalesce, incorporate suberin deposits, and develop external cutin coverings. These cells are called **macrosclereid** cells.

In other species, such as white mustard and spinach, mucilaginous layers inside and outside the seed coats are produced particularly under high moisture conditions, which also function to restrict gaseous exchange (Figure 3–7B).

In most families the inner seed coat becomes membranous but remains alive and semipermeable. In the Compositae, for instance, this layer coalesces with the remnant layers of the endosperm. These layers of integument and remnants of the endosperm and nucellus remain physiologically active during ripening and for a period of time after the seed is separated from the plant (see Figure 3–7C). Such physiologically active layers play a major role in maintaining primary dormancy, mainly because this semipermeable nature restricts aeration and inhibitor movement.

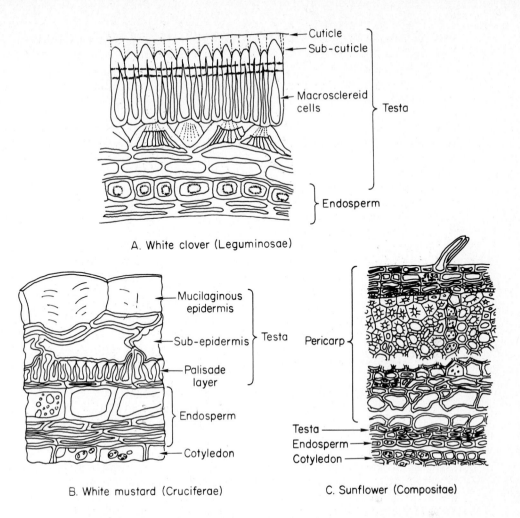

A. White clover (Leguminosae)

B. White mustard (Cruciferae)

C. Sunflower (Compositae)

**Figure 3-7**    Three types of seed coat covering that influence seed dormancy conditions. (A) White clover (*Melilotus alba*) LEGUMINOSEAE: The outer layer of cells becomes hard and impervious because of vertically oriented macrosclereid cells that are covered by a cuticle layer. (B) White mustard (*Brassica hirta*) CRUCIFERAE: Outer seed coats develop a mucilaginous layer when soaked in water. (C) Sunflower (*Helianthus annuus*) COMPOSITAE: Pericarp hardens into a fibrous layer coalescing with the seed coat. Endosperm layer is thin and more or less membranous, and in some species and cultivars can function in controlling dormancy. Courtesy Bewley and Black (*3*).

## Ripening and Dissemination

Specific physical and chemical changes that take place during maturation and ripening of the fruit lead to fruit senescence and dissemination of the seed. One of the most obvious changes is the drying of the fruit tissues. In certain fruits, this leads to dehiscence and the discharge of the seeds from the fruit. Changes may take place in the color of the fruit and the seed coats, and softening of the fruit may occur.

Seed dispersal is accomplished by many agents. Fish, birds, rodents, and bats consume and carry seeds in their digestive tracts. Fruits with spines or hooks become attached to the fur of animals and are often moved considerable distances. Wind dispersal of seed is facilitated in many plant groups by "wings" on dry fruits; tumbleweeds can move long distances by rolling in the wind. Seeds carried by moving water—streams or irrigation canals—can be taken great distances and often become a source of weeds in cultivated fields. Some plants themselves have mechanisms for short-distance dispersal, such as explosive liberation of spores from fern sporangia. The activities of man in purposeful shipment of seed lots all over the world are, of course, of considerable effectiveness in seed dispersal (52).

## THE MATURE SEED

Botanically, in the angiosperms the seed is a matured ovule enclosed within the mature ovary, or fruit. Seeds and fruits of different species vary greatly in appearance, size, shape, and location and structure of the embryo in relation to storage tissues (Figure 3–8). These points are not only useful for identification (25, 35) but have a major effect on germination requirements. From the standpoint of seed handling, it is not always possible to separate the fruit and seed, since they are sometimes joined in a single unit. In such cases, the fruit itself is treated as the "seed," as in wheat or corn.

### Parts of the Seed

The seed has three basic parts: (a) embryo, (b) food storage tissues, and (c) seed coverings.

#### EMBRYO

The embryo is a new plant resulting from the union of a male and female gamete during fertilization. Its basic structure is an embryo **axis** with growing points at each end, one for the shoot and one for the root, and one or more seed leaves (**cotyledons**) attached to the embryo axis. Plants are classified by the number of cotyledons. Monocotyledonous plants (such as the coconut palm or grasses) have a single cotyledon, dicotyledonous plants (such as the bean or peach) have two, and gymnosperms (such as the pine or ginkgo) may have as many as fifteen.

Embryos differ in size, reflecting the extent to which the embryo has developed within the seed. Seeds can be separated into basic types: **nonendospermic** in which the embryo is dominant, and **endospermic** in which the embryo is reduced in size in proportion to the rest of the seed. In the latter, embryos are classified into four basic types characteristic of specific plant families.

#### STORAGE TISSUES

The nonendospermic seeds store foods in the cotyledons, which are the dominant seed parts. In the endospermic seeds the storage material is in the **en-**

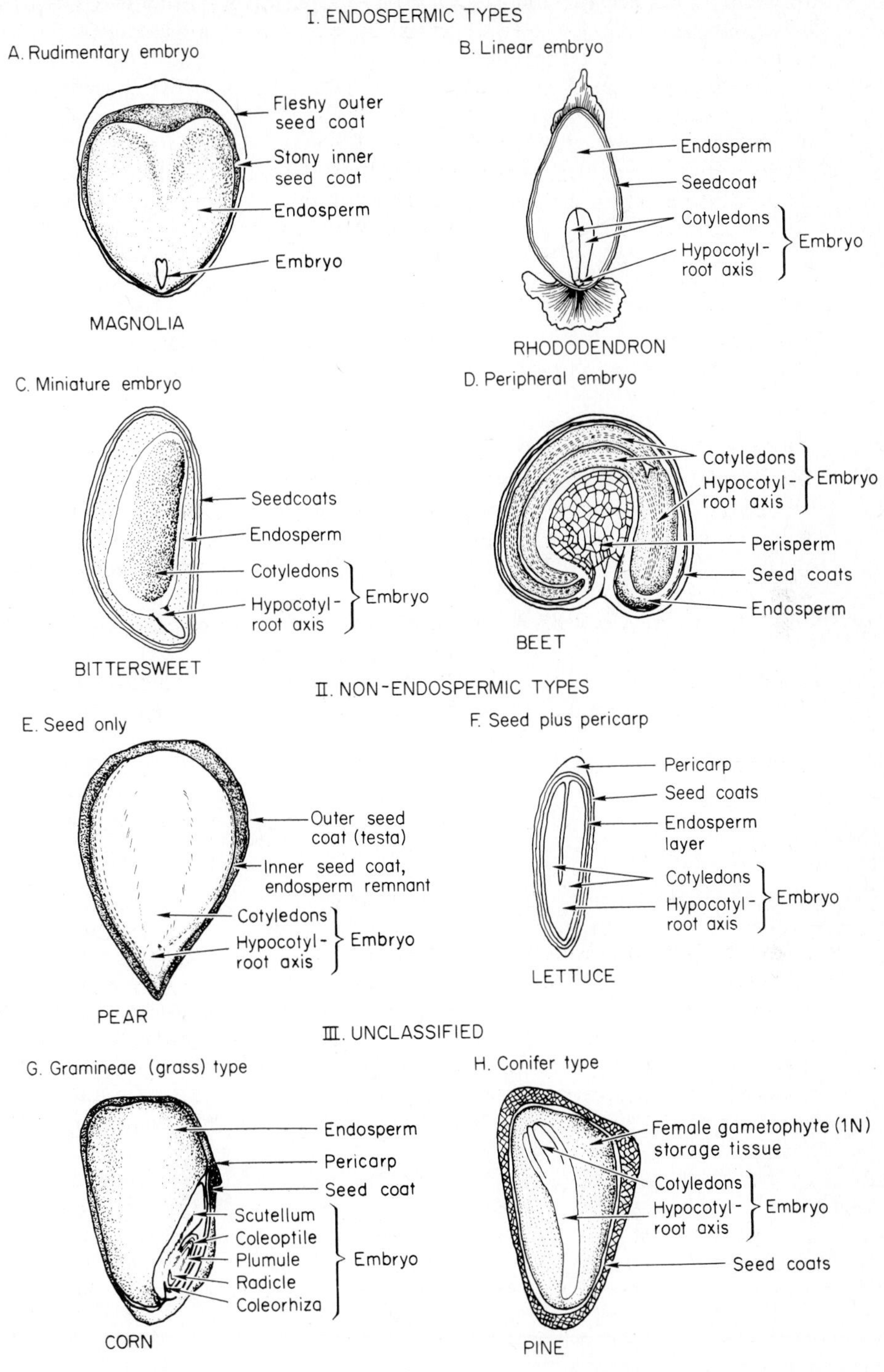

**Figure 3-8** Morphological types of seeds as described in text. *Rhododendron* and *Celastris* Redrawn from (*46*).

The Development of Seeds and Spores

dosperm, **perisperm,** or in the case of gymnosperms, the **haploid female gametophyte.**

SEED COVERINGS

The seed coverings may consist of the seed coats, the remains of the nucellus and endosperm, and sometimes parts of the fruit. The seed coats, or *testa,* usually one or two (rarely three) in number, are derived from the integuments of the ovule. During development the seed coats become modified so that at maturity they present a characteristic appearance. The properties of the seed coat may be highly characteristic of the plant family. Usually the outer seed coat becomes dry, somewhat hardened and thickened, and brownish in color. In particular families it becomes hard and impervious to water. On the other hand, the inner seed coat is usually thin, transparent, and membranous. Remnants of the endosperm and nucellus are found within the inner seed coat, sometimes making a distinct, continuous layer around the embryo.

In some plants, parts of the fruit remain attached to the seed so that the fruit and seed are commonly handled together as the "seed." In certain kinds of fruits, e.g., achenes, caryopsis, samaras, and schizocarps, the fruit and seed layers are contiguous. In others, such as the acorn, the fruit and seed coverings separate but the fruit covering is indehiscent. In still others, such as the "pit" of stone fruits or the shell of walnuts, the covering is a hardened portion of the pericarp, but it is dehiscent and can usually be removed without much difficulty.

The seed coverings provide mechanical protection for the embryo, making it possible to handle seeds without injury, and thus permitting transportation for long distances and storage for long periods of time. The seed coverings can also play an important role in influencing germination, as discussed in chapter 6.

# APOMIXIS

In some species the embryo is not produced as a result of meiosis and fertilization, but from a cell in the embryo sac or surrounding nucellus, which does not undergo meiosis but develops to form a zygote of the same genetic makeup as the female parent. (Nonrecurrent apomixis is an exception.) The occurrence of this asexual reproductive process—in place of the sexual reproductive processes of reduction, division, and fertilization to produce an embryo—is known as **apomixis.** Seedling plants produced in this manner are known as **apomicts.** Plants that produce only apomictic embryos are known as **obligate apomicts;** those that produce both apomictic and sexual embryos are **facultative apomicts** (*4, 26, 34, 51*).

## Types of Apomixis

RECURRENT APOMIXIS

An embryo sac (female gametophyte) develops from the egg mother cell (or from some adjoining cell, the egg mother cell disintegrating), but complete meiosis does not occur. Consequently, the egg has the normal diploid number of chromosomes, the same as the mother plant. The embryo subsequently develops directly from the egg nucleus without fertilization.

The following classification of seeds is based upon morphology of embryo and seed coverings. It includes, as examples, families of herbaceous plants (after Atwater [1]).

I. Seeds with dominant endosperm (or perisperm) as seed storage organs **(endospermic).**

A. Rudimentary embryo. Embryo is very small and undeveloped but undergoes further increase at germination (see Figure 3–8A, Magnolia).

RANUNCULACEAE (*Aquilegia, Delphinium*), PAPAVERACEAE (*Eschscholtzia, Papaver*), FUMARIACEAE (*Dicentra*), ARALIACEAE (*Fatsia*), MAGNOLIACEAE (*Magnolia*), AQUIFOLIACEAE (*Ilex*)

B. Linear embryo. Embryo is more developed than those in A and enlarges further at germination.

UMBELLIFERAE (*Daucus*), ERICACEAE (*Calluna, Rhododendron*), PRIMULACEAE (*Cyclamen, Primula*), GENTIANACEAE (*Gentiana*), SOLANACEAE (*Datura, Solanum*), OLEACEAE (*Fraxinus*)

C. Miniature embryo. Embryo fills more than half the seed.

CRASSULACEAE (*Sedum, Heuchera, Hypericum*), BEGONIACEAE (Begonia), SOLANACEAE (*Nicotiana, Petunia, Salpiglossis*), SCROPHULARIACEAE (*Antirrhinum, Linaria, Mimulus, Nemesia, Penstemon*), LOBELIACEAE (*Lobelia*)

D. Peripheral embryo. Embryo encloses endosperm or perisperm tissue.

POLYGONACEAE (*Eriogonum*), CHENOPODIACEAE (*Kochia*), AMARANTHACEAE (*Amaranthus, Celosia, Gomphrena*), NYCTAGINACEAE (*Abronia, Mirabilis*)

II. Seeds with embryo dominant **(nonendospermic);** classified according to the type of seed covering.

A. Hard seed coats restricting water entry.

LEGUMINOSAE, GERANIACEAE (*Pelargonium*), ANACARDIACEAE (*Rhus*), RHAMNACEAE (*Ceanothus*), MALVACEAE (*Abutilon, Altea*), CONVOLVULACEAE (*Convolvulus*)

B. Thin seed coats with mucilaginous layer.

CRUCIFERAE (*Arabis, Iberis, Lobularia, Mathiola*), LINACEAE (*Linum*), VIOLACEAE (*Viola*), LABIATAE (*Lavandula*)

C. Woody outer seed coats with inner semipermeable layer (see Figure 3–8E).

ROSACEAE (*Geum, Potentilla*), ZYGOPHYLLACEAE (*Larrea*), BALSAMINACEAE (*Impatiens*), CISTACEAE (*Cistus, Helianthemum*), ONAGRACEAE (*Clarkia, Oenothera*), PLUMBAGINACEAE (*Armeria*), APOCYNACEAE, POLEMONIACEAE (*Phlox*), HYDROPHYLLACEAE (*Nemophila, Phacelia*), BORAGINACEAE (*Anchusa*), VERBENACEAE (*Lantana, Verbena*), LABIATEAE (*Coleus, Moluccela*), DIPSACAEAE (*Dipsacus, Scabiosa*)

D. Fibrous outer seed coat with more or less semipermeable membranous layer, including endosperm remnant. (see Figure 3–8F).

COMPOSITAE (many species)

**III.** Unclassified

    **A.** Rudimentary embryo with no food storage.

        ORCHIDACEAE (orchids, in general)

    **B.** Modified miniature embryo located on periphery of seed.

        GRAMINEAE (grasses) (see Figure 3–8G)

    **C.** Axillary miniature embryo surrounded by gametophyte tissue (see Figure 3–8H)

        Gymnosperms, in particular, conifers.

This series of events is known to occur in some species of *Crepis, Taraxacum* (dandelion), *Poa* (bluegrass), and *Allium* (onion) without the stimulus of pollination; in others, e.g., species of *Parthenium* (guayule), *Rubus* (raspberry), *Malus* (apple), some *Poa* species, and *Rudbeckia,* pollination appears to be necessary, either to stimulate embryo development or to produce a viable endosperm.

ADVENTITIOUS EMBRYONY

**Adventitious embryony** is also known as **nucellar embryony** or **nucellar budding.** In this type of apomixis the embryos rise from a cell or group of cells either in the nucellus (usually) or in the integuments, as shown in Figure 3–9. It differs from recurrent apomixis in that such embryos develop outside of the embryo sac and in addition to the regular embryo. In some plants, *Citrus* for instance, fertilization takes place in the usual manner, and a sexual plus several apomictic embryos may develop. In others, e.g., some species of *Opuntia,*

**Figure 3–9**    The development of nucellar embryos in *Citrus. Left:* stage of development just after fertilization showing zygote and remains of pollen tube. Note individual active cells (those shaded) of the nucellus which are in the initial stages of nucellar embryony. *Right:* a later stage showing developing nucellar embryos. The large one may be a sexual embryo. Redrawn from Gustafsson (*26*).

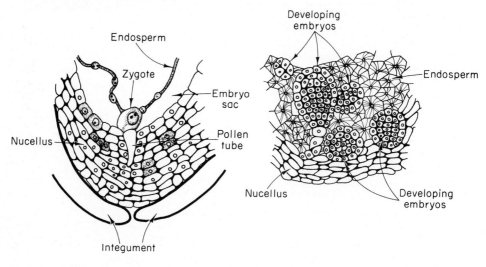

apomictic embryos develop spontaneously, no pollination or fertilization apparently being needed.

### NONRECURRENT APOMIXIS

In **nonrecurrent apomixis** an embryo arises directly from the egg nucleus without fertilization. Since the egg is haploid, the resulting embryo will also be haploid. This case is rare and primarily of genetic interest. It does not consistently occur in any particular kind of plant as do recurrent apomixis and adventitious embryony.

### VEGETATIVE APOMIXIS

In some cases vegetative buds or **bulbils** are produced in the inflorescence in place of flowers. This occurs in *Poa bulbosa* and some *Allium, Agave,* and grass species.

## Polyembryony

The phenomenon in which two or more embryos are present within a single seed is called **polyembryony** (*4, 34*). It may result from one of several causes. Nucellar embryony, as described for *Citrus,* is one cause (see Figure 3-10). The occasional development of more than one nucleus within the embryo sac (in addition to the egg nucleus) is another. Cleavage of the proembryo during the early stages of development, a common occurrence in conifers, leads to multiple embryos.

**Figure 3-10**     Polyembryony in trifoliate orange (*Poncirus trifoliata*) seeds as shown by the several seedlings arising from each seed. One seedling, usually the weakest, is of sexual origin, while the others are nucellar.

### Significance of Apomixis

Apomixis occurs in nature in many different plant families (*26, 39*). Usually polyploids or complex hybrids are involved. Apomixis occurs in some kinds of plant cultivars and can be produced by breeding.

The ability to reproduce plant cultivars apomictically has significant uses in horticulture. Because the seedling plants produced result from an asexual process, the procedure is a means of producing genetically uniform seedling populations (see chapter 4). Such apomictic seedling plants follow the same seedling cycle as shown by sexually produced seedlings, including a juvenile-to-mature transition, except that both the upper and lower cycles are asexual.

Consequently, apomictic seedling production would be most useful for reproducing plants in which juvenility is an important part of their development, such as annuals, biennials, or forest trees. Apomictic seedlings would not normally be suitable for reproducing fruit and nut cultivars in which the juvenile phase involves a delay in onset of bearing and undesirable growth characteristics (see chapter 8). Apomictic seedlings, however, are useful for producing uniform rootstock cultivars in some species.

Many *Citrus* species and cultivars produce apomictic seedlings that are used as rootstocks (*11*). These invariably are uniform and vigorous in contrast to the occasional weaker, variable hybrid seedlings that may also occur. *Malus toringoides* and *M. sikkimensis* are apple species that are apomictic if self-pollinated, although they produce large proportions of hybrid seeds if cross-pollinated (*12*). Some grass species are facultative apomicts. Kentucky bluegrass (*Poa pratensis*) plants consist of both apomicts and sexually reproduced individuals; in one study (*9*) 85 percent of such plants were found to be apomictically reproduced. Certain pasture grasses, such as 'King Ranch' bluestem (*Andropogon*), 'Argentine' Bahiagrass (*Paspalum notatum*), and 'Tucson' side oats grama (*Bouteloua curtipendula*), are examples of apomictic cultivars (*10*).

Apomictic or nucellar seedlings are suitable for screening out pathogens such as viruses, which in many cases are not transmitted by seed. Again this process produces a juvenile plant which must then be consecutively repropagated to attain the mature phase (chapter 8).

## SPORE DEVELOPMENT

Reproduction by spores occurs in ferns (*43*). Two separate stages or generations are involved as shown in Figure 3–11. One is the asexual, or **sporophyte** generation, in which the plant has conspicuous roots, stems, and leaves (fronds). The other is the sexual, or **gametophyte,** generation, in which the plant is small; inconspicuous; without roots, stems, or leaves; and is called a **prothallium.** Spores are produced on the underside of fronds in clusters of **sporangia,** or spore cases, that appear as brownish ''dots.'' Groups of sporangia are called **sori** and sometimes have a cover known as the **indusium.** Within each sporangium 16 **spore mother cells** form, each of which undergoes meiosis to produce four **spores.** Each spore is haploid, with half the normal chromosome number of the species.

Spores are discharged and under favorable temperature and moisture conditions ''germinate'' to produce the **prothallus,** a flat green plate of cells with

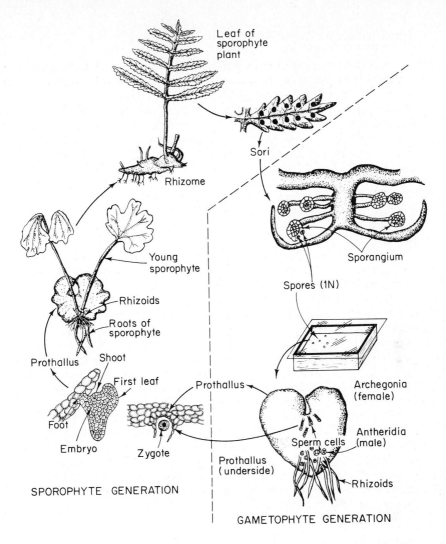

**Figure 3–11**     Development of spores in the reproduction cycle of a fern.

small rootlike structures **(rhizoids).** This grows to about ¼ in. in diameter in about three months. Male (**antheridia**) and female (**archegonia**) structures are produced on the underside of the same prothallus (except in two economically unimportant families). Sperm cells are discharged and, in the presence of water, are attracted into the archegonium to fuse with the egg and subsequently produce a zygote.

The **zygote** develops into an embryo, which grows to produce the fern plant (sporophyte), which has the diploid chromosome number for the species. In its initial development, the embryo develops a **foot,** through which it absorbs moisture and nutrients from the prothallus. A root is produced which grows downward into the soil. Also produced is the primary leaf, which acts as a temporary photosynthetic organ, and the stem, which develops into the rhizome from which fronds and permanent roots arise.

Propagation procedures for ferns are given in chapter 20.

## REFERENCES

1. Atwater, B. R. 1980. Germination, dormancy and morphology of the seeds of herbaceous ornamental plants. *Seed Sci. and Tech.* 8:523–73.

2. Austin, R. B. 1972. Effects of environment before harvesting on viability. In *Viability of seeds,* E. H. Roberts, ed. New York: Syracuse Univ. Press, pp. 114–49.

3. Bewley, J. D., and M. Black.1978. *Physiology and biochemistry of seeds in relation to germination. Vol. 1 Development, Germination, and Growth.* Berlin: Springer-Verlag.

4. Bhatnager, S. P., and B. M. Johri. 1972. Development of angiosperm seeds. In *Seed biology,* Vol. 1, T. T. Kozlowski, ed. New York, Academic Press, pp. 77–149.

5. Black, M. 1980/1981. The role of endogenous hormones in germination and dormancy. *Israel Jour. Bot.* 29:181–92.

6. Bohart, G. E., and T. W. Koerber. 1972. Insects and seed production. In *Seed biology,* Vol. 3, T. T. Kozlowski, ed. New York: Academic Press, pp. 1–54.

7. Brink, R. A. 1962. Phase change in higher plants and somatic cell heredity. *Quart. Rev. Bio.* 37(1):1–22.

8. Brink, R. A., and D. C. Cooper. 1947. The endosperm in seed development. *Bot. Rev.* 13:423–541.

9. Brittingham, W. H. 1943. Type of seed formation as indicated by the nature and extent of variation in Kentucky bluegrass, and its practical application. *Jour. Agr. Res.* 67:255–64.

10. Burton, G. W., and G. F. Sprague. 1961. Use of hybrid vigor in plant improvement. In *Germ plasm resources,* R. E. Hodgson, ed. Washington, D.C.: Amer. Assoc. Adv. Sci. Publ. No. 66, pp. 191–203.

11. Cameron, J. W., R. K. Soost, and H. B. Frost. 1959. The horticultural significance of nucellar embryony in citrus. In *Citrus virus diseases,* J. Wallace, ed. Berkeley: Univ. of Calif., Division of Agr. Sci., pp. 191–96.

12. Campbell, A. J., and D. Wilson. 1962. Apomictic seedling rootstocks for apples: Progress report, III. *Ann. Rpt. Long Ashton Hort. Res. Sta. (1961):* 68–70.

13. Cooper, D. C., and R. A. Brink. 1940. Somatoplastic sterility as a cause of seed failure after interspecific hybridization. *Genetics* 25:593–617.

14. ———. 1945. Seed collapse following matings between diploid and tetraploid races of *Lycopersicon pimpinellifolium. Genetics* 30:375–401.

15. Craig, R. 1976. Flower seed industry. In *Bedding plants; A manual on the culture of bedding plants as a greenhouse crop,* J. W. Mastalerz, ed. University Park, Pa.: Pennsylvania Flower Growers, pp. 25–46.

16. Davis, L. D. 1939. Size of aborted embryos in the Phillips Cling peach. *Proc. Amer. Soc. Hort. Sci.* 37:198–202.

17. Delouche, J. C. 1980. Environmental effects on seed development and seed quality. *HortScience* 15(6):13–18.

18. Dickson, M. H. 1980. Genetic aspects of seed quality. *HortScience* 15(6):771–74.

19. Dure, L. S., III. 1975. Seed formation. *Ann. Rev. Plant Phys.* 26:259–78.

20. ———. 1979. Role of stored messenger RNA in late embryo development and germination. In *The plant seed: Development, preservation, and germination,* I. Rubenstein et al., eds. New York: Academic Press, pp. 113–28.

21. Dure, L. S., III, G. A. Galau, and S. Greenway. 1980/1981. Changing protein patterns during cotton cotyledon embryogenesis and germination as shown by *in vivo* and *in vitro* synthesis. *Israel Jour. Bot.* 29:293–306.

22. Edwards, D. G. W. 1980. Maturity and quality of tree seeds: A state-of-the-art review. *Seed Sci. and Tech.* 8:625–57.

23. Ehrenberg, C., A. Gustafsson, C. P. Forshell, and M. Simak. 1955. Seed quality and the principles of forest genetics. *Hereditas* 41:291–366.

24. Goldbach, H., and C. Michael. 1976. Abscisic acid content of barley grains during ripening as affected by temperature and variety. *Crop Sci.* 16:797–99.

25. Gunn, C. R. 1972. Seed collecting and identification. In *Seed biology,* Vol. 3, T. T. Kozlowski, ed. New York: Academic Press, pp. 55–144.

26. Gustafsson, A. 1946–1947. Apomixis in higher plants, Parts I–III. *Lunds Univ. Arsskrift,* N.F. Avid. 2 Bd 42, Nr. 3:42(2); 43(2); 43(12).

27. Hackett, W. P. 1982. Phase change and intra-clonal variability. *HortScience:* (in press).

28. Hesse, C. O., and D. E. Kester. 1955. Germination of embryos of *Prunus* related to degree of embryo development and method of handling. *Proc. Amer. Soc. Hort. Sci.* 65:251–64.

29. Jones, H. A. 1927. Pollination and life history studies of the lettuce (*Lactuca sativa* L.). *Hilgardia* 2:425–79.

30. Kester, D. E. 1982. The clone in horticulture. *HortScience:* (in press).

31. Kester, D. E., and C. O. Hesse. 1955. Embryo culture of peach varieties in relation to season of ripening. *Proc. Amer. Soc. Hort. Sci.* 65:265–73.

32. Lammerts, W. E. 1942. Embryo culture, an effective technique for shortening the breeding cycle of deciduous trees and increasing germination of hybrid seed. *Amer. Jour. Bot.* 29:166–71.

33. Luckwill, L. C. 1948. The hormone content of the seed in relation to endosperm development and fruit drop in the apple. *Jour. Hort. Sci.* 24:32–44.

34. Maheshwari, P., and R. C. Sachar. 1963. Polyembryony. In *Recent advances in the embryology of angiosperms,* P. Maheshwari, ed. Delhi, India: Univ. of Delhi, Int. Soc. of Plant Morph., pp. 265–96.

35. Martin, A. C. 1946. The comparative internal morphology of seeds. *Amer. Midland Nat.* 36:512–660.

36. Mayer, A. M., and A. Poljakoff-Mayber. 1975. *The germination of seeds.* Oxford: Pergamon Press.

37. Nitsch, J. P. 1971. Perennation through seeds and other structures. *Plant physiology,* Vol. 4A, F. C. Steward, ed. New York: Academic Press.

38. Norstog, K. 1979. Embryo culture as a tool in the study of comparative and developmental morphology. In *Plant cell and tissue culture: Principles and applications,* W. R. Sharp et al., eds. Columbus, Ohio: Ohio State Univ. Press, pp. 179–202.

39. Nygren, A. 1954. Apomixis in the angiosperms II. *Bot. Rev.* 20:577–649.

40. Pollock, B. M., and E. E. Roos. 1972. Seed and seedling vigor. In *Seed biology,* Vol. 1, T. T. Kozlowski, ed. New York: Academic Press, pp. 314–88.

41. Radley, M. 1979. The role of gibberellin, abscisic acid, and auxin in the regulation of developing wheat grains. *Jour. Exp. Bot.* 30(116):381–89.

42. Rietsema, J., S. Satina, and A. F. Blakeslee. 1955. Studies on ovule and embryo growth in *Datura*. I. Growth analysis. *Amer. Jour. Bot.* 42:449–54.

43. Roberts, D. J. 1965. Modern propagation of ferns. *Proc. Inter. Plant Prop. Soc.* 15:317–22.

44. Rolston, M. P. 1978. Water impermeable seed dormancy. *Bot. Rev.* 44:365–96.

45. Sanders, M. E. 1948. Embryo development in four *Datura* species following self and hybrid pollinations. *Amer. Jour. Bot.* 35:525–32.

46. Schopmeyer, C. S., ed. 1974. *Seeds of woody plants in the United States.* U.S. Dept. Agr. Handbook No. 450. Washington, D.C.: U.S. Government Printing Office.

47. Singh, H., and B. M. Johri. 1972. Development of gymnosperm seeds. In *Seed biology,* Vol. 1, T. T. Kozlowski, ed. New York: Academic Press, pp. 22–77.

48. Shepherd, P. H. 1955. The kernel told the tale. *Amer. Fruit Grower* 75:37.

49. Stephens, W. 1969. The mangrove. *Oceans* 2(5):51–55.

50. Taylorson, R. B., and S. B. Hendrick, 1977. Dormancy in seeds. *Ann. Rev. Plant Phys.* 28:331–54.

51. Tisserat, B., E. B. Esan, and T. Murashige. 1979. Somatic embryogenesis in angiosperms. *Hort. Rev.* 1:1–78.

52. Van der Pijl, L. 1969. *Principles of dispersal in higher plants.* Berlin: Springer-Verlag.

53. Varner, J. E. 1965. Seed development and germination. In *Plant biochemistry,* J. Bonner and J. E. Varner, eds. New York: Academic Press.

54. Walton, D. C. 1980. Biochemistry and physiology of abscisic acid. *Ann. Rev. Plant Phys.* 31:453–89.

55. Wardlaw, C. W. 1965. *Physiology of embryonic development in cormophytes.* In *Encyclopedia of Plant Physiology,* Vol. 15, pp. 844–965.

56. Wellington, P. S., and V. W. Durham. 1958. Varietal differences in the tendency of wheat to sprout in the ear. *Empire Jour. Exp. Agr.* 26:47–54.

57. Zimmerman, R. H., ed. 1976. Symposium on juvenility in woody perennials. *Acta Hort.* 56:1–317.

## SUPPLEMENTARY READING

HESLOP-HARRISON, J. 1972. Sexuality of angiosperms. Chapter 9 in *Plant physiology,* Vol. 6C, F. C. Steward, ed. New York: Academic Press, pp. 134–289.

KOZLOWSKI, T. T., ed. 1972. *Seed biology,* Vol. 1. *Importance, Development and Germination.* New York: Academic Press.

MARTIN, A. C., and W. D. BARKLEY. 1961. *Seed identification manual.* Berkeley: Univ. of Calif. Press.

MUSIL, A. F. 1963. *Identification of crop and weed seeds.* U.S. Dept. Agr. Handbook 219. Washington, D.C.: U.S. Govt. Printing Office.

NITSCH, J. P. 1971. Perennation through seeds and other structures. Chapter 4 in *Plant physiology,* Vol. 6A, F.C. Steward, ed. New York: Academic Press.

RUBENSTEIN, I., R. L. PHILLIPS, C. E. GREEN, and G. GENGENBACH, eds. 1979. *The plant seed: Development, preservation and germination.* New York: Academic Press.

U.S. DEPT. OF AGRICULTURE. 1961. *Seeds: Yearbook of agriculture.* Washington, D.C.: U.S. Govt. Printing Office.

In nature higher plants reproduce primarily by seeds. The characteristic genetic variability that occurs among groups of seedlings is important to allow continued adaptation of a particular species to possible changes in the environment. Those individuals within each generation that are best adapted to their environment tend to survive, produce the next generation, and reproduce the species.

On the other hand, propagation of cultivated plants requires that genetic variability during seed reproduction be controlled or the value of a cultivar may be lost. Characteristics selected as important in agriculture and horticulture may not be consistently perpetuated into the next generation unless specific principles and procedures are followed.

## USES OF SEEDLINGS IN PROPAGATION

### Mass Propagation

Seedling propagation is the most efficient and economical method of propagation as long as genetic variability can be controlled within acceptable limits. Many millions of seeds can be produced in bulk, stored for long periods, transported all over the world, and mass propagated in a highly efficient manner.

Annuals and biennials that must be grown from seed account for essentially all the agronomic crops (cereals, forages, grasses, fiber, and oil plants), most vegetable crops, and many garden and florist crops (25). The bedding plant industry depends largely on seeds for the production of flower and vegetable plants (33). In general, seed propagation is made possible through development of cultivars that are true breeding lines in which genetic variability is controlled during seed production.

Seed propagation of herbaceous and woody perennials is also important for many crops. For example, seeds are used to reproduce species and botanical varieties of various bulb-producing plants, such as tulips and lilies, as well as cultivars of various herbaceous perennials, such as cyclamen and tuberous begonias.

4

# Production
# of Genetically Pure Seed

Forest trees and many landscape trees and shrubs are primarily grown from seeds because vegetative methods are either not available or uneconomical in the quantities required. Furthermore, the juvenile growth habit obtained from seedlings is desirable and, in addition, some variability among seedlings is useful. Seed selection procedures to minimize variability are an important part of propagation of woody perennials.

### Rootstocks

Most cultivars of fruit, nut, and shade tree plants are vegetatively propagated clones. Seedlings are used extensively in nurseries to provide rootstocks upon which these cultivars are grafted or budded.

### Plant Breeding

Growing seedlings is the most important means of developing new cultivars because of the variability that results from segregation of chromosomes attending sexual reproduction.

## POLLINATION REQUIREMENTS OF PLANTS

Pollination is the transfer of pollen from an anther to the stigma of a flower. A pollen tube grows down the style into the ovule, where fertilization takes place. Seed development may result from (a) **self-pollination,** in which the pollen may come from the same flower, from different flowers on the same plant, or from flowers on different plants of the same clone; (b) **cross-pollination,** in which pollen comes from a different plant or from a different clone; or (c) **apomixis,** in which an asexual process is substituted for a sexual one. In addition, in some plants, either self- or cross-pollination can occur, the amount of either varying with the kind of plant and environmental factors. In other cases, seed production is partially sexual and partially apomictic (*18*).

### Self-Pollinated Plants

Cultivars of self-pollinated crop plants can be propagated by seed with little variability occurring even when the crop is grown in close proximity to other closely related cultivars. Cross-pollination is usually less than 4 percent (*18*).

Self-pollinated cultivars can be maintained readily by seed propagation because they are largely homozygous. With self-pollination, offspring of homozygous plants are also homozygous and inherit the characteristics of the parent. Heterozygous plants, if self-pollinated, segregate (for instance, like the tall and dwarf offspring of garden peas as shown in Figure 1–3). With continued generations of self-pollination, the proportion of homozygous plants increases while the heterozygous plants decrease by a factor of one-half for each generation. After six to ten generations, the group of descendants from the original parents have segregated into a mixture of more or less true-breeding lines (*6*).

The **line** is one of the major categories of cultivars. It is a population of self-pollinated plants that have been selected to a standard by eliminating "off-type" plants and "fixing" the characteristics of the cultivar through con-

**Table 4–1**   Minimum isolation requirements for seed production of certain species of field and vegetable crops (3).

| Type of Pollination | Species | Seed Class | | | |
| --- | --- | --- | --- | --- | --- |
| | | Foundation | Registered | Certified | |

| Type of Pollination | Species | Foundation | Registered | Certified |
| --- | --- | --- | --- | --- |
| Self-pollinated | barley, oats, wheat, rice, peanut, soybean, lespedeza, field pea, garden bean, cowpea, flax grasses (self-pollinated and apomictic species) | Fields should be separated by a definite boundary adequate to prevent mechanical mixture | | |
| | | 18 m (60 ft) | 9 m (30 ft) | 4.5 m (15 ft) |
| Self-pollinated but to a lesser degree than those in the above list | cotton (upland type) | 30 m (100 ft) from cultivars which differ markedly | | |
| | cotton (Egyptian type) | 400 m (1320 ft) | 400 m (1320 ft) | 200 m (660 ft) |
| | pepper | 60 m (200 ft) | 30 m (100 ft) | 9 m (30 ft) |
| | tomato | 60 m (200 ft) | 30 m (100 ft) | 9 m (30 ft) |
| | tobacco | 45 m (150 ft) or by four border rows of each cultivar. Isolation between cultivars of different types should be 400 m (1320 ft) | | |
| Cross-pollinated by insects | alfalfa, birdsfoot trefoil, red clover, white clover, sweet clover | 182 m (600 ft) or 273 m (900 ft)* | 91 m (300 ft) or 136 m (450 ft)* | 50 m (165 ft) |
| | millet | 400 m (1320 ft) | 400 m (1320 ft) | 200 m (660 ft) |
| | onion | 1600 m (5280 ft) | 800 m (2640 ft) | 400 m (1320 ft) |
| | watermelon | 800 m (2640 ft) | 800 m (2640 ft) | 400 m (1320 ft) |
| Cross-pollinated by wind | hybrid field corn | 200 m (660 ft) (may be reduced if field is surrounded by specified numbers of border rows and the cultivars nearby are of same color and texture). | | |
| | grasses | 273 m (900 ft) | 91 m (300 ft) | 50 m (165 ft) |

*First number if plot is less than 0.2 ha (5 acres); second number if plot is more than 0.2 ha (5 acres).

secutive generations of self-pollinations. If all descendants are from a single homozygous plant, it is termed a **pure line.**

Natural self-pollination in various plants (Table 4–1) results from a particular flower structure. Some of the earliest and most important food crops, such as wheat, barley, and rice, are naturally self-pollinated. This characteristic has undoubtedly made possible their maintenance in cultivation, leading to the beginnings of agriculture some 10,000 years ago.

## Cross-Pollinated Plants

Most species and cultivars are naturally cross-pollinated. In many cases either self- or cross-pollination is possible but some plants have mechanisms that prevent self-pollination. These mechanisms include the following types:

**Dioecy** Pistillate (female) and staminate (male) flowers are borne on separate plants (holly, pistachio, date palm, asparagus).

**Monoecy** Pistils (female) and stamens (male) develop in separate flowers of the same plant (corn, walnut, conifers).

**Dichogamy** Pollen is shed at a different time than when the pistil is receptive (includes many walnut and pecan cultivars).

**Self-Sterility** Specific genes prevent normal formation of either male (pollen) or female (pistil) reproductive structure. Male (pollen) sterile lines of corn (Figure 4–1) or onions have been used to produce hybrid plants.

**Incompatibility** Inability of pollen to grow into pistils of the same flower or flowers of the same plant even though the pollen is itself viable. Incompatibility also occurs between flowers on different plants of the same vegetatively propagated cultivar (sweet cherry, almond, lily, petunia, cabbage).

Cross-pollination is accomplished by various agencies.

**Wind pollination** is the rule with many plants having inconspicuous flowers. Examples are grasses, conifers, the olive, and catkin-bearing trees such as the walnut, oak, alder, and cottonwood. The pollen produced from such plants is generally light and dry, and in some cases is carried long distances in wind currents.

**Insect pollination** is the rule with plants with white, or brightly colored, fragrant, and otherwise conspicuous flowers which attract insects. The honey bee is one of the most important pollinating insects, although wild bees, butterflies, moths, and flies also obtain pollen and nectar from the flower. Generally, pollen is heavy and sticky, adhering to the insect.

Controlled **hand-pollination** by humans is an important practice in breeding programs and is also used in producing seed of some plants.

Other **pollinating agents**, such as birds, bats, snails, and water, are effective with certain kinds of plants.

IN NATURE

Cross-pollination appears to be the most desirable condition for most species of plants. It maintains a level of variability and heterozygosity that provides the opportunity for evolutionary change if environments change. In nature, **species** or **botanical varieties** tend to be relatively uniform phenotypically. On the other hand, subgroups within species (**ecotypes**) may occur that are specifically adapted to a particular ecological location although not morphologically distinct. Variations that occur along a particular gradient are called **clines** (*31*). When seeds of these natural groups are selected for propagation they tend to reproduce the particular plants from which they were taken.

Sharp increases in variability may develop in seedling populations in certain cases. For example, if one species is brought into close proximity to a second closely related species so that cross-pollination occurs, one may find hybrid offspring widely differing in appearance from either parent. Likewise, if enforced self-pollination occurs, then poor seed production and weak, variable seedlings often result. Either situation may develop when individual plants of a species are brought into an arboretum (*50, 53*).

IN CULTIVATION

Seed propagation of cross-pollinated cultivars poses more production problems than that of self-pollinated cultivars since cross-pollinated plants tend to be heterozygous to some degree, and groups of seedlings are often variable. Most herbaceous cross-pollinated cultivars are the result of breeding procedures in which the potential for variation is controlled. Such control must be maintained if the seed production process is to be successful, but the procedures used vary with the kind of cultivar.

Several categories of cross-pollinated cultivars are grown:

**Inbred lines** are true-breeding cultivars that result from enforced self-pollination of selected parent plants followed by continued selection to a desired type in succeeding generations. Once a desired cultivar of this type has been selected, it is maintained by growing plants in isolation and allowing them to crosspollinate or self-pollinate naturally. This practice is called **open pollination.** In some species, hand pollination may be required.

**Hybrid line** describes the $F_1$ hybrid of two inbred lines (*12, 21, 22*). Enforced self-pollination of plants of many species results in a decline in vigor and size but an increase in uniformity. Some inbred lines die out; others stabilize. If two inbred lines are cross-pollinated, the $F_1$ hybrid seedling generation is likely to be a population of uniform, vigorous heterozygous plants. Such $F_1$ hybrid cultivars, first developed for corn, provide a classic example of the advantage of hybrid vigor (see Figure 4–1).

Hybridization may be made between two inbred lines (**single-cross**), two single-crosses (**double-cross**), an inbred line and an open pollinated cultivar (**top-cross**), or between a single-cross and an inbred line (**three-way cross**) (*45*).

Hybrid lines have also become important for various flower and vegetable crops used in bedding plants (petunia, geranium, zinnia, marigold, snapdragon, onion, tomato, and asparagus) and in some grain crops such as rice (*51*).

Seed production for hybrid lines requires that the parents be maintained as (a) *inbred lines* or as (b) *vegetatively propagated clones.* Cross-pollination between these parents must take place so that the $F_1$ generation is reconstituted for each group of seeds. Seeds from the $F_1$ hybrid plants normally cannot be used to produce the next generation because this $F_2$ generation will include a wide range of variable offspring plants and the desirable characteristics will be lost.

**Nonuniform assemblages** (seedling populations) involve controlled variability for particular characteristics. **Mixtures** may include a combination of genotypes that vary in a certain character such as flower color but are similar in all other characteristics (*12*).

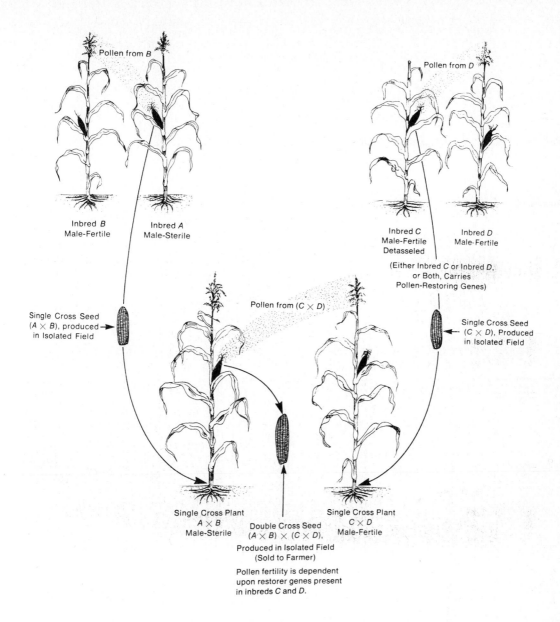

**Figure 4-1** Hybrid seed corn production by the utilization of cytoplasmic male sterility in the production of a single cross and double cross. In this example only one inbred, *A,* is male sterile. The cytoplasmic male sterility is transmitted to the single cross *A* × *B*. Pollen-restoring genes carried by the inbreds *C* or *D* give pollen fertility to the double cross (*A* × *B*) × (*C* × *D*) also. *Source:* USDA.

A **synthetic cultivar** (*6*) is composed of seedling plants produced by combining a number of genetically distinct but phenotypically similar lines or clones, which have been allowed to cross-pollinate at random. The first generation of seeds is called *Syn-1*, the second generation, *Syn-2*, and so on. The original stocks are maintained and allowed to increase only a limited number of seed generations from the original, *Syn-O*. The cultivars are also known as **composites** (*48*).

**F$_2$ strains** are produced for some flower crops such as petunia, pansy, and snapdragon (*12*). Relatively uniform populations have been produced by plant breeding in which the range of variability is acceptable for the purpose intended. Since the seed is grown open-pollinated, or self-pollinated, from F$_1$ hybrids, such seed is less costly to produce than the true F$_1$ hybrid seed.

# SEED PRODUCTION
# OF HERBACEOUS PLANTS

Control of the seed source and production procedure is necessary to maintain the genetic identity of a seed-propagated cultivar. Contamination may result from chance-pollination with plants of a different genotype, or from a mixture with other crop seed of lesser value or with weed seed. Foreign seed similar to the pure crop seed cannot be removed easily with seed-cleaning equipment.

## Methods of Maintaining
## Genetic Identity

### ISOLATION

Isolation is necessary (1) to prevent contamination by cross-pollination with a different but related cultivar, and (2) to prevent mechanical mixing of the seed during harvest. It is primarily achieved through distance, but it also can be attained by enclosing plants or groups of plants in cages, enclosing individual flowers, or removing male flower parts and then employing artificial pollination.

Separation of different cultivars is needed in production of self-pollinated plants to prevent mechanical mixing of seed during harvest. The minimum distance usually specified between plots is 3 m (10 ft) (see Table 4–1). Seed of different cultivars must be kept separate during harvest. Careful cleaning of the harvesting equipment is required when a change is made from one cultivar to another. Likewise, sacks and other containers used to hold the seed must be cleaned carefully to remove any seed that has remained from previous lots.

Considerably more separation for isolation is needed in production of plants cross-pollinated by wind or insects. The minimum distance between cultivars depends upon a number of factors, such as the degree of natural cross-pollination, the pollination agent, direction of prevailing winds, and the number of insects present. The minimum distance for insect-pollinated plants is 0.4 km (1/4 mi) to 1.6 km (1 mi).

The distance that should separate cultivars of wind-pollinated plants varies with the kind of plants. The distance usually specified for corn is 0.2 km (⅛ mi) but this distance may be reduced by planting border rows of the pollinator cultivar. With beets, 0.8 km (½) to 1.6 km (1 mi) is recommended and, to produce stock seed, 3.2 km (2 mi) is preferable (*26, 39*). Seed-producing fields of two different cultivars should not be in the line of prevailing winds.

Cross-pollination takes place between certain cultivars and not between others. In general, any cultivar can contaminate any other of the same species; it may or may not contaminate cultivars of a different species but in the same genus; and rarely will it contaminate cultivars belonging to another genus. Since the horticultural classification may not indicate taxonomic relationships, the seed producers should be familiar with the botanical relationships among the cultivars they grow.

## ROGUING

Off-type plants, plants of other cultivars, and weeds in the seed production field should be eliminated by **roguing.** A low percentage of such plants may not seriously affect the performance of any one lot of seed, but their continued presence will lead to deterioration of the cultivar over a period of time.

Off-type plants may arise because recessive genes are present in a heterozygous condition even in highly homozygous cultivars. Recessive genes may arise by mutation, a process continually occurring at a low rate. The effect of a mutant recessive gene controlling a given plant characteristic may not be immediately observed in the plant in which it arises. The plant becomes heterozygous for that gene, and in a later generation the gene segregates and the character appears in the offspring. Some cultivars have mutable genes that continuously produce specific off-type individuals (*38*). Off-type individual plants should be rogued out of the seed production fields before pollination occurs. Regular supervision of the seed producing fields by trained personnel is required.

Volunteer plants arising from accidentally planted seed or from seed produced by earlier crops is another source of contamination. Fields for producing seed of a particular cultivar should not have grown a potentially contaminating cultivar for a number of preceding years.

## TESTING

Cultivars being grown for seed production should be grown periodically in test plantings to make sure that the cultivar remains true to its characteristics. Seed companies maintain test gardens for this purpose under the supervision of trained personnel.

## CONTROLLING GENERATION SEQUENCE

The distribution process for a seed-propagated cultivar begins by designating a limited supply of seed as the nucleus of future propagations of that cultivar. Plants producing this seed, which have been carefully selected in the process of cultivar development, establish an initial standard for the cultivar. This initial seed material is referred to as **breeder's seed** and is made available to seed producers for commercial distribution.

Commercial seed production not only requires (a) isolation and inspection but also (b) a controlled generation sequence combined with (c) a limited supply of primary or nuclear plants under rigid controls to maintain genetic purity in a few cases. The primary source might be maintained by vegetative propagation, as in asparagus or brussels sprouts. Most primary seed sources are maintained as **foundation seed stock** under control of a seed certifying agency or by a commerical seed company that supplies **stock seed** to contracted growers for production of commercial seed.

## Seed Certification

Commercial seed production of various agronomic crops is regulated through a system of voluntary **seed certification** (*3, 11*). Such programs exist in most states of the United States through the cooperative efforts of public research, extension, and regulatory agencies in agriculture and a state seed-certifying agency often known as a Crop Improvement Association. The agency is usually designated by law to certify seeds. Its members include growers as well as other interested individuals involved in the production of "certified seed." These individual state organizations are coordinated through the Association of Official Seed Certifying Agencies (AOSCA) (*3*) in the United States and Canada. International certification is regulated through the Organization for Economic Cooperation and Development (OECD).

The principal objective of seed certification is to protect the genetic qualities of a cultivar, but requirements of seed quality also may be enforced. The seed-certifying agency determines the eligibility of particular cultivars; sets up production standards for isolation, presence of off-type plants, and quality of harvested seed; and makes regular inspections of the production fields to see that the standards are maintained. Supervision of seed processing is also involved. Recommended minimum standards exist for many crops. Since specific requirements for certification may vary from state to state, the local, state, and national regulations should be examined.

Genetic purity is maintained by utilizing certain classes of seed that designate the generations allowed away from the original source. Their purposes are to allow an increase in seed supplies under conditions to preserve genetic identity and purity.

**Breeder's seed** is that which originates with the sponsoring plant breeder or institution and provides the initial source of all the certified classes.

**Foundation seed** is the progeny of breeder's seed and is so handled as to maintain the highest standard of genetic identity and purity. It is the source of all other certified seed classes but can also be used to produce additional foundation seed plants. **Select seed** is a comparable seed class used in Canada.

**Registered seed** is the progeny of foundation seed (or sometimes breeder's seed or other registered seed) that is produced under specified standards approved and certified by the certifying agency and designed to maintain satisfactory genetic identity and purity.

**Certified seed** is the progeny of registered seed (or sometimes breeder's, foundation, or other certified seed) and is that which is produced in the largest volume and sold to growers. It is produced under specified standards designed to maintain a satisfactory level of genetic identity and purity and is approved and certified by the certifying agency.

Bags of the different seed classes are identified by different color tags attached so that they are difficult to remove without defacing the bags.

Bags of certified seed have a *blue* tag distributed by the seed certifying agency as evidence of the genetic identity and purity of the seed contained therein. Bags of registered seed are labeled with a blue tag marked with the word "registered" or with a *purple* tag. Likewise, foundation seed is labeled with a regular certified seed tag with the word "foundation" or a *white* tag. The International OECD scheme includes **basic** (equivalent to either foundation or registered) seed. It also allows for **certified first generation** (blue tag) and **second generation** (red tag) seed.

Subsequent production of seeds for commercial purposes requires that the seed supply be increased to a large volume to supply potential demand. During this period of massive increase and consecutive seed generations, control over the process may be relaxed to reduce the cost and make production feasible. During this process of increase, there is danger that the genetic variability of seed populations may undergo subsequent change because of environmental selection or differential roguing by different seed producers. With time, characteristics of the same cultivar produced by different lines of production may undergo some change, not because of effects on the genotype of individual plants, but because of changes in proportions of different genotypes within the population. Certain seedlings survive better than others and contribute more to the next generation. The result might be a **genetic shift** (*48*). This potential for change is compounded if chance mutation or contamination from other cultivars occurs.

A potentially serious example has been described for certain forage crops. Cultivars were originated in a cold-winter area but the commerical seed was produced in a warm-winter area (*20*). Over a period of time the cultivar was found to become less winter hardy. To minimize the opportunity for such genetic shifts to occur, only one generation of seed production is allowed in regions of mild climate for certain crops.

### Production of Seed for Hybrid Cultivars

Hybrid cultivars have become an increasingly important category of cultivated plants. These are the $F_1$ progeny produced by the repetitive crossing of two or more parental lines that are maintained either (a) by seed, such as inbred lines, or (b) asexually as clones. Mass production of hybrid seed requires a system to prevent self-pollination and to enforce cross-pollination.

Control by **hand pollination** involves emasculation of pollen bearing parts of the flower and applying pollen by hand. It is used in breeding operations and except for certain cultivars of flower crops, such as petunia, is too expensive to use in production. Hand pollination for seed production is usually carried out commercially in greenhouses in countries where a supply of low-cost labor is available.

Other methods of hybrid production require the availability of plant characteristics that prevent production of pollen from the seed-bearing parental line. These procedures include the following:

**Male (pollen) sterility**
*vegetables:* onion, carrot, leek, radish, tomato
*flowers:* petunia, marigold,* dianthus,* geranium,* snapdragon, zinnia*
*field crops:* corn, sugar beets, sorghum, pearl millet, wheat, barley (*6, 18*)

**Self-Incompatibility**
*vegetables:* cabbage (*37*), broccoli, brussels sprouts
*flowers:* ageratum, *Bellis*
*trees:* almond-peach hybrids
*field crops:* some grasses (*7*)

---

* Pollen applied manually

## Hand Pollination Techniques

Hands and instruments are washed with alcohol before handling pollen or flowers to eliminate foreign pollen. Pollen should be collected from flower buds immediately preceding bloom and before the anthers open (dehisce). Anthers are extracted by pulling them from the flowers with tweezers, by squeezing the flower between the fingers, or by rubbing the buds across a wire screen. Anthers are dried on a sheet of paper until one can observe their opening under a magnifying glass. Pollen and anthers can be further screened through a fine sieve to remove extraneous material; they are then stored in a glass vial or bottle. Catkins (walnut, birch, aspen) or staminate cones (pine) will open on drying and will shed large quantities of pure, dry pollen if placed on a sheet of paper in a warm room.

Most kinds of pollen will remain viable for only a few days or weeks at warm temperatures, but many kinds can be preserved for several months to several years if stored at low temperatures and relatively low humidity. Effective storage conditions are a combination of 10 to 50 percent relative humidity and a temperature of 0° to 10°C (32° to 50°F). Moisture content of the pollen can be controlled by storing over a desiccant, such as calcium chloride or sulfuric acid. Some pollen—that of grasses, for instance—is best stored at 90 to 100 percent relative humidity. Pollen can be stored effectively at about $-18°C$ (0°F), as in a home freezer.

The stigma at the tip of the pistil must be exposed to apply the pollen. If the plant is self-fertile, the stamens must be removed (emasculated). Petals and stamens can be removed at the bud stage either with tweezers or by cutting the base of the flower between the fingernails. If the flower is to be self-fertilized or is a self-sterile cultivar, the flower need not be emasculated but only enclosed to keep out pollinating insects or wind-blown pollen.

Pollen is applied to the sticky, receptive stigma of insect-pollinated flowers with a fine-hair brush, a glass rod, a pencil eraser, or the tip of the finger. Removal of all the flower parts with the exception of the pistil reduces or eliminates the likelihood of further chance pollination by insects. Complete protection can be obtained by covering the individual flower, the pistil, or the branch with cellophane, plastic, paper, or a cloth bag.

For wind-pollinated plants, such as grasses, pines, and walnut, female flowers (or cones) must be carefully sealed throughout the flowering period. They can be covered with cellophane, paper, or a tightly woven cloth bag. Pollen can be transferred by inverting a paper bag containing pollen over the exposed flower; or dry pollen can be blown into the bag with an atomizer or inserted with a hypodermic needle.

The flower, or the branch on which it is borne, should be carefully labeled to maintain a record of the pollen used. The usual method is to identify the cross as follows: *"seed parent × pollen parent."*

**Dioecy**

*vegetables:* spinach, asparagus

*field crops:* some grasses ( *7* )

**Monoecy**

*vegetables:* cucurbits (special genotypes) ( *10, 40, 41* )

*field crops:* field and sweet corn (manual removal of tassels)

*trees:* Paradox hybrid walnuts (natural cross-pollination)

**Chemical Control**

*induction of male sterility:* cotton ( *13* )

*female flower production by gibberellic acid:* cucurbits ( *49* )


## PLANT VARIETY PROTECTION ACT

In the United States, breeders of a new seed-reproduced plant variety (cultivar) may retain exclusive propagation rights through the protection given by the Plant Variety Protection Act, which became effective in 1970 ( *4, 42* ). The breeder applies to the U.S. Department of Agriculture for a Plant Variety Protection Certificate. For one to be granted, the cultivar must be "novel." It must differ from all known cultivars by one or more morphological, physiological, or other characteristics. It must be uniform; any variation must be describable, predictable, and acceptable. It must be stable; i.e., essential characteristics must remain unchanged during propagation. A certificate is good for 17 years. The applicant may designate that the cultivar be certified and that reproduction continue only for a given number of seed generations from the Breeder or Foundation stock. If designated that the cultivar be certified, it becomes unlawful under the Federal Seed Act to market seed by cultivar name unless it is certified.


## SEED SOURCES OF WOODY PERENNIALS

Most trees and shrubs are heterozygous and cross-pollinated and have considerable potentiality for genetic variability. Two general problems are involved. One is the variability that might occur within particular seed lots. The second is suitability of a particular seed source and how close one can predict the characteristics of the offspring from the characteristics of the parents.

Seedling-grown woody plants are used for many purposes, and the amount of allowable variability will differ with intended use. Where the characteristics of the individual plant are important, as in landscaping, careful seed selection is essential or vegetative propagation methods become necessary. Uniformity among plants of a given rootstock is important for fruit-tree growing. In both cases, identifying seed sources that produce seedlings of the desired type is necessary.

On the other hand, some seedling variation is beneficial in reforestation where initial seed planting density is high. Close spacing promotes good stem form; competition among plants gradually eliminates the weakest trees. Modern forestry practices, however, emphasize upgrading tree production through seed selection to reproduce trees having high timber quality, rapid growth habit, and other desirable characteristics.

## Principles of Selection

IDENTIFICATION OF SEED ORIGIN

The climatic and geographical locality where tree seed is produced is referred to as **seed origin** in the United States or **provenance** in Europe (*16, 23, 31, 43,*).

Knowledge of seed origin is important when seed collections are made from a single species that grows over a wide range of ecologically different areas in nature. Variation can occur among groups of plants associated with latitude, longitude, and elevation (*2, 17*). Differences may be shown by morphology, physiology, adaptation to climate and to soil, and in resistance to diseases and insects. If seeds of a given species are collected in one locality, plants may be produced that are completely unadapted to another locality. Seed collected from trees in warm climates or at low altitudes is likely to produce seedlings that will not stop growing sufficiently early in the fall to escape freezing when grown in colder regions. Although the reverse situation—collecting seed from colder areas for growth in warmer regions—would be more satisfactory, it might result in a net reduction in growth resulting from the inability of the trees to utilize the growing season fully because of differences in response to photoperiod.

One application of the seed origin concept is to use only "local" seed whenever the adaptation of a seed source is not known for a given location. "Local seed" means seed from an area subject to similar climatic influences. This is usually considered to mean within 160 m (100 mi) of the planting and within 1600 m (1000 ft) of its elevation. A second application is to use seed from a region having as nearly as possible the same climatic characteristics. A third application is to use seed from a known source shown to have the same or improved adaptability for that area. Important environmental characteristics to consider are length of growing season, mean temperature during the growing season, frequency of summer droughts, and latitude.

**Seed-collection zones,** designating particular climatic and geographical areas, have been established in most of the forest tree growing areas in the world (*1, 43*). California, for example, is divided into six physiographic and climatic regions, 32 subregions, and 85 seed-collection zones (*44*). Similar zones are established in Washington and Oregon and in the central region of the United States (*32*).

Distinct ecotypes have been identified in many native forest tree species, including Douglas fir, ponderosa pine, lodgepole pine, red pine, eastern white pine, slash pine, loblolly pine, shortleaf pine, and white spruce (*43*). Some ecotypes have been shown to be superior seed sources and, in fact, may be preferable to local sources—for example, the East Baltic race of Scotch pine, the Hartz Mountain source of Norway spruce, the Sudeten (Germany) strain of European larch, the Burmese race of teak, Douglas fir from the Palmer area in Oregon, ponderosa pine from the Lolo Mountains in Montana, and white spruce from the Pembroke, Ontario (Canada) area.

Seed source is also important in growing conifers and various hardwood species of trees in the nursery industry. Plants are grown for individual use in the landscape or for special purposes, such as Christmas trees (*15, 27*), in which specific tree characteristics are important. Some characteristics can be demonstrated in nursery tests (*5*), for example Figure 4–2. Douglas fir has at least three recognized races, *viridis, caesia,* and *glauca,* with various geographical strains within them (*23*). The *viridis* strain from the U.S. west coast was not

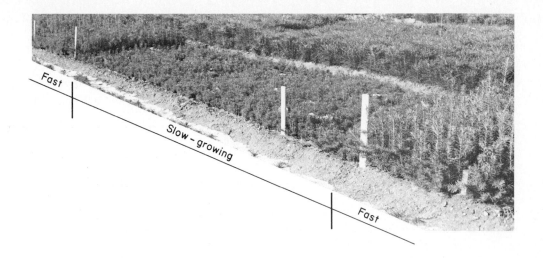

**Figure 4–2** Two-year Douglas fir (*Pseudotsuga menziesii*) seedlings from different seed sources growing side by side in nursery plots. *Left:* two fast-growing green strains from Washington and British Columbia. *Center:* four slow-growing strains from Montana and Wyoming, mostly gray-green in color. *Right:* fast-growing strain from Arizona, deep blue in color. Seedlings from different sources vary in height from 5 to 10 cm (2 to 4 in.) up to 25 to 43 cm (10 to 16 in.). Courtesy C. E. Heit.

winter-hardy in New York tests but it is well suited to Western Europe (*23*). Strains collected further inland were winter-hardy and vigorous; those from Montana and Wyoming were very slow growing. Trees of the *glauca* (blue) strain from the Rocky Mountain region were winter-hardy but varied in growth rate and appearance. Similarly, tree differences due to source occurred in Scotch pine, mugho pine, Norway spruce, and others.

SELECTION OF SEED TREE OR SHRUB SOURCES

Selection of special seed tree or shrub sources may be necessary to reproduce especially desired characteristics and to improve quality in various characteristics of the species (*35*) over that of "local seed." There is evidence that the *phenotype* of the individual seed tree is a good indicator of the *phenotype* of the offspring that will be produced (*32*). Consequently, parent tree selection can be effective for determining many characteristics, such as stem form, branching habit, growth rate, resistance to diseases and insects, presence of surface defects, and other qualities. This procedure is known as **phenotype selection.**

When individual trees in native stands show a superior phenotype, foresters call them "plus" trees. Such trees are significant in natural reseeding and are often used as seed sources. In nature such dominant seed trees may contribute the bulk of natural reseeding in any given area (*43*).

On the other hand, collecting seed from an isolated plant, particularly if it is growing near others of related species, as it might in an arboretum (*50, 53*), is usually undesirable. Such plants are likely to be self-pollinated with seed production poor, seeds of inferior quality, and seedlings weak and variable. Moreover, cross-pollination with an undesirable pollen source, such as a different species, can nullify the potential value of a given seed source and produce excessive and unpredictable variability.

Before about 1915, "Old French" pear seedlings were grown in the United States from seed produced in Europe by *Pyrus communis* trees. These seedlings were used as rootstocks for pear cultivars. Much variability in the seedlings resulted from seed produced by hybridization of the *P. communis* trees with nearby *P. nivalis* trees. Similarly, "Oriental pear" rootstock seedlings used in California about the same time originated from seeds obtained from Asia with little knowledge about the characteristics of either the seed tree or the pollen source. Plants grown from these sources have turned out to be a heterogeneous group with mixed parentage of many species and hybrids (*8, 24*).

Differences among flame eucalyptus (*Eucalyptus ficifolia*) trees arising from separate seed collections taken from a particular seed source in Australia have been explained by variable seed collection. Seeds collected from the center of the block of trees reproduced the true species with upright, uniform trees, while seeds from trees at the edge of the block were hybrids with surrounding *E. calophylla* trees, and consequently produced many dwarfed and off-type plants (*46*).

In selecting an individual seed tree or shrub source, three steps should be followed:

**1.** Evaluate the phenotypic characteristics of the seed tree or trees.

**2.** Evaluate the characteristics of any surrounding trees that could provide cross-pollination.

**3.** Where possible, conduct a seedling progeny test of the trees selected.

Evaluating the phenotypic characteristics of the seed trees is important because plants possessing the desired characteristics are most likely to pass them on to their offspring. This assumption cannot be made, however, without consideration of pollen sources and progeny testing. For most woody tree species, cross-pollination will produce, in general, seeds of better quality (*14*).

The most desirable procedure is to collect seeds from groups of the same kind of plants with desired characteristics growing in a **pure stand,** where cross-pollination can come from a number of plants of the same type. Seedling plants grown from such sources, although not necessarily homozygous, would be most likely to reproduce the characteristics observed among the parent seed trees.

The true genetic value of a particular seed source can be identified by a **seedling progeny** test (*5*). A representative sample of seeds is planted, and the resulting progeny are grown under test conditions that will identify essential characteristics or demonstrate superiority to other sources, or both. This procedure is known as **genotype selection.**

Foresters refer to seed-source trees with a superior genotype, as demonstrated by a progeny test, as "elite" trees. Nursery tests can identify and characterize specific seed sources for landscape and Christmas tree uses (*27*). Progeny testing is an essential phase of identifying seed sources of superior "elite" forest trees (*5*).

Control of both seed and pollination source can be achieved by collecting from **seed orchards** with known pollination arrangements. This procedure utilizes improved seed sources in forest tree improvement programs (*43*).

Seedling rootstocks of fruit trees are usually produced from seed collected from known vegetatively propagated cultivars (or clones) whether as by-products of fruit production or from specially selected rootstock-producing clones. For example, to grow *Pyrus communis* seedlings, the so-called "French pear," seeds are obtained from 'Winter Nelis' pear trees grown for cross-pollination purposes in 'Bartlett' ('Williams Bon Cretien') pear orchards. All seedlings are 'Winter Nelis' × 'Bartlett' hybrids. Likewise, peach seeds or apricot seeds collected from fruit-drying yards come from commercial fruit orchards where, for example, the cultivars 'Lovell' peach and 'Royal' apricot are clones grown in a solid block of trees in which self-pollination occurs.

Progeny testing is used to identify clones that will transmit high yielding characteristics—as in tung (*Aleurites*) (*36*)—or important rootstock qualities, such as nematode resistance (*9*). 'Nemaguard' peach, for example, is a cultivar whose seedling offspring are resistant to nematodes. Seeds of such cultivars are collected from seed orchards.

### HYBRID TREE SEED PRODUCTION

Production of $F_1$ trees may result from crosses between species or within species (*16, 28, 52*). Hybridization between the parental plants to produce the desired $F_1$ offspring may be achieved by various means.

**1.** Hybrids may be produced by hand pollination. This procedure is used mostly to produce $F_1$ hybrids for testing purposes. Mass production of seeds by this method would normally be too expensive, although hybrids of *Pinus rigida* × *P. taeda* have been produced this way in Korea.

**2.** Hybrids may be produced in seed orchards by interplanting the parental trees. Hybrids of *Larix decidua* and *L. leptolepsis* have been produced in seed orchards in Europe. In Idaho, a western white pine (*Pinus monticola*) seed orchard was established in the late 1950s containing 13 parental clones selected by progeny testing (*52*). Their $F_1$ offspring showed an average of 30 percent resistance to blister rust. The $F_2$ generation showed an even higher average level of resistance in tests and are therefore utilized as a seed source. Hybrids of almond × peach have been produced by interplanting the parental peach and almond cultivars and then growing only the seeds of the almond, which is self-incompatible (*30*). A high percentage of seedlings produced are interspecies hybrids which must be identified in the nursery row.

**3.** Production of Paradox hybrid walnut seed is obtained by collecting seed from individual Northern California black walnut (*Juglans hindsii*) seed trees where natural cross-pollination is known to take place with pollen coming from nearby English walnut (*Juglans regia*) trees (*47*). Seed-source trees are identified by progeny testing. Individual Paradox seedlings are identified in the nursery by increased vigor and by leaf characteristics, as shown in Figure 4–3. Other seedlings are rogued out.

**4.** Seed from $F_1$ seed trees gives $F_2$ or second generation hybrids. Collection of such seed where interspecific hybrids are involved produces a group of seedlings that are highly variable in regard to vigor and other characteristics. As a group these may fail to reproduce the desirable attributes of the $F_1$ hybrid generation, although some individuals may be highly desirable. Such variability is undesirable if the hybrid

**Figure 4–3** Paradox hybrid (*Juglans hindsii* × *J. regia*) seedlings (*far left*) sometimes appear among Northern California black walnut (*J. hindsii*) seedlings (*center*) being grown for rootstocks. Identification of the hybrid seedlings in the nursery row is by their lighter bark color and larger leaves. Leaves of Persian walnut (*top right*), Paradox hybrid (*center right*), and Northern California black walnut (*below right*).

plants are to be used for rootstocks or for landscaping where uniformity and predictability of the final form is important. On the other hand, in forest trees, some variability in vigor commonly associated with second-generation hybrids may have advantages (*16*). In tree plants with relatively high initial seedling density, competition among plants will favor the most vigorous and tend to eliminate the weaker, less desirable plants. Consequently, planting seeds from $F_1$ hybrids (the $F_2$) may be an economical way to obtain a natural stand of vigorous hybrids. $F_2$ hybrids of selected lines within species may be quite uniform and desirable as in the blister-rust-resistant, white pine hybrids described above (*52*). The value of any $F_1$ hybrid tree as a seed source should be established by progeny testing, as should that of any other seed source.

USE OF APOMICTIC SEEDLINGS

The seedlings produced by many citrus cultivars originate from apomictic embryos in the nucellus. Various citrus rootstock clones have been selected that are known to produce many apomictic seedlings. Such seedlings are particularly useful as rootstocks because they are highly uniform and vigorous and, being produced asexually, maintain the desirable qualities of the female parent.

**Procedures
in Selection
of Woody Plant Seed**

SEED-COLLECTION ZONE

A **seed-collection zone** (*43*) is an area with defined boundaries and altitudinal limits in which soil and climate are sufficiently uniform to indicate high probability of reproducing a single ecotype. Such zones have been mapped and indicated by numbered codes that are recognized by seed certifying agencies for production of **source-identified seed.**

### SEED-PRODUCTION AREAS

A **seed-production area** (*43*) is an area containing a group of trees that has been identified or set aside specifically as a seed source. Such seed trees within the area are selected for their desirable characteristics. The value of the area can be improved by removing off-type trees, those that do not meet minimum standards, and other trees or shrub species that would interfere with the operations. It is desirable to eliminate additional trees to provide adequate space for tree development and seed production. An isolation zone at least 120 m (400 ft) wide from which off-type trees are removed should be established around the area. Trees may or may not have been progeny tested prior to establishment of the area.

### SEED ORCHARDS

Orchards or plantations may be established specifically for seed production, propagated from seed trees of particular origin and quality, and preferably from progeny-tested seed trees. Seed orchards are used directly or indirectly in the production of fruit-tree rootstock seeds (*9*) and in the production of forest trees (*28, 32, 43*). Seed orchards are useful for maintaining rootstock seed trees involved in virus-control programs.

There are three general types of seed orchards (*32, 48*): (a) seedling trees produced from selected parents through natural or controlled pollination; (b) clonal seed orchards in which selected clones are propagated by grafting, budding, or rooting cuttings; and (c) seedling-clonal seed orchards in which certain clones are grafted into branches of some of the trees. The choice of which type should be used will depend upon the species that is being grown and upon the desired outcome of the seed improvement program.

Details of procedures for setting up a seed orchard vary with the species. A site should be selected for maximum seed production. For forest trees and most other native species enough different selections should be included in a suitable arrangement to insure cross-pollination and to decrease effects of possible inbreeding. Various fruit-tree rootstock clones are self-pollinated and are planted in solid blocks (*9*). An isolation zone at least 120 m (400 ft) wide should be established around the orchard to reduce pollen contamination from other sources. The size of this zone can be reduced if a buffer area of the same kind of tree is present around the orchard.

### NURSERY SELECTION

Desired individuals in variable seedling populations can be identified through selection in the nursery. Variability involves identifiable characters of vigor or appearance, or both. For example, Paradox hybrid walnut seedlings can be identified in nursery plantings of *Juglans hindsii* seedlings (see Figure 4–3). ''Blue'' seedlings of the Colorado spruce (*Picea pungens*) appear among others having the usual green form. Differences in fall coloring of *Liquidambar* and *Pistacia* trees among seedlings necessitate fall selection of individual trees for landscaping. Variation among seedlings grown for rootstocks may be reduced by grading to a size and eliminating the weak, small seedlings, a usual practice in U.S. nurseries.

## Tree-Certification Programs

Certification of tree seeds is available in some states and European countries similar to that for crop seed (*11, 29, 43*). Recommended minimum standards are given by the Association of Official Seed Certifying Agencies (*3*).

These programs seek to provide for improved, progeny-tested sources of forest-tree seed that are maintained and produced under standards of a regulatory body. Categories of seed developed for this program include the following:

**Source-identified tree seed** Seed is collected from natural stands where the geographic origin (source and elevation) is known and specified or from seed orchards or plantations of known provenance, specified by seed-certifying agencies. These seeds carry a *yellow* tag.

**Selected tree seed** Seed is collected from trees that have been rigidly selected for promising phenotypic characteristics but have not been progeny tested. The source and elevation must be stated. These seeds are given a *green* label.

**Certified tree seed** Seed shall be collected from trees of proven genetic superiority, as defined by a certifying agency, and produced under conditions that assure genetic identity. These could come from trees in a seed orchard, or from superior (''plus'') trees in natural stands with controlled pollination. These seeds carry a *blue* tag.

## REFERENCES

1. Aldous, J. R. 1975. Nursery practice. London: *Forestry Commission Bull. No. 43.*

2. Anonymous. 1961. *Forest tree seed directory.* Rome: Food and Agriculture Organization of the United Nations.

3. Assoc. Off. Seed Cert. Agencies. 1971. *AOSCA certification handbook.* Publ. No. 23.

4. Barton, J. H. 1982. The international breeder's rights system and crop plant innovation. *Science* 216:1071–75.

5. Barker, S. C. 1964. Progeny testing forest trees for seed certification programs. *Ann. Rpt. Inter. Crop Imp. Assoc.* 46:83–87.

6. Briggs, F. N., and P. Knowles. 1967. *Introduction to plant breeding.* New York: Reinhold.

7. Burton, G. W., and G. F. Sprague. 1961. Use of hybrid vigor in plant improvement. In *Germplasm resources,* R. E. Hodgson, ed. Washington, D.C.: Amer. Assoc. Adv. Sci., pp. 191–204.

8. Catlin. P. B., and E. A. Olsson. 1966. Identification of some *Pyrus* species after paper chromatography of leaf and bark extracts. *Proc. Amer. Soc. Hort. Sci.* 88:127–44.

9. Cochran, L. C., W. C. Cooper, and E. C. Blodgett. 1961. Seed for rootstocks of fruit and nut trees. In *Seeds: Yearbook of agriculture.* Washington, D.C.: U.S. Govt. Printing Office, pp. 233–39.

10. Conner, L. J., and E. C. Martin. 1971. Staminate: Pistillate flower ratio best suited to the production of gynoecious hybrid cucumbers for machine harvest. *HortScience* 6(4):337–39.

11. Cowan, J. R. 1972. Seed certification. In *Seed biology,* Vol. 3, T. T. Kozlowski, ed. New York: Academic Press, pp. 371–97.

12. Craig, R. 1976. Flower seed industry. In J. W. Mastalerz, *Bedding plants: A manual on the culture of bedding plants as a greenhouse crop.* University Park, Pa.: Pennsylvania Flower Growers, pp. 25–46.

13. Eaton, F. M. 1957. Selective gametocide opens way to hybrid cotton. *Science* 126:74–75.

14. Ehrenberg, C., A. Gustafsson, G. P. Forshell, and M. Simak. 1955. Seed quality and the principles of forest genetics, *Heredity* 41:291–366.

15. Flint, H. 1970. Importance of seed source to propagation. *Proc. Inter. Plant Prop. Soc.* 20:171–78.

16. Fowells, H. A. 1961. Making better forest trees available. In *Seeds: Yearbook of agriculture.* Washington, D.C.: U.S. Govt. Printing Office, pp. 378–82.

17. ———. 1965. *Silvics of forest trees of the United States.* U.S. Dept. of Agric. Handbook No. 271. Washington, D.C.: U.S. Govt. Printing Office, pp. 1–762.

18. Fryxell, P. A. 1957. Mode of reproduction of higher plants. *Bot. Rev.* 23:135–233.

19. Gabelman, W. H. 1956. Male sterility in vegetable breeding. In *Genetics in plant breeding.* Brookhaven Symposia in Biology No. 9, pp. 113–22.

20. Garrison, C. S., and R. J. Bula. 1961. Growing seeds of forages outside their regions of use. In *Seed: Yearbook of agriculture.* Washington, D.C.: U.S. Govt. Printing Office, pp. 401–6.

21. Goldsmith, G. A. 1968. Current developments in the breeding of $F_1$ hybrid annuals. *HortScience* 3(4):269–71.

22. ———. 1976. The creative search for new $F_1$ hybrid flowers. *Proc. Inter. Plant Prop. Soc.* 26:100–103.

23. Haddock, P. G. 1968. The importance of provenance in forestry. *Proc. Inter. Plant Prop. Soc.* 17:91–98.

24. Hartman, H. 1961. Historical facts pertaining to root and trunkstocks for pear trees. *Oregon State Univ. Agr. Exp. Sta. Misc. Paper 109,* 1–38.

25. Hartmann, H. T., W. J. Flocker, and A. M. Kofranek. 1981. *Plant science: Growth, development, and utilization of cultivated plants.* Englewood Cliffs, N.J.: Prentice-Hall.

26. Hawthorn, L. R., and L. H. Pollard. 1954. *Vegetable and flower seed production.* New York: Blakiston Co.

27. Heit, C. E. 1964. The importance of quality, germinative characteristics and source for successful seed propagation and plant production. *Proc. Inter. Plant Prop. Soc.* 14:74–85.

28. Hoekstra, P. A., E. P. Merkel, and H. R. Powers, Jr. 1961. Production of seeds of forest trees. In *Seeds: Yearbook of agriculture.* Washington, D.C.: U.S. Govt. Printing Office, pp. 227–32.

29. Horne, F. R. 1964. Forest tree seeds in Europe and the OECD proposals. *Ann. Rpt. Inter. Crop Imp. Assoc.* 46:90–94.

30. Kester, D. E., and R. N. Asay. 1975. Almonds. In *Advances in plant breeding,* J. Janick and J. D. Moore, eds. Columbus, Ohio: Ohio State Univ. Press.

31. Langlet, O. 1962. Ecological variability and taxonomy of forest trees. In *Tree growth,* T. T. Kozlowski, ed. New York: Ronald Press, pp. 357–69.

32. Linstrom, G. A. 1965. Interim forest tree improvement guides for the central states. *U.S. For. Ser. Res. Paper CS-12,* pp. 1–63.

33. Mastalerz. J. W. 1976. *Bedding plants: A manual on the culture of bedding plants as a greenhouse crop.* University Park, Pa.: Pennsylvania Flower Growers.

34. McMillan-Browse, P. D. A. 1979. *Hardy, woody plants from seed.* London: Growers Books.

35. Mergen, F. 1962. Selection of superior forest trees. In *Tree growth,* T. T. Kozlowski, ed. New York: Ronald Press, pp. 327–44.

36. Merrill, S., Jr., et al. 1954. Relative growth and yield of budded and seedling tung trees for the first seven years in the orchard. *Proc. Amer. Soc. Hort. Sci.* 63:119–27.

37. Odland, M. L., and C. L. Noll. 1950. The utilization of cross-incompatibility and self-incompatibility in the production of $F_1$ hybrid cabbage. *Proc. Amer. Soc. Hort. Sci.* 55:391–402.

38. Pearson, O. H. 1968. Unstable gene systems in vegetable crops and implications for selection. *HortScience* 3(4):271–74.

39. Pendleton, R. A., H. R. Fennell, and F. C. Reimer. 1950. Sugar beet seed production in Oregon. *Ore. Agr. Exp. Sta. Bul. 437.*

40. Peterson, C. E. 1960. A gynoecious inbred line of cucumber. *Mich. Agr. Exp. Sta. Quart. Bul.* 43:40–42.

41. Pike, L. M., and M. A. Mulkey. 1971. Use of hermaphrodite cucumber lines in development of gynoecious hybrids. *HortScience* 6(4):339–40.

42. Rollins, S. F. 1971. The plant variety protection act. *Seed World* 109(9):8–9.

43. Schopmeyer, C. S., ed. 1974. *Seeds of woody plants in the United States.* U.S. Dept. Agr. Handbook No. 450. Washington, D.C.: U.S. Govt. Printing Office.

44. Schubert, G. H., and R. S. Adams. 1971. *Reforestation practices for conifers in California.* Sacramento: Calif. State Div. of Forestry.

45. Sprague, G. F. 1950. Production of hybrid corn. *Iowa Agr. Exp. Sta. Bul. P48,* pp. 556–82.

46. Stoutemyer, V. T. 1960. Seed propagation as a nursery technique. *Proc. Plant Prop. Soc.* 10:251–55.

47. Stuke, W. 1960. Seed and seed handling techniques in production of walnut seedlings. *Proc. Plant Prop. Soc.* 10:274–77.

48. Thomson, J. R. 1979. *An introduction to seed technology.* New York: John Wiley.

49. Weaver, R. J. 1972. *Plant growth substances in agriculture.* San Francisco: W. H. Freeman & Co. Publishers.

50. Westwood, M. N. 1966. Arboretums—a note of caution on their use in agriculture. *HortScience* 1:85–86.

51. Wills, A. B., and C. North. 1978. Problems of hybrid seed protection. *Acta Hort.* 83:31–36.

52. Wright, J. W. 1963. New forest tree varieties. *Agr. Sci. Rev.* 1:27–37.

53. Wyman, D. 1953. Seeds of woody plants. *Arnoldia* 13:41–60.

# SUPPLEMENTARY READING

BRIGGS, F. N., and P. KNOWLES. 1967. *Introduction to plant breeding.* San Francisco: W. H. Freeman & Co. Publishers.

COPELAND, L. O. 1976. *Principles of seed science and technology.* Minneapolis: Burgess.

SCHOPMEYER, C. S., ed. 1974. *Seeds of woody plants in the United States.* USDA Agr. Handbook No. 450. Washington, D.C.: U.S. Govt. Printing Office.

SEEDS: YEARBOOK OF AGRICULTURE. 1961. Washington, D.C.: U.S. Govt. Printing Office.

The production of high-quality seed is of prime importance to propagators, whether they collect or produce the seed themselves or obtain the seed from others. In the production of any crop, the cost of the seed is usually minor compared to other production costs. Yet no single factor is as important in determining the success of the operation.

## SOURCES OF SEED

### Commercial Seed Production

The commercial production of seeds is a large, specialized industry that produces cereal and forage crop seed, vegetable seed, and annual, biennial, and perennial flower seeds for both commercial growers and home gardeners.

Commercial seed is raised primarily in areas where environmental conditions are suited for such production. Unfavorable climatic conditions before and during harvest might impair viability or increase disease (2, 4). Much vegetable and flower seed is produced in somewhat limited areas that are especially suitable, although climate may not necessarily correspond with that of the locality in which the crops will be grown. Low humidity and lack of summer rainfall are desirable conditions for seed crops that must be dried for harvesting. On the other hand, low humidity may be undesirable for some seed crop species since it can cause premature shattering of the seed pods and cracking of seeds during harvesting. For this reason, much of the U.S. flower seed production is located in a limited area along the west coast where moist air from the Pacific Ocean and frequent night and morning fogs tend to prevent the pods from dehiscing during harvest.

Conditions of low atmospheric moisture make it relatively easy to control fungus and bacterial diseases. For instance, two seedborne diseases, anthracnose and bacterial blight, are serious problems in bean seed production in all but

5

# Techniques of Seed Production and Handling

the drier parts of the country. The mountain states and the central valley of California are particularly desirable for bean seed production because of low humidity. Furrow irrigation may be more desirable than sprinklers in seed production fields because it allows better disease control.

Adequate isolation of cross-pollinated plants also is important and may influence the choice of seed-production areas. For example, little sweet corn seed is produced in the Midwestern states of the United States because of the large amount of field corn grown there whose pollen would contaminate the seed production procedures.

## Seed Collecting

Tree and shrub seeds are usually collected from plants not grown specifically for their seeds. Seeds of native species of plants for forest planting and similar purposes may be obtained from natural stands in the forests and other wild areas. Such seed may be collected from trees felled in logging operations, from standing trees, or from squirrel caches. Seed-production areas have been designated in certain regions. Other sources may include parks, roadways, streets, or woodlots. Seed selection for genetic adaptability to a growing area is essential.

## Seed Orchards

Seed orchards or plantations are used to maintain a certain seed source of a particularly valuable genotype. They can also maintain a source of plant material that otherwise might be difficult to locate, produce higher seed yields than would be found in nature, and prevent virus contamination. Seed orchards are used to produce virus-tested fruit tree rootstocks. The procedure has considerable importance in forestry, where scions of genetically superior trees are grafted into plantings for seed production and become an important aspect of forest tree improvement.

## Fruit-Processing Industries

Seeds of fruit cultivars used for producing rootstocks may be obtained as by-products of fruit-processing industries. For instance, pear seed may be obtained from canneries, apple seed from canneries or cider presses, and some cultivars of apricot and peach seed from fruit-drying yards. An important example of this is the 'Lovell' peach, which has been a useful drying peach cultivar in California and also an important rootstock for a number of stone fruits. Peach pits are collected in large numbers at drying yards, cleaned, and sold to nurserymen. This kind of source can be undesirable, however, if one does not know the genetic identity or the virus status of the seed source.

# HARVESTING AND PROCESSING SEEDS

## Maturity and Ripening

A seed becomes mature when it has reached a stage at which it can be removed from the plant without impairing the seed's germination. Usually it has reached a stage on the plant when no further increase in dry weight will occur (*2*). If the fruit ripens or the seed is harvested when the embryo is insufficiently developed, the seed is apt to be thin, light in weight, shriveled, poor in quality, and short-lived (*30*). In addition, various changes in the seed occur during maturation, including dehydration and changes in the physiology of the seed and seed covering.

Early seed harvest may actually be desirable for seeds of some species of woody plants that produce a hard seed covering in addition to a dormant embryo. If seeds become dry and the seed coats harden, the seeds may not germinate until the second spring (*38*), whereas otherwise they would have germinated the first spring. On the other hand, if harvesting is delayed too long, the fruit may dehisce or "shatter," drop to the ground, or be eaten or carried off by birds or animals. Thus, a balance must be made between late and early harvest in order to obtain the maximum number of high-quality seed.

Plants can be divided into three types according to the way their fruit ripens (*30*):

**1.** Plants with dry fruits that do not dehisce and do not disseminate seeds immediately upon maturity. This type includes most of the crop plants (such as corn, bean, wheat, and other grains) whose seeds are used for food or other human needs and which have undergone considerable selection during domestication for ease of handling.

**2.** Plants producing dry fruits that dehisce readily at maturity and those that distribute single dry seeds or fruits. This type includes follicles, pods, capsules, siliques, and cones of conifers. Species of Type 2 differ from Type 1 in that they are either native plants or are used for purposes other than food production.

**3.** Plants with fleshy fruits. Includes important fruit and vegetable species used for food (such as berries, pomes, and drupes) as well as many related tree and shrub species used in landscaping or forestry.

## Harvesting Procedures for Seeds of Herbaceous Plants

*Type 1*    Seeds of field-grown crops that produce indehiscent fruits, such as cereals, grasses, and corn, can be harvested by a combine, a machine that cuts and threshes the standing plant in a single operation. Other plants having a tendency to fall over or "lodge" are cut, piled, or windrowed for drying and curing before the dry fruits are separated out. Low humidity is important during harvest to facilitate drying and curing. Rain damage results in seeds that show low vigor. Seeds of Type 1 may be somewhat difficult to separate from the plant, and the force required to dislodge them may result in mechanical damage. Such injury can reduce viability and result in abnormal seedlings (see Figure 5–1). Some of these injuries are internal and not noticeable, but they result in low viability after storage (*1, 22, 30*). Damage is a potential factor in any operation involving beating or flailing of the seeds, and is most likely to occur if the machinery used is not properly adjusted. Usually less injury occurs if the seeds are somewhat moist at harvest (around 12 to 15 percent).

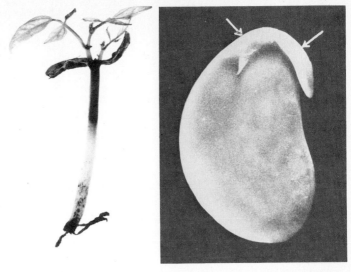

**Figure 5–1**    Seed injury during harvest can affect viability and produce seedling abnormalities. A break in the radicle below the cotyledons (*right arrow*) can prevent germination. Baldhead in lima beans results from a break in the stem of the embryo immediately above the cotyledons (*left arrow*). The slower development of the shoots (*left*) arising from the cotyledon axils results in delayed maturity and lower yields. From R. W. Allard, "Production of Dry, Edible Lima Beans in California," *California Agriculture Experimental Station Circular 423.*

*Type 2*    Plants with dry seeds and fruit that dehisce or "shatter" readily are cut (often by hand) and placed on a canvas or tray to dry for one to three weeks. To avoid seed losses plants of this group must be harvested before the fruits are fully ripe and then cured or dried before the seeds are extracted. The collector or harvester must know the criteria that indicate the optimum yield and quality for that particular kind of seed. A problem with Type 2 seeds is that those on a single plant may not develop uniformly and at harvest a portion of the seeds will be immature. Much of this poor seed can be eliminated by screening or blowing to remove the lightweight seed. This procedure is used for many flower crops, such as delphinium, pansy, and petunia, a number of vegetables, such as onion, cabbage and other cruciferous plants, and okra. Where only a few plants from a backyard garden are involved, they may be cut and hung upside down in a paper bag to dry.

In threshing, the plants are beaten, flailed, or rolled to loosen the seed from the plant. Special seed harvesting machines that are modified for individual crops are used to minimize mechanical damage. The essential parts are a revolving cylinder, which acts as a beater to loosen the seeds, and devices to separate the good seeds from the rest of the plant along with chaff, dirt, and other debris. Threshing may also be accompanied by pulling large rollers across the plants. For small lots of seeds, beating, flailing, or screening by hand may be utilized. Figure 5–2 shows a threshing box that may be convenient for a small amount of seed. Following threshing further cleaning may be required to remove dirt, chaff, extraneous plant parts, and weed and other crop seeds of different shapes and sizes. Small lots of seed can be cleaned by screening or by pouring back and forth from one container to another, letting the wind blow away the lighter materials. Commercial seed cleaning and processing utilizes various kinds of specialized equipment, such as screens of different sizes, air blasts, and gravity separators (*36, 37*). These take advantage of differences between the seeds and the materials to be removed in such physical characteristics as size, thickness, weight, friction, and color.

**Figure 5-2** Threshing box useful for cleaning small lots of dry seed.

*Type 3*    Herbaceous plants with fleshy fruits include tomato, pepper, eggplant, and the various kinds of cucurbits. Seeds tend to lose moisture during maturation even though surrounded by the flesh of the fruit. Fruits are harvested ripe, or in some cases (e.g., cucumber and eggplant) overripe. For small lots of seeds, the fruits may be cut open and the seeds within scooped out, cleaned, and dried. Commercial harvesting of these crops utilizes machines that macerate the fruit. The pulp and seeds are then separated by fermentation, mechanical means, or washing through screens.

Fermentation is used in extracting tomato seed. The macerated fruits are placed in large barrels or vats and allowed to ferment for about four days at about 21° C (70° F), with occasional stirring. If the process is continued too long, sprouting of the seeds may result. Higher temperatures during fermentation shorten the required time. As the pulp releases the seeds, the heavy, sound seeds sink to the bottom of the vat, and the pulp remains at the surface. Following extraction, the seeds are washed and dried either in the sun or in dehydrators. Additional cleaning is sometimes necessary to remove dried pieces of pulp and other materials. Extraction by fermentation is particularly desirable for tomato seed, because it controls bacterial canker.

Special machines have been developed to extract and clean the seeds from the pulp of cucumber and other vine crops. Following separation, the seeds are washed and dried as in the fermentation process.

---

**Drying**

Most kinds of seeds must be dried following harvest. If left in bulk for even a few hours, seeds that have more than 20 percent moisture will heat; this impairs viability. Drying may occur either naturally in open air if the humidity is low, or artificially with heat or other devices (*21*). Drying temperatures should not exceed 43° C (110° F); if the seeds are quite wet 32° C (90° F) is better. Too-rapid drying can cause shrinkage and cracking and can sometimes produce hard seed coats. The minimum safe moisture content for most seeds is in the range of 8 to 15 percent.

---

## Harvesting Procedures for Tree and Shrub Seed

Both dry and fleshy fruits can be collected from standing trees by shaking them onto canvas, by knocking with poles, by using cone hooks attached to long poles (as for conifers), or by hand picking. Collection may be made from trees felled in logging operations. In some cases squirrel caches yield high-quality seeds. Seeds of some street trees—elm and hackberry, for instance—can be swept up from the street. Seeds on low trees and shrubs can be harvested by hand picking, clipping seed stalks, or knocking. Cones from tall conifer trees can be removed by mechanical tree shakers.

### DRY FRUITS

Seeds are extracted from dry, dehiscent fruits as pods and capsules of such plants as certain woody legumes (honey locust), *Caragana, Ceanothus,* poplar, and willow. The fruits of these plants are spread out to dry in shallow layers on canvas, on cloth, on the floor or shelves of open sheds, or in screen-bottom trays. Air-drying takes one to three weeks if the relative humidity is low.

**Figure 5–3**    Power-operated seed cleaner for fleshy fruits. The whole fruits are placed in top of cleaner with side door closed. A stream of water from a hose washes over the fruit. A rapidly rotating plate with low vertical flanges is raised slightly from the bottom of the cleaner and removes the flesh from the fruit. The flesh washes out through the bottom of the cleaner leaving inside the cleaned seeds, which can then be removed through the side door.

Extraction may be accomplished by beating the pods with a flail, treading them under foot, or rubbing them through screens. For larger operations, commercial threshers are more suitable. A macerator has been developed by the USDA Forest Service to extract seed from both dry and fleshy fruits (*32*). Made of metal, it is sufficiently watertight that running water can be used in it when macerating fleshy fruits. Fruits and seeds pass through the hopper and are macerated by means of a revolving cylinder like that of a threshing machine. Such machines will extract and clean 500 lb of seed per hour. Following extraction, the seeds may require additional cleaning to remove extraneous materials, using conventional screening or fanning mills.

### CONES

Extraction of conifer seeds from cones requires special procedures. Cones of some species will open if they are dried in the open air for two to twelve weeks. Others must be force-dried at higher temperatures in special heating kilns. Under such conditions, the cones will open within several hours or at most two days. The temperature of artificial drying should be 46° to 60° C (115° to 140° F) depending upon the species, although a few require even higher temperatures. Jack pine (*Pinus banksiana*) and red pine (*P. resinosa*), for example, need temperatures of 77° C (170° F) for five to six hours. Caution must be used with high temperatures; overexposure will damage seeds. After the cones have been dried, the scales open, exposing the seeds. The cones must then be shaken by tumbling or raking to remove the seeds. A revolving wire tumbler or a metal drum is used when large numbers of seeds are to be extracted. The seeds should be removed immediately upon drying since the cones may close. Conifer seeds have wings, which are removed except in species whose seed coats are easily injured, such as incense cedar (*Calocedrus*). Fir (*Abies*) seed is easily injured, but the wings can be removed if the operation is done gently. Redwood (*Sequoia* and *Sequoiadendron*) seed have wings that are an inseparable part of the seed. For small lots of seed, dewinging can be done by rubbing the seeds between moistened hands or trampling or beating the seeds packed loosely in sacks. For larger lots of seeds, special dewinging machines are used. The seeds are cleaned after extraction to remove the wings and other light chaff. As a final step, separation of heavy, filled seed from light seed is accomplished by gravity or pneumatic separators.

### FLESHY FRUITS

In nature, many kinds of fleshy fruits are eaten by birds and the seeds disseminated through their digestive tracts. Such fruits must become sufficiently ripe to facilitate removal of the fleshy portion. Overly ripe fruit that fall to the ground may produce injury to the seed, through either heating or the effects of microorganisms. Fruits which are allowed to dry around the seed may contribute to a very hard covering that increases dormancy problems.

Fleshy fruits include berries (grape), drupes (peach, plum), pomes (apple, pear), aggregate fruits (raspberry, strawberry), and multiple fruits (mulberry). The flesh must be removed promptly to prevent spoilage and injury to the seeds. Cleaning by hand, treading in tubs, and rubbing through screens are suitable methods for small lots of seed. Relatively large fruits can be cleaned conveniently by placing the fruits in a wire basket and washing them with water from a high-pressure spray machine. For larger lots of seed, a **macerator** is convenient.

A macerator is constructed with a watertight feeder; water is passed through it along with the fleshy fruits, and the pulverized mass is diverted into a tank where the pulp and seeds can be separated by flotation.

A **flotation** process involves placing the seeds and pulp in water so that the heavy, sound seeds will sink to the bottom and the lighter pulp, empty seeds, and other extraneous materials will float off the top. This procedure can also be used for removing the poor seeds and other materials from dry fruits, such as acorn fruits infested with weevils.

Fruit-tree seeds collected from the wastes of drying yards, canneries, or juice presses should be separated from the pulp, washed as quickly as is convenient, and not allowed to ferment or heat in the piles. Such seeds can be handled by flotation or washing with high-pressure spray machines.

In the production of rootstock seedlings in oranges, separation of the seed from the surrounding fruit pulp is facilitated by the addition of a commercial pectinase enzyme (5).

The small berries of some species, such as *Cotoneaster, Juniperus,* and *Viburnum,* are somewhat difficult to process because of small size and the difficulty in separating the seeds from the pulp. One way to handle such seeds is to crush the berries with a rolling pin, soak them in water for several days, and then remove the pulp by flotation. Another device that removes seeds from small-seeded fleshy fruits is an **electric mixer** or **blender** (34). To avoid injuring the seeds, the metal blade of the latter machine can be replaced with a piece of rubber, 1½ in. square, cut from a tire casing. It is fastened at right angles to the revolving axis of the machine (40). A mixture of fruits and water is placed in the mixer and stirred for about two minutes. When the pulp has separated from the seed, the pulp is removed by flotation. This procedure is satisfactory for fruits of *Amelanchier* (serviceberry), *Berberis* (barberry), *Crataegus* (hawthorn), *Fragaria* (strawberry), *Gaylussacia* (huckleberry), *Juniperus* (juniper), *Rosa* (rose), and others (34).

# SEED STORAGE

Seeds are usually stored for varying lengths of time after harvest. Viability at the end of any storage period is the result of the initial viability at harvest, as determined by factors of production and methods of handling, and the rate at which deterioration takes place. This rate of physiological change, or aging, varies with the kind of seed and the environmental conditions of storage, primarily temperature and humidity.

Seeds of certain species are **short-lived** if they are not allowed to germinate immediately in their natural habitat (21, 40). The period of viability may be as short as a few days, months, or at most a year. This group particularly includes certain spring-ripening seeds of temperate zone trees such as poplar (*Populus*), some maple (*Acer*) species, willow (*Salix*), and elm (*Ulmus*). Their seeds drop to the ground and normally germinate immediately. Seeds of many tropical plants grown under high temperature and humidity conditions are short-lived. The group includes such plants as sugar cane, rubber, jackfruit, macadamia, avocado, loquat, citrus, many palms, litchi, mango, tea, choyote, cocoa, coffee, tung, and kola. Another group with short-lived seeds includes many aquatic plants of the temperate zone, such as wild rice (*Zizania aquatica*), pondweeds, arrowheads, and rushes. Many tree nut and similar species produce seeds with large fleshy cotyledons and are relatively short-lived, particularly if allowed to dry out—hickories and pecan (*Carya*), birch (*Betula*), hornbeam (*Carpinus*), hazel and filbert (*Corylus*), chestnut (*Castanea*), beech (*Fagus*), oak (*Quercus*), walnut (*Juglans*), and buckeye (*Aesculus*). Seed longevity in many of these species can be increased significantly with proper handling and storage.

Seeds that may be considered **medium-lived** are those that remain viable for periods of two or three up to perhaps 15 years, providing the seeds are stored at low humidity and, preferably, at low temperatures. Seeds of most conifers and commercially grown vegetables, flowers, and grains fall in this group.

Seeds that are **long-lived,** even at warm temperatures, generally have hard seed coats that are impermeable to water. If the hard seed coat remains undamaged, such seeds should remain viable for at least 15 to 20 years. The maximum life can be as long as 75 to 100 years and perhaps more. Records exist of seeds being kept in museum cupboards for 150 to 200 years still retaining viability. Indian lotus (*Nelumbo nucifera*) seeds that had been buried in a Manchurian peat bog for an estimated 1000 years germinated perfectly when the impermeable seed coats were cracked (*29, 39*). Some weed seeds retain viability for many years (50 to 70 years or more) while buried in the soil, even though they have imbibed moisture (*31*). Longevity seems related to dormancy induced in the seeds by environmental conditions deep in the soil.

## Seed Storage Factors Affecting Viability

The storage conditions that maintain seed viability are those that slow respiration and other metabolic processes without injuring the embryo. The most important conditions are low moisture content of the seed, low storage temperature, and modification of the storage atmosphere. Of these, the moisture temperature relationships have the most practical significance.

### MOISTURE CONTENT

Many kinds of the short-lived seeds are **desiccation sensitive** and lose viability if the moisture content becomes low. For instance, in silver maple (*Acer saccharinum*) seeds, the moisture content was 58 percent in the spring when the seeds matured. Viability was lost when the moisture content dropped below 30 to 34 percent (*25*). Citrus seeds can withstand only slight drying (*8, 14*) without loss of viability. The same is true for seeds of some water plants, such as wild rice, which can be stored directly in water at low temperatures (*27*). The large fleshy seeds of oaks (*Quercus*), hickories (*Carya*), and walnut (*Juglans*) lose viability if allowed to dry after ripening. They are normally stored moist for no more than one year (*32*).

Most species with medium- to long-lived seeds are **desiccation tolerant** and must be dry to survive long periods of storage. A 4 to 6 percent moisture content is favorable for prolonged storage (*16*), although a somewhat higher moisture level is allowable if the temperature is reduced (*35*). For example, for tomato seed stored at 4.5° to 10°C (40° to 50°F), percent moisture should be no more than 13 percent; if 21°C (70°F), 9 percent; and if 26.5°C (80°F), 9 percent. However, if the moisture content of the seed is too low (1 to 2 percent), loss in viability and reduced germination rate can occur in some kinds of seeds (*13*). Some seeds can be stored at these low moisture levels but must be rehydrated with water vapor before planting (*28*). Moisture in seeds is in equilibrium with the relative humidity of the storage atmosphere and increases if the relative humidity increases and decreases if it is reduced (*21*). Thus moisture percentage varies with the kind of seed, depending on the kind of storage reserves within the seed (6). Longevity of seed is maximum if stored in a relative humidity range of 20 to 25 percent (21).

Fluctuations in seed moisture during storage reduce seed longevity (7). Consequently the ability to store seeds exposed to the open atmosphere varies greatly in different climatic areas. Dry climates are conducive to increased longevity; in areas with high relative humidity, seed life is shorter. In open storage in tropical areas, seed viability is particularly difficult to maintain.

Various storage problems arise with increasing seed moisture (21). At 8 or 9 percent or more insects are active and reproduce; above 12 to 14 percent (65 percent relative humidity or more) fungi are active; above 18 to 20 percent heating may occur; and above 40 to 60 percent germination occurs.

Storage in sealed, moisture-resistant containers is advantageous for long storage, but the moisture content of the seed must be low at the time of sealing. In fact, a seed moisture content in a sealed container of 10 to 12 percent (in contrast to 4 to 6 percent) is worse than storage in an unsealed container (16).

TEMPERATURE

Reduced temperature invariably lengthens the storage life of seeds and, in general, can offset the adverse effect of a high moisture content. Harrington has given two "rules of thumb" (21): (1) for seeds not adversely affected by low moisture conditions, each 1 percent decrease in seed moisture, between 5 and 14 percent, doubles the life of the seed; and (2) each decrease of 5° C (9° F), between 0° and 44.5° C (32° and 112° F) in storage temperature, also doubles seed storage life. On the other hand, seeds stored at low temperature but at a high relative humidity may lose viability rapidly when moved to a higher temperature (9).

Subfreezing temperatures, at least down to $-18°$ C (0° F), will increase storage life of most kinds of seeds but moisture content should be in equilibrium with 70 percent relative humidity or lower, or the free water in the seeds may freeze and cause injury (21). Such storage is particularly useful for conifer seeds (32, 33). Refrigerated storage should be combined with dehumidification or with sealing dried seeds in moisture-proof containers.

ULTRA FREEZING

**Cryopreservation** of seeds immersed in liquid nitrogen ($LN_2$) at a very low temperature, $-196°$ C, can be used for many seeds providing the moisture content is relatively low, about 8 to 15 percent, and the seeds are in sealed containers (3, 11, 18). Regulating freezing and thawing rates may be important in the operation.

Special types of equipment are required in order to reduce the temperature to $-196°$C in liquid nitrogen (18). This procedure is useful for very long storage of valuable seed stocks as germplasm (11, 18).

## Types of Seed Storage

OPEN STORAGE WITHOUT MOISTURE
OR TEMPERATURE CONTROL

Many kinds of seeds that are used in large commercial volumes are stored in bins or in sacks or other containers. Under these conditions, seed longevity depends on the relative humidity and temperature of the storage atmosphere, although it also depends upon the kind of seed and their condition at the beginning of storage. Retention of viability consequently varies with the climatic factors of the area in which storage occurs. Poorest conditions are found in warm, humid climates; best storage conditions occur in dry, cold regions. Fumigation or insecticidal treatments may be necessary to control insect infestations.

Uncontrolled storage can be used for many kinds of commercial seeds for at least a year—i.e., to hold seeds from one season to the next. Seeds of many species, including most agricultural, vegetable, and flower seeds, will retain viability for longer periods up to four to five years (*12, 19, 26*),except under the most adverse conditions.

Seeds that have a water-impervious seed coat will retain viability in open storage for many years (10 to 20 years or more) once they have been dried. Open storage is adequate. Some woody plants whose seeds are handled in this manner are:

*Acacia* spp.

*Albizia* spp. (albizzia)

*Amorpha fruticosa* (indigo bush)

*Caragana arborescens*
   (Siberian pea shrub)

*Elaeagnus angustifolia* (Russian
   olive)

*Eucalyptus* spp.

*Koelreuteria paniculata* (golden rain
   tree)

*Rhus ovata* (sumac)

*Robina pseudoacacia* (locust)

*Tilia* (linden)

WARM STORAGE
WITH HUMIDITY CONTROL

Improved seed storage can be achieved by drying the seeds, then storing them in humidity-controlled rooms. Toole (*35*) recommends for vegetable seeds: (1) seeds exposed to 27° C (80° F) for more than a few days should be in air with relative humidity no higher than 45 percent; (2) seeds exposed to 21° C (70° F) should be at no higher humidity than 60 percent; (3) very short-lived seed (onion, peanut), old seed, or those contaminated with fungi should be kept at an even lower humidity.

Dry seeds may be stored in sealed moisture-resistant containers. Containers vary in durability and strength, cost, protective capacity against rodents and insects, and ability to retain or transmit moisture. Different materials vary in moisture transmitting qualities (*20*). Those completely resistant to moisture transmission include tin cans (if properly sealed), aluminum cans, hermetically sealed glass jars, and aluminum pouches. Those almost as good (80 to 90 percent effective) are polyethylene (3 mil or thicker) and various types of aluminum-paper laminated bags. Somewhat less desirable, in regard to moisture transmission, are asphalt and polyethylene laminated paper bags and friction-top tin cans. Paper and cloth bags give no protection against moisture change.

Seeds kept in sealed containers for long periods should have a low moisture content of no more than 5 to 8 percent, depending on the species.

Almost without exception, seed longevity in the species listed in the two previous categories would be enhanced by reducing the storage temperature to 10° C (50° F) or less. Toole (*35*) recommends that for vegetable seed stored at 4.5° to 10° C (40° to 50° F), the relative humidity be no higher than 70 percent, preferably no higher than 50 percent. When removed from storage at a relative humidity of more than 50 percent, seeds should be dried to a safe moisture content, unless they are planted immediately.

Low storage temperatures, down to freezing or lower, might be desirable if the need justifies the added cost. Below-freezing temperatures can be used for very long storage. In such low-temperature storage, relative humidity should be 70 percent or lower (*21*). Sealed containers should be used to keep out moisture, which could form ice crystals that damage seeds. Such seeds may deteriorate quickly unless brought out of storage.

The most effective storage procedure is to dry seeds to a low moisture content (3 to 8 percent), place them in sealed containers, and store them at a very low temperature. Such a procedure is best for seeds that deteriorate rapidly or when the maximum possible storage life is needed for a particular reason, for instance, to maintain seeds of breeding stocks.

Cold storage of tree and shrub seed used in nursery production is generally advisable if the seeds are to be held for longer than one year (*23, 24, 32, 33*), except for hard-coated seeds listed previously. Seed storage is useful in forestry because of the uncertainty of good seed-crop years. Seeds of many species are best stored under cold, dry conditions (*40*).

### COLD MOIST STORAGE

The storage temperature should be 0° to 10°C (32° to 50°F). Seeds should not be dried but should be either stored in a container that will maintain their high moisture content, or mixed with moisture-retaining material. The relative humidity in storage should be 80 to 90 percent. The procedure is similar to moist-chilling (stratification). Acorns and large nuts can be dipped in paraffin or

---

**U.S. National Seed Storage Laboratory**

The National Seed Storage Laboratory (*11*) was built in 1958 on the Colorado State University campus, Fort Collins, Colorado, to preserve seed stocks of valuable germplasm. Seeds are acquired from public agencies, seed companies, and individuals engaged in plant breeding or seed research. Descriptive material is received at the same time. Seed samples are tested for viability and, if suitable for long term storage, are dried to 5 to 7 percent moisture and stored at −10°C to −12°C in hermetically sealed metal cans. Seed lots are tested every five years for germinability. If viability drops, a new generation of seed is produced and stored. Seeds are made available to research workers and plant breeders on request. This facility also conducts seed storage research.

sprayed with latex paint before storage to preserve their moisture content (*24*).

Examples of species whose seeds require this storage treatment are: *Acer saccharinum* (silver maple), *Aesculus* spp. (buckeye), *Carpinus caroliniana* (American hornbeam), *Carya* spp. (hickory), *Castanea* spp. (chestnut), *Corylus* spp. (filbert), *Citrus* spp. (citrus), *Eriobotrya japonica* (loquat), *Fagus* spp. (beech), *Juglans* spp. (walnut), *Litchi, Nyssa silvatica* (tupelo), *Persea* spp. (avocado), and *Quercus* spp. (oak).

# REFERENCES

1. Asgrow. 1959. *A study of mechanical injury to seed beans.* Asgrow Monograph No. 1. New Haven: Associated Seed Growers, 1949.

2. Austin, R. B. 1972. Effects of environment before harvesting on viability. In *Viability of seeds,* E. H. Roberts, ed. Syracuse: Syracuse Univ. Press.

3. Bajaj, Y. P. S. 1979. Establishment of germplasm banks through freeze storage of plant tissue culture and their implications in agriculture. In *Plant cell and tissue culture principles and applications,* W. R. Sharp et al., eds. Columbus: Ohio State Univ. Press, pp. 745–74.

4. Baker, K. F. 1980. Pathology of flower seeds. *Seed Sci. and Tech.* 8:575–89.

5. Barmore, C. R., and W. S. Castle. 1979. Separation of citrus seed from fruit pulp for rootstock propagation using a pectolytic enzyme. *HortScience* 14(4):526–27.

6. Barton, L. V. 1941. Relation of certain air temperatures and humidities to viability of seeds. *Contrib. Boyce Thomp. Inst.* 12:85–102.

7. ———. 1943. Effect of moisture fluctuations on the viability of seeds in storage. *Contrib. Boyce Thomp. Inst.* 13:35–45.

8. ———. 1943. The storage of some citrus seeds. *Contrib. Boyce Thomp. Inst.* 13:47–55.

9. ———. 1953. Seed storage and viability. *Contrib. Boyce Thomp. Inst.* 17:87–103.

10. ———. 1954. Storage and packeting of seeds of Douglas fir and Western hemlock. *Contrib. Boyce Thomp. Inst.* 18:25–37.

11. Bass, L. N. 1979. Physiological and other aspects of seed preservation. In *The plant seed: Development, preservation, and germination,* I. Rubenstein, et al., eds. New York: Academic Press, pp. 145–70.

12. ———. 1980. Flower seed storage. *Seed Sci. and Tech.* 8:591–99.

13. ———. 1980. Seed viability during long term storage. *Hort. Rev.* 2:117–41.

14. Childs, J. F. L., and G. Hrnciar. 1948. A method of maintaining viability of citrus seeds in storage. *Proc. Fla. State Hort. Soc.* 64:69.

15. Craig, R. 1976. Flower seed industry. In *Bedding plants: A manual on the culture of bedding plants as a greenhouse crop,* J. W. Mastalerz, ed. University Park, Pa.: Pennsylvania Flower Growers, pp. 25–46.

16. Crocker, W., and L. V. Barton. 1953. *Physiology of seeds.* Waltham, Mass.: Chronica Botanica.

17. Delouche, J. C. 1980. Environmental effects on seed development and seed quality. *HortScience* 15:775–80.

18. Dougall, D. K. 1978. Preservation of germ plasm. In *Propagation of higher plants through tissue culture: A bridge between research and applications,* K. W. Hughes, R. Henke, and M. Constantin, eds. U. S. Dept. Energy Tech. Infor. Center, Springfield, Va.

19. Goss, W. L. 1937. Germination of flower seeds stored for ten years in the California state seed laboratory. *Calif. Dept. Agr. Bul.* 26:326–33.

20. Harrington, J. F. 1963. The value of moisture-resistant containers in vegetable seed packaging. *Calif. Agr. Exp. Sta. Bul. 792,* pp. 1–23.

21. ———. 1972. Seed storage and longevity. In *Seed biology,* T. T. Kozlowski, ed. New York: Academic Press, pp. 145–245.

22. Hawthorn, L. R., and L. H. Pollard. 1954. *Vegetable and flower seed production.* New York: Blakiston Co.

23. Heit, C. E. 1967. Propagation from seed, Part 10: Storage methods for conifer seed. *Amer. Nurs.* 126(20):14–15.

24. ———. 1967. Propagation from seed, Part 11: Storage of deciduous tree and shrub seed. *Amer. Nurs.* 126(21):12–13, 86–94.

25. Jones, H. A. 1920. Physiological study of maple seeds. *Bot. Gaz.,* 69:127–52.

26. MacGillivray, J. H. 1953. *Vegetable production.* New York: Blakiston Co.

27. Muenscher, W. C. 1936. Storage and germination of seeds of aquatic plants. *New York (Cornell Univ.) Agr. Exp. Sta. Bul.* 652:1–17.

28. Nutile, G. E. 1964. Effect of desiccation on viability of seeds. *Crop Sci.* 4:325–28.

29. Ohga, I. 1926. The germination of century old and recently harvested Indian lotus with special reference to the effect of oxygen supply. *Amer. Jour. Bot.* 13:754–59.

30. Pollock, B. M., and E. E. Roos. 1972. Seed and seedling vigor. In *Seed biology,* Vol. 1. T. T. Kozlowski, ed. New York: Academic Press.

31. Roberts, E. H. 1972. Dormancy: A factor affecting seed survival in the soil. In *Viability of seeds,* E. H. Roberts, ed. London: Chapman and Hall, pp. 320–59.

32. Schopmeyer, C. S., ed. 1974. *Seeds of woody plants in the United States.* USDA Agr. Handbook No. 450. Washington, D.C.: U.S. Govt. Printing Office.

33. Schubert, G. H., and R. S. Adams. 1971. *Reforestation Practices for Conifers in California.* State of Calif. Div. of Forestry.

34. Smith, B. C. 1950. Cleaning and processing seeds. *Amer. Nurs.* 92(11):13–14, 33–35.

35. Toole, E. H. 1958. Storage of vegetable seeds. *USDA Leaflet No. 220* (rev.).

36. Van der Berg, H. H., and R. Hendricks. 1980. Cleaning flower seeds. *Seed Sci. and Tech.* 8:505–22.

37. Vaughn, C. E., B. R. Gregg, and J. C. Delouche. 1968. *Seed processing and handling.* State College, Miss.: Miss. State Univ. Seed Technology Library.

38. Wells, J. S. 1955. *Plant propagation practices.* New York: Macmillan.

39. Wester, H. V. 1973. Further evidence on age of ancient viable lotus seeds from Tulantien deposit, Manchuria. *HortScience* 8:371–77.

40. Wyman, D. 1953. Seeds of woody plants. *Arnoldia* 13:41–60.

# SUPPLEMENTARY READING

BARTON, L. V. 1967. *Bibliography of seeds.* New York: Columbia Univ. Press.

CANTLIFFE, D. J., ed. 1979. Symposium on seed quality: An overview of its relationship to horticulturists and physiologists. *HortScience* 15:764–89.

HAWTHORN, L. R., and L. H. POLLARD. 1954. *Vegetable and flower seed production.* New York: Blakiston Co.

JAMES, E. 1961, 1963. *An annotated bibliography on seed storage and deterioration.* U.S. Dept. of Agr., Agr. Res. Serv. No. 34-15-1 and No. 34-15-2.

JUSTICE, O. L., and L. N. BASS. 1978. *Principles and practices of seed storage.* USDA. Agr. Handbook No. 506. Washington, D.C.: U.S. Govt. Printing Office.

KOZLOWSKI, T. T., ed. 1972. *Seed biology,* Vols. 1, 2, 3. New York: Academic Press.

ROBERTS, E. H., ed. 1972. *Viability of seeds.* London: Chapman and Hall.

*Seed World,* published monthly. 434 S. Wabash Ave., Chicago, Ill. 60605.

SCHOPMEYER, C. S., ed. 1974. *Seeds of woody plants in the United States.* USDA. Agr. Handbook No. 450. Washington, D.C.: U.S. Govt. Printing Office.

THOMSON, J. R. 1979. *An introduction to seed technology.* New York: John Wiley.

U.S. DEPT OF AGRICULTURE. 1961. *Seeds: Yearbook of agriculture.* Washington, D.C.: U.S. Govt. Printing Office.

A seed consists of an embryo and its stored food supply, surrounded by protective seed coverings. At the time the seed separates from the parent plant, the moisture content of most seeds is low, metabolism is at a low level, and no apparent growth activity occurs. In this dry state, seeds can be stored for long periods, particularly at low temperatures, transported all over the world, and used for propagation at a time and under conditions chosen by the propagator.

## THE GERMINATION PROCESS

The initiation of germination requires that three conditions must be fulfilled (25, 65):

First, the seed must be viable; that is, the embryo must be alive and capable of germination.

Second, the seed must be nondormant and the embryo quiescent. There must be no dormancy-inducing physiological, physical, or chemical barrier to germination.

Third, the seed must be subjected to the appropriate environmental conditions: available water, proper temperature, a supply of oxygen, and sometimes light. Because of the complex interactions between the environment and specific dormancy conditions in the seed, these environmental requirements may change with time and methods of seed handling. Thus, the second requirement, avoiding dormancy, may sometimes be met by providing the appropriate environmental condition.

### Stages of Germination

The process of germination can be divided into several separate consecutive but overlapping stages.

6

# Principles
# of Propagation
# by Seeds

## Terminology

**Seed**    A seed is a ripened ovule that consists of an embryo, its stored food supply, and protective coverings. The term *seed* as commonly used also includes the ovules of one-seeded, dry, indehiscent fruits, such as caryopses, achenes, and nuts; a better term for these seeds might be **seed dispersal units.**

**Germination**    Germination is the process of reactivation of the metabolic machinery of the seed and the emergence of the radicle (root) and plumule (shoot), leading to the production of a seedling. Physiologically, germination begins with the initial stages of biochemical reactivation and ends with emergence of the radicle (*65*). Morphologically and for seed testing and plant propagation, the definition must include the production of a normal seedling plant (*123*).

**Dormancy**    The term *dormancy* has broad applications in plant physiology to refer to *lack of growth in any plant part resulting from either internally or externally induced factors* (*124*). Seed technologists, on the other hand, define dormancy in a somewhat narrower sense as the result of conditions within the seed (other than nonviability) that prevent germination (*108, 123, 125*). In this sense, a **dormant seed** is one that *fails to germinate even though it has absorbed water and is exposed to favorable temperature and oxygen levels.*

In the "Rules for Testing Seeds," the Association of Official Seed Analysts further distinguishes between hard seeds and dormant seeds (*4*). **Hard seeds** include those that cannot absorb moisture because of their impermeable seed coat. **Dormant seeds** include those that fail to germinate even though the embryo is alive and moisture is absorbed.

**Primary dormancy**    includes conditions that exist within the seed to prevent germination (*29*) at the time it matures on the plant and in the immediate period afterward.

**Secondary dormancy**    refers to dormancy that develops within the moist seed *after* it is removed from the plant and if it is subjected to adverse environmental conditions (*21, 67, 74*).

**After-ripening**    is a term long used in horticultural literature to describe those physiological changes that occur within the seed after harvest that enable germination to take place. The term is applied both to changes that occur in the dry seed, perhaps physical in nature, and to changes occurring in imbibed seeds during chilling, which are largely biochemical.

**Quiescence**    describes the condition in which the seed can germinate immediately upon the absorption of water in the absence of any internal germination barriers. The embryo (or the seed) is said to be **quiescent** (*21, 65, 125*).

STAGE 1:
AWAKENING OR ACTIVATION

*Imbibition of water*    Water is absorbed by the dry seed and the moisture content increases rapidly at first, then levels off (see Figure 6–1). Initial absorp-

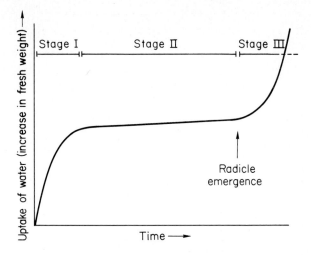

**Figure 6–1** Water uptake of a dry seed. Radicle emergence marks the first visible evidence of germination (*16*). Reprinted by permission with revision from J. D. Bewley and M. Black. 1978. Physiology and biochemistry of seeds in relation to germination. Vol. 1. *Development, germination, and growth.* Berlin: Springer-Verlag.

tion involves the imbibition of water by colloids of the dry seed, which softens the seed coverings and causes hydration of the protoplasm. The seed swells and the seed coats may break. Imbibition is a physical process and can take place even in dead seeds.

*Synthesis of enzymes*　Enzyme activity begins very quickly after the start of germination as the seed becomes hydrated (*16*). Activation results in part from reactivation of stored enzymes previously formed during development of the embryo and in part from synthesis of new enzymes as germination starts (*105*).

Seed development and germination represent a biological system in which the metabolic machinery of the cell is turned ''on'' and ''off'' by control of the flow of genetic information from DNA in the cell. Simply stated, the process involves two basic steps. One is the *transcription* of genetic instructions from the DNA to form specific *messenger RNA* molecules (mRNA). The second step is the *translation* of this information by means of another group of *transfer RNA* molecules (tRNA) to synthesize specific proteins. These proteins are enzymes that control the complex biochemical reactions involved in metabolism and growth. Energy for these processes is obtained from the high energy phosphate bonds in adenosine triphosphate (ATP) present in mitochondria.

Synthesis of enzymes during germination then requires RNA molecules, which are present in the embryo axis. The concept has been advanced that the initiation of germination results from the activation of unique germination controlling mRNA molecules in which genetic information for germination was transcribed during seed development but the translation is delayed until imbibition of the seed (*40, 41*).

Some of the ATP to provide energy was formed during seed development, preserved in the dormant seed, and then reactivated with cell hydration.

*Cell elongation and emergence of the radicle*　The first visible evidence of germination is emergence of the radicle, which results from the elongation of cells rather than from cell division (*14, 53*). In a nondormant seed emergence of the radicle can occur within a few hours or a few days after planting and is usually considered to mark the end of Stage 1.

STAGE 2:
DIGESTION AND TRANSLOCATION

Fats, proteins, and carbohydrates are stored in the endosperm, cotyledons, perisperm, or female gametophyte (conifers). The compounds are digested to simpler chemical substances, which are translocated to the growing points of the embryo axis.

The metabolic patterns in seeds of different species differ with the type of chemical reserves in the seed. Fats and oils—the major food constituent in the seeds of most higher plants—are converted enzymatically to fatty acids and eventually to sugar. Storage proteins, present in most seeds, are a source of amino acids and nitrogen essential to the growing seedling. Starch, present in many seeds as an energy source, is converted to sugar. The metabolic patterns that occur during germination involve the activation of specific enzymes in the proper sequence and the regulation of their activity. Control may be exercised by various biochemical processes within cells and may depend on the presence of specific chemicals or hormones. Water uptake and respiration now continue at a steady rate. The existing cell systems have been activated and the protein-synthesizing system is functioning to produce new enzymes, structural materials, regulator compounds, and nucleic acids to carry on the cell functions and synthesize new materials.

STAGE 3:
SEEDLING GROWTH

In the third stage, development of the seedling plant results from continued *cell division* in the separate growing points of the embryo axis, followed by the expansion of the seedling structures. The initiation of cell division in the growing points appears to be independent of the initiation of cell elongation (*14, 53*). Once growth begins from the embryo axis, fresh weight and dry weight of the new seedling plant increase but total weight of storage tissue decreases. The respiration rate, as measured by oxygen uptake, increases steadily with advance in growth. Storage tissues of the seed eventually cease to be involved in metabolic activities except in plants of which the cotyledons emerge from the ground and become active in photosynthesis. Water absorption increases steadily as new roots explore the germination medium and the fresh weight of the seedling plant increases.

As germination proceeds, the structure of the seedling soon becomes evident. The embryo consists of an **axis** bearing one or more seed leaves, or **cotyledons.** The growing point of the root, the **radicle,** emerges from the base of the embryo axis. The growing point of the shoot, the **plumule,** is at the upper end of the embryo axis, above the cotyledons. The seedling stem is divided into the section below the cotyledons—the **hypocotyl**—and the section above the cotyledons—the **epicotyl.**

The initial growth of the seedling follows one of two patterns. In one type—**epigeous** germination—the hypocotyl elongates and raises the cotyledons above the ground. In the other type—**hypogeous** germination—the lengthening of the hypocotyl does not raise the cotyledons above the ground, and only the epicotyl emerges (Figures 6-2 and 6-3).

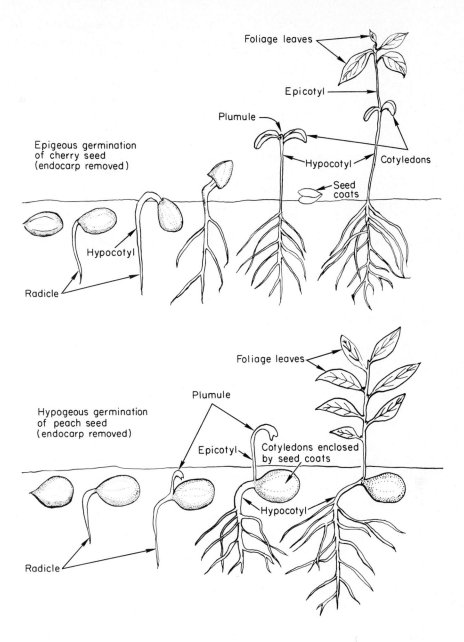

**Figure 6–2**  Seed germination in dicotyledonous plants. *Top:* epigeous germination of cherry. The cotyledons are above ground. *Below:* hypogeous germination of peach. The cotyledons remain below ground.

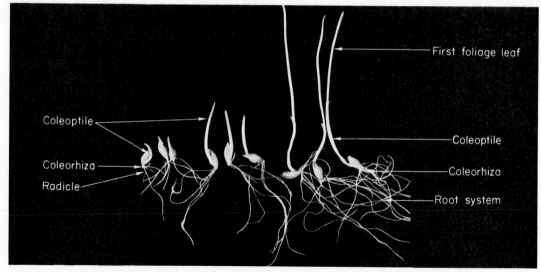

**Figure 6–3**     Germination of barley seeds illustrates the special pattern shown by the grass (Gramineae) family, a monocotyledon. See Figure 3–11 for seed structure. The **coleoptile** and **coleorhiza** enclose the **plumule** and **radicle** respectively and appear first from the seed. The **first foliage leaf** and **radicle** then emerge through these structures to produce the stems and roots. Source: Hartmann, H. T., W. J. Flocker, and A. M. Kofranek. 1981. *Plant science.* Englewood Cliffs, N.J.: Prentice-Hall.

## QUALITY OF SEEDS

A method of judging viability is essential in successful seed propagation. The dead or dying seed is characterized by a gradual decline in vigor, and necrosis or injuries may appear in localized areas of the seed coat. But the difference between a live seed and a dead one is not always distinct (*59, 98*). **Viability** can be expressed by the **germination percentage,** which indicates the number of seedlings produced by a given number of seeds. Additional characteristics of high quality seed are prompt germination, vigorous seedling growth, and normal appearance (*2*). **Vigor** of seed and seedling are important attributes of quality but may be somewhat difficult to measure (*88*). Low germination percentage, low germination rate. and low vigor are often associated. Low germination can be due to genetic properties of certain cultivars (*38*), incomplete seed development on the plant, injuries during harvest, improper processing (*36*) and storage (*101*), disease, and aging. Loss in viability is usually preceded by a period of declining vigor (*122*).

Seeds with low vigor may not be able to withstand unfavorable conditions in the seed bed, such as attacks by disease organisms. The seedlings may lack the strength to emerge if the seeds are planted too deeply or if the soil surface is crusted. Field survival of low-vigor seeds is apt to be less than a laboratory germination percentage test would indicate.

*Measurement of seed quality*     If one measures the time sequence of germination of a given lot of seeds, or the emergence of seedlings from a seed bed, one

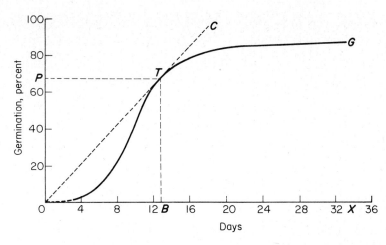

**Figure 6–4**    Typical germination curve of a sample of germinating seeds. After an initial delay, the number of seeds germinating increases, then decreases. Such a curve can be used to measure **germination value.** Redrawn from Czabator (*32*).

usually finds a pattern like the germination curve shown in Figure 6-4. There is an initial delay in the start of germination, then a rapid increase in the number of seeds that germinate, followed by a decrease in their rate of appearance. When viability is less than 100 percent, the end point may not be exact.

Germination is measured on two parameters—the **germination percentage** and the **germination rate.** Vigor may be indicated by these measurements, but seedling growth rate and morphological appearance also must be considered. Sometimes abnormally growing seedlings result from low seed quality (*59*).

Statements of germination percentage should involve a time element, indicating the number of seedlings produced within a specified length of time. Germination rate can be measured by several methods. One determines the number of days required to produce a given germination percentage. Another method calculates the average number of days required for radicle or plumule emergence as follows:

$$\text{Mean days} = \frac{N_1 T_1 + N_2 T_2 + \ldots N_x T_x}{\text{total number of seeds germinating}}$$

$N$ values are the numbers of seeds germinating within consecutive intervals of time; $T$ values indicate the times between the beginning of the test and the end of the particular interval of measurement. Kotowski (*78*) has used the reciprocal of this formula multiplied by 100 to determine a **coefficient of velocity.** Gordon (*51*) has suggested the term **germination resistance** as the time (hours or days) to average germination, based on seeds that germinate.

Czabator (*32*) has suggested another measurement for seeds of woody perennials in which germination may be slow: the **germination value (*GV*).** It includes both the germination rate and percentage. To calculate *GV*, a germination curve, as shown in Figure 6-4, must be obtained by periodic counts of radicle or plumule emergence. The important values on the curve are **T**—the point at which the germination rate begins to slow down—and **G**—the final germination percentage. These points divide the curve into two parts—a rapid phase and a slow phase. Peak value ($PV$) is the germination percentage at $T$, divided by the days to reach that point. Mean daily germination ($MDG$) is the final germination percentage divided by number of days in the test. For example:

$$
\begin{aligned}
GV &= PV \times MDG \\
GV &= \frac{68}{13} \times \frac{85}{34} \\
&= 5.2 \times 2.5 \\
&= 13.0
\end{aligned}
$$

## DORMANCY: REGULATION OF GERMINATION

Maturation of seeds includes the development of internal dormancy mechanisms that control the onset of germination (see chapter 3). One method of control is reduction in moisture to a level below that required for germination. Most freshly harvested seeds, however, have additional regulating mechanisms that prevent germination even if water is absorbed.

*In nature*    The effect of these controls is to preserve seeds and to regulate germination so it will coincide with periods during the year having environmental conditions that favor survival of the seedlings in nature. Consequently, germination-controlling mechanisms exist as adaptations for natural survival of the species. The particular requirements for germination are related to the environment where the plant species has evolved (*77, 120, 136*). Knowledge of ecological requirements can aid in establishing treatments to induce germination.

Such mechanisms are particularly important in plants growing where extremes of environment occur, such as deserts or cold regions where environmental conditions may not be favorable for germination immediately following dissemination of the seed (*13, 99, 137*).

*In cultivation*    The domestication of seed-propagated cultivars of many cultivated crop plants, such as grains and vegetables, has included selection for easier propagation so that dormancy controls have disappeared during the dry storage of normal handling.

Testing seed viability in seed labs, especially when attempted shortly after the seed is harvested, is particularly difficult. Such seed testing is a major occupation and requires a knowledge of dormancy problems, which are common in seeds of many tree and shrub species, particularly those of the temperate zones. Dormancy problems are found in seeds of both native plant species and species recently introduced into cultivation.

Propagators of cultivated plants have long recognized these germination-delaying phenomena and have learned to cope with them through the adoption of appropriate pregermination and handling procedures discovered by trial and error.

## Categories of Seed Dormancy

Various types of seed dormancy result from different germination-controlling mechanisms within the seed. Classifications have been developed to explain the biological mechanisms involved and suggest ways to overcome dormancy. The first system was formulated by Crocker (29), who described seven types of seed dormancy. More recently Nicolaeva (94) has defined a system based predominantly on the physiological causes of dormancy. Atwater (5) has shown that the morphological characteristics, in terms of embryo development and types of seed coverings, are associated with specific dormancy categories. Seed morphology is characteristic of taxonomically established plant families. The interaction of morphology and physiology is the basis for the various kinds of seed dormancy that enable certain members of those families to adapt to specific environments.

The following categories are based on both physiological and morphological differences, primarily following Nicolaeva's classification (94).

### SEED COAT DORMANCY

**Physical dormancy** is characteristic of a large number of plant families in which the seed coats and sometimes hardened sections of other seed coverings are impermeable to water. These include Leguminoseae, Malvaceae, Cannaceae, Geraniaceae, Chenopodiaceae, Convolvulaceae, and Solanaceae. The embryo is quiescent (nondormant) but is sealed inside a water impermeable covering that can preserve the seed at low moisture contents for many years even at warm temperatures. Among cultivated crops hardseededness is found chiefly in the herbaceous legumes, including clover, alfalfa, etc., as well as many woody legumes (*Robinia, Acacia,* etc.). Hardseededness depends on the genetic nature of the species and cultivar, environmental conditions during seed maturation, and environmental conditions during seed storage. Drying at high temperatures during ripening will increase hardseededness. Harvesting slightly immature seeds and preventing them from drying can reduce or overcome hardseededness in some cases.

Impermeability of the seed coat is due to a layer of palisade-like **macrosclereid** cells, especially thick-walled on their outer surfaces and coated with a layer of waxy, cuticular substances (100) (see Figure 6-5). Disintegration of the "caps" of such cells, or mechanical stress separating the cells, may allow water to enter and produce germination (20). In some legume species, the point of attachment (**hilum**) of the seed acts as a one-way valve during ripening by opening to allow water to escape in a dry atmosphere but closing in a moist atmosphere to prevent water uptake (64). In such species as *Albizia lophantha* (35) a small opening (**strophiole**) near the hilum is sealed with a cork-like plug, which can be dislodged with vigorous shaking or impaction (54) or by exposure to dry heat as in a fire (35).

In nature, seed coverings are softened by various agents of the environment, including mechanical abrasion, alternate freezing and thawing, attack by soil microorganisms, passage through the digestive tracts of birds or mammals, or fire. Softening of seed coverings by fungi and bacteria is most effective at temperatures above 10°C (50°F) or when seeds are planted in a nonsterile medium. Chapter 7 lists various methods by which seed coats can be modified artificially (see also Figure 6-5).

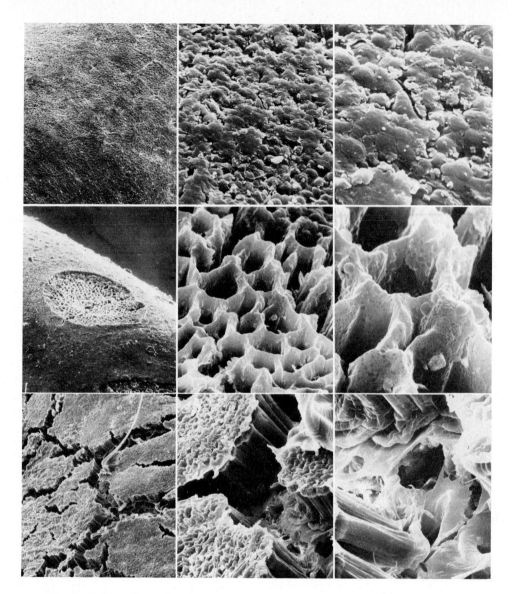

**Figure 6–5**  Scanning electron micrographs of 'Penngift' crownvetch seed showing scarification effects on seed coats. *Top row:* unscarified seed. Left to right: 170×, 1700×, and 4250× magnification. Note intact platy surface covering. *Center row:* after 30 min acid scarification. Left to right: 170×, 1700×, and 4250× magnification. Caps of macroscoereid cells have been destroyed, exposing the lumens of the cells. *Bottom row:* after seed immersion in boiling water for 30 sec with agitation. Left to right: 170×, 850×, 1700× magnification. Note columnar structure of exposed macroscoereid cells. Courtesy R. E. Brant, G. W. McKee, and R. W. Cleveland (*20*).

**Mechanical dormancy**     refers to seed coverings that are too hard to allow the embryo to expand during germination. This factor probably is not the sole cause of dormancy in any species, but combines with other factors to delay germination. In general, once water is absorbed by the seed, the expanding force of germination ruptures the seed coats and breaks apart any outer covering. The shells of walnuts and the pits of stone fruits or other hard fruit coverings can delay germination, however. The hard endocarp (pit) of peach seeds slows water absorption and delays the leaching of germination inhibitors out of the seed (*42*). It has been difficult to demonstrate in many species that such mechanical effects are a primary cause of dormancy (*94*). Any effects of mechanically resistant coats can be overcome by the same treatments used for impermeable seed coats in native or cultivated plants.

**Chemical dormancy**     Chemicals that act as seed **germination inhibitors** have been extracted from various plant parts and identified (*48*). Such chemicals are produced and accumulate in the fruit as well as in the seed coverings.

Fleshy fruits, or juices from them, can strongly inhibit seed germination. This occurs in citrus, cucurbits, stone fruits, apples, pears, grapes, and tomatoes. Likewise, dry fruits and fruit coverings, such as the hulls of guayule, *Pennisetum ciliare*, wheat, as well as the capsules of mustard (*Brassica*) can inhibit germination. When the fruit coverings remain with the seed, such inhibition can persist into the germination period. Some of the substances associated with inhibition are various phenols, coumarin, and abscisic acid.

Establishing a direct relationship between specific chemicals and germination is difficult. Nevertheless, in some specific cases where inhibitors are present, germination can be improved or stimulated by leaching with water, or removing the seed coverings, or both (*95*). For example, specific seed germination inhibitors play a role in the ecology of certain desert plants (*77, 133, 134*). Inhibitors are leached out of the seeds by heavy soaking rains which would in turn provide sufficient soil moisture to insure survival of the seedlings. Since a light rain shower is insufficient to cause leaching, such inhibiting substances have been referred to as "chemical rain gauges." Germination inhibitors are reported to be widespread in seeds of tropical species (*94*). Dormancy in iris seeds is due to a water- and ether-soluble germination inhibitor in the endosperm, which can be leached from the seeds with water or avoided by embryo excision (*3*).

These examples show direct effects of inhibitors externally located in seed coverings that can be removed by leaching. Inhibitors, however, may be involved in other kinds of dormancy in a more complex manner. Inhibitors have been cited as the cause of rudimentary embryos (*5*), and their effect must be overcome before embryo elongation can occur. Beet "seeds" contain substances in their coverings giving off ammonia, which interferes with germination in laboratory seed testing; but if such seeds are planted in soil, leaching or absorption of the ammonia by soil particles overcomes this factor (*123*). Thus seed inhibitors may be a more serious problem in seed testing laboratories than in the field.

Inhibitors have been found in the seeds of such families as Polygonaceae, Chenopodiaceae (*Atriplex*), Portulaceae (*Portulaca*), and other species in which the embryo is peripherally located (see beet seed, Figure 3–8). Likewise, seeds of a group of such families as Cruciferae (mustard), Linaceae (flax), Violaceae (violet), Labitae (*Lavendula*) have a thin seed coat with a mucilaginous inner layer that contains inhibitors (*5*).

Control of dormancy resulting from internally located inhibitors is more complex than simple leaching and is considered in more detail under subsequent headings.

MORPHOLOGICAL DORMANCY

Morphological dormancy is found in plant families that characteristically have seeds in which the embryo is not fully developed at the time of ripening. Additional embryo growth is necessary after separation of the seed from the plant and prior to germination. As a general rule embryo enlargement is favored by warm temperatures, but the response may be complicated by the presence of other dormancy mechanisms. There are two groups within this category:

**Rudimentary embryo**     Plants with rudimentary embryos (*5*) produce seeds with little more than a proembryo embedded in a massive endosperm (like the magnolia in Figure 3–8) at the time of fruit maturation. This type occurs in various families, including some that are important as herbaceous flower crops, such as Ranunculaceae (anemone, ranunculus), Papaveraceae (poppy, *Romneya*), and Araliaceae (ginseng, fatsia). Germination inhibiting chemicals also occur in the endosperm, and become particularly active at high temperatures. Effective aids for inducing germination include: (a) exposure to temperatures of 15°C (59°F) or below; (b) exposure to alternating temperatures; (c) treatment with chemical additives such as potassium nitrate or gibberellic acid.

Certain temperate zone woody plants such as holly (*Ilex*) and snowberry (*Symphoricarpus*) have rudimentary embryos but, in addition, have other types of dormancy, such as hard seed coats and dormant embryos, which must be overcome (*94*) for germination to occur.

Orchids have both rudimentary embryos and undeveloped seeds. They are not considered dormant in the same sense as others in this category and are prepared for germination by special aseptic methods as discussed in chapter 17.

**Undeveloped embryos**     At fruit maturity some seeds have partially developed, torpedo-shaped embryos that may attain a size up to one half that of the seed cavity. Further embryo enlargement occurs prior to germination. Some important families and species in this category include Umbelliferae (carrot), Ericaceae (rhododendron, heather), Primulaceae (cyclamen, primula), and Gentianaceae (gentian). Other conditions, such as semipermeability of the inner seed coats and internal germination inhibitors, may be involved in freshly harvested seed. A temperature of about 20°C (68°F) is favorable for germination, as are gibberellic acid treatments.

In seeds of some species, subsequent chilling is also required for germination after the warm embryo development period. Various temperate zone trees fall into this category, including *Fraxinus* (ash) and *Euonymous* species (*94*).

Various tropical species, many of which are monocots, have seeds with undeveloped embryos that require an extended period at high temperatures for germination to take place. For example, seeds of various palm species ordinarily require storage for several years to germinate, but this period can be reduced to three months by holding the seeds at temperatures of 38 to 40°C (100 to 104°F), or to 24 hours by excising the embryos and germinating them aseptically. Gibberellic acid (1000 ppm) has accelerated germination in palm seed, but a seed coat treatment is needed to assure penetration of the chemical (*91*). Other examples include *Actinidia,* whose seed requires two months' warmth, and *Anona squamosa* seed, which requires three months' (*94*).

INTERNAL DORMANCY

Several categories include seeds in which dormancy is internally controlled within the living tissues of the seed. Two separate phenomena are involved in the internal control of germination. The first is the control exerted by the *semipermeability* of seed coverings. The second is a *dormancy* present within the embryo that is overcome by exposure to moist-chilling.

**Physiological dormancy**    Freshly harvested seed of many if not most herbaceous plants of the temperate zone have a **physiological dormancy** that tends to disappear with dry storage (*16, 87, 94, 117*). For most cultivated cereals, grasses, vegetables, and flower crops, the dormant period may last for one to six months but disappears with dry storage during normal handling procedures. For many noncultivated plants such dormancy usually lasts longer, particularly if the seeds remain moist, as when they are buried in the ground.

When they are dormant, seeds tend to have more specific environmental requirements for germination than when they are nondormant. Freshly harvested seeds of cocklebur or amaranth, for example, germinate only at high temperatures, about 30°C (86°F). In freshly harvested seeds of other species, such as some cultivars of lettuce and celery, germination is inhibited at temperatures above 25°C (77°F). Heat sensitivity is called **thermodormancy.**

Dormant seeds of many species that have temperature sensitivity also have light sensitivity (**photodormancy**). Seeds of some plants, including lettuce and many flower crops, require light to germinate, whereas others require darkness.

The controlling mechanisms of physiological dormancy apparently lie within the living, physiologically active seed coverings (*49*). Removal of the seed coats, puncturing them, or increasing the oxygen level can bring about germination. This controlling mechanism is due to the semipermeability of the seed coats, which allows water to enter the seed but restricts movement of gases (oxygen entering, carbon dioxide escaping), and prevents leaching of germination inhibitors from within the seed.

In general, the outer layers, such as the outer integument (seed coat) and the fruit tissues, are made up of nonliving cells, but those of the inner integument, the nucellus, and the endosperm are physiologically active. The most important controlling layers seem to be the ones nearest the embryo.

**Intermediate internal dormancy**    Another type of dormancy is exercised primarily by the seed coverings and the surrounding storage tissues. **Intermediate internal dormancy** is characteristic of conifers—pine, spruce, and the like (*108*). The embryo itself is not dormant and is capable of more or less normal germination if it is excised from the seed and placed on aseptic nutrient agar. This type of dormancy is further characterized by the fact that chilling may not be an absolute requirement of germination, but greatly accelerates the germination rate.

**Embryo dormancy**    Embryo dormancy is characterized principally by the requirement of a period of **moist-chilling** for germination and the inability of the excised embryo to germinate normally (*26, 30, 81, 94, 125*). Plant propagators have known since early times that moist seeds of many fall-ripening tree and shrub species of the temperate zone must be chilled as they overwinter in the ground before they will germinate in the spring. This requirement led to the horticultural practice termed **stratification,** in which seeds are placed between layers of moist sand or soil in boxes (or in the ground) and exposed to chilling temperatures either out-of-doors or in refrigerators.

In this situation two separate but interacting conditions have been identified as controlling germination (*94*). The first is the semipermeability of the seed covering membranes that exert effects on aeration and inhibitor retention similar to those described for seeds having physiological dormancy. The second is the internal dormancy condition within the embryo itself, particularly in the growing points of the embryo axis. Influences of surrounding endosperm or female gametophyte storage tissue also may be involved. Modification of all of these effects takes place gradually with time during chilling so that finally the seed can germinate promptly when given higher temperatures.

Seeds of plants in this group generally have an absolute chilling requirement of one to four months in order for germination to take place. Furthermore, when excised and placed on nutrient agar the embryo does not germinate normally but exhibits various degrees of dormancy symptoms. A range of responses appears that can be utilized as a test of viability for these dormant seeds (*30, 94*). Such responses include enlargement and greening of the cotyledons; thick, short radicle growth with no epicotyl development; or development of normal root systems but with various degrees of dwarfed abnormal epicotyls. The latter may remain in such a condition until the plants are given a chilling treatment, after which some growing points resume normal growth. Such seedlings are termed **physiological dwarfs** (*30, 50*) (Figure 6–6).

Dormant embryos are most common in seeds of trees and shrubs and some herbaceous plants of the temperate zones and of even colder climates, where seeds in nature overwinter in the ground and germinate in the spring.

Conditions required to overcome embryo dormancy include:

1. Imbibition of moisture by the seed
2. Exposure to chilling, but not necessarily freezing, temperatures
3. Aeration
4. A certain amount of time.

**Figure 6–6**    Effect of chilling 'Lovell' peach seed for different lengths of time on subsequent growth and appearance of the seedlings. Seed stored before planting under moist conditions at 5°C (41°F) for 102 days (*left*) and 68 days (*right*). Note physiologically dwarfed and abnormal seedlings in the latter.

*Moisture*    After-ripening requires that the seed have imbibed water. A dry seed with a dormant embryo initially absorbs moisture by imbibition to around 50 percent moisture (*81, 94*). If the seed is enclosed by a hard endocarp (*42*) or other seed layers, water absorption may be slow so that the required after-ripening period becomes much longer. Initiation of after-ripening is delayed so that the entire stratification time is greatly lengthened. During the ensuing chilling period, the moisture content of the seed remains relatively constant or it may even gradually increase. As dormancy ends the embryo absorbs water rapidly with the beginning of germination processes so that the demand for water may become quite high (*81*). Seeds swell and seed coverings split.

Reduction in moisture content can have various effects. Drying near the end of stratification can cause injury to the embryo. Drying of seeds during stratification stops the after-ripening process (*57*) and secondary dormancy can develop (*97*). On the other hand, several experimental reports describe procedures whereby partially after-ripened seeds were partially dried and then successfully stored at low temperatures for later planting (*33, 43, 66, 115*). When reimbibed with water, germination occurred promptly without loss of germinability.

*Temperature*    Temperature is the most important factor affecting the rate of after-ripening.

Temperatures somewhat above freezing, 2° to 7°C (35° to 45°F), are generally most effective; higher and lower temperatures produce changes at a slower rate. The minimum effective temperature is reported to be about −5°C (23°F) (*34*), and the maximum 17°C (62°F) (*1*). At higher temperatures the seeds not only fail to germinate but revert to secondary dormancy (*1, 107, 108, 127*). For apple seeds (*1*) the processes involved in after-ripening and those leading to induction of secondary dormancy have been found to be in equilibrium at 17°C (62°F), which is called a *compensation temperature*. This point has been determined also in dormant seeds of other plants (*115*), but the exact temperature varies with different species and different stages of after-ripening (*109*). The response of partially after-ripened apple seeds to germination temperatures varies with temperature. At low temperatures the seeds germinate slowly but the percentage is high. At higher temperatures, germination rates are faster, but the germination percentage decreases in proportion to the increase in temperature (*34, 107*). Nongerminating seeds revert to secondary dormancy and may not germinate until the following spring after again going through a chilling treatment.

The temperatures most effective for germination are similar to those in the natural environment, where soil temperatures are relatively cool in early spring but gradually increase as the season advances. Secondary dormancy in the seed, resulting in inhibition of germination during an early hot period in the spring, can be advantageous to species survival in nature by limiting germination to early season periods that allow good establishment before hot, possibly dry, summer weather occurs.

Physiological dwarfing in excised nonafter-ripened embryos has been shown to result from exposure of the apical meristem to warm germination temperatures (*97*). In peaches, a temperature of 23° to 27°C (73° to 81°F) produced symptoms of physiological dwarfing, but at lower temperatures the seedlings grew normally. In almonds, exposing incompletely stratified seed to high temperatures subsequently induced physiological dwarfing symptoms in the seedling.

Pinching out the apex can circumvent dwarfing by forcing lateral growth from nondwarfed lower nodes. Dwarfing has also been offset by exposing seedlings to long photoperiods or continuous light (*50, 80, 129*), provided that this action is taken before the apical meristem becomes fully dormant. Repeated application of gibberellic acid has also overcome dwarfing (*11, 12, 50*).

*Aeration*    Good aeration that provides oxygen during seed stratification is necessary for proper after-ripening. The amount of oxygen needed depends on temperature (*27, 124*). The moist seed coverings of dormant, imbibed seeds have been shown to restrict oxygen uptake because of (a) low oxygen solubility in water, and (b) oxygen fixation by phenolic substances in the seed coats. At chilling temperatures, however, the embryo's oxygen requirement is low and oxygen is generally adequate. But if the temperature is increased, the oxygen requirement of the embryo increases, the solubility of oxygen is less, and the amount fixed by the phenols increases. Secondary dormancy that develops at high temperatures can be attributed to an interaction with seed coat permeability and temperature. Since seed coats have also been shown to contain other inhibitors, such as abscisic acid (*37*), the precise explanation is uncertain. Nevertheless, as previously stated, removal of seed coats can greatly reduce seed dormancy in these groups of plants.

*Time*    The time required to after-ripen dormant seeds of most woody perennial species is from one to three months although, for certain species, five to six months are necessary. The time required is characteristic of the species and a general correlation exists between the chilling requirements of the buds of the plant and of the seeds (*68, 69, 136*).

Germination time of the seed can be influenced by the environmental exposure of the mother plant during seed development (*125*). Consequently, much variation in the chilling requirements of the seeds can occur, not only among different species but also among different seed sources of the same species; this, in turn, depends upon the year the seed was produced and the location of the parent plants. Variation among individual seeds in a single seed lot can produce uneven germination in the lot.

**Epicotyl dormancy**    Some seeds have separate stratification requirements for the radicle, hypocotyl, and epicotyl (*30, 94*). These species fall into two subgroups.

**1.** Seeds that initially germinate during a warm period of one to three months to produce root and hypocotyl growth but then require one to three months' chilling to enable the epicotyl to grow. This group includes various lily (*Lilium*) species, *Viburnum* spp., and peony (*Paeonia*).

**2.** Seeds that require a chilling period to after-ripen the embryo, followed by a warm period for the root to grow, then a second cold period to stimulate shoot growth. In nature, such seeds require two full growing seasons to complete germination. Examples include *Trillium* and certain other native perennials of the temperate zone.

## DOUBLE DORMANCY

**Double dormancy** is characterized by having both seed coat dormancy (lack of water permeability) and a dormant embryo. To produce germination both blocking conditions must be eliminated in proper sequence. Seed coats must be modified to allow water to penetrate to the embryo; after-ripening of the embryo can then take place. Warm followed by cold stratification generally overcomes these situations.

This type of dormancy is characteristic of species of trees and shrubs in families having seeds with hard seed coats but whose plants grow in cold winter areas. In nature, various agents of the environment—those that affect physical dormancy—soften the seed coat when the seed falls to the ground, then the seeds are chilled as they overwinter.

## SECONDARY DORMANCY

Secondary dormancy is acquired after the seed has been separated from the plant ( *29* ). Although the term might be applied to seed coat effects such as hard-seededness that could develop by dehydration during storage ( *123* ), the term is best applied to a kind of embryo dormancy that can develop gradually if the intact seeds are exposed to environmental conditions that allow imbibition but prevent germination ( *67, 73* ). This type of dormancy develops after imbibition of water by the seed and before emergence of the radicle. Such conditions can include high temperature, low oxygen supply, and lack of light ( *72, 120* ).

Induction of secondary dormancy is well illustrated by experiments with freshly-harvested seeds of lettuce that show physiological dormancy ( *74* ). If germinated at 25°C (77°F) the seeds require light, but if imbibed with water for two days in the dark, excised embryos germinate immediately, illustrating that only primary dormancy was present. If imbibition continues for as long as eight days, however, excised embryos will not germinate since they have then developed secondary dormancy.

Release from secondary dormancy can be induced by chilling, sometimes by light, and in various cases, treatment with germination-stimulating hormones, particularly gibberellic acid.

Prevention of secondary dormancy can be achieved by dehydration and dry storage.

Secondary dormancy is an important ecological adaptation for plants in nature. In cultivation seeds separated from the plant remain viable for long periods if dried and lose their primary dormancy in storage. When imbibed with water, they germinate promptly. However, if seeds retain a high moisture content after ripening, they must either germinate promptly or may lose viability. The induction of secondary dormancy allows imbibed seeds to retain viability for long periods. Furthermore, with induction of a chilling requirement, seasonal patterns of germination can be established. Secondary dormancy is particularly significant for seeds buried in soil, which then do not germinate until the soil is disturbed and the seeds are exposed to light.

Physiological changes that occur during stratification and after-ripening of dormant embryos are very complex. Nevertheless, they have been associated with changing levels of growth-promoting substances.

High levels of growth-inhibiting substances, identified usually as abscisic acid (ABA), have been found in dormant seeds of plum (Figure 6–7), walnut (*Juglans* spp.) (*84*), peach (*Prunus persica*) (*37, 83*), and filbert (*Corylus avellana*) (*138*). These substances occur in highest concentrations in the seed coverings in the freshly harvested seed. The level invariably decreases during chilling, as shown in Figure 6–7 in plum, and sometimes they can be leached out with water. Their disappearance does not necessarily coincide with the beginning of germination, however. On the other hand, application of ABA to chilled seed ready to germinate invariably prevents germination.

Growth-promoting substances, identified usually as gibberellin, are at a low level in dormant seeds of a number of species, such as plum (*82*) and peach (*86*). Gibberellin levels increase during chilling (*86*), as shown in Figure 6–7 for plum seed (*82*).

**Figure 6–7**     Changes in endogenous levels of abscisic acid and gibberellin in plum seeds during stratification. Redrawn from Lin and Boe (*82*).

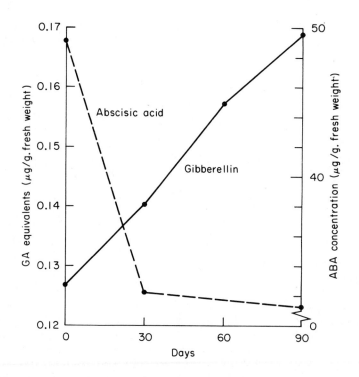

Filbert (*Corylus avellana*) seeds illustrate separation of growth inhibiting and growth-promoting hormonal systems in control of germination. At the time of ripening, the intact seed is dormant but the embryo is quiescent. A significant amount of abscisic acid can be detected in the seed covering (*138*) and a detectable amount of gibberellin in the embryo (*103*). When the seed is dried following harvest, the embryo becomes dormant, during which time the gibberellin level decreases significantly (*102*). Stratification of several months is then required for germination. During this chilling period the gibberellin level remains low but increases after the seeds are placed at warm temperature and germination begins (Figure 6–8, *right*).

Gibberellic acid applied to the dormant seed (*19*) can replace the chilling requirement (Figure 6–8, *left*). ABA applied at the same time as gibberellin will offset the effect of GA and prevent germination (*104*).

**Figure 6–8**    Interactions of gibberellin, stratification, and germination in filbert seeds. Reproduced by permission from A. W. Galston and P. S. Davies, *Control mechanisms in plant development.* Englewood Cliffs, N.J.: Prentice-Hall, Inc., 1970.

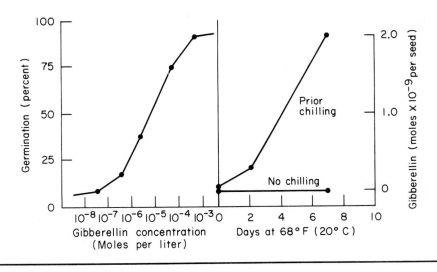

## Hormonal Control of Dormancy and Germination

Much experimental evidence supports the concept that control of dormancy and germination is mediated by specific endogenous growth-promoting and growth inhibiting hormones (*17, 71*), such as *gibberellins, cytokinins, ethylene,* and inhibitors, primarily *abscisic acid.* Such evidence comes from biochemical and physiological studies of the various germination stages and correlated changes in hormone concentrations in various parts of the seed. However, these correlations do not necessarily prove their role in nature (*17*). Nevertheless, applied synthetic hormones can act as **primary germination agents** in stimulating and inhibiting germination of various seeds (*73*). Hormone applications are very useful as experimental probes, or germination aids. Responses do not automatically follow, however, and may vary with the species, cultivar, and the dormancy or germination stage. Response also depends on the ability to introduce the hormone into the right tissue of the seed such that special methods as **infusion** must be used to insure hormone availability (*73*).

### Gibberellins

Gibberellins (GA) comprise the class of hormones most directly implicated in the control and promotion of seed germination (see Figure 6–9). While there are many molecular variations of gibberellin, the one most widely used experimentally and commercially is **gibberellic acid** ($GA_3$). These compounds occur at relatively high levels in developing seeds but usually drop to a lower level in mature dormant seeds, particularly in dicotyledonous plants.

Applied gibberellins can function in relieving many types of dormancy, including physiological dormancy, photodormancy, and thermodormancy. Gibberellins appear to play a role at two different stages of germination. In Stage 1, it has been suggested that gibberellins act at the initial stage of enzyme induction in their transcription from the chromosomes.

**Figure 6–9**     Chemical structure of various plant regulators involved in germination or germination control: gibberellic acid, abscisic acid, kinetin, and ethylene.

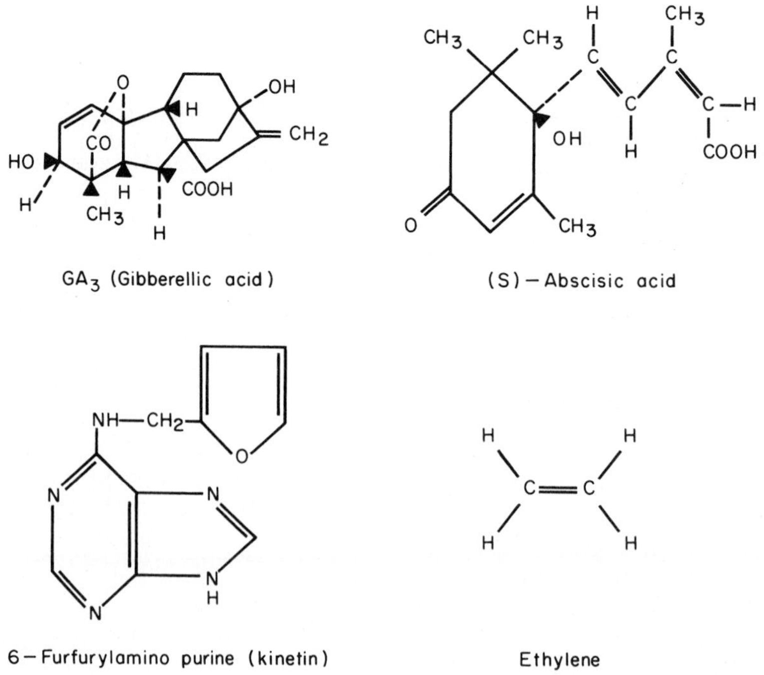

GA$_3$ (Gibberellic acid)

(S) — Abscisic acid

6 — Furfurylamino purine (kinetin)

Ethylene

A later stage in which gibberellin is effective is the activation of enzymes involving the food mobilizing system. Experiments (28) have been carried out with barley seeds. When the quiescent dry seed imbibes water, gibberellin appears in the embryo and is translocated to the *aleurone* (a layer three to four cells thick surrounding the endosperm), where it causes new production of $\alpha$-amylase. This enzyme moves to the endosperm, where it converts starch to sugar, which is in turn translocated to the growing points of the embryo to provide energy for growth. In barley seeds GA also promotes the induction or stimulation of other specific enzymes.

## Abscisic Acid (ABA)

Of the germination inhibitors discussed earlier in this chapter, the most significant is **abscisic acid.** This naturally occurring compound is one of the most important growth-regulating compounds not only in seed germination but in plant growth in general (129, 130). In cotton, ABA appears to play a role in preventing ''precocious germination'' of the developing embryo in the ovule (40). High inhibitor levels have been considered responsible for the lack of development in rudimentary embryos (5).

ABA tends to increase with maturation of the fruit and may be involved with prevention of vivipary and the induction of dormancy. It has been isolated from dormant seeds, as in the seed coats of peach, walnut, apple, rose, and plum, but it disappears during stratification (Figure 6–7).

Application of ABA can inhibit germination of nondormant seed and offset the effects of applied gibberellic acid. In general, these effects are temporary and disappear when seeds are shifted to an ABA-free solution.

## Cytokinin

A number of naturally occurring compounds classed as cytokinins have been found in plants (119). These have a basic chemical structure of an $N^6$-substituted adenine (see Figure 6–9). Synthetic cytokinins available for experimental use include benzyladenine, kinetin, and others. In addition, cytokinin activity is shown by such compounds as thiourea (46) and diphenylurea (119).

Cytokinin activity tends to be high in developing fruits and seeds, but decreases and becomes difficult to detect as the seeds mature. In seed germination, cytokinin is believed to offset the effect of inhibitors, notably ABA. It has been described, therefore, as playing a ''permissive'' role in germination in allowing gibberellic acid to function (71). It is believed to be active, therefore, at a different germination stage than gibberellins.

Reversal of dormancy by cytokinins has been demonstrated in some seeds. This reversal may involve seeds in which inhibition is due to naturally occurring inhibitors, as in dormant cocklebur (see Figure 6–10), or it may occur where inhibition has been imposed by ABA, as in light-sensitive lettuce. In both of these cases it is noted that GA does not overcome the inhibition. Dormant sycamore maple (*Acer pseudoplatanus*) seeds contain inhibitors that can be removed by leaching. In this case, cytokinin applications improve germination but gibberellins do not. Similarly, germination in light-sensitive celery seed is controlled, in part, by inhibitors. Seed germination in some cultivars can be

**Figure 6-10** Germination of cocklebur (*Xanthium*). Two burs are shown at A, each of which contains two seeds (B and C); the smaller one is dormant but the other is not. Early experiments showed that low permeability to gases was a dormancy inducing factor. Later experiments (*132*) showed that the smaller, dormant, seed contains two water-insoluble inhibitors that prevent germination. If these are removed by leaching, or if the seed is subjected to high oxygen pressure, then germination will occur. The two seeds also differ in seed coat strength and the germinating forces required to rupture them. Treatment with kinetin or ethylene (*72*) will stimulate germination of both seeds, while abscisic acid will inhibit it. From Khan et al. (*76*).

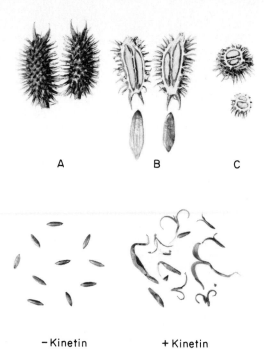

A    B    C

−Kinetin    + Kinetin

stimulated by GA alone, but in others with higher inhibitor levels both GA and cytokinin is required to obtain a response.

## Ethylene

Ethylene gas (*70*) is an important, naturally occurring hormone involved in many aspects of plant growth. Response to ethylene treatment of dormant seeds of snowberry (*Symphoricarpos*), honeysuckle (*Lonicera*), and similar species, as well as seeds of corn and other cereals, was demonstrated many years ago. Ethylene production from germinating bean and pea seed was shown in 1935. Later work demonstrated that ethylene is a natural germination-promoting agent for certain kinds of seeds.

Ethylene is given off by seeds of subterranean clover (*Trifolium subterranean*), which produces seeds underground. It has been suggested that this ethylene stimulates germination of seed clusters under conditions of soil crusting (*47*). Similarly, ethylene is produced and dormancy is overcome in seeds of Virginia-type peanut (*Arachis hypogaea*), also produced underground. Likewise, ethylene produces 100 percent germination of witchweed (*Striga asiatica*) seeds (*45*). This annual plant is parasitic on many species. The witchweed seeds are caused to germinate by a naturally occurring stimulant coming from the roots of the host plants. Ethylene has been used experimentally to overcome high temperature dormancy in lettuce seeds. It is particularly effective if used in combination with kinetin (*110*) (Figure 6–11), carbon dioxide (*92, 93*), or light treatments (*96*). Dormant cocklebur (*Xanthium*) seeds, both intact seeds and excised axis and cotyledon segments, also were found to respond to ethylene.

**Figure 6-11** Germination of lettuce seed at high temperature (35°C; 95°F). No germination occurred in water, but 80 percent germination took place if seeds were presoaked in water for 24 hours at 25°C (77°F) (No. 1). Presoaking could be partially replaced by ethylene or by kinetin treatments and completely by a combination of both (No. 2). Treatments were 3 minutes in 100 mg/l kinetin, followed by germination on filter paper soaked with 100 mg/l ethephon, an ethylene-releasing chemical. From data of Sharples (*110*).

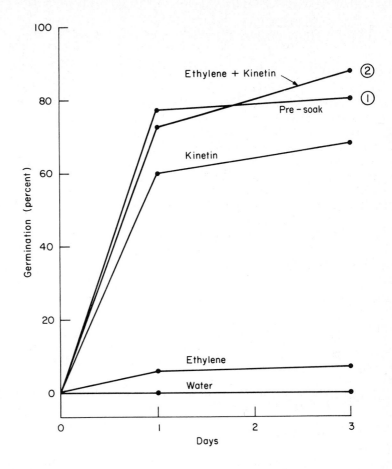

Carbon dioxide levels higher (0.5 to 5.0 percent) than atmospheric levels (0.03 percent) have been shown to stimulate germination of subterranean clover (*Trifolium subterranean*) seed, as well as some other legumes (*77*). At such higher levels carbon dioxide interacts with ethylene to overcome thermodormancy in lettuce seeds (*96*).

### Other Compounds

Certain other compounds are known to stimulate seed germination but their role is not clear. Use of **potassium nitrate** has been an important seed treatment in seed-testing laboratories for many years without a good explanation for its action. **Thiourea** overcomes certain types of dormancy, such as the seed coat inhibiting effect of deep embryo dormant *Prunus* seeds as well as the high temperature inhibition of lettuce seeds (*119*). The effect of thiourea may be due to its cytokinin activity in overcoming inhibition (*46*). Two other naturally occurring substances, *fusicoccin* and *cotylenin* (*72*), have been reported to mimic the combination of GA plus cytokinin.

# ENVIRONMENTAL FACTORS
# AFFECTING SEED GERMINATION

## Water

Water content is a very important factor in controlling seed germination. Below 40 to 60 percent of water in the seed (fresh weight basis) germination does not occur. A water absorption curve for dry seeds has three parts: (a) an initial rapid uptake, which is mostly imbibitional, (b) a slow period, and (c) a second rapid increase as the radicle emerges and the seedling develops (see Figure 6–1).

Dry seeds have great absorptive power for water during imbibition because of their colloidal nature. This absorptive power is present both in storage (see chapter 5) and in the germination medium. It varies with the nature of the seed and the permeability of the seed covering, but absorption also depends on the availability of water in the surrounding medium (*112, 113*). Higher temperatures increase water uptake so that the higher temperatures of the respiring seed may cause increased water uptake (*77*). Once the seed germinates and the radicle emerges, the water supply to the seedling depends on the ability of the root system to grow into the germination medium and the ability of the new roots to absorb water.

The amount of moisture supplied to the imbibing seed before the radicle emerges is particularly important because it can affect both the germination percentage and the germination rate (*39, 55, 63*). Most kinds of seed germinate equally well over most of the range of available soil moisture, from field capacity (FC) to permanent wilting percentage (PWP) (*6*). However, germination of other seeds, particularly those with dormancy problems, such as beet, lettuce, endive, or celery, is inhibited at low moisture levels. Such seeds apparently contain inhibitors and require leaching at high moisture levels. Activated charcoal treatments have proved beneficial for lettuce seed germination (*111*). Addition of Ethrel (2.0 mM), which releases ethylene, plus kinetin (0.3 mM) to lettuce and red clover seed also has been proved beneficial to moisture-stressed seeds (*58*). Seeds of other species, such as spinach, when exposed to excess water, produce extensive mucilage that restricts oxygen supply to the embryo (*60*); an inhibitor is also present (*5*). In these cases germination improves with less moisture.

The rate of seedling emergence from a seed bed, in general, is strongly reduced by a decrease in moisture supply. As the available moisture decreases to a level approximately halfway through the range from FC to PWP, a decline in emergence rate occurs (Figure 6–12) (*6, 39, 55*).

Maintaining an adequate continuous moisture supply to the seeds can be difficult because germination takes place in the upper surface of the germination medium, which is subject to fluctuations of moisture and to rapid water loss. The problem is greater with the necessarily shallow planting of small seeds, or where the germination rate of the particular seeds is low. Two properties of the germination medium affect availability of water: **matric potential** and **osmotic potential.**

The **matric potential** is the ability of the water to move by capillarity through the pores of the soil to the seed. Rate of movement depends upon (a) the pore structure (texture) of the germination medium, (b) the soil packing, and (c) the closeness and distribution of the soil-seed contact. As moisture is re-

**Figure 6–12** The effect of different amounts of available soil moisture on the germination (emergence) of 'Sweet Spanish' onion seed in Pachappa fine sandy loam. From Ayres (6).

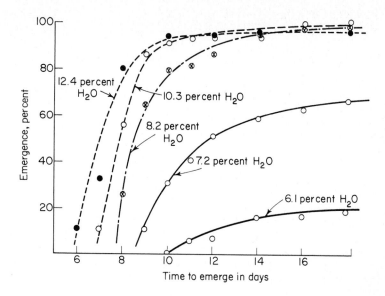

moved from the soil by the imbibing seed, the area nearest the seed becomes dry and must be replenished by water in pores farther away. Consequently a firm, fine-textured seed bed closely compacted to the seed is important in maintaining a uniform moisture supply.

**Osmotic potential** depends upon the presence of solutes (salts) in the soil solution. Excess soluble salts (high salinity) in the germination medium may inhibit germination and reduce seedling stands (6). Such salts originate in the soil and other materials used in the germination medium, the irrigation water, or in excessive fertilization. Since the effects of salinity become more acute when the moisture supply is low and the concentration of salts thereby increased, it is particularly important to maintain a high moisture supply in the seed bed where the possibility of high salinity exists. Surface evaporation from subirrigated beds can result in the accumulation of salts at the soil surface even under conditions in which salinity would not be expected. Planting seeds several inches below the crown of a sloping seed-bed can minimize this hazard (15).

*Seed soaking*    Seeds are sometimes soaked in water before planting to speed up germination and to overcome certain dormancy conditions. Seeds of most herbaceous species benefit from eight hours of soaking but may be injured by soaking periods of 24 hours or more. Dormant seeds of woody plants can be leached for longer periods without injury (95). Large seeds are particularly vulnerable to prolonged soaking. Excess water may be trapped between cotyledons and suffocate the embryo (60). Most harmful results have been attributed to the effects of microorganisms and to a reduced oxygen supply (9, 10). If soaking is to be prolonged, the water should be changed at least once every 24 hours.

Seeds of some species, such as rice, can tolerate low oxygen and germinate readily under water.

*Seed priming*    This process involves procedures that initiate germination to various stages by imbibition prior to planting. Some of these procedures include **fluid drilling** (*52, 62, 106*), in which germinated seeds are planted by special machines, and **osmoconditioning,** in which biochemical processes are initiated but radicle emergence prevented by high osmotic conditions (see chapter 7) (*18, 22, 60, 61, 75*).

## Temperature

Temperature is, perhaps, the most important environmental factor that regulates germination and controls subsequent seedling growth. Dry, unimbibed seeds can withstand extremes of temperature. Seeds can be placed in boiling water for short periods without killing them. In nature, brush fires are sometimes effective in overcoming dormancy without damaging the seeds.

### EFFECTS ON GERMINATION

Temperature affects both germination percentage and germination rate (*78*).

*Germination rate* is usually low at low temperature but increases continuously as temperature rises, similar to a chemical rate-reaction curve (*77*). Above an optimum level, where the rate is most rapid, a decline occurs as the temperatures approach a lethal limit and seed is injured (*56*).

*Germination percentage,* unlike the germination rate, may remain relatively constant, at least over the middle part of the temperature range, if sufficient time is allowed for germination to occur.

Three temperature points (minimum, optimum, and maximum), varying with the species, are usually designated for seed germination (*44*). **Minimum** is the lowest temperature for effective germination. **Maximum** is the highest temperature at which germination occurs. Upper limits may be determined by either the lethal limit or the effects of dormancy induction. **Optimum** temperatures for seed germination fall within the range at which the largest percentage of seedlings are produced at the highest rate. The optimum for non-dormant seeds of most plants is between 25° and 30°C (77° and 86°F). Seeds of different species, whether cultivated or native, can be categorized into temperature-requirement groups.

*Cool-temperature tolerant*    Seeds of many kinds of plants, mostly native to temperate zones, will germinate over a wide temperature range from about 4.5°C (40°F) (or sometimes near freezing) up to the lethal limit—from 30°C (86°F) to about 40°C (104°F) (*56*). Exposure to higher temperatures for very short periods is possible, however, for disease control. The optimum germination temperature is usually about 25° to 30°C (77° to 86°F). Examples include broccoli, carrot, cabbage, alyssum, and others (*85*).

*Cool-temperature requiring*    Seeds of some cool-season plants require low temperatures and fail to germinate at temperatures higher than about 25°C (77°F) (*56*). The inability to germinate at high temperatures is common in freshly harvested seeds of many species, as described earlier in this chapter. This category involves **thermodormancy,** which is present in many species with physiologically dormant seeds but tends to disappear with after-ripening at dry storage. Examples include celery, lettuce, onion, coleus, cyclamen, freesia, primula, delphinium, and others (*5, 85*).

Thermodormancy is found also in seeds of some species with embryo dormancy (see Table 7-1). High temperature, especially in conjunction with reduced aeration, appears to be a major agent of dormancy induction.

*Warm temperature requiring*   Seeds of another broad group of species, originating primarily in subtropical or tropical regions, have a minimum germination requirement of about 10°C (50°F) (asparagus, sweet corn, and tomato) or 15°C (60°F) (beans, eggplant, pepper, and cucurbits). Seeds of species such as lima bean, cotton, soybean, and sorghum are susceptible to "chilling injury." Exposure of the seed to temperatures of 10° to 15°C (50° to 60°F) during initial imbibition, as could occur if they are planted in a cold soil, can injure the embryo axis and result in abnormal seedlings (*98*). Chilling injury is more severe if the seed is very dry at the start of imbibition, or if the oxygen supply is limited.

*Alternating temperatures*   Fluctuating day-night temperatures give better results than constant temperatures for both seed germination and seedling growth. Use of fluctuating temperatures is a standard practice in seed-testing laboratories, even for seeds not requiring it. The alternation should be a 10°C (18°F) difference (*123*). This requirement is particularly important with dormant, freshly harvested seeds (*4*). Seeds of a few species will not germinate at all at constant temperatures. It has been suggested that one of the reasons imbibed seeds deep in the soil do not germinate is that soil temperature fluctuations disappear with increasing soil depth (*99*).

EFFECTS ON
SEEDLING GROWTH

The optimum temperature may shift after germination begins since seedling growth tends to have different temperature requirements than seed germination. In the nursery or laboratory the usual practice is to shift the seedlings to a somewhat lower temperature regime following germination in order to prepare the plants for transplanting and to reduce disease problems in the seed bed.

If seeds are germinated and the seedlings grown at high temperatures, it is important that other environmental conditions be favorable. Plants should have increased light, preferably long photoperiods, adequate fertilization, and sterile conditions to eliminate disease pathogens. Increased carbon dioxide is also a useful component of this system.

## Aeration

Good exchange of gases between the germination medium and the embryo is essential for rapid and uniform germination. Oxygen ($O_2$) is essential for the respiratory processes in the germinating seed. Oxygen uptake can be measured shortly after imbibition of water begins. Rate of oxygen uptake is an indicator of germination progress and has been suggested as a measure of seed vigor (78). In general, $O_2$ uptake is proportional to the amount of metabolic activity taking place.

The supply of oxygen to the embryo can be limited by the condition of the soil medium or by the restrictions imposed by the seed coverings.

Oxygen supply is limited where there is excessive water in the soil medium. Poorly drained outdoor seed beds, particularly after heavy rains or irrigation, can have the pore spaces of the soil so filled with water that little oxygen is available to the seeds. The amount of oxygen in the germination medium is affected by its low solubility in water and its slow diffusability into the medium. Thus, gaseous exchange between the germination medium and the atmosphere, where the $O_2$ concentration is 20 percent, is reduced significantly by soil depth and, in particular, by a hard crust on the surface, which can limit oxygen diffusion (55).

Carbon dioxide ($CO_2$) is a product of respiration and under conditions of poor aeration can accumulate in the soil. At lower soil depths increased $CO_2$ may inhibit germination to some extent but probably plays a minor role, if any, in maintaining dormancy. In fact, high levels of $CO_2$ can be effective in overcoming dormancy in some seeds (77).

SEED COAT RESTRICTION
OF AERATION

In most kinds of seeds there is probably some physical restriction in the movement of gases (particularly oxygen) to the imbibed embryo, due either to the inner membranous seed coat or to the enclosing nucellus or endosperm (21). This situation has been used to explain the germination control by dormant embryos of some seeds (27, 79, 122, 127). For example, freshly harvested dormant seeds that respond to light and are temperature sensitive will germinate if the embryo is excised, the seed coat altered, or the seed subjected to an oxygen level higher than that found in the atmosphere. The effect of after-ripening in dry storage has been attributed to an increase in permeability of the seed coverings. These permeability effects are operable only during the initial stages of germination because once germination occurs and the seed coat is ruptured, the gas exchange capacity is completely altered.

GERMINATION
IN WATER

Seeds of different species vary in their ability to germinate at very low oxygen levels, as occurs under water (89, 90). Seeds of some water plants germinate readily under water, with germination inhibited in air. Rice seeds can germinate in a shallow layer of water. At low oxygen levels, rice seedlings, however, develop differently than those of other monocots. Shoot development is stimulated and the plumule grows to extend up through the water into the air; root growth is suppressed and poor anchorage results unless the water layer is drained away (24).

## Light

It has been known since the mid-nineteenth century that light can affect seed germination (*31*) of certain epiphytic plants, such as mistletoe (*Viscum album*) and strangling fig (*Ficus aurea*). Such seeds have an absolute requirement for light, and lose viability in a few weeks without it. There is another large group of species (most grasses, many herbaceous vegetable and flower species, many conifers, and many native plants) in which seed germination is promoted by light.

Germination in seeds is inhibited by light in a smaller group including some species of *Phacelia, Nigelia, Allium, Amaranthus,* and *Phlox.* Seeds of still other species respond to length of day (photoperiod). Eastern hemlock (*Tsuga canadensis*) (*114*) and birch (*Betula*) (*113*), for instance, have "long-day" seeds, but germination is also promoted by low temperatures. Photoperiodic control can partially replace a seed-chilling requirement, and vice versa. Seed germination in some other species, such as *Nemophila, Nigella, Veronica persica,* and *Eschscholzia californica,* is inhibited by long days (*87*). The light requirement in some cases is also associated with diurnal temperature requirements.

### MECHANISM

The basic mechanism of light sensitivity in seeds involves a photochemically reactive pigment called **phytochrome** widely present in plants (*117*). Exposure of the imbibed seed to **red light** (660 to 760 nm) causes the seeds' phytochrome to change to phytochrome$_{fr}$ (or $\mathbf{P}_{fr}$), which stimulates germination. Exposure of the seed to **far-red** light (760 to 800 nm) causes a change to the alternate form ($\mathbf{P}_r$), which inhibits germination. Both of these changes are instantaneous and can be repeated indefinitely, the last treatment being the one that is effective. In darkness a slow change to $P_r$ occurs and prevents germination. Of some 200 species tested (*117*) the seeds in about half responded to a single light exposure, the seeds of about one-fourth required repeated exposures, and in the remainder germination was inhibited by high light exposure.

Two reaction steps are involved. The first one can occur at moisture levels lower than the level essential for germination. The second step, in which the photoreaction is linked to metabolic reactions controlling germination, requires full seed imbibition. Light control appears to involve the seed coats or the endosperm layer, and the light requirement disappears if the embryos are excised and germinated in aseptic culture. Light control involves hormone regulation, and various hormones can modify the light effect (see Figure 6–13). Furthermore, light and temperature controls are interactive and light response requires a certain optimum temperature.

One can list red light, gibberellic acid, cytokinin, ethylene, moderate to low temperatures (less than 20°C, 68°F) as germination-promoting agents, and far-red light, abscisic acid, moisture stress, and high temperatures as germination-inhibiting agents.

### IN CULTIVATION

Light sensitivity is primarily a property of freshly-harvested, physiologically dormant seeds and tends to disappear in dry storage as these seeds lose their

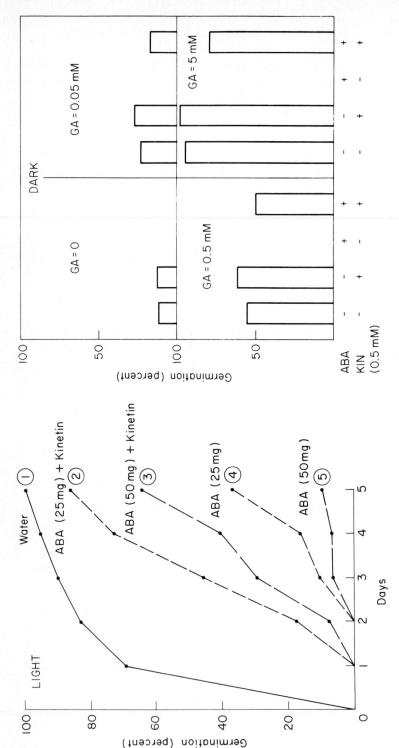

**Figure 6-13**  *Left:* Germination of 'Grand Rapids' lettuce seed occurs in continuous light, as in water control (No. 1). Abscisic acid (ABA) inhibits this light-promoted germination in proportion to concentration (Nos. 4 and 5). Germination is partially restored by treatment with kinetin (K) (Nos. 2 and 3). Redrawn from Khan et al. (*76*). *Right:* Germination percentages of 'Grand Rapids' lettuce seed placed in darkness and promoted by increasing concentrations of gibberellic acid (GA). Abscisic acid (ABA) inhibits gibberellin-promoted germination. Kinetin (K) overcomes the effect of ABA. Kinetin only slightly stimulates gibberellin-promoted germination in the absence of ABA. Redrawn from Khan et al. (*76*).

primary dormancy. For most cultivated crops, light sensitivity is a problem encountered only in seed testing laboratories, where it may cause secondary dormancy.

Light sensitivity can be induced in nonsensitive seeds by exposing the imbibed seeds to conditions inhibiting germination, such as high temperature, high osmotic pressure, or germination-inhibiting gases (135). Light requirements must be considered also in seed priming procedures.

### IN NATURE

Light sensitivity of seeds is a major ecological factor in adaptation of plant species to particular environments. In natural sunlight, red wavelengths dominate over far-red at a ratio of 2:1 so that the phytochrome tends to remain in the active $P_{fr}$ form. Under a foliage canopy, far-red is dominant and the red/far-red ratio may be as low as 0.12:1.00 to 0.70:1.00, which, therefore, mimics darkness, thus inhibiting seed germination (99).

Light-sensitive seeds are often small and thus germination is favored by placement near the soil surface where the seedlings can emerge quickly and begin photosynthesis. Such seeds would not have the capacity to emerge from greater depths. Red light penetrates less deeply into the soil than far-red so that the red/far-red ratio becomes lower with depth until eventually darkness is complete (118). Imbibed light-sensitive seeds buried in the soil will remain dormant until such time as the soil is cultivated or disturbed so as to expose them to light. Similarly, seedling survival is not favored if the seed germinates in close proximity to other plants where there would be intense competition for light, nutrients, and water by the already established plant population.

In contrast, plants producing light-inhibited seeds tend to be found in dry desert environments where seedling survival would be enhanced if the seeds germinated at somewhat greater soil depths where there is less heat and more moisture.

### LIGHT AND SEEDLING GROWTH

Light of a relatively high intensity is desirable to produce sturdy, vigorous plants, particularly if transplanting is involved. Low light intensity results in etiolation and reduced photosynthesis and poor seedling survival if transplanted.

High light intensity, on the other hand, often results in high temperatures that produce heat injury to the seedling, particularly at the soil level, in a manner resembling "damping-off" fungi attacks. Shading is desirable for many kinds of plants during their early seedling growth out-of-doors to avoid heat injury.

Supplementary artificial light to maintain high light intensity and to increase the photoperiod can increase seedling growth, particularly during winter when days are short and light intensity is low. Artificial light can be used as the sole illumination source in special growth rooms (23, 85). Suggested light intensities are 500 to 1200 foot candles or more at the top of the plant, with a photoperiod of at least 16 hours. Combining high light intensity, high temperature, high humidity, supplementary fertilizers, and enrichment with $CO_2$ can be used to produce maximum seedling growth (85).

A balanced light spectrum should be used, including both red and blue light to produce stocky seedlings; red light alone produces tall, "leggy" plants. Cool-white fluorescent lamps provide an adequate, well balanced light source for plant growth. Incandescent lights produce much red and infra-red, which results in heating. Combinations of incandescent and fluorescent often are used at a wattage ratio of 30:100 (*85*).

## Disease Control During Seed Germination

The control of disease during seed germination is one of the most important tasks of the propagator. The most universally destructive pathogens are those resulting in "damping-off," which may cause serious loss of seeds, seedlings, and young plants. In addition, there are a number of fungus, virus, and bacterial diseases that are seedborne and may infect certain plants (*7*). In such cases, specific methods of control are required during propagation. (See chapter 2—Sanitation.)

### DAMPING-OFF

**Damping-off** is a term long used to describe the death of small seedlings resulting from attacks by certain fungi, primarily *Pythium ultimum* and *Rhizoctonia solani,* although other fungi—for example, *Botrytis cinerea* and *Phytophthora* spp.—may also be involved. Mycelia from these organisms occur in soil, in infected plant tissues, or on seeds, from which they contaminate clean soil and infect clean plants. *Pythium* and *Phytophthora* produce spores that are moved about in water.

Damping-off occurs at various stages during seed germination and subsequent seedling growth (*8*):

1. The seed may decay or the seedling may rot before emergence from the soil (**pre-emergence damping-off**).
2. The seedling may develop a stem rot near the surface of the medium and fall over (**post-emergence damping-off**).
3. The seedling may remain alive and standing but the stem will become girdled and the plant stunted, eventually dying (**wire-stem**).
4. Rootlets of larger plants may be attacked; the plants will become stunted and eventually die (**root rot**).

The environmental conditions prevailing during the germination period will affect the growth rate of both the attacking fungi and the seedling. For instance, the optimum temperature for the growth of *Pythium ultimum* and *Rhizoctonia solani* is between approximately 20° and 30°C (68° and 86°F), with a decrease in activity at both higher and lower temperatures. Seeds that have a high minimum temperature for germination (warm-season plants) are particularly susceptible to damping-off, because at lower or intermediate temperatures (less than 23°C or 75°F) their growth rate is low at a time when the activity of the fungi is high. At high temperatures, not only do the seeds germinate faster, but also the activity of the fungi is less. Field planting of such seeds should be delayed until the soil is warm. On the other hand, seeds of cool-season plants germinate (although slowly) at temperatures of less than 13°C

(55°F), but since there is little or no activity of the fungi, they can escape the effects of damping-off. As the temperature increases, their susceptibility increases, because the activity of the fungi is relatively greater than that of the seedling.

The *moisture content* of the germination medium is of great importance in determining the incidence of damping-off (see Figure 6-14). Conditions usually associated with damping-off include over-watering, poor drainage, lack of ventilation, and high density planting.

The control of damping-off involves two separate procedures: (a) the complete elimination of the pathogens during propagation and (b) the control of plant growth and environmental conditions, which will minimize the effects of damping-off or give temporary control until the seedlings have passed their initial vulnerable stages of growth.

If damping-off begins after seedlings are growing, it may sometimes be controlled by treating that area of the medium with a fungicide. The ability to control attacks depends on their severity and on the modifying environmental conditions. (See chapter 2.)

Symptoms resembling damping-off are also produced by certain unfavorable environmental conditions in the seedbed. Drying, high soil temperatures, or high concentrations of salts (see Figure 6-14) in the upper layers of the germination medium can cause injuries to the tender stems of the seedlings near the ground level. The collapsed stem tissues have the appearance of being "burned off." These symptoms may be confused with those caused by pathogens. Damping-off fungi can grow in concentrations of soil solutes high enough to inhibit the growth of seedlings. Where salts accumulate in the germination medium, damping-off can thus be particularly serious.

**Figure 6-14**  Interaction of pathogens and environment in affecting seed germination and seedling growth (*8*). Redrawn from K. F. Baker, ed. 1957. The U.C. system for producing healthy container-grown plants. *Calif. Agr. Exp. Sta. Man. 23.*

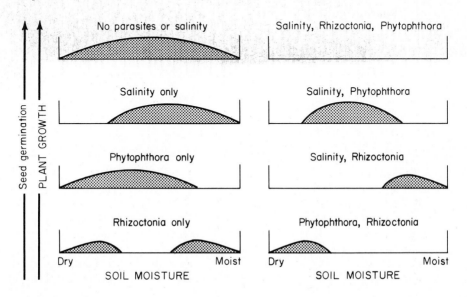

# REFERENCES

1. Abbott, D. L. 1955. Temperature and the dormancy of apple seeds. *Rpt. 14th Inter. Hort. Cong.* Vol. 1, pp. 746–53.

2. Abdul-Baki, A. A. 1980. Biochemical aspects of seed vigor. *HortScience* 15 (6):765–71.

3. Arditti, J., and P. R. Pray. 1969. Dormancy factors in iris (Iridaceae) seeds. *Amer. Jour. Bot.* 56(3):254–59.

4. Assoc. Off. Seed Anal. 1975. Rules for seed testing. *Jour. Seed Tech.* 3(3):1–126.

5. Atwater, B. R. 1980. Germination, dormancy and morphology of the seeds of herbaceous ornamental plants. *Seed Sci. and Tech.* 8:523–73.

6. Ayers, A. D. 1952. Seed germination as affected by soil moisture and salinity. *Agron. Jour.* 44:82–84.

7. Baker, K. F. 1972. Seed pathology. In *Seed biology*, Vol. 2, T. T. Kozlowski, ed. New York, Academic Press.

8. Baker, K. F., and P. A. Chandler. 1957. Development and maintenance of healthy planting stock. In *The U.C. system for producing healthy container-grown plants*, K. F. Baker, ed. Calif. Agr. Exp. Sta. Man. 23:217–36.

9. Barton, L. V. 1950. Relation of different gases to the soaking injury of seeds. *Contrib. Boyce Thomp. Inst.* 16(2):55–71.

10. ———. 1952. Relation of different gases to the soaking injury of seeds, II. *Contrib. Boyce Thomp. Inst.* 17(1):7–34.

11. ———. 1956. Growth response of physiologic dwarfs of *Malus arnoldiana* Sarg. to gibberellic acid. *Contrib. Boyce Thomp. Inst.* 18:311–17.

12. Barton, L. V. and C. Chandler. 1957. Physiological and morphological effects of gibberellic acid on epicotyl dormancy of tree peony. *Contrib. Boyce Thomp. Inst.* 19:201–14.

13. Baskin, J. M., and C. C. Baskin. 1977. Dormancy and germination in seeds of common ragweed with reference to Beal's buried seed experiment. *Amer. Jour. Bot.* 64(9):1174–76.

14. Berlyn, G. P. 1972. Seed germination and morphogenesis. In *Seed biology*, Vol. 3, T. T. Kozlowski, ed. New York: Academic Press.

15. Berstein, L., A. J. MacKenzie, and B. A. Krantz. 1955. The interaction of salinity and planting practice on the germination of irrigated row crops. *Proc. Soil Science Soc. Amer.* 19:240–43.

16. Bewley, J. D., and M. Black. 1978. *Physiology and biochemistry of seeds in relation to germination, Vol. 1, Development, germination, and growth.* Berlin: Springer-Verlag.

17. Black, M. 1980/1981. The role of endogenous hormones in germination and dormancy. *Israel Jour. Bot.* 29:181–92.

18. Bodsworth, S., and J. D. Bewley. 1981. Osmotic priming of seeds of crop species with polyethylene glycol as a means of enhancing early and synchronous germination at cool temperatures. *Can. Jour. Bot.* 59:672–76.

19. Bradbeer, J. W., and N. J. Pinfield. 1967. Studies in seed dormancy. III. The effects of gibberellin on dormant seeds of *Corylus avellana* L. *New Phytol.* 66:515–23.

20. Brant, R. E., G. W. McKee, and R. W. Cleveland. 1971. Effect of chemical and physical treatment on hard seed of Penngift crown vetch. *Crop Sci.* 11:1–6.

21. Brown, R. Germination. 1972. In *Plant physiology,* Vol. 7C, F. C. Steward, ed. New York: Academic Press.

22. Cantliffe, D. J., K. D. Shuler, and A. C. Guedes. 1981. Overcoming seed thermodormancy in a heat sensitive romaine lettuce by seed priming. *HortScience* 16(2):196–98.

23. Cathey, H. M., and L. E. Campbell. 1980. Light and lighting systems for horticultural plants. *Hort. Rev.* 2:491–537.

24. Chapman, A. L., and M. L. Peterson. 1962. The seedling establishment of rice under water in relation to temperature and dissolved oxygen. *Crop. Sci.* 2:391–95.

25. Ching, Te May. 1972. Metabolism of germinating seeds. In *Seed biology,* Vol. 2, T. T. Kozlowski, ed. New York: Academic Press.

26. Comé, D. 1980/1981. Problems of embryonal dormancy as exemplified by apple embryo. *Israel Jour. Bot.* 29:145–57.

27. Comé, D., and T. Tissaoui. 1973. Interrelated effects of imbibition, temperature, and oxygen on seed germination. In *Seed ecology,* W. Heydecker, ed. University Park, Pa.: Pennsylvania State Univ. Press.

28. Chrispeals, M. J., and J. E. Varner. 1967. Hormonal control of enzyme synthesis: On the mode of action of gibberellic acid and abscisin in aleurone layers of barley. *Plant Phys.* 42:1008–16.

29. Crocker, W. 1916. Mechanics of dormancy in seeds. *Amer. Jour. Bot.* 3:99–120.

30. ———. 1948. *Growth of plants.* New York: Reinhold.

31. ———. 1930. Effect of the visible spectrum upon the germination of seeds and fruits. In *Biological effects of radiation.* New York: McGraw-Hill, pp. 791–828.

32. Czabator, F. 1962. Germination value: An index combining speed and completeness of pine seed germination. *For. Sci.* 8:386–96.

33. DeHaas, P. G. 1955. Neue Beitrage zur Samlingsanzucht bei Kernobst. *Rpt. 14th Inter. Hort. Cong.* Vol. 1: 1185–95.

34. DeHaas, P. G., and H. Schander. 1952. Keimungsphysiologische Studien an Kernobst. I. Kamen and Keimung, *Z. f. Pflanz.* 31(4):457–512.

35. Dell, B. 1980. Structure and function of the strophiolar plug in seed of *Albizia lophantha. Amer. Jour. Bot.* 67(4):556–61.

36. Delouche, J. C. 1980. Environmental effects on seed development and seed quality. *HortScience* 15(6):13–18.

37. Diaz, D. H., and G. C. Martin. 1972. Peach seed dormancy in relation to endogenous inhibitors and applied growth substances. *Jour. Amer. Soc. Hort. Sci.* 97(5): 651–54.

38. Dickson, M. H. 1980. Genetic aspects of seed quality. *HortScience* 15(6):771–74.

39. Doneen, L. D., and J. H. MacGillivray. 1943. Germination (emergence) of vegetable seed as affected by different soil conditions. *Plant Phys.* 18:524–29.

40. Dure, L. S., III. 1975. Seed formation. *Ann. Rev. Plant Phys.* 26:259–78.

41. Dure, L. S., G. A. Galau, and S. Greenway. 1980/1981. Changing protein patterns during cotton cotyledon embryogenesis and germination as shown by *in vivo* and *in vitro* synthesis. *Israel Jour. Bot.* 29:293–306.

42. du Toit, H. J., G. Jacobs, and D. K. Strydom. 1979. Role of the various seed parts in peach seed dormancy and initial seedling growth. *Jour. Amer. Soc. Hort. Sci.* 104(4):490–92.

43. Edwards, D. G. W. 1981. Improving seed germination in *Abies. Proc. Inter. Plant Prop. Soc.* 31:69–78.

44. Edwards, T. J. 1932. Temperature relations of seed germination. *Quart. Rev. Biol.* 7:428–43.

45. Eplee, R. E. 1975. Ethylene, a witchweed seed germination stimulant. *Weed Sci.* 23(5):433–36.

46. Erez, A. 1978. Thiourea, a growth promoter of callus tissues. *Jour. Exp. Bot.* 29:159–65.

47. Esashi, Y., and A. C. Leopold. 1969. Dormancy regulation in subterranean clover seeds by ethylene. *Plant Phys.* 44:1470–72.

48. Evenari, M. 1949. Germination inhibitors. *Bot. Rev.* 15:153–94.

49. Evenari, M., and G. Newman. 1952. The germination of lettuce seed. II. The influence of fruit coats, seed coat and endosperm upon germination. *Bul. Res. Council, Israel* 2:75–78.

50. Flemion, F., and E. Waterbury. 1945. Further studies with dwarf seedlings of non-after-ripened peach seeds. *Contrib. Boyce Thompson Inst.* 13:415–422.

51. Gordon, A. G. 1973. The rate of germination. In *Seed ecology,* W. Heydecker, ed., University Park, Pa.: Pennsylvania State Univ. Press, pp. 391–410.

52. Gray, D. 1981. Fluid drilling of vegetable seeds. *Hort. Rev.* 3:1–27.

53. Haber, A. H., and H. J. Luippold. 1960. Separation of mechanisms initiating cell division and cell expansion in lettuce seed germination. *Plant Phys.* 35:168–73.

54. Hamly, D. H. 1932. Softening the seeds of *Melilotus alba. Bot Gaz.* 93:345–75.

55. Hanks, R. S., and F. C. Thorp. 1956. Seedling emergence of wheat as related to soil moisture content, bulk density, oxygen diffusion rate and crust strength. *Proc. Soil Sci. Soc. Amer.* 20:307–10.

56. Harrington, J. F. 1962. The effect of temperature on the germination of several kinds of vegetable seeds. *XVIth Inter. Hort. Cong.* Vol. 2:435–41.

57. Haut, I. C. 1932. The influence of drying on after-ripening and germination of fruit tree seeds. *Proc. Amer. Soc. Hort. Sci.* 29:371–74.

58. Hegarty, T. W., and H. A. Ross. 1980/1981. Investigations of control mechanism of germination under water stress. *Israel Jour. Bot.* 29:83–92.

59. Heydeker, W. 1972. Vigour. In *Viability of seeds,* E. H. Roberts, ed. Syracuse, N.Y.: Syracuse Univ. Press.

60. ———. 1977. Stress and seed germination: An agronomic view. In *The physiology and biochemistry of seed dormancy and germination,* A. A. Khan, ed. Amsterdam: North-Holland Publishing Co.

61. Heydecker, W., and B. M. Gibbins. 1978. Attempts to synchronize seed germination. *Acta Hort.* 72:79–92.

62. Hiron, R. W. P., and R. C. Balls. 1978. The development and evaluation of an air pressurized fluid drill. *Acta Hort.* 72:109–20.

63. Hunter, J. B. and A. E. Erickson. 1952. Relation of seed germination to moisture tension. *Agron. Jour.* 44:107–9.

64. Hyde, E. O. C. 1956. The function of some Papilionaceae in relation to the ripening of the seed and permeability of the testa. *Ann. Bot.* 18:241–56.

65. Jann, R. C., and R. D. Amen. 1977. What is germination? In *The physiology and biochemistry of seed dormancy and germination,* A. A. Khan, ed. Amsterdam: North-Holland Publishing Co., pp. 7–28.

66. Kamininski, W., and R. Rom. 1973. Secondary dormancy in stratified peach embryos. *HortScience* 8(5):401.

67. Karssen, C. M. 1980/1981. Environmental conditions and endogenous mechanisms involved in secondary dormancy of seeds. *Israel Jour. Bot.* 29:45–64.

68. Kester, D. E. 1969. Pollen effects on chilling requirements of almond and almond hybrid seeds. *Jour. Amer. Soc. Hort. Sci.* 94:318–21.

69. Kester, D. E., P. Raddi, and R. Asay. 1977. Correlations of chilling requirements for germination, blooming and leafing within and among seedling populations of almond. *Jour. Amer. Soc. Hort. Sci.* 102(2):145–48.

70. Ketrick, D. L. 1977. Ethylene and seed germination. In *The physiology and biochemistry of seed dormancy and germination,* A. A. Khan, ed. Amsterdam: North-Holland Publishing Co., pp. 156–78.

71. Khan, A. A. 1971. Cytokinins: Permissive role in seed germination. *Science* 171:853–59.

72. ———. 1977. Preconditioning, germination and performance of seeds. In *The physiology and biochemistry of seed dormancy and germination,* A. A. Khan, ed. Amsterdam: North-Holland Publishing Co.

73. ———. 1978. Incorporation of bioactive chemicals into seeds to alleviate environmental stress. *Acta Hort.* 83:225–34.

74. ———. 1980/1981. Hormonal regulation of primary and secondary dormancy. *Israel Jour. Bot.* 29:207–24.

75. Khan, A. A., N. H. Peck, and C. Samiry. 1980/1981. Seed osmoconditioning: Physiological and biochemical changes. *Israel Jour. Bot.* 29:133–44.

76. Khan, A. A., C. E. Heit, E. C. Waters, C. C. Anojulu, and L. Anderson. 1971. Discovery of a new role for cytokinins in seed dormancy and germination. In *Search.* New York Agr. Exp. Sta. (Geneva) 1(9):1–12.

77. Koller, D. 1972. Environmental control of seed germination. In *Seed biology,* Vol. 2. T. T. Kozlowski, ed. New York: Academic Press.

78. Kotowski, F. 1926. Temperature relations to germination of vegetable seeds. *Proc. Amer. Soc. Hort. Sci.* 23:176–84.

79. Kozlowski, T. T., and A. C. Gentile. 1959. Influence of the seed coat on germination, water absorption and oxygen uptake of eastern white pine seed. *For. Sci.* 5:389–95.

80. Lammerts, W. E. 1943. Effect of photoperiod and temperatures on growth of embryo-cultured peach seedlings. *Amer. Jour. Bot.* 30:707–11.

81. Lewak, S., and R. M. Rudnicki. 1977. After-ripening in cold requiring seeds. In *The physiology and biochemistry of seed dormancy and germination,* A. A. Khan, ed. Amsterdam: North-Holland Publishing Co.

82. Lin, C. F., and A. A. Boe. 1972. Effects of some endogenous and exogenous growth regulators on plum seed dormancy. *Jour. Amer. Soc. Hort. Sci.* 97:41–44.

83. Lipe, W., and J. C. Crane. 1966. Dormancy regulation in peach seeds. *Science* 153:541–42.

84. Martin, G. C., H. Forde, and M. Mason. 1969. Changes in endogenous growth substances in the embryo of *Juglans regia* during stratification. *Jour. Amer. Soc. Hort. Sci.* 94:13–17.

85. Mastalerz, J. W. 1976. *Bedding plants.* University Park, Pa.: Pennsylvania Flower Growers.

86. Mathur, D. D., G. A. Couvillon, H. M. Vines, and C. H. Hendershott. 1971. Stratification effects of endogenous gibberellic acid (GA) in peach seeds. *HortScience* 6:538–39.

87. Mayer, A. M., and A. Poljakoff-Mayber. 1975. *The germination of seeds* (2nd ed.). New York: Macmillan.

88. McDonald, M. B., Sr. 1980. Assessment of seed quality. *HortScience* 15:784–88.

89. Morinaga, T. 1926. Germination of seeds under water. *Amer. Jour. Bot.* 13:126–31.

90. ———. 1926. The favorable effect of reduced oxygen supply upon the germination of certain seeds. *Amer. Jour. Bot.* 13:150–65.

91. Nagao, M. A., K. Kanegawa, and W. S. Sakai. 1980. Accelerating palm seed germination with gibberellic acid, scarification, and bottom heat. *HortScience* 15(2):200–201.

92. Negm, F. B., O. E. Smith, and J. Kumamoto. 1972. Interaction of carbon dioxide and ethylene in overcoming thermodormancy of lettuce seeds. *Plant Phys.* 49:869–72.

93. ———. 1973. The role of phytochrome in an interaction with ethylene and carbon dioxide in overcoming lettuce seed thermodormancy. *Plant Phys.* 51:1089–94.

94. Nikolaeva, M. G. 1977. Factors affecting the seed dormancy pattern. In *The physiology and biochemistry of seed dormancy and germination,* A. A. Khan, ed. Amsterdam: North-Holland Publishing Co., pp. 51–76.

95. Norton, C. R. 1980. Deleterious metabolic and morphological changes resulting from seed soaking prior to sowing. *Proc. Inter. Plant Prop. Soc.* 30:132–34.

96. Olatoye, S. T., and M. A. Hall. 1973. Interaction of ethylene and light on dormant weed seeds. In *Seed ecology,* W. Heydecker, ed. University Park, Pa.: Pennsylvania State Univ. Press.

97. Pollock, B. M. 1962. Temperature control of physiological dwarfing in peach seedlings. *Plant Phys.* 37:190–97.

98. Pollock, B. M., and E. E. Roos. 1972. Seed and seedling vigor. In *Seed biology,* Vol. 3, T. T. Kozlowski, ed. New York: Academic Press.

99. Roberts, E. H. 1972. Dormancy: A factor affecting seed survival in the soil. In *Viability of seeds,* E. H. Roberts, ed. Syracuse, N. Y.: Syracuse Univ. Press.

100. Rolston, M. P. 1978. Water impermeable seed dormancy. *Bot. Rev.* 44:365–96.

101. Roos, E. E. 1980. Physiological, biochemical and genetic changes in seed quality during storage. *HortScience* 15:781–84.

102. Ross, J. D., and J. W. Bradbeer. 1968. Concentrations of gibberellin in chilled hazel seeds. *Nature* 220:85–86.

103. ———. 1971. Studies in seed dormancy. V. The concentrations of endogenous gibberellins in seeds of *Corylus avellana* L. *Planta* (Berl.) 100:288–302.

104. ———. 1971. Studies in seed dormancy. VI. The effects of growth retardants on the gibberellin content and germination of chilled seeds of *Corylus avellana* L. *Planta* (Berl.) 100:303–8.

105. Rubenstein, I., R. L. Phillips, G. E. Green, and B. G. Gengenbach. 1979. *The plant seed: Development, preservation and germination.* New York: Academic Press.

106. Salter, P. J. 1978. Techniques and prospects for 'fluid' drilling of vegetable crops. *Acta Hort.* 72:101–8.

107. Schander, H. 1955. Keimungsphysiologische Studien an Kernobst. III. Sortenvergleichende Untersuchungen über die Temperature-ansprüche stratifizierten Saatgutes von Kernobst und über die Reversibilitat der Stratifikationsvorgange, *Z. f. Pflanz.* 35:89–97.

108. Schopmeyer, C. S., ed. 1974. *Seeds of woody plants in the United States.* USDA Handbook No. 450. Washington, D.C.: U.S. Govt. Printing Office.

109. Semeniuk, P., and R. N. Stewart. 1962. Temperature reversal of after-ripening of rose seeds. *Proc. Amer. Soc. Hort. Sci.* 80:615–21.

110. Sharples, G. C. 1973. Stimulation of lettuce seed germination at high temperatures by ethephon and kinetin. *Jour. Amer. Soc. Hort. Sci.* 98(2):207–9.

111. ———. 1978. Interaction of moisture potential and activated carbon on lettuce seed germination. *Jour. Amer. Soc. Hort. Sci.* 103(1):135–37.

112. Shull, C. A. 1916. Measurement of the surface forces in soils. *Bot. Gaz.* 62:1–29.

113. ———. 1920. Temperature and rate of moisture uptake in seeds. *Bot. Gaz.* 69:361–90.

114. Stearns, F., and J. Olson. 1958. Interactions of photoperiod and temperature affecting seed germination in *Tsuga canadensis. Amer. Jour. Bot.* 45:53–58.

115. Stewart, R. N., and P. Semeniuk. 1965. The effect of the interaction of temperature with after-ripening requirement and compensating temperature on germination of seed of five species of *Rosa. Amer. Jour. Bot.* 52:755–60.

116. Suszka, B. 1978. Germination of tree seed stored in a partially after-ripened condition. *Acta Hort.* 83:181–88.

117. Taylorson, R. B., and S. B. Hendricks. 1977. Dormancy in seeds. *Ann. Rev. Plant Phys.* 28:331–54.

118. Thomas, H. 1972. Control mechanisms in the resting seed. In *Viability of seeds,* E. H. Roberts, ed. Syracuse, N.Y.: Syracuse Univ. Press.

119. Thomas, T. H. 1977. Cytokinins, cytokinin-active compounds and seed germination. In *The physiology and biochemistry of seed dormancy and germination,* A. A. Khan, ed. Amsterdam: North-Holland Publishing Co., pp. 111–24.

120. Thompson, P. A. 1973. Geographical adaptation of seeds. In *Seed ecology,* W. Heydecker, ed. University Park, Pa.: Pennsylvania State Univ. Press.

121. Thornton, N. C. 1945. Importance of oxygen supply on secondary dormancy and its relation to the inhibiting mechanism regulating dormancy. *Contrib. Boyce Thomp. Inst.* 13:497–500.

122. Toole, E. H., V. K. Toole, and E. A. Gorman. 1948. Vegetable seed storage as affected by temperature and relative humidity. *USDA Tech. Bul. 972.*

123. USDA. 1952. *Manual for testing agricultural and vegetable seeds.* USDA Agr. Handbook 30. Washington, D.C.: U.S. Govt. Printing Office.

124. Vegis, A. 1964. Dormancy in higher plants. *Ann. Rev. Plant Phys.* 15:185–224.

125. Villiers, T. A. 1972. Seed dormancy. In *Seed biology,* Vol. 2, T. T. Kozlowski, ed. pp. 220–82. New York: Academic Press.

126. Visser, T. 1956. The role of seed coats and temperature in after-ripening, germination and respiration of apple seeds. *Proc. Koninkl. Nederl. Akad. van Wetens,* Series C 59:211–22.

127. ———. 1956. Some observations on respiration and secondary dormancy in apple seeds. *Proc. Koninkl. Akad. van Wetens,* Series C 59:314–24.

128. ———. 1956. The growth of apple seedlings as affected by after-ripening, seed maturity and light. *Proc. Koninkl. Nederl. Akad. van Wetens,* Series C 59:325–34.

129. Walton, D. C. 1980. Biochemistry and physiology of abscisic acid. *Ann. Rev. Plant Phys.* 31:453–89.

130. ———. 1980/1981. Does ABA play a role in seed germination? *Israel Jour. Bot.* 29:168–80.

131. Wareing, P. F. 1963. The germination of seeds. In *Vistas in botany,* III. New York: Macmillan, pp. 195–227.

132. Wareing, P. F., and H. A. Foda. 1957. Growth inhibitors and dormancy in *Xanthium* seed. *Phys. Plant.* 10(2):266–80.

133. Went, F. W. 1949. Ecology of desert plants. II. The effect of rain and temperature on germination and growth. *Ecology* 30:1–13.

134. Went, F. W., and M. Westergaard. 1949. Ecology of desert plants. III. Development of plants in the Death Valley National Monument, California. *Ecology* 30:26–38.

135. Wesson, G., and P. F. Wareing. 1969. The induction of light sensitivity in weed seeds by burial. *Jour. Exp. Bot.* 20(63):414–25.

136. Westwood, M. N., and H. O. Bjornstad. 1948. Chilling requirement of dormant seeds of fourteen pear species as related to their climatic adaptation. *Proc. Amer. Soc. Hort. Sci.* 92:141–49.

137. Willemsen, R. W. 1975. Effect of stratification temperature and germination temperature on germination and the induction of secondary dormancy in common ragweed seeds. *Amer. Jour. Bot.* 62(1):1–5.

138. Williams, P. M., J. D. Ross, and J. W. Bradbeer. 1973. Studies in seed dormancy. VII. The abscisic acid content of the seeds and fruits of *Corylus avellana* L. *Planta* (Berl.) 110:303–10.

## SUPPLEMENTARY READING

AMEN, R. D. 1968. A model of seed dormancy. *Bot. Rev.* 34:1–31.

BARTON, L. V. 1967. *Bibliography of seeds.* New York: Columbia Univ. Press.

BEWLEY, J. D., and M. BLACK. 1978. *Physiology and biochemistry of seeds in relation to germination, Vol. 1, Development, germination, and growth.* Berlin: Springer-Verlag.

CROCKER, W., and L. V. BARTON. 1953. *Physiology of seeds.* Waltham, Mass.: Chronica Botanica.

HEYDECKER, W., ed. 1973. *Seed ecology: Proceedings of the nineteenth Easter School in Agricultural Science, University of Nottingham, 1972.* University Park: The Pennsylvania State Univ. Press.

KHAN, A. A., ed. 1977. *The physiology and biochemistry of seed dormancy and germination.* Amsterdam: North-Holland Publishing Co.

KOZLOWSKI, T. T., ed. 1972. *Seed biology,* Vols. 1, 2, 3. New York: Academic Press.

MAYER, A. M. 1980/1981. Control mechanisms in seed germination. *Israel Jour. Bot.* 29:1–4.

Mayer, A. M., and A. Poljakoff-Mayber. 1975. *The germination of seeds* (2nd ed.). New York: Macmillan.

Roberts, E. H., ed. 1972. *Viability of seeds.* Syracuse, N.Y.: Syracuse Univ. Press.

Rubenstein, I., R. L. Phillips, G. E. Green, and B. G. Gengenbach. 1979. *The plant seed: Development, preservation and germination.* New York: Academic Press.

Schopmeyer, C. S., ed. 1974. *Seeds of woody plants in the United States.* USDA Handbook No. 450. Washington, D.C.: U.S. Govt. Printing Office.

Taylorson, R. B., and S. B. Hendricks. 1977. Dormancy in seeds. *Ann. Rev. Plant Phys.* 28:331–54.

Seed propagation involves careful management of germination conditions and facilities and a knowledge of the requirements of individual kinds of seeds. Its success depends upon the degree to which the following conditions are fulfilled:

(a) The seed must maintain the particular cultivar or species which the propagator wishes to grow. This can be accomplished by obtaining seed from a reliable dealer, buying certified seed, or—if producing one's own—following the principles of seed selection described in chapter 4.

(b) The seed must be viable and capable of germination. It should germinate rapidly and vigorously to withstand possible adverse conditions in the seedbed. Viability can be determined by seed tests, but seed vigor is difficult to predict.

(c) Any dormancy condition of the seed which would inhibit germination must be overcome by applying any necessary pregermination treatments. The propagator should know the requirements of the seed with which he is concerned. A germination test will be helpful in indicating the necessity for any pregermination treatment. In the absence of specific knowledge, the propagator should try to duplicate the natural environmental conditions associated with germination of seed of this particular kind of plant.

(d) Assuming that the seed is capable of prompt germination, propagation success depends upon providing the proper environment—moisture, temperature, oxygen, and light or darkness—to the seed and the resulting seedling until it is well established. A proper environment also includes control of diseases and insects.

## SEED TESTING

Good-quality seed has the following characteristics: It is genetically true to species or cultivar; capable of high germination; free from disease and insects; and free from mixture with other crop seeds, weed seeds, and inert and ex-

7

# Techniques of Propagation by Seeds

traneous material. The germination capability and purity of the seed can be determined by conducting a seed test on a small representative sample drawn from the seed lot in question (*4, 38, 71*).

In the United States, state laws regulate the shipment and sale of agricultural and vegetable seeds within that state. Seeds entering interstate commerce or those sent from abroad are subject to the Federal Seed Act adopted in 1939 (*3*). Such regulations require labeling by the shipper of commercially produced seeds as to: name and cultivar; origin; germination percentage; and the percentage of pure seed, other crop seed, weed seed, and inert material. The regulations may set minimum standards of quality, germination percentage, and freedom from weed seeds. Shipment and sale of tree seed is regulated by law in some states (*59*) and in most European countries.

Seed testing provides information to meet legal standards, determines seed quality, and establishes the rate of sowing for a given stand of seedlings. It is desirable to retest seeds that have been in storage for a prolonged period.

Procedures for testing agriculture and vegetable seed in reference to the Federal Seed Act are given by the U.S. Department of Agriculture (*3, 71*). The Association of Official Seed Analysts also publishes procedures for testing these seeds in addition to procedures for testing seeds of many flower, tree, and shrub species (*4*). International rules for testing seeds of many tree, shrub, agricultural, and vegetable species are published by the International Seed Testing Association (*35*). The Western Forest Tree Seed Council also has published testing procedures for tree seed (*76*).

## Sampling

The first step in seed testing is to obtain a uniform sample representing the entire lot under consideration. Equally sized **primary** samples are taken from evenly distributed parts of the seed lot, such as a sample from each of several sacks in lots of less than five sacks or from every fifth sack with larger lots. The seed samples are thoroughly mixed to make a **composite** sample. A representative portion is used as a **submitted sample** for testing. This sample is further divided into smaller lots to produce a **working sample,** i.e., the sample upon which the test is actually to be run. The amount of seed required for the working sample varies with the kind of seed and is specified in the Rules for Seed Testing.

## Purity Determination

**Purity** is the percentage by weight of the ''pure seeds'' present in the sample. **Pure seed** refers to the principally named kind, cultivar, or type of seed present in the seed lot. After the working sample has been weighed, it is divided visually into (a) the pure seed of the kind under consideration; (b) other crop seed; (c) weed seed; and (d) inert material, including seed-like structures, empty or broken seeds, chaff, soil, stones, and other debris (see Figure 7–1). In some cases, it is possible to check the genuineness of the seed or trueness to cultivar or species by visual inspection. Often, however, identification cannot be made except by growing the seeds and observing the plants. At the time of making the purity test, the number of pure seeds per pound can be calculated. These data are necessary as a guide to seeding rates.

**Figure 7–1**   Purity of seeds is determined by visual examination of individual seeds in a weighed sample taken from the larger lot in question. Impurities may include other crop seed, weed seed, and inert, extraneous material. Courtesy E. L. Erickson Products, Brookings, S.D.

## Moisture Determination

Moisture content is found by the loss of weight when a sample is dried under standardized conditions (*67*). Oven drying at 130°C (266°F) for one to four hours is used for many kinds of seeds. For oily seeds 103°C (217°F) for 17 hours is used (*67*) and for some seeds which lose oil at that temperature (e.g., fir, cedar, beech, spruce, pine, hemlock) a toluene distillation method is used. Various kinds of electronic meters can be used for quick moisture tests (*9*).

## Viability Determination

Viability can be determined by several tests, the **direct germination, excised embryo,** and **tetrazolium** tests being the most important. In the **direct germination test** the **germination percentage** is determined by the percent of normal seedlings produced by the pure seed (the kind under consideration). To produce a good test, it is desirable to use at least 400 seeds picked at random and divided into lots of one hundred each. If any two of these lots differ by more than 10 percent, a retest should be carried out. Otherwise, the average of the four tests becomes the germination percentage (see Figure 7–2).

### DIRECT GERMINATION TESTS

In a standard germination test the seeds are placed under optimum environmental conditions of light and temperature to induce germination (Figure 7–3). The conditions required to meet legal standards are specified in the rules for seed testing, which may include type of test, environmental conditions, and length of test.

**Figure 7-2**    Germination testing of seeds. *Top:* 100 seeds from the sample to be tested are placed on a moistened blotter. In this case placing the seeds evenly and quickly is made possible by an automatic vacuum counter. *Below:* after one or more weeks in a germinator the number of germinated seeds is counted. Note that this test consists of four lots of 100 seeds each. Courtesy E. L. Erickson Products, Brookings, S.D.

**Figure 7-3**    Commercial seed germinator with light and temperature control for testing viability. Courtesy C. E. Heit.

Various techniques are used for germinating seed in seed-testing laboratories. Small seeds are placed on germination trays (not galvanized steel, which contains toxic zinc salts). Plastic boxes, paraffined cardboard boxes, or covered glass Petri dishes also are useful containers. Absorbent paper is cut into small pieces (*blotters*) and small seeds are placed on top or between two layers. Other media are absorbent cotton, paper toweling (five thicknesses), filter paper (five layers), and for large seeds, sand, vermiculite, perlite, or soil (16 mm, ⅝ in.). Containers are placed in special germinators in which temperature, moisture, and light are controlled. To discourage the growth of microorganisms, all materials and equipment should be kept scrupulously clean, sterilized when possible, and the water amount carefully regulated. No water film should form around the seeds; neither should the germination medium be so wet that a film of water appears when that medium is pressed with a finger. Relative humidity in the germinator should be 90 percent or more to prevent drying. Containers with sand should be kept tightly closed. Water should not be added during the test.

The **rolled towel** test is commonly used for testing cereal grains. Several layers of moist paper towelling, about 2.8 by 3.6 cm (11 by 14 in.) in size, are folded over the seeds, then rolled into cylinders and placed vertically in a germinator (*1*).

A germination test usually runs from one to four weeks but could continue for three months for some slow-germinating tree seeds with dormancy. A *first count* may be taken at one week and germinated seeds discarded with a formal count taken later. At end of the test seeds are divided into (a) normal seedlings, (b) hard seeds, (c) dormant seeds, and (d) abnormal seedlings plus dead or decaying seeds. A normal seedling generally should have a well-developed root and shoot, although the criteria for a "normal seedling" vary with different kinds of seeds. "Abnormal seedlings" can be caused by age of seed or poor storage conditions; insect, disease, or mechanical injury; overdoses of fungicides; frost damage; mineral deficiencies (manganese and boron in peas and beans); or toxic materials sometimes present in metal germination trays, substrata, or tap water. Any ungerminated seed should be examined to determine the possible reason. "Hard seeds" have not absorbed water. Dormant seeds are those that are firm, swollen, and free from molds, but show erratic sprouting, or none.

Dormancy of freshly harvested seeds produces difficulties in direct testing by prolonging the testing period and interfering with the reliability of the test.

Under seed testing rules specific environmental requirements to overcome dormancy may be specified routinely for many kinds of agricultural, vegetable, and flower seeds (*4, 35*). Tree and shrub seeds often require special pregermination treatments before tests are run (Table 7–1).

---

**Table 7–1**    Types of germination requirements for tree and shrub seeds (with examples) when tested in the laboratory.*

---

**Group 1.** Seeds that germinate within a wide temperature range and without light exposure.

| | |
|---|---|
| Beefwood (*Casuarina glauca*) | some spruce species (*Picea abies, P. asperata, P. polita*) |
| Italian cypress (*Cupressus semipervirens*) | Chinese and Siberian elm (*Ulmus parvifolia, U. pumila*) |
| many species of eucalyptus | |
| Honeylocust (*Gleditia triacanthos*)† | |

**Group 2.** Seeds that have specific temperature requirements but do not require light.

**20° to 30°C (68° to 86°F) diurnally alternating:**
Catalpa
Ailanthus
Red pine ( *Pinus resinosa*)
or 25°C (77°F) constant

**10° to 30°C (50° to 86°F) diurnally alternating:**
Mountain mahogany ( *Cercocarpus ledifolius*)
Cliffrose ( *Cowania stansburiana*)
Antelope bitterbush ( *Purshia tridentata*)

**20°C (68°F) constant:**
several pine species ( *Pinus cembroides, P. halepensis, P. pinea*)
Lilac ( *Syringa vulgaris*)
Arborvitae ( *Thuja orientalis*)

**Group 3.** Seeds that germinate in 7 to 12 days within a wide temperature range if exposed to artificial light.
several spruce species ( *Picea engelmannii, P. mariana, P. omerika*)
several pine species ( *Pinus banksiana, P. nigra, P. mugo var. mughus, P. rigida, P. sylvestris, P. ponderosa scopulorum*)

**Group 4.** Seeds that germinate in 14 to 28 days if exposed to artificial light and to warm alternating temperatures of 20° to 30°C (68° to 86°F). Seeds may also respond to moist-chilling.

Birch ( *Betula*)
Elm ( *Ulmus americana*)
Larch ( *Larix sibirica*)
Mulberry ( *Morus alba, M. nigra*)
*Liquidambar styraciflua*
some spruce series ( *Picea glauca, P. orientalis, P. rubens, P. sitchensis*)

some pine species ( *Pinus densiflora, P. echinata, P. elliotti, P. taeda, P. thunbergii, P. virginiana*)
*Rhododendron*
*Sequoia* and *Sequoiadendron*
*Thuja plicata, T. occidentalis*

**Group 5.** Seeds that require 3 to 4 weeks moist-chilling at 3°C (37°F) before germination at alternating 20° to 30°C (68° to 86°F) (except as noted) temperatures in light for 2 to 4 weeks.

Fir species ( *Abies balsamea, A. fraseri, A. grandis, A. homolepsis, A. procera*)
*Cedrus* species 20°C (68°F)
some pine species ( *Pinus flexilis, P. glabra, P. leucodermis, P. strobus*)

some sources of Douglas fir ( *Pseudotsuga menziesii*)
*Rosa multiflora* 10° to 30°C (50° to 86°F)
Eastern hemlock ( *Tsuga canadensis*) 15°C (59°F)
Sumac ( *Rhus aromatica*)†

**Group 6.** Seeds that require 2 to 6 months moist-chilling as a minimum requirement prior to germination. Some also have other dormancy problems. Embryo excision or a tetrazolium test may be useful in determining germinative capacity.

**Sensitive to high germination temperature 20°C (68°F)**
Maple ( *Acer* spp.)
Apple ( *Malus* spp.)
Pear ( *Pyrus* spp.)
Peach, cherry, etc. ( *Prunus* spp.)
Yew ( *Taxus* spp.)

**Not sensitive to high germination temperature:**
Some pine species ( *Pinus cembra, P. lambertiana, P. monticola, P. peuce*)

* From C. E. Heit ( *34*).
† Must be treated also for hard seed coats.

The excised-embryo test is used to test the seed viability of woody shrubs and trees whose dormant embryos require long periods of after-ripening before true germination will take place (*20, 30*). In this test the embryo is excised from the seed and germinated alone (see Figure 7–4).

The seeds are soaked for one to four days by one of the following methods until they are completely swollen: (a) in slowly running water, (b) in standing water below 15°C (59°F), or (c) in standing water at about 20°C (68°F), with at least two changes of water daily.

Storing seeds in moist peat for three days to two weeks at cool temperatures is also satisfactory in preparing seeds for excision. The excision must be done carefully to avoid injury to the embryo. Any hard, stony seed coverings, such as the endocarp of stone fruit seeds, must first be removed.

The moistened seed coats are cut with a sharp scalpel, razor blade, or knife, under clean but nonsterile conditions with sterilized instruments, preferably under a sheet of glass. The embryo is carefully removed. If a large endosperm is present, the seed coats may be slit and the seeds covered with water, and after about a half hour the embryo will float out or can easily be removed.

Procedures for germinating excised embryos are similar to those for germinating intact seeds. Petri dishes with a moist substratum, such as blotting or filter paper, are used. The embryos are placed on the filter paper so that they do not touch. The dishes are kept in the light at a temperature of 18° to 22° C (64° to 74° F). At higher temperatures, molds may develop and interfere with the test. The time required for the test varies from three days to three weeks.

Nonviable embryos become soft, turn brown, and decay within two to ten days; viable embryos remain firm and show some indication of viability, depending upon the species. Types of response that occur include spreading of the cotyledons, development of chlorophyll, and growth of the radicle and plumule. The rapidity and degree of development gives some indication of the vigor of the seed.

**Figure 7–4**    The excised embryo method of testing seed germination. *Top:* germination of apple seeds showing the range of vigor from strong to weak to dead. Actual germination percentages for the four lots were (left to right): 100, 70, 44, and 0. *Below:* seed vigor and germination of four peach stocks (left to right): strong seed, vigorous growth (80 percent viability); good seed, fair vigor (52 percent viability); old seed, weak (18 percent viability); and dead seed. Courtesy C. E. Heit.

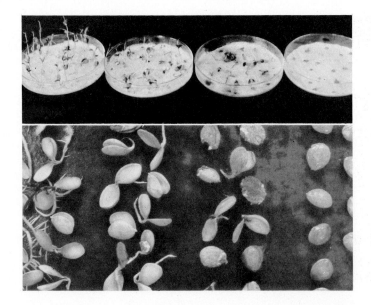

## TETRAZOLIUM TEST

The tetrazolium test is a biochemical method in which viability is determined by the red color appearing when the seeds are soaked in a 2,3,5-triphenyltetrazolium chloride (TTC) solution. Living tissue changes the TTC to an insoluble red compound (chemically known as formazan); in nonliving tissue the TTC remains uncolored. The test is positive in the presence of dehydrogenase enzymes. This test was developed in Germany by Lakon (43) who referred to it as a topographical test, since loss in embryo viability begins to appear at the extremity of the radicle, epicotyl, and cotyledon tips. The reaction takes place equally well in dormant and nondormant seed. Results can be obtained within 24 hours, sometimes in two or three hours. TTC is soluble in water, making a colorless solution. Although the solution deteriorates with exposure to light, it will remain in good condition for several months if stored in a dark bottle. The solution should be discarded if it becomes yellow. A 0.1 to 1.0 percent concentration is commonly used. The pH should be 6 or 7.

The TTC test is used primarily to obtain rapid results for both nondormant and dormant seeds.

The test distinguishes between living and dead tissues within a single seed and can indicate weakness before germination is actually impaired. Necrotic areas may be attacked by pathogenic organisms, and seeds with such dead tissues may decay during stratification or give reduced germination under unfavorable soil conditions. In the hands of a skilled technologist this test can be used for seed-quality evaluation and as a tool in seed research (47, 57).

On the other hand, this test may not adequately measure certain types of injury which could lead to seedling abnormality—for instance, an overdose of chemicals, seedborne diseases, frost or heat injury. Standardized procedures and skills are required for evaluating results.

Although details vary with different kinds of seeds, in general, the following procedures must be followed (4, 35, 57):

1. Any hard covering such as an endocarp, wing, or scale must be removed. Tips of dry seeds of some plants, such as *Cedrus,* should be clipped.

2. Seeds should first be soaked in water in the dark; moistening activates enzymes and facilitates cutting or removal of seed coverings. Seeds with fragile coverings, such as snap beans or citrus, must be softened slowly on a moist medium to avoid fracturing.

3. Most seeds require preparation for TTC absorption. Embryos with large cotyledons, such as *Prunus,* apple, and pear, often comprise the entire seed, requiring only seed coat removal. Other kinds of seed are cut longitudinally to expose the embryo (corn and large seeded grasses, larch, some conifers); or transversely ¼ to ⅓ at the end away from the radicle (small seeded grasses, juniper, *Carpinus, Cotoneaster, Crataegus, Rosa, Sorbus, Taxus*). Seed coats can be removed, leaving the large endosperm intact (some pines, *Tilia*). Some seeds (legumes, timothy) require no alteration prior to the tests.

4. Seeds are soaked in the TTC solution for two to 24 hours. Cut seeds require a shorter time; those with exposed embryos somewhat longer; intact seeds 24 hours or more.

5. Interpretation of results depends upon the kind of seed and its morphological structure. Completely colored embryos indicate good seed. Conifers must have both the megagametophyte and embryo stained. In grass and grain seeds only the

embryo itself colors, not the endosperm. Seeds with declining viability may have uncolored spots, or be unstained at the radicle tip and the extremities of the cotyledons. Nonviability depends upon the amount and location of necrotic areas, and correct interpretation depends upon standards worked out for specific seeds (*47*).

**6.** If the test continues too long, even tissues of known dead seeds become red due to respiration activities of infecting fungi and bacteria. Likewise, the solution itself can become red because of such contamination.

X-RAY ANALYSIS

X-ray photographs of seeds (*39, 63*) can be used as a rapid test for seed soundness (*2*). X-ray photographs do not normally measure seed viability but provide an examination of the inner structure for mechanical disturbance, absence of vital tissues, such as embryo or endosperm, insect infestation, cracked or broken seed coats, and shrinkage of interior tissues (possibly a sign of age). (See Figure 7–5.)

Standard X-ray equipment is used. A convenient Plexiglass seed holder with 100 seed compartments can be made by drilling 100 holes 15 mm in diameter in a 5 mm thick Plexiglass plate. Then glue a 0.1 mm thick Mylar sheet to the underside. Place one dry seed in each compartment and expose for ½ to 3 minutes at 15 to 20 kilovolts tube potential. The plate can be processed immediately and the status of seeds determined. Seeds with dimensions less than 2 mm are too small to show details. Since X-rays do not injure the seed, further tests for viability can be conducted on the same batch (*2*).

Other tests for seed viability have been developed utilizing contrast agents, such as solutions of certain salts or heavy metals (*63*), followed by X-ray photography. In one test, for example, seeds are soaked in water for 16 hours, then transferred to a concentrated (20 to 30 percent) barium chloride solution for one to two hours, washed to remove all excess material, and dried. The salts will not enter living cells because of their semipermeability but can penetrate the dead cells of damaged seeds because of membrane destruction and thus produce an exposure on the film. This penetration differentiates between living and dead seeds, or portions of seeds. Fairly consistent and reliable correlations with germination percentages can be obtained with fresh conifer seeds, although consistent results may not occur with stored seed (*16*).

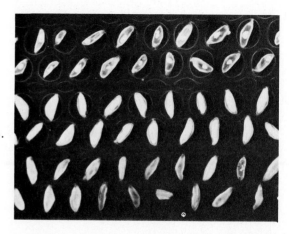

**Figure 7–5** X-radiograph of *Abies procera* seed. Seed has been segregated to illustrate empty seed (top two rows), normally filled seed with distinct embryos (center two rows), and seed infested with chalcid fly larvae (bottom two rows). Courtesy Jay Allison (*2*).

Several other kinds of seed tests are used in addition to those already described (*67*).

*Verification of species and cultivar,* in some cases, is possible by visual observation of the external seed characteristics; more accurate verification requires observations of seedlings produced either in the laboratory or in field plots.

*Seed health* tests are used to obtain observations on the presence of pathogens. These specialized tests require training in plant pathology methods.

*Estimating relative seed vigor* is important, but standard tests are available only for special situations (*12, 50*). **Germination speed** provides a general index: A seedling count at one week is compared to the final germination count (*12*). Previously, tree seed technologists have used the term **germination energy,** which is the germination percentage achieved at the maximum germination rate (*70*). The **germination value,** as described in chapter 6, is a useful method that includes vigor measurements.

# TREATMENTS TO OVERCOME SEED DORMANCY

## Softening Seed Coats and Other Coverings

**Scarification** is any process of breaking, scratching, mechanically altering, or softening the seed coverings to make them permeable to water and gases.

### MECHANICAL SCARIFICATION

Mechanical scarification is simple and effective with seeds of many species if suitable equipment is available (Figure 7–6). The seeds are dry after such treatment and may be stored or planted immediately by mechanical seeders. Scarified seeds are more susceptible to injury from pathogenic organisms, however, and will not store as well as comparable nonscarified seeds.

Chipping hard seed coats by rubbing with sandpaper, cutting with a file, or cracking with a hammer or a vise are simple methods useful for small amounts of relatively large seed. For large-scale mechanical operations, special scarifiers

**Figure 7–6**    Disk scarifier used to modify hard seed coats. Construction based on one described by U.S. Department of Agriculture for use with tree seeds. Scarifier consists of five disks covered with abrasive paper mounted on a shaft and enclosed within a metal cylinder also lined with abrasive paper. The bottom third of the cylinder is filled with seeds and the disks are rotated at a speed of 500 to 900 rpm. Duration of treatment must be established for the particular seed lot and kind of seed.

are used. Small seeds of legumes, such as alfalfa and clover, are often treated in this manner to increase germination (*12*). Seeds may be tumbled in drums lined with sandpaper or in concrete mixers, combined with coarse sand or gravel (*70*) (Figure 7-6). The sand or gravel should be of a different size than the seed to facilitate subsequent separation.

Scarification should not proceed to the point at which the seeds are injured. To determine the optimum time, a test lot can be germinated, the seeds may be soaked to observe swelling, or the seed coats may be examined with a hand lens. The seed coats generally should be dull but not so deeply pitted or cracked as to expose the inner parts of the seed.

### HOT WATER SCARIFICATION

Drop the seeds into four to five times their volume of hot water 77° to 100° C (170° to 212°F). The heat source is immediately removed, and the seed soaked in the gradually cooling water for 12 to 24 hours. Following this the unswollen seeds can be separated from the swollen ones by suitable screens and either re-treated or subjected to some other treatment. The seeds should usually be planted immediately after the hot-water treatment; some kinds of seed have been dried and stored for later planting without impairing the germination percentage, although the germination rate was reduced. (See Figure 6–5.)

### ACID SCARIFICATION

Dry seeds are placed in containers and covered with concentrated sulfuric acid (specific gravity 1.84) in a ratio of about one part seed to two parts acid. The amount of seed treated at any one time should be restricted to no more than about 10 kg (22 lbs) (*54*) to avoid uncontrollable heating. Containers should be glass, earthenware, or wood—not metal or plastic. The mixture should be stirred cautiously at intervals during the treatment to produce uniform results and to prevent accumulation of the dark, resinous material from the seed coats that is sometimes present. Since stirring tends to raise the temperature, vigorous agitation of the mixture should be avoided to prevent injury to the seeds. The time of treatment may vary from as little as ten minutes for some species to six hours or more for other species. Since treatment time may vary with different seed lots, making a preliminary test on a small lot is recommended prior to treating large lots (*33, 54*). With thick-coated seeds that require long periods, the progress of the acid treatment may be followed by drawing out samples at intervals and checking the thickness of the seed coat. When it becomes paper thin, the treatment should be terminated immediately.

At the end of the treatment period the acid is poured off, and the seeds are washed to remove the acid. Glass funnels are useful in removing the acid from small lots of seed. Placing seeds in a large amount of water with a small amount of baking soda (sodium bicarbonate) will neutralize any adhering acid; or the seeds can be washed for 10 minutes in running water. The acid-treated seeds can either be planted immediately when wet, or dried and stored for later planting.

Large seeds of most legume species respond to the simple sulfuric acid treatment, but variations are required for some species (*54*). Some roseaceous seeds (*Cotoneaster, Rosa*) have hard pericarps that are best treated partially with acid and then given warm stratification. A third group, such as *Hamamelis* and *Tilia*, have very "tough" pericarps that may first need to be treated with nitric acid and then with sulfuric acid.

### WARM MOIST SCARIFICATION

Keeping seeds in a nonsterile, moist, warm medium (e.g., in nonpasteurized sandy soil) for several months can soften seed coats through microorganism activity. This treatment can be provided by planting hard seeds in summer or early fall while soil temperatures are warm. Usually it is most effective for seeds having double dormancy where the warm moist scarification precedes cold stratification during the winter.

### HIGH TEMPERATURE SCARIFICATION

Seeds of certain species of native plants with hard seed coats germinate extensively after a forest or range fire. Seed coats are modified by the high temperatures. Closed cone species of pine, for example, *Pinus radiata,* also respond to high temperatures in fires by melting the enclosing resins that seal the cone. This action allows release of the seeds which are then able to germinate (*70*).

### HARVESTING IMMATURE FRUITS

Extracting seeds from immature fruits improves germination in certain tree species by avoiding the development of hard seed coats. Such seeds must be planted immediately without drying.

## Stratification

**Stratification** is a method of handling dormant seeds in which the imbibed seeds are subjected to a period of chilling to after-ripen the embryo. The term originated because nurserymen placed seeds in stratified layers interspersed with a moist medium, such as soil or sand, in out-of-doors pits during winter. The term *moist-chilling* has been used as a synonym for stratification.

### REFRIGERATED STRATIFICATION

Dry seeds should be fully imbibed with water prior to refrigerated stratification. Twelve to 24 hours of soaking at warm temperatures may be sufficient for seeds without hard seed coats or coverings. Longer periods are required, with aeration, for seeds enclosed in hard endocarp or pericarp; soaking for three days to a week or more may be necessary. Leaching seeds in running water also can be used, or alternately soaking seeds for 12 hours, then draining for 12 hours (*44*).

After soaking, seeds are usually mixed with a moisture-retaining medium for the stratification period. Almost any medium that holds moisture, provides aeration, and contains no toxic substances is suitable. These include well-washed sand, peat moss, chopped or screened (.6 to 1.0 cm [¼ to ⅜ in.] with smaller parts discarded) sphagnum moss, vermiculite, and composted sawdust. Fresh sawdust may contain toxic substances. A good material is a mixture of one part sand and one part peat moss, moistened and allowed to stand 24 hours before use. Any medium used should be moist but not so wet that water can be squeezed out.

Seeds are mixed with one to three times their volume of the medium or they may be stratified in layers, alternating with similarly sized layers of the medium. Suitable containers are boxes, cans, glass jars with perforated lids, or other containers that provide aeration, prevent drying, and protect against rodents. Polyethylene bags are excellent containers either with or without media. Stratification of seeds in a plastic bag without a surrounding medium has been called *naked chilling*. A fungicide may be added as a seed protectant.

The usual stratification temperature is 0° to 10°C (32° to 50°F). At higher temperatures seeds often heat or sprout prematurely. Lower temperatures (just above freezing) delay sprouting.

The time required for stratification depends on the kind of seed, and sometimes upon the individual lot of seed as well (Figure 7–7). For seeds of most species, one to four months is sufficient for low-temperature stratification. During this time the seeds should be examined periodically; if they are dry, the medium should be remoistened. When sprouting begins, the seeds should be planted or moved to lower storage temperatures. The seeds to be planted are removed from the containers and separated from the medium, using care to prevent injury to the moist seeds. A good method is to use a screen that allows the medium to pass through while retaining the seeds. The seeds are usually planted without drying to avoid injury and reversion to secondary dormancy. Some success has been reported for partially drying previously stratified seeds, holding them for a time at low temperatures, then planting them "dry" without injury or loss of after-ripened condition. Beech and mahaleb cherry seeds were successfully dried to 10 percent, then held near freezing (*66*). Similarly, conifer seed has been dried to 20 to 35 percent, then stored for a year at low temperatures (*17*) after previous stratification.

**Figure 7–7** Walnut (*Juglans hindsii*) seeds after stratification ready for planting. Seeds on left may not have enough chilling to sprout. Seeds in center are in a favorable condition. Seeds on right are too advanced and radicles are likely to be injured in planting.

### OUTDOOR STRATIFICATION

Where refrigerated storage is not available, stratification may be done by storing outdoors, either in pits several feet deep or in raised beds enclosed in wooden frames (1, 70). Essentially the same seed preparation procedures are used but outdoor winter chilling and natural rainfall provide the required chilling temperature and moisture. However, the seeds need to be protected against freezing, drying, and rodents (65).

When seeds are stratifying in a trench, a layer of wire netting should be placed on the bottom, sides, and top of the trench to protect the seeds from rodents. A layer of clean sharp sand or gravel (10 cm or 4 in.) is placed on the bottom, then the seeds are mixed with the medium or placed in alternating layers of seed and medium. The trench is then filled and covered with the netting.

### OUTDOOR PLANTING

As an alternative to container stratification, seeds requiring a cold treatment may be planted out-of-doors directly in the seedbed, cold frame, or nursery row at a time of the year when the natural environment provides the necessary conditions for after-ripening. Several different categories of seeds can be handled in this way with good germination in the spring following planting.

Seeds must be planted early enough in the fall to allow them to become imbibed with water and to get the full benefit of the winter chilling period. The seeds generally germinate promptly in the spring when the soil begins to warm up, but while the soil temperature is still low enough to inhibit damping-off organisms and to avoid high-temperature inhibition.

Seeds with a hard endocarp, such as *Prunus* species (the stone fruits, including cherries, plums, and peaches), show increased germination if planted early enough in the summer or fall to provide one to two months of warm temperatures prior to the onset of chilling (44). Seeds that require high temperatures followed by chilling can thus be planted in late summer to fulfill their warm-temperature requirements, followed by the subsequent winter period that satisfies the chilling requirement.

For other seeds having hard coverings (such as juniper and some *Magnolia* species), germination can be facilitated if the fruit is harvested when ripe and the seeds planted immediately without drying (54, 73). Once the seeds become dry, the seed coats harden and germination may be delayed, perhaps until the second spring. Seeds that ripen early in the growing season and lose viability rapidly should be collected and planted in spring or summer as soon as they mature. Where effective treatments are not known, the propagator should attempt to reproduce the natural seeding habits of the plant and provide the germinating conditions of its natural environment. When seeds remain for a long period in an outdoor nursery or seedbed prior to germination, they must be protected from drying, adverse weather conditions, rodents, birds, diseases, and competition from weeds. Herbicides can be used for weed control.

## Leaching

The purpose of leaching is to remove inhibitors by soaking seeds in running water or by placing them in frequent changes of water. The length of leaching time is 12 to 24 hours. If longer periods are involved, the water should be changed every 12 hours to provide oxygen to the submerged seeds (although dormant seeds of woody plants are unlikely to be injured).

## Combinations of Treatments

Many woody plant species have more than one type of dormancy, for example, a hard seed coat plus a dormant embryo. For such species multiple treatments are required. Treatments must be given in sequence, first, one to soften the seed coat to allow uptake of water and, second, a chilling period to overcome embryo dormancy.

## Laboratory Treatments to Overcome Dormancy

Seed laboratories use various methods to initiate germination in viability testing. The same treatments might apply in propagation, although in many cases, as with seeds of cereals, vegetables, and flower crops, the seeds lose their dormancy during dry storage prior to planting.

*Prechilling*   For prechilling, place imbibed seeds at 5° to 10°C (37° to 42°F) for five to seven days before germination is attempted.

*Predrying*   In predrying, dry seeds are subjected to 37° to 40°C (99° to 104°F) for five to seven days prior to germination.

*Daily alternation of temperature*   When alternate temperatures are required, the usual combinations are 15° to 30°C (59° to 86°F), or 20° to 30°C (68° to 86°F), with seeds held at the lower temperatures for 16 hours and at the higher temperature for eight hours. Such temperature fluctuations are found normally out-of-doors at certain seasons of the year and often greatly promote germination.

*Light exposure*   Exposure to light can stimulate germination of many kinds of seeds, depending upon age, previous handling, and accompanying temperature. Light sensitivity is strongest just following harvesting and tends to disappear with dry storage. Consequently light influence is quite important in seed testing. The Rules for Seed Testing (*4*) specify light exposure for testing seeds of 42 percent of the 493 species cited. These include most grasses and conifers, many flowers, and a number of vegetables, including lettuce and endive.
Light should be provided by cool-white fluorescent lamps at an intensity of at least 75 to 125 ft-c (800 to 1345 lux) for at least eight hours daily. Seeds must have imbibed moisture at the time of light exposure and should be placed on top of the medium.

*Potassium nitrate*   Many freshly harvested dormant seeds germinate better after soaking in a potassium nitrate solution. Seeds are placed in germination trays or petri dishes and the substratum moistened with 0.2 percent potassium nitrate. For Kentucky bluegrass (*Poa pratensis*) or Canada bluegrass (*P. compressa*) a 0.1 percent solution should be used. If they are rewatered, tap or distilled water is used rather than additional nitrate solution.

# SEED TREATMENTS
# TO FACILITATE GERMINATION

## Hormones and
## Other Chemical Stimulants

*Gibberellins*　　Gibberellins (see Figure 6–10) comprise a group of plant hormones that have significant activity in seed physiology. While there are many kinds of natural gibberellins in plants, **gibberellic acid** (GA$_3$) is the one most widely used for exogenous applications. Gibberellic acid is produced by fungus cultures and is available commercially (*75*). GA$_3$ treatments can overcome physiological dormancy in various kinds of seeds and stimulate germination in seeds with dormant embryos. In seed testing laboratories, the germination medium is usually moistened with 500 ppm GA$_3$, although ranges of 200 to 1000 ppm are used. At concentrations above 800 ppm, addition of a buffer solution[1] is recommended.

For large seeds, a 12-hour soak of 500 to 1000 ppm GA$_3$ is recommended.

*Cytokinins*　　**Cytokinins** (see Figure 6–10) are natural growth hormones that stimulate germination of some kinds of seed by overcoming germination inhibitors. A commercial preparation, **kinetin** (6-furfurylamino purine) is available. Dissolve first in a small amount of HCl, then dilute with water. Other available synthetic cytokinins are **BA** (6-benzylamino purine) and **PBA** (6-benzylamino)-9-(2-tetrahydropyranyl)-9H-purine); these are more active for higher plants than is kinetin (*75*). These materials stimulate germination and overcome high temperature dormancy of certain kinds of seeds, such as lettuce, where seeds were soaked in 100 ppm kinetin solutions for three minutes. Large-scale treatments should be preceded by trials at varying concentrations. Cytokinins are sometimes effective in promoting germination when used in combination with gibberellic acid and with ethylene-producing compounds.

*Ethylene*　　Ethylene (see Figure 6–10) occurs naturally in plants and has growth-regulating properties. Applied to some kinds of seed, such as cocklebur, ethylene stimulates germination. Commercial ethylene-generating chemicals, such as ethephon (*75*), are available to supply ethylene.

*Thiourea*　　Thiourea, $CS(NH_2)_2$, at a ½ to 3 percent solution, has been used to stimulate germination of some kinds of dormant seeds, particularly those that do not germinate well in darkness or at high temperatures, and those that require a moist-chilling treatment. Since thiourea is somewhat inhibitory to growth, it is best to soak the seeds no longer than 24 hours and then rinse in water.

*Sodium hypochlorite*　　Sodium hypochlorite stimulates germination of rice seed, apparently by overcoming a water-soluble inhibitor in the hull (*56*). Use one gallon of commercial concentrate to 100 gal of water.

---

[1] Dissolve 1779.9 mg of $NaHPO_4 \bullet 2H_2O$ and 1379.9 mg of $NaH_2PO_4 \bullet H_2O$ in 1 liter of distilled water (*35*).

## Seed Priming

Various procedures that initiate germination prior to planting are used to improve speed and uniformity of seedling establishment and to overcome various dormancy problems in freshly harvested seed. These methods are called **seed priming.**

*Osmoconditioning*    In osmoconditioning (*41*) seeds are placed in a shallow layer in a container of 20 to 30 percent polyethylene glycol (PEG 6000) solution, which may include other types of chemicals, such as hormones or fungicides (*55*). Seeds are incubated at 15° to 20°C (59° to 68°F) for seven to 21 days. At the end of this period the seeds are washed with distilled water, air dried at 25°C (77°F), and stored until use. The seeds are then planted directly in their permanent location.

*Infusion*    Organic solvent infusion is a method for incorporating chemicals, such as growth regulators, fungicides, insecticides, antibiotics, and herbicidal antidotes (*40*) into seeds by means of organic solvents. Seeds are immersed for one to four hours in an acetone and dichloromethane solution which contains the chemical to be infused. The solvent is then removed by evaporation, and the seeds are spread out in a shallow container for drying in a vacuum desiccator for one to two hours. The incorporated chemical is then absorbed directly into the embryo with subsequent soaking in water.

*Fluid drilling*    Fluid drilling (*24*) is an integrated system involving the treatment and pre-germination of seeds followed by their sowing suspended in a special gel. Various methods are utilized to carry out the procedure. Seeds must be pregerminated under conditions of aeration, light, and optimum temperatures for the species. Among the procedures that can be used are (a) germinating seeds in trays on absorbent blotters covered with paper, or (b) placing seeds in water in glass jars or plastic columns through which air is continuously bubbled and fresh water continuously supplied. Osmoconditioning or addition of growth regulators can potentially be incorporated into the system. Pregerminated seeds of various vegetables have been stored for seven to 15 days at temperatures of 1° to 5°C (34° to 41°F) in air or aerated water.

Various kinds of gels are commercially available. Among the materials used are sodium alginate, hydrolyzed starch-polyacrylonitrile, guar gum, synthetic clay, and others. Special machines are needed to deposit the seeds and gel into the seed bed.

## Protection of Seeds against Pathogens

Three types of seed treatments are used to control diseases: disinfestation, disinfection, and seed protection (*7, 26, 45, 49, 72*).

**Disinfestants** elminate organisms present on the surface of the seed. Materials that have this sole action are useful if the seeds or embryos are to be grown in aseptic culture or in some type of sterile medium. Materials listed in this chapter as disinfectants or protectants are also likely to be good disinfestants.

**Calcium hypochlorite** is an effective disinfestant (*79*). In this treatment, 10 gm of calcium hypochlorite are placed in 140 ml of water and shaken for 10 minutes or allowed to stand for an hour. The filtrate containing about 2 percent hypochlorite is usually used, although this solution is sometimes diluted by half. The time of contact that will disinfest and still avoid injury varies with different kinds of seeds, usually being between 5 and 30 minutes. Adjusting the pH to 8 to 10 produces the most consistent results. Commercial bleaching preparations, such as Clorox (containing 5.25 percent sodium hypochlorite), also have been used, diluted with water 1:9.

**Disinfectants** eliminate organisms within the seed itself. Treatments of this type include hot water, formaldehyde, and aerated steam.

**Protectants** are materials applied to the seeds to protect them from pathogenic fungi in the soil. These materials may also be applied as a soil drench either before or after seed planting. Numerous materials for seed treatment are available commercially under various trade names. The manufacturer's directions should be followed carefully, since improper use can cause injury or death to the seedlings of some plants. And since some protectants are hazardous to the person using them, they should be handled carefully to avoid contact with the skin or breathing of the dust. In many cases, special machines must be used for treating the seeds.

### Thermotherapy

**Hot water treatment** can be used as a disinfectant procedure. Dry seeds are immersed in hot water (49° to 57°C, or 120° to 135°F) for 15 to 30 minutes, depending upon the species (*5, 6*). After treatment, the seeds are cooled and spread out in a thin layer to dry. To prevent injury to the seeds, temperature and timing must be regulated precisely; a seed protectant should subsequently be used, and old, weak seeds should not be treated. Hot water is effective for specific seedborne diseases of vegetables and cereals, such as *Alternaria* blight in broccoli and onion, and loose smut of wheat and barley.

**Aerated steam** (see chapter 2) is an alternate method that is less expensive, easier to handle, and less likely to injure seeds (*6*). Seeds are treated in special machines in which steam and air are mixed and drawn through the seed mass to raise the temperature of the seeds rapidly (about two minutes) to the desired temperature. The treatment temperature and time vary with the organism to be controlled and kind of seed. Usually the treatment is 30 minutes, but may be as little as 10 or 15 minutes. Temperatures range from 46° to 57°C (83° to 143°F). At the end of treatment, temperatures must be lowered rapidly to 32°C (60°F) by evaporative cooling continuing until dry. Holding seeds in moisture saturated air at room temperature for one to three days prior to treatment will improve effectiveness.

The two basic methods of application are (a) the dry or dust method and (b) the "slurry" method. In the latter the seed is immersed in a thick water suspension of the chemicals.

Other seed treatments include the application of insecticides, or combinations of insecticides and fungicides, plus bird and rodent repellents.

*All state and federal regulations governing the use of such materials on crop plants should be followed carefully.*

## DIRECT SEEDING OUT-OF-DOORS

In direct seeding, seeds are planted in the location where the plant is to remain throughout its entire life cycle. This is the primary commercial method of mass propagating all kinds of field crops (grains, forages, fiber crops, oil crops, and the like), as well as vegetable crops and lawn grasses. Direct seeding is important for home vegetable and flower gardeners, although the availability of bedding plant transplants (both vegetables and flowers) has made their use for "instant gardening" much more attractive. Trees and shrubs are directly seeded into permanent locations for reforestation and planting natural areas (*14*), for planting in permanent places in the landscape (*11*), and occasionally for planting in orchards.

### Direct Seeding of Herbaceous Plants

SEEDBED PREPARATION

Good seedbed preparation is essential for successful field and garden sowing. A good seedbed should have a loose but fine physical texture that produces close contact between seed and soil so that moisture can be supplied continuously to the seed. Such a soil should provide good aeration, but not too much or it dries too rapidly. The surface soil should be free of clods and trashy crop residues. The subsoil should be permeable to air and water with good drainage and aeration. Adequate soil moisture should be available to carry the seeds through the germination and early seedling growth stages but the soil should not be water-logged nor anaerobic (without oxygen). A medium loam texture, not too sandy and not too fine, is best. A good seedbed is one in which three-fourths of the soil particles (*aggregates*) range from 1 to 12 mm in diameter (*37*).

Preparation of seedbeds involves several consecutive operations. The implements used depend on the scale of operations but the principle is the same. In field operations the seedbed is first plowed to incorporate crop residues and break up the soil layer to a depth of 6 to 10 inches. Plowing is followed by disking and harrowing, which pulverize the surface layers into smaller particles and compact them together. In garden or nursery operations the seedbed is prepared by a rototiller-type implement, or by hand spading, followed by raking.

Adding organic matter to the soil improves texture. The soil may be conditioned by plowing under a green manure (growing crop) or animal manure (after allowing time for decomposition) or incorporating peat moss or other organic matter. These materials should be mixed uniformly with the soil so that no stratified layers exist that will impede movement of water. Seedbed preparation may include fumigation and other soil treatments to control harmful insects, disease organisms, and weed seeds.

## TIME FOR PLANTING

Planting time is determined by germination temperature requirements of the seed and the need to meet production schedules. Temperature requirements vary with the particular kind of seed. Too early plantings of seeds requiring warm soil temperatures can cause slow and uneven germination, disease problems, and injury to the seedlings resulting in abnormalities of growth. Too high soil temperatures can result in excessive drying, injury or death to the seedlings, or induction of thermodormancy in the case of heat sensitive seeds such as lettuce.

## DEPTH OF PLANTING

Depth of planting is a critical factor that determines the rate of emergence and perhaps stand density. If too shallow, the seed may be in the upper surface that dries out rapidly; if too deep, emergence of the seedling is delayed. Depth varies with the kind and size of seed and, to some extent, the condition of the seedbed and the environment at the time of planting. When exposure to light is necessary, seeds should be planted shallowly. *A rule of thumb is to plant seeds to a depth approximately three to four times their diameter.*

## RATE OF SOWING AND PLANT DENSITY

Establishing the rate of sowing is critical in direct sowing in order to produce a desired plant density. If final plant density is too low, yields will be reduced because the number of plants per unit area is low; if too high, size and quality of the finished plants may be reduced by competition among plants for available space, sunlight, water, and nutrients. Rate of sowing can be estimated by the following formula:

$$\text{Pounds (ounces) of seed required per unit area} = \frac{\text{Number of plants per unit area desired}}{\text{Number of seeds per pound (ounces)}} \times \text{percentage germination} \times \text{percentage purity}$$

This rate is a minimum and should be adjusted to account for expected losses in the seedbed, determined by previous experience at that site.

Rates will vary with the spacing pattern. Field crops or lawn seeds may be *broadcast,* i.e., spaced randomly over the entire area, or *drilled* at given spaces. Other field crops, particularly vegetables, are planted in *rows* so that the rate per linear distance in the row must be determined.

Some crops are grown in rows on *raised beds,* particularly in areas of low rainfall, where irrigation is practiced and excess soluble salts may accumulate to toxic levels through evaporation. Overhead sprinkling and planting seed below the crest of sloping seedbeds may eliminate or reduce this problem. With row planting, excess seed is planted, then the plants are thinned to the desired spacing. **Thinning** is an expensive, time-consuming operation that could be reduced or eliminated by better methods of precision planting. A low rate of sowing requires seeds of very high quality that produce not only very high germination rates but also vigorous, uniform, healthy seedlings.

**Seed tapes** are available for flower and vegetable cultivars. Properly spaced seeds are attached to a plastic tape. The tapes are buried in shallow furrows and disintegrate from moisture as the seeds germinate.

**Pelleted seeds** are seeds coated with an adhesive material, such as clay, which makes them uniformly round for use in precision planters. Nutrients, hormones, activated charcoal (63), and other chemicals are often included in the pelleting material. Seed emergence and final stands are sometimes reduced with pelleting, however, apparently because the coating restricts available oxygen (18). Placing vermiculite over the seeds in the planting row has improved emergence in precision planting (77); activated carbon added to the vermiculite protects the seeds against pre-emergence weed sprays (78).

## Direct Seeding of Trees and Shrubs

Direct seeding of forest trees is accomplished in reforestation through either natural seed dissemination or planting. Costs and labor requirements of direct seeding are lower than those for transplanting seedlings, provided soil and site conditions favor the operation (14). The major difficulty is the very heavy losses of seeds and young plants that result from predation by insects, birds, and animals and from drying, hot weather, and disease (42, 61). A proper seedbed is essential and an open mineral soil with competing vegetation removed is best. The soil may be prepared by burning, disking, or furrowing. Seeds may be broadcast by hand or by special planters, or drilled with special seeders. Seeds should be coated with a bird and rodent repellent. An available mixture for this purpose is a combination of thiram and endrin.

In certain situations, trees and shrubs can be directly seeded into a natural setting for landscape purposes (11). Dig a 4- to 5-inch hole, carefully pulverizing soil if compacted. On a rocky slope, place seeds in a crevice or pocket of soil. If on a slope, make a slight backslope to avoid erosion and to prevent the seed hole from being covered with loose soil from above. Place 1 gram of slow-release fertilizer in the bottom of the hole. Replace soil, leaving a slight depression for the seed. Plant 2 to 20 seeds, and cover with pulverized soil to 3.2 to 12.7 cm ($\frac{1}{8}$ to $\frac{1}{2}$ in.) deep depending upon size. Contact herbicides or mulches around the seed site are essential to remove weed competition. Mulches can be coarse organic materials (wood chips, coarse bark) or sheet materials (uncoated and polyethylene-coated mulching paper, black polyethylene film, asphalt roofing paper, or kraft building paper). With the coarse organic mulches use a tin can (size $2\frac{1}{2}$), milk carton, or asphalt or kraft paper collar around the seed site. Cut a hole in the center of the sheet mulches to allow the seedlings to emerge. Small wire cages may be needed to cover the site for rodent and bird protection.

# GROWING SEEDLINGS IN FIELD NURSERIES

Outdoor seed propagation of trees and shrubs utilizes two basic nursery operations:

SEEDBED CULTURE

Seeds are planted relatively close together in **seedbeds,** sometimes raised, commonly 1.1 to 1.2 cm ($3\frac{1}{2}$ to 4 ft) wide and of various lengths (Figure 7–8).

**Figure 7–8**    Growing seedlings in seedbeds. Olive ( *Olea europaea* ) seedlings in Italy planted thickly in unshaded beds.

Here the seedlings remain for one to three years and then are dug for transplanting to a permanent location. For some species the plants may be shifted to a **transplant bed** after one year and then grown for a period of time at wider spacing. This basic procedure is used to propagate millions of forest tree seedlings, both conifer and deciduous species.

Seedlings produced in a seedbed nursery are often designated by numbers to indicate the length of time in a seedbed and the length of time in a transplant bed. For instance, a designation of 1–2 means a seedling grown one year in a seedbed and two years in a transplant bed or field. Similarly, a designation of 2–0 means a seedling produced in two years in a seedbed and none in a transplant bed.

In other cases, seedbeds may be used to produce **liners,** which are small plants dug after one year and then ''lined out'' in a nursery row, used for growing woody ornamentals to a salable size or rootstocks of deciduous fruit trees. This procedure is used particularly where the growing season is short, or for apple, pear, and cherry, which grow relatively slowly and are not large enough for budding in one year.

NURSERY ROW CULTURE

Planting directly in nursery rows about four feet apart (Figure 7–9) is the primary method used to propagate rootstocks of most fruit and nut tree species. Cultivars are budded or grafted to the seedlings in place. The method is also used to propagate shade trees and ornamental shrubs, either as seedlings or on rootstocks as budded selected cultivars.

**Figure 7–9**    Direct seeding in nursery row, illustrated by peach seedlings grown for rootstocks. These seedling plants will be budded to peach cultivars.

### Nursery Soil Preparation

Nursery production requires a fertile, well-drained soil of medium texture. Preparation for planting may include rotation with other crops and incorporation of a green manure crop or animal or chicken manure (*64*). Preplant fumigation and weed control are essential aspects of most nursery operations.

Weed control can be facilitated by careful seedbed preparation, cultivation, and chemical sprays. Three types of chemical controls are available. **Preplanting** fumigation with methyl bromide, chloropicrin, D–D, Vapam, steam, or the like is effective and also kills disease organisms and nematodes. **Pre-emergence** herbicides are applied to the soil before the weed seeds emerge. **Post-emergence** herbicides are applied as foliage sprays after the weed seedlings have emerged. A wide range of selective and nonselective commercial products are available. Such materials should be used with caution, however, since improper use can cause injury to the young nursery plants. Not only should directions of the manufacturer be followed, but preliminary trials should be made before large-scale use (*19, 74*).

### Time of Planting

Seeds of the different species to be propagated are planted in the nursery in the fall, spring, or summer, depending on the dormancy conditions of the seed, the temperature requirements for germination, the management practices at the nursery, and the location of the nursery (in a cold-winter or a mild-winter area). Planting time varies for several general categories of seed (*34, 53, 61*):

**1.** Certain species (e.g., apple, pear, *Prunus,* yew) require moist-chilling (stratification) and are adversely affected by high germination temperatures (Table 7–1), which produce secondary dormancy. Germination temperatures of 10° to 17°C (50° to 62°F) are optimum. Seeds of these species could be planted in the fall with germination taking place in late winter or early spring. This procedure is particularly desirable in mild-winter areas. Or seeds can be stratified during winter storage and planted as early in the spring as permitted by climate. This procedure is more likely to be used in cold-winter areas where fall-planted germinating seed might be injured by very low temperatures.

A modification of the fall planting procedure is useful for seeds that have hard seed coats plus dormant embryos. Seeds could be planted in summer or early fall to allow 6 to 8 weeks of warm stratification in the seedbed prior to the winter chilling (*44, 48*). Another modification is collecting seeds before they are fully ripe and planting without allowing them to dry (*48*).

**2.** Many kinds of seeds, including most conifers (pine, fir, spruce) and many deciduous hardwood species, benefit from moist-chilling, do not germinate until soil temperatures have warmed up, and are not inhibited by high soil temperatures. Optimum germination temperatures are 20° to 30°C (68° to 86°F). Seeds of these species would not germinate in the cool fall season or in early spring. Such seeds can be handled either by fall planting or by spring planting following the required chilling treatments (stratification).

**3.** Seeds of some species (e.g., black locust, Colorado and Norway spruce, Chinese and Siberian elm, European larch, bristlecone pine, and Douglas fir) are able to germinate over a wide range of temperatures 15° to 32°C (60° to 90°F). Seeds of some of these species require moist-chilling or have hard coats requiring special treatment. If fall-planted in cold areas, seeds of this group may germinate prematurely when the seedlings can be injured by winter cold. Spring planting is recommended, although the actual time is not as critical as that for the other two groups.

**4.** Seeds of some species (maple, poplar, aspen, elm, cottonwood) ripen in spring or early summer. Such seeds should be planted immediately as their viability declines rapidly.

## Seedbed Culture

A common size of seedbed is 1.1 to 1.2 m (3½ to 4 ft) wide with the length varying according to the size of the operation. Beds may be raised to insure good drainage, and in some cases, sideboards are added after sowing to maintain the shape of the bed and to provide support for glass frames or lath shade. Beds are separated by walkways.

Seed may be either broadcast over the surface of the bed or drilled into closely spaced rows with seed planters. For economy, seeds should be planted as close together as feasible without overcrowding, which increases damping-off and reduces vigor and size of the seedlings (*32*), resulting in thin, spindly plants and small root systems. Seedlings with these characteristics do not transplant well (*31*).

The optimum seed density depends primarily on the species but also depends on the nursery objectives. If a high percentage of the seedlings are to reach a desired size for field planting, low densities might be desired, but if the seedlings are to be transplanted into other beds for additional growth, higher densities (with smaller seedlings) might be more practical. Once the actual density is determined, the necessary rate of sowing can be calculated from data obtained from a germination test and from experience at that particular nursery.

The following formula is used (*21, 70*):

$$W = \frac{A \times S}{D \times P \times G \times L}$$

where $W$ = weight of seeds in pounds to sow per given area.
  $A$ = area of seedbed in square feet.
  $S$ = number of living seedlings desired per square foot.
  $D$ = average number of seeds per pound at the moisture content at which the seeds are sown.
  $P$ = average purity percentage of seeds (expressed as a decimal).
  $G$ = average effective germination percentage in the laboratory (expressed as a decimal).
  $L$ = average percentage of germinating seeds that will be living plants at the end of the season (expressed as a decimal). This is a correction factor that takes into account the expected losses which experience at that nursery indicates will occur with that species.

Seeds of a particular lot should be thoroughly mixed before planting to be sure that the density in the seedbed will be uniform. Treatment with a fungicide for control of damping-off is desirable. Small conifer seeds may be pelleted for protection against disease, insects, birds, and rodents. Seeds are planted either by hand or by machine. Depth of planting varies with the kind and size of the seed. In general, a depth of three to four times the diameter of the seed is satisfactory.

A mulch applied to the seedbed after planting helps to protect against drying, crusting, and cold, and discourages weed growth (*64*).

During the first year in the seedbed, the seedlings should be kept growing continuously without any check in development. A continuous moisture supply, cultivation or herbicides to control weeds, and proper disease and insect control contribute to successful seedling growth. Fertilization with nitrogen is usually necessary, particularly when a mulch has been applied, since decomposition of organic material can produce nitrogen deficiency. In the case of tender plants, glass frames can be placed over the beds, although for most species a lath shade is sufficient. With some species, shade is necessary throughout the first season; with others, shade is necessary only during the first part of the season. Sprinkling with water to reduce ground temperature during the hot part of the day is sometimes useful (*15*).

### Nursery Row Culture

Deciduous fruit, nut, and shade tree propagation usually begins by planting seeds or "liners" in nursery rows. Where plants are to be budded or grafted in place, the width between rows is about 1.2 m (4 ft) and the seeds are planted 7.6 to 10 cm (3 to 4 in.) apart in the row (see Figure 7–9). Seeds known to have low germinability must be planted closer together to get the desired stand of seedlings. Large seed (walnut) can be planted 10 to 15 mm (4 to 6 in.) deep, medium-sized seed (apricot, almond, peach, and pecan) about 7.6 cm (3 in.), and small seed (myrobalan plum), about 3.8 cm (1½ in.). Spacing may vary with soil type. If germination percentage is low and a poor stand results, the surviving trees may, because of the wide spacing, grow too large to be suitable for budding. Plants to be grown to a salable size as seedlings without budding could be spaced at shorter intervals and in rows closer together.

Fall planting of fruit and nut tree seeds is commonly used in mild-winter areas such as California (*25*). Seeds are planted 2.5 to 3.6 cm (1 to 1½ in.) deep and 10 to 15 cm (4 to 6 in.) apart depending on size, and then covered with a ridge of soil 15 to 20 cm (6 to 8 in.) deep, in which the seeds remain to stratify over winter. The soil ridge is removed in the spring just before seedling emergence. Herbicide control of weeds and protection of the seeds from rodents become important considerations during these procedures.

# INDOOR SEEDLING PRODUCTION IN CONTAINERS

Indoor seedling production for later transplanting to a permanent location out-of-doors is extensively used in the production of bedding plants, both flowers and vegetables (*8, 51*), and in the production of woody plants for reforestation (*69*), landscaping, and sometimes orcharding (*22, 23, 36*).

## Types of Systems

Two general sequences of operation are followed in the initial establishment of the seedling plants:

### TRANSPLANT METHOD

Seeds are planted in a special **germination flat** or container, and later the germinated seedlings are "pricked-out" and transplanted to develop either in a **transplant flat** at wider spacing or in individual containers where they remain until transplanted to the out-of-doors. This procedure makes it possible to optimize environmental conditions for seed germination, which may be different than for seedling growth. It allows for selection of the seedling plants to improve uniformity and spacing. The transplant method is mostly used with small-sized seeds where direct seeding would not be practical.

On the other hand, transplanting is labor-intensive and therefore costly. Each transplant operation causes a "transplant shock" that may injure roots, delay growth, and increase total production time.

### DIRECT SEEDING

Seed may be placed at the desired spacing in a seed flat or directly into an individual container. Direct seeding is used mostly with medium- or large-sized seeds, such as those of most woody plants, or those for which transplanting is particularly harmful, such as cucurbits. Special mechanical seeding devices have been developed for small-sized seeds. Pelleted seed, seed tapes, or pregerminated seeds are alternatives that allow omission of the transplanting step. The procedure requires high quality seed that germinates at nearly 100 percent with uniform rates of germination.

## Containers

The most common container is the standard greenhouse flat, either wood or plastic, available in various sizes. The second general type is the individual container in which each seedling plant has its own root ball. These come in various sizes, made from various materials, including plastic, peat moss, paper, and wood. A common type used in bedding plant production is the Cell-Pak in which individual plastic ''cells'' are joined for convenience in handling. Chapter 2 describes various types of containers.

Larger individualized containers are needed for tree seedlings and many kinds are available (*29*). The shape and size can affect the type of root system and the ultimate survival of the plant. In round containers lateral roots can grow around the edge in a circular pattern and encircle the tap root. If not removed these roots can eventually girdle the plant later when it has been planted into the permanent location (*68*) (see Figure 2–19). A better container is one with square corners, such as a paraffined milk carton (*13*), since roots tend to grow down the corners. A depth of 23 to 30 cm (9 to 12 in.) with a width of 6.5 × 6.5 cm (2½ × 2½ in.) is optimum for several species (*12*). Various other containers are available with vertical grooves that direct root growth downward (see Figure 2–13).

An open base on a wire support is useful to ''air prune'' the tap root and stimulate lateral root branching.

## Facilities

Indoor seedling production occurs in several types of structures, including greenhouses, cold frames, and hot beds, as described in chapter 2.

Some bedding plant operations have special **growth rooms** where seed flats are placed on carts or shelves in an enclosed area and subjected to controlled environments. These facilities are used during the germination period, which usually lasts for the first two to three weeks. The flat is then moved to the greenhouse or growing area, where the small seedlings are usually transplanted.

Growth rooms need controlled lighting (daylength and intensity) and temperature (duration, level) and may include control of relative humidity, carbon dioxide fertilization, irrigation, and fertilization (*58*). Under proper conditions a grower may produce not only rapid uniform germination but also healthy seedlings with excellent ability to transplant without any check in plant growth.

## Propagating Herbaceous Bedding Plants

### MEDIA

Germination media for herbaceous bedding plants must retain moisture, yet provide good drainage and aeration. The pH should be neutral and soluble salts should be low. Some nutrients must be present, particularly phosphorus and calcium. Regular soil mixes, for example, equal parts soil, sand, and peat, might be used. These have been supplanted to a large extent by special nonsoil mixes made up of such components as peat moss, perlite, ground bark, sawdust, and vermiculite, fortified by mineral nutrients or slow-release fertilizers (see chapter 2). These mixes are available commercially and are used in large quantities by bedding plant producers. The mix should have excellent drainage to inhibit damping-off and reduce crusting. About 6 mm (¼ in.) of sterile sand, gently worked into the upper surface of the medium, is sometimes used to avoid

**Figure 7–10**    Steps in greenhouse seedling production. *Top left:* the soil medium is firmed to eliminate air pockets. *Top right:* the seed is planted into shallow furrows (on surface for very fine seed). *Below left:* when seedlings have produced the first true leaves, they are pricked off into a transplant flat at wider spacing or into individual containers. The peg board in the background is convenient for producing uniform spacing. *Below right:* after hardening, the seedlings are transplanted into the garden or other permanent location. A cube of soil from the flat is removed with the plant to avoid disturbing the root system.

damping-off. Sphagnum moss inhibits damping-off and is an effective germination medium. Small seeds should have a finer and more compact medium than is used for larger seeds. Seed flats and media should be pasteurized, and seeds should be treated for disease organisms (see chapter 2).

Seed flats should be filled completely, the medium worked carefully into the corners, and the excess medium removed with a straight board drawn across the top (Figure 7–10). The medium is then tamped with a block to provide a uniformly firm seedbed to a level about 12 mm (½ in.) below the top of the flat. The flat should be watered from above or presoaked and drained. Some media, such as vermiculite, should not be compacted.

Seeds may be broadcast over the surface or planted in rows. Seeding equipment may be used for larger seeds. Advantages of row planting are reduced damping-off, better aeration, easier transplanting, and less drying out.

Planting at too high a density encourages damping-off, makes transplanting more difficult, and produces weaker, nonuniform seedlings. Suggested rates are 1000 to 1200 seeds per 29 × 54 cm (11 × 22 in.) flat for small-seeded species (e.g., petunia) and 750 to 1000 for larger seeds. Small seeds are dusted on the surface; medium seeds are covered lightly to about the diameter of the seed. Larger seeds may be planted at a depth of two to three times their minimum diameter.

MOISTURE

Maintaining proper moisture conditions during the germination period is essential so that the medium neither dries out at any time nor remains so wet that damping-off becomes a problem. The seed flats may be held under polyethylene tents (see Figure 2–10), or in small operations, covered with glass or plastic to keep the surface from drying out. An alternate method is to use a fine intermittent mist spray set for about eight seconds of water every ten minutes during the day. Covered flats should not be exposed directly to sunlight, as excessive heat buildup injures the seedling.

TEMPERATURE

Maintaining the proper germination temperature in the medium is the most important control at this stage. The optimum temperature depends on the particular species and sometimes the cultivar. This is discussed fully in chapter 6. In general, a daily alternating temperature 20° to 30°C (68° to 86°F) is optimum for seeds of *warm temperature-requiring* and *cool temperature-tolerant* species. The alternations are 20°C for 16 hours and 30°C for eight hours. For seeds of *cool temperature-requiring* species, the germination temperature should be about 20°C (68°F) or lower.

Light should be provided for germination of some kinds of seed.

TRANSPLANTING

The time required for germination can vary from about one to three weeks. When the first true leaves have appeared, the seedlings should be transplanted. Direct seeded flats should be shifted to a lower temperature for the seedlings of most kinds of plants to harden properly.

Transplanting flats or containers are filled with a growing medium and handled in the same manner as the seed flats. Holes are made in the medium at the correct spacing with a small **dibble.** The roots of each small seedling are inserted into a hole, and medium is pressed around them to provide good contact. Dibble boards are often used to punch holes for an entire flat at once. As soon as the flat is filled, it is thoroughly watered.

SEEDLING GROWING

Once seeds have germinated and the seedlings have been transplanted, the principal objectives of production are to prevent damping-off and to develop stocky, vigorous plants capable of further transplanting with little check in growth. The usual procedure in production is to shift the flats to lower temperatures (10°C or less) and to expose them to good light. High temperatures and low light tend to produce spindly, elongated plants that will not survive transplanting (see Figure 7–11). Such growth is termed ''stretching.''

Once root systems have developed sufficiently to grow into the medium, irrigation can be scheduled to keep the medium somewhat dry on the surface but

**Figure 7–11** Tomato
plants ready for transplanting.
The two plants on the left are
spindly and undesirable for
transplanting. The three on
the right are healthy, well-
formed plants suitable for
transplanting. The latter show
the effects of sufficient space
and desirable growing conditions. From *Calif. Agr. Ext.
Cir. 167.*

moist underneath. Such irrigation helps prevent damping-off and produces
sturdy seedlings.

## Production of Woody Plant Seedlings

Production of seedling trees and shrubs in containers involves two main activities, the production of landscape or fruit or nut tree plants and the mass propagation of small seedlings for reforestation.

Landscape and fruit tree seedlings are started in germination flats, then
transplanted to either a transplant flat or a small pot. Later they are moved to
slightly larger containers, or transplanted directly into the containers where they
will remain until transplanted out-of-doors.

With this system root pruning is essential to induce a desirable root system.
Root pruning should be done at the first transplanting, soon after the roots reach
the bottom of the flat (*27, 28*). Further root pruning should take place when the
liner flat is transplanted, which should be done when the roots protrude through
the pot about an inch. These roots are removed. Any delay in transplanting at
either stage, or omission of pruning, will increase the incidence of poor root
systems. Copper or plastic screen-bottomed flats (*22*) or supports stimulate formation of branch roots.

The second activity involving woody plant production is the mass propagation of small seedlings for direct planting in reforestation (*46, 69*) or for planting
into outdoor beds for an additional one or two years' growth (*60*). Seeds are invariably planted directly into various kinds of small individualized containers—usually narrow and relatively deep. These may be plastic containers
from which the seedling plug is removed and planted (see Figure 2–13) or they
may be containers made of substances such as peat or fiber blocks, which are
planted with the seedling (see Figure 2–14).

## Transplanting to Permanent Location

The final step in indoor production is transplanting seedlings to their permanent location (*51, 52*). The seedlings may be transplanted either with bare roots, as is customary with many vegetable transplants, or with a soil ball containing the roots (see Figure 7–12).

Bare-root transplants invariably undergo *root damage* and *transplant shock*, both checking growth. The use of a soil ball minimizes transplant shock, how much depending on the degree of root disturbance. In both cases success in transplanting depends to a large extent on the previous handling of the plants. The operation requires that the plant be "hardened" prior to its shift to the open field. **Hardening** involves a checking of growth resulting in the accumulation of carbohydrates, which makes the plant better able to withstand adverse environmental conditions. This process can be brought on by temporarily withholding moisture, reducing the temperature, and gradually shifting from

**Figure 7–12**    Mechanized seedling production utilizing the Speedling system. *Upper left:* mechanized seeder. *Upper right:* open greenhouse with automatic watering. *Lower left:* transplant with characteristic root ball. *Lower right:* mechanized transplanting from special flats. Courtesy George Todd, Sr. From The Speedling system. 1981. *Proc. Inter. Plant Prop. Soc.* 31:612–16.

the greenhouse or hotbed to the environment of the permanent location. Flats of seedlings can be moved into a cold frame or a lathhouse and allowed to remain for seven to ten days prior to placing in the field.

Tree and shrub plants should be allowed to stop growth, develop dormancy, and develop a terminal bud. Any supplementary lighting can be reduced to short days, and water can be withheld, but not eliminated.

Before being moved into the field, the plants should be watered thoroughly. Planting is done in the field by hand or, in most cases, by transplanting machines. During transplanting, it is desirable to retain as much soil about the roots as practical to avoid disturbing the root system. Afterward, the plants should be thoroughly watered. If practical, temporary shade should be provided for the first few days, until the plants have become established. They should be watered as necessary to prevent wilting.

The use of "booster" or "starter" solutions containing nitrogen, phosphorus, and potassium shortly before or after transplanting is sometimes beneficial in establishing plants in the field. Such solutions should be used with caution, since if the soil is low in moisture at transplanting time, high concentrations of fertilizer will develop and injure the plants.

# REFERENCES

1.  Aldous, J. R. 1972. Nursery practice. *For. Comm. Bul. 43.* London: Her Majesty's Stationery Office.

2.  Allison, C. J. 1980. X-ray determination of horticultural seed quality. *Proc. Inter. Plant Prop. Soc.* 30:78–86.

3.  Anonymous. 1975. Seed testing regulations under the Federal Seed Act. *U. S. Dept. Agr., Agr. Mark. Ser.,* 184 pp.

4.  Assoc. Off. Seed Anal. 1978. Rules for seed testing. *Jour. Seed Tech.* 3(3):1–126.

5.  Baker, K. F. 1962. Thermotherapy of planting material. *Phytopathology* 52:1244–55.

6.  ———. 1972. Seed pathology. In *Seed biology,* Vol. 2, T. T. Kozlowski, ed. New York: Academic Press.

7.  Baker, K. F., and P. A. Chandler. 1957. Development and maintenance of healthy planting stock. In *The U.C. system for producing healthy container-grown plants,* K. F. Baker, ed. Calif. Agr. Exp. Sta. Man. 23, pp. 217–36.

8.  Ball, V., ed. 1972. *The Ball red book* (12th ed.). Chicago: Geo. J. Ball, Inc.

9.  Bonner, F. T. 1974. Seed testing. In *Seeds of woody plants in the United States,* C. S. Schopmeyer, ed. U.S. Dept. Agr. Handbook No. 450, pp. 136–52.

10. Carville, L. 1978. Seed bed production in Rhode Island. *Proc. Int. Plant Prop. Soc.* 28:114–17.

11. Chan, F. J., R. W. Harris, and A. T. Leiser. 1971. Direct seeding of woody plants in the landscape. *Univ. of Calif. Ext. Ser. AXT-n27.*

12. Copeland, L. O. 1976. *Principles of seed science and technology.* Minneapolis: Burgess.

13. Davis, R. E., and C. E. Whitcomb. 1975. Effects of propagation container size on development of high quality seedlings. *Proc. Inter. Plant Prop. Soc.* 25:448–53.

14. Deer, H. J., and W. F. Mann, Jr. 1971. *Direct seedling pines in the South.* U.S. Agr. Handbook No. 391. Washington, D.C.: U.S. Govt. Printing Office.

15. Eden, C. J. 1962. Conifer seed—from cone to seed bed. *Proc. Plant Prop. Soc.* 12:208–14.

16. ———. 1965. Use of X-ray technique for determining sound seed. *U.S. Forest Service, Tree Planters' Notes* 72:25–27.

17. Edwards, D. G. W. 1981. Improving seed germination in *Abies*. *Proc. Inter. Plant Prop. Soc.* 31:69–78.

18. Ellis, J. E. 1965. Prevention of stand losses in tomato due to soil crust formation. *Proc. Amer. Soc. Hort. Sci.* 87:433–37.

19. Elmore, C. 1971. Management of undesirable plants in ornamental containers and ground covers. *Proc. Inter. Plant Prop. Soc.* 21:184–90.

20. Flemion, F. 1938. A rapid method for determining the viability of dormant seeds. *Contrib. Boyce Thomp. Inst.* 9:339–51.

21. Fordham, D., 1976. Production of plants from seed. *Proc. Inter. Plant Prop. Soc.* 26:139–45.

22. Frolich, E. F. 1971. The use of screen bottom flats for seedling production. *Proc. Inter. Plant Prop. Soc.* 21:79–80.

23. Garner, R. S., and S. A. Chaudhri. 1976. The propagation of tropical fruit trees. *Hort. Rev.* No. 4. East Malling, England: Commonwealth Bureau of Hort. and Plant Crops.

24. Gray, D. 1981. Fluid drilling of vegetable seeds. *Hort. Rev.* 3:1–27.

25. Hall, T. 1975. Propagation of walnuts, almonds and pistachios in California. *Proc. Inter. Plant Prop. Soc.* 25:53–57.

26. Hansen, E. W., E. D. Hansing, and W. T. Schroeder. 1961. Seed treatments for control of disease. In *Seeds: Yearbook of agriculture.* Washington, D.C.: U.S. Govt. Printing Office, pp. 272–80.

27. Harris, R. W., W. B. Davis, N. W. Stice, and D. Long. 1971. Root pruning improves nursery tree quality. *Jour. Amer. Soc. Hort. Sci.* 96:105–9.

28. ———. 1971. Influence of transplanting time in nursery production. *Jour. Amer. Soc. Hort. Sci.* 96:109–10.

29. Hart, J. W., and J. W. Hanover. 1979. Practical requirements for controlled environment nursery stock production. *Proc. Inter. Plant Prop. Soc.* 29:304–13.

30. Heit, C. E. 1955. The excised embryo method for testing germination quality of dormant seed. *Proc. Assn. Off. Seed Anal.* 45:108–17.

31. ———. 1964. The importance of quality, germinative characteristics and source for successful seed propagation and plant production. *Proc. Inter. Plant Prop. Soc.* 14:74–85.

32. ———. 1967. Propagation from seed, Part 5. Control of seedling density. *Amer. Nurs.* 125(8):14–15, 56–59.

33. ———. 1967. Propagation from seed, Part 6. Hardseededness, a critical factor. *Amer. Nurs.* 125(10):10–12, 88–96.

34. ———. n.d. Thirty five years testing of tree and shrub seed. *New York Agr. Exp. Sta. (Geneva) Mimeographed leaflet.*

35. International Seed Testing Association. 1976. International rules for seed testing. *Seed Sci. and Tech.* 4:114–77.

36. Joley, L. E., and K. W. Opitz. 1971. Further experiments with propagation of *Pistacia*. *Proc. Inter. Plant Prop. Soc.* 21:67–76.

37. Hoyle, B. J., H. Yamada, and T. D. Hoyle. 1972. Aggresizing—to eliminate objectionable soil clods. *Calif. Agr.* 26(11):3–5.

38. Justice, O. L. 1972. Essentials of seed testing. In *Seed biology*, Vol. 3, T. T. Kozlowski, ed. New York: Academic Press.

39. Kamra, S. K. 1964. The use of x-rays in seed testing. *Proc. Inter. Seed Testing Assn.* 29:71–79.

40. Khan, A. A. 1978. Incorporation of bioactive chemicals into seeds to alleviate environmental stress. *Acta Hort.* 83:225–34.

41. Khan, A. A., K. Tao, J. S. Knypl, and B. Berkowska. 1978. Osmoconditioning of seeds: Physiological and biochemical changes. *Acta Hort.* 83:267–77.

42. Kozlowski, T. T. 1971. *Growth and development of trees,* Vol. 1. New York: Academic Press.

43. Lakon, G. 1949. The topographical tetrazolium method for determining the germinating capacity of seeds. *Plant Phys.* 24:389–94.

44. Lawyer, E. M. 1978. Seed germination of stone fruits. *Proc. Inter. Plant Prop. Soc.* 28:106–9.

45. Leukel, R. W. 1953. Treating seeds to prevent diseases. In *Plant diseases: Yearbook of agriculture.* Washington, D.C.: U.S. Govt. Printing Office, pp. 134–45.

46. Maclean, N. M. 1968. Propagation of trees by tube technique. *Proc. Inter. Plant Prop. Soc.* 18:303–9.

47. MacKay, D. B. 1972. The measurement of viability. In *Viability of seeds,* E. H. Roberts, ed. Syracuse, N.Y.: Syracuse Univ. Press.

48. Mastalerz, J. W. 1976. *Bedding plants.* University Park, Pa.: Pennsylvania Flower Growers.

49. Maude, R. B. 1978. Seed treatments for pest and disease control. *Acta Hort.* 83:205–12.

50. McDonald, M. B., Jr. 1980. Assessment of seed quality. *HortScience* 15:784–88.

51. McKee, J. M. T. 1981. Physiological aspects of transplanting vegetables and other crops. I. Factors which influence re-establishment. *Hort. Abstracts* 51(5):265–72.

52. ———. 1981. Physiological aspects of transplanting vegetables and other crops. II. Methods used to improve transplant establishment. *Hort. Abstracts* 51(6):355–68.

53. McMillan-Browse, P. D. A. 1978. Stratification—a detail of technique. *Proc. Inter. Plant Prop. Soc.* 28:191–92.

54. ———. 1979. *Hardy woody plants from seed.* London: Grower Books.

55. Mechel, B. E., and M. R. Kaufmann. 1973. The osmotic potential of polyethylene glycol 6000. *Plant Phys.* 51:914–16.

56. Mikkelson, D. S., and M. N. Sinah. 1961. Germination inhibition in *Oryza sativa* and control by preplanting soaking treatments. *Crop Sci.* 1:332–35.

57. Moore, R. P. 1973. Tetrazolium staining for assessing seed quality. In *Seed ecology,* W. Heydecker, ed. London: Butterworth, pp. 347–66.

58. Poynter, M. J. 1978. Building and using a growing room for seed germination of bedding plants. *Proc. Inter. Plant Prop. Soc.* 28:109–14.

59. Rudolf, P. O. 1965. State tree seed legislation. *U.S. Forest Service, Tree Planters' Notes* 72:1–2.

60. Schalla, S. L., and K. Doughton. 1978. Transplanting the Douglas fir plug. *Proc. Inter. Plant Prop. Soc.* 28:177–84.

61. Schubert, G. H., and R. S. Adams. 1971. *Reforestation practices for conifers in California.* Sacramento: Div. of Forestry.

62. Sharples, G. C. 1981. Lettuce seed coating for enhanced seedling emergence. *HortScience* 16:661–62.

63. Simak, M. 1957. The x-ray contrast method for seed testing. *Medd. f. Stat. skogsforssr. Inst.* 47(4):1–22.

64. Steavenson, H. 1979. Maximizing seedling growth under midwest conditions. *Proc. Inter. Plant Prop. Soc.* 29:66–71.

65. Stuke, W. 1960. Seed and seed handling techniques in production of walnut seedlings. *Proc. Plant Prop. Soc.* 10:274–77.

66. Suszka, B. 1978. Germination of tree seed stored in a partially after-ripened condition. *Acta Hort.* 83:181–88.

67. Thomson, J. R. 1979. *An introduction to seed technology.* New York: John Wiley.

68. Tinus, R. W. 1978. Root system configuration is important to long tree life. *Proc. Inter. Plant Prop. Soc.* 28:58–62.

69. Tinus, R. W., and S. E. McDonald. 1979. *How to grow tree seedlings in containers in greenhouses.* USDA For. Ser. Gen. Tech. Rpt. RM-60.

70. U.S. Dept. of Agriculture, Forest Service. 1974. *Seeds of woody plants in the United States.* C. S. Schopmeyer, ed. Agr. Handbook No. 450.

71. ———. 1952. *Manual for testing agriculture and vegetable seeds.* USDA Agr. Handbook No. 30. Washington, D.C.: U.S. Govt. Printing Office.

72. Vaartaja, O. 1964. Chemical treatment of seed beds to control nursery diseases. *Bot. Rev.* 30:1–91.

73. Vanstone, D. E. 1978. Basswood (*Tilia americana* L.) seed germination. *Proc. Inter. Plant Prop. Soc.* 28:566–69.

74. Wakeley, P. C. 1954. Planting the southern pines. *USDA Monograph No. 18.* Washington, D.C.: U.S. Govt. Printing Office.

75. Weaver, R. J. 1972. *Plant growth substances in agriculture.* San Francisco: W. H. Freeman & Co. Publishers.

76. Western Forest Tree Seed Council. 1966. *Sampling and service testing western conifer seeds.* Portland, Ore.: Western Forestry and Conservation Association, 36 pp.

77. Wilcox, G. E., and P. E. Johnson. 1971. An integrated tomato seedling system. *HortScience* 6:214–16.

78. Williams, R. D., and R. R. Romanowski. 1972. Vermiculite and activated carbon adsorbents protect direct-seeded tomatoes from partially selective herbicides. *Jour. Amer. Soc. Hort. Sci.* 97:245–49.

79. Wilson, J. K. 1915. Calcium hypochlorite as a seed sterilizer. *Amer. Jour. Bot.* 2:420–27.

## SUPPLEMENTARY READING

ALDOUS, J. R. 1972. Nursery practice. *Forestry Commission Bulletin 43.* London: Her Majesty's Stationery Office.

ASSOCIATION OF OFFICIAL SEED ANALYSTS. Proceedings of Annual Meetings.

BALL, V., ed. 1976. *The Ball red book* (13th ed.). Chicago: Geo. J. Ball, Inc.

BEDDING PLANTS, INC. Proceedings of Annual Conference.

GRAY, D. 1981. Fluid drilling of vegetable seeds. *Hort. Rev.* 3:1–27.

INTERNATIONAL PLANT PROPAGATORS' SOCIETY. Proceedings of Annual Meetings.

INTERNATIONAL SEED TESTING ASSOC. *Jour. Seed Technology.* Annual.

MASTALERZ, J. W. 1976. *Bedding plants.* University Park, Pa.: Pennsylvania Flower Growers.

McMILLAN-BROWSE, P. D. A. 1979. *Hardy woody plants from seed.* London: Grower Books.

SCHOPMEYER, C. S., ed. 1974. *Seeds of woody plants of the United States.* U.S. For. Ser. Agr. Handbook No. 450.

STOECKELER, J. H., and P. E. SLABAUGH. 1965. *Conifer nursery practice in the prairie states.* U.S. Dept. Agr. Handbook No. 279.

THOMSON, J. R. 1979. *An introduction to seed technology.* New York: John Wiley.

TINUS, R. W., and S. E. McDONALD. 1979. *How to grow tree seedlings in containers in greenhouses.* USDA For. Ser. Gen. Tech. Rpt. RM–60.

U.S. DEPT. OF AGRICULTURE. 1952. *Manual for testing agricultural and vegetable seeds.* U.S. Dept. Agr. Handbook No. 30.

———. 1961. *Seeds: Yearbook of agriculture.* Washington, D.C.: U.S. Govt. Printing Office.

Asexual propagation, reproduction from vegetative parts of the original plant, is possible because every cell of the plant contains the genetic information necessary to regenerate the entire plant. Reproduction can occur through the formation of **adventitious** roots and shoots or through the uniting of vegetative parts by grafting or budding (see chapter 11). Stem cuttings and layers have the ability to form adventitious roots, and root cuttings can regenerate a new shoot system. Leaves can regenerate both new roots and new shoots. A stem and a root can be grafted together to form a single plant.

New plants can start from a single cell, either adventitiously on intact plants or in aseptic culture systems. The property of living vegetative cells of plants to have all the genetic information needed to regenerate the complete organism is called **totipotency** (*92*). Entire plants have been regenerated in aseptic cultures from individual cells of tobacco pith (*110*) and carrot root (*91*). These new plants were identical with the one from which the original cells were taken.

## REASONS FOR USING
## VEGETATIVE PROPAGATION

*Maintenance of clones*   Vegetative propagation is asexual in that it involves mitotic cell divisions that duplicate the genotype of the plant. Such genetic duplication is known as **cloning** and the population of offspring plants is known as a **clone** (*88, 90, 96, 115*). In cloning the unique characteristics of any single plant are perpetuated during propagation. Cloning is particularly important in horticulture because the genotype of most fruit and ornamental cultivars is highly heterozygous and the unique characteristics of such plants are immediately lost if propagated by seed.

*Propagation of seedless plants*   Asexual propagation is necessary to maintain cultivars that produce no viable seeds, such as some kinds of bananas, figs, oranges, and grapes.

# 8

# General Aspects
# of Asexual Propagation

*Avoidance of long juvenile periods*    Plants grown from seed go through a juvenile period in which flowering does not occur. Some woody plants and certain herbaceous perennials, such as orchids and bulb species, may require 5 to 10 years before flowering begins. Once it has reached the flowering stage, the plant flowers regularly. Vegetative propagation then retains this flowering capacity and avoids the nonflowering juvenile phase.

*Control of growth form*    During the juvenile period the seedling plant not only fails to produce flowers and fruit, but often exhibits distinct morphological features. Some of these are highly undesirable traits (e.g., thorniness or excessive vigor) that can be avoided by vegetatively propagating the adult form. Some plants, such as English ivy (*Hedera helix*), are more desirable in the juvenile phase than in the adult form. Selection of cutting material in the juvenile phase facilitates propagation of difficult-to-root species and cultivars since cuttings taken from juvenile material root more easily than those from adult material.

*Combination of clones*    The possibility of combining two or more clones in the same plant by grafting or budding is an important aspect of vegetative propagation. This process is described in chapters 11, 12, and 13.

*Economics*    In general, mass propagation by vegetative means is not more economical than comparable propagation by seedlings but its use is justified by the superiority and uniformity of specific clones. The major economy in vegetative propagation comes from the elimination of the juvenile phase and shortening the time to reach reproductive maturity.

# THE CLONE

## History

One of the major advances during primitive agriculture was the discovery that the best individuals of important food plants, such as grape, olive, or fig, could be reproduced simply by thrusting pieces of their woody stems (cuttings) into the ground and letting them produce roots (*124*). Bulb, tuber, and rhizome-producing plants, such as potato, onion, garlic, sugar cane, banana, pineapple, and bamboo, can likewise be reproduced by dividing the vegetative structures and planting the pieces (see chapters 14 and 15). Similarly, the discovery of budding and grafting techniques showed that one can reproduce exactly specific superior individuals of important fruit crops, such as apple, pear, and peach that are difficult to propagate by cuttings.

Later, the development of greenhouses and other propagation aids made possible large-scale vegetative propagation by small leafy cuttings of an even wider range of plant species used either for food or as ornamentals (chapter 10).

## Significance

Cloning in horticulture makes possible the exploitation of a single superior plant with a unique genotype. Such plants may be selected from a variable population of seedling plants or from a single variant tree or branch within a group of vegetatively propagated plants. With the modern developments in plant breeding, cloning is even more important as a powerful tool of selection (63).

Another significance of cloning is that, since all members of the clone have the same basic genotype, the population tends to be highly uniform phenotypically. All members of the clone generally have the same appearance, size, ripening time, blooming time, and so on. The standardization of production and other uses of the cultivar are thus possible.

Many clones of horticultural interest have been discovered and perpetuated by man as named cultivars (71). The 'Bartlett' ('Williams Bon Chrétien') pear clone originated from a seedling in England about 1770 and has been maintained asexually ever since. The 'Delicious' apple clone originated about 1870 as a chance seedling in Jesse Hiatt's apple orchard in Peru, Iowa. Buds taken from this tree and budded to rootstocks produced new trees having tops of the same genetic constitution as the original seedling tree. By continuing this process of asexual propagation, millions of trees have since been grown whose budded tops, collectively, make up the 'Delicious' apple clone.

Clones exist in nature, reproducing naturally by such structures as bulbs, rhizomes, runners, stolons, and tip layers (see chapters 14 and 15). Apomixis in some species of Rosaceae, Gramineae, and Compositeae produces seedling plants that can be considered a type of clone. A clone may perpetuate itself successfully in nature, sometimes better than can sexually propagated plants, as long as the environment remains reasonably constant. If the environment changes drastically, however, a clonally reproduced species will be at a disadvantage because it has no opportunity to evolve forms better adapted to the new environment.

Similarly, use of a single or only a few cultivated clones in any crop industry has a disadvantage in being genetically uniform. Such industries are examples of **monoculture** (32). An adverse situation, such as disease or insect attack, can affect all members of the clone equally and may even destroy it. If a major food crop in a country is based on only one or two cultivars, the results of such attacks can be disastrous. Genetic diversity among different cultivars would be more desirable.

### Variability

The actual appearance and behavior of a plant, its **phenotype,** results from the interaction of the **genotype** with the **environment** in which the plant is growing. Among the plants of a clone, environmental variation can occur without changing the genotype of the clone. The appearance of the plants, or of fruits and flowers on different plants, can vary because of the effects of climate, soil, disease, and the like. In many years 'Bartlett' pear trees grown in California tend to produce round, apple-shaped fruit, whereas 'Barlett' pears grown in Washington and Oregon are relatively long and narrow; this difference is believed to be due to different climatic factors (105). Some plants produce leaves of different appearance when growing in shade than when growing in sun. Certain water plants produce leaves quite different in appearance on the part of the plant submerged than on the part in the air. Within a single orchard, fruit trees of the same cultivar often differ noticeably because of differences in soil, water availability, rootstock, or competition from surrounding plants.

**Figure 8-1** Zonal variations in location of different maturation phases on a mature seedling plant of a woody perennial.

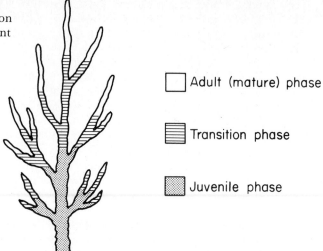

☐ Adult (mature) phase

▤ Transition phase

▨ Juvenile phase

Horticulturists of the seventeenth and eighteenth centuries believed that a clone would deteriorate with age and could be rejuvenated only by seed propagation (*56, 72*). Later evidence has shown, however, that the life of the clone is theoretically unlimited if it is grown in a suitable environment and continually renewed by vegetative shoots. Some horticultural cultivars, such as 'Cabernet Sauvignon' and 'Sultana' grapes, have been grown for 2000 years or more, producing millions of plants with little, if any, basic change in the cultivar.

Variability can, however, occur among plants of a clone and changes may lead to deterioration. Exposure to a continually *unfavorable environment* may result in progressive deterioration of a clone. For example, lack of sufficient winter chilling for certain strawberry cultivars (*12*) or an insufficiently long postbloom vegetative period for some bulbous plants may lead to gradual decline in vigor and productivity, but the basic genotype is not altered.

The effect of *pathogens* is probably the most significant cause of deterioration. These are mostly virus and viruslike organisms (*104*). Virtually any clone that is repeatedly propagated for long periods is likely to become infected, its survival being determined by its ability to tolerate the pathogens.

*Genetic changes* (mutations) in the clone, although not always degenerative in a strict sense, can produce off-type individuals that reduce the value of the clone (*86*).

The variability between *juvenile* to *adult* growth and development phases, described in chapter 3, manifested in differences in appearance within a single seedling plant (Figure 8–1), represents another kind of variation associated with vegetative propagation. Propagators must recognize these variability factors and know how to control them.

## Phase Variation in Seedlings and Clones

Most clones originate from buds taken from some part of a seedling plant and then reproduced consecutively by some form of vegetative propagation.

The ontogenetic development of the seedling plant through its life cycle (see chapter 3) occurs in distinct *phases* (*13*), designated as **juvenile** and **adult** (mature) separated by a **transition** phase. These phases can be manifested in three basic ways (*48, 54, 55, 114, 122, 123*):

**1.** The potential to shift from vegetative growth to **reproductive maturation** is controlled in the shoot tips (meristem) which in the juvenile phase are unable to initiate flowers even when given appropriate flower inductive conditions. "Maturation" is not the same as "aging."

**2.** Variations may occur in specific morphological and physiological traits, including leaf shape, vigor, and thorniness which are associated with different phases.

**3.** Differences occur in the ability of the plant parts in the different phases to regenerate shoots or roots; the juvenile phase is more likely to regenerate than the mature.

These phase change phenomena have two significant aspects in propagation. One is that juvenile, transition, and adult phases may be exhibited simultaneously in different parts of the seedling plant (*42, 84, 114, 123*). As illustrated in Figure 8–1, a zonal pattern occurs in many seedling trees, such as apple, pear, citrus, and conifers, in which flowering begins in the outer and upper extremities of the plant, sometimes after many years of juvenile growth. Similar but less conspicuous patterns are shown in herbaceous plants, in which the first nodes are vegetative and nodes produced later are reproductive.

Phase changes may also be expressed in other traits in seedling plants. Leaf shape, for example, is often an important "marker" for the change from juvenile to mature phase, as illustrated in Figure 8–2.

A second aspect of phase change is that vegetative propagation of a meristem reproduces the physiological and morphological features of the particular phase that it had on the source plant. This means vegetative offspring reproduced from different locations on the source plant can show much varia-

**Figure 8–2** The juvenile phase in a seedling plant often is indicated by leaf characteristics, as illustrated in *Eucalyptus.* Juvenile leaves at the base are large, broad, and have no petiole (*sessile*); mature leaves on the upper part of the shoots are elongated and have a distinct petiole. *Left:* variation within tree. *Right:* close-up of leaf variation in single shoot.

**Figure 8–3**    Two different types of expression of juvenile → mature phase changes. *Left: Acacia melanoxylon.* In this seedling plant the bipinnate leaves at the base are characteristic of the juvenile condition; the "leaves," which are actually expanded petioles (phyllodia), at the top of the seedling are characteristic of the adult form. Note the gradual change in the leaves from base to top. *Right:* adult (*middle*) and juvenile (*right*) forms of a variegated (chimeral) ivy (*Hedera helix*). Note flowering in adult and vine-like growth of juvenile with adventitious roots along stem.

tion both among the plants produced and between source plant and vegetative offspring. Foresters use the term **ortet** to designate the original seedling tree and the term **ramet** to refer to its vegetative offspring. These terms are useful to indicate the sequence of changes that may occur during vegetative propagation where a strong juvenile → adult shift occurs.

Plants propagated with buds taken from the adult phase of the plant invariably begin to flower at a younger age than did the original seedling plant. In one example, grafted sugar maple ramets flowered in two years in the greenhouse and seven to nine years in the open, but comparable seedling plants did not flower until 19 to 21 years (*43*).

Figure 8–3 shows the striking difference in growth form of plants propagated from the juvenile and adult phase in English ivy (*Hedera helix*). Figure 8–4 shows citrus nursery trees, the very thorny plant on the left budded from the thorny juvenile phase and the plant on the right from the nonthorny adult phase of the same seedling plant.

**Figure 8-4**    Two nursery trees of
'Kara' mandarin propagated from single
buds (arrows) taken from the same tree, a
nucellar seedling. Thorny tree on left
grew from a bud taken from the thorny,
juvenile portion of the tree; the thornless
tree on right grew from a bud taken from
the thornless, adult part of the tree. Courtesy H. B. Frost (*42*). From H. J. Webber
and L. D. Batchelor, *The citrus industry,*
vol. 1. Berkeley: Univ. of California
Press.

## TOPOPHYSIS

The phenomenon in which different parts of the plant show phase variation
and from which meristems perpetuate these different phases in their vegetative
offspring has been called **topophysis** (*9, 72, 120*).

Topophysis is also manifested in the persistence of growth form in certain
plants following vegetative propagation. Figure 8–5 shows coffee plants started
as rooted cuttings taken from an upright (**orthotrophic**) shoot (left) and from a
lateral (**plagiotrophic**) branch (right) on the same source plant. New shoots
from cuttings will grow either in an upright direction or in a horizontal orientation depending from which branch on the original plant the cutting was obtained. The same phenomenon is shown strongly by many conifers where prostrate or horizontal forms can be obtained if cuttings are taken from horizontally
growing shoots. Norfolk Island pine (*Araucaria*) is another plant with a strong

**Figure 8–5**    *Left:* coffee plant propagated from a vertically oriented cutting
maintains an upright (orthotropic)
growth habit. *Right:* coffee
plant propagated from a lateral
shoot produces a horizontal (plagiotropic) growth habit. Plants in many
species exhibit this growth phenomenon, requiring care in selecting
cutting material.

tendency to maintain such different growth forms in cuttings. This phenomenon is also referred to as **determination** in meristems (*9*).

CONTROL MECHANISM

The mechanism involved in the cellular control of phase changes is not well understood. It is clear that the control is inherent in the cells of the individual meristems and involves a system that determines embryonic, juvenile, and mature states. These also involve control of various other morphological traits associated with these phases. Expression of the maturation (flowering) state is complex, however; experiments have shown that an interaction exists between the inherent state of the meristem with hormonal influences from the surrounding leaves, and perhaps from the roots and cellular environments of the growing point (*48*).

The term **epigenetic** is used to describe persistent changes in phenotype that involve the expression of particular genes. Such changes can develop gradually but once induced the changed phenotype persists more or less indefinitely after an initial stimulus to change is removed (*18, 68, 69*). Epigenetic changes are thus different from purely **physiological changes** that depend upon the continued presence of a particular stimulus. Epigenetic changes differ from genetic changes in that in the latter case the DNA is altered. Such altered genes can be demonstrated by inheritance and segregation studies of the traits affected in subsequent generations (*70*). Although the mechanism of epigenetic control is not understood, it is believed to involve control of the RNA transfer system where genetic information is relayed from chromosomes to appropriate enzyme systems.

Epigenetic variation is evidently widespread in plants as part of their normal growth and developmental patterns. The development of the seedling beginning with the zygote and embryo can be interpreted as a series of persistent epigenetic changes resulting in variation in separate meristems but leading eventually to a relatively stable but reversible mature state (*54*). In the formation of the seed, epigenetic reversion to the embryonic and subsequent juvenile state occurs to repeat the entire cycle. Both sexual (Figure 3–1) and apomictic seedlings (*42*) go through the same juvenile → adult cycle indicating that no genetic change is involved and that the epigenetic changes occurring in the vegetative cells are not dependent on sexual reproduction.

EFFECTS OF PHASE CHANGES
ON PLANT BEHAVIOR

Different species, and different seedling plants within those species, vary greatly in the effect of phase changes on the juvenile → adult cycle of their development. Strong or weak juvenile effects or early and late bearing tendencies are inherited, so that the process is under an overall genetic control. Various traits that are affected fall into one of the following three categories:

**Reproductive maturation** refers specifically to the ability to flower, which is the primary manifestation of the juvenile → adult transition. Induction of flowering depends on both the maturation state of the meristems and the environment to which the plant is subjected. In plants propagated from seed, location of flowering may be zonal (Figure 8–1) and delayed. In plants vegetatively propagated from different parts of the seedling plant, variation in age of flowering may occur in the first vegetative generation because of topophysis, depending on the location of the buds. Once the plant starts to flower, whether seedling or vegetatively propagated, it should continue to flower from then on.

Most horticultural cultivars that are clones have been repeatedly propagated vegetatively so that the entire clone is in the adult phase. *Consequently, plants propagated with buds taken from different parts of grafted or own-rooted plants, young or old, are not likely to exhibit the variation associated with juvenile and adult characteristics.* The growth cycle of the clonally propagated plant (*53*) has only two parts, **vegetative** (but nonjuvenile) and **reproductive.** Some meristems remain vegetative and continue to generate new branches or are used to propagate new plants. All parts of the plant, both vegetative and reproductive, are adult and can respond to flowering stimuli.

**Morphological and physiological traits** in the juvenile and adult zones in different parts of a seedling plant may differ distinctly. In many plants leaf shape can identify phase changes (*1*), as illustrated in *Eucalyptus* in Figure 8–2. For other examples, seedling plants of citrus (*16*), pear (*113*), and apple (*9, 80, 113*) are characterized by delayed flowering, excessive vigor, and thorniness. As parts of the plant attain the adult phase, flowers and fruit are produced, vigor decreases, and thorns tend to disappear in that part of the plant. The juvenile phase of English ivy (*Hedera helix*) (Figure 8–3), which is a classic example of this phenomenon, is a trailing vine with alternate palmate leaves. The mature, flowering form is an erect or semierect shrub with entire, ovate leaves, produced oppositely around the stem. In some conifers, the juvenile phase produces needle-shaped leaves, but the adult produces scale-like leaves.

Other differences between juvenile and mature phases are shown physiologically (*13, 48, 84*). Leaf persistence in the autumn may be greater in the juvenile zones of the seedling plant. Differences may be shown also by stem pigmentation, disease resistance, and cold resistance. Juvenile shoots are often less palatable to browsing animals.

**Regeneration** is one of the major manifestations of the juvenile to mature phase changes. It has been known for a long time that cuttings made from plant material taken from the juvenile phase of woody perennial plants tend to produce adventitious roots and shoots more easily than cuttings taken from material in the mature phase. It has been more recently learned that regeneration of new organs (roots, shoots, or embryos) on explants and propagules in tissue culture systems is strongly influenced by the source of tissue in relation to maturation (see chapter 16) (*6, 65*). Thus, manipulation of the juvenile state may be a major factor in extending clonal propagation by cuttings to hard-to-root species.

CONTROL OF PHASE CHANGES

Learning to control variability associated with juvenile → adult phase changes adds an important dimension to vegetative propagation techniques. The epigenetic changes can be *progressive, arrested,* or *reversed.*

Epigenetic changes are **progressive** in the ontogenetic development of the plant. Lateral meristems behind the continually dividing apical meristem maintain their independent pattern of epigenetic change. There is evidence to indicate that the rate of epigenetic change in each of these meristems is related to the number of cell divisions (as measured by node number). Thus the basal (proximal) part of the seedling plant tends to remain juvenile through the life of the plant. Although this zone is the oldest part of the plant in terms of chronological age, it is the most juvenile in terms of maturation. On the other hand, the upper or outer (distal) extremities of the seedling plant would be the most adult and, once flowering starts, would continue to flower.

Continued rapid juvenile → adult phase changes then are important in obtaining early flowering in breeding programs or in the initial establishment of a clone as a cultivar (*83, 98, 121*). Thus, procedures have been developed to speed up flowering. Likewise, selection of propagating material from the distal (mature) portion of the plant is useful to stabilize the clone (*16, 106*) particularly if undesirable vegetative characters, such as thorniness, are associated with the juvenile phase.

Loss of rooting potential in newly developed clonal apple rootstocks has occurred between the time of original selection of the rootstocks and the time that the rootstock was distributed for commercial distribution. Changes took place in the juvenile → mature phase in the intervening consecutive propagations (*76*). This is an example of an undesirable phase change.

**Arresting** development in the juvenile phase may be desirable in order to maintain a particular growth habit, as in ivy (Figure 8–3), or to retain ease of rooting. Selection of cutting material from the basal juvenile parts of seedling plants is widely recognized as a method to propagate hard-to-root plants. If such plants are grown in a hedge and repeatedly pruned, it is possible to arrest the juvenile → mature change and retain good rooting potential (*5, 61*) (see Figure 9–19).

**Reversions** from adult to the juvenile phase occurs in the normal process of sexual reproduction with the subsequent development of the embryo and seedling plant. Other types of reversion in vegetative material may also occur.

**1.** Reversions develop with the production of adventitious shoots either in callus (*99*) on roots (*48, 85*) or in tissue culture systems (see chapter 16).

**2.** Reversions have taken place with consecutive grafting to seedling rootstocks in rubber (*75*) and eucalyptus (*37*) and on juvenile ivy (*36*) if accompanied by high temperature. Consecutive aseptic reculturing using an ancient grape cultivar, 'Cabernet Sauvignon' (*73*), has induced reversion.

**3.** Hormone treatment, primarily gibberellic acid on ivy (*48*), has produced reversions under experimental conditions.

## GENETIC VARIATION IN ASEXUALLY PROPAGATED PLANTS

### Mutations

Genetic changes can occur in cells and may lead to permanent changes in parts of the clone. This situation requires that the genetic change be followed by cell divisions in which the daughter cells occupy a substantial portion of the meristem.

Chromosomes are located in the nucleus and are composed of long chains of **deoxyribonucleic acid** (DNA) with repeating units (**nucleotides**) containing sugar, phosphate, and combinations of four chemicals (bases): cytosine, guanine, adenine, and thymine. The arrangements of these bases form the genetic code that makes up the **genome** (i.e., the genes) for that cell. DNA replicates itself during cell division, but occasionally a change occurs during this replication that results in a **mutation.** A plant that arises as a result of a mutation is known as a **mutant.**

Genetic changes may result from a rearrangement of the four bases (**point mutation**) or from **deletions, duplications,** or **inversions** of parts of some chromosome. Genetic changes may result also from the addition or subtraction of individual chromosomes (**aneuploidy**) or from the multiplication of entire sets of chromosomes (**polyploidy**).

The cytoplasm also has replicable units that contain DNA, including **plastids** and **mitochondria.** These systems, referred to as the **plastome** and **chondrianome,** are independently involved in determining various plant characteristics (*111*). The most conspicuous plastome mutants involve loss of chlorophyll, which results either in production of **albino** plants or, more commonly, **variegated** plants having both albino and green tissue (*93, 111*). These variegated forms occur in essentially all species.

The effect of mutation on the variability of a given clone depends on the rate at which mutations occur and the extent that the daughter cells from the original mutant cell occupy the shoot meristems (*15, 27, 93, 103*). Mutations can occur spontaneously at a more or less set rate but the rate may vary with the location of the cell in the meristem. Cells in the apical meristem appear to be relatively stable and less likely to mutate than those in less organized cell systems, as in the inner layers of the meristem or in callus (see chapter 16). This is particularly true for changes in polyploidy. Mutation may be increased by the use of various mutagenic agents (*15*).

The location of a mutated cell in a meristem determines whether it produces any permanent change in the plant. If it occurs near the base of the apical meristem, it will likely be bypassed by the growth of the other nonmutated cells; if, however, it is among the dividing cells at the tip of the shoot meristem, it will likely produce a sector of mutant tissue in the shoot and result in significant change. If the new mutant tissue involves a conspicuous trait, such as color (chlorophyll or pigmentation), it may be detected early. Usually such changes go unnoticed until the mutant cells occupy a large part of a lateral meristem and produce an entire branch (**bud-sport** or bud mutation) (Figure 8–6) or are used for propagation. Such plants are composed of more than one kind of genetic tissue and are known as **chimeras.**

Most mutants are defective and in time may reduce productivity or produce other deleterious effects in a clone. Long-term propagation of clones may lead to

**Figure 8–6**　An almond (*Prunus dulcis*) tree in which one branch has conspicuously later bloom. This represents a ''bud-sport'' where a mutation in genes controlling time of bloom has evidently occurred.

variations among propagation sources that are not readily detected without extensive testing (*109, 116*). If the change is inferior, it can give rise to inferior off-type plants that should be avoided (*34, 87*). Citrus, for instance, is noted for having mutations in an inferior direction (*89*).

On the other hand, some bud-sports have resulted in important new clones in both fruit and ornamental species (*86, 108*). Pink-fleshed grapefruits were found on a single branch of a tree in a grove of thousands of trees producing white-fleshed grapefruits: an obvious mutation. The seedless 'Washington Navel' orange probably arose as bud mutation of the Brazilian orange, 'Laranja Selecta' (*71*).

### Chimeras

**Chimeras** are plants composed of two or more genetically different tissues growing separately but adjacent to each other in one plant (*7*). These tissues are commonly arranged in layers or sectors in the stem. Most chimeras originate by a mutation in one of the cells of the shoot apex. Some, however, appear as variable seedlings as a result of hybridization. Chimeras tend to be unstable in propagation but the degree of stability depends on their structure, as described later, and the genotype of the plant (*55, 111*).

Chimeras may be horticultural curiosities or economically important plants. Plants with variegated foliage, as found in *Citrus, Vitis, Pelargonium, Chrysanthemum, Hydrangea, Dahlia, Coleus, Euonymus, Bouvardia, Sansevieria,* and others, are examples of chimeras. In these plants the plastids in part of the leaf tissue lack the capacity to produce chlorophyll, whereas other leaf cells are normal. The resulting pattern shows distinct green and white (or yellow) areas in the leaf.

Fruit chimeras like the orange in Figure 8–7 are frequently observed. Apple fruits that have mixtures of sweet and sour flesh and peach fruits that have both fuzzy and smooth (nectarine) surfaces have been discovered (*27*). Red forms of apple (*25, 86, 108*) and potato cultivars are chimeras in which the red skin color mutation is present in the outer layer or epidermis, the inner tissue being the original unmutated form. Likewise, thornless blackberry cultivars exist in which the thornlessness is restricted to the outer stem layer. Chromosomal chimeras (*cytochimeras*) may occur in which cells in part of the tissue of the stem are diploid, i.e., they have the normal chromosome number for the species, and cells in other parts of the stem are polyploid, i.e., they have some larger number of chromosomes (*26, 27, 33, 78, 79*).

Chimeras usually originate from a mutation (spontaneous or induced) in one of the dividing cells in the meristem (*27, 93, 103*). Figure 8–8 shows that a

**Figure 8–7** Mericlinal chimera in 'Washington Navel' orange. Left portion of the fruit has developed from the mutation, producing a thicker rind that is yellow instead of orange in color. This example is an undesirable mutant.

shoot apex is composed of two (or more) distinct layers (**tunica**) over a mass of less organized cells (**corpus**). Angiosperm plants commonly have three distinct cell layers; gymnosperms and monocots have two. For convenience these layers have been called L–I, L–II, and L–III.

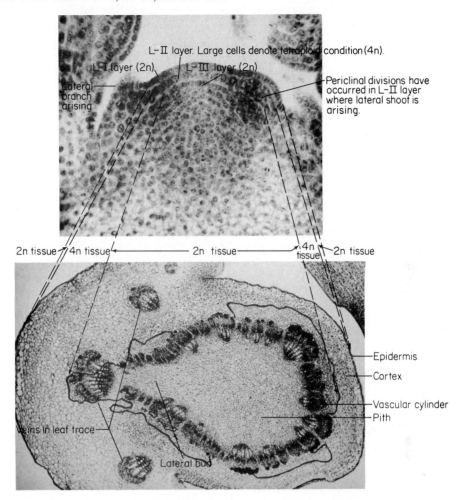

**Figure 8–8** Chimeras may originate by *mutation in one of the histogenic layers* of the apical meristem. Photomicrographs illustrate a chimera in peach showing an enlarged, longitudinal section of the shoot tip (top) and a cross section of the stem under the tip (below). *Top:* three histogenic cell layers in the tip are identified as L–I, L–II, and L–III (see text). In this example the L–I is diploid (2n), the L–II is tetraploid (4n), and the L–III is diploid (2n). Note that the L–I layer is a single row of cells, but that the L–II layer has increased in thickness on each side because of periclinal divisions at the point of leaf initiation. *Below:* the epidermis of the stem is 2n because it was derived from the L–I layer. The cortex and part of the vascular system are 4n, being derived from the L–II layer. The remainder of the vascular system and the pith are 2n, being derived from the L–III layer. From Derman (*27*).

These layers maintain their integrity because cell divisions occur in a particular direction. A cell in the growing point divides either at right angles to the outer surface (**anticlinally**) or parallel (**periclinally**) to the outer surface. Cells in L-I usually divide anticlinally and generally a distinct, more or less stable, outer cell layer results. Cells in L-II divide most often anticlinally near the apex but divide randomly farther away from the apex. Cells in L-III divide in both directions.

Occasionally a cell produced in one of the three layers becomes displaced into an adjoining layer. If this shift occurs relatively close to the growing tip so that the displaced cell, as it divides, produces a significant part of the new stem or leaf, a change in chimeral structure will result. For example, a shift could occur from a 2-4-2 cytochimera to one of 2-4-4 (or from 4-2-2 to 4-4-2), although the change might not be readily observed externally. A change from 2-4-2 to 4-4-2 would probably be more readily detected (since it would effect a change in the epidermis), but this change occurs less commonly (*27*).

Cells of a given layer give rise to a more or less specific part of the stem and are consequently referred to as **histogenic layers.** L-I produces the one-cell layered epidermis but sometimes becomes several layers thick. L-II produces the outer cortex and some portion of the vascular cylinder. The reproductive cells in the anthers and ovules are usually produced by this layer. L-III layer usually gives rise to the inner cortex, vascular cylinder, and pith. However, as seen in Figure 8-8, the parts of the stem arising from the L-II and L-III layers are not always consistent. Initiation of a chimera depends on the location of the mutant cell within these layers. A mutation that occurs in a cell in one of these layers will, in general, affect only that part of the stem arising from that layer. Thus, if a tetraploid cell arises in the L-II layer of a diploid plant, the resulting plant is described as a 2-4-2 cytochimera.

The further development of a chimera depends on the location of lateral buds in relation to the mutated areas. If a mutation occurs in a cell on one side of the growing point, only a small part of that stem may be affected. A lateral shoot that develops from the mutated part of the stem will show the mutant trait; a lateral shoot developing from the opposite side of the original stem will show the original (non-mutated) trait. If the bud contains tissue from both the original and the mutant sides, then various types of chimeral arrangement might occur.

### Periclinal Chimera

**In the periclinal** type, a tissue with one genotype occurs as a relatively thin layer, one or several cell layers in thickness, completely covering a genetically different "core." This type is common and is the most stable during propagation if stem cuttings or grafts are used. The genus *Rubus*, for example, has numerous thornless blackberry forms in which the epidermal layer lacks the gene for thorniness (*21*). Such plants will usually retain this characteristic if propagated by stem cuttings or by tip layering. In propagating such thornless chimeras by stem cuttings, it is significant that the adventitious roots arise endogenously beneath the mutated tissue, with the result that the root system of such thornless plants is composed entirely of nonmutated cells. If such roots are subsequently used as a source of root cuttings, the resulting plants are thorny. Likewise, seedlings developing from seeds taken from the thornless plant are thorny because the gametes are produced from cells of the tissue which originated from the L-II (nonmutated) layer. Some thornless forms, however, do involve the L-II layer in the mutation, and these transmit the altered characteristics to the seedling offspring (*22*).

Some potato cultivars are periclinal chimeras. Their nature can [be] demonstrated experimentally by removing the buds ("eyes") from tubers, forcing adventitious buds to arise from internal tissues. Removing the buds f[rom] 'Noroton Beauty', which has mottled tubers, gives 'Triumph', which has [solid] tubers (2). 'Golden Wonder', which has tubers with a thick brown russet skin, is also a chimera, having inner tissues characteristic of 'Langworthy', whose tubers have a thin, white, smooth skin (20).

Propagation of a periclinal chimera by some types of leaf cuttings, as in variegated *Sansevieria,* may also result in reversion to the nonmutated form, since adventitious shoots and roots, in this case, arise from the inner, nonmutated tissues.

### Mericlinal Chimera

The mericlinal type is similar to the periclinal chimera except that the mutated tissue occupies only part of one of the layers and does not extend completely around the shoot (see Figure 8–7 and Figure 8–9). This type of chimera is one of the most frequent to occur naturally, since a mutation in a single cell of a growing point would be apt to produce this type. It is not stable, however, and with continued vegetative propagation tends to change to a periclinal chimera or reverts to a non-chimeral shoot. A lateral bud arising from the chimeral portion would become a periclinal chimera, whereas lateral buds from the normal part produce normal shoots. Only in shoots from the terminal growing point and from lateral shoots at the edge of the mutated portion would the mericlinal chimera continue.

### Sectorial Chimera

In the sectorial type (Figure 8–9), the shoot is composed of two genetically different

**Figure 8–9**    Some variegated plants originate as seedlings in which the initial structure is a *sectorial chimera.* This type can develop if a mutation occurred very early in embryo development or if a "mixed" embryo or egg cell is produced with both normal and mutated plastids. Such chimeras are unstable and buds arising at different positions may produce shoots consisting entirely of mutated cells or entirely of nonmutated cells; or the shoots may be a sectorial, mericlinal, or periclinal chimera, depending upon their location. Adapted from W. N. Jones, *Plant chimeras and graft hybrids.* Courtesy Methuen & Co., London, Publishers.

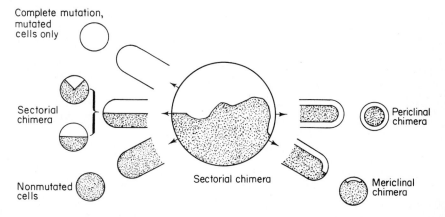

tissues situated side by side to occupy distinct sectors of the entire stem. Leaves and lateral buds arising from such a shoot may be composed of two tissues combined in various ways, depending upon their location. Because of the nature of the shoot apex, this type is uncommon in shoots but usual in roots. It must occur early in the development of the embryo before distinct layers develop in the growing point. It is unstable, and with continued growth, different parts of the plant develop either as normal, as mutated, or as a periclinal chimera.

**Variegation** in plants (Figure 8–10) is usually caused by mutant plasmagenes affecting chlorophyll development (*94, 111*). The green-white color patterns result in part from the cellular arrangements shown in Figure 8–8 but also depend on the unique cell development patterns in leaves. The leaf axis, or vein, grows like a stem, mostly from L–III with epidermal and subepidermal layers from L–I and L–II, respectively. The leaf blade grows laterally from each of the layers with L–I producing only the single-layered epidermis. The leaf blade is made up of layers of cells with varying proportions of L–II and L–III cells, thus producing the typical green-white (albino) variegation patterns.

Some of the variegation may be caused by segregation of green and white plastids within the cell itself during mitosis. This produces a more variable pattern than the more or less organized distribution of cells in the normal chimera (*93*).

Variability may also be due to causes other than the typical chimera. In some plants, genetic mutations occur at high rates in a set pattern controlled by the genotype of the plant. For example, certain unusual types of color variation are produced by unstable gene systems (*14, 36, 66*) involving specific units on the chromosome called **controlling elements** (*36*) or **insertion mutants** (IS) (*46*). These units are capable of shifting positions on the chromosome and produce variation in spots and sectors with patterned regularity. In a rare type of

**Figure 8–10**    A variegated pink 'Eureka' lemon, a type of chimera. From Hartmann, H. T., W. J. Flocker, and A. M. Kofranek. 1981. *Plant science.* Englewood Cliffs, N.J.: Prentice-Hall.

variation called **somatic crossing-over** (*35*), genetic material is exchanged between chromosomes during mitosis causing "twin spots" to occur as two kinds of tissue develop, both different from the parent tissue. **Mosaic** patterns can be caused by certain viruses. For example, "color breaks" in some kinds of tulips are caused by a virus rather than by a mutation. Likewise, some variations are caused by "pattern" genes that continue to reproduce that pattern (*111*).

### GRAFT CHIMERAS

While most chimeras occur naturally, they can also be established artificially by grafting (*8, 57*). In a young, grafted plant, if the scion is cut back severely, almost to the stock, an adventitious bud may sometimes arise from the callus at the junction of the stock and scion. Shoots resulting from such an origin are chimeras in which the cells of the two graft components remain genetically independent regardless of how intermingled they become.

Various graft chimeras have been reported in the horticultural literature. Many years ago Winkler (*118, 119*) artificially produced numerous graft chimeras of tomato (*Lycopersicon esculentum*) on black nightshade (*Solanum nigrum*) and black nightshade on tomato, both of which are easily grafted and produce callus readily (see Figure 8–11). Winkler gave such mixed shoots the name "chimera" after the mythological monster that was part lion and part dragon.

Natural graft chimeras have been known to exist and have been perpetuated vegetatively. The "bizzaria" orange, a fruit half-orange and half-citron, was discovered in 1644 in Florence, Italy. It apparently originated from an adventitious bud arising from callus at the junction of a sour orange (*Citrus aurantium*) graft on a citron (*Citrus medica*) rootstock (*41, 101*). Similarly, medlar (*Mespilus germanica*) grafted on hawthorn (*Crataegus monogyna*) has led to several chimeras known as hawmedlars (*8, 47*).

The possibility of developing chimeral plants through *mixed callus* cultures has been demonstrated in tobacco, adventitious shoots being produced in a chimeral arrangement from two separate species (*93*).

**Figure 8–11** Stages in Winkler's method of producing a graft-chimera shoot between the nightshade and tomato. Adapted from W. N. Jones, *Plant chimeras and graft hybrids.* Courtesy Methuen & Co., London, Publishers.

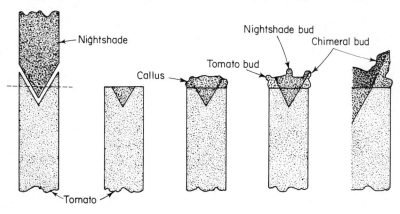

# PATHOGENS AND VEGETATIVE PROPAGATION

Propagation procedures, either seedling or vegetative, may be affected adversely by various pathogens, such as fungi, bacteria, insects, and mites, which need to be controlled for successful propagation. Control programs must be included as an essential part of the propagation operation (see chapters 2 and 7).

Equally important, certain pathogens may be disseminated along with the propagating material either on the external surface of the plant or internally within the plant. Consequently, procedures need to be included that either eliminate the pathogen if present, or detect those plants or plant parts free of the pathogen, which can then be utilized as stock material for propagation. Pathological problems vary with the different plant species and, for specific crops, publications need to be consulted for details on control measures. On the other hand, certain principles of control apply to all species, depending on the kind of problem involved. Important groups of plant pathogens that become involved in plant propagation are described as follows:

## Fungi

Fungi are organisms that do not photosynthesize but must obtain organic nutrients either from dead organic matter (**saprophytes**) or from living tissue (**parasites**). In some cases a fungus can live on both dead and living tissue (*facultative*) as compared to living on only dead or living tissue (*obligate*).

Fungi produce long, branching, threadlike **mycelia** made up of **hyphae,** which invade the substrate to absorb food and water. Some fungi also produce **sclerotia,** which are dense compact masses of hyphae, by which they can resist periods of unfavorable environments.

Some of the many detrimental fungi associated with propagation are *Phytophthora* spp., *Pythium* spp., and *Rhizoctonia* spp. (*52*). Control measures include fumigation, pasteurization, and fungicide treatments coupled with strict sanitation procedures (see chapter 2).

Certain beneficial fungi grow in a symbiotic relationship in root cells of many higher plants in a combination known as **mycorrhiza.** These combinations have been found in certain crop plants, such as corn and soybeans, and in apple, citrus, conifers, aspens, beeches, and other trees. Their role is beneficial and in some cases essential (*62, 64, 112*). Elimination of mycorrhiza by fumigation has resulted in zinc deficiency in certain instances (*60*).

## Bacteria

A bacterium is a one-celled organism with a cell wall. Bacteria occur universally and include both beneficial and pathogenic forms. They usually invade the plant through open wounds or natural openings where free moisture exists. Dormant spores of nonpathologic forms allow bacteria to survive dry periods and to be carried on the surface of propagating material or tools. The major sources of the pathogens are contaminated stock plants, soil, and potting mix. Bacteria may exist in latent infections until a plant or propagation part is placed in an environment favorable for their growth. Infection is often associated with wounds;

bacteria can be particularly harmful during propagation procedures where cuts into the plant tissues are necessary.

Bacterial pathogens that cause serious propagation problems include *Erwinia chrysanthemi, Erwinia carotovora,* and *Pseudomonas* spp. (*52*). Crown gall (*Agrobacterium tumefasciens*) attacks many nursery crops by producing large tumerous galls that make the plants unmarketable. A segment of chromosome from the organism is transmitted onto the host chromosome causing the plant to continue to form galls even in absence of the bacteria (*53*).

Bacteria are difficult to control. Eliminating sources of contamination, sterilizing tools, and pasteurizing soils are important sanitation procedures. The use of antibiotics or other bactericides is sometimes essential for economic control.

## Viruses

Viruses are submicroscopic organisms made up of nucleic acids (RNA or DNA) encased in an outer sheath of protein. The organisms live within plant cells as obligate parasites multiplying and growing in collaboration with the chromosomes of the cell, or in the bodies of insect vectors. Viruses are very small but can be identified in some cases by use of the electron microscope. Since viruses have a protein coat, serology tests also have been used for identification. In a susceptible host viruses disrupt some of the normal functions of the plant, causing various external symptoms. Viruses are characterized primarily by the name of the host and a description of symptoms, for example, tobacco mosaic virus (TMV) and *Prunus* ring spot virus (PRSV). Viruses may move from plant to plant in vectors, mostly aphids, leaf hoppers, thrips, mites, nematodes, and sometimes in pollen and infected seeds, but the most important means of spread is in propagation material. Viruses move within the plant in the phloem and can cross a graft union to infect a scion cultivar grafted onto an affected stock, or vice versa. Any clone that has been propagated for a long period of time probably has become infected with one or more viruses. These may influence growth, appearance, and production, not only of the infected plant but also of plants propagated from it. These organisms may remain with the clone, as would a genetic change, but they are also infectious. Nursery propagation practices using budwood or seedling stocks infected with viruses can be held partly responsible for the widespread distribution of viruses. Once present a virus remains more or less indefinitely with the host. Methods have been devised, however, to detect such plants and free them from viruses.

How severely a virus affects the clone depends upon the characteristics of the virus, the tolerance of the particular clone to the virus, and sometimes the accompanying environmental conditions. For example, *Prunus* ring spot virus (PRSV) produces a "shock" symptom that can sharply reduce bud survival during budding in the nursery (*44*). Recovery then occurs, and subsequent symptoms are much less noticeable. The severity of PRSV "shock" differs with species, plum and apricot being less affected than almond, peach, or cherry (*107*). There are a number of recognized virus diseases of strawberry, and there is a great difference in the tolerance of particular cultivars to them (*38*).

Rather commonly, plant viruses produce distinct symptoms in the plant at low temperatures, but these are not detectable when it is grown at higher temperatures (*40*). This is a reason, for instance, for the production of potato propagation stock in cooler climates where diseased plants, particularly those affected by mosaic virus, can be easily detected.

Multiple infections of several virus components may be involved in many (if not most) virus diseases. For instance, three latent viruses—A, B, and C—if present individually in strawberry may cause slight or no symptoms in a given clone, but will produce a severe reaction when present in combinations of two or three. This situation indicates a danger of distributing clones carrying a single "nonpathogenic" virus, since later infections by additional viruses can have serious consequences.

## Mycoplasma-like Organisms

Intermediate between bacteria and viruses in size, mycoplasmas are extremely small parasitic organisms that will live in plant or animal cells and may be found in the phloem tissues of plants. They have no constant shape, consisting only of a membrane enclosing the living protoplasm. These organisms multiply rapidly, utilizing the plant's food for their own growth, and they also disrupt translocation of food materials in the phloem, causing the plant to wilt and start yellowing. In some diseases due to these organisms, hormonal disturbances in affected plants result in *witches'-brooms* and other malformations, such as conversion of flower parts to leafy structures (*29, 49, 51*).

A number of other plant diseases formerly thought to be due to viruses are now assumed to be caused by spiroplasma mycoplasma. These include *aster yellows* in strawberries and certain vegetables and ornamentals, *blueberry stunt disease, pear decline,* and *X disease* in stone fruits. Differential susceptibility to mycoplasma-like pathogens between the scion cultivar and the rootstock have caused serious graft incompatibility disorders, such as *pear decline.* These organisms may be moved from plant to plant by insect vectors, usually leafhoppers or psylla, and are carried in propagating material.

## Rickettsia-like Organisms

Rickettsia-like organisms are associated with certain plant diseases, such as *Pierce's disease* of grapes (*45*) and *Phony disease* of peach (*77*). These organisms are transmitted by insect vectors from plant to plant or carried in seed and vegetative propagating material. Rickettsiae are considered to be highly modified bacteria, parasitic principally in arthropods, but man and other mammals are known hosts as are certain homopterous insects (*24*).

## Viroids

Viroids are extremely small agents that cause such infectious diseases as *spindle-tuber* of *potato, citrus exocortis, chrysanthemum mottle,* and *chrysanthemum stunt,* and probably *avocado sunblotch* (*100*) and *cadang-cadang* in coconut. Some of these diseases were originally attributed to viruses. Viroid structure appears to be a single, free RNA molecule of low-molecular weight that becomes involved in some manner with the chromosomes of the cell to produce symptoms. A viroid is about 1/10 the size of the smallest virus molecule. Some are easily transmitted mechanically as well as by grafting. Special procedures and sensitive indicator plants may be needed in studying viroids because symptoms are often slow to develop and difficult to detect. Viroids are carried in vegetative propagating material, and some are transmitted by seeds. Characteristically some of them multiply most rapidly at high temperatures.

Detection methods involve extraction and purification of nucleic acids of the cell, followed by detection of unique RNA viroid species by electrophoresis (*100*). Viroids cannot be identified by serological tests or by most virus testing procedures.

### Nematodes

Plant parasitic nematodes are microscopic (0.5 mm to 3 mm; 0.02 to 0.125 in.), eel-like worms that attack roots, stems, foliage, and inflorescences. Nematodes are significant in propagation because of their presence in roots of plants and in bulbs and other structures.

Propagators control nematodes by using nematode-free material and by sanitation, fumigation, or pasteurization of propagation material (see chapter 2).

## PRODUCTION AND MAINTENANCE OF PATHOGEN–FREE, TRUE–TO–TYPE CLONES

Vegetative propagation of clones needs to involve procedures that maintain genetic identity, detect genetic modifications, and exclude pathogens, such as fungi, bacteria, nematodes, viruses, and mycoplasmalike organisms. Historically, many propagation and production practices had their basis in minimizing the effects of disease. Clones (sometimes highly desirable ones) subject to deterioration due to any of these causes have tended to disappear. Clonally propagated cultivars that have survived are those that grow and produce in a given environment despite the fact they may be carrying latent viruses or are exposed to various agents of disease. However, when a clone satisfactory in one area is transferred to a different environment, it may become exposed to new disease agents for which it has no resistance. Or it may contribute agents of disease that it is carrying to other species in the new environment. Graft incompatibilities are sometimes caused by infectious agents to which either the stock or scion is hypersensitive. Bringing one diseased plant into proximity to other clean stock can result in infection of all. Quarantine procedures are designed to minimize disease spread that is due to movement of planting and propagating stock carrying harmful pathogens.

### Maintaining Genetic Identity

*True-to-type* involves two somewhat different concepts that are often used synonymously. One is **true-to-name,** which means that the propagation source is the cultivar intended and not from some other incorrectly identified plant. *Propagation stock may become mechanically mixed, or labels changed, and the mistake not discovered until after thousands of new plants have been propagated.*

The second concept—**trueness-to-type**—means not only that the plant is the correct cultivar, but also that it does not represent any significant variant from the expected phenotype that would affect performance and production (*55*).

Visual inspection of the source plants, preferably when in flowering or fruiting or at a mature stage of development, is the usual test for trueness-to-type. Visual inspections have limitations, however, in that a variant may not be sufficiently conspicuous for detection, or it may be masked by the environment or by the handling of the source block. The variant may be a chimera that involves only limited sectors of tissue. Buds arising from those sectors will produce variant plants. Growing test plants (**progeny testing**) may be desirable to select specific sources of a cultivar (*55, 116*).

Some biochemical tests exist to aid in separating different species and cultivars. A useful test is to identify banding patterns in electrophoresis gels which are unique to specific species and cultivars. These are used for biochemical "finger-printing" (*58*).

## Excluding Pathogens

Control of pathogens in propagation stock (*95*) involves two different concepts of disease control (*4, 5*):

The term **disease-free stock** indicates the absence of disease symptoms or effects, even though a pathogen may be present. Various environmental conditions, cultural practices, or tolerances of the cultivar may mask any visual disease symptoms, and plants may be propagated containing pathogens that might later, under particular conditions, show visible symptoms of the disease.

The term **pathogen-free stock** indicates the absence of pathogens on or in the propagation stock and the prevention of infection of plants grown during propagation (*117*). Absolute freedom from all pathogens is not, however, necessarily achieved and may not be a practical ideal (*97*). In the absence of visible disease symptoms, the existence of a pathogen is known only through specific indexing tests. Consequently, pathogens might be present but not detected by the tests used. Thus, a preferred term would be **specific pathogen-tested** or SPT, meaning "tested for specific pathogens as determined by specific index tests" (*74*). If specific viruses are involved, the term **SVT** might be used. If a single meristem or plant is the source, then the source would be an **SPT** (or **SVT**) **clone**.

Any program to utilize SPT, true-to-type sources or clones for propagation involves the following three phases:

1. Initial selection of a source of propagation stock
2. Maintenance of such stock in a block with adequate safeguards against infection
3. A system of propagation and distribution whereby such stock is disseminated without infection, and cultivar and source identity is maintained. The propagation system itself should not introduce variability (*10*).

## Initial Selection

Selection of a suitable propagation source begins with the identification of plants that are true-to-type. A full-grown flowering or fruiting plant, preferably one with a history of superior performance, is usually the best starting place. Individual plants should be visually inspected for absence of virus symptoms or other pathogenic diseases, and for trueness-to-type. Information about the performance of progeny plants propagated from the source is of value. It is essential that the source plant and the propagating material taken from it be labeled with its correct name. Color-coding is a useful nursery practice when large numbers of plants are involved.

The second step is indexing the plant source for latent viruses, fungi, or other pathogens, using a recommended minimum indicator host range and prescribed procedures (*39*).

**Culture indexing** to identify material free of fungi and bacteria is illustrated in Figure 8–12 (*5*). The principle is to place pieces of the plant in aseptic culture conditions using a medium favoring the growth of the pathogen. This method is used commercially in production systems for various ornamentals such as chrysanthemum, carnation, and *Pelargonium*.

**Virus indexing** is used to detect viruses and other systemic organisms by transmitting them to a sensitive indicator plant, which then develops symptoms (*95*). In some cases virus indexing is done mechanically by transferring sap into leaves of specific herbaceous indicator hosts. In other cases, used mostly with woody plants, a piece of the suspect plant, i.e., bud, scion, leaf, or bark, is

**Figure 8–12**    Testing propagating sources for disease organisms (fungi, bacteria) as illustrated for *Verticillium* wilt in chrysanthemum cuttings. From Baker and Chandler (*5*), *California Division of Agricultural Sciences Manual 23.*

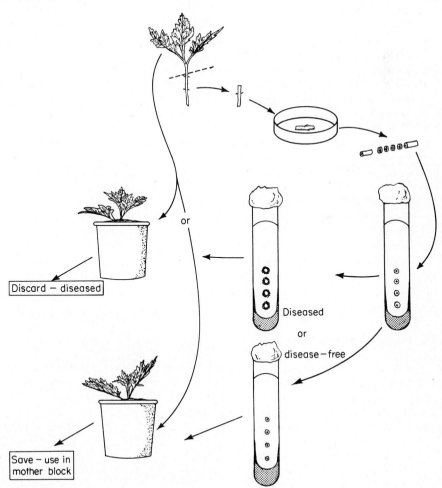

**Figure 8-13**    Shirofugen cherry indexing method. Buds are taken from suspect tree and inserted into branch of Shirofugen cherry. If no Prunus ring-spot virus (PRSV) is present, buds will heal normally (*right*). If PRSV is present, buds will die, gum will appear, and necrotic dead tissue will be present in bud patch. Test requires one month.

grafted to the **indicator** plant. Figures 8-13 and 8-14 show two techniques. In another procedure used with fruit plants, buds of both the test plant and the indicator plants are placed into a seedling plant grown in a greenhouse under prescribed environmental conditions favoring the development of virus symptoms. The virus must move not only across the union but through an intervening section of neutral tissue (*39*).

Other techniques for detecting viruses include **electron microscopy** and **serology.** Serology tests have important uses since these tests can identify unique proteins associated with particular pathogens. A particularly useful method is the ELISA test (**enzyme-linked immunosorbent assay**), which can be developed for essentially any pathogen or virus with a unique protein pattern. Purified extract of the pathogen is injected into a rabbit or other animal. Specific antibodies produced in the animal's blood react against the protein of the virus. When an extract containing the same protein is tested against these antibodies, no reaction will occur; only a foreign protein will cause a reaction. This method is very sensitive and highly specific (*19, 95*). Identification of cer-

**Figure 8-14**    Grafting procedure used in excised leaf method of transferring strawberry viruses. *Left·* terminal leaf removed, petiole split, and scion (excised leaf) inserted. *Center:* union partly wrapped with latex tape. *Right:* enlarged view showing healed union 40 days after grafting. Courtesy R. S. Bringhurst.

tain viroids is possible by extracting and purifying the unique RNA species and observing characteristic bands on electrophoresis gels (*100*).

## Procedures for Eliminating Pathogens from Plant Parts

If pathogen-free source plants are not found, procedures to eliminate the pathogen from some part of the plant are useful. A single small part, such as a tip cutting, a bulb scale, a lateral bud or the meristem tip can be the starting point of SPT propagation stock. In this case the SPT source becomes an SPT clone.

*a. Selection of uninfected parts*    Some parts of a plant may be infected and others not. Soilborne organisms, such as *Phytophthora,* can be avoided by taking only tip cuttings from tall growing stock plants that are off the ground. Likewise, using the apical portion of vegetative shoots and discarding lower portions can often avoid tissue infected with organisms that cause vascular wilt (*Fusarium, Verticillium,* and *Phytophthora*) (*5*).

*b. Shoot apex culture*    The terminal growing point of a plant is often free of virus and other pathogens even if the rest of the plant is infected. Excision and aseptic culture of this small segment can be the start of an SPT clone. Examples of plants where shoot apex culture has been used to eliminate viruses include (*74, 81*):

> *flower crop:* chrysanthemum, carnation, dahlia, geranium, orchid spp., amaryllis, freesia, gladiolus, iris, lily, narcissus, nerine
> *vegetable:* garlic, cauliflower, brussels sprouts, rhubarb, potato, sweet potato
> *fruit crop:* apple, stone fruits, gooseberry, strawberry, raspberry, grape, citrus, pineapple, banana
> *other:* sugar cane, hops, cassava, taro

For woody plant cultivars difficult to start as cuttings, in vitro micro-grafting possibly can be used. In all cases, however, shoot-tip culture may not eliminate all pathogens; follow-up index tests should be done for verification (*10, 52, 81, 95*).

*c. Heat treatments of short duration*    This procedure has been used widely in freeing many kinds of plants or plant parts, bulbs, and seeds from fungi, bacteria, and nematodes (*3, 59*). The plant is subjected to temperatures high enough and of long enough duration to destroy the pathogen but not so high as to kill the plant. Only one plant need survive to provide a beginning of pathogen-free material. (Less drastic treatments are needed where all the planting stock is to be treated, as, for instance, in commercial planting of gladiolus corms.) Treatment may vary with different plants from 43.5° to 57° C (110° to 135° F) for one-half to four hours. Methods include hot water soaking and exposure to hot air or to aerated steam. For example, the causal agent for *Pierce's disease* in grapevines can be eliminated by immersion of the propagating wood in hot water for three hours at 45° C (113° F), thus preventing the possibility of moving this disease to new areas by diseased wood (*45*). Hardening-off vegetative material or reducing the moisture content in seeds is generally desirable prior to such heat treatment.

*d. Heat treatments of low intensity and long exposure*    This procedure will free many kinds of plants from virus diseases (*3*). Plants are grown in containers until they are well established and have ample carbohydrate reserves; then they are held in a heat chamber at 37° to 38°C (98° to 100° F) for two to four weeks or longer. Buds may be taken from the treated plants and inserted into "virus-free" rootstocks, or cuttings may be taken and rooted. Another procedure is to insert the bud from the infected plant into a rootstock prior to treatment.

*e. Combinations of heat treatment and shoot apex culture*    In various plants certain viruses are heat tolerant and are not eliminated by the usual heat treatments, and shoot apex culture alone will not eliminate viruses from all plants. A combination of the two methods has been successful in eliminating the viruses. Infected plants are heat-treated at 38° to 40° C (100° to 103° F) for four to six weeks then apical shoot tips about 0.33 mm long are aseptically removed from the heat-treated plants and grown in in vitro test tubes containing a nutrient medium. Subsequent growth from these shoot tips results in mother plants that become the source for SVT clones.

*f. Chemical treatment of propagation stock*    This procedure can sometimes be used to eradicate externally carried pathogens. For example, white calla lily can be freed of the *Phytophthora* root fungus by soaking the rhizomes for an hour in a weak formaldehyde solution.

*g. Growing seedlings (apomictic and nonapomictic)*    Since many viruses are not transmitted through seeds (although there are exceptions), seedling growing may be used to produce new SVT clones to replace older cultivars that have deteriorated from virus infections. Growing apomictic seedlings, in species where they occur, provides a means of both preserving an existing clone and eliminating viruses from the clone. This procedure has been utilized in citrus. Nucellar seedlings are the basis of improved SVT clones of old cultivars (*16*). In citrus species that do not naturally produce nucellar embryos, aseptic culture of nucellar tissue has induced nucellar embryos to form (*82*).

## Maintenance of Propagation Stock

Once a *specific pathogen-tested* (SPT), true-to-type source has been obtained, plants should be multiplied and maintained under conditions that will prevent infection and at the same time allow detection of any significant genetic change from the original source. A planting either in the greenhouse or in the field or orchard in which propagation stock is maintained under control especially as a source for propagation is termed a **stock block.** In the case of fruit and nut trees, the stock block may be a **scion orchard** used mainly as a source for propagating material. Such blocks allow the propagator to manage the plants for maximum scion yield and to control potential pathogen and genetic problems.

Preservation of an SPT propagation source has three aspects: *isolation, sanitation,* and *periodic inspection and testing.* Isolation is necessary to separate the plants from infective agents. A minimum requirement is to separate the propagation source block from the propagation area. It can best be achieved by growing plants in containers and keeping them in an insect-proof screenhouse or a special greenhouse with restricted entry. Tree crops should be at least one-half mile from potential sources of virus diseases. Strawberry nurseries in California are grown in isolated mountain valleys far from production areas.

Sanitation is essential to eliminate disease agents in or on unclean equipment, tools, soil mixes, and so on (see chapter 2).

Source plants should be tested periodically to detect evidence of contamination and to see that the plants conform to their original standards. Visual inspections are used in conjunction with indexing and culturing methods. Specific procedures must be established for individual crops.

Special programs are sometimes used to maintain and distribute SPT true-to-type plants. Registration and Certification programs are available for some crops. The U.S. Dept. of Agriculture maintains repositories for fruit and nut crops, both for established cultivars (39) and for germplasm purposes (102).

### Distribution Systems for Propagation Materials

The aim of distribution systems is to provide nursery plants that meet some minimum standard of cleanliness and genetic purity. These standards may be imposed by the propagator or established by a legally designated authority.

---

**Registration and Certification
of Specific Pathogen-Tested (SPT) Clones**

Programs to select and distribute SPT true-to-type clones have been developed for various clonally propagated species based on similar programs for seed-propagated crops (67). Examples are potato, strawberry, citrus, grape, apple, pear, and stone fruits. The details vary depending on the crop and the specific pathogens to be controlled.

The basic program involves some general terms and principles. **Nuclear stock,** usually from a single plant or meristem, is selected from within the cultivar and placed into a **foundation planting** (orchard, vineyard, field, test tube, greenhouse). Individual plants are **registered** by location. Usually only a limited number of plants are grown in this step since these are maintained to supply material to propagators for further increase, not to supply material for commercial use. If clones of more than one SPT source are utilized, separate identity should be maintained, at least until progeny tests show them to be identical.

An individual nursery will establish a separate **mother block** or **scion orchard** from registered stock coming from a foundation planting. These also have prescribed regulations for isolation, management, and inspection.

To multiply the supply of propagating material to produce the commercial crop, an **increase block,** with less stringent requirements, can be utilized in order to reduce the size of the mother blocks, which are more expensive to maintain.

**Certified stock** is produced for commercial sale. Propagation material comes primarily from mother blocks or increase blocks, but could come directly from foundation blocks. By convention, certified stock is identified by a *blue* tag label, foundation stock by a *white* tag, and increase block material by a *purple* tag.

---

**Terms Used to Designate Categories
of Clonally Propagated Cultivars**

**Clone:** a genetically uniform cultivar propagated by vegetative propagation originating from a single plant, either a seedling or a bud-mutation of a single member of the clone.

**Variant:** individual plant or part of a plant differing by some significant trait from other members of the clone. If the variant is shown to be due to genetic causes, it is called a **mutant.** This variant could give rise to a **bud-sport** and become the origin of a new cultivar if it is useful and stable after vegetative propagation.

**Source:** a term that refers to a specific origin of propagation material within a cultivar. For example, fruit tree and grape propagators may maintain a specific source through consecutive vegetative generations. Such sources have been called **budlines, strains,** or **subclones** (*109*). These may or may not represent actual differences in pathogen or genetic potential.

**SPT (specific pathogen-tested) or SVT (specific virus-tested) source:** a group of plants within a clone selected as free of specific pathogens, as determined by prescribed index tests. If this group has originated from a single bud, cutting, shoot tip or meristem, it should be called an **SPT (or SVT) clone.** The SPT clone should have a designated label to indicate its unique identity.

**Ortet:** a term used primarily in forestry to designate the original seedling plant from which vegetatively propagated plants are derived.

**Ramet:** a term that refers to the vegetatively propagated offspring of an ortet. One can refer to consecutive propagation from an original ortet as scion generations.

Propagators should be familiar with quarantine regulations and controls on disease pathogens that govern movement or sale of propagated plants. The development, maintenance, and distribution of SPT source material may increase production costs, but such added expense can be justified by protection to the propagator and consumer and the increased value of the propagated material. Furthermore, increased production from the nursery material may result.

If different source materials are utilized and many cultivars are propagated, maintaining both **cultivar** and **source identity** is important, not only in the nursery operation, but also in its distribution to the ultimate user. If a question arises as to the final quality of the stock, one would be able to trace the origin of the problem to the proper step in the propagation sequence and correct it promptly in the propagation sequence.

## THE PLANT PATENT LAW

An amendment to the United States Patent Law (*30*) was enacted in 1930 which enabled the originators of new plant forms in the U.S. to obtain patents on them. This amendment added impetus to the development and introduction of new and improved plants by establishing the possibility of monetary rewards to private plant breeders and alert horticulturists for valuable plant introductions. It has stimulated growers of fruit and ornamentals to be on the lookout for improved plant forms and has handsomely paid numerous fortunate or observant persons for their finds.

Essentially, what can be patented as stated in the statute is "any distinct and new variety of plant, including cultivated sports, mutants, hybrids, and newly found seedlings, other than a tuber-propagated plant or a plant found in an uncultivated state." For a new plant to be patentable, it must be one that has been asexually reproduced and can be so propagated commercially, as by cuttings, layering, budding, or grafting. The plant patent law also includes microorganisms. Protection is also given to some seed-propagated cultivars.

Characteristics which would cause a plant to be "distinct and new" and thus patentable include such things as growth habit; immunity from disease; resistance to cold, drought, heat, wind, or soil conditions; the color of the flower, leaf, fruit, or stem; flavor; productivity, including everbearing qualities in the case of fruits; storage qualities; form; and ease of reproduction.

The applicant for a plant patent must be the person who invented or discovered and subsequently reproduced the new cultivar of plant for which the patent is sought. If a person who was not the inventor applied for a patent, the patent, if it were obtained, would be void; in addition, such a person would be subject to criminal penalties for committing perjury. A plant found growing wild in nature is not considered patentable.

A plant patent issued to an individual is a grant consisting of the right to exclude others from propagation of the plant or selling or using the plant so reproduced. Essentially, it is a grant by the United States Government, acting through the Patent Office, to the inventor (or his heirs or assigns) of certain exclusive rights to his invention for a term of 17 years throughout the United States and its territories and possessions. A U.S. plant patent affords no protection in other countries. The mere fact that a patent has been issued on a new plant does not imply any endorsement by the government of high quality or merit. The only implication of a patent on a plant is that it is "distinct and new."

The U.S. Plant Variety Protection Act, which became effective in 1970, extends plant patent protection to certain sexually propagated cultivars which can be maintained as "lines," such as those of cotton, alfalfa, soybeans, marigolds, bluegrass, and others.

# REFERENCES

1. Ashby, E. 1949. Leaf shape. *Scient. Amer.* 181:22–24.

2. Asseyera, T. 1927. Bud mutations in the potato and their chimerical nature. *Jour. Gen.* 19:1–26.

3. Baker, K. F. 1962. Thermotherapy of planting material. *Phytopathology* 52:1244–55.

4. ———. 1972. Disease-free propagation in relation to standardization of nursery stock. *Proc. Inter. Plant Prop. Soc.* 21:191–98.

5. Baker, K. F., and P. A. Chandler. 1957. Development and maintenance of healthy planting stock. In *The U.C. system for producing healthy container-grown plants,* K. F. Baker, ed. Calif. Agr. Exp. Sta. Man. 23, pp. 217–36.

6. Banks, M. S. 1979. Plant regeneration from callus from two growth phases of English ivy, *Hedera helix* L. *Z. Pflanzenphysiol.* 92:349–53.

7. Bateson, W. 1916. Root cuttings, chimeras and "sports." *Jour. Gen.* 6:75–80.

8. Baur, E. 1909. Pfropfbastarde Periclinal Chimaeren and Hyperchimaeren. *Ber. Deuts. Bot. Ges.* 27:603–5.

9. Bonga, J. M. 1982. Vegetative propagation in relation to juvenility, maturity, and rejuvenation. In Bonga, J. M., and D. J. Durzan, eds., *Tissue culture of forest trees.* Amsterdam: Elsevier.

10. Boxus, P., and P. Druart. 1980. Micropropagation, an industrial propagation method of quality plants true-to-type at a reasonable price. In *Plant cell cultures; Results and perspectives,* F. Sala et al., eds. Amsterdam: Elsevier/North Holland: Biomed. Press, pp. 265–69.

11. Bringhurst, R. S., and V. Voth. 1956. Strawberry virus transmission by grafting excised leaves. *Plant Disease Rpt.* 40(7):596–600.

12. Bringhurst, R. S., V. Voth, and D. van Hook. 1960. Relationship of root starch content and chilling history to performance of California strawberries. *Proc. Amer. Soc. Hort. Sci.* 75:373–81.

13. Brink R. A. 1962. Phase change in higher plants and somatic cell heredity. *Quart. Rev. Bio.* 37(1):1–22.

14. ———. 1973. Paramutation. *Ann. Rev. Genetics* 7:129–52.

15. Broertjes, C., and A. M. van Harten. 1978. *Application of mutation breeding methods in the improvement of vegetatively propagated plants.* Amsterdam: Elsevier.

16. Cameron, J. W., R. K. Soost, and H. B. Frost. 1957. The horticultural significance of nucellar embryony in citrus. In *Citrus virus diseases,* J. M. Wallace, ed. Berkeley, Calif.: Univ. of Calif. Div. of Agr. Sci., pp. 191–96.

17. Chacko, E. K., P. R. Kohli, R. Doreswamy, and G. S. Randhawa. 1974. Effect of (2-chloroethyl) phosphoric acid on flower induction in juvenile mango (*Mangifera indica*) seedlings. *Phys. Plant.* 32:188–90.

18. Chaleff, R. S. 1981. *Genetics of higher plants: Applications of cell culture.* Cambridge: Cambridge Univ. Press.

19. Clark, M. F., and A. N. Adams. 1977. Characteristics of the microplate method of enzyme-linked immunosorbent assay for the detection of plant viruses. *Jour. Gen. Virol.* 34:475–83.

20. Crane, M. B. 1936. Note on a periclinal chimera in the potato. *Jour. Gen.* 32:73–77.

21. Darrow, G. M. 1928. Notes on thornless blackberries. *Jour. Hered.* 19:139–42.

22. ———. 1931. A productive thornless sport of the Evergreen blackberry. *Jour. Hered.* 22:404–6.

23. Darrow, G. M., R. A. Gibson, W. E. Toenjes, and H. Dermen. 1948. The nature of giant apple sports. *Jour. Hered.* 39:45–51.

24. Davis, R. E., and R. F. Whitcomb. 1971. Mycoplasmas, rickettsiae, and chlamydiae: Possible relation to yellows diseases and other disorders of plants and insects. *Ann. Rev. Phytopath.* 9:119–54.

25. Dayton, D. F. 1969. Genetic heterogeneity in the histogenic layers of apple. *Jour. Amer. Soc. Hort. Sci.* 94(6):592–95.

26. Dermen, H. 1954. Colchiploidy in grapes. *Jour. Hered.* 45:159–72.

27. ———. 1960. Nature of plant sports. *Amer. Hort. Mag.* 39:123–73.

28. Diener, T. O. 1979. Viroids and viroid diseases. New York: John Wiley.

29. Doi, Y., M. Teranaka, K. Yora, and H. Asuyama. 1967. Mycoplasma or PLT grouplike microorganisms found in the phloem elements of plants infected with mulberry dwarf, potato witches' broom, aster yellows, or paulownia witches' broom. *Ann. Phytopath. Soc. Jap.* 33:259–66.

30. Donahue, L. J. 1980. Plant patents and legalities. *Proc. Inter. Plant Prop. Soc.* 30:414–21.

31. Doorenbos, J. 1954. "Rejuvenation" of *Hedera helix* in graft combinations. *Proc. Koninkl. Nod. Akad. Wetenschap. Ser. C* 57:99–102.

32. Duvick, D. N. 1978. Risks of monoculture via clonal propagation. In *Propagation of higher plants through tissue culture; A bridge between research and application,* K. W. Hughes et al., eds. U.S. Dept. of Energy, Tech. Information Center, pp. 73–84.

33. Einset, J., and C. Pratt. 1954. "Giant" sports of grapes. *Proc. Amer. Soc. Hort. Sci.* 63:251–56.

34. ———. 1959. Spontaneous and induced apple sports with misshapen fruit. *Proc. Amer. Soc. Hort. Sci.* 73:1–8.

35. Evans, D. A., and E. T. Paddock. 1979. Mitotic crossing-over in higher plants. In *Plant cell and tissue culture: Principles and applications,* W. R. Sharp et al., eds. Columbus: Ohio State Univ. Press, pp. 315–51.

36. Fincham, J. R. S., and G. R. K. Sastry. 1974. Controlling elements in maize. *Ann. Rev. Genetics* 8:15–50.

37. Franclet, A. 1979. Rajeunissement des Arbres adultes en vue de leur propagation vegetative. In *Micropropagation d'Arbres Forestiers.* AFOCEL, etudes et recherches no. 12, pp. 3–31.

38. Frazier, N. W., ed. 1970. *Virus diseases of small fruits and grapevines.* Berkeley: Div. of Agricultural Sciences, Univ. of Calif.

39. Fridlund, P. R. 1980. Maintenance and distribution of virus-free fruit trees. In *Proc. conf. on nursery production of fruit plants through tissue culture,* R. H. Zimmerman, ed. USDA Sci. and Education Administration, ARR–NE–11, pp. 11–22.

40. ———. 1970. Temperature effects on virus disease symptoms in some *Prunus, Malus,* and *Pyrus* cultivars. *Wash. Agr. Exp. Sta. Bul. 726.*

41. Frost, H. B. 1926. Polyembryony, heterozygosis, and chimeras in *Citrus. Hilgardia* 1:365–402.

42. ———. 1938. Nucellar embryony and juvenile characters in clonal varieties of citrus. *Jour. Hered.* 29:423–32.

43. Gabriel, W. J. 1979. Early flowering and seed production in plantations of sugar maple. *Tree planters notes* 30(3):11–13.

44. Gilmer, R. M., K. D. Brase, and K. G. Parker. 1957. Control of virus diseases of stone fruit nursery trees in New York. *New York Agr. Exp. Sta. Bul. 779.*

45. Goheen, A. C., G. Nyland, and S. K. Lowe. 1973. Association of a rickettsia-like organism with Pierce's disease of grapevines and alfalfa dwarf and heat therapy of the disease in grapevines. *Phytopathology* 63:341–45.

46. Green, M. M. 1978. Insertion mutants and the control of gene expression in *Drosophila melanogaster.* In *The clonal basis of development,* S. Subtelny et al., eds. New York: Academic Press, pp. 239–46.

47. Haberlandt, G. 1930. Das Wesen der *Crataegomespili* Sitzber. *Preuss. Akad. Wiss.* 20:374–94.

48. Hackett, W. P. 1983. Phase change and intra-clonal variability. *HortScience* 18:(in press).

49. Hampton, R. O. 1972. Mycoplasmas as plant pathogens: Perspectives and principles. *Ann. Rev. Plant Phys.* 23:389–418.

50. Hood, J. V., and W. J. Libby. 1978. Continuing effects of maturation state in radiata pine and a general maturation model. In *Propagation of higher plants through tissue culture; A bridge between research and application,* K. W. Hughes et al., eds. U.S. Dept. of Energy, Tech. Information Center, pp. 220–32.

51. Hooper, G. R., and M. L. Lacy. 1971. Mycoplasma: New causes for old diseases in Michigan. *Mich. Agr. Ext. Bul. E-644.*

52. Joiner, N., ed. 1981. *Foliage plant production.* Englewood Cliffs, N.J.: Prentice-Hall.

53. Kester, D. E. 1976. The relationship of juvenility to plant propagation. *Proc. Inter. Plant Prop. Soc.* 26:71–84.

54. ———. 1983. The clone in horticulture. *HortScience* 18:(in press).

55. Kirk, J. T. O., and R. A. E. Tilney-Bassett. 1978. *The plastids.* 2nd ed. San Francisco: W. H. Freeman & Company Publishers.

56. Knight, T. A. 1795. Observations on the grafting of trees. *Phil. Trans. Roy. Soc., London* 85:290.

57. Krenke, N. P. 1933. *Wundkompensation Transplantation und Chimären bei Pflanzen* (translated from Russian). Berlin: Springer.

58. Kuhns, L. J., and T. A. Fretz. 1978. Distinguishing rose cultivars by polyacrilamide gel electrophoresis. II. Isozyme variation among cultivars. *Jour. Amer. Soc. Hort. Sci.* 103:509–16.

59. Kunkel, L. O. 1936. Heat treatments for the cure of yellows and other virus diseases of peach. *Phytopathology* 26:809–30.

60. Lambert, D. H., R. F. Stouffer, and H. Cole, Jr. 1979. Stunting of peach seedlings following soil fumigation. *Jour. Amer. Soc. Hort. Sci.* 104:433–35.

61. Libby, W. J., and J. V. Hood. 1976. Juvenility in hedged radiata pine. *Acta Hort.* 56:91–98.

62. Linderman, R. G. 1978. Mycorrhizae in relation to rooting cuttings. *Proc. Inter. Plant Prop. Soc.* 28:128–32.

63. Maliga, P. 1980. Isolation, characterization, and utilization of mutant cell lines in higher plants. In *Perspectives in plant cell and tissue culture,* I. K. Vasil, ed. Int. Rev. of Cyt. Supp. 11A. New York: Academic Press.

64. Maronek, D. M., and J. W. Hendrix. 1978. Mycorrhizal fungi in relation to some aspects of plant propagation. *Proc. Inter. Plant Prop. Soc.* 28:506–14.

65. Martin, B., and G. Quillet. 1974. Bouturage des arbres forestiers au Congo. *Rev. Bois et Forets des Tropiques* 154:41–57; 155:15–33; 156:39–60; 157:21–39.

66. McClintock, B. 1978. Development of the maize endosperm as revealed by clones. In *The clonal basis of development,* S. Subtelny et al., eds. New York: Academic Press, pp. 217–38.

67. Mather, S. M. 1961. Nursery stock registration and certification in California. *Calif. Dept. of Agr. Quart. Bul.* L(3):173–84.

68. Meins, F., Jr., and A. N. Binns. 1979. Cell determination in plant development. *Bio-Science* 29:221–25.

69. ———. 1978. Epigenetic clonal variation in the requirement of plant cells for cytokinin. In *The clonal basis for development,* S. Subtelny et al., eds. New York: Academic Press, pp. 185–201.

70. Meredith, C. P., and P. S. Carlson. 1978. Genetic variation in cultured plant cells. In *Propagation of higher plants through tissue culture: A bridge between research and application,* K. W. Hughes et al., eds. U.S. Dept. of Energy, Tech. Information Center, pp. 166–76.

71. Miller, E. V. 1954. The natural origins of some popular varieties of fruit. *Econ. Bot.* 8:337–48.

72. Molisch, H. 1938. *The longevity of plants* (1928). Lancaster, Pa.: E. Fulling (English translation).

73. Mullins, M. G., Y. Nair, and P. Sampet. 1979. Rejuvenation *in vitro:* Induction of juvenile characters in an adult clone of *Vitis vinifera. Ann. Bot.* 44:623–27.

74. Murashige, T. 1978. The impact of plant tissue culture in agriculture. In *Frontiers of plant tissue culture,* T. A. Thorpe, ed. Calgary: Univ. of Calgary Press, pp. 15–26.

75. Musik, T. J., and H. J. Cruzado. 1958. Transmission of juvenile rooting ability from seedlings to adults of *Hevea braziliensis. Nature* 181:1288.

76. Nelson, S. H. 1977. Loss of productivity in clonal apple rootstocks. *Proc. Inter. Plant Prop. Soc.* 27:350–55.

77. Nyland, G., A. C. Goheen, S. K. Lowe, and H. C. Kirkpatrick. 1973. The ultrastructure of a rickettsialike organism from a peach tree affected with phony disease. *Phytopathology* 63(10):1275–78.

78. Olmo, H. P. 1952. Breeding tetraploid grapes. *Proc. Amer. Soc. Hort. Sci.* 59:285–90.

79. ———. 1936. Bud mutation in the vinifera grape. II. Sultanina gigas. *Proc. Amer. Soc. Hort. Sci.* 33:437–39.

80. Passecker, F. 1949. Zur Frage der jugend Formen der Apfel. *Zuchter* 19:311.

81. Quak, F. 1977. Meristem culture and virus-free plants. In *Applied and fundamental aspects of plant cell, tissue, and organ culture,* J. Reinert and Y. P. S. Bajaj, eds. Berlin, Springer-Verlag, pp. 598–615.

82. Rangan, T. S., T. Murashige, and W. P. Bitters. 1968. *In vitro* initiation of nucellar embryos in monoembryonic *Citrus. HortScience* 3:226–27.

83. Robinson, L. W., and P. F. Wareing. 1969. Experiments on the juvenile-adult phase change in some woody species. *New Phytol.* 68:67–78.

84. Schaffelitzky de Muckadell, M. 1959. Investigations on aging of apical meristems in woody plants and its importance in silviculture. *Forstl. Forsogov. Danm.* 25:310–455.

85. Schwabe, W. W. 1976. Applied aspects of juvenility and some theoretical considerations. *Acta Hort.* 56:45–56.

86. Shamel, A. D., and C. S. Pomeroy. 1936. Bud mutations in horticultural crops. *Jour. Hered.* 27:487–94.

87. Shamel, A. D., C. S. Pomeroy, and R. E. Caryl. 1929. Bud selection in the Washington Navel orange; progeny tests of limb variations. *USDA Tech. Bul. 123.*

88. Shull, C. H. 1912. "Phenotype" and "clone." *Science* N. S. 35:182–83.

89. Soost, R. K., J. W. Cameron, W. P. Bitters, and R. G. Platt. 1961. Citrus bud variation. *Calif. Citrograph* 46:176, 188–93.

90. Stern, W. T. 1943. The use of the term "clone." *Jour. Roy. Hort. Soc.* 74:41–47.

91. Steward, F. C., L. M. Blakely, A. E. Kent, and M. A. Mapes. 1963. Growth and organization in free cell colonies. In *Brookhaven symp. in bio.* No. 16, pp. 73–88.

92. Steward, F. C., and A. D. Krikorian. 1978. Problems and potentialities of cultured plant cells in retrospect and prospect. In *Plant cell and tissue culture: Principles and applications,* W. R. Sharp et al., eds. Columbus: Ohio State Univ. Press, pp. 221–62.

93. Stewart, R. N. 1978. Ontogeny of the primary body in chimeral forms of higher plants. In *The clonal basis of development,* S. Subtelny et al., eds. New York: Academic Press, pp. 131–60.

94. ———. 1965. The origin and transmission of a series of plastogene mutants in *Dianthus* and *Euphorbia. Genetics* 52:925–47.

95. Stone, O. M. 1978. The production and propagation of disease free plants. In *Propagation of higher plants through tissue culture: A bridge between research and application,* K. W. Hughes et al., eds. U.S. Dept. of Energy, Tech. Information Center, pp. 25–34.

96. Stout, A. B. 1940. The nomenclature of cultivated plants. *Amer. Jour. Bot.* 27:339–47.

97. Stout, G. L. 1962. Maintenance of "pathogen-free" planting stock. *Phytopathology* 52:1255–58.

98. Stoutemyer, V. T. 1962. The control of growth phases and its relation to plant propagation. *Proc. Plant Prop. Soc.* 12:260–64.

99. ———. 1937. Regeneration in various types of apple wood. *Iowa Agr. Exp. Sta. Res. Bul.* 220:308–52.

100. Symons, R. H. 1980. Rapid indexing of sunblotch viroid in avocados and of exocortis viroid in citrus. *Proc. Inter. Plant Prop. Soc.* 30:578–83.

101. Tanaka, T. 1927. Bizzarria—a clear case of periclinal chimera. *Jour. Gen.* 18:77–85.

102. Thompson, M. M. 1981. Utilization of fruit and nut germplasm. *HortScience* 16:132–35.

103. Tilney-Bassett, R. A. 1963. The structure of periclinal chimeras. *Heredity* 18:265–85.

104. Tincker, M. A. H. 1945. Propagation, degeneration, and vigor of growth. *Jour. Roy. Hort. Soc.* 70:333–37.

105. Tufts, W. P., and C. J. Hansen. 1931. Variation in shape of Bartlett pears. *Proc. Amer. Soc. Hort. Sci.* 28:627–33.

106. Tolley, I. S. 1975. A technique for the accelerated production of commercially acceptable citrus clones from seed. *Proc. Inter. Plant Prop. Soc.* 25:294–97.

107. U.S. Dept. Agr. 1951. *Virus and other disorders with virus-like symptoms of stone fruits in North America.* USDA Agr. Handbook No. 10. Washington, D.C.: U.S. Govt. Printing Office.

108. Upshall, W. H. 1970. *North American apples: Varieties, rootstocks, outlook.* E. Lansing, Mich.: Michigan State Univ. Press.

109. van Oosten, H. J., and H. H. van der Borg, eds. 1977. Symposium on clonal variation in apple and pear. *Acta Hort.* 75:1–185.

110. Vasil, V., and A. C. Hildebrandt. 1965. Differentiation of tobacco plants from single, isolated cells in microculture. *Science* 150:889–92.

111. Vaughn, K. 1983. Chimeras: Problems in propagation. *HortScience* 18:(in press).

112. Verkade, S. D., and D. F. Hamilton. 1980. Mycorrhizae and their uses in the nursery. *Proc. Inter. Plant Prop. Soc.* 30:353–62.

113. Visser, T. 1966. Juvenile phase and growth of apple and pear seedlings. Meded. 224. *Inst. voor de Vereh, vontuinb. Wageningen* (Nederland).

114. Wareing, P. F. 1960. Problems of juvenility and flowering in trees. *J. Linn. Soc. Bot.* 56:282–89.

115. Webber, H. J. 1903. New horticultural and agricultural terms. *Science* N.S., 18:501–2.

116. Whiting, J. R., and W. J. Hardie. 1981. Yield and compositional differences between selections of grapevine cv. Cabernet Sauvignon. *Amer. Jour. Enol. and Vit.* 32:212–18.

117. Wilhelm, S. 1962. Symposium on pathogen-free stock. *Phytopathology* 52:1234–35.

118. Winkler, H. 1907. Uber pfropfbastarde und pflanzliche Chimaren. *Ber. Deuts. Bot. Ges.* 25:568–76.

119. ———. 1910. Uber die Nachkommenschaft der *Solanum* Pfropfbastarde und die Chromosomenzahlen ihrer Keimzellen. *Zeits. f. Bot.* 2:1–38.

120. Wright, J. W. 1976. *Introduction to forest genetics.* New York: Academic Press.

121. Zimmerman, R. H. 1971. Flowering in crabapple seedlings; methods of shortening the juvenile phase. *Jour. Amer. Soc. Hort. Sci.* 96(4):404–11.

122. ———. 1972. Juvenility in woody plants: A review. *HortScience* 7:447–55.

123. ———. 1973. Juvenility and flowering in fruit trees. *Acta Hort.* 34:139–42.

124. Zohary, D., and P. Spiegel-Roy. 1975. Beginnings of fruit growing in the old world. *Science* 187(4174):319–27.

## SUPPLEMENTARY READING

BAKER, K. F., ed. 1957. The U.C. system for producing healthy container-grown plants. *Calif. Agr. Exp. Sta. Man. 23.*

DOORENBOS, J. 1965. Juvenile and adult phases in woody plants. In *Handbuch der Pflanzenphysiologie,* Vol. 15 (Part I), pp. 1222–35. Berlin: Heidelberg; New York: Springer-Verlag.

FRAZIER, N. W., ed. 1970. *Virus diseases of small fruits and grapevines.* Berkeley: Div. of Agr. Sci., Univ. of Calif.

KENNETH, N., and J. KATAN, eds. 1970. Symposium on production of heathy plants by therapeutic and other methods and their maintenance and use. *Proc. XVIII Inter. Hort. Cong. 3.*

KIRK, J. T. O., and R. A. E. TILNEY-BASSETT. 1978. *The plastids.* Amsterdam: Elsevier.

KNEEN, O. H. 1948. Patent plants enrich our world. *National Geographic* 93:357–78.

KOENIG, R., ed. 1980. Fifth international symposium on virus diseases of ornamental plants. *Acta Hort.* 110:1–334.

MARSTON, M. E. 1955. The history of vegetative propagation. *Report 14th International Horticultural Congress,* Vol. 2, pp. 1157–64.

NEILSON-JONES, W. 1969. *Plant chimeras* (2nd ed.). London: Methuen & Company.

NYLAND, G., and A. C. GOHEEN, 1969. Heat therapy of virus diseases of perennial plants. *Ann. Rev. Phytopathol.* 7:331-54.

POSNETTE, A. F., ed. 1963. Virus diseases of apples and pear. *Tech. Comm. 30.* Bucks, England: Commonwealth Agricultural Bureaux, Farnham Royal.

Symposium on the clone in horticulture. 1983. *HortScience* 18:(in press).

Symposium on viruses in fruit crops. 1977. *HortScience* 12(5):463-90.

WEISS, F. E. 1930. The problem of graft hybrids and chimaeras. *Biol. Rev.* 5:231-71.

In propagation by **stem** and **leaf-bud cuttings,** it is only necessary that a new adventitious root system[1] be formed, since a potential shoot system—a bud—is already present. **Root cuttings** must initiate a new shoot system—from an adventitious bud[2]—as well as an extension of the existing root piece, often by the production of adventitious roots. In **leaf cuttings,** both a new root and a new shoot system must be regenerated.

This capacity for regenerating the entire plant structure, a property of essentially all the plant's living cells, is demonstrated in the various cell and tissue systems described in chapters 16 and 17. This capacity depends on two fundamental characteristics of plant cells. One is **totipotency,** which means that each living plant cell contains the genetic information necessary for reconstituting all the plant parts and functions (*268*). The second is **dedifferentiation,** the capability of mature cells to return to a meristematic condition and develop a new growing point. Since these two characteristics are more pronounced in some cells and plant parts than in others, the propagator must do some manipulation to provide the proper conditions for rooting.

## ADVENTITIOUS ROOT FORMATION

Adventitious roots form naturally on various kinds of plants. For example, ''brace'' roots develop on corn, pandanus, and other monocots, arising from the

[1] **Adventitious roots** are those arising from any plant part other than the seedling root and its branches.

[2] **Adventitious buds** (and shoots) are those arising from any plant part other than terminal, lateral, or latent buds on stems.

# 9

# Anatomical and Physiological Basis of Propagation by Cuttings

**Figure 9-1** The ultimate in adventitious root production is shown on this banyan tree (*Ficus benghalensis*) at Hilo, Hawaii. Such roots arise from branches, extend downward, and grow into the soil.

intercalary regions at the base of internodes. Banyan trees (Figure 9-1) produce huge aerial roots that grow to and into the ground. Many plants that develop from rhizomes, bulbs, and other such structures develop adventitious roots.

Adventitious roots are of two types: **preformed roots** and **wound roots.** Preformed roots develop naturally on stems while they are still attached to the parent plant but do not emerge until the stem piece is severed. Wound roots develop only after the cutting is made, a response to the wounding effect in preparing the cutting. When a cutting is made, living cells at the cut surfaces are injured and the dead conducting cells of the xylem are opened and exposed. In the subsequent healing and regeneration process three steps occur:

First, as the outer injured cells die, a necrotic plate forms and seals the wound with a corky material (suberin), and plugs the xylem with gum. This plate protects the cut surfaces from desiccation.

Second, living cells behind this plate begin to divide after a few days and a layer of parenchyma cells (callus) may form.

Third, certain cells in the vicinity of the vascular cambium and phloem begin to initiate adventitious roots.

The anatomical changes that can be observed in the stem during root initiation can be divided into four stages:

1. **Dedifferentiation** of specific mature cells.

2. **Formation of root initials** from certain cells near vascular bundles, or vascular tissue, which have become meristematic by dedifferentiation.

3. Subsequent development of these root initials into organized **root primordia.**

4. Growth and **emergence** of the root primordia outward through other stem tissue plus the formation of vascular connections between the root primordia and the conducting tissues of the cutting itself.

The precise location inside the stem where adventitious roots originate has intrigued plant anatomists for centuries. Probably the first study of this phenomenon was made by a French dendrologist, Duhamel du Monceau, in 1758 (*53*); a great many subsequent studies have covered a wide range of plant species (*78*).

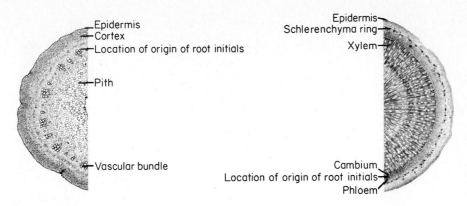

**Figure 9–2** Stem cross sections showing the usual location of origin of adventitious roots. *Left:* young, herbaceous, dicotyledonous plant. *Right:* young, woody plant.

In **herbaceous plants** adventitious roots usually originate just outside and between the vascular bundles (*204*), but the tissues involved at the site of origin can vary widely depending upon the kind of plant (Figure 9–2). For example, in the tomato, pumpkin (*199*), and mung bean (*15*), adventitious roots arise in the phloem parenchyma; in *Crassula* they arise in the epidermis (*175*); and in coleus they originate from the pericycle (*25*).

Adventitious roots in castor bean (*Ricinus communis*) cuttings arise between the vascular bundles, as shown in Figure 9–3. Root initials in carnation cuttings arise in a layer of parenchymatous cells inside a fiber sheath; the developing root tips, upon reaching this band of impenetrable fiber cells, do not push through it, but turn downward, emerging from the base of the cutting (*236*).

**Figure 9–3** Adventitious root primordia (arrows) arising laterally and adjacent to vascular bundles in the castor bean (*Ricinus communis*), a herbaceous plant. Vascular connections will develop between the adventitious roots and the plant's vascular bundles. Epidermis is at top, pith at bottom. From Priestley and Swingle (*204*).

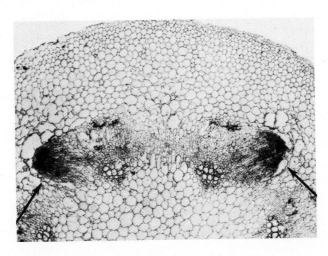

**Figure 9–4** Tissues involved in adventitious root formation in 'Brompton' plum hardwood cuttings. cx = cortex; ppf = primary phloem fibers; rp = root primordium; p = phloem; r = rays; c = cambium; x = xylem. Courtesy A. Beryl Beakbane, East Malling Research Station (*10*).

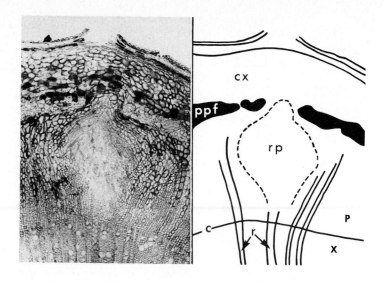

In **woody perennial plants,** where one or more layers of secondary xylem and phloem are present, adventitious roots in stem cuttings usually originate from living parenchyma cells, primarily in the young, secondary phloem (Figure 9–4), but sometimes from such other tissues as vascular rays, cambium, phloem, lenticels, or pith (*37, 42, 77, 172, 180, 272*).

Generally, the origin and development of adventitious roots takes place next to and just outside the central core of vascular tissue. Upon emergence from the stem (Figure 9–5), the adventitious roots have developed a root cap and the usual root tissues, as well as a complete vascular connection with the originating stem (*62*). Adventitious roots (and shoots) usually arise within the stem (endogenously) near the vascular cylinder, just outside the cambium.

The time for root initials to develop after cuttings are placed in the propagating bed varies widely. In one study (*236*) they were first observed microscopically after three days in the chrysanthemum, five days in the carna-

**Figure 9–5** Emergence of adventitious roots in plum stem cuttings. Observe the tendency of the roots to form in longitudinal rows, which appear directly below buds.

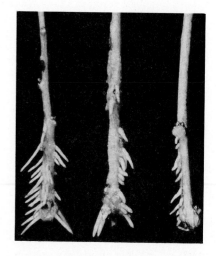

**Figure 9-6** Burr knots developing from preformed root initials at the base of shoots of the 'Colt' cherry rootstock. Photographed at East Malling Research Station, England.

tion (*Dianthus caryophyllus*), and seven days in the rose (*Rosa*). Visible roots emerged from the cuttings after ten days for the chrysanthemum, but three weeks were required for the carnation and rose.

*Preformed, or latent, root initials (27, 160, 263)* generally lie dormant until the stems are made into cuttings and placed under environmental conditions favorable for further development and emergence of the primordia as adventitious roots. In *Populus* × *robusta* they form in the stems in mid-summer and then emerge from cuttings made the following spring (*232*). In some species they develop into aerial roots on the intact plant and become quite prominent. Such preformed root initials occur in a number of easily rooted genera, such as willow (*Salix*), hydrangea (*Hydrangea*), poplar (*Populus*), jasmine (*Jasminum*), currant (*Ribes*), citron (*Citrus medica*), and others (*78*). The position of origin of these preformed root initials is essentially the same as that of other adventitious roots (*26, 159*). In some of the clonal apple and cherry rootstocks and in old trees of some apple and quince cultivars, these preformed latent roots cause swellings, called **burr knots** (*215, 251*), as shown in Figure 9-6. Species with preformed root initials generally root rapidly and easily, but cuttings of many species without such root initials root just as easily.

In willow, latent root primordia can remain dormant, embedded in the inner bark for years if the stems remain intact on the tree (*26*). Their location can be observed by peeling off the bark and noting the protuberances on the wood, with corresponding indentations on the inside of the removed bark.

## CALLUS

Some amount of callus usually develops at the basal end of the cutting when it is placed under environmental conditions favorable for rooting. **Callus** is an irregular mass of parenchyma cells in various stages of lignification. Callus growth proliferates from young cells at the base of the cutting in the region of the vascular cambium, although cells of the cortex and pith may also contribute to its formation. Frequently the first roots appear through the callus, leading to the belief that callus formation is essential for rooting. In most kinds of plants, the formation of callus and the formation of roots are independent of each other;

**Figure 9–7** Adventitious root (arrow) forming in callus tissue at base of a *Hedera helix* (adult phase) cutting, as seen in longitudinal section.
P = phloem; $X^1$ = primary xylem; $X^2$ = secondary xylem; V = vessels. Courtesy R. M. Girouard (*79*).

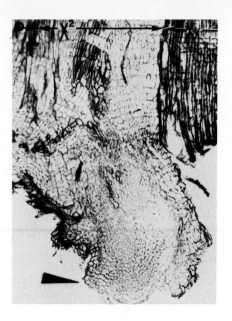

that they often occur simultaneously is due to their dependence upon similar internal and environmental conditions.

In some species, however, callus formation apparently is a precursor of adventitious root formation. For example, in *Pinus radiata* (*24*), *Sedum* (*296*), and *Hedera helix* (*79*) (adult phase) adventitious roots originate in the callus tissue that has formed at the base of the cutting (Figure 9–7).

There is evidence that the **pH**[3] of the rooting medium can influence the type of callus produced, which in turn can affect emergence of newly formed adventitious roots. In studies (*38, 39*) using stem cuttings of balsam poplar, at pH 6.0 the callus cells were large and somewhat soft and the cuttings rooted readily. With increasing alkalinity, the callus masses became smaller until at pH 11.0, the callus cells were small and compactly arranged with a firm calcareous structure. Such cuttings did not root although upon sectioning, well-formed root primordia were found beneath the callus.

## STEM STRUCTURE AND ROOTING

The development of a continuous **sclerenchyma ring** (Figure 9–8) between the phloem and cortex, exterior to the point of origin of adventitious roots, which is often associated with maturation, possibly constitutes an anatomical barrier to rooting. In a study of olive stem cuttings (*32, 33*), such a ring was associated with types of cuttings difficult to root, while easily rooted types were characterized by discontinuity of this sclerenchyma ring. Leafy cuttings of difficult-to-root types having a continuous schlerenchyma ring, when placed under mist for rooting, showed active proliferation of parenchyma ray cells,

---

[3] A measure of relative acidity or alkalinity: A pH of 7.0 is neutral; values below 7.0 indicate increasing acidity; values above 7.0 indicate increasing alkalinity.

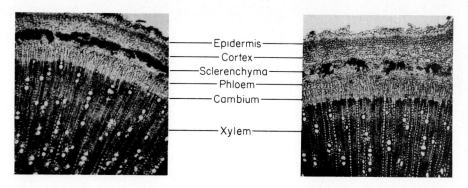

Epidermis
Cortex
Sclerenchyma
Phloem
Cambium

Xylem

**Figure 9–8**    Transverse sections of olive stem cuttings, × 30. *Left:* the continuous sclerenchyma ring is believed by some to constitute a barrier to emergence of adventitious roots. *Right:* the sclerenchyma ring is broken into groups of fibers separated by parenchyma cells, which would permit emergence of roots without difficulty. Courtesy Ciampi and Gellini (*32*).

resulting in breaking of the continuity of the sclerenchyma ring, making rooting possible in anatomically unsuitable stems. Instances of such associations between shy-rooting cultivars and the presence of a heavily lignified continuous sclerenchyma ring have been noted in cuttings of other species (*10*).

While a sheath of lignified tissue in stems may in some cases act as a mechanical barrier to root emergence, there are so many exceptions that this certainly cannot be a primary cause of rooting difficulty. Moreover, auxin treatments and rooting under mist (*170*) cause considerable cell expansion and proliferation in the cortex, phloem, and cambium, resulting in breaks in continuous sclerenchyma rings, yet in shy-rooting cultivars of several fruit species there is still no formation of root initials.

Difficult-to-root cuttings of the mature stage of English ivy, *Hedera helix,* show intense groups of discontinuous schlerenchyma fibers in the cortex, but adventitious roots have no difficulty growing through them; again in this case, some other factor must be the cause for the low rooting performance (*80*).

Easily rooted carnation cultivars have a band of sclerenchyma present in the stems, yet the developing root primordia emerge from the cuttings by growing downward and out through the base (*236*). In other plants in which an impenetrable ring of sclerenchyma could block root emergence this same possibility is open. Rooting is more likely to be related to the actual formation of root initials than to the mechanical restriction of a sclerenchyma ring barring root emergence.

Certain types of stem structure or tissue relationships within the stem seem to be more favorable to the initiation of root primordia than others. This is shown by studies (*176*) with the easily rooted citron (*Citrus medica*), which produces roots profusely from preformed root initials along the entire stem after a short time in the rooting medium, and the sour orange (*C. aurantium*), which forms only a few roots at the base of the cutting after several weeks. When the difficult-to-root bark piece of the sour orange was grafted on the citron stem—where it would presumably be able to obtain any translocatable essential rooting factors from the leaves of the easily rooted citron—it still failed to form roots more readily than it did normally.

Adventitious root formation may be dependent on certain inherent non-translocatable factors as determined by the genotype of the individual cells of the tissue. It is likely, however, that interactions between certain fixed or nonmobile factors located within the cells—perhaps certain enzymes—and easily conducted nutrients and endogenous rooting factors take place to establish conditions favoring root initiation.

## LEAF CUTTINGS

Many plant species, including both monocots and dicots, can be propagated by leaf cuttings (*88*). Although the origin of new shoots and new roots in leaf cuttings is quite varied, they generally develop from the so-called primary or secondary meristems, the latter type being the most common.

**Primary (preformed) meristems** are groups of cells directly descended from embryonic cells that have never ceased to be involved in meristematic activity.

**Secondary (wound) meristems** are groups of cells that have differentiated and functioned in some mature tissue system and then again resumed meristematic activity.

### Leaf Cuttings with Primary Meristems

In detached leaves of *Bryophyllum*, small plants arise from the notches around the leaf margin (see Figure 10–16). These small plants originate from so-called foliar "embryos," formed in the early stages of leaf development from small groups of cells at the edges of the leaf. As the leaf expands, a foliar embryo develops until it consists of two rudimentary leaves with a stem tip between them, two root primordia, and a "foot" that extends toward a vein (*108, 295*). As the leaf matures, cell division in the foliar embryo ceases, and it remains dormant. If the leaf is detached and placed in close contact with a moist rooting medium, the young plants rapidly break through the leaf epidermis and become visible in a few days. Roots extend downward, and after several weeks many new independent plants form while the original leaf dies. The new plants thus develop from latent primary meristems—from cells that have not assumed mature characteristics. Such production of new plants from leaf cuttings by the renewed activity of primary meristems is found, too, in other species, as the piggy-back plant (*Tolmiea*) and walking fern (*Camptosorus*).

### Leaf Cuttings with Secondary Meristems

In leaf cuttings of such plants as *Begonia rex, Sedum,* African violet (*Saintpaulia*), snake plant (*Sansevieria*), *Crassula,* and lily, new plants may develop from secondary meristems arising from mature cells at the base of the leaf blade or petiole as a result of wounding.

In *Lilium longiflorum* and *L. candidum,* the bud primordium originates in parenchyma cells in the upper side of the bulb scale, whereas the root primordium arises from parenchyma cells just below the bud primordium. Although the original scale serves as a source of food for the developing plant, the vascular system of the young bulblet is independent of that of the parent scale, which eventually shrivels and disappears (*275*).

In African violet new roots and shoots arise by the formation of meristematic cells from mature cells in the leaves. The roots are produced endogenously from thin-walled cells lying between the vascular bundles. The new shoots arise from cells of the epidermis and the cortex immediately below the epidermis. The roots emerge, form branch roots, and continue to grow for several weeks before the shoots appear. Although the original leaf supplies nutrient materials to the young plant, it does not become a part of the new plant (*189*).

In several species, for example, sweet potato, *Peperomia,* and *Sedum,* new roots and new shoots on leaf cuttings arise in callus tissue which develops over the cut surface through activity of secondary meristems. The petiole of leaf cuttings of *Sedum* forms a considerable pad of callus within a few days after the cuttings are made. Root primordia are organized within the callus tissue, and shortly thereafter four or five roots develop from the parent leaf. Following this, stem primordia arise on a lateral surface of the callus pad and develop into new shoots (*296*).

Adventitious roots form on leaves much more readily than do adventitious buds. In some plants, such as the India rubber fig (*Ficus elastica*) and the jade plant (*Crassula argentea*), the cutting must include a portion of the old stem containing an axillary bud because although adventitious roots may develop at the base of the leaf, an adventitious shoot is not likely to form. In fact, rooted leaves of some species will survive for years without producing an adventitious shoot. Treatments with a cytokinin such as benzyladenine may, however, initiate buds and shoots (*17*).

## ROOT CUTTINGS

Development of adventitious shoots and, in many cases, adventitious roots must take place when new plants are regenerated from root pieces (root cuttings) (*213*). In some plants, adventitious buds form readily on roots of intact plants, producing suckers. When roots are dug up, removed from plants, and cut into pieces, buds are even more likely to form as a response to wounding. In young roots, such buds may arise in the pericycle near the vascular cambium (*224, 291*). The developing buds first appear as groups of thin-walled cells having a prominent nucleus and a dense cytoplasm (*58, 269*). In old roots buds may arise exogenously in a calluslike growth from the phellogen; or they may appear in a calluslike proliferation from ray tissue (*62*). Bud primordia may develop also from wound callus tissue that proliferates from the cut ends or injured surfaces of the roots (*204*) or they may just arise at random in the cortex parenchyma (*214*).

Regeneration of new root meristems on root cuttings is often more difficult than the production of adventitious buds. New roots may not be adventitious but develop from latent root initials contained in old branch roots and present on the root piece (*227*). Generally, such branch roots arise from mature cells of the pericycle or endodermis, or both, adjacent to the central vascular cylinder (*14, 62*). Adventitious root initials have been observed to arise in the region of the vascular cambium in roots.

Regeneration of new plants from root cuttings takes place in different ways, depending upon the species. In the most common type, the root cutting first produces an adventitious shoot, and later produces roots, often from the base of the new shoot rather than from the original root piece itself. Sometimes these adven-

> ### Root Cutting Propagation of Chimeral Plants
>
> One of the chief advantages claimed for asexual propagation is the faithful reproduction of all characteristics of the parent plant. With root cuttings, however, this generalization does not always hold true. In periclinal chimeras, in which the cells of the outer layers are of a different genetic makeup from those of the inner tissues, the production of a new plant by root cuttings results in a plant that is different from the parent. This is well illustrated in the thornless boysenberry, and the 'Thornless Evergreen' trailing blackberry, in which stem or leaf-bud cuttings produce plants that retain the thornless condition, but root cuttings develop into thorny plants.

titious shoots can be removed and rooted as stem cuttings when treated with a rooting hormone (*214*). In other plants, a well developed root system has formed by the time the first shoots appear. Root cuttings of some species form a strong adventitious shoot, but no new roots develop, and the cutting eventually dies. In certain species, root cuttings produce a strong new root system, but no adventitious shoot arises, so the cutting finally dies (*131*).

Root cuttings taken from very young seedling trees are much more successful than those taken from older trees. Failures in the latter case are apparently due to the inability of the root pieces to regenerate a new root system. This inability is probably related to the phenomenon of juvenility, which is also involved in root formation in stem cuttings.

Not all underground portions of plants are roots—some, like rhizomes, are stem tissue. New plants developing from shoots arising from such underground stem tissue could not, therefore, be said to have been propagated by root cuttings.

A list of plants commonly propagated by root cuttings is given in chapter 10.

## POLARITY

The polarity inherent in shoots and roots is dramatically shown in rooting cuttings (Figures 9–9 and 9–10). Stem cuttings form shoots at the **distal** end (nearest the shoot tip), and roots at the **proximal** end (nearest the crown of the plant). Root cuttings form roots at the distal end and shoots at the proximal end. Changing the position of the cuttings with respect to gravity does not alter this tendency (*16*) (Figure 9–10).

In Vöchting's early studies (*274*) on the polarity of regeneration in plants, he pointed out that stem tissue is strongly polarized. The theory was then advanced that this property could be attributed to the individual cellular components, since no matter how small the piece, regeneration was consistently polar. When a root piece is cut into two segments, the two surfaces at the cut are similar in all respects, yet upon regeneration of roots and shoots, one surface of the cut produces a shoot and the other produces roots. Vöchting also concluded that the intensity of the polarity effect varied considerably among the different plant organs. Stems showed strong regeneration polarity, with somewhat weaker polarity by roots and much weaker regeneration polarity by leaves. It is

**Figure 9–9**  Results of planting a cutting of red currant (*Ribes sativum*) upside down (reversed polarity). *Left:* several months after starting. *Right:* one year later. The shoot from the center bud with new roots at its base has become the main plant and is growing with correct polarity. The shoot from the top bud, while still alive, has failed to develop normally.

commonly observed in leaf cuttings that roots and shoots arise at the same position, usually the base of the cutting, showing that little, if any, polarity influence is present (see Figure 10–12).

When tissue segments are cut, the physiological unity is disturbed. This must cause a redistribution of some substance, probably auxin, thus accounting for the different responses observed at previously adjacent surfaces. The correlation of polarity of root differentiation with auxin movement has been noted in several instances (*173, 215, 231, 253, 276*). It is also known that the polarity in auxin transport varies in intensity among different tissues, being particularly weak in leaf petioles, The polar movement of auxins is an active transport process and apparently is a secretion activity, with its basis found within the structural features of the individual phloem cells (*162*).

**Figure 9–10**  Polarity of root regeneration in grape hardwood cuttings. Cuttings at left were placed for rooting in an inverted position, but roots still developed from the morphologically basal (proximal) end. Cuttings on right were placed for rooting in the normal, upright orientation with roots forming at the basal end.

# PHYSIOLOGICAL BASIS OF ADVENTITIOUS ROOT AND SHOOT INITIATION

## Plant Growth Substances

Certain concentrations of naturally occurring materials having hormonal properties are more favorable than others for adventitious root initiation. Much study has been given to determining these relationships. In distinguishing between **plant hormones**[4] and **plant growth regulators**[5] it can be said that all hormones regulate growth but not all growth regulators are hormones. Various classes of growth regulators, such as the auxins, cytokinins, gibberellins, abscisic acid, and ethylene, influence root initiation. Of these, auxins have the greatest effect on root formation in cuttings. In addition to these groups, other naturally occurring materials that are not well defined, such as various inhibitors and promoters, may have a less direct part in adventitious root initiation.

### AUXINS

In the mid-1930s and later, studies of the physiology of auxin action showed that auxin was involved in such varied plant activities as stem growth, root formation, lateral bud inhibition, abscission of leaves and fruits, and activation of cambial cells.

**Indole-3-acetic acid (IAA)** was identified in 1934 as a naturally occurring compound having considerable auxin activity (*149, 254*) and was soon found to promote adventitious root formation (*152, 256, 258, 286*). This action of IAA was originally shown by a biological test using etiolated pea epicotyls under a set of standard conditions (*285, 287, 288*).

Synthetic indoleacetic acid (see Figure 9–20) was subsequently tested for its activity in promoting roots on stem segments, and in 1935 several investigators (*34, 155, 256*) demonstrated the practical use of this material in stimulating root formation on cuttings. About the same time it was shown (*253, 301*) that two similar materials, **indolebutyric acid (IBA)** and **napththaleneacetic acid (NAA)** (see Figure 9–20)—although not naturally occurring—were even more effective than the naturally occurring indoleacetic acid for this purpose. It has been confirmed many times that auxin, natural or artificially applied, is a requirement for initiation of adventitious roots on stems (*75*) and, indeed, it has been shown that the division of the first root initial cells are dependent upon either applied or endogenous auxin (*90*).

Physiological studies (*59, 60, 181*) with pea cuttings have shown the role of auxins in the intricate developmental process of root initiation. Root formation and development in pea cuttings was found to occur in two basic stages:

---

[4] *Plant hormones* are organic compounds, other than nutrients, produced by plants which, in low concentrations, regulate plant physiological processes. They usually move within the plant from a site of production to a site of action.

[5] *Plant growth regulators* are either synthetic compounds or plant hormones that modify plant physiological processes. They regulate growth by mimicking hormones, by influencing hormone synthesis, destruction, or translocation, or (possibly) by modifying hormonal action sites.

**1. An initiation stage** in which root meristems are formed. This stage could be further divided into:

    **a. An auxin-active stage,** lasting about four days, during which auxin must be supplied continuously for roots to form, coming either from a terminal bud or from applied auxin (if the cutting has been decapitated) (*60, 181*). This stage is followed by

    **b. An auxin-inactive stage.** Withholding auxin during this stage (which lasts about four days) does not adversely affect root formation.

**2. A root elongation and growth stage,** during which the root tip grows outward through the cortex, finally emerging from the epidermis of the stem (see Figure 9–5). A vascular system then develops in the new root primordia and becomes connected to adjacent vascular bundles. At this stage there is no response to applied auxin.

CYTOKININS

Cytokinins are plant growth hormones involved in cell growth and differentiation. Various natural and synthetic materials such as **zeatin, kinetin,** and **6-benzyl adenine** have cytokinin activity. Cuttings of species with high native cytokinin levels have been more difficult to root than those with low cytokinin levels (*190*). Generally, applied synthetic cytokinins have inhibited root initiation in stem cuttings (*132, 225*). However, cytokinins at very low concentrations, when applied to decapitated pea cuttings at an early developmental stage (*61*) or to begonia leaf cuttings (*107*) promoted root initiation, while higher concentrations inhibited initiation. Application to pea cuttings at a later stage in root initiation did not show such inhibition. The influence of cytokinins in root initiation may thus depend upon the particular stage of initiation and the concentration (*61*). Cytokinins relate to auxins in controlling organ differentiation (*89, 231*), as shown in Figure 9–11 for studies with tobacco stem segments.

**Figure 9–11**    Effects of adenine sulfate (a cytokinin) and indoleacetic acid (auxin) on growth and organ formation in tobacco stem segments. *Far left:* control. *Central left:* adenine sulfate, 40 mg per liter. Bud formation with decrease in root formation. *Center right:* indoleacetic acid, 0.02 mg per liter. Root formation with prevention of bud formation. *Far right:* adenine sulfate, 40 mg per liter plus indoleacetic acid, 0.02 mg per liter. Growth stimulation but without organ formation. Courtesy Folke Skoog.

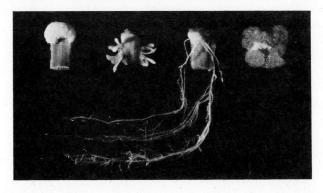

Cytokinins strongly promote bud initiation. For example, in root cuttings of *Isatis tinctoria* (a biennial herb) grown in a sterile medium, after several subcultures, shoot formation on the root pieces did not occur unless the medium was supplied with kinetin (*43*). Likewise, cytokinin treatments of *Convolvulus* (bindweed) root segments in a sterile nutrient medium induced bud initiation, especially in the light (*18*).

Leaf cuttings provide good test material for studying auxin-cytokinin relationships, since such cuttings must initiate both roots and shoots. In a study (*107*) with *Begonia* leaf cuttings under aseptic conditions, cytokinin at relatively high concentrations (about 13 ppm) promoted bud formation and inhibited root formation. Auxins, at high concentrations, gave the opposite effect. There were interacting relationships, however, between auxins and cytokinins. At low concentrations (about 2 ppm), IAA promoted bud formation, enhancing the cytokinin influence. Also, at low concentrations (about 0.8 ppm), kinetin stimulated the effect of IAA on root promotion.

It seems that the considerable seasonal changes in the regenerative ability of *Begonia* leaf cuttings are due to a complex interaction of temperature, photoperiod, and light intensity, controlling the levels of endogenous auxins and other growth regulators (*110*).

Buds are initiated in the leaf notches of *Bryophyllum* at an early stage in leaf formation. Shoots arising from the leaves originate from these buds. Cytokinin applications have a stimulatory effect on bud development while auxin applications inhibit bud development but stimulate root formation (*108*).

### GIBBERELLINS
(GA)

The gibberellins (Figure 6–10) are a group of closely related, naturally occurring compounds first isolated in Japan in 1939 and known principally for their effects in promoting stem elongation. At relatively high concentrations (i.e., $10^{-3}$ M) they have consistently inhibited adventitious root formation. There is evidence that this inhibition is a direct local effect that prevents the early cell divisions involved in transformation of mature stem tissues to a meristematic condition (*22*). Gibberellins have a function in regulating nucleic acid and protein synthesis and may be suppressing root initiation by interfering with these processes (*148*). At lower concentrations ($10^{-11}$ to $10^{-7}$M), however, gibberellin has promoted root initiation in pea cuttings, especially when the stock plants were grown at low irradiances (*94*).

In *Begonia* leaf cuttings, gibberellic acid was noted (*111*) to inhibit both adventitious bud and root formation, probably by blocking the organized cell divisions that initiate formation of bud and root primordia. There is evidence (*90*) in willow stem cuttings that applied gibberellin blocks auxin activity in root primordium development subsequent to the earlier initiation phase.

Lowering the natural levels of gibberellin in the tissues should stimulate adventitious root formation in cuttings. In fact, promotion of rooting has been done experimentally by various chemical substances that interfere with gibberellin activity such as Alar® (SADH) (*209, 294*), abscisic acid (*9, 29*), gonadotropins (*163*), and EL 531 (Arest® ) [α cycloprophyl α (4 methoxyphenyl-5-pyrimidine methanol], a gibberellin antagonist (*142*).

## ABSCISIC ACID
(ABA)

Reports on the effect of abscisic acid, a naturally occurring inhibitor in plants, on adventitious root formation are contradictory (*9, 29, 109, 208*), apparently depending upon the concentration and upon the nutritional status of the stock plants from which the cuttings are taken.

## ETHYLENE
($C_2H_4$)

Ethylene, a gaseous material, is produced by plants and has a number of hormonal effects although it does not exactly fit the definition of a hormone (*1*). In 1933 Zimmerman and Hitchcock (*299*) showed that applied ethylene at about 10 ppm causes root formation on stem and leaf tissue as well as the development of preexisting latent roots on stems. These and other workers (*301*) also showed about the same time that auxin applications can regulate ethylene production, and suggested that auxin-induced ethylene may account for the ability of auxin to cause root initiation. Centrifuging *Salix* cuttings in water, or just soaking them in hot or cold water, stimulates ethylene production in the tissues as well as root development, suggesting a possible causal relationship between ethylene production and subsequent root development (*142, 144*). Centrifugation seems to increase the water content of the cutting, which may block ethylene diffusion out of the cutting, the increased ethylene thus stimulating root formation (*145*).

Studies (*186*) of root initiation in mung bean cuttings showed that ethylene, from 0 to 1000 ppm, decreased root initiation.

Still other studies (*153*) showed that applications of **ethephon,** an ethylene-generating compound, to mung bean cuttings *did* stimulate root formation. The relationships between auxin, ethylene, and adventitious root formation apparently are complex, involving more than a simple alteration in ethylene concentration.

## The Effects of Buds and Leaves

Duhamel du Monceau (*53*) explained in 1758 the formation of adventitious roots on stems on the basis of the downward movement of sap. In extending this concept, Sachs, the German plant physiologist, postulated (*219*) in 1882 the existence of a specific root-forming substance manufactured in the leaves, which moves downward to the base of the stem, where it promotes root formation. It was shown (*159, 161*) by van der Lek in 1925 that strongly sprouting buds promote the development of roots just below the buds in cuttings of such plants as willow, poplar, currant, and grape. It was assumed that hormone-like substances were formed in the developing buds and transported through the phloem to the base of the cutting, where they stimulated root formation.

The existence of a specific root-forming factor was first determined by Went (*285*) in 1929 when he found that if leaf extracts from the *Acalypha* plant were applied back to *Acalypha* or to *Carica* tissue, they would induce root formation. Bouillenne and Went (*19*) in 1933 found substances in cotyledons, leaves, and buds that stimulated rooting of cuttings; they called this material "rhizocaline."

In Went's pea test (*287*) for root-forming activity of various substances, it is significant that the presence of at least one bud on the pea cutting was essential for root production. A budless cutting would not form roots even when treated with an auxin-rich preparation. This finding indicated again that a factor other than auxin, presumably one produced by the bud, was needed for root formation. In 1938, Went postulated that specific factors other than auxin were manufactured in the leaves and were necessary for root formation. Later studies (*59, 182*) with pea cuttings confirm this observation. For roots to form, the presence of an actively growing shoot tip (or a lateral bud) is necessary during the first three or four days after the cuttings are made. But after the fourth day the shoot terminals and buds could be removed without interfering with subsequent root formation.

That the amount of some naturally occurring root-forming substance(s) other than auxin, as yet unidentified but essential for root initiation, may be abundant in some plants and slight or even lacking in others, was shown long ago. In Cooper's studies (*35*) in 1938 apple and lemon cuttings were treated with auxin, and after analysis for auxin in the two groups of cuttings, it was found that there was little difference in the amount recovered, yet none of the apple cuttings rooted, whereas the lemon cuttings did. It was assumed that the apple cuttings were lacking in certain unidentified natural substances necessary for root formation, while the lemon cuttings had this substance(s) in abundance.

From their studies on the rooting of coniferous evergreen cuttings, Thimann and Delisle (*255*) agreed that some unknown factor(s), other than auxin, is involved in root initiation. They believed that this factor might exist in larger quantities in young plants, such as one-year-old seedlings, thus accounting for the comparative ease of rooting cuttings taken from young plants—the "juvenility effect" (*119*).

Removal of the buds from cuttings in certain plants will stop root formation almost completely, especially in species without preformed root initials (*159, 285*). In some plants, if a ring of bark is removed down to the wood just below a bud, root formation is reduced, indicating that some influence travels through the phloem from the bud to the base of the cutting, where it is active in promoting root initiation. It has been shown (*65, 161*) that if hardwood cuttings are taken in midwinter when the buds are in the rest period,[6] they have no stimulating effect on rooting, but if the cuttings are made in early fall or in the spring, when the buds are active and not in the "rest" influence, they show a strong root-promoting effect.

It has been shown too, with cuttings of apple and plum rootstocks, that the capacity of the shoots to regenerate roots increased during the winter, reaching a high point just before bud-break in the spring; this is believed to be associated with a decreasing level of bud dormancy following winter chilling (*127*).

Studies (*212*) with Douglas-fir cuttings showed a pronounced relationship between bud activity and rooting of cuttings. Rooting was least in September

---

[6] The "rest period" is a physiological condition of the buds of many woody perennial species beginning shortly after the buds are formed. While in this condition, they will not expand into flowers or leafy shoots even under suitable growing conditions. After exposure to sufficient cold, however, the "rest" influence is broken, and the buds will develop normally with the advent of favorable growing temperatures.

and October when bud dormancy was highest. Rooting was greatest in December and January (if auxin was applied), and in February and March (when no auxin was added), after winter cold had removed bud dormancy.

In rooting hardwood carob (*Ceratonia siliqua*) cuttings at monthly intervals throughout the year, bud removal changed the rooting pattern, chiefly by delaying time of peak rooting from mid-spring to mid-summer. This is shown in Figure 9–18 (*66*).

LEAF EFFECTS ON ROOTING

It has long been known, and there is considerable supporting experimental evidence (*35, 207, 285*), that the presence of leaves on cuttings exerts a strong stimulating influence on root initiation (see Figures 9–12 and 9–13).

**Figure 9–12** Effect of leaves on cuttings of 'Lisbon' lemon cuttings. Both groups were rooted under intermittent mist and were treated with indolebutyric acid at 4000 ppm by the concentrated-solution-dip method.

**Figure 9–13** Effect of leaves, buds, and applied auxin on adventitious root formation in leafy 'Old Home' pear cuttings. *Top:* cuttings treated with auxin (indolebutyric acid at 4000 ppm for 5 sec). *Bottom:* untreated cuttings. Left to right: with leaves; leaves removed; buds removed; one-fourth natural leaf area. Courtesy W. Chantarotwong.

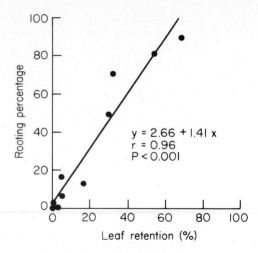

**Figure 9–14** Correlation between leaves retained on cuttings of different avocado clones and the rooting percentage. From data of Reuveni and Raviv (*212*).

$y = 2.66 + 1.41 x$
$r = 0.96$
$P < 0.001$

The stimulatory effect of leaves on rooting in stem cuttings is nicely shown by studies (*210*) with the avocado. Cuttings of difficult-to-root cultivars under mist soon shed their leaves and die, whereas leaves on the cuttings of cultivars that rooted are retained as long as nine months. The positive correlation between percent of leaf retention and percent of cuttings rooting is shown in Figure 9–14. In this study, after five weeks in the rooting bed, there was five times more starch in the base of the easily rooted cuttings than there was at the beginning of the tests.

Carbohydrates translocated from the leaves undoubtedly contribute to root formation. However, the strong root-promoting effects of leaves and buds are probably due to other, more direct factors (*21*). Leaves and buds are known to be powerful auxin producers, and the effects are observed directly below them, showing that polar apex-to-base transport is involved.

Cuttings of certain clones are easily rooted but cuttings of other, closely related, clones root with considerable difficulty. It would seem that grafting a leafy portion of the easily rooted clone onto a basal stem portion of the difficult-to-root clone and then preparing the combination as a cutting, would cause it to root readily. The rooting factors provided by the leaves or buds of the easily rooted clone could perhaps stimulate rooting of the difficult-to-root basal part. Such experiments have been tried but with conflicting results. Van Overbeek and others (*270, 271*) rooted the difficult-to-root 'Purity' (white) hibiscus by previously grafting onto it sections of the easily rooted 'Brilliant' (red) hibiscus, then applying auxins. Auxin-treated 'Purity' cuttings, ungrafted, did not root, the leaves soon dropping. Auxin-treated ungrafted 'Brilliant' cuttings rooted readily and retained their leaves.

Later experiments with the same hibiscus cultivars by Ryan and others (*217*) gave different results. With cuttings held for six weeks, they obtained 100 percent rooting with 'Brilliant', 90 percent with 'Purity', and 84 percent with cuttings made from 'Brilliant' on 'Purity' grafts. It is possible that in van Overbeek's studies, difficulty in rooting 'Purity' cuttings was due primarily to the early loss of leaves under the conditions of their experiments. Leaves of the 'Purity' cultivar may stimulate rooting just as well as those of the 'Brilliant' if they can be retained long enough.

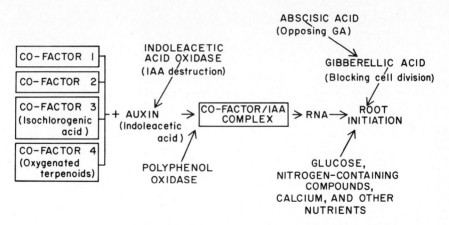

**Figure 9–15** Hypothetical relationships of various components leading to adventitious root initiation. In addition, specific root-inhibiting factors may be present, which interfere with root development.

Bouillenne and Bouillenne-Walrand (*19*) in 1955 proposed that "rhizocaline" be considered as a complex of three components: (1) a specific factor, translocated from the leaves, and characterized chemically as an orthodihydroxy phenol; and (2) a nonspecific factor (auxin), which is translocated and is found in biologically low concentrations; and (3) a specific enzyme located in cells of certain tissues (pericycle, phloem, cambium), which is probably of the polyphenol-oxidase type.

They further proposed that the ortho-dihydroxy phenol reacts with auxin wherever the required enzyme is present, giving rise to the complex "rhizocaline," which may be considered one step in a chain of reactions leading to root initiation (Figure 9–15). Libbert (*164*) in 1956 also developed evidence showing that auxin, forming a complex with a mobile factor "X," would result in root initiation, but he rejected the concept that a non-mobile enzyme was involved. He asserted that auxin itself acts to cause dedifferentiation of cells, determining the site of root formation.

### ROOTING CO-FACTORS (AUXIN SYNERGISTS)

Hess (*113, 114, 115, 117*) in 1962 isolated various **rooting co-factors** from cuttings, using chromatography together with mung bean (*Phaseolus aureus*) bioassay techniques. He worked with cuttings of the easily rooted juvenile form and the difficult-to-root mature form of English ivy (*Hedera helix*). He also used easy and difficult-to-root cultivars of chrysanthemum and of the red-flowered and white-flowered forms of *Hibiscus rosa-sinensis*. These co-factors are naturally occurring substances that appear to act synergistically with indoleacetic acid in promoting rooting. The easily rooted forms of plants he has worked with have a larger content of such co-factors than the difficult ones.

One of these cofactors (No. 4) represents a group of active substances, tentatively characterized as oxygenated terpenoids. Another (No. 3) was identified in 1965 as chlorogenic acid (*116*). Further work showed that there were three lipidlike root-promoting compounds in juvenile ivy tissue that contain functional alcohol and nitrile groups. All three were colorless and unstable, breaking down to orange-yellow compounds and losing their root-promoting activity (120).

In testing the biological activity of compounds structurally related to cofactor 4, Hess (114) found that the phenolic compound catechol reacts synergistically with indoleacetic acid in root production in the mung bean bioassay. Since, as he points out, catechol is readily oxidized to a quinone, and since the mung bean itself is a good source of phenolase, it may be that oxidation of an ortho-dihydroxy phenol is one of the first steps leading to root initiation, as suggested earlier by Bouillenne and Bouillenne-Walrand (17). In addition, various other compounds have been found to react synergistically with auxin in promoting rooting (84).

One of the postulated rooting co-factors could possibly be abscisic acid which can promote root initiation (29), perhaps by antagonizing gibberellic acid which, at certain concentrations, inhibits root formation.

Rooting co-factors were found in hardwood cuttings of 'Crab C' and 'Malling 26' apple rootstocks by the mung bean bioassay (30). Increased co-factor activity was found when cuttings of 'Malling 26' were subjected to elevated temperatures 18.5°C (65°F) during the winter storage period, a practice known to increase rooting. Hardwood stem tissue of the easily rooted 'Malling 106' was found (4) to show strong root promoting factors in the mung bean bioassay, whereas in the difficult-to-root 'Malling 2' such factors were present in lower amounts and rooting inhibitors appeared.

Fadl and Hartmann (64, 65) isolated an endogenous root-promoting factor from basal sections of hardwood cuttings of an easily rooted pear cultivar ('Old Home'). This highly active root-promoting material appeared only in cuttings having buds and treated with indolebutyric acid. It was found in largest amounts about ten days after the cuttings had been made, treated with IBA, and placed in the rooting medium. Extracts from basal segments of similar cuttings of a difficult-to-root cultivar ('Bartlett'), treated with IBA, did not show this rooting factor. It did not appear either in 'Old Home' cuttings which had the buds removed or at a time of year when the buds were in a deep "rest" condition. Tests using UV spectrum analysis and infrared spectroscopy indicated that this rooting factor is a complex structure of high molecular weight and possibly is a condensation product between the applied auxin and a phenolic substance produced by the buds.

The action of these phenolic compounds in root promotion could be, at least partly, in protecting the root-inducing, naturally occurring auxin—indoleacetic acid—from destruction by the enzyme, indoleacetic acid oxidase (50). This possibility was suggested also in rooting (87) juvenile *Hedera helix,* in which the phenol, catechol, showed remarkable synergism with IAA in promoting rooting. However, when naphthaleneacetic acid (NAA), which is not affected by IAA oxidase, was used as the auxin, NAA alone was as effective as NAA plus catechol. This finding implies that catechol was protecting IAA from destruction by the enzyme.

Using in vitro techniques, phloroglucinol (1,3,5-trihydroxybenzene), a phenolic material, was shown (136) to act synergistically with indolebutyric acid in stimulating adventitious root initiation in 'Malling 9' and other, but not all, apple rootstocks (302). In these studies, shoot cultures developing in the presence of phloroglucinol subsequently rooted better than shoots grown in its absence. However, the interaction of IBA and phloroglucinol in promoting rooting in vitro, depends in the apple at least, upon the growth phase of the cutting material (juvenile or adult), as well as the concentrations of each compound (*282*). Phloroglucinol has also been shown to stimulate rooting in *Rubus* (*134, 135*) and in *Prunus* (*141*) species.

The role played by phenolic rooting co-factors in adventitious root initiation is controversial, however, with some studies (*166*) failing to show any correlation between rooting cofactors and rooting response.

### Endogenous Rooting Inhibitors

Cuttings of certain difficult-to-root plants may fail to root because of naturally occurring rooting inhibitors. This was found many years ago (*235*) to be the case with grapes, in which chromatographic studies suggested the presence of two inhibitors associated with rooting response. Leaching the cuttings with water enhanced the quantity and quality of roots. An inhibitor released into the water during leaching had a detrimental effect on rooting cuttings of the easily rooted *Vitis vinifera*. Shy-rooting cuttings of *V. berlandieri* seemed to possess a high inhibitor content.

'Bartlett' pear hardwood cuttings are difficult to root under treatments which give good rooting of 'Old Home' pear hardwood cuttings (*64*). Extracts taken from cuttings of both cultivars 20 days after being treated with IBA and placed in a rooting medium showed distinctly different amounts of inhibitors and promoters.

Other evidence implicating rooting inhibitors is given in studies (*41, 194*) in Australia with the difficult-rooting adult tissues of *Eucalyptus grandis,* which contained compounds that blocked adventitious root formation. In these studies three such inhibitors were found and considerable information was developed concerning their structure and physical properties. They were determined to be naturally occurring derivatives of the 2, 3-dioxa-bicycle [4,4,0] decane system. These inhibitors were not present in the easily rooted juvenile tissue of *E. grandis*.

From studies (*11, 13*) with cuttings taken from dahlia cultivars it was determined that in plants whose cuttings were difficult to root, inhibitors formed in the roots and moved upwards, accumulating in the shoots, subsequently interfering with root formation. In cultivars whose cuttings rooted easily, inhibitor levels were low.

### Classes of Plants in Respect to Ease of Rooting

Plants can be divided into three classes in regard to their relation to materials involved in adventitious root initiation:

**a.** Those in which the tissues provide all the various native substances, including auxin, essential for root initiation. When cuttings are made and placed under proper environmental conditions, rapid root formation occurs.

**b.** Those in which the naturally occurring co-factors are present in ample amounts but auxin is limiting. With the application of auxin, rooting is greatly increased.

**c.** Those that lack the activity of one or more of the internal co-factors although natural auxin may or may not be present in abundance. External application of auxin gives little or no response, owing to lack of the effects of one or more of the naturally occurring materials essential for root formation.

Concerning the last group, Haissig (91) postulates that lack of root initiation in response to applied auxin (or even to native auxin) may be due to one or more of the following:

1. Lack of necessary enzymes to synthesize the root-inducing auxin-phenol conjugates
2. Lack of enzyme activators
3. Presence of enzyme inhibitors
4. Lack of substrate phenolics
5. Physical separation of enzyme reactants due to cellular compartmentalization.

## Biochemical Changes in the Development of Adventitious Roots

Once adventitious roots have been initiated in cuttings, considerable metabolic activity takes place as new root tissues are developed and the roots grow through and out of the surrounding stem tissue to become external functioning roots. Protein synthesis and RNA production were both shown indirectly to be involved in adventitious root development in etiolated stem segments of *Salix tetrasperma* (133). The fact that auxin action requires the presence of nutritional factors (glucose) is due to the requirement of a carbon source for the biosynthesis of nucleic acids and proteins.

Some significant studies (183, 184) have been made of the biochemical changes taking place during the *development* of preformed root initials in hydrangea into emerging roots. These studies followed, in particular, the changing patterns of DNA and enzyme levels as the roots developed.

Root initials were found to originate in the phloem ray parenchyma, with roots emerging 10 to 12 days after the cuttings were made. The total protein content of the root initials increased over 100 percent in the first four days after the cuttings were made but there was no pronounced increase in the DNA content of the cells until the sixth day. Apparently, then, considerable protein synthesis preceded large-scale replication of nuclear DNA.

Pronounced increases in enzyme activity took place during adventitious root formation in hydrangea (183); the enzymes peroxidase, cytochrome oxidase, succinic dehydrogenase, and starch-hydrolyzing enzymes were observed in the phloem and xylem ray cells of the vascular bundles. Then, during root development, enzyme activity shifted from the vascular tissues to the periphery of the bundles. Such increases in enzyme activity occurred two to three days after the cuttings were made.

Starch was found to disappear from the endodermis, phloem and xylem rays, and pith in the region of the developing root primordia, apparently being utilized as a carbohydrate food source. Starch evidently plays an important nutritional role in adventitious root development.

In the development of adventitious roots on IBA-treated plum cuttings, it was determined (20), using radioactive $CO_2$ applied to leaves, that as soon as callus and roots started forming, pronounced sugar increases—and starch losses—occurred at the base of the cuttings, the $^{14}C$ appearing in sucrose, glucose, fructose, and sorbitol. The developing callus and roots apparently act as a "sink" for the movement of soluble carbohydrates from the top of the cutting.

# FACTORS AFFECTING REGENERATION OF PLANTS FROM CUTTINGS

Great differences exist among plants of different species and cultivars in the rooting ability of cuttings taken from them. Empirical trials with each cultivar are necessary to determine these differences. This has already been done with most plants of economic importance. Stem cuttings of some cultivars root so readily that the simplest facilities and care give high rooting percentages. On the other hand, cuttings of many cultivars or species have yet to be rooted. Cuttings of some "difficult" cultivars can be rooted only if various influencing factors are taken into consideration and maintained at the optimum condition. Environmental factors to be discussed in this section are of great importance to this group, and the attention given to them makes the difference between success or failure in obtaining satisfactory rooting. These factors are:

**A.** *Selection of cutting material*
  **1.** Physiological condition of the stock plant
  **2.** Juvenility factor (age of the stock plant)
  **3.** Type of wood selected
  **4.** Presence of viruses
  **5.** Time of year cuttings are taken
**B.** *Treatment of cuttings*
  **1.** Growth regulators
  **2.** Mineral nutrients
  **3.** Fungicides
  **4.** Wounding
**C.** *Environmental conditions during rooting*
  **1.** Water relations
  **2.** Temperature
  **3.** Light
    **a.** Intensity
    **b.** Day length
    **c.** Light quality
  **4.** Rooting medium

The remainder of this chapter discusses these factors in detail.

## Selection of Cutting Material

PHYSIOLOGICAL CONDITION
OF THE STOCK PLANT

*Water stress*    Plant propagators often emphasize the desirability of taking cuttings early in the morning when the plant material is in a turgid condition. There is experimental evidence to support this view. Studies with both cacao (*63*) and pea (*206*) cuttings showed reduced rooting when the cuttings were taken from stock plants having a water deficit.

*Carbohydrates*    The nutrition of the stock plant can exert a strong influence on the development of roots and shoots from cuttings (*197, 203, 222*). This effect, which may relate to the particular physiological state of the tissue, can be associated with certain carbohydrate/nitrogen relationships. For example, Kraus and Kraybill (*151*) observed long ago in making tomato cuttings that those with yellowish stems, high in carbohydrates but low in nitrogen, produced many roots but only feeble shoots, whereas those with greenish stems, containing ample carbohydrates but higher in nitrogen, although producing fewer roots gave stronger shoots. Green, succulent stems, very low in carbohydrates but high in nitrogen, all decayed without producing either roots or shoots.

Internal factors, such as auxin levels, rooting cofactors, and carbohydrate storage, can, of course, influence root initiation of cuttings. In a study (*240*) in which all these factors were determined in easy and difficult-to-root cultivars of chrysanthemum, the only correlation that could be obtained was in the carbohydrate storage in the stems, the easily rooted cultivars having the highest storage levels. Softwood cuttings of the hop plant, when held under low light intensities—at about the compensation point—responded to sugar pretreatments with large increases in rooting, illustrating the need for ample carbohydrates for root production (*128*). The necessity for applying sucrose (*81*) or glucose (*75, 200*) for root formation on stem segments has been demonstrated by in vitro tests (see Figure 9–25), although excessive concentrations may decrease rooting (*95, 188*).

Selecting suitable cutting material, as far as the carbohydrate content is concerned, can often be determined by stem firmness. Those undesirably low in carbohydrates are soft and flexible, while those higher in carbohydrates are firm and stiff, and break with a snap rather than bending. This desirable condition may be confused, however, with firmness that is due to the maturing of the tissues, caused by thickening and lignification of the cell walls.

A simple method of selecting cutting material having the desirable high starch content is the iodine test. Freshly cut ends of a bundle of cuttings are immersed for one minute in a 0.2 percent solution of iodine in potassium iodide. Cuttings having the highest starch content stain the darkest color. A rough grouping of cuttings high, medium, and low in starch can then be made. In such tests with grapes (*292*), 63 percent of the high-starch cuttings rooted, 35 percent of the medium-starch cuttings rooted, and 17 percent of the low-starch cuttings rooted.

Propagation studies with both raspberry (*171*) and apple (*215*) root cuttings have shown greatest success with cuttings taken in the fall when carbohydrate content of the roots was highest. Survival was poorest with cuttings taken during the summer when carbohydrate storage in the roots was very low.

Easily-rooted avocado cuttings retained their leaves and accumulated starch in the cutting bases, whereas cuttings of difficult-to-root cultivars dropped their leaves and died (210).

*Mineral nutrition*    Just as for carbohydrates, a moderate level of nitrogen in the tissue is best for optimum rooting. Very low nitrogen leads to reduced vigor, whereas very high nitrogen causes excessive vigor. Either extreme is unfavorable for good rooting. Tissues with high nitrogen content are likely to be rank-growing, soft, and succulent, with low carbohydrate storage. Such rapidly growing shoots may also be low in other components necessary for rooting.

The low-nitrogen–high-carbohydrate balance in stock plants, which in many cases seems to favor rooting, can be achieved in several ways:

**1.** Reduce nitrogen fertilizers to the stock plants, thus reducing shoot growth and allowing for carbohydrate accumulation. Any type of root restriction of the stock plants, such as occurs when they are grown in containers or close together in hedge rows, tends to prevent excessive vegetative growth and permits the accumulation of carbohydrates.

**2.** For cutting material, select portions of the plant that are in the desirable nutritive stage. For example, take lateral shoots in which rapid growth has decreased and carbohydrates have accumulated, rather than succulent terminal shoots. (But for plants showing a plagiotropic growth pattern [see p. 205], use of lateral shoots should be avoided.)

**3.** Select regions of the shoot that are known to have a high carbohydrate content. In a chemical analysis (264) of rose shoots of the type used for cuttings, the nitrogen content increased uniformly from the base of the shoot to the tip. Basal portions of such a shoot would therefore have the low-nitrogen–high-carbohydrate balance favorable for good rooting.

It cannot be said, however, that a high carbohydrate content in cuttings is invariably associated with ease of rooting. Other, stronger influencing factors may be present.

Studies (197, 198) with grapes showed that when stock plants were grown under phosphorus, potassium, magnesium, or calcium deficiency, root formation in cuttings was poorer than in cuttings from full-nutrient plants; but with reduced nitrogen in the stock plants, root formation by the cuttings was increased. However, extreme nitrogen deficiency of the stock plants lowered rather than increased rooting. This was substantiated by tests (104) with geranium in which stock plants were grown at three levels of nitrogen, phosphorus, and potassium. The nitrogen nutrition of the stock plants had a greater effect on rooting response of the cuttings than did phosphorus or potassium, with the low and medium levels of nitrogen resulting in higher percentages of rooted cuttings than the high level.

For root initiation to take place, however, nitrogen is necessary for nucleic acid and protein synthesis. Below a cut-off level of nitrogen availability, root initiation is impaired; in such cases added nitrogen promotes rooting.

Grape cuttings taken from vines fertilized with zinc rooted in higher percentages and were of better quality than cuttings taken from untreated vines (222). This effect could be due to an increase in native auxin production resulting from the increased level of tryptophan (an auxin precursor) found in

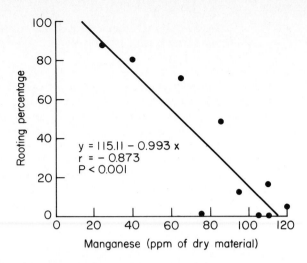

**Figure 9–16**
Correlation between manganese content of leaves of different avocado clones and rooting percentage. From data of Reuveni and Raviv (*210*).

$y = 115.11 - 0.993 x$
$r = -0.873$
$P < 0.001$

the treated plants. (Zinc is required for tryptophan production.) In fact, applications of synthetic tryptophan have increased rooting of vine cuttings. Beneficial effects of zinc applications to stock plants have been noted also in South Africa in the propagation of 'Marianna' plum cuttings.

High levels of manganese were found (*210*) in leaves of cuttings taken from difficult-to-root avocado cultivars, whereas cuttings from easy-to-root cultivars had a much lower manganese level (Figure 9–16). Of eleven elements for which analyses were made in this study, only manganese showed any correlation with rooting. Manganese is known (*259*) to be an activator of the enzyme, indoleacetic acid oxidase, which destroys auxin and could lead to a reduced amount of natural auxin at the base of the cuttings, thus causing poor rooting.

For plants difficult to root, several treatments can be used to alter the physiological or nutritional condition of the stock plant or portions of the plant. These treatments, which often result in increased rooting of cuttings taken from such plants, and include such practices as etiolation (see p. 278) and/or girdling the shoots some time before they are made into cuttings.

*Girdling*    Girdling, or otherwise constricting the stem, blocks the downward translocation of carbohydrates, hormones, and other possible root-promoting factors and can result in an increase in root initiation. Using this technique on shoots prior to their removal for use as cuttings should improve rooting. This practice has been remarkably successful in some instances. For example, rooting of citrus and hibiscus cuttings was stimulated by girdling or binding the base of the shoots with wire several weeks before taking the cuttings (*137, 239*). In cuttings from mature trees of the water oak (*Quercus nigra*), a three-fold improvement in rooting was obtained when cuttings were taken from shoots that had been girdled six weeks previously, especially if a powder consisting of 1 percent each of IBA, PPZ (1-phenyl-3-methyl-5-pyrazolone), 20 percent sucrose, and 5 percent captan in talc, was rubbed into the girdling cuts (*97*). Girdling (*238*) also caused substantial increase in a rooting cofactor above the girdle in an easily rooted hibiscus clone. Girdling just below a previously etiolated stem section was particularly effective in promoting rooting in apple cuttings (*45*).

## Juvenility Factor or Phase Change
## (Age of the Stock Plant)

In plants difficult to root the age of the stock plant can be an overriding factor in root formation. Either stem or root cuttings taken from plants in the **juvenile** development growth phase, as found in young seedlings, often form new roots much more readily than those taken from plants in the **adult** growth phase, whether seedlings or vegetatively propagated (*73, 122, 223*) (see chapter 8). Experiments with apple, pear, eucalyptus, live oak, Douglas fir, and many other species have shown that the ability of cuttings to form adventitious roots decreased with increasing age of the plants from seed; in other words, when the stock plant changed from the juvenile to the adult phase.

In a study (*255*) of rooting cuttings of certain coniferous and deciduous species known to root only with extreme difficulty, it was concluded that the most important single factor affecting root initiation was the age of the tree from which the cuttings were taken.

Some old English garden books list great numbers of plants that were rooted from cuttings—plants now considered very difficult, if not impossible, to propagate by cuttings. It is likely that in earlier days, such species were brought to England from distant countries as seeds; cuttings were subsequently made from young seedlings, the juvenility factor accounting for ease of rooting.

In some species such as apple, English ivy, and olive, differences in certain morphological characteristics, such as leaf size and shape, make it easy to distinguish between the adult and the lower, juvenile portions of the plant. Cuttings should be taken from the latter type for best rooting (*201*). In some kinds of trees, such as oak and beech, leaf retention late into the fall occurs on the basal parts of the tree and indicates the part still in the juvenile stage. Cuttings should be taken from this type of wood (*85*).

Any treatment that maintains the juvenile growth phase would be of value in preventing the decline in rooting potential as the stock plant ages. The hedging or shearing treatments given *Pinus radiata* trees (Figure 9–17) was quite effec-

**Figure 9–17** Stock plants for *Pinus radiata* cuttings maintained in a hedge form by shearing (compare with unsheared trees in background). Such "hedged" trees can yield high numbers of cuttings which maintain the high rooting percentages, root quality, and growth potential normally associated with the younger stock trees (*165*). Photo courtesy W. J. Libby.

tive in maintaining the rooting potential of cuttings taken from them as the trees aged, compared to nonhedged trees (165). Such benefits in maintaining rooting potential by hedging are probably explained by the prevention of the normal phase change from the juvenile to the adult form. In England it has been a practice for many years to use as stock plants for hardwood cuttings of fruit-tree rootstocks, special plants that are maintained as hedges rather than allowed to grow to a tree form (76).

Juvenility in relation to rooting may possibly be explained by the increasing production of rooting inhibitors as the plant grows older. Stem cuttings taken from young seedlings of a number of eucalyptus species root easily, but as the stock plants become older rooting decreases dramatically. Studies (194) in Australia showed that there was a direct and quantitative association between such decreased rooting and the production of a rooting inhibitor in the tissues at the base of the cuttings. In easily rooted young seedling stems this inhibitor was absent, as it was absent in adult stem tissue of the easily rooted *Eucalyptus deglupta.*

Reduced rooting potential as plants age may possibly be a result of lowering phenolic levels. Phenols are postulated as acting as auxin cofactors or synergists in root initiation. In certain plants, lower phenolic levels were noted in mature forms than in the juvenile forms (80).

In rooting cuttings of difficult species it would be useful to be able to induce rejuvenation of the easily rooted juvenile stage from plants in the adult form. This has been done in several instances by the following methods:

Juvenile forms of apple can be obtained from mature trees by causing adventitious shoots to develop from root pieces, which are then made into softwood stem cuttings and rooted (241).

By removing terminal and lateral buds and spraying stock plants of *Pinus sylvestris* with a mixture of cytokinin, tri-iodobenzoic acid, and Alar, many interfascicular shoots can be forced out. With proper subsequent treatment, high percentages of these shoots can be rooted (289).

In some plants juvenile wood can be obtained from mature plants by forcing juvenile growth from **sphaeroblasts** (wartlike protuberances containing meristematic and conductive tissues sometimes found on trunks or branches). These are induced to develop by disbudding and heavily cutting back stock plants. These juvenile shoots then may be easily rooted under the usual conditions (283). Using the mound-layering (stooling) method on these rooted sphaeroblast cuttings produces rooted shoots that continue to possess the juvenile characteristics. In the stooling method of propagation the juvenile phase may be maintained continually.

Grafting adult forms of ivy onto juvenile forms has induced a change of the adult to the juvenile stage, provided the plants are held at fairly high temperatures (51, 247); such transmission of the juvenile rooting ability from seedlings to adult forms by grafting has also been accomplished in rubber trees (*Hevea brasiliensis*) (187).

Gibberellin sprays on English ivy plants in the adult growth phase have caused substantial stimulation of growth and reversion of some of the branches to the juvenile stage (211) as well as improved rooting of cuttings taken from the sprayed plants (247).

It is important to remember that the juvenile condition is found only in such tissues as those originating from young seedlings, those on the lower juvenile portion of mature adult trees, those arising from adventitious (not latent) buds on stems, or those caused to revert to juvenility either by gibberellin treatments or by grafting to juvenile wood. Tissues from young plants propagated from material taken from plants in the adult phase are not in a truly juvenile state.

## Type of Wood Selected

There are many choices of the type of material to use, ranging (in woody perennials) from the very succulent terminal shoots of current growth to large hardwood cuttings several years old. Here, as with most of the other factors affecting rooting of cuttings, it is impossible to state any one type of cutting material that would be best for all plants. What may be ideal for one plant would be a failure for another. What has been found to hold true for certain species or cultivars, however, often may be extended to other related species or cultivars.

### DIFFERENCES BETWEEN INDIVIDUAL SEEDLING PLANTS

In rooting cuttings taken from individual plants of a species which ordinarily is propagated by seed, experience has shown that wide differences exist among individuals in the ease with which cuttings taken from them form roots. Just as seedlings differ in many respects, it is not surprising that this difference in root-forming ability should also exist. Differences in the rooting ability of various clones are, of course, recognized. Likewise, such differences should be anticipated when woody plants usually propagated by seed—such as most forest tree species—are propagated by cuttings. In rooting cuttings taken from old seedling trees of Norway spruce (*Picea abies*), white pine (*Pinus strobus*), and red maple (*Acer rubrum*) marked differences occurred in the rooting capacity of shoots taken from the individual trees (*48, 233*).

### DIFFERENCES BETWEEN LATERAL AND TERMINAL SHOOTS

Experiments in rooting plum stem cuttings compared the rooting of different types of softwood cuttings taken in the spring. The results showed a marked superiority in rooting of lateral shoots; the terminals had only 10 percent rooting; laterals in active growth, 19 percent; and laterals that had ceased active growth, 35 percent.

Similarly, lateral branches of white pine and Norway spruce gave consistently higher percentages of rooted cuttings than did terminal shoots (*48, 67*). In rhododendrons, too, thin cuttings made from lateral shoots consistently give higher rooting percentages than these taken from vigorous, strong terminal shoots. In certain species, however, plants propagated as rooted cuttings taken from lateral branches may have an undesirable growth habit (see Figure 8–5).

### DIFFERENCES BETWEEN VARIOUS PARTS OF THE SHOOT

In some woody plants, hardwood cuttings are made by sectioning shoots several feet long and obtaining four to eight cuttings from a single shoot. Marked differences are known to exist in the chemical composition of such shoots from base to tip. Variations in root production on cuttings taken from different portions of the shoot are often observed, with the highest rooting, in many cases, found in cuttings taken from the basal portions of the shoot. Cuttings prepared from shoots of three cultivars of the highbush blueberry (*Vaccinium corymbosum*) were significantly more successful if taken from the basal portions of the shoot rather than from terminal portions (*192*). The number of preformed root initials in woody stems has been determined (in some plants at least) as distinctly decreasing from the base to the tip of the shoot (*160*). Consequently the rooting capacity of basal portions of such shoots would be considerably higher than that of the apical parts.

In studies with a different type of wood, however, cherries (*Prunus cerasus, P. avium, P. mahaleb*) rooted under mist by softwood cuttings prepared from succulent new growth gave the following percentages of rooted cuttings: 'Stockton Morello', basal—30 percent, tip—77 percent; 'Bing', basal—0 percent, tip—100 percent; and 'Montmorency', basal—10 percent, tip—90 percent (*98*).

In woody stems a year or more in age, where carbohydrates have accumulated at the base of the shoots and, perhaps, where some root initials have formed in the basal portions, possibly under the influence of root-promoting substances from buds and leaves, the best cutting material is the basal portions of such shoots. An entirely different physiological situation exists in deciduous plants in which succulent shoots are used for softwood cuttings. Here carbohydrate storage and preformed root initials are not present. The better rooting of shoot tips may be explained by the possibility of higher concentrations of endogenous root-promoting substances arising in the terminal bud. There is also less differentiation in the terminal cuttings, with more cells capable of becoming meristematic.

This factor is of little importance, however, in cuttings of easily rooted species in which entirely satisfactory rooting is obtained regardless of the position of the cutting on the shoot.

### FLOWERING OR VEGETATIVE WOOD

In most plants, cuttings could be made from shoots that are in either a flowering or a vegetative condition. Again, with easily rooted species it makes little difference which is used, but in difficult-to-root species this can be an important factor. For example, in a blueberry species (*Vaccinium atrococcum*), hardwood cuttings from shoots bearing flower buds did not root nearly so well as those bearing only leaf buds (*191*). When vegetative wood was used, 39 percent of the cuttings rooted, but when cuttings contained one or more flower buds, not one rooted. Removing the flower buds prior to rooting did not increase the rooting percentage, indicating that it is not the actual presence of flower buds that inhibits rooting, but rather some previous physiological or anatomical condition associated with the presence of flower buds. In the same manner, dahlia cuttings bearing flower buds are more difficult to root than cuttings having only vegetative buds (*13*).

In the rhododendron, early removal of potential flower buds increased rooting, presumably through the elimination of a flower-promoting stimulus antagonistic to rooting; later flower bud removals still enhanced rooting, possibly by eliminating competition for materials necessary for rooting (*138*).

Some antagonism apparently exists between vegetative regeneration and flowering. The basis for this is probably found in auxin relationships, since it is known that high auxin levels, which are favorable for adventitious root formation in stem cuttings, tend to inhibit flower initiation (*260*).

It has been consistently noted for both root and stem cuttings of many species that better regeneration takes place when cuttings are taken either before or after, rather than during, the flowering period (*52*). In some instances cuttings taken any time while the stock plants were in the vegetative state rooted very well, but as soon as the stock plants started initiating flower buds, the cuttings failed to root (*221*).

In order to have the highest regenerative capacity, stock plants should be showing active vegetative growth (and not be entering the flowering stage). In England (*76*) the usual sources for hardwood cuttings of fruit tree rootstocks are hedges that are severely pruned in winter in collecting the cutting material, thus maintaining a state of vigorous nonflowering growth for the following year.

"HEEL" VS. "NONHEEL" CUTTINGS

In preparing cuttings, it is often recommended that a "heel" (a small slice of older wood) be retained at the base of the cutting in order to obtain maximum rooting. For hardwood cuttings of some plants, this may be true. In quince (*Cydonia oblonga*), considerably better rooting was obtained with the heel type of cutting, probably owing in this case to the presence of preformed root initials in the older wood. Narrow-leaved evergreen cuttings often, but not always, root more readily if a heel of old wood is retained at the base of the cuttings.

## Presence of Viruses

Since procedures have been developed for eliminating viruses from clones by heat treatment, it has been possible to show the depressing effects of viruses on adventitious root initiation. This was found to be true for both apple (*129*) and gooseberry (*266*) cuttings, where those from virus-free clones rooted considerably better than those from infected stock. The presence of viruses reduces not only rooting percentages but also numbers of roots forming on the cuttings. Poor results often obtained in rooting cuttings may possibly be traced to the use of virus-infected cutting material and may account for the variable results often obtained in different rooting tests with the same cultivar.

## Time of Year

Season of the year can have, in some instances, dramatic influences on the results obtained in rooting cuttings, and may provide the key to highly successful rooting. It is possible, of course, to make cuttings at any time during the year. In propagating deciduous species, hardwood cuttings could be taken during the dormant season, or leafy softwood or semi-hardwood cuttings could be prepared during the growing season, using succulent or partially matured wood. The narrow- and broad-leaved evergreen species have one or more flushes of growth during the year, and cuttings could be obtained at various times in relation to these flushes of growth.

Certain species, such as the privets, can be rooted readily if cuttings are taken almost any time during the year; on the other hand, for example, excellent rooting of leafy olive cuttings under mist can be obtained during late spring and summer, whereas rooting drops almost to zero with similar cuttings taken in midwinter (*101*). Cuttings of *Ficus infectoria* root best during spring and summer, when cambial activity is highest, rooting dropping to zero during the fall and winter when cambial activity stops (*2*). Softwood cuttings of deciduous woody species taken during spring or summer usually root more readily than hardwood cuttings procured in the winter. For plants difficult to root, it is thus often necessary to resort to the use of softwood cuttings. In tests (*98*) with cherries, not one hardwood cutting taken in winter was induced to root, whereas softwood cuttings made in the spring gave satisfactory rooting with most cultivars. This situation is also well illustrated in the lilacs (*Syringa* spp.). About the only way cuttings of these plants can be rooted is to make softwood cuttings during a short period in the spring when the shoots are several inches long and in active growth. The Chinese fringe tree (*Chionanthus retusus*) is notoriously difficult to root, but by taking cuttings during a short period in mid-spring, high rooting percentages can be obtained (*242*).

The effect of timing is also strikingly shown by difficult-to-root deciduous azalea cuttings. These root readily if the cuttings are taken from succulent growth in early spring; by late spring, however, the rooting percentages rapidly decline (*150*).

Figure 9–18 shows the importance of time of collection in rooting hardwood carob (*Ceratonia siliqua*) cuttings. No rooting was obtained from early fall through late winter, but cuttings rooted well from early spring through midsummer. For any given plant, empirical tests are required to determine the optimum time of taking cuttings, which is more related to the physiological condition of the plant than to any given calendar date.

It is possible for hardwood cuttings of deciduous species to be taken at any time from just before leaf fall until the buds start development in the spring. For

**Figure 9–18**    Seasonal fluctuations in rooting of budded and disbudded carob (*Ceratonia siliqua*) hardwood cuttings. All cuttings treated with indolebutyric acid at 7500 ppm. Untreated cuttings did not root. Redrawn from Fadl, El-Deen, and El-Mahadi (*66*).

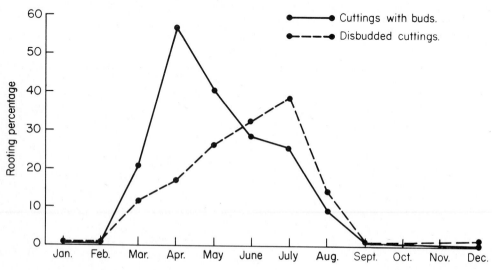

easily rooted species, it makes little difference when the cuttings are made during the dormant season. Rapidly developing buds sometimes tend to promote root formation, whereas buds in the "rest period" may inhibit root development (*161*).

Often the effects of timing are merely a reflection of the response of the cuttings to environmental conditions at the different times of the year. When hardwood cuttings of deciduous species are taken and planted in the nursery in early spring after the rest period of the buds has been broken by winter chilling, the results are quite often a complete failure, since the buds quickly open with the onset of warm days. The newly developing leaves will start transpiring and remove the moisture from the cuttings before they have the opportunity to form roots; thus they soon die. If cuttings can be taken and planted in the fall while the buds are still in the rest period, roots may form and be well established by the time the buds open in the spring. A preplanting moist, warm (15° to 21°C; 60° to 70°F) storage period is often helpful in starting adventitious root initiation (*99, 100, 102, 103, 127*).

For softwood cuttings of deciduous species, best results are generally secured if the cuttings are taken in the spring just as soon as the leaves are fully expanded and the shoots have attained some degree of maturity. Good results are sometimes obtained when softwood cuttings are taken from potted plants that have started growth early in the spring after being moved into a greenhouse.

Broad-leaved evergreens usually root most readily if the cuttings are taken after a flush of growth has been completed and the wood is partially matured. This occurs, depending upon the species, from spring to late fall.

In rooting cuttings of narrow-leaved evergreens, best results may be expected if the cuttings are taken during the period from late fall to late winter. For some species (Figure 9–19), day length—changing during the seasons—

**Figure 9–19**    Rooting of 'Andorra' juniper cuttings taken at monthly intervals throughout the year at Lafayette, Indiana. Note that the photoperiod under which the cuttings were rooted had very little effect. Courtesy Lanphear and Meahl (*157*).

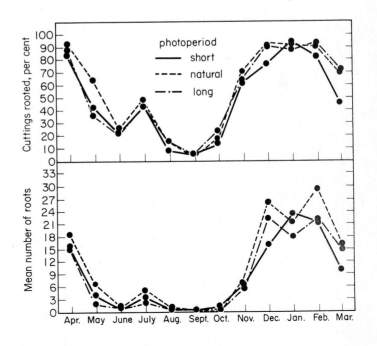

apparently is not a factor accounting for this striking seasonal variability in rooting. Juniper cuttings held under long and short days and under natural day lengths rooted with equal ease or difficulty. Rooting was lowest during the season of active vegetative growth and highest during the dormant period. Furthermore, the low temperatures occurring at the time when such coniferous evergreens root best apparently is not a requirement. Juniper stock plants held in a warm greenhouse from early fall to midwinter produced cuttings that rooted better than those taken from out-of-door stock plants (*158*).

The season of the year in which root cuttings are taken is very important with some species; in others it makes no difference. For example, in the red raspberry (*Rubus idaeus*), regeneration from root cuttings taken from autumn to spring was almost 100 percent successful. Cuttings taken during the summer months failed to survive. On the other hand, root cuttings of horseradish were completely successful regardless of the time of year they were taken (*131*).

## Treatment of Cuttings

### GROWTH REGULATORS

*To induce adventitious roots*    Before the use of synthetic root-promoting growth regulators (auxins) in rooting stem cuttings, many other chemicals were tried with varying degrees of success. In some of the earlier studies (*42*) of this type, now of historical interest, treatments of tomato and privet cuttings with sugar as well as with compounds of manganese, iron, and phosphorus were tried. Improved rooting often resulted, especially with potassium permanganate. The effect of such compounds, however, was so slight and variable that they are no longer used.

Zimmerman (*299*) showed in 1933, before auxins were discovered, that certain unsaturated gases, such as ethylene, carbon monoxide, and acetylene, stimulate initiation of adventitious roots as well as development of latent, pre-existing root initials. Cuttings of many herbaceous plants respond to these gases with increased rooting. Many studies some 30 years later showed that ethylene is a naturally-occurring plant hormone.

The discovery in 1934 and 1935 that auxins, such as **indoleacetic acid (IAA)** and **indolebutyric acid (IBA),** were of real value in stimulating the production of adventitious roots in stem and leaf cuttings was a major milestone in propagation history. The response, however, is not universal; cuttings of some difficult-to-root species still root poorly after treatment with auxin. It is believed that certain naturally occurring materials (rooting cofactors) limit rooting in such cases.

A practice of some European and Middle Eastern gardeners in early days was to embed grain seeds into the split ends of cuttings to promote rooting. This seemingly odd procedure had a sound physiological basis, for it is now known that germinating seeds are good producers of auxin, which, of course, would materially aid in root formation in the cuttings.

Some of the phenoxy compounds having auxin activity promote root formation even at very low concentrations (*123, 125, 246*). The weed-killer, **2,4-dichlorophenoxyacetic acid (2,4–D),** is quite potent at low concentrations in inducing rooting of certain species, but it has the disadvantage of inhibiting shoot development and is very toxic to plants at higher concentrations.

Mixtures of root-promoting substances are sometimes more effective than either component alone. For example, equal parts of indolebutyric acid (IBA) and naphthaleneacetic acid (NAA), when used on a number of widely diverse species, were found to induce a higher percentage of cuttings to root and more roots per cutting than either material alone (124).

Adding a small percentage of certain of the phenoxy compounds to either IBA or NAA has, in some instances, caused excellent rooting (125) and produced root systems qualitatively better than those obtained with phenoxy compounds alone. Likewise, other tests (57) have shown that a combination of IBA, NAA, and 2,4–D gave much better rooting than the use of any of these auxins alone.

The acid form of most of these compounds is relatively insoluble in water, but can be dissolved in a few drops of alcohol or ammonium hydroxide before adding to the water. The salts of some of the growth regulators may be more desirable than the acid in some instances, because of their comparable activity and greater solubility in water (300). The aryl esters of both IAA and IBA have been reported (92) to be more effective than the acids in promoting rooting initiation.

For general use in rooting stem cuttings of the majority of plant species, indolebutyric acid, or sometimes, naphthaleneacetic acid, is recommended (Figure 9–20). To determine the best material and optimum concentration for rooting any particular species under a given set of conditions, empirical trials are necessary.

**Figure 9–20**    Structural formulas of some growth regulators (auxins) active in promoting adventitious root initiation in cuttings.

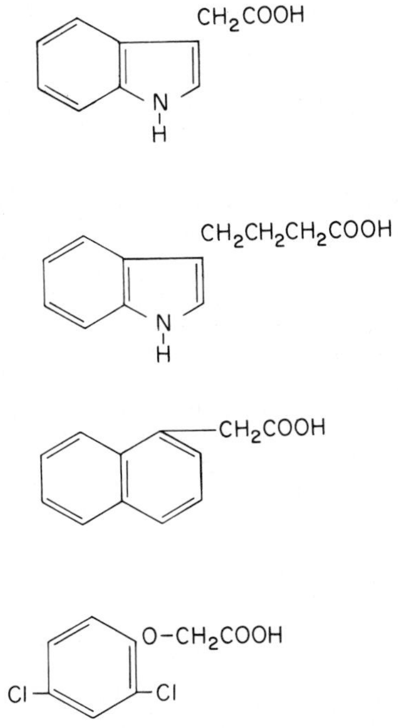

3—Indoleacetic acid
beta—Indoleacetic acid
(IAA)

3—Indolebutyric acid
gamma—(Indole–3)–butyric acid
(IBA)

1—Naphthaleneacetic acid
alpha—naphthaleneacetic acid
(NAA)

2,4–Dichlorophenoxyacetic acid
(2,4–D)

Application of synthetic auxins to stem cuttings at high concentrations can inhibit bud development, sometimes to the point at which no shoot growth will take place even though root formation has been adequate. Also, applications of rooting substances to root cuttings may inhibit the development of shoots from such root pieces.

There is often a question of how long the various root-promoting preparations will keep without losing their strength. Bacterial destruction of IAA occurs readily in unsterilized solutions. A 9-ppm concentration was found to disappear in 24 hours and a 100-ppm concentration in 14 days. In sterile solutions, this material remains active for several months. A widely distributed species of *Acetobacter* destroys IAA, but the same organism has no effect on IBA. Uncontaminated solutions of NAA and 2,4–D maintained their strength for as long as a year.

IAA is sensitive to light, strong sunlight destroying a 10-ppm concentration in about 15 minutes. IBA is much more light-stable than IAA, a 20-hour exposure to strong sunlight causing only a slight change in concentration. Both NAA and 2,4–D seem to be entirely light-stable. With their resistance to bacterial decomposition and light destruction, the latter two compounds are more likely to maintain their effectiveness over a long period of time than the indole compounds. When dilute indoleacetic acid solutions are prepared, they should be used immediately, owing to their rapid deterioration (*177*). In the plant, too, the enzyme system, indoleacetic acid oxidase, will break down IAA but it has no effect on IBA.

The natural auxin flow in stem tissue is in a basipetal direction (apex to base). In early work, synthetic auxin applications were made to cuttings at the upper end to conform to the natural downward flow. As a practical matter, however, it was soon found that basal applications gave better results. Sufficient movement apparently resulted to carry the applied auxin into the parts of the cutting where it stimulated root production. In tests using radioactive IAA (4000 ppm for 5 seconds) for rooting leafy plum cuttings, it was noted (*249*) that IAA was absorbed and distributed throughout leafy cuttings in 24 hours, whether application was at the apex or base. However, with basal application, most of the radioactivity remained in the basal portion of the cuttings, this being the case from 24 hours after treatment until rooting occurred. Leafless cuttings absorbed the same amount of IAA as leafy cuttings, indicating that transpiration "pull" was not the chief cause of absorption and translocation. Twenty-eight days after the cuttings were made and placed under mist there was still considerable radioactivity throughout the cuttings and in the adventitious roots which subsequently developed.

In respiration studies of tissues at the basal ends of IBA-treated and of control cuttings it was found that by the time roots had formed on treated cuttings, their respiration rate was four times as great as in untreated cuttings. In addition, IBA-treated cuttings had a considerably higher level of amino acids at their bases 48 hours after treatment than untreated cuttings. This pattern continued, with nitrogenous substances accumulating in the basal part of treated cuttings, apparently mobilized in the upper part and translocated as asparagine (*250*).

Tests have been conducted by research workers all over the world, using most of the species of plants whose propagation is of interest, to determine the value of the various synthetic root-promoting substances in inducing roots on cuttings. (For application methods of root-promoting growth regulators, see chapter 10.)

*To induce adventitious shoots*     In propagation by leaf or root cuttings the production of adventitious buds and shoots is necessary, along with the development of adventitious roots. Treatments with one of the **cytokinins,** such as **kinetin, BA,** or **PBA** may be helpful (*44, 107, 290*) (see Figure 6–9).

MINERAL NUTRIENTS

The rooting of cuttings has been distinctly promoted by the addition of **nitrogenous compounds** in a number of different plants. The addition of several nitrogen compounds, both organic and inorganic, was found (*49*) to have a beneficial effect on the rooting response of rhododendron cuttings. Also, in experiments (*257*) in rooting bean leaf cuttings, treatments with two organic forms of nitrogen—asparagine and adenine—were found to be very effective. Leafless hibiscus cuttings, when treated with arginine or ammonium sulfate in combination with sucrose, were stimulated markedly in initiating roots, provided that they were also treated with auxin (*271*). The fact that concentrations as low as 0.05 ppm are sometimes quite effective supports the hypothesis that the role of these nitrogenous materials may be in hormonal interactions.

**Boron** stimulates root production in cuttings, at least for some plants, owing, generally, to its promotion of root growth rather than to an effect on root initiation. Experiments (*105*) in rooting cuttings prepared from bean hypocotyls showed that when cuttings were placed in nutrient solutions completely lacking in boron, visible roots failed to appear, although in complete nutrient solutions or in solutions lacking in certain other trace elements, adequate rooting took place. The use of boron—in combination with IBA—increased the rooting percentage, the number and length of roots, and the speed of rooting of English holly cuttings taken in the fall. Apparently a synergistic reaction with IBA occurred, since boron alone had no effect (*279*). While the mechanism of the action of boron in stimulating rooting is not precisely known, there is evidence (*280*) to support the belief that boron is acting in oxidative processes, possibly increasing mobilization of oxygen-rich citric and iso-citric acids into the rooting tissues.

FUNGICIDES

Adventitious root initiation followed by survival of the rooted cuttings are two different phases. Often cuttings root but do not survive for very long. During the rooting and immediate postrooting period, cuttings are subject to attacks by various microorganisms. Treatments with fungicides should give some protection and result in both better survival and improved root quality. Such benefits have been demonstrated in a number of instances, but there is a question whether improvement is due to protection from fungus attacks, or stimulation of root initiation by the fungicide, or both. Studies (*93*) indicate, as shown in Figure 9–21, that captan[7] may act by protecting the newly formed roots from fungal attacks and giving increased cutting survival. Other reports (*267, 284*) also indicate considerable improvement in cutting survival and in the quality of the rooted cuttings resulting from the use of captan. Captan may be used as a powder dip following an IBA treatment, or the IBA in talc may be mixed with the captan powder. Captan is especially suitable for treating cuttings, since it does not decompose easily and has a long residual action.

[7] N-trichloro methyl mercapto-4-cyclohexene 1,2-dicarboximide

**Figure 9-21**  Typical peach hardwood cuttings placed for rooting in sterile peat moss. Treatments, left to right: control; captan alone, 25 percent; indolebutyric acid alone, 4000 ppm, 5 sec dip; IBA, 4000 ppm, 5 sec dip plus 25 percent captan. In this case the captan was not responsible for stimulation of root initiation.

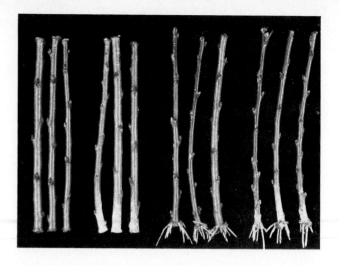

Benomyl[8] is a very effective systemic fungicide, controlling many fungi on a wide range of host plants and its use as a preplanting soak for cuttings has been shown to promote cutting survival (*68*). In tests (*252*) with cuttings of *Pinus strobus,* treatments with a mixture of benomyl (5 percent) and captan (25 percent) in talc gave maximum rooting. It has been noted (*126*), too, that untreated rhododendron cuttings became heavily infected by the fungus, *Cylindrocladium,* and did not root, while treatments with benomyl inhibited development of the fungus and greatly improved rooting.

WOUNDING

Basal wounding is beneficial in rooting cuttings of certain species, such as rhododendrons and junipers, especially cuttings with older wood at the base. Following wounding, callus production and root development frequently are heavier along the margins of the wound. Evidently, wounded tissues are stimulated into cell division and production of root primordia. This is due, perhaps, to a natural accumulation of auxins and carbohydrates in the wounded area and to an increase in the respiration rate. In addition, injured tissues from wounding would be stimulated to produce ethylene, which is known to promote adventitious root formation (*153, 299*).

Wounded cuttings probably absorb more water from the medium than unwounded, and wounding would also permit greater absorption of applied growth regulators by the tissues at the base of the cuttings. In stem tissue of some species there is a sclerenchymatic ring of tough fiber cells in the cortex external to the point of origin of adventitious roots. There is some evidence (*10, 32*) that newly formed roots may have difficulty penetrating this band of cells. A shallow wound would cut through these cells and perhaps permit outward penetration of the developing roots more readily.

[8] 1-(butylcarbamoyl)-2-benzimidazole carbamic acid, methyl ester

# Environmental Conditions during Rooting

WATER RELATIONS

Although the presence of leaves on cuttings is a strong stimulus to root initiation, loss of water from the leaves may reduce the water content of the cuttings to such a low level as to cause them to die before root formation can take place. For good rooting of leafy cuttings it is essential that they maintain their turgidity and have a high leaf water potential.[9] Rooting procedures should be aimed at accomplishing this goal. Measurements (*167*) have shown that low water potentials (well below −10 bars) often occur, and this low level is related to poor rooting.

In cuttings, the natural water supply to the leaf from the roots has been cut off, yet the leaf still carries on transpiration. In cuttings of species that root rapidly, quick root formation soon permits water uptake to compensate for that removed by the leaves, but in more slowly rooting species, water loss from the leaves must be reduced to a very low rate to keep the cuttings alive until roots form. To reduce transpiration of the leaves on cuttings to a minimum, the vapor pressure of the water in the atmosphere surrounding the leaves should be maintained nearly equal to the water vapor pressure in the intercellular spaces within the leaf.

There are different methods by which water loss from the leaves of cuttings during rooting can be reduced (*71*).

**1.** Rooting the cuttings in a glass (or polyethylene) covered propagating frame, either within a greenhouse or out-of-doors (Figure 2–10). Frequent watering, automatically or by hand, maintains a high-humidity environment within the frame. Such closed structures require close temperature control, usually by shading, otherwise they become excessively hot during sunny days. Sometimes the entire greenhouse becomes the high-humidity structure.

An adaptation of this method is the added feature of a humidification system incorporating ventilation. Automatic rotating humidifiers finely atomize large quantities of water, which are dispersed into forced-air ventilation to give fog-laden air across the cuttings (*178, 179*).

**2.** Laying a thin (0.03 mil) layer of polyethylene film over the bed of leafy cuttings, the film directly touching the leaves, and tucking in the film around the edges. If the bed is well watered before the film is laid down and is situated in a shaded lathhouse, cuttings will root, especially in cool climate areas. This method is widely and successfully used in Holland and Belgium. (See Figure 10–26.)

**3.** Use of mist propagation in which water loss from the leaves is reduced by water mist sprays directly on the leaves. This spray reduces leaf temperature and maintains a high-humidity environment around the leaves.

---

[9] **Water potential** refers to the difference between the activity of water molecules in pure distilled water at atmospheric pressure and 30°C (standard conditions), and the activity of water molecules in any other system in the plant. Pure water under these conditions has a water potential of zero. Since the activity of water in a cell is usually less than that of pure water, the water potential in a cell is usually a negative number. The magnitude of water potential is expressed in ''bars'' (1 bar = 0.987 atmosphere).

## Mist Propagation

A major advance in propagation of plants, starting about 1940, occurred with the development of techniques for rooting leafy cuttings under *mist* (*72, 193, 205, 245*). (See Figure 10–29.) Such sprays maintain a film of water on the leaves, which not only results in a high relative humidity surrounding the leaf, but also lowers the air and leaf temperature—all factors tending to lower the transpiration rate. In tests in which leaf temperatures were recorded by thermocouples, leaves under mist were found to be 5.5° to 8.5°C (10° to 15°F) cooler than leaves not under mist (*156*). In other comparisons, the air temperature in a mist bed in the greenhouse was very uniform, averaging about 21°C (70°F) during the hottest part of the day. Cooling from water sprays is so effective that propagating beds can be placed in full sun in summer without an appreciable temperature increase of the leaves.

A distinction should be made between humidification and mist. Under systems that only increase the relative humidity, the water vapor pressure in the area around the leaves is increased. Under mist systems this increase also occurs, but the leaf itself is covered with a film of water, which has the additional benefit of reducing leaf temperature due to cooling by the water and to evaporation of water from the leaf surfaces, processes which absorb heat. This cooling reduces the internal water vapor pressure in the leaf and consequently the transpiration rate.

Under mist, conditions are ideal for rooting leafy cuttings. Transpiration is reduced to a low level, but the light intensity can be high, thus promoting photosynthetic activity (although there is evidence (*168, 278*) that best rooting is not always obtained at the highest light intensity). The temperature of the entire cutting is relatively low, thereby reducing the respiration rate. On the other hand, in the closed propagating bed, temperatures tend to build up, and some ventilation and shading are necessary. Otherwise the cuttings would "burn up." Under these conditions of higher temperature, the transpiration rate increases and the higher temperature increases respiration throughout the cutting. It can use up its reserve food faster than it is being manufactured, which, of course, will soon result in its death. Cuttings under mist, conversely, can be synthesizing food in excess of that used in respiration, such nutrients being very important in promoting the initiation and development of new roots (*118*). (See Table 9–1.) The leaves of certain species, however, will not tolerate the continual wetting which occurs under mist and bud deterioration can occur with such high moisture levels.

Although mist nozzles should maintain a film of water on the leaves at all times during the daylight hours, the application of additional water as mist is of no advantage; in fact, it is likely to be harmful so the mist is turned off at night. The large amounts of water used in mist systems operating continuously reduce the temperature of the tissues of the cuttings, as well as that of the rooting medium, to that of the water, which may be too low for optimum rooting. Under the universally used *intermittent mist* systems, in which water is applied at frequent, short intervals, relatively little water is used and the temperature of the rooting medium is not adversely low. Continuous mist systems are no longer used.

Studies (*170, 265*) with radioactive isotopes have shown that subjecting leaves to natural or artificial rains, or just soaking in water, will remove nutrients, both organic and inorganic, although considerable variability exists among plant species in regard to their susceptibility to losses of minerals by leaching (*5, 83*). Nutrients added to the mist can replenish, to some extent,

**Table 9–1**    Some comparisons between *Cornus florida* (dogwood) cuttings rooted under mist and in a glass-covered frame.*

|  | *Mist* | *Glass frame* |
|---|---|---|
| *Leaf temperature* | 24°C (75°F) | 30°C (86°F) |
| *Water vapor pressure within leaf (mm mercury)* | 23.76 | 31.82 |
| *Water vapor pressure of surrounding air (mm mercury)* | 17.50 | 17.50 |
| *Vapor pressure differential* | 6.26 | 14.32 |
| *Light intensity (foot-candles)* | 7000 | 240 |
| *Increase in photosynthetic rate under mist (mg $CO_2$/hr/cm$^2$)* | 6.93 | —— |
| *Increase in respiration rate over the photosynthetic rate (mg $CO_2$/hr/cm$^2$)* | —— | 5.26 |
| *Increase in carbohydrates (mg per cutting)* | 138.0 | 17.2 |
| *Percentage of cuttings rooted* | 96 | 22 |

* From data of Hess and Snyder (*118*).

those lost by leaching (*293*). Root initiation itself is not increased under nutrient mist but root quality is better and subsequent growth of the rooted cuttings is enhanced with some plants. In other plants, use of nutrient mist may be harmful, causing foliar toxicity (*146*).

Development of disease organisms under mist might be expected, but such has not been the case; in fact, the reverse has been true. For example, in greenhouse roses, no mildew was found to develop on leaves under mist, but mildew did develop on those not under mist. This is explained by the failure of mildew spores to germinate in water (*156*). Water sprays inhibit the development of powdery mildew (*Sphaerotheca pannosa*), and it may be that other disease organisms are held in check in the same manner (*297*).

With mist, softwood cuttings may be taken early in the season, at a stage most favorable for rooting. Under conventional equipment, such immature cuttings would be difficult to maintain without wilting. As illustrated in Figure 9–22, the reduction in transpiration under mist permits the rooting of large cuttings with a considerable leaf area.

*(For details in setting up mist propagating equipment, see chapter 10.)*

**Figure 9–22**    Propagation under mist enables the use of large cuttings with greater leaf area. This often causes stronger rooting, which results in a larger rooted plant. These cuttings of the olive were treated with indolebutyric acid at 4000 ppm by the concentrated-solution-dip method.

## TEMPERATURE

Daytime air temperatures of about 21° to 27°C (70° to 80°F) with night temperatures about 15°C (60°F) are satisfactory for rooting cuttings of most species, although some root better at lower temperatures. Excessively high air temperatures tend to promote bud development in advance of root development and to increase water loss from the leaves. It is important to have root development ahead of shoot development. In cutting beds, some type of thermostatically controlled heat applied below the cuttings is beneficial in maintaining the temperature at the base of the cutting higher than that of the buds, which materially promotes rooting in most instances.

## LIGHT

In all types of plant growth and development light is of major importance as the source of energy in photosynthesis. In rooting cuttings, the products of photosynthesis are important for root initiation and growth. Light effects in rooting can be due to intensity (irradiance), photoperiod (day length), and light quality. Such effects can be exerted either on the stock plants from which the cutting material is obtained or on the cuttings themselves during the rooting process.

*Intensity (irradiance)*    There is strong evidence from rooting studies with cuttings of many plants, including dahlia (*12*), pea (*94, 95*), rhododendron (*139*), hibiscus (*140*), *Hedera helix* (*202*), and pine (*96, 248*), that a relatively **low** light intensity (irradiance)[10] of *the stock plant* gives better rooting than is obtained with cuttings taken from stock plants grown at a **high** light intensity. In apple, one study (*31*) showed that cuttings would respond to auxin treatments only if the stock plants were grown at low irradiance. On the other hand, with chrysanthemum cuttings, increasing irradiance of stock plants was shown (*69*) to increase rooting of cuttings taken from them.

The level of irradiance under which the *cuttings* themselves are being rooted also can influence rooting. Evidence that cuttings of some plants root best under relatively low irradiance (Figure 9–23) was shown many years ago (*243*) in studies with fluorescent lamps and confirmed in more recent studies with blueberry (*278*), weigela, forsythia, viburnum, and hibiscus cuttings (*168*). However, cuttings of certain herbaceous plants such as chrysanthemum, geranium, and poinsettia rooted better in tests when the irradiance over the cuttings during rooting was increased to 116 w/m$^2$ during trials in winter months. Very high irradiance, however (174 w/m$^2$), damaged leaves on the cuttings, delayed rooting, and reduced root growth.

Reasons for the association of enhanced rooting of cuttings with reduced irradiance given to both the stock plants and the cuttings themselves is not clear, but several theories have been advanced (*94*). It has been shown (*54, 261*) that levels of certain natural growth inhibitors are higher in plant tissue grown in light than in etiolated tissues. These inhibitors may reduce rooting, although

---

[10] **Irradiance** is the relative amount of light as measured by radiant energy per unit area. It is best expressed technically as *watts per square meter of photosynthetically active radiation (PAR)* (about 400 to 750 nanometers). These values can be determined by the use of a calibrated thermopile and a galvanometer. Measurements of spectral distribution can be obtained with a quanta spectrometer.

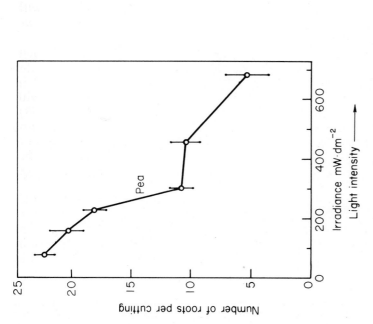

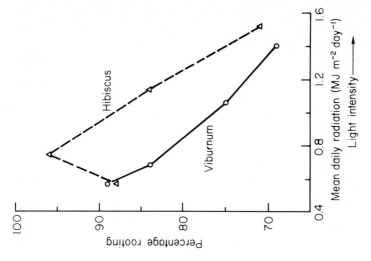

**Figure 9-23** *Left:* Effect of irradiance (light intensity) given the stock plants on the root number subsequently produced on pea cuttings. From data of Hansen and Ericson (*95*). *Right:* Effect of irradiance of hibiscus and viburnum cuttings, during rooting under mist, on rooting percentage. From data of Loach (*168*). Note in both cases that reduced irradiance enhanced rooting, in one instance as affecting the stock plants prior to rooting, in the other as affecting the cuttings themselves during rooting.

this effect has not been proved. A second theory suggests that although actively growing plants under high light often contain more native auxins (*261*), the auxin remains at the growing points, leaving a low level at the basal, root-producing tissues. A third theory is based on relationships known to exist between the carbohydrate content of plant tissue and root initiation (*86, 188*), there being an optimum carbohydrate level for auxin to promote rooting. Carbohydrate levels higher than optimum—which would occur in plants under high irradiance—would not interact efficiently with auxin in root formation.

In conclusion, in working with cuttings of some plants that are difficult to root, possible enhancement of rooting may be obtained by growing the stock plants and rooting the cuttings at reduced irradiance levels, either by placing shade cloth over outdoor-grown stock plants, or by placing container-grown stock plants in a shade house.

*Etiolation*     The practice of **etiolation,**[11] in which the irradiance given the stock plants, or given the tissue which is to be the site of adventitious root initiation, is reduced to zero or near zero, has long been known to be remarkably effective in increasing adventitious root formation in stem tissue (*70*). This effect was shown by Sachs in 1864 (*218*) and has been confirmed many times (*46*). Light has a negative effect on root initiation. Experimental irradiation with white light of the basal part of pea cuttings (*55*) and of aspen and willow cuttings (*56*) during the rooting period caused a strong inhibition in rooting. The mechanism for this inhibition is not known although it is possible that light interferes with the root-promoting action of endogenous auxin. Applications of the synthetic auxin, indolebutyric acid, could overcome the inhibitory effect of direct light on the tissues in pea cuttings (*55*).

The procedures of trench and mound layering (see chapter 14) in which shoot bases are kept in the dark, covered by soil, is based upon etiolation of stem tissue. Even rooting of stem cuttings where the bases are in the dark, within the rooting medium, is probably promoted by stimulation from etiolation effects.

In studies (*112*) with bean cuttings, etiolated tissue was found to react strongly to indolebutyric acid in rooting, implying the existence of some other materials in the tissues that act synergistically as a rooting cofactor. Cuttings under etiolation were found to have a higher level of endogenous auxin (IAA) at the etiolation site during the period of root initiation (*143*).

Adult English ivy (*Hedera helix*) stem tissue is known to be difficult to root. In a study (*87*) in rooting adult shoot apices under aseptic conditions, excellent rooting was obtained under low light intensities or darkness, especially with IAA plus catechol treatments. These same treatments, but under high light intensity, gave little rooting. Other studies (*200*) showed root formation in stem segments of rhododendron cultivars maintained under aseptic conditions to be completely inhibited by continuous light.

A significant study (*154*) with bean hypocotyls implicates the destruction of root-promoting factors by light. DNP (2,4-dinitrophenol) (known to be an uncoupler of oxidative phosphorylation), applied to the tissues, which were then

---

[11] **Etiolation** is the development of plants or plant parts in the absence of light. This results in such characteristics as small, unexpanded leaves, elongated shoots, and lack of cholorophyll resulting in a yellowish or whitish color.

maintained in darkness by covering with opaque rubber tubing, caused root primordia to form along the stem, increasing in amount with concentration. No roots formed on treated but uncovered areas or on those treated but covered by transparent tubing. Adding an auxin (IAA) greatly stimulated the DNP effect.

There are theories (64, 115) that phenolic compounds, interacting with auxin, promote adventitious root formation and, too, certain auxin effects have been shown (259) to be enhanced by phenolic compounds. In the presence of ascorbate and chlorophyll a (naturally occurring materials)—and in the presence of light—DNP is reduced to 2-amino, 4-nitrophenol (174), a compound which is inactive in causing roots to form on bean hypocotyls (154) either in the light or dark. Thus, the promotive influence of the absence of light on tissues during root initiation (the etiolation effect) may possibly be explained by photoinactivation of an essential component in a complex required for root initials to form.

In stimulating root formation by etiolation (74), shoots which later are to be made into cuttings may be allowed to develop during their initial stages in complete darkness, thus assuming the elongated, whitish characteristics of etiolation. The basal portion of the shoots, where the root initials develop, can be retained under this etiolated condition by being covered with black adhesive tape; the terminal leafy portion of the shoots is allowed to develop further in the light, thus assuming a normal type of growth. After the shoots attain sufficient length, they can be cut off, either immediately or at a later time, and placed in a rooting bed.

*Day length (photoperiod)*    There is some evidence that the **photoperiod** under which the stock plants are grown may exert an influence on the rooting of cuttings taken from them (7). This may be related to carbohydrate accumulation, the best rooting being obtained under photoperiods promoting carbohydrate increase, although in some cases (237) stock plants held under short photoperiods have produced the best rooting cuttings. In some species, the photoperiod under which the cutting is actually rooted may affect root initiation, long days or continuous illumination generally being more effective than short days (243), although in other species photoperiod has no influence (234, 237).

This situation can become quite complex, however, since photoperiod can involve shoot development as well as root initiation. For example, in propagation by leaf cuttings there must be development of both adventitious roots and shoots. Using *Begonia* leaf cuttings (106), where the light intensity was adjusted so that the total light energy was about the same under both long days and short days, it was found that short days and relatively low temperatures promoted adventitious bud formation on the leaf pieces, whereas short days suppressed adventitious root formation. Roots formed best under long days with relatively high temperatures.

In rooting cuttings of 'Andorra' juniper, pronounced variations in rooting occurred during the year, but the same variations took place whether the cuttings were maintained under long days, short days, or the natural day length (157). A number of tests have been made of the effect of photoperiod on root formation in cuttings, but the results are conflicting; it is difficult to make any generalization (6, 157, 277).

In some plants photoperiod will control growth after the cuttings have been rooted. Certain plants cease active shoot growth in response to natural changes in day length. This is the case with spring cuttings of deciduous azaleas and dwarf rhododendrons which had rooted and were potted in late summer or early fall. Considerably improved growth of such plants was obtained during the winter in the greenhouse if they were placed under continuous supplementary light, in comparison with similar plants subjected only to the normal short winter days. The latter plants, without added day length, remained in a dormant state until the following spring (*40, 82, 281*).

*Light quality*    Radiation in the orange-red end of the spectrum seems to favor rooting of cuttings more than that in the blue region (*243*), but in one test (*244*), when stock plants were exposed for six weeks to light sources of different quality before taking the cuttings, those from plants exposed to blue light rooted most readily.

The preformed root primordia found in the stems of Lombardy poplar (*Populus nigra,* var. *italica*) develop and emerge if the cuttings are placed in darkness, but they fail to emerge if the cuttings are exposed to light each day. Red light (about 680 nm) is more inhibitory than blue, green, or far-red light (*226*).

ROOTING MEDIUM

The rooting medium has three functions:

**a.** to hold the cutting in place during the rooting period;
**b.** to provide moisture for the cutting; and
**c.** to permit penetration of air to the base of the cutting.

An ideal rooting medium provides sufficient porosity to allow good aeration, has a high water-holding capacity yet is well drained, and is free from harmful pathogens.

The rooting medium can affect the type of root system arising from cuttings. Cuttings of some species, when rooted in sand, produce long, unbranched, coarse, and brittle roots, but when rooted in a mixture, such as sand and peat moss, or perlite and peat moss, develop roots that are well branched, slender, and flexible, a type much more suited for digging and repotting.

Experiments (*169*) to determine which of the characteristic differences between peat moss and sand were responsible for the different types of root system produced indicated that it was the difference in the moisture content. Determinations of the air and moisture content of peat moss and sand when each was at a point considered optimum for rooting showed that, on a volume basis, peat moss contained over twice as much air and three times as much moisture as sand.

The pH level of the rooting medium can be an important consideration in the production of adventitious roots. Studies with *Thuja occidentalis* cuttings rooted in perlite saturated with solutions maintained at various pH values gave best rooting at pH 7. Increasing acidity of the medium markedly inhibited rooting but high alkaline levels did not significantly reduce rooting (*23*). There is evidence, too, in tests (*303*) with mung bean that the pH at which the stock plants are grown and the cuttings rooted can have a pronounced influence on root initiation. In these tests a pH of 6.5 gave much better rooting than 4.5 or 7.5.

The level of exchangeable calcium in peat moss, used as a rooting medium for chrysanthemum cuttings, had an influence on the number of roots per cutting and on root length. Increasing the exchangeable calcium (complementary ion—hydrogen) from 0 to 100 percent caused root numbers per cutting to decrease linearly from 15 to 4, respectively. Root length per cutting was at a maximum with 37.5 percent calcium, decreasing at higher or lower calcium levels. In a rooting medium for chrysanthemum cuttings, between 37.5 and 75 percent calcium saturation should give the best results (195).

In an extension of such studies to several kinds of woody plants, it was noted (196) that cuttings of different species did not respond the same to different levels of exchangeable calcium in the sphagnum peat rooting medium, so generalizations would be difficult. Some plants, such as euonymus and pyracantha, were strongly affected by the calcium level, while osmanthus and rhododendron were not.

Available oxygen in the rooting medium is essential for root production, although the requirement varies with different species. Willow cuttings form roots readily in water with an oxygen content as low as 1 ppm, but English ivy requires about 10 ppm for adequate root growth (298). Rooting of carnation and chrysanthemum cuttings increased markedly as the water in which they were rooted was aerated with increasing amounts of oxygen, from 0 to 21 percent (262). Adding supplementary oxygen to rooting bins in which apple cuttings were being rooted gave pronounced increases in rooting (130). When roots are produced only near the surface of the rooting medium, it is likely that the oxygen supply in the medium is inadequate (see Figure 9–24).

The different kinds and mixtures of rooting media commonly used are discussed in chapter 10.

**Figure 9–24**    Necessity of oxygen for adventitious root development in willow cuttings. Started October 24 and photographed November 16. Water depths in tubes are 2, 8, 15 in. *Left:* water not aerated. *Right:* water aerated with oxygen from a commercial gas cylinder. Roots on cuttings in tubes at the left developed only at the water surface, where oxygen is available. Cuttings in oxygenated water at the right produced roots throughout their length. Courtesy P. W. Zimmerman.

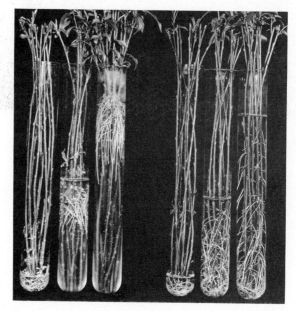

## A Study of Rooting Factors under Aseptic Conditions

A significant study (*200*) using in vitro aseptic conditions showed the effects of many factors on rooting. Small rhododendron explants were used, and different factors affecting rooting were modified one at a time. Several conclusions were:

1. Rooting took place only in segments taken from young, soft shoots.
2. Removal of a bark strip (wounding) promoted rooting.
3. An oxygen supply was essential for rooting.
4. Darkness was necessary for roots to form.
5. There was an optimum temperature for rooting, i.e., 25°C (77°F).
6. Root formation occurred only when sugar was included in the medium (Figure 9–25).
7. The various macroelements, boric acid, and the pH of the medium were not important in rooting.
8. Auxin was an absolute requirement for rooting to take place.
9. A cytokinin, benzylaminopurine, at low concentrations increased the amount of roots produced.
10. Addition of either gibberellic acid or abscisic acid decreased rooting.
11. Variability occurred among the three rhododendron cultivars tested under identical conditions—one was difficult to root, and two were easy—which shows that another, unknown, naturally-occurring factor was playing a regulatory role.

**Figure 9–25**
Necessity for a carbohydrate source for root initiation is shown by in vitro studies with stem segments of rhododendrons. *Left column:* sucrose added. *Right column:* glucose added. *Top row:* 3 percent. *Center row:* 2 percent. *Bottom row:* 1 percent. *Center, below:* Same as above but no sugar added. Courtesy R. L. M. Pierik and H. H. M. Steegmans.

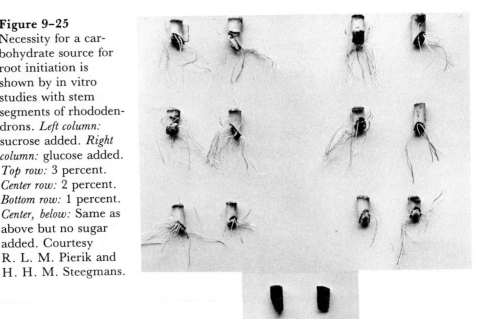

# REFERENCES

1. Abeles, F. B. 1973. *Ethylene in plant biology.* New York: Academic Press.

2. Anand, V. K., and G. T. Heberlein. 1975. Seasonal changes in the effects of auxin on rooting in stem cuttings of *Ficus infectoria. Phys. Plant.* 34:330–34.

3. Ashiru, G. A., and R. F. Carlson. 1968. Some endogenous rooting factors associated with rooting of East Malling II and Malling-Merton 106 apple clones. *Proc. Amer. Soc. Hort. Sci.* 92:106–12.

4. Aung, L. H. 1972. The nature of root-promoting substances in *Lycopersicon esculentum* seedlings. *Phys. Plant.* 26:306–9.

5. Bahn, K. C., A. Wallace, and O. R. Lunt. 1959. Some mineral losses from leaves by leaching. *Proc. Amer. Soc. Hort. Sci.* 73:289–93.

6. Baker, R. L., and C. B. Link. 1963. The influence of photoperiod on the rooting of cuttings of some woody ornamental plants. *Proc. Amer. Soc. Hort. Sci.* 82:596–601.

7. Barba, R. C., and F. A. Pokorny. 1975. Influence of photoperiod on the propagation of two rhododendron cultivars. *Jour. Hort. Sci.* 50:55–59.

8. Bassuk, N. L., and B. H. Howard. 1981. Seasonal rooting changes in apple hardwood cuttings and their implications to nurserymen. *Proc. Inter. Plant Prop. Soc.* 30:289–93.

9. Basu, R. N., B. N. Roy, and T. K. Bose. 1970. Interaction of abscisic acid and auxins in rooting of cuttings. *Plant and Cell Phys.* 11:681–84.

10. Beakbane, A. B. 1969. Relationships between structure and adventitious rooting. *Proc. Inter. Plant Prop. Soc.* 19:192–201.

11. Biran, I., and A. H. Halevy. 1973. Endogenous levels of growth regulators and their relationship to the rooting of dahlia cuttings. *Phys. Plant* 28:436–42.

12. ———. 1973. Stock plant shading and rooting of dahlia cuttings. *Sci. Hort.* 1:125–31.

13. ———. 1973. The relationship between rooting of dahlia cuttings and the presence and type of bud. *Phys. Plant.* 28:244–47.

14. Blakely, L. M., S. J. Rodaway, L. B. Hollen, and S. G. Croker. 1972. Control and kinetics of branch root formation in cultured root segments of *Haplopappus ravenii. Plant Phys.* 50:35–49.

15. Blazich, F. A., and C. W. Heuser. 1979. A histological study of adventitious root initiation in mung bean cuttings. *Jour. Amer. Soc. Hort. Sci.* 104(1):63–67.

16. Bloch, R. 1943. Polarity in plants. *Bot. Rev.* 9:261–310.

17. Boe, A. A., R. B. Steward, and T. J. Banko. 1972. Effects of growth regulators on root and shoot development of *Sedum* leaf cuttings. *HortScience* 74(4):404–5.

18. Bonnett, H. T., Jr., and J. G. Torrey. 1965. Chemical control of organ formation in root segments of *Convolvulus* cultured in vitro. *Plant Phys.* 40:1228–36.

19. Bouillenne, R., and M. Bouillenne-Walrand. 1955. Auxines et bouturage. *Rpt. 14th Inter. Hort. Cong.* Vol. 1, pp. 231–38.

20. Breen, P. J., and T. Muraoka. 1973. Effect of indolebutyric acid on distribution of $^{14}C$ photosynthate in softwood cuttings of Marianna 2624 plum. *Jour. Amer. Soc. Hort. Sci.* 98(5):436–39.

21. ———. 1974. Effect of leaves and carbohydrate content and movement of $^{14}C$-assimilate in plum cuttings. *Jour. Amer. Soc. Hort. Sci.* 99(4):326–32.

22. Brian, P. W., H. G. Hemming, and D. Lowe. 1960. Inhibition of rooting of cuttings by gibberellic acid. *Ann. Bot. n.s.* 24:407–9

23. Bruckel, D. W., and E. P. Johnson. 1969. Effects of pH on rootability of *Thuja occidentalis. The Plant Propagator* 15(4):10–12.

24. Cameron, R. J., and G. V. Thomson. 1969. The vegetative propagation of *Pinus radiata:* Root initiation in cuttings. *Bot. Gaz.* 130(4):242–51.

25. Carlson, M. C. 1929. Origin of adventitious roots in *Coleus* cuttings. *Bot. Gaz.* 87:119–126.

26. ———. 1950. Nodal adventitious roots in willow stems of different ages. *Amer. Jour. Bot.* 37:555–61.

27. Carpenter, J. B. 1961. Occurrence and inheritance of preformed root primordia in stems of citron ( *Citrus medica* L. ). *Proc. Amer. Soc. Hort. Sci.* 77:211–18.

28. Carpenter, W. J., G. R. Beck, and G. A. Anderson. 1973. High intensity supplementary lighting during rooting of herbaceous cuttings. *HortScience* 8(4):338–40.

29. Chin, T. Y., M. M. Meyer, Jr., and L. Beevers. 1969. Abscisic acid stimulated rooting of stem cuttings. *Planta* 88:192–96.

30. Challenger, S., H. J. Lacey, and B. H. Howard. 1965. The demonstration of root promoting substances in apple and plum rootstocks. *Ann. Rpt. E. Malling Res. Sta. for 1964,* pp. 124–28.

31. Christiansen, M. V., E. N. Eriksen, and A. S. Andersen. 1980. Interaction of stock plant irradiance and auxin in the propagation of apple rootstocks by cuttings. *Sci. Hort.* 12:11–17.

32. Ciampi, C., and R. Gellini. 1958. Anatomical study on the relationship between structure and rooting capacity in olive cuttings. *Nuovo Giorn. Bot. Ital.* 65:417–24.

33. ———. 1963. Formation and development of adventitious roots in *Olea europaea* L.: Significance of the anatomical structure for the development of radicles. *Nuovo Giorn. Bot. Ital.* 70:62–74.

34. Cooper, W. C. 1935. Hormones in relation to root formation on stem cuttings. *Plant Phys.* 10:789–94.

35. ———. 1938. Hormones and root formation. *Bot. Gaz.* 99:599–614.

36. ———. 1944. The concentrated-solution-dip method of treating cuttings with growth substances. *Proc. Amer. Soc. Hort. Sci.* 44:533–41.

37. Corbett, L. C. 1897. The development of roots from cuttings. *W. Va. Agr. Exp. Sta. Ann. Rpt.* 9(1895–96):196–99.

38. Cormack, R. G. H. 1965. The effect of calcium ions and pH on the development of callus tissue on stem cuttings of balsam poplar. *Can. Jour. Bot.* 43:75–83.

39. Cormack, R. G. H., and P. L. Lemay. 1966. A further study of callus tissue development on stem cuttings of balsam poplar. *Can. Jour. Bot.* 44:47–50.

40. Crossley, J. H. 1965. Light and temperature trials with seedlings and cuttings of *Rhododendron molle. Proc. Inter. Plant Prop. Soc.* 15:327–34.

41. Crow, W. D., W. Nicholls, and M. Sterns. 1971. Root inhibitors in *Eucalyptus grandis:* Naturally occurring derivatives of the 2,3-dioxabicyclo (4,4,0) decane system. *Tetrahedron Letters 18.* London: Pergamon Press, pp. 1353–56.

42. Curtis, O. F. 1918. Stimulation of root growth in cuttings by treatment with chemical compounds. *Cornell Univ. Agr. Exp. Sta. Mem. 14.*

43. Danckwardt-Lillieström, C. 1957. Kinetin-induced shoot formation from isolated roots of *Isatis tinctoria*. *Phys. Plant.* 10:794–97.

44. Davies, F. T., Jr., and B. C. Moser. 1980. Stimulation of bud and shoot development of Rieger begonia leaf cuttings with cytokinins. *Jour. Amer. Soc. Hort. Sci.* 105(1):27–30.

45. Delargy, J. A., and C. E. Wright. 1978. Root formation in cuttings of apple (cv. Bramley's Seedling) in relation to ringbarking and to etiolation. *New Phytol.* 81:117–127.

46. ———. 1979. Root formation in cuttings of apple in relation to auxin application and to etiolation. *New Phytol.* 82:341–47.

47. Delisle, A. L. 1942. Histological and anatomical changes induced by indoleacetic acid in rooting cuttings of *Pinus strobus* L. *Va. Jour. Sci.* 3:118–24.

48. Deuber, C. G. 1940. Vegetative propagation of conifers. *Trans. Conn. Acad. Arts and Sci.* 34:1–83.

49. Doak, B. W. 1940. The effect of various nitrogenous compounds on the rooting of rhododendron cuttings treated with naphthaleneacetic acid. *New Zealand Jour. Sci. and Tech.* 21:336A–43A.

50. Donoho, C. W., A. E. Mitchell, and H. N. Sell. 1962. Enzymatic destruction of $C^{14}$ labelled indoleacetic acid and naphthaleneacetic acid by developing apple and peach seeds. *Proc. Amer. Soc. Hort. Sci.* 80:43–49.

51. Doorenbos, J. 1954. Rejuvenation of *Hedera helix* in graft combinations. *Proc. Kon. Ned. Akad. Wet., Series C.* 57:99–102.

52. Dore, J. 1953. Seasonal variation in the regeneration of root cuttings. *Nature* 172:1189.

53. Duhamel du Monceau, H. L. 1758. *La physique des arbres,* Vols. 1 and 2. Paris: Guerin and Delatour.

54. Eliasson, L. 1971. Growth regulators in *Populus tremula*. II. Effect of light on inhibitor content in root suckers. *Phys. Plant.* 24:205–8.

55. ———. 1980. Interaction of light and auxin in regulation of rooting in pea stem cuttings. *Phys. Plant.* 48:78–82.

56. Eliasson, L., and L. Brunes. 1980. Light effects on root formation in aspen and willow cuttings. *Phys. Plant.* 48:261–65.

57. Ellyard, R. K. 1981. Effect of auxin combinations on the rooting of *Persoonia chameapitys* and *P. pinifolia* cuttings. *Proc. Inter. Plant Prop. Soc.* 31:251–55.

58. Emery, A. E. H. 1955. The formation of buds on roots of *Chamaenerion angustifolium* (L.) Scop. *Phytomorphology* 5:139–45.

59. Ericksen, E. N. 1973. Root formation in pea cuttings. I. Effects of decapitation and disbudding at different development stages. *Phys. Plant.* 28:503–6.

60. ———. 1974a. Root formation in pea cuttings. II. The influence of indole-3-acetic acid at different development stages. *Phys. Plant.* 30:158–62.

61. ———. 1974b. Root formation in pea cuttings. III. The influence of cytokinin at different development stages. *Phys. Plant.* 30(2):163–67.

62. Esau, K. 1965. *Plant anatomy* (2nd ed.). New York: John Wiley, pp. 513–14.

63. Evans, H. 1952. Physiological aspects of the propagation of cacao from cuttings. *Proc. 13th Inter. Hort. Cong.* 2:1179–90.

64. Fadl, M. S., and H. T. Hartmann. 1967. Isolation, purification, and characterization of an endogenous root-promoting factor obtained from the basal sections of pear hardwood cuttings. *Plant Phys.* 42:541–49.

65. ———. 1967. Relationship between seasonal changes in endogenous promoters and inhibitors in pear buds and cutting bases and the rooting of pear hardwood cuttings. *Proc. Amer. Soc. Hort. Sci.* 91:96–112.

66. Fadl, M., S. A. S. El-Deen, and M. A. El-Mahady. 1979. Physiological and chemical factors controlling adventitious root initiation in carob (*Ceratonia siliqua*) stem cuttings. *Egyptian Jour. Hort.* 6(1):55–68.

67. Farrar, J. H., and N. H. Grace. 1942. Vegetative propagation of conifers. XI. Effects of type of cutting on the rooting of Norway spruce cuttings. *Can. Jour. Res., Sect. C.* 20:116–21.

68. Fiorino, P., J. N. Cummins, and J. Gilpatrick. 1969. Increased production of rooted *Prunus besseyi* Bailey softwood cuttings with pre-planting soak in benomyl. *Proc. Inter. Plant Prop. Soc.* 19:320–36.

69. Fischer, P., and J. Hansen. 1977. Rooting of chrysanthemum cuttings: Influence of irradiance during stock plant growth and of decapitation and disbudding of cuttings. *Scient. Hort.* 7:171–78.

70. Frolich, E. F. 1961. Etiolation and the rooting of cuttings. *Proc. Inter. Plant Prop. Soc.* 11:277–83.

71. Gaffney, J. J. 1978. Humidity: Basic principles and measurement techniques. *HortScience* 13(5):551–55.

72. Gardner, E. J. 1941. Propagation under mist. *Amer. Nurs.* 73(9):5–7.

73. Gardner, F. E. 1929. The relationship between tree age and the rooting of cuttings. *Proc. Amer. Soc. Hort. Sci.* 26:101–4.

74. ———. 1937. Etiolation as a method of rooting apple variety stem cuttings. *Proc. Amer. Soc. Hort. Sci.* 34:323–29.

75. Gautheret, R. J. 1969. Investigations on the root formation in the tissues of *Helianthus tuberosus* cultured in vitro. *Amer. Jour. Bot.* 56(7):702–17.

76. Garner, R. J., and E. S. J. Hatcher. 1962. Regeneration in relation to vegetative growth and flowering. *Proc. 16th Inter. Hort. Cong.*, pp. 105–11.

77. Ginzburg, C. 1967. Organization of the adventitious root apex in *Tamarix aphylla*. *Amer. Jour. Bot.* 54:4–8.

78. Girouard, R. M. 1967. Anatomy of adventitious root formation in stem cuttings. *Proc. Inter. Plant Prop. Soc.* 17:289–302.

79. ———. 1967. Initiation and development of adventitious roots in stem cuttings of *Hedera helix*. *Can. Jour. Bot.* 45:1883–86.

80. ———. 1969. Physiological and biochemical studies of adventitious root formation. Extractible rooting co-factors from *Hedera helix*. *Can. Jour. Bot.* 47(5):687–99.

81. Greenwood, M. S., and G. P. Berlyn. 1973. Sucrose-indole-3-acetic acid interactions on root regeneration by *Pinus lambertiana* embryo cuttings. *Amer. Jour. Bot.* 60(1):42–47.

82. Goddard, W. 1963. Forestalling dormancy and inducing continuous growth of *Azalea molle* with supplementary light for winter propagation. *Proc. Inter. Plant Prop. Soc.* 13:276–78.

83. Good, G. L., and H. B. Tukey, Jr. 1964. Leaching of nutrients from cuttings under mist. *Proc. Inter. Plant Prop. Soc.* 14:138–42.

84. Gorter, C. J. 1969. Auxin-synergists in the rooting of cuttings. *Phys. Plant.* 22:497–502.

85. Grace, N. H. 1939. Rooting of cuttings taken from the upper and lower regions of a Norway spruce tree. *Can. Jour. Res.* 17(C):172–80.

86. Greenwood, M. S., and G. P. Berlyn. 1973. Sucrose: Indoleacetic acid interactions on root regeneration by *Pinus lambertiana* embryo cuttings. *Amer. Jour. Bot.* 60:42–47.

87. Hackett, W. P. 1970. The influence of auxin, catechol, and methanolic tissue extracts on root initiation in aseptically cultured shoot apices of the juvenile and adult forms of *Hedera helix. Jour. Amer. Soc. Hort. Sci.* 95(4):398–402.

88. Hagemann, A. 1932. Untersuchungen an Blattstecklingen. *Gartenbauwiss* 6:69–202.

89. Haissig, B. E. 1965. Organ formation *in vitro* as applicable to forest tree propagation. *Bot. Rev.* 31:607–26.

90. ———. 1972. Meristematic activity during adventitious root primordium development. Influences of endogenous auxin and applied gibberellic acid. *Plant Phys.* 49:886–92.

91. ———. 1973. Influence of hormones and auxin synergists on adventitious root initiation. In *Proc. I.U.F.R.O. Working Party on Reprod. Processes,* Rotorua, New Zealand.

92. ———. 1979. Influence of aryl esters of indole-3-acetic and indole-3-butyric acids on adventitious root primordium initiation and development. *Phys. Plant.* 47:29–33.

93. Hansen, C. J., and H. T. Hartmann. 1968. The use of indolebutyric acid and captan in the propagation of clonal peach and peach-almond hybrid rootstocks by hardwood cuttings. *Proc. Amer. Soc. Hort. Sci.* 92:135–40.

94. Hansen, J. 1976. Adventitious root formation induced by gibberellic acid and regulated by irradiance to the stock plants. *Phys. Plant.* 36:77–81.

95. Hansen, J., and E. N. Ericksen. 1974. Root formation of pea cuttings in relation to the irradiance of the stock plants. *Phys. Plant.* 32:170–73.

96. Hansen, J., L. H. Strömquist, and A. Ericsson. 1978. Influence of the irradiance on carbohydrate content and rooting of cuttings of pine seedlings (*Pinus sylvestris* L.). *Plant Phys.* 61:975–79.

97. Hare, R. C. 1977. Rooting of cuttings from mature water oak (*Quercus nigra*). *Southern Jour. Appl. For.* 1(2):24–25.

98. Hartmann, H. T., and R. M. Brooks. 1958. Propagation of Stockton Morello cherry rootstock by softwood cuttings under mist sprays. *Proc. Amer. Soc. Hort. Sci.* 71:127–34.

99. Hartmann, H. T., and C. J. Hansen. 1958. Effect of season of collecting, indolebutyric acid, and pre-planting storage treatments on rooting of Marianna plum, peach, and quince hardwood cuttings. *Proc. Amer. Soc. Hort. Sci.* 71:57–66.

100. ———. 1958. Rooting pear and plum rootstocks. *Calif. Agr.* 12(10):4, 14, 15.

101. Hartmann, H. T., and F. Loreti. 1965. Seasonal variation in the rooting of olive cuttings. *Proc. Amer. Soc. Hort. Sci.* 87:194–98.

102. Hartmann, H. T., W. H. Griggs, and C. J. Hansen. 1963. Propagation of own-rooted Old Home and Bartlett pears to produce trees resistant to pear decline. *Proc. Amer. Soc. Hort. Sci.* 82:92–102.

103. Hatcher, E. S. J., and R. J. Garner. 1957. Aspects of rootstock propagation. IV. The winter storage of hardwood cuttings. *Ann. Rpt. E. Malling Res. Sta. for 1956,* pp. 101–8.

104. Haun, J. R., and P. W. Cornell. 1951. Rooting response of geranium (*Pelargonium hortorum,* Bailey, var. Ricard) cuttings as influenced by nitrogen phosphorus, and potassium nutrition of the stock plant. *Proc. Amer. Soc. Hort. Sci.* 58:317–23.

105. Hemberg, T. 1951. Rooting experiments with hypocotyls of *Phaseolus vulgaris* L. *Phys. Plant.* 11:1–9.

106. Heide, O. M. 1965. Photoperiodic effects on the regeneration ability of *Begonia* leaf cuttings. *Phys. Plant.* 18:185–90.

107. ———. 1965. Interaction of temperature, auxin, and kinins in the regeneration ability of *Begonia* leaf cuttings. *Phys. Plant.* 18:891–920.

108. ———. 1965. Effects of 6-benzylamino-purine and 1-naphthaleneacetic acid on the epiphyllous bud formation in *Bryophyllum*. *Planta* 67:281–96.

109. ———. 1968. Stimulation of adventitious bud formation in *Begonia* leaves by abscisic acid. *Nature* 219(5157):960–61.

110. ———. 1968. Auxin level and regeneration of *Begonia* leaves. *Planta* 81:153–59.

111. ———. 1969. Non-reversibility of gibberellin-induced inhibition of regeneration in *Begonia* leaves. *Phys. Plant.* 22:671–79.

112. Herman, D. E., and C. E. Hess. 1963. The effect of etiolation upon the rooting of cuttings. *Proc. Inter. Plant Prop. Soc.* 13:42–62.

113. Hess, C. E. 1962. A physiological analysis of root initiation in easy and difficult-to-root cuttings. *Proc. 16th Inter. Hort. Cong.*, pp. 375–81.

114. ———. 1962. Characterization of the rooting co-factors extracted from *Hedera helix* L. and *Hibiscus rosa-sinensis* L. *Proc. 16th Inter. Hort. Cong.*, pp. 382–88.

115. ———. 1963. Naturally-occurring substances which stimulate root initiation. *Col. Int. du Centre Nat. Recherche Sci.* No. 123, pp. 517–27. Paris.

116. ———. 1965. Phenolic compounds as stimulators of root initiation. *Plant Phys.* (suppl.) 40 XLV.

117. ———. 1968. Internal and external factors regulating root initiation. Root Growth (Proc. 15th Easter Sch. Agr. Sci., U. of Nott.). London: Butterworth.

118. Hess, C. E., and W. E. Snyder. 1955. A physiological comparison of the use of mist with other propagation procedures used in rooting cuttings. *Rpt. 14th Inter. Hort. Cong.*, pp. 1133–39.

119. Heuser, C. W. 1976. Juvenility and rooting co-factors. *Acta Hort.* 56:251–61.

120. Heuser, C. W., and C. E. Hess. 1972. Isolation of three lipid root-initiating substances from juvenile *Hedera helix* shoot tissue. *Jour. Amer. Soc. Hort. Sci.* 97(5):571–74.

121. Hiller, Charlotte. 1951. A study of the origin and development of callus and root primordia of *Taxus cuspidata* with reference to the effects of growth regulator. Master's thesis, Cornell Univ., Ithaca, N.Y.

122. Hitchcock, A. E., and P. W. Zimmerman. 1932. Relation of rooting response to age of tissue at the base of green wood cuttings. *Contrib. Boyce Thomp. Inst.* 4:85–98.

123. ———. 1937. A sensitive test for root formation. *Amer. Jour. Bot.* 24:735–36.

124. ———. 1940. Effects obtained with mixtures of root-inducing and other substances. *Contrib. Boyce Thomp. Inst.* 11:143–60.

125. ———. 1942. Root inducing activity of phenoxy compounds in relation to their structure. *Contrib. Boyce Thomp. Inst.* 12:497–507.

126. Hoitink, H. A. J., and A. F. Schmitthenner. 1970. Disease control in rhododendron cuttings with benomyl or thiabendazole mixtures. *Plant Dis. Rpt.* 54:427–30.

127. Howard, B. H. 1965. Increase during winter in capacity for root regeneration in detached shoots of fruit tree rootstocks. *Nature* 208:912–13.

128. Howard, B. H., and J. Y. Sykes. 1966. Regeneration of the hop plant (*Humulus lupulus* L.) from softwood cuttings. II. Modification of the carbohydrate resources within the cutting. *Jour. Hort. Sci.* 41:155–63.

129. ———. 1972. Depressing effects of virus infection on adventitious root production in apple hardwood cuttings. *Jour. Hort. Sci.* 47:255–58.

130. ———. 1975. Improved rooting of cuttings by diffusion of oxygen through the rooting medium. *Jour. Hort. Sci.* 50:173–74.

131. Hudson, J. P. 1955. The regeneration of plants from roots. *Proc. 14th Inter. Hort. Cong.*, Vol. 2, pp. 1165–72.

132. Humphries, E. C. 1960. Inhibition of root development of petioles and hypocotyls of dwarf bean (*Phaseolus vulgaris*) by kinetin. *Phys. Plant.* 13:659–63.

133. Jain, M. K., and K. K. Nanda. 1972. Effect of temperature and some antimetabolites on the interaction effects of auxin and nutrition in rooting etiolated stem segments of *Salix tetrasperma*. *Phys. Plant.* 27:169–72.

134. James, D. J. 1979. The role of auxins and phloroglucinol in adventitious root formation in *Rubus* and *Fragaria* grown *in vitro*. *Jour. Hort. Sci.* 54:273–77.

135. James, D. J., V. H. Knight, and I. J. Thurbon. 1980. Micropropagation of red raspberry and the influence of phloroglucinol. *Scient. Hort.* 12:313–19.

136. James, D. J., and I. J. Thurbon. 1981. Shoot and root initiation *in vitro* in the apple rootstock M9 and the promotive effects of phloroglucinol. *Jour. Hort. Sci.* 56:15–20.

137. Jauhari, O. S., and S. F. Rahman. 1959. Further investigations on rooting in cuttings of sweet lime (*Citrus limettoides*) Tanaka. *Sci. and Cult.* 24:432–34.

138. Johnson, C. R. 1970. The nature of flower bud influence on root regeneration in the *Rhododendron* shoot. Ph. D. Dissertation. Ore. State Univ., Corvallis, Ore.

139. Johnson, C. R., and A. N. Roberts. 1971. The effect of shading rhododendron stock plants on flowering and rooting. *Jour. Amer. Soc. Hort. Sci.* 96:166–68.

140. Johnson, C. R., and D. F. Hamilton. 1977. Rooting of *Hibiscus rosa-sinensis* L. cuttings as influenced by light intensity and ethephon. *HortScience* 12(1):39–40.

141. Jones, O. P., and M. E. Hopgood. 1979. The successful propagation *in vitro* of two rootstocks of *Prunus:* the plum rootstock Pixy (*P. insititia*) and the cherry rootstock F 12/1 (*P. avium*). *Jour. Hort. Sci.* 54:63–66.

142. Kawase, M. 1964. Centrifugation, rhizocaline, and rooting in *Salix alba. Phys. Plant.* 17:855–65.

143. ———. 1965. Etiolation and rooting in cuttings. *Phys. Plant.* 18:1066–76.

144. ———. 1971. Causes of centrifugal root promotion. *Phys. Plant.* 25:64–70.

145. ———. 1976. Centrifugation and rooting of cuttings. *Revista dell' Ortoflorofrutticoltura Italiana* N. 2.

146. Keever, G. I., and H. B. Tukey, Jr. 1979. Effect of nutrient mist on the propagation of azaleas. *HortScience* 14(6):755–56.

147. Kefford, N. P. 1973. Effect of a hormone antagonist on the rooting of shoot cuttings. *Plant Phys.* 51:214–16.

148. Key, J. L. 1969. Hormones and nucleic acid metabolism. *Ann. Rev. Plant Phys.* 20:449–74.

149. Kögl, F., A. J. Haagen-Smit, and H. Erxleben. 1934. Über ein neues Auxin ("Heteroauxin") aus Harn. XI. *Mitteilung. Z. physiol. Chem.* 228:90–103.

150. Kraus, E. J. 1953. Rooting azalea cuttings. *Nat. Hort. Mag.* 32:163–64.

151. Kraus, E. J., and H. R. Kraybill. 1918. Vegetation and reproduction with special reference to the tomato. *Ore. Agr. Exp. Sta. Bul. 149.*

152. Kraus, E. J., N. A. Brown, and K. C. Hamner. 1936. Histological reactions of bean plants to indoleacetic acid. *Bot. Gaz.* 98:370–420.

153. Krishnamoorthy, H. N. 1970. Promotion of rooting in mung bean hypocotyl cuttings with Ethrel, an ethylene releasing compound. *Plant & Cell Phys.* 11:979–82.

154. Krul, W. R. 1968. Increased root initiation in Pinto bean hypocotyls with 2,4-dinitrophenol. *Plant Phys.* 43(3):439–41.

155. Laibach, F., and O. Fischnich. 1935. Künstliche Wurzelneubildung mittels Wuchsstroffpaste. *Ber. Deuts. bot. Ges.* 53:528–39.

156. Langhans, R. W. 1955. Mist for growing plants. *Farm Res.* (Cornell Univ.) 21(3).

157. Lanphear, F. O., and R. P. Meahl. 1961. The effect of various photoperiods on rooting and subsequent growth of selected woody ornamental plants. *Proc. Amer. Soc. Hort. Sci.* 77:620–34.

158. ———. 1966. Influence of the stock plant environment on the rooting of *Juniperus horizontalis* 'Plumosa'. *Proc. Amer. Soc. Hort. Sci.* 89:666–71.

159. Lek, H. A. A., van der. 1925. Root development in woody cuttings. *Meded. Landbouwhoogesch. Wageningen* 38(1).

160. ———. 1930. Anatomical structure of woody plants in relation to vegetative propagation. *Proc. IX. Inter. Hort. Cong.,* pp. 66–76.

161. ———. 1934. Over den invloed der knoppen op de wortelvorming der stekken, *Meded. Landbouwhoogesch. Wageningen* 38(2):1–95.

162. Leopold, A. C. 1964. The polarity of auxin transport, in Meristems and Differentiation. *Brookhaven Symposia in Biology, Rpt. No. 16,* pp. 218–34. Upton, N.Y.: Brookhaven Natl. Lab.

163. Lesham, Y., and B. Lunenfield. 1968. Gonadotropin in promotion of adventitious root production on cuttings of *Begonia semperflorens* and *Vitis vinifera. Plant Phys.* 43:313–17.

164. Libbert, E. 1956. Untersuchungen über die Physiologie der Adventivewurzelbildung, I. Die Wirkungsweise einiger Komponenten des 'Rhizokalinkomplexes.'' *Flora* 144:121–50.

165. Libby, W. J., A. G. Brown, and J. M. Fielding. 1972. Effects of hedging Radiata pine on production, rooting, and early growth of cuttings. *New Zealand Jour. For. Sci.* 2(2):263–83.

166. Lipecki, J., and F. G. Dennis. 1972. Growth inhibitors and rooting cofactors in relation to rooting response of softwood apple cuttings. *HortScience* 7(2):136–38.

167. Loach, K. 1977. Leaf water potential and the rooting of cuttings under mist and polythene. *Phys. Plant.* 40:191–97.

168. ———. 1979. Mist propagation: Past, present, future. *Proc. Inter. Plant Prop. Soc.* 29:216–229.

169. Long, J. C. 1933. The influence of rooting media on the character of roots produced by cuttings. *Proc. Amer. Soc. Hort. Sci.* 29:352–55.

170. Long, W. G., D. V. Sweet, and H. B. Tukey. 1956. The loss of nutrients from plant foliage by leaching as indicated by radioisotopes. *Science* 123:1039–40.

171. MacKenzie, J. A. 1957. The regeneration of plants from roots: Seasonal variations in *Rubus idaes* L. var. Malling Promise. Ph. D. Dissertation, Univ. Nottingham.

172. Mahlstede, J. P., and D. P. Watson. 1952. An anatomical study of adventitious root development in stems of *Vaccinium corymbosum*. *Bot. Gaz.* 113:279–85.

173. Maini, J. S. 1968. The relationship between the origin of adventitious buds and the orientation of *Populus tremuloides* root cuttings. *Bul. Ecol. Soc. Amer.* 49:81–82.

174. Massini, P., and G. Voorn. 1967. The effect of ferrodoxin and ferrous ion on the chlorophyll sensitized photoreduction of dinitrophenol. *Photochem.-Photobiol.* 6:851–56.

175. McVeigh, I. 1938. Regeneration in *Crassula multicava*. *Amer. Jour. Bot.* 25:7–11.

176. Mes, M. G. 1951. Cuttings difficult to root. *Plants and Gardens* 7(2):95–97.

177. ———. 1951. Plant hormones. *Prog. Rpt. Plant Phys. Res. Inst.,* 1950–1951, Univ. of Pretoria.

178. Milbocker, D. C. 1977. Propagation in a humid chamber. *Proc. Inter. Plant Prop. Soc.* 27:455–61.

179. Milbocker, D. C., and R. Wilson. 1979. Temperature control during high humidity propagation. *Jour. Amer. Soc. Hort. Sci.* 104:123–26.

180. Mittempergher, L. 1964. Indagini sull' origine delle radici avventizie in talee legnose di pero. *Ortoflorofrutticoltura Ital.* 48:39–44.

181. Mohammed, S., and E. N. Ericksen. 1974. Root formation in pea cuttings. IV. Further studies on the influence of indole-3-acetic acid at different development stages. *Phys. Plant.* 32:94–96.

182. Mohammed, S. 1975. Further investigations on the effects of decapitation and disbudding at different development stages of rooting in pea cuttings. *Jour. Hort. Sci.* 50:271–73.

183. Molnar, J. M., and L. J. LaCroix. 1972. Studies of the rooting of cuttings of *Hydrangea macrophylla:* Enzyme changes. *Can. Jour. Bot.* 50(2):315–22.

184. ———. 1972. Studies of the rooting of cuttings of *Hydrangea macrophylla:* DNA and protein changes. *Can. Jour. Bot.* 50(3):387–92.

185. Morgan, D. L., E. L. McWilliams, and W. C. Parr. 1980. Maintaining juvenility in live oak. *HortScience* 15(4):493–94.

186. Mullins, M. G. 1972. Auxin and ethylene in adventitious root formation in *Phaseolus aureus* (Roxb.). In *Plant growth substances—1970,* D. J. Carr, ed. Berlin: Springer-Verlag.

187. Muzik, T. J., and H. J. Cruzado. 1958. Transmission of juvenile rooting ability from seedlings to adults of *Hevea brasiliensis. Nature* 181:1288.

188. Nanda, K. K., M. K. Jain, and S. Malhotra. 1971. Effect of glucose and auxins in rooting etiolated stem segments of *Populus nigra. Phys. Plant.* 24:387–91.

189. Naylor, E. E., and B. Johnson. 1937. A histological study of vegetative reproduction in *Saintpaulia ionantha. Amer. Jour. Bot.* 24:673–78.

190. Okoro, O. O., and J. Grace. 1978. The physiology of rooting *Populus* cuttings. II. Cytokinin activity in leafless hardwood cuttings. *Phys. Plant.* 44:167–70.

191. O'Rourke, F. L. 1940. The influence of blossom buds on rooting of hardwood cuttings of blueberry. *Proc. Amer. Soc. Hort. Sci.* 40:332–34.

192. ———. 1944. Wood type and original position on shoot with reference to rooting in hardwood cuttings of blueberry. *Proc. Amer. Soc. Hort. Sci.* 45:195–97.

193. ———. 1949. Mist humidification and the rooting of cuttings. *Mich. Agr. Exp. Sta. Quart. Bul. 32,* pp. 245–49.

194. Paton, D. M., R. R. Willing, W. Nichols, and L. D. Pryor. 1970. Rooting of stem cuttings of eucalyptus: A rooting inhibitor in adult tissue. *Austral. Jour. Bot.* 18:175–83.

195. Paul, J. L., and L. V. Smith. 1966. Rooting of chrysanthemum cuttings in peat as influenced by calcium. *Proc. Amer. Soc. Hort. Sci.* 89:626–30.

196. Paul, J. L., and A. T. Leiser. 1968. Influence of calcium saturation of sphagnum peat on the rooting of five woody species. *Hort. Res.* 8(1):41–50.

197. Pearse, H. L. 1943. The effect of nutrition and phytohormones on the rooting of vine cuttings. *Ann. Bot.* n.s., 7:123–32.

198. ———. 1946. Rooting of vine and plum cuttings as affected by nutrition of the parent plant and treatment with phytohormones. *Sci. Bul. 249, Dept. of Agr. Union of S. Afr.*

199. Petri, P. S., S. Mazzi, and P. Strigoli. 1960. Considerazione sulla formazione delle radici avventizie con particolare riguardo a: *Cucurbita pepo, Nerium oleander, Menyanthes trifoliatae, Solanum lycopersicum, Nuovo Giorn. Bot. Ital.* 67:131–75.

200. Pierik, R. L. M., and H. H. M. Steegmans. 1975. Analysis of adventitious root formation in isolated stem explants of *Rhododendron. Scient. Hort.* 3:1–20.

201. Porlingis, I. C., and I. Therios. 1976. Rooting response of juvenile and adult leafy olive cuttings to various factors. *Jour. Hort. Sci.* 51:31–39.

202. Poulsen, A., and A. S. Anderson. 1980. Propagation of *Hedera helix:* Influence of irradiance to stock plants, length of internode and topophysis of cutting. *Phys. Plant.* 49:359–65.

203. Preston, W. H., J. B. Shanks, and P. W. Cornell. 1953. Influence of mineral nutrition on production, rooting, and survival of cuttings of azaleas. *Proc. Amer. Soc. Hort. Sci.* 61:499–507.

204. Priestley, J. H., and C. F. Swingle. 1929. Vegetative propagation from the standpoint of plant anatomy. *USDA Tech. Bul. No. 151.*

205. Raines, M. A. 1940. Some uses of a spray chamber in experimentation with plants. *Amer. Jour. Bot.,* Suppl. to Vol. 27, No. 10, p. 185.

206. Rajagopal, V., and A. S. Andersen. 1980. Water stress and root formation in pea cuttings. *Phys. Plant.* 48:144–49.

207. Rappaport, J. 1940. The influence of leaves and growth substances on the rooting response of cuttings. *Natuurw Tijdschr.* 21:356–59.

208. Rasmussen, S., and A. S. Andersen. 1980. Water stress and root formation in pea cuttings. II. Effect of abscisic acid treatment of cuttings from stock plants grown under two levels of irradiance. *Phys. Plant.* 48:150–54.

209. Read, P. E., and V. C. Hoysler. 1969. Stimulation and retardation of adventitious root formation by application of B-Nine and Cycocel. *Jour. Amer. Soc. Hort. Sci.* 94:314–16.

210. Reuveni, O., and M. Raviv. 1981. Importance of leaf retention to rooting avocado cuttings. *Jour. Amer. Soc. Hort. Sci.* 106(2):127–30.

211. Robbins, W. J. 1960. Further observations on juvenile and adult *Hedera. Amer. Jour. Bot.* 47:485–91.

212. Roberts, A. N., and L. H. Fuchigami. 1973. Seasonal changes in auxin effect on rooting of Douglas-fir stem cuttings as related to bud activity. *Phys. Plant.* 28:215–21.

213. Robinson, J. C. 1975. The regeneration of plants from root cuttings with special reference to the apple. *Hort. Abst.* 45(6):305–15.

214. Robinson, J. C., and W. W. Schwabe. 1977. Studies on the regeneration of apple cultivars from root cuttings. I. Propagation aspects. *Jour. Hort. Sci.* 52:205–20.

215. ———. 1977. Studies on the regeneration of apple cultivars from root cuttings. II. Carbohydrate and auxin relations. *Jour. Hort. Sci.* 52:221–33.

216. Rom, R. C., and S. A. Brown. 1979. Factors affecting burrknot formation on clonal *Malus* rootstocks. *HortScience* 14(3):231–32.

217. Ryan, G. F., E. F. Frolich, and T. P. Kinsella. 1958. Some factors influencing rooting of grafted cuttings. *Proc. Amer. Soc. Hort. Sci.* 72:454–61.

218. Sachs, J. 1865. Ueber die Neubildung von Adventivwurzelin durch Dunkelheit. *Verhandlingen des naturhistorischen Vereines der preussischen Rheinlande und Westphalens,* pp. 110–11. Abs. in *Bul. Soc. Bot. de France,* 12, Part 2, p. 221.

219. ———. 1880 and 1882. Stoff und Form der Pflanzenorgane. I and II. *Arb. bot. Inst. Würzburg* 2:452–88 and 4:689–718.

220. Sachs, R. M., F. Loreti, and J. DeBie. 1964. Plant rooting studies indicate sclerenchyma tissue is not a restricting factor. *Calif. Agr.* 18(9):4–5.

221. Selim, H. A. A. 1956. The effect of flowering on adventitious root formation. *Meded. Landbouwhoogesch. Wageningen* 56(6):1–38.

222. Samish, R. M., and P. Spiegel. 1957. The influence of nutrition of the mother vine on the rooting of cuttings. *Ktavim* 8:93–100.

223. Sax, K. 1962. Aspects of aging in plants. *Ann. Rev. Plant Phys.* 13:489–506.

224. Schier, G. A. 1973. Origin and development of aspen root suckers. *Can. Jour. For. Res.* 3:39–44.

225. Schraudolf, H., and J. Reinert. 1959. Interaction of plant growth regulators in regeneration processes. *Nature* 184:465–66.

226. Shapiro, S. 1958. The role of light in the growth of root primordia in the stem of Lombardy poplar. In *The physiology of forest trees,* K. V. Thimann, ed. New York: Ronald Press.

227. Siegler, E. A., and J. J. Bowman. 1939. Anatomical studies of root and shoot primordia in 1-year apple roots. *Jour. Agr. Res.* 58:795–803.

228. Sircar, P. K., and S. K. Chatterjee. 1973. Physiological and biochemical control of meristemation and adventitious root formation in *Vigna* hypocotyl cuttings. *The Plant Propagator* 19(1):17–26.

229. ———. 1974. Physiological and biochemical changes associated with adventitious root formation in *Vigna* hypocotyl cuttings. II. Gibberellin effects. *The Plant Propagator* 20(2):15–22.

230. ———. 1980. Effect of foliar applications of kinetin and Ethrel on adventitious root formation at the base of *Vigna* hypocotyl cuttings. *The Plant Propagator* 26(4):3–5.

231. Skoog, F., and C. Tsui. 1948. Chemical control of growth and bud formation in tobacco stem and callus. *Amer. Jour. Bot.* 35:782–87.

232. Smith, N. G., and P. F. Wareing. 1972. The distribution of latent root primordia in stems of *Populus* × *robusta* and factors affecting emergence of preformed roots from cuttings. *Forestry* 45:197–210.

233. Snow, A. G., Jr. 1939. Clonal variation in rooting response of red maple cuttings. *USDA Northeastern For. Exp. Sta. Tech. Note 29*.

234. Snyder, W. E. 1955. Effect of photoperiod on cuttings of *Taxus cuspidata* while in the propagation bench and during the first growing season. *Proc. Amer. Soc. Hort. Sci.* 66:397–402.

235. Spiegel, P. 1954. Auxins and inhibitors in canes of *Vitis. Bul. Res. Coun., Israel* 4:176–83.

236. Stangler, B. B. 1949. An anatomical study of the origin and development of adventitious roots in stem cuttings of *Chrysanthemum morifolium* Bailey, *Dianthus caryophyllus* L., and *Rosa dilecta* Rehd. Ph.D. dissertation, Cornell Univ., Ithaca, N.Y.

237. Steponkus, P. L., and L. Hogan. 1967. Some effects of photoperiod on the rooting of *Abelia grandiflora* Rehd. 'Prostrata' cuttings. *Proc. Amer. Soc. Hort. Sci.* 91: 706–15.

238. Stoltz, L. P., and C. E. Hess. 1966. The effect of girdling upon root initiation: Auxin and rooting co-factors. *Proc. Amer. Soc. Hort. Sci.* 89:744–51.

239. ———. 1966. The effect of girdling upon root initiation: Carbohydrates and amino acids. *Proc. Amer. Soc. Hort. Sci.* 89:734–43.

240. ———. 1966. Factors influencing root initiation in an easy- and difficult-to-root chrysanthemum. *Proc. Amer. Soc. Hort. Sci.* 92:622–26.

241. Stoutemyer, V. T., 1937. Regeneration in various types of apple wood. *Iowa Agr. Exp. Sta. Res. Bul.* 220:309–52.

242. ———. 1942. The propagation of *Chionanthus retusus* by cuttings. *Nat. Hort. Mag.* 21(4):175–78.

243. Stoutemyer, V. T., and A. W. Close. 1946. Rooting cuttings and germinating seeds under fluorescent and cold cathode light. *Proc. Amer. Soc. Hort. Sci.* 48:309–25.

244. ———. 1947. Changes of rooting response in cuttings following exposure of the stock plants to light of different qualities. *Proc. Amer. Soc. Hort. Sci.* 49:392–94.

245. Stoutemyer, V. T., and F. L. O'Rourke. 1943. Spray humidification and the rooting of greenwood cuttings. *Amer. Nurs.* 77(1):5–6, 24–25.

246. ———. 1945. Rooting of cuttings from plants sprayed with growth-regulating substances. *Proc. Amer. Soc. Hort. Sci.* 46:407–11..

247. Stoutemyer, V. T., O. K. Britt, and J. R. Goodin. 1961. The influence of chemical treatments, understocks, and environment on growth phase changes and propagation of *Hedera canariensis. Proc. Amer. Soc. Hort. Sci.* 77:552–57.

248. Strömquist, L., and J. Hansen. 1980. Effects of auxin and irradiance on the rooting of cuttings of *Pinus sylvestris. Phys. Plant.* 49:346–50.

249. Strydom, D. K., and H. T. Hartmann. 1960. Absorption, distribution, and destruction of indoleacetic acid in plum stem cuttings. *Plant Phys.* 35:435–42.

250. ———. 1960. Effect of indolebutyric acid on respiration and nitrogen metabolism in Marianna 2624 plum softwood stem cuttings. *Proc. Amer. Soc. Hort. Sci.* 76:124–33.

251. Swingle, C. F. 1927. Burr knot formation in relation to the vascular system of the apple stem. *Jour. Agr. Res.* 34:533–44.

252. Thielges, B. A., and H. A. J. Hoitink. 1972. Fungicides and rooting of eastern white pine cuttings. *For. Sci.* 18(1):54–55.

253. Thimann, K. V. 1935. On an analysis of activity of two growth-promoting substances on plant tissues. *Proc. Kon. Ned. Akad. Wet.* 38:896–912.

254. ———. 1935. On the plant growth hormone produced by *Rhizopus suinus. Jour. Bio. Chem.* 109:279–91.

255. Thimann, K. V., and A. L. Delisle. 1939. The vegetative propagation of difficult plants. *Jour. Arnold Arb.* 20:116–36.

256. Thimann, K. V., and J. B. Koepfli. 1935. Identity of the growth-promoting and root-forming substances of plants. *Nature* 135:101–2.

257. Thimann, K. V., and E. F. Poutasse. 1941. Factors affecting root formation of *Phaseolus vulgaris. Plant Phys.* 16:585–98.

258. Thimann, K. V., and F. W. Went. 1934. On the chemical nature of the root-forming hormone. *Proc. Kon. Ned. Akad. Wet.* 37:456–59.

259. Thomaszewski, M., and K. V. Thimann. 1966. Interactions of phenolic acids, metallic ions, and chelating agents on auxin-induced growth. *Plant Phys.* 41:1443–54.

260. Thurlow, J., and J. Bonner. 1947. Inhibition of photoperiodic induction in *Xanthium. Amer. Jour. Bot.* 34:603–4.

261. Tillburg, E. 1974. Levels of indole-3-acetic acid and acid inhibitors in green and etiolated bean seedlings (*Phaseolus vulgaris*). *Phys. Plant.* 31:106–11.

262. Tinga, J. H. 1952. The effect of five levels of oxygen on the rooting of carnation cuttings in tap water culture. M.S. Thesis, Cornell Univ., Ithaca, N.Y.

263. Trécul, A. 1846. Recherches sur l' origine des racines. *Ann. Sci. Nat. Bot. Ser.* 3:340–50.

264. Tukey, H. B., and E. L. Green. 1934. Gradient composition of rose shoots from tip to base. *Plant Phys.* 9:157–63.

265. Tukey, H. B., Jr., H. B. Tukey, and S. H. Wittwer. 1958. Loss of nutrients by foliar leaching as determined by radioisotopes. *Proc. Amer. Soc. Hort. Sci.* 71:496–506.

266. Vander der Meer, F. A. 1965. Nerfvergelingsmozaick bij kruisbessen. *Fruitteelt.* 55:245–46.

267. Van Doesburg, J. 1962. Use of fungicides with vegetative propagation. *Proc. 16th Inter. Hort. Cong.,* Vol. 4, pp. 365–72.

268. Vasil, V., and A. C. Hildebrandt. 1965. Differentiation of tobacco plants from single, isolated cells in microcultures. *Science* 150:889–92.

269. Vasilevskaya, V. K. 1957. The anatomy of bud formation on the roots of some woody plants. *Russian Vest. Leningr. Univ., Ser. Bio. Bul.* 1:3–21.

270. Van Overbeek, J., and L. E. Gregory. 1945. A physiological separation of two factors necessary for the formation of roots on cuttings. *Amer. Jour. Bot.* 32:336–41.

271. Van Overbeek, J., S. A. Gordon, and L. E. Gregory. 1946. An analysis of the function of the leaf in the process of root formation in cuttings. *Amer. Jour. Bot.* 33:100–107.

272. Van Tieghem, P., and H. Douliot. 1888. Recherches comparatives sur l'origine des membres endogènes dans les plantes vasculaires. *Ann. Sci. Nat. Bot.* VII. 8:1–160.

273. Venverloo, G. J. 1976. The formation of adventitious organs. III. A comparison of root and shoot formation on *Nautilocalyx* explants. *Z. Pflanzenphysiol.* 80:310–22.

274. Vöchting, H. 1878. *Uber Organbildung im Pflanzenreich.* Bonn: Verlag Max Cohen, pp. 1–258.

275. Walker, R. I. 1940. Regeneration in the scale leaf of *Lilium candidum* and *L. longiflorum. Amer. Jour. Bot.* 27:114–17.

276. Warmke, H. E., and G. L. Warmke. 1950. The role of auxin in the differentiation of root and shoot primordia from root cuttings of *Taraxacum* and *Cichorium. Amer. Jour. Bot.* 37:272-80.

277. Waxman, S., and J. P. Nitsch. 1956. Influence of light on plant growth. *Amer. Nurs.* 104(10):11-12.

278. Waxman, S. 1965. Propagation of blueberries under fluorescent light at various intensities. *Proc. Inter. Plant. Prop. Soc.* 15:154-158.

279. Weiser, C. J., and L. T. Blaney. 1960. The effects of boron on the rooting of English holly cuttings. *Proc. Amer. Soc. Hort. Sci.* 75:704-10.

280. ———. 1967. The nature of boron stimulation to root initiation and development in beans. *Proc. Amer. Soc. Hort. Sci.* 90:191-99.

281. Weiser, C. J. 1963. Rooting and night-lighting trials with deciduous azaleas and dwarf rhododendrons. *Amer. Hort. Mag.* 42:95-100.

282. Welander, M., and I. Huntrieser. 1981. The rooting ability of shoots raised *in vitro* from the apple rootstock A2 in juvenile and in adult growth phase. *Phys. Plant.* 53(3):301-6.

283. Wellensiek, S. J. 1952. Rejuvenation of woody plants by formation of sphaeroblasts. *Proc. Kon. Ned. Akad. Wet.* 55:567-73.

284. Wells, J. S. 1963. The use of captan in rooting rhododendrons. *Proc. Inter. Plant Prop. Soc.* 13:132-35.

285. Went, F. W. 1929. On a substance causing root formation. *Proc. Kon. Ned. Akad. Wet.* 32:35-39.

286. ———. 1934. A test method for rhizocaline, the root-forming substance. *Proc. Kon. Ned. Akad. Wet.* 37:445-55.

287. ———. 1934. On the pea test method for auxin, the plant growth hormone. *Proc. Kon. Ned. Akad. Wet.* 37:547-55.

288. ———. 1935. Hormones involved in root formation. *Proc. 6th Inter. Bot. Cong.* 2:267-69.

289. Whitehill, S. J., and W. W. Schwabe. 1975. Vegetative propagation of *Pinus sylvestris. Phys. Plant.* 35:66-71.

290. Wildren, J. A., and R. A. Criley, 1975. Cytokinins increase shoot production from leaf cuttings of begonia. *The Plant Propagator* 20(4) + 21(1):7-9.

291. Wilkinson, R. E. 1966. Adventitious shoots on saltcedar roots. *Bot. Gaz.* 127:103-4.

292. Winkler, A. J. 1927. Some factors influencing the rooting of vine cuttings. *Hilgardia* 2:329-49.

293. Wott, J. A., and H. B. Tukey, Jr. 1967. Influence of nutrient mist on the propagation of cuttings. *Proc. Amer. Soc. Hort. Sci.* 90:454-61.

294. Wylie, A. W., K. Ryugo, and R. M. Sachs. 1970. Effects of growth retardants on biosynthesis of gibberellin precursors in root tips of peas, *Pisum sativum* L. *Jour. Amer. Soc. Hort. Sci.* 95(5):627-30.

295. Yarborough, J. A. 1932. Anatomical and developmental studies of the foliar embryos of *Bryophyllum calycinum. Amer. Jour. Bot.* 19:443-53.

296. ———. 1936. Regeneration in the foliage leaf of *Sedum. Amer. Jour. Bot.* 23:303-7.

297. Yarwood, C. E. 1939. Control of powdery mildews with a water spray. *Phytopath.* 29:288-90.

298. Zimmerman, P. W. 1930. Oxygen requirements for root growth of cuttings in water. *Amer. Jour. Bot.* 17:842-61.

299. ——. 1933. Initiation and stimulation of adventitious roots caused by unsaturated hydrocarbon gases. *Contrib. Boyce Thomp. Inst.* 5:351–69.

300. ——. 1937. Comparative effectiveness of acids, esters, and salts as growth substances and methods of evaluating them. *Contrib. Boyce Thomp. Inst.* 8:337–50.

301. Zimmerman, P. W., and F. Wilcoxon. 1935. Several chemical growth substances which cause initiation of roots and other responses in plants. *Contrib. Boyce Thomp. Inst.* 7:209–29.

302. Zimmerman, R. H., and O. C. Broome. 1981. Phloroglucinol and *in vitro* rooting of apple cultivar cuttings. *Jour. Amer. Soc. Hort. Sci.* 106(5):648–52.

303. Zucconi, F., and A. Pera. 1978. The influence of nutrients and pH effects on rooting as shown by mung bean cuttings. *Acta Hort.* 79:57–62.

## SUPPLEMENTARY READING

Argles, G. K. 1969. Root formation by stem cuttings. *Nurseryman and garden centre,* Vol. 148, Nos. 18, 19; Vol. 149, No. 3.

Dore, J. 1965. Physiology of regeneration in cormophytes. *Handbuch der Pflanzenphysiologie,* Vol. 15 (Part 2), pp. 1–91. Berlin: Springer-Verlag.

Esau, K. 1977. *Anatomy of seed plants* (2nd ed.). New York: John Wiley.

Fernquist, I. 1966. Studies on factors in adventitious root formation. *Lantbrukshogskolans Annaler* (Annals of Agricultural College of Sweden, Uppsala), Vol. 32. pp. 109–244.

Galston, A. W., P. J. Davies, and R. L. Satler. 1980. Hormonal control of rate and direction of growth. Chapter 9 in *The life of the green plant* (3rd ed.). Englewood Cliffs, N.J.: Prentice-Hall.

Haissig, B. E. 1974. Origins of adventitious roots, Vol. 4, No. 2; Influences of auxins and auxin synergists on adventitious root primordium initiation and development, Vol. 4, No. 2; Consideration of metabolism during adventitious root primordium initiation and development, Vol. 4, No. 2. *New Zealand Jour. For. Sci.*

International Plant Propagators' Society. *Proceedings of Annual Meetings.*

Klein, R. M., and D. T. Klein. 1970. *Research methods in plant science.* New York: Natural History Press.

Komissarov, D. A. 1964. *Biological basis for the propagation of woody plants by cuttings* (translated from Russian). Springfield, Va.: U.S. Dept. Commerce, Clearinghouse Fed. Sci.

In propagation by cuttings, a portion of a stem, root, or leaf is cut from the parent plant, after which this plant part is placed under certain favorable environmental conditions and induced to form roots and shoots, thus producing a new independent plant which, in most cases, is identical with the parent plant.

## THE IMPORTANCE AND ADVANTAGES OF PROPAGATION BY CUTTINGS

Cuttings are the most important means of propagating ornamental shrubs—deciduous species as well as the broad- and narrow-leaved types of evergreens. Cuttings are also used widely in commercial greenhouse propagation of many florists' crops and are commonly used in propagating several fruit species.

For species that can be propagated easily by cuttings, this method has numerous advantages. Many new plants can be started in a limited space from a few stock plants. It is inexpensive, rapid, and simple, and does not require the special techniques necessary in grafting or budding. There is no problem of compatibility with rootstocks or of poor graft unions. Greater uniformity is obtained by absence of the variation which sometimes appears as a result of the variable seedling rootstocks of grafted plants. The parent plant is usually reproduced exactly, with no genetic change.

It is not always desirable, however, to produce plants on their own roots by cuttings even if it is possible to do so. It is often advantageous or necessary to use a rootstock resistant to some adverse soil condition or soil-borne organism, or to utilize available dwarfing or invigorating rootstocks.

## TYPES OF CUTTINGS

Cuttings are made from the vegetative portions of the plant, such as **stems, modified stems** (rhizomes, tubers, corms, and bulbs), **leaves,** or **roots.** Cut-

# 10

# Techniques of Propagation by Cuttings

tings can be classified according to the part of the plant from which they are obtained:

Stem cuttings
   Hardwood
     Deciduous
     Narrow-leaved evergreen
   Semihardwood
   Softwood
   Herbaceous
Leaf cuttings
Leaf-bud cuttings
Root cuttings

Many plants can be propagated by several different types of cuttings with satisfactory results. The preferred type depends on individual circumstances; the least expensive and easiest method is usually selected.

For easy-to-root woody perennial plants, hardwood stem cuttings in an outdoor nursery are ordinarily used because of the simplicity and low cost. For more tender herbaceous species, or for those more difficult to propagate, it is necessary to resort to the more expensive, and more elaborate facilities required for rooting the leafy types of cuttings. Root cuttings of some species are also satisfactory, but cutting material may be difficult to obtain in large quantities.

In selecting cutting material it is important to use stock plants that are free from diseases, moderately vigorous, and of known identity. Propagators should avoid stock plants that have been injured by frost or drought, defoliated by insects, stunted by excessive fruiting, or by lack of soil moisture or proper nutrition, and plants that have made rank, overly vigorous growth.

A commendable practice for the propagator is the establishment of stock blocks as a source of propagating material, where uniform, true-to-type, pathogen-free mother plants can be maintained and held under the proper nutritive condition for the best rooting of cuttings taken from them.

**Stem Cuttings**

The stem cutting, which is the most important type, can be divided into four groups, according to the nature of the wood used: **hardwood, semihardwood, softwood,** and **herbaceous.**

In propagation by stem cuttings, segments of shoots containing lateral or terminal buds are obtained with the expectation that under the proper conditions adventitious roots will develop and thus produce independent plants.

The type of wood, the stage of growth used in making the cuttings, the time of year when the cuttings are taken, and several other factors can be very important in securing satisfactory rooting of some plants. Information concerning these factors is given in chapter 9, although much of this knowledge can be obtained by actual experience over the years in propagating plants.

**Hardwood cuttings** are those made of matured, dormant hardwood after leaves have dehisced and before new shoots emerge in the spring. The use of hardwood cuttings is one of the least expensive and easiest methods of vegetative propagation. Hardwood cuttings are easy to prepare, are not readily perishable, may be shipped safely over long distances if necessary, and require little or no special equipment during rooting.

Because of the low cost of hardwood cutting propagation, this method makes feasible high-density meadow orchards, consisting of precocious dwarfed fruit trees planted many thousands to an acre, in which there is considerable interest. Some peach cultivars, for example, can be propagated easily on a large scale from rooted hardwood cuttings (*11, 16*).

Hardwood cuttings are prepared during the dormant season—late fall, winter, or early spring—usually from wood of the previous season's growth, although with a few species, such as the fig, olive, and certain plum cultivars, two-year-old or older wood can be used. Hardwood cuttings are most often used in propagation of deciduous woody plants, although some broad-leaved evergreens, such as the olive, can be propagated by leafless hardwood cuttings. Many deciduous ornamental shrubs are started readily by this type of cutting. Some common ones are privet, forsythia, wisteria, honeysuckle, crape myrtle, and spiraea. Rose rootstocks, such as *Rosa multiflora,* are propagated in great quantities by hardwood cuttings, to be used as the roots for rose cultivars. A few fruit species are propagated commercially by this method—for example, fig, quince, olive, mulberry, grape, currant, gooseberry, pomegranate, and some plums. Certain trees, such as the willow and poplar, are propagated by hardwood cuttings.

The propagating material for hardwood cuttings should be taken from healthy, moderately vigorous stock plants growing in full sunlight. The wood selected should not be from extremely rank growth with abnormally long internodes, or from small, weakly growing interior shoots. Wood of moderate size and vigor is the most desirable. The cuttings should have an ample supply of stored foods to nourish the developing roots and shoots until the new plant becomes self-sustaining. Tip portions of a shoot, which are usually low in stored foods, are discarded. Central and basal parts make the best cuttings.

Hardwood cuttings vary considerably in length—from 10 to 76 cm (4 to 30 in.). Long cuttings, when they are to be used as rootstocks for fruit trees, permit the insertion of the cultivar bud into the original cutting following rooting, rather than into a smaller new shoot arising from the original cutting.

At least two nodes are included in the cutting; the basal cut is usually just below a node and the top cut 1.3 to 2.5 cm ($\frac{1}{2}$ to 1 in.) above a node. However, in preparing stem cuttings of plants with short internodes, little attention is ordinarily given to the position of the basal cut, especially when quantities of cuttings are prepared and cut to length, many at a time, by a band saw.

The diameter of the cuttings may range from 0.6 to 2.5 or even 5 cm ($\frac{1}{4}$ in. to 1 or 2 in.), depending upon the species. Three different types of cuttings are shown in Figure 10–1: the "**mallet**," the "**heel**," and the "**straight**" cutting. The mallet includes a short section of stem of the older wood, whereas the heel cutting includes only a small piece of the older wood. The straight cutting, not including any of the older wood, is the most common and gives satisfactory results in most instances.

**Figure 10–1** Types of hardwood cuttings. *Left:* straight—the type ordinarily used. *Center:* heel cuttings. A small piece of older wood is retained at the base. *Right:* mallet cuttings. An entire section of the branch of older wood is retained.

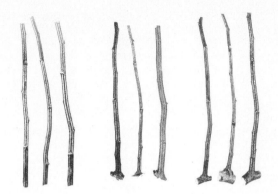

Hardwood cuttings of quince, freshly prepared and after a summer's growth in the nursery, are shown in Figure 10–2.

Where it is difficult to distinguish between the top and base of the cuttings, it is advisable to make one of the cuts at a slant rather than at right angles. In large-scale operations, bundles of cutting material are cut to the desired lengths by band saws or other types of mechanical cutters rather than individually by hand (Figure 10–3). For large-scale commercial operations planting the cuttings is mechanized, using equipment as illustrated in Figure 10–4; but for planting a limited number of cuttings the method shown in Figure 10–5 is satisfactory.

**Figure 10–2** *Left:* hardwood cuttings of quince (*Cydonia oblonga*) prepared and ready for planting in the outdoor nursery in early spring. *Right:* rooted cuttings after one summer's growth. Source: Univ. of Calif. Div. Agr. Sci. Leaflet 21103.

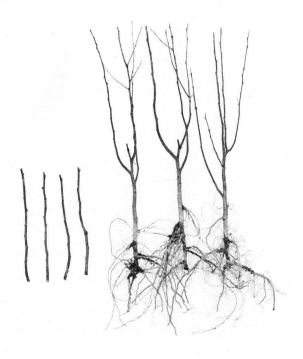

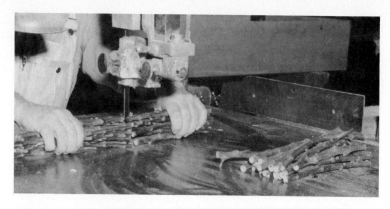

**Figure 10-3**    Sawing hardwood cuttings to length with a band saw. This method is much faster than preparing each cutting individually and in most cases gives equally good results.

**Figure 10-4**    Machine for large-scale planting of hardwood cuttings. Developed at the Tree Nursery Division, P. F. R. A., Indian Head, Saskatchewan, Canada, primarily for propagation of willow and poplar for shelter belt use. This four-unit machine will plant 10 to 12 thousand cuttings per hour. Coutesy Canada Department of Regional Economic Expansion.

**Figure 10–5**     Steps in making and planting hardwood cuttings. *Top left:* preparing the cuttings from dormant and leafless one-year-old shoots. A common length is 15 to 20 cm (6 to 8 in.) and the basal cut is generally made just below a node. *Top right:* treating the cuttings with a root-promoting substance. On the left a bundle of cuttings is dipped in a commercial talc preparation. On the right the basal ends of the cuttings are soaked for 24 hours in a dilute solution of the chemical. With easily rooted plants such treatments are unnecessary. *Middle left:* the cuttings may be planted immediately, but with some plants it is helpful to callus the cuttings for several weeks in a box of moist shavings or peat moss before planting. *Middle right:* planting the cuttings in the nursery row. A *dibble* (heavy, pointed, flat-bladed knife) is a useful tool for inserting the cutting and at the same time firming the soil around the previously planted cutting. *Bottom left:* the cuttings should be planted 7.6 or 10 cm (3 or 4 in.) apart and deep enough so that just one bud shows above the ground. A loose, sandy loam is best for starting hardwood cuttings. *Bottom right:* several weeks after planting, the cuttings start to grow. They must be watered frequently if rains do not occur, and weeds must be controlled.

Several methods are commonly used for preparing and handling hardwood cuttings before planting:

*Winter callusing*    During the dormant season, make the cuttings of uniform length, tie them with heavy rubber bands into convenient-sized bundles, placing the tops all one way, and store them under cool, moist conditions until spring. The bundles of cuttings may be buried out-of-doors in sandy soil, sand, or sawdust in a well-drained location. They may be buried horizontally or in a vertical position, but upside down with the basal end of the cuttings several inches below the surface of the soil. The basal ends are somewhat warmer and better aerated than the terminal ends. This procedure tends to promote root initiation at the base, while retarding bud development at the top. At planting time in the spring, the bundles of cuttings are dug up and the cuttings planted right side up. In regions with mild winters, the bundles of cuttings are often stored during this callusing period in large boxes of moist sand, sawdust, peat moss, or shavings, either in an unheated building or out-of-doors. This probably would not be enough protection for the cuttings, however, in regions where severe, subzero winter temperatures are experienced. A cool, but above-freezing, cellar would be satisfactory for such climates. If refrigerated rooms are available, the cuttings can be safely stored during the callusing period at temperatures of about 4.5° C (40° F) until they are ready to plant.

*Direct spring planting*    It is often sufficient with easily rooted species to gather the cutting material during the dormant season, wrap it in heavy paper or polyethylene with slightly damp peat moss, and store at 0° to 4.5° C (32° to 40°F) until spring. The cutting material should not be allowed to dry out or to become excessively wet during storage. At planting time, the cuttings are made into proper lengths and planted in the nursery.

Stored cutting material should be examined frequently. If signs of bud development appear, lower storage temperatures should be used or the cuttings should be made and planted without delay. If the buds are far developed when the cuttings are planted, leaves will form before the roots appear, and the cuttings will die, because of water loss from the leaves.

*Direct fall planting*    In regions with mild winters, cuttings can be made in the autumn and planted immediately in the nursery. Callusing, and perhaps rooting, may take place before the dormant season starts, or the formation of roots and shoots may occur simultaneously the following spring. Hardwood cuttings of peach and peach × almond hybrids have been successfully rooted in the nursery by this method provided they were treated prior to planting with indolebutyric acid and captan (*16*). Fall-planted cuttings may be injured by rodents and, unless herbicides are used, weed growth may be considerable.

*Warm temperature callusing*    Take the cuttings in the fall while the buds are in or entering the ''rest period,'' treat them with a root-promoting chemical, then store under moist conditions at relatively warm temperatures—18° to 21°C (65° to 70° F)—for three to five weeks to stimulate root initiation. After this, plant the cuttings in the nursery (in mild climates) or hold in cold storage (2° to 4.5° C; 35° to 40° F) until spring. Experimentally, good rooting of hardwood pear cuttings occurred when the cuttings were allowed to callus (and initiate roots) while the buds were under the ''rest'' influence and did not start growth and compete for food reserves in the cuttings (*1*).

*Bottom heat callusing*     This method has been successful for difficult-to-root subjects such as some apple, pear, and plum rootstocks. Cuttings are collected in either the fall or late winter, the basal ends treated with root-promoting

**Figure 10–6**     Steps in propagation by hardwood cuttings using the bottom-heat technique for difficult-to-root materials. *Upper left:* removing 'M 26' apple cuttings from hard-pruned, vigorous hedges by cutting one-year shoots at their base. *Upper right:* 60 cm (2 ft) cuttings inserted (after IBA treatment) to a 25 cm (10 in.) depth in insulated bins filled with rooting compost (one-half coarse peat and one-half grit—½ cm [3/16 in.] gravel and washed sand) maintained at 21°C (70°F) by bottom heat. Bins are situated in a cool building to retard bud development. *Lower left:* root development after six weeks (shown here for plum cuttings). *Lower right:* apple rootstock hardwood cuttings ('M 26', 'M 106', 'MM 111') after one season's growth in nursery. Photos courtesy East Malling Research Station, England.

chemicals (IBA at 2500 to 5000 ppm) then placed upright for about four weeks in damp packing material over bottom heat at 18° to 21° C (65° to 70° F) but with the top portion of the cuttings left exposed to the cool outdoor temperatures. It is best to do this in a covered open shed for protection against excessive moisture from rains. The East Malling Research Station in England has developed (25, 26, 27) commercial procedures, as shown in Figure 10–6, for propagating difficult subjects by this method. Cuttings must be transplanted before buds commence growth; this is usually done as roots begin to emerge. It is important to prevent decay in the cuttings by avoiding excessive application of water to the rooting compost. As long as the correct stimulation has been given, it is not essential to await root emergence before transplanting.

This procedure is probably best suited for regions having relatively mild winters (17, 18, 25). When soil or weather conditions are not suitable for planting after roots become visible, it has been satisfactory to leave the cuttings undisturbed in the rooting bed, shut off the bottom heat, then plant them in the nursery when conditions become suitable (5).

*Plastic bag storage*   The hardwood cuttings are taken during the dormant season, the bases dipped into a root-promoting material—for example IBA at 2000 ppm, for a few seconds—then sealed in polyethylene bags which are placed in the dark at a temperature of about 10° C (50° F). Studies with this technique using peach hardwood cuttings showed 85 to 100 percent rooting after about 50 days (46). While high rooting can be obtained by this method it may be difficult to obtain survival of the cuttings following transplanting.

### HARDWOOD CUTTINGS
### (NARROW-LEAVED EVERGREEN SPECIES)

**Narrow-leaved evergreen cuttings** must be rooted under moisture conditions that will prevent excessive drying as they usually are slow to root, sometimes taking several months to a year. Some species root much more readily than others. In general *Chamaecyparis*, *Thuja*, and the low-growing *Juniperus* species root easily and the yews (*Taxus* spp.) fairly well, whereas the upright junipers, the spruces (*Picea* spp.), hemlocks (*Tsuga* spp.), firs (*Abies* spp.), and pines (*Pinus* spp.) are more difficult. In addition, there is considerable variability among the different species in these genera in regard to the ease of rooting of cuttings. Cuttings taken from young seedling stock plants root much more readily than those taken from older trees because of the juvenility factor. Treatments with root-promoting substances, particularly indolebutyric acid at relatively high concentrations, are usually beneficial in increasing the speed of rooting, the percentage of cuttings rooted, and obtaining heavier root systems.

Narrow-leaved evergreen cuttings ordinarily are best taken between late fall and late winter (see Figure 9–21). Rapid handling of the cuttings after the material is taken from the stock plants is important. The cuttings are usually best rooted in a greenhouse with relatively high light intensity and under conditions of high humidity or very light misting but without heavy wetting of the leaves. A bottom heat temperature of 24° to 26.5° C (75° to 80° F) has given good results. Dipping the cuttings into a fungicide helps prevent disease attacks. Sand alone is a satisfactory rooting medium, as is a 1:1 mixture of perlite and peat moss. Some individual cuttings take longer to root than others. The slower-

**Figure 10–7**    Narrow-leaved evergreen cuttings. *Above:* juniper cuttings ready for sticking in rooting medium. *Below:* cuttings after rooting. Source: Hartmann, H. T., W. J. Flocker, and A. M. Kofranek. 1981. *Plant science.* Englewood Cliffs, N.J.: Prentice-Hall.

rooting ones can be inserted again in the rooting medium, and often will root eventually.

The type of wood to use in making the cuttings varies considerably with the particular species being rooted. As shown in Figure 10–7, the cuttings are made 10 to 20 cm (4 to 8 in.) long with all the leaves removed from the lower half. Mature terminal shoots of the previous season's growth are usually used. In some instances, as with *Juniperus chinensis* 'Pfitzeriana', older and heavier wood also can be used, thus resulting in a larger plant when it is rooted. On the other hand, some nurserymen use small tip cuttings, 5.0 to 6.6 cm (2 to 3 in.) long, placed very close together in a flat for rooting. In some species, as *Juniperus excelsa,* older growth taken from the sides and lower portion of the stock plant roots better than the more succulent tips. Cuttings of *Taxus* root best if they are taken with a piece of old wood at the base of the cutting; such cuttings seem less subject to fungus attacks (*58*). In certain of the narrow-leaved evergreen species, some type of basal wounding is often beneficial in inducing rooting.

SEMIHARDWOOD CUTTINGS

**Semihardwood cuttings** are those made from woody, broad-leaved evergreen species, but leafy summer cuttings taken from partially matured wood of deciduous plants could also be considered as semihardwood. Cuttings of broad-leaved evergreen species are generally taken during the summer from new shoots just after a flush of growth has taken place and the wood is partially matured. Many ornamental shrubs, such as camellia, pittosporum, rhododen-

**Figure 10–8**
Semihardwood cuttings as shown by euonymus. *Left:* cuttings as prepared for rooting from partially matured wood in midsummer. *Right:* cuttings after rooting.

dron, euonymus, the evergreen azaleas, and holly, are commonly propagated by semihardwood cuttings. A few fruit species, such as citrus and olive, can also be propagated in this manner.

The cuttings are made 7.5 to 15 cm (3 to 6 in.) long with leaves retained at the upper end, as shown in Figure 10–8. If the leaves are very large, they should be reduced in size to lower the water loss and to allow closer spacing in the cutting bed. The shoot terminals are often used in making the cuttings, but the basal parts of the stem will usually root also. The basal cut is usually just below a node. The cutting wood should be obtained in the cool, early morning hours when the stems are turgid, and kept wrapped in clean moist burlap or put in large polyethylene bags. Keep out of the sun at all times until the cuttings are made.

It is necessary that leafy cuttings be rooted under conditions that will keep water loss from the leaves at a minimum; commercially, they are ordinarily rooted under intermittent mist sprays or, in cool, moist climates, under polyethylene sheets laid over the cuttings. Bottom heat and growth-regulator treatments are also beneficial. Rooting media, such as 1:1 mixture of perlite and peat moss, or perlite and vermiculite, give satisfactory results.

SOFTWOOD CUTTINGS

Cuttings prepared from the soft, succulent, new spring growth of deciduous or evergreen species may properly be classed as **softwood cuttings.** Many ornamental woody shrubs can be started by softwood cuttings. Typical examples are the hybrid French lilacs, forsythia, magnolia, weigela, and spiraea. Other examples are shown in Figure 10–9. Some deciduous ornamental trees, such as the maples, also can be started in this manner. Although fruit tree species are not commonly propagated by softwood cuttings, those of apple, peach, pear, plum, apricot, and cherry will root, especially under mist.

Softwood cuttings generally root easier and quicker than the other types but require more attention and equipment. This type of cutting is always made with leaves attached. They must, consequently, be handled carefully to prevent drying, and be rooted under conditions which will avoid excessive water loss from the leaves. Temperature should be maintained during rooting at 23° to 27° C

**Figure 10–9**
Softwood cuttings of
several ornamental
species. *Top:* cuttings
made in late spring
from young shoots.
*Below:* cuttings after
rooting. Left to right:
*Myrtus, Pyracantha,
Oleander,* and *Hebe.*

(75° to 80° F) at the base and 21° C (70° F) at the leaves for most species. Soft-wood cuttings produce roots in two to four or five weeks in most cases. In general, they respond well to treatments with root-promoting substances.

It is important in making softwood cuttings to obtain the proper type of cutting material from the stock plant. Such material will vary greatly, however, with the species being propagated. Extremely fast-growing, soft, tender shoots are not desirable, as they are likely to deteriorate before rooting. At the other extreme, older woody stems are slow to root or may just drop their leaves and not root. The best cutting material has some degree of flexibility but is mature enough to break when bent sharply. Weak, thin, interior shoots should be avoided as well as vigorous, abnormally thick, or heavy ones. Average growth from portions of the plant in full light is the most desirable to use. Some of the best cutting material is the lateral or side branches of the stock plant. Heading back the main shoots will usually force out numerous lateral shoots from which cuttings can be made. Softwood cuttings are 7.5 to 12.5 cm (3 to 5 in.) long with two or more nodes. The basal cut is usually made just below a node. The leaves on the lower portion of the cutting are removed, with those on the upper part retained. Large leaves should be reduced in size to lower the transpiration rate and to occupy less space in the propagating bed. All flowers or flower buds should be removed. In some nurseries where quantities of cuttings are prepared, bundles of cutting material are rapidly cut into uniform lengths by paper cutters.

The cutting material is best gathered in the early part of the day and should be kept moist, cool, and turgid at all times by wrapping in damp, clean burlap or placing in large unsealed polyethylene bags (kept out of the sun). Laying the cutting material or prepared cuttings in the sun for even a few minutes will cause serious damage. Soaking the cutting material or cuttings in water for prolonged periods to keep them fresh is undesirable.

**Figure 10-10**     Typical herbaceous cuttings. Left to right: chrysanthemum, begonia, and geranium. It is often necessary with large-leaved plants, such as the begonia, to trim back some of the leaves to prevent wilting and to conserve space in the propagating bench.

HERBACEOUS CUTTINGS

**Herbaceous cuttings** are made from such succulent, herbaceous plants as geraniums, chrysanthemums, coleus, or carnations. They are 7.5 to 12.5 cm (3 to 5 in.) long with leaves retained at the upper end, as shown in Figure 10–10, or without leaves (Figure 10–11). Most florists' crops are propagated by herbaceous cuttings which root easily. They are rooted under the same conditions as softwood cuttings, requiring high humidity. Bottom heat also is helpful. Under proper conditions, rooting is rapid and in high percentages. Although root-promoting substances are usually not required, they are often used to gain uniformity in rooting and development of heavier root systems. Herbaceous cuttings of some plants that exude a sticky sap, such as the geranium, pineapple, or cactus, do better if the basal ends are allowed to dry for a few hours before they are inserted in the rooting medium. This practice tends to prevent the entrance of decay organisms.

**Figure 10–11**     A type of stem cutting consisting only of a leafless stem piece; used here in propagating the monocotyledonous plant, *Dieffenbachia picta*. Latent buds develop into shoots along with the formation of adventitious roots.

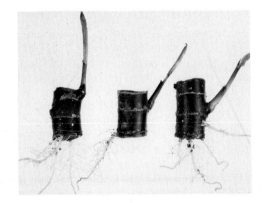

**Figure 10–12**    Leaf cuttings of *Sansevieria. Left:* the thick, leathery leaves are cut into pieces 7.5 to 10 cm (3 to 4 in.) long. To avoid trying to root upside down, the basal end can be marked by cutting on a slant as shown with two of the cuttings. *Right:* development of the plant. The original cutting does not become a part of the new plant

## Leaf Cuttings

In **leaf cuttings,** the leaf blade, or leaf blade and petiole, is utilized in starting new plants. Adventitious roots and an adventitious shoot form at the base of the leaf and develop into the new plant; the original leaf does not become a part of the new plant.

One type of propagation by leaf cuttings is illustrated by *Sansevieria.* The long tapering leaves are cut into sections 7.5 to 10 cm (3 to 4 in.) long as shown in Figure 10–12. These leaf pieces are inserted three-fourths of their length into sand, and after a period of time a new plant forms at the base of the leaf piece, the original cutting disintegrating. The variegated form of *Sansevieria, S. trifasciata laurenti,* is an example of a periclinal chimera that will not reproduce true to type from leaf cuttings; to retain its characteristics, it must be propagated by division of the original plant.

In starting plants with fleshy leaves, such as *Begonia rex,* by leaf cuttings, the large veins are cut on the undersurface of the mature leaf, which is then laid flat on the surface of the propagating medium. The leaf is pinned or held down in some manner, with the natural upper surface of the leaf exposed. After a period of time under humid conditions, new plants form at the point where each vein was cut. The old leaf blade gradually disintegrates.

Another method, sometimes used with fibrous-rooted begonias, is to cut large, well-matured leaves into triangular sections, each containing a piece of a large vein. The thin outer edge of the leaf is discarded. These leaf pieces are then inserted upright in sand with the pointed end down. The new plant develops from the large vein at the base of the leaf piece.

New plants arise from leaves in a variety of ways. An interesting example is illustrated in Figure 10–13, where the new plant develops at the junction of the leaf blade and petiole, even while the leaf is still growing on the mother plant.

The African violet (*Saintpaulia*) is typical of leaf cuttings that can be made of an entire leaf (leaf blade plus petiole), the leaf blade only, or just a portion of the

**Figure 10-13** Leaf cuttings of piggy-back plant (*Tolmiea menziesi*). The large parent leaf (underneath) is set in a moist rooting medium in a humid location for rooting. The new plants (arrows) arise at the junction of the leaf blade and petiole.

leaf blade. The new plant forms at the base of the petiole or midrib of the leaf blade. (See Figures 10-14 and 10-15.)

An unusual type of leaf cutting is illustrated in Figure 10-16, where many new plants arise at the margins of the leaf. The leaf itself eventually deteriorates.

Leaf cuttings should be rooted under the same conditions of high humidity used for softwood or herbaceous cuttings. Root-promoting chemicals are usually helpful.

**Figure 10-14** Leaf cuttings of African violet (*Saintpaulia*). *Left:* each cutting consists of a leaf blade and petiole. *Right:* leaf cuttings after rooting. One or more new plants will form at the base of the petiole. The original leaf can be cut off and used again for rooting.

**Figure 10-15** Inserting African violet leaf cuttings in flats of rooting medium.

**Figure 10–16** Leaf cuttings of *Kalanchoe pinnata (Bryophyllum pinnata)*, air plant. *Left:* new plants developing from foliar "embryos" in the notches at the margin of the leaf. *Right:* leaves ready to lay flat on the rooting medium. They should be partially covered or pegged down to hold the leaf margin in close contact with the rooting medium.

## Leaf-Bud Cuttings

A **leaf-bud cutting** consists of a leaf blade, petiole, and a short piece of the stem with the attached axillary bud (Figure 10–17).

Such cuttings are of particular value in plants where adventitious roots but not adventitious shoots are initiated from detached leaves; the axillary bud at the base of the petiole provides for the new shoot. A number of plant species such as the black raspberry (*Rubus occidentalis*), blackberry, boysenberry, lemon, camellia, and rhododendron are readily started by leaf-bud cuttings, as well as many tropical shrubs and most herbaceous greenhouse plants usually started by stem cuttings. Red raspberries (*Rubus idaeus*) apparently will not reproduce in this manner.

Leaf-bud cuttings are particularly useful when propagating material is scarce, because they will produce at least twice as many new plants from the same amount of stock material as can be started by stem cuttings. Each node can be used as a cutting. For plants with opposite leaves, two leaf-bud cuttings can be obtained from each node. Leaf-bud cuttings are best made from material having well-developed buds and healthy, actively growing leaves.

Treatment of the cut surfaces with one of the root-promoting substances should stimulate root production. The cuttings are inserted in the rooting

**Figure 10–17** Leaf-bud cutting used in the propagation of rhododendrons. *Left:* cutting when made. *Center:* root development after several weeks. *Right:* appearance of root ball and new shoot after five months. Courtesy H. T. Skinner.

medium with the bud 1.3 to 2.5 cm (½ to 1 in.) below the surface. High humidity is essential, and bottom heat is desirable for rapid rooting. Sand, or sand and peat moss, 1:1, are satisfactory rooting media for leaf-bud cuttings.

## Root Cuttings

Best results with **root cuttings** are likely to be attained if the root pieces are taken from young stock plants in late winter or early spring when the roots are well supplied with stored foods but before new growth starts. Taking the cuttings during the spring when the parent plant is rapidly making new shoot growth should be avoided. Root cuttings of the Oriental poppy (*Papaver orientale*) should be taken in midsummer, the dormant period for this species.

Securing cutting material in quantities for root cuttings can be quite laborious unless it can be obtained by trimming roots from nursery plants as they are dug.

It is important with root cuttings to maintain the correct polarity when planting. To avoid planting them upside down, the proximal end (nearest the crown of the plant) may be made with a straight cut and the distal end (away from the crown) with a slanting cut. The proximal end of the root piece should always be up. In planting, insert the cutting vertically so that the top is at about soil level. With many species, however, it is satisfactory to plant the cuttings horizontally 2.5 to 5 cm (1 to 2 in.) deep (Figure 10–18), avoiding the possibility of planting them upside down.

In using root cuttings to propagate chimeras with variegated foliage, such as some *Aralias* and *Pelargoniums,* the new plants lose their variegated form.

Propagation by root cuttings is very simple, but the root size of the plant being propagated may determine the best procedure.

**Figure 10–18**    Propagation by root cuttings. *Left:* horseradish (*Armoracia rusticana*) root pieces planted horizontally. *Right:* apple (*Malus pumila*) root pieces set vertically. Adventitious buds arising from the root piece form the new shoot system.

## PLANTS WITH SMALL, DELICATE ROOTS

Root cuttings of plants with small, delicate roots should be started in flats of sand or finely screened soil in the greenhouse or hotbed. The roots are cut into short lengths, 2.5 to 5 cm (1 to 2 in.) long, and scattered horizontally over the surface of the soil. They are covered with a 1.2 cm (½-in.) layer of fine soil or sand. After watering a polyethylene cover or a pane of glass should be placed over the flat to prevent drying until the plants are started. The flats are set in a shaded place. After the plants become well formed, they can be transplanted to other flats or lined-out in nursery rows for further growth.

## PLANTS WITH SOMEWHAT FLESHY ROOTS

Cuttings of plants with fleshy roots are best started in a flat of sandy soil in the greenhouse or hotbed. The root pieces should be 5 to 7.5 cm (2 to 3 in.) long and planted vertically, observing correct polarity. New adventitious shoots should form rapidly, and as soon as the plants become well established with good root development they can be transplanted.

## PLANTS WITH LARGE ROOTS, PROPAGATED OUT-OF-DOORS

Large root cuttings are made 5 to 15 cm (2 to 6 in.) long. They are tied in bundles, care being used to keep the same ends together to avoid planting upside down later. The cuttings are packed in boxes of damp sand, sawdust, or peat moss for about three weeks and held at about 4.5° C (40° F). After this they should be planted 5 to 7.5 cm (2 to 3 in.) apart in a well-prepared nursery soil with the tops of the cuttings level with, or just below, the top of the soil.

Table 10–1 lists many of the species which can be propagated by root cuttings (*9, 12, 42, 50*).

**Table 10–1**    Some species that can be propagated by root cuttings.

| | |
|---|---|
| *Actinidia chinensis* (Kiwifruit) | *Plumbago* spp. (leadwort) |
| *Aesculus paviflora* (bottle-brush buckeye) | *Populus alba* (white poplar) |
| *Ailanthus altissima* (tree-of-heaven) | *Populus tremula* (European aspen) |
| *Albizia julibrissin* (silk tree) | *Populus tremuloides* (quaking aspen) |
| *Aralia spinosa* (devil's walking stick) | *Prunus glandulosa* (dwarf flowering almond) |
| *Artocarpus altilis* (breadfruit) | *Pyrus calleryana* (oriental pear) |
| *Broussonetia papyrifera* (paper mulberry) | *Rhus copallina* (shining sumac) |
| *Campsis radicans* (trumpet vine) | *Rhus glabra* (smooth sumac) |
| *Celastrus scandens* (American bittersweet) | *Rhus typhina* (staghorn sumac) |
| *Chaenomeles japonica* (Japanese flowering quince) | *Robinia hispida* (rose acacia) |
| *Clerodendrum trichotomum* (glory-bower) | *Robinia pseudoacacia* (black locust) |
| *Comptonia peregrina* (sweet fern) | *Rosa blanda* (rose) |
| *Daphne genkwa* (daphne) | *Rosa nitida* (rose) |
| *Eschscholzia californica* (California poppy) | *Rosa virginiana* (rose) |
| *Koelreuteria paniculata* (goldenrain tree) | *Rubus* spp. (blackberry, raspberry) |
| *Ficus carica* (fig) | *Sassafras albidum* (sassafras) |
| *Malus* spp. (apple, flowering crabapple) | *Sophora japonica* (Japanese pagoda tree) |
| *Myrica pennsylvanica* (bayberry) | *Syringa vulgaris* (lilac) |
| *Papaver orientale* (oriental poppy) | *Ulmus carpinifolia* (smooth-leaved elm) |
| *Phlox* spp. (phlox) | |

# STOCK PLANTS:
## SOURCES OF CUTTING MATERIAL

In cutting propagation, the source of the cutting material is very important (*3*). The stock plants, from which the cutting material is obtained, should be:

1. True-to-name and type
2. Free of disease and insect pests
3. In the proper physiological state so that cuttings taken from them are likely to root.

Several sources are possible for obtaining cutting material:

1. From plants growing in the landscape in parks, around houses or buildings, or in the wild. For nurserymen propagating plants for sale this can be a dangerous practice. While the species may be known, identification as to cultivar may be pure guesswork. In addition, such plants may be infected with virus, fungal, or bacterial diseases, which may appear later in the rooted cuttings or in the nursery plants.

2. Prunings from young nursery plants as they are trimmed and shaped. Many nurseries use prunings as the primary source of their cutting material. Sometimes, however, the trimming is not done at the proper time to root the cuttings—and the unrooted cuttings must be stored. Since young nursery plants are often difficult to identify accurately, a mix-up in labeling may cause a great many new plants to be propagated and improperly identified without anyone being aware of the mix-up until the plants have matured.

3. Stock plants especially maintained as a source of cutting material. Although such plants may occupy valuable land space, this is probably the ideal source of cutting material. The history and identity of each stock plant can be determined accurately. The health status can be controlled, and the plants can be maintained in the proper degree of nutrition and vigor.


# ROOTING MEDIA

Cuttings of many species root easily in a variety of rooting media but those more difficult to root may be greatly influenced by the kind of rooting medium used, not only in the percentage of cuttings rooted, but in the quality of root system formed (*29*).

Combinations of some of the materials listed below often give better results than any one used alone. It is advisable to experiment with the plants being propagated under the actual environmental conditions at hand to determine the best rooting mixture. These are discussed in detail in chapter 2.

**Soil** is ordinarily used for planting deciduous hardwood cuttings and root cuttings. A well-aerated sandy loam is preferable to a heavy clay soil, a higher percentage of the cuttings forming roots, which are usually of better quality. Also, in the lighter, sandy soils, cuttings may be planted and—after rooting—dug much sooner following rains than when the heavier soils are used. The nursery soil should be free from nematodes, verticillium, and crown gall. Nematodes can be eliminated effectively by treating the soil prior to planting with some fumigant such as D–D (dichloropropene-dichloropropane). This requires a three-week period or more following application for the fumigant to dissipate (*36*). Soil is not a suitable rooting medium for the more succulent softwood and semihardwood types of cuttings, although some commercial nurserymen have used it successfully. Cuttings of certain easily rooted plants, such as chrysanthemums and geraniums, are sometimes started directly in small containers or plant bands, using a mixture of 2 parts coarse sand to 1 part soil. This mixture should be heat-treated or fumigated before using.

**Sand** was a widely used rooting medium for cuttings in the past. It is inexpensive and readily available. Clean, sharp plaster sand, free from organic matter and soil, as usually supplied to the building trade, was the type used. Sand is not as retentive of moisture, however, as most other rooting media, necessitating more frequent watering—and it is heavy. The sand should be fine enough to retain some moisture around the cuttings yet coarse enough to allow water to drain freely through it. Used alone, very fine particle sand or very coarse sand does not give good results with cuttings of most woody ornamentals. As with other rooting media, it is best to use the sand only once for rooting cuttings unless it can be sterilized.

For such evergreens as yews, junipers, and arborvitaes, sand is a satisfactory rooting medium. With some species, however, cuttings rooted in sand produce a long, unbranched, brittle root system in contrast to the more desirable fibrous and branched systems developed in other media (*29*).

**Peat moss** is often added to perlite in varying proportions, mainly to increase the water-holding capacity of the mixture. This combination makes a good rooting medium for cuttings of many species. Mixtures used vary from 2 parts perlite and 1 of peat moss to 1 part perlite and 3 of peat moss.

Including peat moss in a rooting medium considerably increases the mixture's water-holding capacity, and consequently the danger of overwatering. High proportions of peat moss in the mixture, if kept wet, as in a mist bed, will sometimes cause deterioration of the roots soon after they are formed.

**Shredded sphagnum moss** is sometime used as a rooting medium when mixed with an equal part of sand (*8*).

**Vermiculite** is often used as a rooting medium. A mixture of equal parts of vermiculite and perlite usually gives better results than either material used alone.

**Perlite** is widely used as a rooting medium for leafy cuttings, especially under mist, owing to its good drainage properties. It may be used alone but is best when used in combination, in varying proportions, with peat moss or vermiculite.

**Pumice,** and mixtures of pumice and peat moss, are satisfactory rooting media (*28, 35*). Pumice is volcanic rock closely resembling perlite; it is a sterile product and is available in different grade sizes.

**Synthetic rooting blocks** are becoming more widely used in the nursery industry, being well adapted to automation. Other advantages are their light weight, reproducibility, and sterile condition. However, watering has to be controlled carefully to maintain constant moisture, yet provide adequate aeration.

**Water** can be used to root cuttings of easily propagated species. Its great disadvantage is lack of aeration. Artificially aerating water with air or oxygen, can produce excellent rooting of cuttings of some species (*47, 61*). In aerated water, the best roots are produced near the basal end of the cuttings, whereas in nonaerated water, the best roots are produced near the surface of the water where the oxygen content is higher.

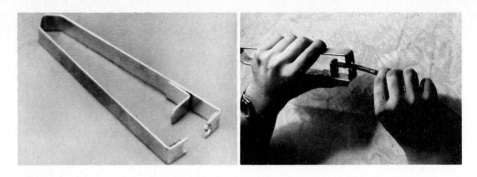

**Figure 10-19** Patented tool designed for making wounding cuts in the base of cuttings to stimulate rooting. Four sharp prongs make the actual cuts as the cutting is pulled through the opening, as shown in the photo on the right.

## WOUNDING

Root production on stem cuttings can be promoted by wounding the base of the cutting. This has proved useful in a number of species such as juniper, arborvitae, rhododendron, maple, magnolia, and holly (*59*). Wounds may be produced in cuttings of narrow-leaved evergreen species, such as arborvitae, by stripping off the lower side branches of the cuttings. A vertical cut with the tip of a sharp knife down each side of the cutting for an inch or two, penetrating through the bark and into the wood, may be enough. A more drastic wound is made with a razor blade device. This consists of four single-edge blades soldered together along their backs. Four wound cuts are then made simultaneously with this equipment.

Larger cuttings, such as magnolias and rhododendrons, may be more effectively wounded by removing a thin slice of bark for about an inch from the base on two sides of the cutting, exposing the cambium but not cutting deeply into the wood. For the greatest benefit, the cuttings should be treated after wounding with one of the root-promoting compounds, either a talc or a concentrated solution dip preparation, working the material into the wounds.

The device shown in Figure 10-19 can be used for wounding cuttings rapidly and uniformly.

## TREATING CUTTINGS
## WITH GROWTH REGULATORS

The purpose of treating cuttings with auxin-type growth regulators ("**hormones**") is to increase the percentage of cuttings that form roots, to hasten root initiation, to increase the number and quality of roots produced per cutting, and to increase uniformity of rooting. Figures 10-20 and 10-21 show examples of the benefits of such materials. Plants whose cuttings root easily may not justify the additional expense and effort of using these materials. Best use of rooting hormones is with plants whose cuttings will root but only with difficulty. The use of these substances, however, does not permit other good practices in cutting propagation, such as the maintenance of proper water relations, temperature, and

**Figure 10-20** Effect of wounding and auxin treatment on the rooting of cuttings of *Juniperus sabina* 'Tamariscifolia' under intermittent mist in the greenhouse. *Top:* wounded. *Below:* not wounded. *Left:* treated with indolebutyric acid at 4000 ppm by the concentrated dip method. *Center:* treated with indolebutyric acid in talc, at 8000 ppm. *Right:* not treated. Cuttings started in early spring and dug six weeks later.

light conditions, to be ignored. The value of these chemicals in propagation is well established, as shown by the tremendous number of reports in scientific and trade journals of tests made on the rooting of cuttings of almost all plant species of economic importance (*39, 52*). Although treatment of cuttings with root-

**Figure 10-21** Effect of indolebutyric acid at four concentrations on rooting of *Escallonia* leafy cuttings under mist. Source: Hartmann, H. T., W. J. Flocker, and A. M. Kofranek. 1981. *Plant science.* Englewood Cliffs, N.J.: Prentice-Hall.

promoting substances is useful in propagating plants, the ultimate size and vigor of such treated plants is no greater than obtained with untreated plants (6).

## Materials

The synthetic root-promoting chemicals that have been found most reliable in stimulating adventitious root production in cuttings are **indolebutyric acid (IBA)** and **naphthaleneacetic acid (NAA)** although others can be used. Indolebutyric acid is probably the best material for general use, because it is nontoxic to plants over a wide concentration range and is effective in promoting rooting of a large number of plant species. These chemicals are available in commercial preparations, dispersed in talc, or liquid formulations that can be diluted with water to the proper strength. The pure chemicals are also available from chemical supply companies, so it is possible for propagators to prepare their own solutions.

## Methods of Application

### COMMERCIAL POWDER PREPARATIONS

Complete directions come with the commercial materials, together with a list of plants that are likely to respond to the particular preparation. Woody, difficult-to-root species should be treated with higher concentration preparations, whereas tender, succulent, and easily rooted species should be treated with lower-strength materials. Fresh cuts should be made at the base of the cuttings shortly before they are dipped into the powder. The operation is faster if a bundle of cuttings is dipped at once rather than each cutting individually, being sure that the inner cuttings in the bundle receive as much powder as those on the outside. The powder adhering to the cuttings after they are lightly tapped is sufficient. If there is little or no natural moisture at the base of the cuttings, they may be pressed against a damp sponge before being dipped in the powder so that more will adhere.

It is advisable in using powder preparations to place a small portion of the stock material into a temporary container, sufficient for the work at hand, and discard any remaining portion after use, rather than dipping the cuttings into the entire stock of powder, which may lead to its early deterioration because of contamination with moisture, fungi, or bacteria.

The cuttings should be inserted into the rooting medium immediately after treatment. To avoid brushing off the powder during insertion, a thick knife may be used to make a trench in the rooting medium before the cuttings are inserted. (See Fig. 10–25.)

Talc preparations have the advantage of being readily available and easy to use. Uniform results may be difficult to obtain, owing to the variability in the amount of the material adhering to the cuttings, influenced by such factors as the amount of moisture at the base of the cutting and the texture of the stem (hairy or smooth).

### DILUTE SOLUTION SOAKING METHOD

In an older procedure, the basal part—2.5 cm (1 in.)—of the cuttings is soaked in a dilute solution of the material for about 24 hours just before they are

**Preparing the Dilute Solution**

To prepare one liter of a 100-ppm solution of a root-promoting chemical, 100 mg of the pure material is dissolved in about 10 ml of alcohol (ethyl, methyl, or isopropyl). This solution is then diluted with water to make one liter. Naphthaleneacetic acid dissolves best in a few drops of ammonium hydroxide before adding to the water. The acid form of these growth substances is not directly soluble in water. The potassium salt of indolebutyric acid, which is water-soluble, is available.

An approximate 100-ppm solution of indolebutyric acid can be prepared by dissolving a level ¼ teaspoon of the chemical in a small amount of alcohol and adding to 3.8 liter (1 gal) of water, stirring thoroughly. If a precipitate or cloudiness forms, add a few drops of ammonium hydroxide. This precipitate may be avoided by using distilled or deionized water.

---

inserted into the rooting medium. The concentrations used vary from about 20 ppm for easily rooted species to about 200 ppm for the more difficult species.

During the soaking period, the cuttings should be held at about 20° C (68° F), but not placed in the sun. The amount of the chemical absorbed by the cuttings depends somewhat upon the surrounding conditions during this period, which may lead to some variation in the results obtained.

CONCENTRATED SOLUTION DIP METHOD

In the dip method, a concentrated solution varying from 500 to 10,000 ppm (0.05 to 1.0 percent) of the root-promoting chemical in 50 percent alcohol is prepared, and the basal ½ to 1 cm (⅕ to ⅖ in.) of the cuttings are dipped in it for a short time (about 5 seconds); then the cuttings are inserted into the rooting medium. Cuttings are best dipped as a bundle, not one by one.

The concentrated solution method of application has a number of advantages over the others. It eliminates the necessity of providing equipment for soaking the cuttings and returning later to insert them in the rooting medium. In addition, more uniform results are likely to be obtained, because the uptake of the chemical by the cuttings is not influenced as much by surrounding conditions as is the case with the other two methods. Some propagators spray the concentrate over the bases of a bundle of cuttings rather than dipping.

---

**Preparing the Concentrated Solution**

To prepare 100 ml of a 4000-ppm solution of a root-promoting substance, weigh out 400 mg of the chemical and dissolve it in 100 ml of 50 percent alcohol (ethyl, methyl, or isopropyl).

An approximate 4000-ppm solution of indolebutyric acid can be prepared by dissolving a level ¼ teaspoon of the pure crystals in 100 ml (3⅓ fluid oz) of 50 percent alcohol.

The same solution can be reused for many thousands of cuttings, but it must be tightly sealed when not in use, because the evaporation of the alcohol will change its concentration. It is best to use only a portion of the material at a time, just sufficient for the immediate needs, discarding it after use rather than pouring it back into the stock solution. The propagator (or his pharmacist) can prepare the solutions using the pure crystals, although liquid concentrates, available commercially, can be obtained and diluted according to directions.

Growth regulators used in excessive concentrations for the species may inhibit bud development (6), cause yellowing and dropping of leaves, blackening of the stem, and eventual death of the cuttings. An effective, nontoxic concentration has been used if the basal portion of the stem shows some swelling, callusing, and profuse root production just above the base of the cutting. A concentration just below the toxic point is considered the most favorable for root promotion.

Some negative results obtained in using growth regulators as an aid in rooting cuttings may be due to the use of old or deteriorated chemicals. A simple test using tomato leaf cuttings makes it possible to determine in a short time whether the preparation planned for use has root-promoting properties (21). Tomato leaf cuttings, of the type shown in Figure 10–22, are treated with the material, then inserted in moist sand in a glass- or polyethylene-covered box together with a group of untreated cuttings for comparison. After about two weeks the cuttings can be observed. Tomato cuttings are sensitive to growth regulators and will give a good indication of the effectiveness of the material by the extent of their root production.

Use fresh preparations whenever possible. Dilute solutions, e.g., 25 ppm, lose their activity within a few days, especially if they become contaminated with foreign material. Solutions used in the concentrated solution dip method of application, which contain a high percentage of alcohol, will retain their activity almost indefinitely if kept clean.

Usually just the base of a leafy cutting is dipped into the root-promoting preparation, but in some instances (54) immersing the entire cutting into the solution, which includes a wetting agent, is more effective in promoting rooting than just dipping the base alone. There is an initial retardation of shoot growth, but this does not seem to be an important disadvantage. Just dipping the foliar portion only of the cutting is effective for some species, provided a high enough

**Figure 10–22**   Tomato leaf cuttings provide a sensitive test for the effectiveness of root-promoting substances. Left to right: no treatment; treated with indolebutyric acid in talc at 1000 ppm, at 3000 ppm, and at 8000. 'Marglobe' cultivar after 12 days in sand.

concentration (2000 to 10,000 ppm) is used (*32*). There is evidence, too, that with hardwood cuttings of some species, dipping just the basal cut surface gives better results than dipping 2 cm or more of the base (*27*).

## TREATMENT OF CUTTINGS WITH FUNGICIDES

As a precaution against fungus infection it may be advisable to give the cutting material a dip into a fungicidal preparation, such as benomyl (0.5 g per liter; 3 oz per 50 gal), either before or after the cuttings are made.

Dipping the cutting bases into a combination fungicide-indolebutyric acid mixture often gives better survival results than an IBA treatment alone. Simple preparations may be made:

**1.** Mix captan (50 percent wettable powder), 1:1 (w/w), with a commercial talc preparation containing 0.8 percent IBA to give a 25 percent captan and a 0.4 percent (4000 ppm) indolebutyric acid concentration, or

**2.** Dilute benomyl (50 percent wettable powder) to a 10 percent concentration by mixing with talc (2 g benomyl plus 8 g talc), then mixing this, 1:1 (w/w), with a commercial talc preparation containing 0.8 percent IBA to give a mixture containing 5 percent benomyl and 0.4 percent indolebutyric acid.

If indolebutyric acid is used as a concentrated solution dip, after this treatment and after the cutting bases are allowed to dry the bases can then be swirled around in a fungicidal powder, either 25 percent captan (50 percent wettable powder diluted 1:1 (w/w) with talc), or 5 percent benomyl (2 g 50 percent wettable powder to 16 g talc), before sticking in the rooting medium.

## ENVIRONMENTAL CONDITIONS FOR ROOTING LEAFY CUTTINGS

For the successful rooting of leafy cuttings, the essential environmental requirements are:

Proper temperature, 18° to 27° C (65° to 75° F)
Atmosphere conducive to low water loss from the leaves
Ample but not excessive light
Clean, moist, well aerated, and well drained rooting medium

Many types of equipment are satisfactory for providing these conditions—ranging from a simple glass jar or polyethylene cover placed over a few cuttings stuck in sand, to elaborate greenhouse benches with automatic mist control and automatically controlled electric heating cables below the cuttings. One of the simplest devices, which is satisfactory for rooting a limited number of cuttings, is a wooden box half-filled with the rooting medium, with a pane of glass or a sheet of polyethylene film over the top. If this box is placed next to a window in a heated room in the winter, cuttings of many species can be rooted in it. Such a box can be used successfully in the summer also for rooting cuttings of

**Figure 10–23** Polyethylene plastic sheeting can be used for starting cuttings of easily rooted species. The basal ends of the cuttings are inserted in damp sphagnum or peat moss and rolled in the polyethylene as shown here. The roll of cuttings should then be set upright in a cool, humid location for rooting.

some plants, if placed out-of-doors in the shade. Cuttings of many plants can be rooted successfully with only artificial light, if placed under a large fluorescent lamp fixture. A simple procedure for rooting a few cuttings is illustrated in Figure 10–23.

In commercial operations, large-scale rooting of leafy cuttings is done in specially prepared rooting beds in hotbeds, cold frames, greenhouses, or in lathhouses, or even out-of-doors in mild climates. In these situations, proper environmental conditions must be established and maintained (see chapter 2).

Most commercial propagators recognize the value of maintaining strict sanitary procedures during all stages of making and rooting leafy cuttings. It is much easier to prevent attacks of disease organisms than to try to stop them. Losses can be considerable from a disease attack when hundreds of thousands of cuttings are involved. See chapter 2 for a full discussion of sanitation in propagation.

## PREPARING THE ROOTING FRAME AND INSERTING THE CUTTINGS

The rooting frames or benches should preferably be raised or, if on the ground, equipped with drainage tile, to assure perfect drainage of excess water.

The frames or flats should be deep enough so that about 10 cm (4 in.) of rooting medium can be used, and a cutting of average length—7.5 to 13 cm (3 to 5 in.)—can be inserted up to half its total length, with the end of the cutting still 2.5 cm (1 in.) or more above the bottom of the frame. The rooting medium should be watered thoroughly before the cuttings are inserted, which should be as soon as possible after they are prepared. It is very important that the cuttings be protected from drying at all stages during their preparation and insertion (Figure 10–24).

After a section of the rooting bench or a flat has been filled with cuttings, it should be well watered to settle the rooting medium around the cuttings (see Figure 10–25).

**Figure 10–24**    Commercial preparation and insertion of cuttings in a large wholesale nursery. Cuttings are prepared on the right and stuck into flats for rooting on the left. Spray bottles (on left) contain a rooting hormone (indolebutyric acid) solution for spraying on the base of bundles of cuttings before inserting them in the rooting medium.

**Figure 10–25**    Steps in placing semihardwood cuttings in mist propagating beds for rooting, shown here with English holly. *Upper left:* cutting row in rooting medium with heavy knife against board. *Upper right:* selecting prepared cuttings for inserting. *Lower left:* dipping cuttings for 5 sec in liquid rooting ''hormone'' (auxin) preparation. *Lower right:* sticking cuttings in rooting medium. Courtesy Klass Ellerbrook, West Oregon Nursery, Portland, Oregon.

**Figure 10–26**
Use of thin, light-weight polyethylene film laid directly on the rooting cuttings to reduce transpiration from leaves.

## Polyethylene Coverings over Rooting Beds

If a well shaded lathhouse or greenhouse is available and cool ambient temperatures (17° to 20° C; 63° to 68° F) can be maintained, a thin (0.003 mil) polyethylene sheeting placed directly on top of and touching the cuttings will provide good rooting conditions. The edges of the poly film should be tucked in to seal in the water vapor and provide high humidity around the cuttings (see Figure 10–26).

A variation of this method is to use rooting beds on the ground out-of-doors in the full sun and to lay sheets of 0.63 cm (¼ in.) microfoam directly on the cuttings, covering this with white 4 mil co-polymer film sealed to the ground by gravel or pieces of pipe (*15*).

## Ventilated High Humidity Propagation

The propagating structure can be equipped with a system of forced air ventilation with foglike moisture injected into the incoming air by a humidifier. Best results are obtained with an oscillating humidifier[1] that can produce a large volume (15 to 30 gal/hr) of fog with droplets in the 20 to 30 micron range. It should operate from sunrise to sunset (*33*).

# MIST SYSTEMS FOR ROOTING CUTTINGS

In the propagation of plants by leafy cuttings, one of the chief problems is to maintain the cuttings without wilting until roots are produced. This can be accomplished by keeping the relative humidity of the air surrounding the cuttings at a high level, the foliage, benches, and floors being sprinkled by hand several times a day during the rooting period. Such procedures are cumbersome, how-

---

[1] Agritech, Inc., Box 33083, Raleigh, N.C. 27606 manufactures humidifiers for this purpose.

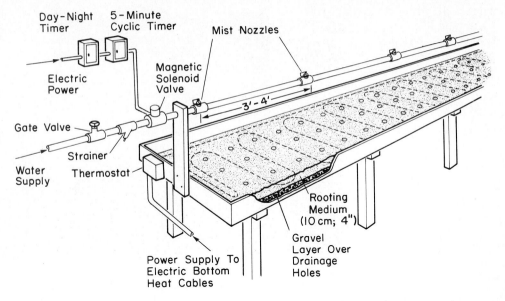

**Figure 10–27** Basic component parts of an intermittent mist propagating installation with electric bottom-heat cable. One timer turns the mist system on in the morning and off at night. The second is a short interval timer to provide the intermittent mist cycles.

ever, and do not avoid heat build-up problems under the necessary glass or plastic covers.

An **intermittent-mist** water spray over the cuttings in the rooting bed is a very effective aid in rooting leafy cuttings of a great many kinds of plants. This system is widely used by propagators throughout the world. Such sprays provide a film of water over the leaves and cuttings; this film lowers their temperature and increases the humidity around the leaves, thus reducing transpiration and respiration. Cuttings of certain plants, however—particularly succulent types with fleshy leaves and some others that show foliage leaching—do not do well under mist, rooting more readily in a closed-frame propagating bed.

This mist technique makes possible the rooting of cuttings of plants previously considered very difficult or impossible to root. It permits the use of soft, succulent, fast-growing cutting material early in the season, which (in some species) is much more likely to root than older, more mature, hardened wood. In addition, intermittent mist keeps slow-rooting cuttings alive for a long period of time, giving them a chance to root before they die from desiccation. By the use of mist propagation techniques, large cuttings with considerable leaf area can be rooted, permitting the production of large-size, salable plants in a short time (*53*).

Mist beds can be set up either in a greenhouse for use in summer and winter, or out-of-doors in a lathhouse or in open sun for use during the warmer months of the year. Over these beds, as shown in Figure 10–27, are placed nozzles that produce a fine fog-like mist spaced so as to give complete coverage of the bed. Figures 10–28 and 10–29 show typical mist propagation installations.

**Figure 10–29**    Rooting cuttings out-of-doors under intermittent mist in a southern California commercial nursery.

## Mist Nozzles

Two basic types of spray nozzles are available with several modifications of each: (a) the oil-burner, whirling action type, and (b) the deflection type (Figure 10–30).

The oil-burner nozzle produces an evenly distributed fine spray and uses a relatively small amount of water. The mist is produced in this nozzle by water passing through small grooves set at an angle to each other. Nozzles have been developed especially for mist propagation. These generally emit a flat, 160-deg angle pattern to give a wide coverage—100 to 120 cm (3½ to 4 ft)—and are designed to operate well at the usual water pressure of about 2.1 kg/cm² (30 lb per sq in.). Water output is relatively low: 9.5 to 19 liters (2½ to 5 gal) per hour.

**Figure 10–30**    Types of nozzles used in mist propagation installations. The two on the left are deflection nozzles. The one on the right (two views) is a whirling action, oil-burner type.

The deflection nozzle develops a mist by a fine stream of water striking a flat surface. The larger aperture used in this type reduces clogging but uses more water. It can operate on a low water pressure more effectively than the oil-burner type. Some types of nozzles can be shut off individually, which facilitates working in the propagating beds.

There are various possible methods of placing the water pipes to which the nozzles are attached. One is to lay the main feeder pipe down the center of the bed, either below, at, or above the surface of the rooting medium, with the nozzles at the end of risers from this pipe. Another method is to place the feeder pipe well above the cuttings down the center of the bed with the nozzles directed downward. Whatever arrangement is used, the nozzles should be placed close enough together and the water pressure should be high enough so that the entire bed is completely under the mist. Unless the mist actually wets the leaves, rooting is likely to be unsatisfactory.

## Controls

Intermittent mist during the daylight hours, which supplies water at intervals frequent enough to keep a film of water on the leaves but no more, gives better results than continuous mist. Since it would be impractical to turn the mist on and off by hand at short intervals throughout the day, automatic control devices are necessary. Several types are available, all operating to control a solenoid (magnetic) valve in the water line to the nozzles.

In a mist installation, especially the outdoor type, the cuttings will be damaged if the leaves are allowed to become dry for very long. Even ten minutes without water on a hot, sunny day can be disastrous. In setting up the control system to provide an intermittent mist, every precaution should be taken to guard against accidental failure of the mist applications. This includes the use of a ''normally open'' solenoid valve—that is, one constructed so that if the electric power becomes disconnected, the valve is open and water passes through it. Application of electricity closes the valve and shuts off the water. If an accidental power failure occurs or any failure in the electrical control mechanism takes place, the mist remains on continuously, and no damage to the cuttings results. On the other hand, in using a ''normally closed'' solenoid, which requires an electric current to open it and allow the passage of water, any failure of the power would mean complete stoppage of the mist and, if not soon detected, possibly total loss of the cuttings.

*Timers* Electrically operated timer mechanisms are available that operate the mist as desired. A successful type uses two timers acting together in series—one turns the entire system on in the morning and off at night; the second, an interval timer, operates the system during the daylight hours to produce an intermittent mist—at any desired combination of timing intervals, such as 6 seconds ON and 90 seconds OFF. This type of control mechanism is almost foolproof, and although it does not automatically compensate for variations in weather conditions, in most situations it can be adjusted closely enough to give satisfactory results. Time clocks for regulating the application of water are preferred by many propagators because of their reliability (*19*). Some electronic timers are very versatile and can operate many banks of mist nozzles in sequence.

The danger of electrical shock should always be kept in mind when installing and using any electrical control unit in a mist bed where considerable water is present. The complete electrical installation should be done by a competent electrician.

*Electronic leaf*    In another type of control mechanism, the so-called "electronic leaf," a small piece of plastic containing two terminals is placed under the mist along with the cuttings (*20, 51*). The alternate wetting and drying of the terminals makes and breaks the electric circuit which, in turn, controls the solenoid valve. There are several variations in this type of control. In one, a piece of filter paper is used as the sensing material connected between two electrodes (*4, 56*). The electronic leaf would theoretically maintain a film of water on the leaves of the cuttings at all times, automatically compensating for changes in the evaporating power of the air. The principal defect of the electronic leaf is the gradual build-up of a mineral deposit between the terminals, which will conduct electricity. The leaf therefore must be cleaned periodically.

*Thermostat and timer*    One type of control is based upon a thermostat placed with the cuttings. When the temperature at the leaf level reaches a certain point, the solenoid is activated, and mist is applied. This lowers the thermostat temperature, and the mist is shut off. A combination of such a thermostat plus a time clock has been used successfully (*14*). The timer operates the mist unless the temperature goes above a certain point, when the thermostat overrides the timer and turns on the mist.

*Screen balance*    Another type of control is based upon the weight of water. A small stainless-steel screen is attached to a lever actuating a mercury switch. When the mist is on, water collects on the screen until its weight trips the mercury switch, shutting off the solenoid. When the water evaporates from the screen, it raises, closing the switch connection, which opens the solenoid, again turning on the mist. This type of control is best adapted to regions where considerable fluctuations in weather patterns may occur throughout the day, from warm and sunny to overcast, cool, and rainy; the unit reacts automatically to changes in the evaporating power of the air.

*Photoelectric cell*    Controls based upon the relationship between light intensity and transpiration rate are available. These contain a photoelectric cell which conducts current in proportion to the light intensity. It activates a magnetic counter, or charges a condenser, so that after a certain period of time the solenoid valve is opened and the mist applied. The higher the light intensity the more frequently mist is applied. At dawn and dusk very little is used, and at night none. During cloudy days less mist is used than during bright, sunny days. Such a control system would not be well suited for outdoor mist beds, where transpiration is affected by wind movement as well as by light intensity (*41, 57*).

## Operation

Difficulties may arise in operating a mist-propagation bed. Lack of sufficient water pressure to operate the nozzles properly can be overcome by installing a small, electrically operated, rotary booster pump between the water source and the solenoid valve. If there is much sand in the water, it is advisable to install filters in the supply line, which will reduce the clogging of the strainers in the nozzles.

Algae growth often develops a green coating on and around mist-propagation installations after an extended period of operation. This coating is not particularly harmful to the cuttings but is unsightly and—because it is very slippery—can be a hazard. This slimy material is principally blue-green (*Oscillatoria, Phormidium,* and *Arthrospira*) and green (*Stichococcus* and *Chlamydomonas*) algae (*7, 10*). Allowing the water to remain off for a period each night so that the mist area can completely dry out will generally hold the algae in check. Sprinkling powdered Bordeaux Mixture on the walks and benches, or using certain proprietary algae inhibitors, such as "Algae-Go 36–20," will help control the algae.

In propagating cuttings under mist, it is essential that a well-drained rooting medium be used and the bed raised, equipped with drainage tile, or otherwise provided for adequate removal of any excess water.

The quality of the water used in the mist—and in watering cuttings during rooting—can influence the rooting obtained. Water relatively high in total salts, if there is enough calcium and magnesium, may be quite satisfactory. But water high in such salts as sodium or potassium carbonates, bicarbonates, or hydroxides, can be very detrimental, especially when coupled with low levels of calcium salts and when the rooting medium contains peat (or other materials of high exchange capacity) (*37, 45*). However, with sand as the rooting medium, high sodium levels in the mist are not as detrimental (*38*). Equivalent amounts of sodium or potassium salts of nitrates, phosphates, or chlorides in the water are not so likely to cause injury. With chrysanthemum cuttings, lack of rooting in media containing peat was associated with mist water having a high sodium to calcium ratio (4.3:0.6). Poor rooting was corrected by adding gypsum ($CaSO_4$) to the rooting medium at the rate of 500 g per sq. meter (*45*). Figure 10–31 shows poor rooting of chrysanthemum cuttings when there is high sodium in the rooting medium. See chapter 2 for a discussion of water quality.

**Figure 10–31** Injurious effects of excess sodium salts in a sand-peat rooting medium on rooting of chrysanthemum cuttings. Left to right: control, 255, 500, and 755 g per sq. meter. Courtesy R. D. Raabe (*45*).

## Nutrient Mist

Mineral nutrients added to the water used for misting may improve root quality and subsequent growth of the rooted cuttings in some plants. In other plants there may be no benefit or even leaf damage. Slow-to-root cuttings held under mist for prolonged periods may have a large amount of their nutrients leached away. A convenient solution to use for nutrient mist can be prepared from a commercial soluble fertilizer, e.g., "Ra-Pid-Gro,"[2] containing nitrogen, phosphorus, and potassium (23–8–14, respectively), applied at the rate of 44 grams per 100 liters (6 oz per 100 gal) of water. This may be made up separately in a large tank and applied to the cuttings as mist through a separate pressure pump at the rate of about 12 sec every 2½ min. (60), or by a proportioning pump connected into the water supply (49).

## Hardening-Off

Moving the rooted cuttings from under mist to a drier environment must be done carefully. With some plants, *Prunus* spp., for example, it is important to remove the cuttings from the mist as soon as they are rooted. Otherwise, rapid defoliation and deterioration of the roots occur.

There are several ways of successfully taking the rooted cuttings from the mist conditions:

1. The cuttings may be left in place in the mist bed but with the duration of the misting periods gradually decreased, either by lessening the ON periods and increasing the OFF periods or by leaving the misting intervals the same but gradually decreasing the time for which the mist is in operation each day.

2. In some out-of-door operations, the rooted cuttings are left in place and allowed to send their roots on through the rooting medium to the soil beneath. The propagating frame is moved to a different location to root another set of cuttings.

3. Another method is to root the cuttings in flats and move the flats after rooting to another mist frame where they are "hardened-off" and then potted into containers. Cuttings may be left in the rooting medium until the dormant season, when they can be dug more safely, to be either lined out in the nursery row for further growth or potted and brought into the greenhouse. If the rooted cuttings are left in the rooting medium for a considerable time, it is advisable to water them at intervals with a nutrient solution.

4. Some propagators root their cuttings directly in small containers set up in flats. Then, after rooting, the plants may be easily moved for transplanting without disturbing the roots. An alternate method is to root the cutting in a solid, block-type rooting medium which, after rooting, permits transplanting without disturbing the roots. Several such products made from wood products and/or compressed peat, plus some added fertilizer, are available.

5. Another method is to pot the cuttings immediately after rooting and hold them for a time in a cool, humid, shaded location, e.g., a fog chamber, closed frame, or greenhouse.

It is a common experience to have the buds of cuttings that are rooted under mist enter an apparent dormant or physiological rest condition in which they do

[2] Ra-Pid-Gro Corp., Dansville, N.Y. 14437.

not continue further growth that season even though they are well rooted. This may possibly be overcome by altering the day-length conditions.

## CARE OF CUTTINGS DURING ROOTING

Hardwood stem cuttings or root cuttings started out-of-doors in the nursery require only the usual care given to other crop plants, such as adequate soil moisture, freedom from weed competition, and insect and disease control. Best results are obtained if the nursery is established in full sun where shading and root competition from large trees or shrubbery do not occur.

Leafy softwood or semihardwood stem cuttings and leaf-bud or leaf cuttings being rooted under high humidity require close attention throughout the rooting period. The temperature should be controlled carefully. The cuttings must not be allowed to show wilting for any length of time. Glass-covered frames, exposed even for a few hours to strong sunlight, will build up excessively high and injurious temperatures, owing to the heat accumulating under the glass. Such equipment should always be protected by cloth screens, whitewash on the glass, or some other method of reducing the light intensity.

If bottom heat is provided, thermometers should be inserted in the rooting medium to the level of the base of the cutting and checked at frequent intervals, especially at first. A temperature of about 24° C (75° F) is desirable. Excessively high temperatures in the rooting medium, even for a short time, are likely to result in death of the cuttings.

It is important to maintain humidity as high as possible in rooting leafy cuttings to reduce water loss from the leaves to a minimum. Without automatic mist equipment, syringing the leaves with a spray nozzle at frequent intervals is necessary especially during hot weather. Although more time-consuming, several light sprinklings with water each day are better than heavy soakings at less frequent intervals. A nozzle should be used that breaks the water into a fine spray. A drop of the humidity to a low level with a consequent pronounced wilting of the cuttings, if prolonged for any length of time, may so injure the cuttings that rooting will not occur, even though high-humidity conditions are subsequently resumed. However, most nursery operations use intermittent-mist propagating beds to overcome such problems.

Adequate drainage must be provided so that excess water can escape and not cause the rooting medium to become soggy and waterlogged. When peat or sphagnum moss is used as a component of the rooting medium, it is especially important to see that it does not become excessively wet.

It is also necessary to maintain sanitary conditions in the propagating frame. Leaves that drop should be removed promptly, as should any obviously dead cuttings. Pathogens find ideal conditions in a humid, closed propagating frame with low light intensity and, if not controlled, can destroy thousands of cuttings in a short time.

Disease problems under mist-propagation conditions have not been serious. Probably frequent washing of the leaves by the aerated water removes spores before they are able to germinate. The greater light intensity possible and the air movement also decrease the incidence of diseases.

If mites, aphids, or mealy bugs appear on the leaves of the cuttings, immediate control measures are necessary.

The more sophisticated techniques, such as the use of carbon-dioxide enriched atmospheres or the application of carbonated mist ($34$), have produced faster and greater rooting by supplying added $CO_2$ for photosynthesis. However, these benefits would be found only in situations (high temperature and light intensity) where the $CO_2$ level surrounding the cuttings is limiting growth. Equipment is available for supplementing the natural $CO_2$ supply. (See chapter 2.)

## HANDLING CUTTINGS AFTER ROOTING

### Hardwood Cuttings

Rooted hardwood cuttings[3] in the nursery row are usually dug during the dormant season after the leaves have dropped. With fast-growing species, the cuttings may be sufficiently large to dig after one season's growth. Slower-growing species may require two or even three years to become large enough to transplant.

The digging should take place on cool, cloudy days, when there is no wind. If possible, digging should not be done when the soil is wet, especially if it has a high clay content. Most of the soil should drop readily from the roots after the plants are removed. After the plants are dug, they should be quickly heeled-in in a convenient location, placed in cold storage, or replanted immediately in their permanent location. **"Heeling-in"** is to place dug, bare-rooted deciduous nursery plants close together in trenches with the roots well covered. This is a temporary provision for holding the young plants until they can be set out in their permanent location. (See Figure 10–32.)

---

[3] The procedures described in this section would apply also to nursery plants propagated as seedlings or as budded or grafted trees.

**Figure 10–32** Temporary storage of nursery stock by "heeling-in" in raised beds filled with damp wood shavings sufficiently deep to cover the roots.

**Figure 10-33** Digging machines specially constructed for undercutting nursery stock in commercial nurseries. Photo on left shows large U-shaped blade which can be lowered to travel 0.3 to 0.6 m (1 to 2 ft) below the soil surface to cut roots. Vibrating "lifter" behind the cutting blade facilitates the movement of the blade through the soil.

Commercial nurseries often store quantities of deciduous plants for several months through the winter in cool, dark rooms with the roots protected by damp wood shavings, shingle tow, or some similar material. Nursery stock to be kept for extended periods should be held under refrigerated conditions at 0° to 2° C (32° to 35° F) (*31*).

If only a few plants are to be removed from the nursery row, they can be dug with shovels, but in large-scale nursery operations some type of mechanical digger, such as is shown in Figure 10-33, is generally used. This digger "undercuts" the plants. A sharp U-shaped blade travels 30 to 60 cm (1 to 2 ft) below the soil surface under the nursery row, cutting through the roots. Sometimes a horizontal, vibrating, "lifting" blade is attached to, and travels behind, the cutting blade. This slightly lifts the plants out of the soil, making them easy to pull by hand.

Unless very small, plants of broad- or narrow-leaved evergreen species usually are not successfully handled bare-root, as is done with dormant and leafless deciduous plants. The presence of leaves on evergreen plants requires the continuous contact of the roots with soil. Therefore, large, salable plants of broad- or narrow-leaved evergreens, and occasionally deciduous plants, are either grown in containers or dug and sold "balled and burlapped." By the latter method, the plants are removed from the soil by carefully digging a trench around each individual. The soil mass around the roots is sometimes tapered at top and bottom, resulting in a ball of soil in which the roots are embedded. It is important that the soil be at the proper moisture level—not too wet and not too dry—otherwise it will fall apart. The ball is tipped gently onto a large square of burlap, which is then pulled tightly around the ball and sewed in place with heavy twine. If properly done, this burlap adequately holds the soil to the roots, and the plant can then be moved safely for considerable distances and replanted successfully.

Prior root pruning is necessary to prepare field-grown nursery stock for transplanting, either by hand ''ball and burlapping'' (Figure 10–34) or by

**Figure 10–34** Steps in ball and burlapping nursery stock, the procedure being used here in digging tree roses. Plant is carefully removed with a ball of soil adhering to the roots (*top*). Soil must be at proper moisture level so it does not fall apart. Burlap is wrapped around soil ball and tied with heavy cord (*middle*). Plants ready to be moved and transplanted (*below*). This method is used for all types of evergreen plants that have not been grown in containers.

machine digging. Pruning to produce a compact, fibrous root system should be started when the young plants (**liners**) are first set in the field. Long or curled and twisted roots should be cut back to a few inches from the crown of the liner. Root pruning should be repeated the second year, either by a tractor-powered U-blade cutter (see Figure 10–33) for large-scale operations, or with a sharp shovel if only a few plants are involved. This procedure helps confine the roots to the soil mass that will be taken with the plant during the digging and balling operation.

The production of balled and burlapped nursery stock is gradually being replaced, especially in areas with mild winters, by the production of plants in containers, primarily because of the lower labor costs involved and the greater opportunities for mechanization.

## Softwood, Herbaceous, Semihardwood, Leaf-Bud, Leaf Cuttings

Cuttings rooted with leaves attached and under conditions of high humidity require considerable care in being removed from the rooting medium. After rooting has started, the humidity should be lowered and ventilation of the bed increased. Cuttings should be dug as soon as a substantial root system with secondary roots has formed. Many propagators have experienced rapid rooting of cuttings only to have them die when they are dug and potted. Sometimes this trouble may be overcome by leaving the cuttings in the rooting bed longer, until after the first-formed primary roots have branched to develop a dense, fibrous secondary root system to which the rooting medium clings in a ball.

## COLD STORAGE OF ROOTED AND UNROOTED LEAFY CUTTINGS

Sometimes it may be convenient to take cuttings at certain times, such as when nursery plants are being sheared and shaped, and store them for later rooting. Storage has been successful with the Kurume type of azaleas overwintered in a cold greenhouse. Softwood cuttings were taken in the spring and held in polyethylene bags at temperatures from −0.5° to 4.5° C (31° to 40° F) for up to ten weeks, with subsequent rooting equal to that from unstored cuttings (*44*).

Unrooted chrysanthemum and carnation cuttings can be stored in sealed plastic bags for several weeks at −0.5° C (31°F) for subsequent rooting. In tests (*2*) on the effects of storage on subsequent performance of the plants, cuttings rooted after storage gave better results than those stored after rooting.

Carnation cuttings, both rooted and unrooted, store well at −0.5 to + 0.5° C (31° to 33° F) for at least five months, if placed in polyethylene-lined boxes with a small amount of moist sphagnum or peat moss. The poly film should not be sealed airtight (*23*).

In some situations cuttings may be rooted in late summer or early fall for planting outdoors the following spring. Several studies have shown that it is possible to hold rooted cuttings of some species in polyethylene bags for prolonged periods under cold storage temperatures of 1.5° to 4.5° C (35° to 40°F) (*55*). In one test, rooted cuttings of certain plants were safely stored for about five months at temperatures of 1° to 4° C (34° to 39° F) when placed in polyethylene bags. Packing material around the roots was of no advantage (*48*). Other studies (*13*) on cold storage of rooted cuttings of 31 different species for six months at two temperatures, 0° and 4.5° C (32° and 40° F), showed better survival with many of the species at 0° C than at 4.5° C, although no difference was noted with other species. Dusting prior to storage with 5 percent captan increased survival of some species but made no difference in others. Apparently some species survive cold-storage treatments much better than others. The storage procedures for all types of nursery plants are, of course, a consideration for commercial nurseries and are an important aspect of nursery management (*21, 30, 31*).

# REFERENCES

*1.* Ali, N., and M. N. Westwood. 1966. Rooting of pear cuttings as related to carbohydrates, nitrogen, and rest period. *Proc. Amer. Soc. Hort. Sci.* 88:145-50.

*2.* Alstadt, R. A., and W. D. Holley. 1964. Effects of storage on the performance of carnation cuttings. *Colo. Flower Growers Assn. Bul.* 173:1-2.

*3.* Baldwin, I., and J. Stanley. 1981. How to manage stock plants. *Amer. Nurs.* 153 (8):16, 74-80.

*4.* Bean, G., E. S. Trickett, and D. A. Wells. 1957. Automatic mist control equipment for the rooting of cuttings. *Jour. Agr. Eng. Res.* 2:44-48.

*5.* Carlson, R. F. 1966. Factors influencing root formation in hardwood cuttings of fruit trees. *Mich. Quart. Bul.* 48:449-54.

*6.* Carlson, R. F., and D. C. Kiplinger. 1938. The effect of synthetic growth substances on the rooting and subsequent growth of ornamental plants. *Proc. Amer. Soc. Hort. Sci.* 36:809-16.

*7.* Coorts, G. D., and C. C. Sorenson. 1968. Organisms found growing under nutrient mist propagation. *HortScience* 3(3):189-90.

*8.* Creech, J. L., R. F. Dowdle, and W. O. Hawley. 1955. Sphagnum moss for plant propagation. *USDA Farmers Bul.* 2085.

*9.* Donovan, D. M. 1976. A list of plants regenerating from root cuttings. *The Plant Propagator* 22(1):7-8.

*10.* Durrell, L. W., and R. Baker. 1959. Algae causing clogging of cooling systems. *Colo. Flower Growers Assn. Bul.* 111 (April-May).

*11.* Erez, A., and Z. Yablowitz. 1981. Rooting of peach hardwood cuttings for the meadow orchard. *Scient. Hort.* 15:137-44.

*12.* Flemer, W., III. 1961. Propagating woody plants by root cuttings. *Proc. Plant Prop. Soc.* 11:42-47.

*13.* Flint, H. L., and J. J. McGuire. 1962. Response of rooted cuttings of several woody ornamental species to overwinter storage. *Proc. Amer. Soc. Hort. Sci.* 80:625-29.

*14.* Floor, J. 1957. Report on experiments with mist propagation of cuttings. *Inst. voor de Vered. van Tuinb. Wageningen. Meded.* 188.

*15.* Gouin, F. R. 1981. Vegetative propagation under thermo-blankets. *Proc. Inter. Plant Prop. Soc.* 30:301-5.

*16.* Hansen, C. J., and H. T. Hartmann. 1968. The use of indolebutyric acid and captan in the propagation of clonal peach and peach-almond hybrid rootstocks by hardwood cuttings. *Proc. Amer. Soc. Hort. Sci.* 92:135-40.

17. Hartmann, H. T., W. H. Griggs, and C. J. Hansen. 1963. Propagation of ownrooted Old Home and Bartlett pears to produce trees resistant to pear decline. *Proc. Amer. Soc. Hort. Sci.* 82:92–102.

18. Hartmann, H. T., C. J. Hansen, and F. Loreti. 1965. Propagation of apple rootstocks by hardwood cuttings. *Calif. Agr.* 19(6):4–5.

19. Hess, C. E., and W. E. Snyder. 1953. A simple and inexpensive time clock for regulating mist in plant propagation procedures. *Proc. Plant Prop. Soc.* 3:56–61.

20. ———. 1954. Interrupted mist found superior to constant mist in tests with cuttings. *Amer. Nurs.* 100(12):11–12, 82.

21. Hitchcock, A. E., and P. W. Zimmerman. 1938. The use of green tissue test objects for determining the physiological activity of growth substances. *Contrib. Boyce Thomp. Inst.* 9:463–518.

22. Hocking, D., and R. D. Nyland. 1971. Cold storage of coniferous seedlings. *AFRI Res. Rpt. No. 6,* Col. Forestry, Syracuse Univ.

23. Holley, W. D., and R. Baker. 1963. *Carnation production.* London: Grower Publications.

24. Howard, B. H. 1971. Propagation techniques. *Scient. Hort.* 23:116–26.

25. Howard, B. H., and R. J. Garner. 1965. High temperature storage of hardwood cuttings as an aid to improved establishment in the nursery. *Ann. Rpt. E. Malling Res. Sta. for 1964,* pp. 83–87.

26. Howard, B. H. 1968. The influence of 4 (indolyl-3) butyric acid basal temperature on the rooting of apple rootstock hardwood cuttings. *Jour. Hort. Sci.* 43:23–31.

27. Howard, B. H., and N. Nahlawi. 1970. Dipping depth as a factor in the treatment of hardwood cuttings with indolybutyric acid. *Ann. Rpt. E. Malling Res. Sta. for 1969,* pp. 91–94.

28. Inose, K. 1971. Pumice as a rooting medium. *Proc. Inter. Plant Prop. Soc.* 21:82–83.

29. Long, J. C. 1932. The influence of rooting media on the character of the roots produced by cuttings. *Proc. Amer. Soc. Hort. Sci.* 352–55.

30. Lutz, J. M., and R. E. Hardenberg. 1968. *The commercial storage of fruits, vegetables, and florist and nursery stocks.* USDA–ARS Agr. Handbook No. 66. Washington, D.C.: U.S. Govt. Printing Office.

31. Mahlstede, J. P., and W. E. Fletcher. 1960. *Storage of Nursery Stock.* Washington, D.C.: Amer. Assn. Nurserymen, pp. 1–62.

32. McGuire, J. J., L. S. Albert, and V. K. Shutak. 1968. Effect of foliar applications of 3-indolebutyric acid on rooting of cuttings of ornamental plants. *Proc. Amer. Soc. Hort. Sci.* 93:699–704.

33. Milbocker, D. C. 1981. Ventilated high humidity propagation. *Proc. Inter. Plant Prop. Soc.* 30:480–82.

34. Molnar, J. M., and W. A. Cumming. 1968. Effect of carbon dioxide on propagation of softwood, conifer, and herbaceous cuttings. *Can. Jour. Plant Sci.* 48:595–99.

35. Noland, D. A., and D. J. Williams. 1980. The use of pumice and pumice-peat mixtures as propagation media. *The Plant Propagator* 26(4):6–7.

36. Osborne, W. W. 1961. Soil sterilization and fumigation. *Proc. Plant Prop. Soc.* 11:57–67.

37. Paul, J. L. 1968. Water quality and mist propagation. *Proc. Inter. Plant Prop. Soc.* 18:183–86.

38. Paul, J. L., and A. T. Leiser. 1968. Influence of sodium in the mist water on rooting of chrysanthemums. *HortScience* 3(3):187–88.

39. Pearse, H. L. 1948. Growth substances and their practical importance in horticulture. *Commonwealth Bur. of Hort. and Plant. Crops, Tech. Comm. No. 20.*

40. Petersen, F. 1961. Current methods in the selection and production of nursery stock. *Proc. Plant Prop. Soc.* 11:235–40.

41. Petersen, H. 1962. A photoelectric timing control for mist application. *Proc. 15th Inter. Hort. Cong.* 3:273–79.

42. Pike, A. V. 1972. Propagation by roots. *Horticulture* 50(5):56, 57–61.

43. Pokorny, F. A. 1965. An evaluation of various equipment and media used for mist propagation and their relative costs. *Ga. Agr. Exp. Sta. Bul.* n.s. 139.

44. Pryor, R. L., and R. N. Stewart. 1963. Storage of unrooted azalea cuttings. *Proc. Amer. Soc. Hort. Sci.* 82:483–84.

45. Raabe, R. D., and J. Vlamis. 1966. Rooting failure of chrysanthemum cuttings resulting from excess sodium or potassium. *Phytopath.* 56:713–17.

46. Scaramuzzi, F. 1965. Nuova technica per stimolare la radicazione delle talle legnose di ramo (A new technique for stimulating rooting in hardwood cuttings). *Riv. Ortoflorofrut-ticoltura Ital.* 49:101–4.

47. Smith, P. F. 1944. Rooting of guayule stem cuttings in aerated water. *Proc. Amer. Soc. Hort. Sci.* 44:527–28.

48. Snyder, W. E., and C. E. Hess. 1956. Low temperature storage of rooted cuttings of nursery crops. *Proc. Amer. Soc. Hort. Sci.* 67:545–48.

49. Sorenson, D. C., and G. D. Coorts. 1968. The effect of nutrient mist on propagation of selected woody ornamentals. *Proc. Amer. Soc. Hort. Sci.* 92:696–703.

50. Stoutemyer, V. T. 1968. Root cuttings. *The Plant Propagator* 14(4):4–5.

51. Templeton, H. M. 1953. The phytotektor method of rooting cuttings. *Proc. Plant Prop. Soc.* 3:51–56.

52. Thimann, K. V., and J. Behnke-Rogers. 1950. *The use of auxins in the rooting of woody cuttings.* Petersham, Mass.: Harvard Forest, M. M. Cabot Foundation.

53. Tinga, J. H., J. J. McGuire, and R. J. Parvin. 1963. The production of pyracantha plants from large cuttings. *Proc. Amer. Soc. Hort. Sci.* 82:557–61.

54. Van Braght, J., H. van Gelder, and R. L. M. Pierik. 1976. Rooting of shoot cuttings of ornamental shrubs after immersion in auxin-containing solutions. *Scient. Hort.* 4:91–94.

55. Vanderbrook, C. 1956. The storage of rooted cuttings. *Amer. Nurs.* 103(9):10, 76–77.

56. Vanstone, F. H. 1959. Equipment for mist propagation developed at the N.I.A.E. *Ann. Appl. Bio.* 47:627–31.

57. Waxman, S., and J. H. Whitaker. 1960. A light-operated interval switch for the operation of a mist system. *Prog. Rpt. No. 40, Storrs (Conn.) Agr. Exp. Sta.*

58. Wells, J. S. 1952. Pointers on propagation: Propagation of *Taxus. Amer. Nurs.* 96(11):13, 37–38, 43.

59. ———. 1962. Wounding cuttings as a commercial practice. *Proc. Plant Prop. Soc.* 12:47–55.

60. Wott, J. A., and H. B. Tukey, Jr. 1967. Influence of nutrient mist on the propagation of cuttings. *Proc. Amer. Soc. Hort. Sci.* 90:454–61.

61. Zimmerman, P. W. 1930. Oxygen requirements for root growth of cuttings in water. *Amer. Jour. Bot.* 17:842–61.

## SUPPLEMENTARY READING

DORAN, W. L. 1957. Propagation of woody plants by cuttings. *Mass. Agr. Exp. Sta. Bul. 491.*

INTERNATIONAL PLANT PROPAGATORS' SOCIETY. Proceedings of Annual Meetings.

JOINER, J. N., R. T. POOLE, and C. A. CONOVER. 1981. Propagation. Chapter 11 in J. N. Joiner, ed., *Foliage plant production.* Englewood Cliffs, N.J.: Prentice-Hall.

LAMB, J. G. D., J. C. KELLY, and P. BOWBRICK. 1975. *Nursery stock manual.* London: Grower Books.

McMILLAN-BROWSE, P. 1979. *Plant propagation.* New York: Simon & Schuster.

ROWE-DUTTON, P. 1959. Mist propagation of cuttings. *Bucks, England: Commonwealth Agricultural Bureaux, Farnham Royal, Digest No. 2.*

WELCH, H. J. 1970. *Mist propagation and automatic watering.* London: Faber and Faber.

The origins of grafting can be traced back to ancient times. There is evidence that the art of grafting was known to the Chinese at least as early as 1000 B.C. Aristotle (384–322 B.C.) discussed grafting in his writings with considerable understanding. During the days of the Roman Empire grafting was very popular, and methods were precisely described in the writings of that era. Paul the Apostle, in his Epistle to the Romans, discussed grafting between the "good" and the "wild" olive trees (Romans 11:17–24).

The Renaissance period (1350–1600 A.D.) saw a renewed interest in grafting practices. Large numbers of new plants from foreign countries were imported into European gardens and maintained by grafting. By the sixteenth century the cleft and whip grafts were in widespread use in England and it was realized that the cambium layers must be matched, although the nature of this tissue was not then understood and appreciated. Propagators were handicapped by a lack of a good grafting wax; mixtures of wet clay and dung were used to cover the graft unions. In the seventeenth century many orchards were planted in England, the trees all being propagated by budding and grafting.

Early in the eighteenth century Stephen Hales, in his studies on the "circulation of sap" in plants, approach-grafted three trees and found that the center tree stayed alive even when severed from its roots. Duhamel, about the same time, studied wound healing and the uniting of woody grafts. The graft union at that time was considered to act as a type of filter changing the composition of the sap flowing through it. Thouin (*167*) in 1821 described 119 methods of grafting and discussed changes in growth habit resulting from grafting. Vöchting (*177*) in the late nineteenth century continued Duhamel's earlier work on the anatomy of the graft union.

Liberty Hyde Bailey in *The Nursery Book* (*6*), published in 1891, described and illustrated the methods of grafting and budding commonly used in the United States and Europe at that time. The methods used today differ very little from those described by Bailey.

# 11

# Theoretical Aspects of Grafting and Budding

# TERMINOLOGY

**Grafting** is the art of connecting two pieces of living plant tissue together in such a manner that they will unite and subsequently grow and develop as one plant. As any technique that will accomplish this could be considered a method of grafting, it is not surprising that innumerable procedures for grafting are described in the literature on this subject. Through the years several distinct methods have become established that enable the propagator to cope with almost any grafting problem at hand. These are described in chapter 12 with the realization that there are many variations of each and that there are other, somewhat different, forms that could give the same results. Figure 11–1 illustrates a grafted plant and the parts involved in the graft.

**Budding** is similar to grafting except that the scion (see below) is reduced in size to contain only one bud. The various budding methods are described in chapter 13.

**Scion** is the short piece of detached shoot containing several dormant buds, which, when united with the stock, comprises the upper portion of the graft and from which will grow the stem or branches, or both, of the grafted plant. It should be of the desired cultivar and free from disease.

**Stock** (*Rootstock, Understock*) is the lower portion of the graft, which develops into the root system of the grafted plant. It may be a seedling, a rooted cutting, or a layered plant. If the grafting is done high in a tree, as in topworking, the stock may consist of the roots, trunk, and scaffold branches.

**Figure 11–1**     In grafted plants the entire shoot system consists of growth arising from one (or more) buds on the scion. The root system consists of an extension of the original rootstock. The graft union remains at the junction of the two parts throughout the life of the plant.

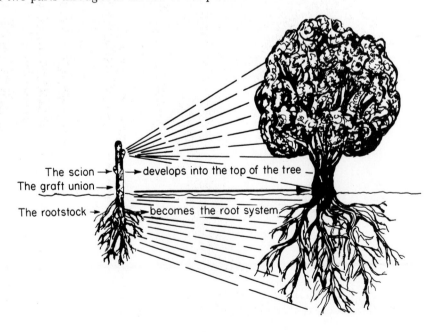

The scion → develops into the top of the tree

The graft union →

The rootstock → becomes the root system

**Interstock** *(Intermediate stock, Interstem)* is a piece of stem inserted by means of two graft unions between the *scion* and the *rootstock.* An interstock is used for several reasons, such as to avoid an incompatibility between the stock and scion, to make use of a winter-hardy trunk, or to take advantage of its growth-controlling properties.

**Cambium** is a thin tissue of the plant located between the bark (phloem) and the wood (xylem). Its cells are meristematic; that is, they are capable of dividing and forming new cells. For a successful graft union, it is essential that the cambium of the scion be placed in close contact with the cambium of the stock.

**Callus** is a term applied to the mass of parenchyma cells that develops from and around wounded plant tissues. It occurs at the junction of a graft union, arising from the living cells of both scion and stock. The production and interlocking of these parenchyma (or callus) cells constitute one of the important steps in the healing process of a successful graft.

# REASONS FOR GRAFTING AND BUDDING

Grafting and budding serve many different purposes:

1. Perpetuating clones that cannot be readily maintained by cuttings, layers, division, or other asexual methods
2. Obtaining the benefits of certain rootstocks
3. Changing cultivars of established plants
4. Hastening the reproductive maturity of seedling selections in hybridization programs
5. Obtaining special forms of plant growth
6. Repairing damaged parts of trees
7. Studying virus diseases.

Each of these reasons is discussed in detail below.

## Perpetuating Clones That Cannot Be Readily Maintained by Cuttings, Layers, Division, or Other Asexual Methods

Cultivars of some groups of plants, including most fruit and nut species and many other woody plants, such as eucalyptus and spruce, are not propagated commercially by cuttings, because they cannot be rooted at all or in satisfactory percentages by the methods now available. Additional individual plants often can be started by layering or division, but for propagation in large quantities, it is necessary to resort to budding or grafting scions of the desired cultivar on rootstock plants with which they are compatible.

Since these plants are highly heterozygous, seed propagation will not maintain the clone. In established cultivars, buds and scions are invariably taken from the *adult* growth phase to avoid effects of juvenility delays in fruiting.

## Rootstock Categories

Rootstocks can be divided into two groups: **seedling** and **clonal.**

**Seedling rootstocks,** those developed from germinated seeds, have certain advantages. Production of seedlings is relatively simple and economical and is well adapted to mass propagation methods (see chapter 7). Most seedling plants do not retain viruses occurring in the parent plant (although some viruses are seed-transmitted). In some instances (e.g., plum rootstocks) the root system developed by seedlings tends to grow deeper and to be more firmly anchored than rootstocks grown from cuttings.

However, seedling rootstocks have the disadvantage of genetic variation, which may lead to variability in the growth and performance of the scion of the grafted plant. Such variation is most likely to occur if the seed is obtained from unknown, unselected sources. The seed source plants may be unusually heterozygous or may have been cross-pollinated with related species. Within the same species some seed sources produce much better rootstock plants than others. With careful testing, individual plants in a species could be selected to be a new mother-tree seed source, or the start of a clonal rootstock.

Variability among seedling rootstocks can be reduced by careful selection of the parental seed source as to identity, and by its protection from cross-pollination. Variation can be reduced if all nursery trees of the same age are dug from the nursery row at one time, and small or obviously off-type seedlings or budded trees discarded. In most nurseries the young trees are graded by size, all those of the same grade being sold together. The practice of retaining slow-growing seedlings or budded trees for an additional year's growth is an undesirable practice as it tends to perpetuate variability. In the United States many fruit orchards grown on seedling rootstocks show no more variability resulting from the rootstock used than from unavoidable environmental differences in the orchard, principally soil variability.

**Clonal rootstocks** have received much attention in European countries, especially England, in regard to their development and use. These rootstocks are propagated vegetatively either by stool layering or as rooted cuttings. Each individual rootstock plant is the same genetically as all the other plants in the clone and can be expected to have identical growth characteristics in a given environment. Clonal rootstocks are often desirable not only to produce uniformity but—equally important—to preserve their special characteristics and the specific influences they have on scion cultivars, such as disease resistance, growth, or flowering habit. To maintain rootstock influence, deep planting of the nursery tree—which may lead to **"scion rooting"**—must be avoided, as illustrated in Figure 11-2. The deeper the graft union below the soil surface, the higher the incidence of scion rooting is likely to be (26).

The combinations of different clonal rootstocks with different scion cultivars allow much refinement in the performance of grafted trees. Each particular scion-stock combination requires thorough testing, however, before its performance is established and can be predicted.

In propagating and using clonal rootstocks it is very important in nursery propagation that only pathogen-free material be obtained, as any diseases present in such stocks are maintained and spread, along with the rootstock material. On such "clean" rootstock material only disease-free scion material should be grafted or budded.

**Figure 11-2** Scion roots of an 'Old Home' pear grafted on quince. (A) Original quince roots. (B) Scion roots arising from the 'Old Home' pear above the graft union. These have assumed the major support of the tree. The dwarfing influence due to the quince roots has disappeared.

## Obtaining the Benefits of Certain Rootstocks

Many plant cultivars selected for their desirable fruit or ornamental qualities do not have comparably suitable root systems but require grafting onto other roots to give satisfactory plants. For many kinds of plants, rootstocks are available that tolerate unfavorable conditions, such as heavy, wet soils, or resist soil-borne insect or disease organisms better than the plant's own roots.[1] Also, for some species size-controlling rootstocks are available that can cause the composite grafted plant to have exceptional vigor or to become dwarfed. Some rootstocks, particularly in citrus species, give better size and quality of the fruit of the scion cultivar than do others (*145*).

Special rootstocks for glasshouse vegetable crops are used in Europe to avoid root diseases such as fusarium and verticillium wilt (*85*). In Holland forcing types of greenhouse cucumbers are grafted onto *Cucurbita ficifolia,* and commercial tomato cultivars are grafted onto vigorous $F_1$ hybrid, disease-resistant rootstocks (*159*).

### DOUBLE WORKING

More than two kinds of plants can be combined together in a vertical arrangement. In addition to the rootstock and scion, one may insert a third kind between them by grafting. Such a section is termed an **intermediate stem section,** an **intermediate stock,** or **interstock.**

There are several reasons for using double-working in propagation.

**1.** The interstock makes it possible to avoid certain kinds of incompatibility.

---

[1] Detailed discussions of the rootstocks available for the various fruit and ornamental species are given in chapters 18 and 19.

**2.** The interstock may possess a particular characteristic (such as disease resistance or cold-hardiness) not possessed by either the rootstock or the scion, which makes the interstock valuable for use as the main framework of the tree.

**3.** A certain scion cultivar may be required for disease resistance in cases where the interstock characteristics are the chief consideration.

**4.** The interstock may have a definite influence on the growth of the tree. For example, when a stem piece of the dwarfing 'Malling 9' apple stock is inserted between a vigorous rootstock and a vigorous scion cultivar, it reduces growth of the composite tree and stimulates flowering in comparison with a similar tree propagated without the interstem piece (*141*) (see Figure 11–24).

Nurserymen supplying trees on seedling or clonal rootstocks, or with a clonal interstock, should identify such stocks on the label just as they do for the scion cultivar.

## Changing Cultivars of Established Plants (Topworking)

A fruit tree, or an entire orchard, may be of an undesirable cultivar. It could be unproductive, or an old cultivar whose fruits are no longer in demand; it could be one with poor growth habits, or possibly one that is susceptible to prevalent diseases or insects. As long as a compatible type is used, the framework of the tree may be regrafted to a more desirable cultivar, if one exists (see Figure 12–31).

In an orchard of a single cultivar of a species requiring cross-pollination, provision for adequate cross-pollination can be obtained by topworking scattered trees throughout the orchard to a proper pollinizing cultivar. Or if a single tree of a species such as the sweet cherry is unfruitful because of lack of cross-pollination, a branch of the tree may be grafted to the proper pollinizing cultivar.

A single pistillate (female) plant of a dioecious (pistillate and staminate flowers borne on separate individual plants) species, such as the hollies (*Ilex*), may be unfruitful because of the lack of a nearby staminate (male) plant to provide proper pollination. This problem can be corrected by grafting a scion taken from a staminate plant onto one branch of the pistillate plant.

Of interest to the home gardener is the fact that several cultivars of almost any fruit species can be grown on a single tree of that species by topworking each primary scaffold branch to a different cultivar. In a few cases, different species can be worked on the same tree. For example, on a single citrus tree it would be possible to have oranges, lemons, grapefruit, mandarins, and limes; or on a peach root system could be grafted plum, almond, apricot, and nectarine. Some different cultivars (or species), however, grow at different degrees of vigor, so careful pruning is required to cut back the most vigorous cultivar on the tree to prevent it from becoming dominant over the others.

## Hastening Reproductive Maturation of Seedling Selections

In various fruit-breeding projects, the young seedling selections, if left to grow on their own roots, may take five to ten or more years to grow out of the juvenile phase and come into bearing. Unless the juvenility influence is too strong, this period can be shortened by grafting the small seedling shoots as scions onto large, established trees (*174*) or onto certain dwarfing rootstocks (*175*). Such grafting does not eliminate seedling juvenility, rather it takes advantage of the existing large root system of the rootstock plant.

It is possible to graft several seedling selections on one mature tree, but this could be inadvisable owing to the danger of virus contamination coming from the old tree, or perhaps from one or more of the seedling scions, and spreading to the others.

Sometimes desirable new seedling plants produced in breeding programs never attain the ability to grow well on their own roots, but when grafted on a vigorous, compatible rootstock, develop into plants of the desired form and stature (*38*). Crosses of *Syringa laciniata* × *S. vulgaris,* for example, produce Chinese lilac hybrids, but many of the seedlings lack sufficient vigor to live more than two or three years. However, if they are budded on the tree lilac, *Syringa amurensis japonica,* they survive and grow vigorously (*149*).

## Obtaining Special Forms of Plant Growth

By grafting certain combinations together it is possible to produce unusual types of plant growth, such as "tree" roses or "weeping" cherries or birches. For this purpose, an upright growing trunk is grafted at a height of 1.2 to 1.8 m (4 to 6 ft) to a cultivar or species that has a drooping or hanging growth habit. (See Figure 11–3.) Cactus is easily grafted to produce unusual plant forms, as shown in Figure 11–4.

**Figure 11–3**   How "weeping" plant forms may be obtained by grafting. An upright growing cultivar is grafted at the top by a side graft with another cultivar having a hanging growth pattern.

**Figure 11-4** Grafted ornamental cactus. An easily-rooted cultivar is used as the rootstock and an unusual attractive type is used as the scion. These grafts are made in large quantities in Japan and shipped to wholesale nurseries in other countries for rooting, potting, and growing until ready for sale in retail outlets.

## Repairing Damaged Parts of Trees

Occasionally the roots, trunk, or large limbs of trees are severely damaged by winter injury, cultivation implements, diseases, or rodents. By the use of bridge-grafting, or inarching, such damage can be repaired and the tree saved. This is discussed in detail in chapter 12.

## Studying Virus Diseases

Virus diseases can be transmitted from plant to plant by grafting. This characteristic makes possible testing for the presence of the virus in plants that may carry the pathogens but show few or no symptoms. By grafting scions or buds from a plant suspected of carrying the virus onto an indicator plant known to be highly susceptible, which shows prominent symptoms, detection is easily accomplished (*110*). This procedure is known as *indexing* (see chapter 8).

In order to detect the presence of a latent virus in a symptomless carrier, it is not necessary to use only combinations which make a permanent, compatible graft union. For example, the 'Shirofugen' flowering cherry (*Prunus serrulata*) is used to detect viruses in peach, plum, almond, and apricot. Cherry does not make a compatible union with these species, but a temporary, incompatible union is a sufficient bridge for virus transfer (*92*).

## NATURAL GRAFTING

One occasionally sees branches that have become grafted together naturally following a long period of being pressed together without disturbance. English ivy (*Hedera helix*) forms such grafts, and detailed studies have been made of translocation in natural grafts of this species (*112*). Not so obvious but of much greater significance and occurrence, particularly in stands of forest species, such as pine, hemlock, oak, and Douglas fir, is the natural grafting of roots (*58, 163*).

Such grafts are most common between roots of the same tree or between roots of trees of the same species. Grafts between roots of trees of different species are rare. In the forest, living stumps sometimes occur, kept alive because their roots have become grafted to those of nearby intact, living trees (*108*). The anatomy of natural grafting of aerial roots has been studied; the initial contact is established by the formation and fusion of epidermal hairs (*139*). Such natural grafting permits transmission of fungi, viruses, and mycoplasms from infected trees to their neighbors. This can be important in closely set orchard and nursery plantings of fruit trees, where numerous root grafts could occur and result in the slow spread of pathogens throughout the planting (*90*). Natural root grafting is a potential source of error in virus indexing procedures where virus-free and virus-infected trees are grown in close proximity (*61*). In addition, fungal pathogens such as those causing oak wilt and Dutch elm disease can be spread by such natural root connections.

## FORMATION OF THE GRAFT UNION

A number of detailed studies have been made of the healing of graft unions, mostly with woody plants (*35, 50, 106, 117, 143, 151, 161, 165*).

Briefly, the usual sequence of events in the healing of a graft union is as follows (Figure 11–5):

**1.** Freshly cut scion tissue capable of meristematic activity is brought into secure, intimate contact with similar freshly cut stock tissue in such a manner that the cambial regions of both are in close proximity. Temperature and humidity conditions must be such as to promote growth activity in the newly exposed, and surrounding, cells.

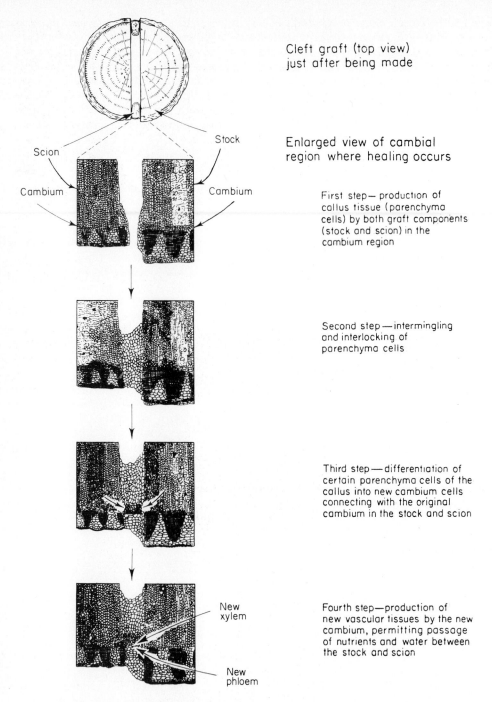

Cleft graft (top view)
just after being made

Enlarged view of cambial
region where healing occurs

Scion

Stock

Cambium

Cambium

First step— production of
callus tissue (parenchyma
cells) by both graft components
(stock and scion) in the
cambium region

Second step—intermingling
and interlocking of
parenchyma cells

Third step—differentiation of
certain parenchyma cells of the
callus into new cambium cells
connecting with the original
cambium in the stock and scion

New
xylem

Fourth step—production of
new vascular tissues by the new
cambium, permitting passage
of nutrients and water between
the stock and scion

New
phloem

**Figure 11–5** Diagrammatic developmental sequence during the healing
of a graft union as illustrated by the cleft graft.

**Figure 11-6**    Callus production from incompletely differentiated xylem, exposed by excision of a strip of bark. × 120. Reprinted with permission from K. Esau, *Plant Anatomy* (New York: John Wiley & Sons, Inc., 1953).

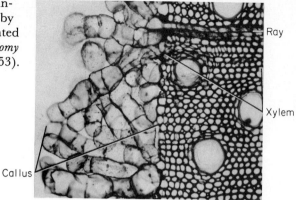

**2.** The outer exposed layers of cells in the cambial region of both scion and stock produce parenchyma cells that soon intermingle and interlock; this is called **callus tissue** (Figures 11-6 and 11-7).

**3.** Certain cells of this newly formed callus in line with the cambium layer of the intact scion and stock differentiate into new cambium cells.

**4.** These new cambium cells produce new vascular tissue, xylem toward the inside and phloem toward the outside, thus establishing vascular connection between the scion and stock, a requisite of a successful graft union.

**Figure 11-7**
Cross section of a *Hibiscus* wedge graft showing the importance of callus development in the healing of a graft union. Cambial activity in the callus has resulted in the production of secondary tissues which have joined the vascular tissues of stock and scion. × 10. Reprinted with permission from K. Esau, *Plant Anatomy* (New York: John Wiley & Sons, Inc., 1953).

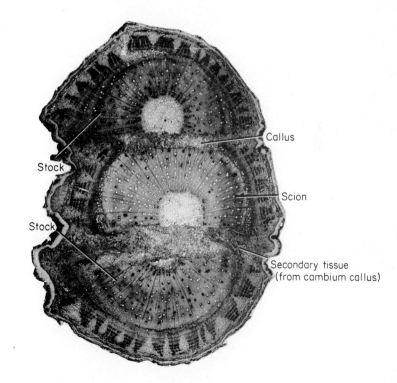

The healing of a graft union can be considered as the healing of a wound (*14*). Such injury to tissue as would occur if the cut end of a branch were split longitudinally would heal quickly if the split pieces were bound tightly together. New parenchyma cells would be produced by abundant proliferation from cells of the cambium region of both pieces, forming callus tissue. Some of the newly produced parenchyma cells differentiate into cambium cells, which subsequently produce xylem and phloem.

If, between the two split pieces, one interposed a third, detached, piece cut so that a large number of its cells in the cambial region could be placed in intimate contact with cells of the cambial region of the two split pieces, proliferation of parenchyma cells from all cambial areas would soon result in complete healing, with the foreign, detached piece joined completely between the two original split pieces. A graft union is essentially a healed wound, with an additional, foreign, piece of tissue incorporated into the healed wound.

This added piece of tissue, the scion, will not resume its growth successfully, however, unless vascular connection has been established so that it may obtain water and mineral nutrients. In addition, the scion must have a terminal meristematic region—a bud—to resume shoot growth and, eventually, to supply photosynthates to the root system.

In the healing of a graft union, the parts of the graft that are originally prepared and placed in close contact do not themselves move about or grow together. The union is accomplished entirely by cells that develop *after* the actual grafting operation has been made.

In addition, it should be stressed that in a graft union *there is no intermingling of cell contents*. Cells produced by the stock and by the scion each maintain their own distinct identity.

Considering in more detail the steps involved in the healing of a graft union, we may say that the first one listed below is a preliminary step, but nevertheless, it is essential and one over which the propagator has control.

**1.** *Establishment of intimate contact of a considerable amount of the cambial region of both stock and scion under favorable environmental conditions.*

Temperature conditions that will cause cell activity are necessary. Usually, temperatures from 12.8° to 32°C (55° to 90°F), depending upon the species, are conducive to rapid growth. Outdoor grafting operations should thus take place at a time of year when such favorable temperatures can be expected and when the plant tissues, especially the cambium, are in a naturally active state. These conditions generally occur during the spring months. Temperature levels under greenhouse and bench grafting situations can, of course, be readily controlled, thereby permitting greater reliability of results.

The new callus tissue arising from the cambial region is composed of thin-walled, turgid cells, which can easily become desiccated and die. It is important for the production of these parenchyma cells that the air moisture around the graft union be kept at a high level. This explains the necessity of thoroughly waxing the graft union or placing root grafts in a moist medium to maintain a high degree of tissue hydration.

It is important, too, that the region of the graft union be kept as free as possible from pathogenic organisms. The thin-walled parenchyma cells, under relatively high humidity and temperature conditions, provide a favorable medium for growth of fungi and bacteria, which are exceedingly detrimental to the successful healing of the union. Prompt waxing of the graft union helps prevent such infection.

It is essential that the two original graft components be held together firmly by some means, such as wrapping, tying, or nailing, or better yet, by wedging (as in the cleft or notch grafts) so that the parts will not move about and dislodge the interlocking parenchyma cells after proliferation has begun.

The statement is often made that for successful grafting the cambium layers of stock and scion must be "matched." Although this is desirable, it is unlikely that complete matching of the two cambium layers is, or ever can be, attained. In fact, it is only necessary that the cambial regions be close enough together so that the parenchyma cells from both stock and scion produced in this region can become interlocked. A slight crossing of the stock and scion cambium layers insures callus interlocking. It is in the region of the cambium that the essential callus production is the highest. Two badly matched cambial layers may delay union or, if extremely mismatched, prevent the graft union from taking place. In studies (*124*) of grafting monocotyledonous plants, it was found that a cambium layer is not necessarily required for a successful graft union but that any meristematic tissue would generate callus tissue to lead to the formation of a union between stock and scion.

**2.** *Production and interlocking of parenchyma cells (callus tissue) by both stock and scion.*

During the grafting operation the cells cut and damaged by the grafting knife turn brown and die, forming a **necrotic plate** that separates the two graft partners. Wound periderm develops and the contact layers then become suberized. Underneath these dead cells the living cells show increased cytoplasmic activity with, in some plants at least, a pronounced accumulation of dictyosomes[2] along the graft interfaces (see Figure 11-8) (*114, 115, 116, 117*). These dictyosomes appear to secrete materials into the cell wall space between the graft components via vesicle migration to the plasmalemma, resulting in a rapid adhesion between parenchymatous cells at the graft interface.

From these living cells new parenchyma cells (**callus**) proliferate in one to seven days from both stock and scion, coming from the parenchyma of the phloem rays and the immature parts of the xylem. The actual cambial layer itself seems to take little part in this first development of the callus (*89, 146, 155*).

In grafting scions on established stocks, the stock produces most of the callus. These parenchyma cells comprising the spongy callus tissue penetrate the thin necrotic layer within two or three days and soon fill the space between the two components of the graft (scion and stock), becoming intimately interlocked and providing some mechanical support, as well as allowing for limited passage of water and nutrients between the stock and the scion. For a time, between the callus arising from the stock and that arising from the scion, there is a more or less continuous brown line consisting of dead and crushed cells remaining between the two tissues of the graft. This line of cells is gradually resorbed, however, and disappears. At the final stages of healing, the cells of the outer layer of callus become suberized (*165*). Previously existing vascular elements (tracheids, vessels) are sealed off with a deposit of gum (*16*).

---

[2] Dictyosomes in plant cells are a series of flattened plates or double lamellae—one of the component parts of the Golgi apparatus.

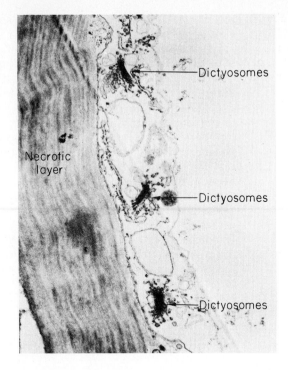

**Figure 11-8** Accumulation of dictyosomes along the cell walls adjacent to the necrotic layer at six hours after grafting in the compatible autograft in *Sedum telephoides.* × 17,500. Photograph courtesy R. Moore and D. B. Walker (*117*).

*3. Production of new cambium through the callus bridge.*

At the edges of the newly formed callus mass, parenchyma cells touching the cambial cells of the stock and scion differentiate into new cambium cells within two to three weeks after grafting. This cambial formation in the callus mass proceeds farther and farther inward from the original stock and scion cambium, and on through the **callus bridge,** until a continuous cambial connection forms between stock and scion.

Figure 11-9 shows the development of a cambial bridge through the callus in a mango "chip bud" graft union.

*4. Formation of new xylem and phloem from the new vascular cambium in the callus bridge.*

The newly formed cambial layer in the callus bridge begins typical cambial activity, laying down new xylem toward the inside and phloem toward the outside, along with the original vascular cambium of the stock and scion, continuing this throughout the life of the plant.

In the formation of new vascular tissues following cambial continuity, it appears that the type of cells formed by the cambium is influenced by the cells of the stock adjacent to the cambium. For example, xylem ray cells are formed where the cambium is in contact with xylem rays of the stock, and xylem elements where they are in contact with xylem elements (*136*). Leaves developing on the scion—but not on the stock—have been observed to exert a strong stimulus to induce the differentiation of vascular tissue across the graft interface (*162*).

**Figure 11-9**

Cambial bridge developing through callus in mango "chip bud" graft union—after 12 days.
Sc = scion; St = stock; Ph = phloem; Xy = xylem; C = cambium; Ca = callus; LC = laticiferous canal. Courtesy J. Soule (*161*).

The new xylem tissue originates from the activities of the scion tissues rather than from that of the stock. This origin is demonstrated in "ring grafting," in which a ring of bark from a young tree is removed and replaced by a ring of bark from another tree (*151*). Researchers using bark rings of the 'Scugog' apple, which has purple xylem tissue, found that subsequent xylem growth of the tree following grafting was entirely purple in color just under the 'Scugog' ring of bark, whereas in the remainder of the tree the xylem remained white (*190*).

Production of new xylem and phloem thus permits vascular connection between the scion and the stock. It is essential that this stage be completed before much new leaf development arises from buds on the scion. Otherwise, the enlarging leaf surfaces on the scion shoots will have little or no water to offset that lost by transpiration, and the scion will quickly become desiccated and die. It is possible, however, even though vascular connections fail to occur, enough translocation can take place through the parenchyma cells of the callus to permit survival of the scion. In grafts of *Vanilla* orchid, a monocot, scions survived and grew for two years with only parenchyma union (*124*).

A somewhat different developmental sequence has been found to take place in tobacco (*37*) and cotton (*84*). Here, xylem tracheary elements or phloem sieve tubes, or both, form directly by differentiation of callus into these vascular elements. A cambium layer subsequently forms between the two vascular elements. Apparently parenchyma cells, which make up the callus, can easily differentiate into tracheidlike elements.

Buds, as would occur on the scions of grafts, are effective in inducing differentiation of vascular elements in the tissues on which they are grafted. This bud influence has been shown by inserting a bud into a piece of *Cichorium* root, consisting only of old vascular parenchyma, and observing, as shown in Figure 11–10, that under the influence of auxin produced by the bud, the old parenchyma cells differentiate into groups of conducting elements (*24, 56*).

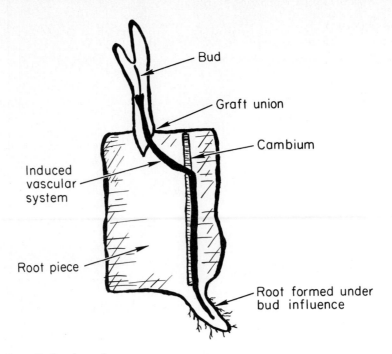

**Figure 11-10** Induction of vascular tissue by grafting a bud on to a piece of *Chicorium* root tissue. Redrawn from Gautheret (*56*).

Induction of vascular tissues in callus, as would be formed in the healing of a graft union, is under the control of materials originating from growing points of shoots. Furthermore, the stimulus from such shoot apices to produce xylem formation can be replaced by appropriate concentrations of such hormones as auxins, cytokinins, and gibberellins, as well as by sugars (*169*). It has been demonstrated (*186*) that sugar concentrations of 2.5 to 3.5 percent, plus IAA or NAA at 0.5 mg/liter, will cause the induction of xylem and phloem in the callus, with a cambium in between.

## THE HEALING PROCESS IN T-BUDDING

In T-budding, the bud piece usually consists of the epidermis, cork layer, cortex, phloem, cambium, and often some xylem tissue. Attached externally to this is a lateral bud subtended, perhaps, by a leaf petiole. In budding, this piece of tissue is laid against the exposed xylem and cambium of the stock, as shown diagrammatically in Figure 11-11.

Detailed studies of the healing process in T-budding have been made for the rose (*20*), citrus (*104*), and apple (*122*). Figure 11-12 shows a longitudinal section through a healed T-bud.

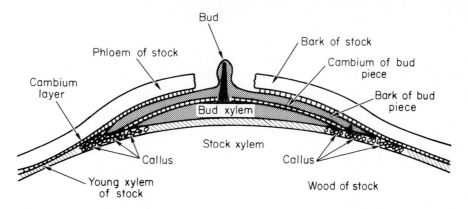

**Figure 11-11** Tissues involved in healing of an inserted T-bud as prepared with the wood (xylem) attached to the bud piece. Healing occurs when callus cells developing from the young xylem of the stock intermingle with callus cells forming from exposed cambium and young xylem of the T-bud piece. As the bark is lifted on the stock for insertion of the bud piece it detaches by separation of the youngest xylem cells.

**Figure 11-12** Healing at the bud union in a three-year-old T-budded almond tree. *Top left:* longisection through bud union. *Top right:* external appearance of union showing slight swelling. *Below:* Detail at union showing location of original inserted bud and the "buried" remains of the budded rootstock. Note clean line at union and good vascular connection.

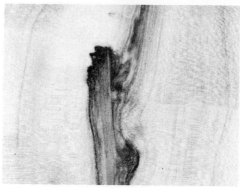

When the bud piece is removed from the budstick, and when the flaps of bark on either side of the "T" incision on the stock are raised, the cambium and newly formed xylem and phloem in these tissues are usually destroyed, owing to their very tender, succulent nature.

In the apple, when the bark of the stock is lifted in the budding operation, the separation occurs in the young, undifferentiated xylem. The entire cambial zone remains attached to the inside of the bark flaps. Very shortly after the bud shield is inserted a necrotic plate of material develops from the cut cells. Next, after about two days, callus parenchyma cells start developing from the rootstock xylem rays and break through the necrotic plate. Some callus parenchyma from the bud scion ruptures through the necrotic area in a similar manner. As additional callus is produced it surrounds the bud shield and holds it in place. The callus originates almost entirely from the rootstock tissue, mainly from the exposed surface of the xylem cylinder. Very little callus is produced from the sides of the bud shield. Callus proliferation continues rapidly for two to three weeks until all internal air pockets are filled. Following this, a continuous cambium is established between the bud and the rootstock. The callus then begins to lignify, and isolated tracheary elements appear. Lignification of the callus is completed about 12 weeks after budding (*122, 179*).

In the rose, about three days after budding, the terminal cells of the broken xylem rays and adjacent cambial derivatives on the exposed surface of the stock begin to enlarge and divide, leading to the production of callus strands. In the same manner, callus strands develop from terminal cells of broken phloem rays and adjacent young secondary phloem cells on the cut surface of the inner side of the bud piece. Within 14 days the space between the stock and the bud piece is completely filled with callus, which has developed mainly from the proliferating immature secondary xylem of the stock and the immature secondary phloem of the bud piece. During the second week, short areas of cambium cells appear in this newly developed callus tissue. By the tenth day, a completed band of cambium tissue extends over the face of the stock and is joined to the uninjured cambium on either side of the bud piece.

After cambial continuity is completed, vascular tissue connection soon becomes established between the bud and stock. In T–budding, then, the primary union is between the surface of the phloem on the inner face of the shield and the meristematic xylem surface of the stock. A secondary type of union may occur, however, at the edges of the shield piece, as it would in chip budding (*16*).

The various stages and time intervals involved in the healing of the union in T–budding of citrus have been determined as follows (*104*):

| Stage of development | Approximate time after budding |
| --- | --- |
| **1.** First cell division | 24 hours |
| **2.** First callus bridge | 5 days |
| **3.** Differentiation of cambium: | |
|     **a.** In the callus of the bark flaps | 10 days |
|     **b.** In the callus of the shield | 15 days |
| **4.** First occurrence of xylem tracheids: | |
|     **a.** In the callus of the bark flaps | 15 days |
|     **b.** In the callus of the shield | 20 days |

**5.** Lignification of the callus completed:

    **a.** In the bark flaps . . . . . . . . . . . . . . . . . . . . . . . . . . . . . . . . . . . . . . . . 25 to 30 days

    **b.** Under the shield . . . . . . . . . . . . . . . . . . . . . . . . . . . . . . . . . . . . . 30 to 45 days

# FACTORS INFLUENCING THE HEALING OF THE GRAFT UNION

As anyone experienced in grafting or budding knows, the results are often inconsistent, an excellent percentage of "takes" occurring in some operations, whereas in others the results are disappointing. A number of factors can influence the healing of graft unions.

## Incompatibility

One of the symptoms of incompatibility in grafts between distantly related plants is a complete lack, or a very low percentage, of successful unions. Grafts between some plants known to be incompatible, however, will initially make a satisfactory union, even though the combination eventually fails. (See Figure 11–17.)

## Kind of Plant

Some plants are much more difficult to graft than others even when no incompatibility is involved. Difficult ones, for example, are the hickories, oaks, and beeches. Nevertheless, such plants, once successfully grafted, grow very well with a perfect graft union. In topgrafting apples and pears, even the simplest techniques usually give a good percentage of successful unions, but in topgrafting certain of the stone fruits, such as peaches and apricots, much more care and attention to details are necessary. Strangely enough, topgrafting peaches to some other compatible species, such as plums or almonds, is more successful than reworking them back to peaches. Many times one method of grafting will give better results than another, or budding may be more successful than grafting, or vice versa. For example, in topworking the native black walnut (*Juglans hindsii*) to the Persian walnut (*Juglans regia*) in California, the bark graft method is more successful than the cleft graft.

Some easily grafted plants, such as apple, form a "wound gum" plugging exposed xylem elements after the grafting operation, thus preventing excessive desiccation and death of tissues. Other plants, such as the walnut, in which graft unions heal with difficulty, form such "wound gum" very slowly, and in these desiccation and death of tissues in the area of the graft union may be extensive.

Some species, such as the Muscadine grape (*Vitis rotundifolia*), mango (*Mangifera indica*), and *Camellia reticulata*, are so difficult to propagate by the usual grafting or budding methods that "approach grafting," in which both partners of the graft are maintained for a time on their own roots, is often used. This variation among plant species and cultivars in their grafting ability is probably related to their ability to produce callus parenchyma, which is essential for a successful graft union.

## Temperature, Moisture, and Oxygen Conditions
## During and Following Grafting

Certain environmental requirements must be met for callus tissue to develop.

**Temperature** has a pronounced effect on the production of callus tissue (Figure 11–13). In apple grafts little, if any, callus is formed below 0°C (32°F) or above about 40°C (104°F). Even around 4°C (40°F), callus development is slow and meager, and at 32°C (90°F) and higher, callus production is retarded, with cell injury becoming more apparent as the temperature increases, until death of the cells occurs at 40°C. Between 4° and 32°C, however, the rate of callus formation increases directly with the temperature (*151*). In such operations as bench grafting, callusing may be allowed to proceed slowly for several months by storing the grafts at relatively low temperatures, 7° to 10°C (45° to 50°F), or if rapid callusing is desired, they may be kept at higher temperatures for a shorter time.

Following bench grafting of grapes, a temperature of 24° to 27°C (75° to 80°F) is about optimum; 29°C (85°F) or higher results in profuse formation of a soft type of callus tissue that is easily injured during the planting operations. Below 20°C callus formation is slow, and below 15°C (60°F) it almost ceases.

Outdoor grafting operations performed late in the spring when excessively high temperatures may occur, often result in failure. Tests (*67*) of walnut top-grafting in California during very hot weather in May showed that white-washing the area of the completed graft union definitely promoted healing of the union. Whitewash reflects the radiant energy of the sun, thus resulting in lower bark temperatures. In addition, in these tests, scions placed on the north and east sides of the stub survived much better than those on the south and west, probably because of the lower temperatures resulting from their shaded position.

**Moisture**     Since the parenchyma cells comprising the important callus tissue are thin-walled and tender, with no provision for resisting desiccation, it is obvious that if they are exposed to drying air for very long, they will be killed.

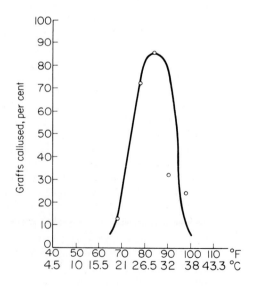

**Figure 11–13**     Influence of temperature on the callusing of walnut (*Juglans*) grafts. Callus formation is essential for the healing of the graft union. Maintaining an optimum temperature following grafting is very important for successful healing of walnut grafts. Adapted from data of Sitton (*158*).

This was found (*157*) to be the case in studies of the effect of humidity on healing of apple grafts. Air moisture levels below the saturation point inhibited callus formation, the rate of desiccation of the cells increasing as the humidity dropped. In fact, the presence of a film of water against the callusing surface was much more conducive to abundant callus formation than just maintaining the air at 100 percent relative humidity.

Highly turgid cells are more likely to give proliferation of callus than those in a wilt condition. In vitro studies (*46*) of stem pieces of ash (*Fraxinus excelsior*) have shown that callus production on the cut surfaces was markedly reduced as the water potential (turgidity) decreased.

Unless the adjoining cut tissues of a completed graft union are kept by some means at a very high humidity level, the chances of successful healing are rather poor. With most plants, thorough waxing of the graft union, which retains the natural moisture of the tissues, is all that is necessary. Often root grafts are not waxed but stored in a moist packing material during the callusing period. Damp peat moss or wood shavings are good media for callusing, providing adequate moisture and aeration.

**Oxygen**　　It has been shown (*157*) that oxygen is necessary at the graft union for the production of callus tissue. This would be expected since rapid cell division and growth are accompanied by relatively high respiration, which requires oxygen. For some plants, a lower percentage of oxygen than is found naturally in air is sufficient, but for others healing of the union is better if it is left unwaxed but placed in a well-moistened medium. This may indicate that the latter plants have a high oxygen requirement for callus formation. Waxing restricts air movement and oxygen may become a limiting factor, thus inhibiting callus formation. This situation seems to be true for the grape in which, usually, the union is not covered with wax or other air-excluding materials during the callusing period.

There is some evidence (*25*) that light will inhibit callus development. In vitro callus cultures of black cherry (*Prunus serotina*) became much larger in darkness than they did in light.

## Growth Activity of the Stock Plant

Some propagation methods, such as T-budding and bark grafting, depend upon the bark "slipping," which means that the cambium cells are actively dividing, producing young thin-walled cells on each side of the cambium. These newly formed cells separate easily from one another, so the bark "slips."

Initiation of cambial activity in the spring results from the onset of bud activity since, shortly after the buds start growth, cambial activity can be detected beneath each developing bud, with a wave of cambial activity progressing down the stems and trunk. This stimulus is due to production of auxin and gibberellins originating in the expanding buds (*180*).

In budding seedlings in the nursery in late summer it is important that they have an ample supply of soil moisture just before and during the budding operation. If they should lack water during this period, active growth is checked, cell division in the cambium stops, and it becomes difficult to lift the bark flaps to insert the bud.

There is evidence to show that callus proliferation—essential for a successful graft union—occurs most readily at the time of year just before and during "bud break" in the spring, diminishing through the summer and into fall. Increasing callus proliferation takes place again in late winter, but this is not dependent upon breaking of bud dormancy (*164*).

At certain periods of high growth activity in the spring, plants exhibiting strong root pressure (such as the walnut, maple, and grape) show excessive sap flow or "bleeding" when cuts are made preparatory to grafting. Grafts made with such moisture exudation around the union will not heal properly. Such "bleeding" at the graft union can be overcome by making slanting knife cuts below the graft around the tree. They should be made through the bark and into the xylem to permit such exudation to take place below the graft union. When potted rootstock plants to be grafted (for example, in species of *Fagus, Betula,* or *Acer*) show excessive root pressure, they should be put in a cool place with reduced watering until the "bleeding" stops.

On the other hand, potted rootstock plants, such as junipers or rhododendrons, when first brought into a warm greenhouse in winter for grafting, are dormant, and grafting should be delayed until the rootstock plants have been held for several weeks at 15° to 18°C (60° to 65°F) and new roots start to form; then the rootstock plant is physiologically active enough for the union to heal.

When the rootstock plant is physiologically overactive (excessive root pressure and "bleeding"), or underactive (no root growth being made), some form of side-graft, in which the rootstock top is not at first removed, should be used. In situations in which the rootstock is neither overactive nor underactive, one of the many forms of topgrafting, in which the top of the stock is completely removed at the time the graft is made, is likely to be successful (*47*).

## Propagation Techniques

Sometimes the techniques used in grafting are so poor that only a small portion of the cambial regions of the stock and scion are brought together. Although healing occurs in this region and growth of the scion may start, after a sizable leaf area develops and high temperatures and high transpiration rates occur, sufficient movement of water through the limited conducting area cannot take place, and the scion subsequently dies. Other errors in grafting technique, such as poor or delayed waxing, uneven cuts, or use of desiccated scions can, of course, result in grafting failure.

Poor grafting techniques, although they may delay adequate healing for some time—weeks or months—do not in themselves cause any permanent incompatibility. Once the union is adequately healed, growth can proceed normally.

## Virus Contamination, Insects, and Diseases

Using virus-infected propagating materials in nurseries can reduce bud "take" as well as the vigor of the resulting plant (*132*). In stone fruit propagation the use of budwood free of ring spot virus has consistently given improved percentages of "takes" over infected wood.

Topgrafting olives in California is seriously hindered in some years by attacks of the American plum borer (*Euzophera semifuneralis*), which feeds on the soft callus tissue around the graft union, resulting in the death of the scion. In England, nurserymen are often plagued with the red bud borer (*Thomasiniana oculiperda*), which feeds on the callus beneath the bud-shield in newly inserted T-buds causing them to die (*55*).

Sometimes bacteria or fungi gain entrance at the wounds made in preparing the graft or bud unions. For example, it was found that a rash of failures in grafts of *Cornus florida* 'Rubra' on *C. florida* stock was due to the presence of the fungus *Chalaropsis thielavioides* (*31*). Chemical control of such infections materially aids in promoting healing of the unions (*45*).

In South and Central America, rubber (*Hevea*) trees are propagated by a modification of the patch bud. A major cause of budding failures in such warm, humid areas has been infection of the cut surfaces by a fungus, *Diplodia theobromae*, but control of this infection can be obtained by fungicidal treatments (*94*). In topgrafting mangos in Florida, control of the fungus diseases anthracnose and scab is essential for success. This is done by spraying the rootstock trees and the source of scionwood regularly with copper fungicides before grafting is attempted (*126*).

### Growth Substances in Healing of Graft Unions

Trials with the use of growth substances, particularly auxin, applied to tree wounds or to graft unions have not given consistent results in promoting subsequent healing; consequently such materials are not generally used for this purpose (*67, 103*). In tissue culture studies, however, a definite relationship has been found to exist between callus production (which is essential for graft healing) and the levels of certain applied growth substances, particularly kinetin and auxin (*46, 52, 123, 130*). There is some evidence, too, that abscisic acid stimulates callus production, especially when applied to tissues in combination with auxin (*1*), or with kinetin (*15*).

Some practical benefit from the use of combinations of auxin and kinetin or abscisic acid plus auxin in stimulating callus formation and subsequent healing of graft unions may be possible.

## POLARITY IN GRAFTING

Proper polarity is essential if the graft union is to be permanently successful. In all commercial grafting operations correct polarity is strictly observed.

As a general rule, as shown in Figure 11–14, in grafting two pieces of stem

**Figure 11–14**     Polarity in grafting. In top grafting, the proximal end of the scion is attached to the distal end of the stock. In root grafting, however, the proximal end of the scion is joined to the proximal end of the stock.

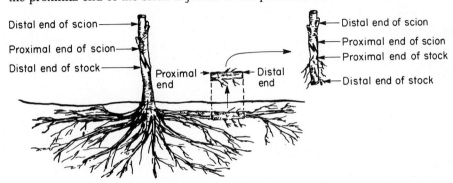

> The **proximal** end of either the shoot or the root is that nearest the stem-root junction of the plant.
>
> The **distal** end of either the shoot or the root is that furthest from the stem-root junction of the plant and nearest the tip of the shoot or root.

tissue together, the morphologically proximal end of the scion should be inserted into the morphologically distal end of the stock. But in grafting a piece of stem tissue on a piece of root, as is done in root grafting, the proximal end of the scion should be inserted into the proximal end of the root piece.

Should a scion be inserted with reversed polarity—"upside down"—in bridge-grafting, for example, it is possible for the two graft unions to be successful and the scion to stay alive for a time. But, as seen in Figure 11–15, the reversed scion does not increase from its original size, whereas the scion with correct polarity enlarges normally.

In **nurse-root grafting,** the rootstock may purposefully be grafted to the scion with reversed polarity. Union will occur, and the root will supply water and mineral nutrients to the scion but the scion is unable to supply necessary organic materials to the rootstock and the stock eventually dies. In nurse-root grafting, the graft union is purposely set well below the ground level, and the scion itself produces adventitious roots, which ultimately become the entire root system of the plant.

**Figure 11–15**    Bridge graft on a pear tree five months after grafting. Center scion was inserted with reversed polarity. Although the scion is alive it has not increased from its original size. The two scions on either side have grown rapidly.

**Figure 11–16**    Two-year-old 'Stayman Winesap' apple budded on 'McIntosh' seedling by inverted T-buds. Note development of wide angle crotches. Courtesy Arnold Arboretum, Jamaica Plain, Mass.

In T-budding or patch budding, the rule for observance of correct polarity is not as exacting. Buds can be inserted with reversed polarity and still make permanently successful unions. As shown in Figure 11–16, inverted T-buds start growing downward, then the shoots curve and grow upward (*148*). In the inverted bud piece, the cambium is capable of continued functioning and growth. In the xylem, phloem, and fibers formed from cambial activity, there is a twisting configuration which apparently allows for normal translocation and water conduction (*32*).

## LIMITS OF GRAFTING

Since one of the requirements for a successful graft union is the close matching of the callus-producing tissues near the cambium layers, grafting is generally confined to the dicotyledons in the angiosperms, and to the gymnosperms, the cone-bearing plants. Both have a vascular cambium layer existing as a continuous tissue between the xylem and the phloem. In the monocotyledonous plants of the angiosperms, which do not have a vascular cambium, grafting is more difficult, with a low percentage of "takes." There are cases of successful graft unions between the stem parts of monocots. By making use of the meristematic properties found in the intercalary tissues (located at the base of internodes), successful grafts have been obtained with various grass species as well as the large tropical monocotyledonous vanilla orchid (*124, 125*).

Before a grafting operation is started, it should be determined that the plants to be combined are capable of uniting and producing a permanently successful union. There is no definite rule that can predict exactly the ultimate outcome of a particular graft combination except that *the more closely the plants are related botanically, the better the chances are for the graft union to be successful.* However, there are numerous exceptions to this rule.

### Grafting within a Clone

A scion can be grafted back on the plant from which it came, and a scion from a plant of a given clone can be grafted to any other plant of the same clone.

For example, a scion taken from an 'Elberta' peach tree could be grafted successfully to any other 'Elberta' peach tree in the world.

## Grafting Between Clones within a Species

In the tree fruit and nut crops different clones within a species can almost always be grafted together without difficulty and produce satisfactory trees. However, in some conifer species, notably Douglas fir (*Pseudotsuga menziesii*), incompatibility problems have arisen in grafting together individuals of the same species, such as selected *P. menziesii* clones onto *P. menziesii* rootstock seedlings (*36*).

## Grafting between Species within a Genus

For plants in different species but in the same genus, grafting is successful in some cases but unsuccessful in others. Grafting between most species in the genus *Citrus,* for example, is successful and widely used commercially. The almond (*Prunus amygdalus*), the apricot (*Prunus armeniaca*), the European plum (*Prunus domestica*), and the Japanese plum (*Prunus salicina*)—all different species—are grafted commercially on the peach (*Prunus persica*), a still different species, as a rootstock. But on the other hand, almond and apricot, both in the same genus, cannot be intergrafted successfully. The 'Beauty' cultivar of Japanese plum (*Prunus salicina*) makes a good union when grafted on the almond, but another cultivar of *P. salicina,* 'Santa Rosa', cannot be successfully grafted on the almond. In another example of this inconsistency, some almond cultivars (which are clones) are highly successful when grafted on the 'Marianna 2624' plum as a rootstock, while other cultivars are not (*91*). Thus, compatibility between species in the same genus depends upon the particular genotype combination of stock and scion.

Reciprocal interspecies grafts are not always successful. For instance, 'Marianna' plum (*Prunus cerasifera* × *P. munsoniana*) on peach (*Prunus persica*) roots makes an excellent graft combination, but the reverse—grafts of the peach on 'Marianna' plum roots—either soon die or fail to develop normally (*2, 101*). In another example, many Japanese plum (*Prunus salicina*) cultivars can be successfully grafted on the European plum (*P. domestica*), but the reverse, the European plum grafted on the Japanese plum, is unsuccessful (*75*).

## Grafting between Genera within a Family

When the plants to be grafted together are in the same family but in different genera, the chances of a successful union become more remote. Cases can be found in which such grafts are successful and used commercially, but in most instances such combinations are failures.

Trifoliate orange (*Poncirus trifoliata*) is used commercially as a dwarfing stock for the orange (*Citrus sinensis*), in a different genus. The quince (*Cydonia oblonga*) has long been used as a dwarfing rootstock for certain pear (*Pyrus communis*) cultivars. The reverse combination, quince on pear, though, is unsuccessful. Intergeneric grafts in the nightshade family, Solanaceae, are quite common. Tomato (*Lycopersicon esculentum*) can be grafted successfully on Jimson weed (*Datura stramonium*), tobacco (*Nicotiana tabacum*), potato (*Solanum tuberosum*), and black nightshade (*Solanum nigrum*). The evergreen loquat (*Eriobotrya japonica*) can be grafted on quince roots (*Cydonia oblonga*), a deciduous tree, giving a dwarfed loquat plant.

### Grafting between Families

Successful grafting between plants of different botanical families is usually considered to be impossible but there are reported instances (*92, 191*) in which it has been accomplished. These are with short-lived, herbaceous plants, though, for which the time involved is relatively short. Grafts, with vascular connections between the scion and stock, were successfully made (*128*) using white sweet clover, *Melilotus alba* (Leguminosae), as the scion and sunflower, *Helianthus annuus* (Compositae), as the stock. Cleft grafting was used, with the scion inserted into the pith parenchyma of the stock. The scions continued growth with normal vigor for more than five months. As far as is known, however, there are no instances in which woody perennial plants belonging to different families have been successfully and permanently grafted together.

## GRAFT INCOMPATIBILITY

The ability of two different plants, grafted together, to produce a successful union and to develop satisfactorily into one composite plant is termed **compatibility.** The opposite, of course, would be **incompatibility.** The distinction between a compatible and an incompatible graft union is not clearcut. On one hand, stock and scion of closely related plants, unite readily and grow as one plant. On the other hand, stock and scion of unrelated plants grafted together are likely to fail completely in uniting. Most graft combinations lie between these two extremes in that they unite initially, with apparent success (*21*) (Figures 11–17 and 11–18), but gradually develop distress symptoms with time, due either to failure at the union or to the development of abnormal growth patterns (*16, 52*). A strong, well-healed union and one showing incompatibility symptoms are illustrated in Figure 11–19.

**Figure 11–17** Breakage at the graft union resulting from incompatibility. *Left:* one-year-old nursery trees of apricot on almond seedling rootstock. *Right:* a 15-year-old 'Texas' almond tree on seedling apricot rootstock which broke off cleanly at the graft union—a case of "delayed incompatibility" symptoms.

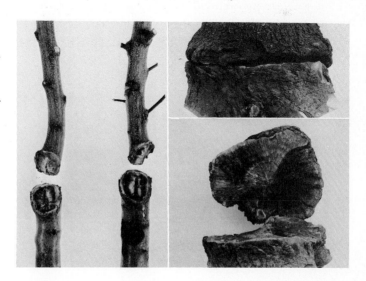

**Figure 11-18** Apple grafts (right portion of tree) five months after being grafted on a pear tree. This is an incompatible combination; the apple grafts eventually died although they initially made a strong growth.

### Symptoms of Incompatibility

Graft union malformations resulting from incompatibility can usually be correlated with certain external symptoms. The following symptoms have been associated with incompatible graft combinations:

**Figure 11-19** Radial sections through two fruit-tree graft unions. *Left:* a strong, well-knit compatible graft union. *Right:* a weak, poorly connected union, showing incompatibility symptoms.

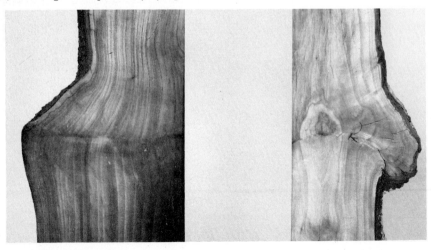

1. Failure to form a successful graft or bud union in a high percentage of cases.

2. Yellowing foliage in the latter part of the growing season, followed by early defoliation. Decline in vegetative growth, appearance of shoot dieback, and general ill health of the tree.

3. Premature death of the trees, which may live only a year or two in the nursery.

4. Marked differences in growth rate or vigor of scion and stock.

5. Differences between scion and stock in the time at which vegetative growth for the season begins or ends.

6. Overgrowths at, above, or below the graft union.

7. Graft components breaking apart cleanly at the graft union.

An isolated case of one or more of the above symptoms (except for the last) does not necessarily mean the combination is incompatible; some of these symptoms can result from unfavorable environmental conditions, such as lack of water or some essential nutrient, attacks by insects or diseases, or poor grafting or budding techniques (*3, 96*).

Incompatibility is clearly indicated by *the breaking off of trees at the point of union, particularly when they have been growing for some years and the break is clean and smooth, rather than rough or jagged.* This break may occur in a year or two after the union is made—for instance, in the apricot on almond roots (see Figure 11-17).

In certain other cases, e.g., some apricot cultivars grafted on myrobalan plum roots, such breakage may not take place until the trees are full-grown and bearing crops (*43, 48, 137*). The occurrence of this situation in a single instance provides justification for stating that the particular combination is incompatible.

Scion overgrowth at the graft union is sometimes associated with incompatibility but it also occurs in compatible unions. This symptom is not a reliable indication of incompatibility (*2, 16*) (Figure 11-20).

**Figure 11-20**
It is possible for the scion to overgrow the stock and yet develop into a large, strong tree. Such overgrowth is shown here for the Caucasian wingnut (*Pterocarya fraxinifolia*) grafted on *Pterocarya stenoptera*.

## Types of Incompatibility

Graft incompatibility in fruit trees has been classified by Mosse (*121*) into two types: **localized incompatibility** and **translocated incompatibility.** Most known cases can be placed in one or the other of these groups.

*Localized incompatibility* includes combinations in which the incompatibility reactions apparently depend upon actual contact between stock and scion. Separation of the components by insertion of a **mutually compatible** interstock overcomes the incompatibility symptoms. In the incompatible combination the union structure is often mechanically weak, with continuity of cambium and vascular tissues broken, although there are cases in which the union is strong and the tissues joined normally. Often external symptoms develop slowly, appearing in proportion to the degree of anatomical disturbance at the graft union. Root starvation eventually results, owing to translocation difficulties across the defective graft union.

An example of an incompatible combination in this category is that of 'Bartlett' ('Williams') pear grafted directly on quince rootstock. With the use of a "compatibility bridge" of 'Old Home' or ('Beurré Hardy') pear as an interstock, the three-part combination is completely compatible, and satisfactory tree growth takes place (*120*).

Masses of parenchymatous tissue, rather than normally differentiated tissues, are commonly found at the union in this type of incompatibility, interrupting the normal vascular connection between stock and scion. Sometimes the conducting vessels make connections around the masses of callus tissue, whereas in other cases the ends of the vessels are separated by the masses of parenchyma. In extreme cases the vascular tissue becomes distorted, but the parenchyma layer between stock and scion is continuous. Apparently the cambial region of either scion or stock or both has failed to produce other than parenchyma cells at the line of union, resulting in a continuous sheet of such cells between scion and stock (*136, 182*).

Inclusions of bark tissue may develop. This occurs in trees of the incompatible combinations of apple on pear (*137*) and plum on cherry (*136*). It was found in these studies that a layer of bark extended almost to the point at which the cambium layers of stock and scion were placed when the graft was made. A small amount of continuous new parenchyma tissue allowed enough movement of water and nutrients to enable the tree to start growth. But the layer of bark laid down by the cambiums of both scion and stock and pinched between the xylem on either side formed a layer of mechanical weakness. Thus, by the second season, with a larger top and greater resistance to wind movement, a break could easily occur at the graft union.

*Translocated incompatibility* includes certain cases in which the incompatible condition is *not* overcome by the insertion of a mutually compatible interstock because, apparently, some labile influence can move across it. This type involves phloem degeneration, and can be recognized by the development of a brown line or necrotic area in the bark. Consequently, restriction of movement of carbohydrates occurs at the graft union—accumulation above and reduction below. Reciprocal combinations may be compatible. In the various combinations in this category the range of bark tissue breakdown can extend from virtually no union at all, to a mechanically weak union with distorted tissues, to a strong union with tissues normally connected.

An example of a combination in this category is that of 'Hale's Early' peach grafted onto 'Myrobalan B' plum rootstock (*76*). This forms a weak union in which distorted tissues occur. Abnormal quantities of starch accumulate at the base of the peach scion. If the mutually compatible 'Brompton' plum is used as an interstock between the 'Hale's Early' peach and the 'Myrobalan B' rootstock, the incompatibility symptoms still persist, with an accumulation of starch in the 'Brompton' interstock. However, in other studies (*78*), in which peach/myrobalan plum grafts were made, but using small seedlings at the cotyledonary stage, no signs of incompatibility appeared even after the grafted trees were 13 years old. In the same study all trees of this combination grafted at the usual stage in the nursery showed incompatibility symptoms one year after grafting. Possibly some factor responsible for the incompatibility symptoms is not present in the juvenile seedling tissues.

In another example of translocated incompatibility, the combination of 'Nonpareil' almond on 'Marianna 2624' plum shows complete phloem breakdown, although the xylem tissue connections are quite satisfactory. However, another cultivar, the 'Texas' almond, on 'Marianna 2624' plum produces a good, compatible combination. Inserting a 6-in. piece of 'Texas' almond as an interstock between the 'Nonpareil' almond and the 'Marianna' plum stock fails to overcome the incompatibility between these two components. Bark disintegration occurs at the normally compatible 'Texas' almond/'Marianna' plum graft union, owing, presumably, to some factor translocated in the phloem from the 'Nonpareil' scion above, through the 'Texas' interstock, to the 'Marianna' plum rootstock (*91*).

Peach grafted on 'Marianna' plum generally forms an incompatible union. Although peach buds unite readily with this plum stock and grow satisfactorily the first season, later an enlargement appears above the graft union, followed by wilting of the peach leaves and death of the tree. Anatomical studies (*101*) showed this to be a case of incompatibility in which good xylem connections developed at the graft union but the phloem tissues failed to unite. Death of the roots resulted, with subsequent wilting and death of the peach top. If leafy shoot growth was retained on the 'Marianna' plum stock to nourish the roots, however, the trees could be kept alive indefinitely, a situation found in other incompatibility cases also (*184*).

There is evidence that compatibility relationships between clones may change with time (*99*). For example, the pear cultivar 'Bristol Cross', when it was introduced in 1932, was known to make a strong union when grafted directly on quince roots. Thirty years later, this cultivar, when worked on quince roots, required a mutually compatible interstock—such as 'Beurré Hardy'—in order to produce an acceptable union. Likewise, 'Conference' pear grafted on quince in 1937 developed strong unions and vigorous trees, but 20 years later many individual nursery trees of this combination showed weak, incompatible unions (*187*). It is possible that such changes result from mutations, or from latent viruses, which have appeared in one or more of the components (*99*).

A mutation in the opposite direction—from incompatibility to compatibility—is offered as the most likely explanation for the appearance of the 'Swiss Bartlett' ('Williams') pear clone, found to be compatible with quince (*133*). Bartlett/quince combinations have long been notorious for their incompatibility.

## Causes and Mechanisms
## of Incompatibility

Although incompatibility is clearly related to genetic differences between stock and scion, the mechanisms by which particular cases are expressed are not clear. With the large numbers of genetically different plant materials that can be combined by grafting, a wide range of different physiological, biochemical, and anatomical systems are brought together, with many possible interactions, both favorable and unfavorable. Several proposals have been advanced in attempts to explain incompatibility, but the evidence supporting most of them is generally inadequate and often conflicting.

One possible mechanism is *physiological and biochemical differences between stock and scion.* This hypothesis is supported by studies with incompatible combinations of certain pear cultivars on quince rootstock (*62, 63, 64*). The experimental evidence supports the following conclusions:

**1.** When certain pear cultivars are grafted onto quince roots, a cyanogenic glucoside, prunasin—normally found in quince, but not in pear, tissues—is translocated from the quince into the phloem of the pear. The pear tissues break down the prunasin in the region of the graft union, with hydrocyanic acid as one of the decomposition products. This enzymatic breakdown is hastened by high temperatures. In addition, different pear cultivars vary in their ability to decompose the glucoside.

**2.** The presence of the hydrocyanic acid leads to a lack of cambial activity at the graft union, with pronounced anatomical disturbances in the phloem and xylem at the union resulting. The phloem tissues are gradually destroyed at and above the graft union. Conduction of water and materials is seriously reduced in both xylem and phloem.

**3.** A reduction of the levels of the sugar reaching the quince roots leads to further decomposition of prunasin, liberating hydrocyanic acid and killing large areas of the quince phloem.

**4.** A water-soluble and readily diffusable inhibitor of the action of the pear enzyme (which breaks down the glucoside) occurs in the various pear cultivars, although they differ in their content of this inhibitor. This may explain why certain pear cultivars are compatible and others incompatible with quince rootstock.

Electron microscope studies (*18, 19*) have been made of compatible and incompatible pear-quince grafts, with detailed observations of cell-wall structure at the graft unions. Examination of the cell walls of compatible graft combinations showed the concentration of lignin at the line of union to be as high as that in cell walls not a part of the union. In incompatible combinations, however, the adjoining cell walls of the two components contained no lignin, with the parts either not connected or interlocked only by cellulose fibers. From these studies it was concluded that the **lignification** processes of cell walls are involved in the formation of strong unions in pear-quince grafts. Reactions that inhibited the formation of lignin and the establishment of a mutual middle lamella between the two components resulted in weak unions.

In a study (*117*) at the cellular level of an obviously incompatible combination, *Sedum telephoides* (CRASSULACEAE) on *Solanum pennellii* (SOLANACEAE), the initial stages of graft union healing were similar to those occurring in a compatible combination (*Sedum* on *Sedum*). In the incompatible *Sedum* on *Solanum* graft,

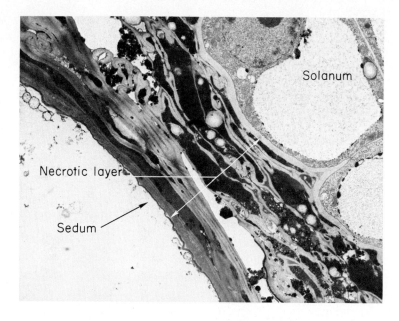

**Figure 11–21**    The graft interface of an incompatible graft between *Sedum telephoides* and *Solanum pennellii* at eight days after grafting. Lethal cellular senescence in *Sedum* has resulted in the formation of a necrotic layer of collapsed cells that separates the two graft partners. × 5000. Photograph courtesy R. Moore and D. B. Walker (*117*).

however, after 48 hours *Sedum* cells at the graft interface deposited an insulating layer of suberin along the cell wall, which later underwent a lethal cellular senescence and collapse, to form a necrotic layer of increasing thickness (*116, 117*). See Figure 11–21. Associated with this cellular senescence in *Sedum* cells a dramatic increase in a hydrolytic enzyme, acid phosphatase, has been observed (*118*). Rather than callus interlocking, cambial formation, and vascular connection, the thick necrotic layer prevented cellular connection, which led to scion desiccation and eventual death. Interestingly, the *Solanum* stock did not show the rejection response that the *Sedum* scion did.

In tests comparing the tensile strength of compatible *Sedum* on *Sedum* grafts with incompatible *Sedum* on *Solanum* grafts, that of the former increased steadily for 11 days after grafting when it leveled off, whereas the tensile strength of the incompatible grafts decreased, being correlated with the continued cellular necrosis of *Sedum* cells bordering the graft interface (*119*).

VIRUS AND MYCOPLASM PROBLEMS

Graft union failures resembling incompatibility symptoms can be due to pathogens. In certain cases abnormalities first attributed to stock/scion incompatibility were later found to be due to latent virus or mycoplasmalike pathogens introduced by grafting from a resistant partner to a susceptible partner (*28, 44, 107*). Figure 11–22 shows such an occurrence in apple.

The pear decline disease, developing first in Italy and later in western North America, killed hundreds of thousands of pear trees in California alone (*127,*

**Figure 11-22** Latent viruses in the scion portion of a graft combination may cause symptoms to appear in a susceptible rootstock following grafting. Here "stem-pitting" virus symptoms have developed in the sensitive 'Virginia Crab' apple rootstock. The wood of the scion cultivar—above the graft union—is unaffected. Courtesy H. F. Winter (*188*).

*154*). Early studies (*7*) related the trouble to the rootstock used and indicated it to be an "induced incompatibility," with the causal factor unknown. 'Bartlett' trees on oriental pear roots, such as *Pyrus pyrifolia*, were highly susceptible to the problem, while trees on *P. communis* roots were not. Subsequent research showed that pear decline was not due to an inherent stock-scion incompatibility but was associated with what was then thought to be a virus (*87*), transmitted by an insect vector, *Psylla pyricola*. The 'Bartlett' scion cultivar and the *P. communis* roots were observed to be resistant to decline, but the rootstock, *P. pyrifolia*, was highly susceptible, showing a phloem degeneration just below the graft union. Later studies (*80, 81*) showed the infective agent to be a mycoplasma-like organism rather than a virus. For the decline condition to appear, the concurrent presence of the infective bodies, pear psylla, and a susceptible rootstock are necessary. The mycoplasma-like organisms were found to be sensitive to tetracycline antibiotics; thus, remission of the symptoms is possible by tree trunk injections (*129*).

Another example of virus involvement is found in citrus. At one time incompatibility was blamed for the difficulty encountered in budding sweet orange (*Citrus sinensis*) onto sour orange (*C. aurantium*) roots in South Africa (1910) and in Java (1928), even though this combination was a commercial success in other parts of the world. The incompatibility was believed to be due to the production of some substance by the scion that was toxic to the stock (*170*). In the light of subsequent studies stimulated by the development of orange "tristeza" or "quick decline" in Brazil and California, it is clear that the toxic substance from the sweet orange scions was a virus, tolerated by the sweet orange, but lethal to the sour orange roots (*12, 183*).

Delayed incompatibility was formerly believed to be the cause of the so-called black-line of walnuts; black-line occurs in certain Persian walnut orchards in Oregon, California, and France where cultivars of *Juglans regia* are grafted onto seedling rootstocks of *J. hindsii* (Northern California black walnut) or onto Paradox roots (*J. hindsii* × *J. regia*) (*109, 153*). Affected trees grow normally, bearing good crops, until they reach maturity—15 to 20 or more years of age. When the trouble appears, a thin layer of cambium and phloem cells dies at the graft union. The dead tissue gradually extends around the tree at the graft union until the tree is girdled; the vertical width of the dead area may reach 30 cm (12 in.). Such girdling kills the tree above the union, but the rootstock usually develops sprouts and remains alive. It is now known that the problem is not due to incompatibility but to a virus that kills the *J. regia* tissue but does not affect the *J. hindsii* root tissue below the graft union (*105*).

### Correcting Incompatible Combinations

If a graft combination known to be incompatible has been made and is discovered before the tree dies or breaks off at the union, it is possible, in graft combinations of the localized type, to correct this condition by bridge grafting with a mutually compatible stock, if such exists. If a tree has been mistakenly propagated on a rootstock known to show symptoms of incompatibility with the scion cultivar, and with the probability that the tree will eventually break off at the graft union, it is also possible to correct this condition by inarching with seedlings of a compatible rootstock. If breakage does not occur until the inarches are strong enough to support the tree, it may thus be possible to save it, the inarched seedlings becoming the main root system.

## SCION–STOCK (SHOOT–ROOT) RELATIONSHIPS

Combining two (or more, in the case of interstocks) different plants (genotypes) into one plant by grafting—one part producing the top and the other part the root—can produce growth patterns that are different from those that would have occurred if each component part had been grown separately. Some of these effects have major horticultural value, while others are detrimental and should be avoided. These altered characteristics may result from (a) incompatibility reactions; (b) the fact that one of the graft partners possesses one or more specific characteristics not found in the other, for example, resistance to certain diseases, insects, or nematodes—or tolerance of certain adverse weather or soil conditions; or (c) specific interactions between the stock and the scion that alter size, growth, productivity, fruit quality, or other horticultural attributes.

In practice it may be difficult to separate which of the three kinds of influencing factors is dominant in any given graft combination growing in a particular environment.

### Effects of the Rootstock on the Scion Cultivar

#### SIZE AND GROWTH HABIT

Size control, and sometimes an accompanying change in tree shape, is one of the most significant rootstock effects. Apparently the rootstock alters the vigor of a given scion cultivar. By proper rootstock selection in apples, the complete range of tree size—from very dwarfed to very large—has been obtained with a given scion cultivar grafted to different rootstocks.

That specific rootstocks can be used to influence the size of trees has been known since ancient times. Theophrastus—and later the Roman horticulturists—made use of dwarfing apple rootstocks that could be easily propagated. The name "Paradise," which refers to a Persian park or garden—"pairi-daeza"—was applied to dwarfing apple stocks about the end of the fifteenth century (*23*).

A wide assortment of size-controlling rootstocks has now been developed for certain of the major tree fruit crops (*17*). Most notable is the series of vegetatively propagated apple rootstocks collected and developed at the East Malling Research Station in England, beginning in 1912. These were classified into four groups according, primarily, to the degree of vigor imparted to the scion cultivar: *dwarf, semidwarf, vigorous,* and *very vigorous* (*70, 113, 134*). Similarly, the size-controlling effects of the rootstock on sweet cherry (*Prunus avium*) scion cultivars has been known since the early part of the eighteenth century (*79*). Mazzard (*P. avium*) seedling rootstocks produce large, vigorous, long-lived trees, whereas *P. mahaleb* seedlings, as a rootstock, tend to produce smaller trees of shorter life (*42*). However, individual seedlings of these species, increased and maintained as clones, can produce distinct rootstock effects different from that of the whole species. Rootstock effects on tree size and vigor are recognized also in citrus, pear, and other species (*10, 53*). A discussion of specific rootstocks for the various fruit and nut crops is given in chapter 18.

Prediction of rootstock effects cannot be made with certainty without considering the entire system in which it is used, including the particular cultivar used as the scion top, which can modify the rootstock influence. Each combination of scion and rootstock must be tested before any conclusion regarding the behavior of the composite tree can be reached (*69*). The environment in which the particular combination is to be grown also must be taken into account. If grafted plants are grown under optimum conditions the differences in performance of those on strong-growing rootstocks vs. those on weaker stocks may be minimized. Good soil and cultural conditions are required for successful tree performance when very dwarfing rootstocks are used.

Alterations in the normal shape of the tree are often associated with the dwarfing effect caused by certain rootstocks. A low and spreading, rather than upright, form is illustrated by grafts of the 'McIntosh' apple on the apomictic, semidwarfing rootstock *Malus sikkimensis* (*148*). Such effects may possibly be due to changes in levels of certain hormones (auxins, gibberellins) in the tree.

Symptomless viruses that occur in plants may exert a dwarfing influence. If no harmful symptoms appear, useful dwarfing could be produced in this manner (*60*). Removing viruses from some dwarfing clonal stocks by heat treatment may decrease their dwarfing influence.

It would be very useful in developing new clonal rootstocks from seedlings to be able to predict whether such stocks would be dwarfing or invigorating. Studies (*8, 102*) in England indicate that apple rootstocks known to produce dwarf trees have a high proportion of bark to wood in the lateral roots, whereas stocks causing increased vigor in the scion cultivar have a lower proportion of bark to wood. Also, much of the functional wood tissue of roots of dwarfing apple stocks was composed of living cells, whereas in nondwarfing, vigorous rootstocks, the wood consisted of a relatively large amount of lignified tissue without living cell contents. However, similar studies (*168*) in Italy with grafted apple and pear trees did not show such a relationship.

FRUITING

*Fruiting precocity, fruit bud formation, fruit set, and yield* of a tree can be influenced by the rootstock used. In general, fruiting precocity is associated with dwarfing rootstocks and slowness to start fruiting with vigorous rootstocks. Long-term yield studies, conducted in England, involving several apple rootstocks showed that the results varied according to tree age and tree spacing (*134, 135*). Trees on 'Malling 9' roots planted 3.6 by 3.6 m (12 by 12 feet) apart showed highest accumulated yield per tree up to ten years of age because of their early bearing. By the tenth year this yield was surpassed by trees on the moderately vigorous 'Malling 4' roots. By the fifteenth year this was surpassed by trees on the vigorous 'Malling 1' roots, but by the twentieth year yield of all trees was superseded by trees grown on the very vigorous 'Malling 16' roots.

Vigorous, strongly growing rootstocks in some cases result in a larger and more vigorous plant that produces greater crops over a long period of years. On the other hand, trees on dwarfing stocks may be more fruitful, and if closely planted, produce higher yields per acre, especially in the early years of bearing.

The presence of the graft union itself may stimulate earlier and perhaps heavier bearing. For instance, in studies with citrus (*83*) five rootstocks—sour orange, sweet orange, trifoliate orange, grapefruit, and rough lemon—all started fruiting two seasons earlier when budded to themselves than when unbudded, although in each case the trees were about the same size.

If there is an imperfect graft union, such as the partial incompatibility that occurs in some combinations, a reduction in translocation at the graft union can have a girdling effect and thus lead to increased fruitfulness.

Rootstock influence can vary greatly with different kinds of plants. In growing Oriental persimmon (*Diospyros kaki*) cultivars the rootstock seems to have a direct effect on flower production and fruit set. In tests (*152*) using the 'Hachiya' persimmon, trees on *D. lotus* roots produced more flowers but matured fewer fruits than similar trees on *D. kaki* roots, while trees of the same cultivar on *D. virginiana* roots produced so few flowers that crops were very poor.

In grapes, where yield is dependent upon the vigor of the current season's growth, the rootstock used can be a strong influencing factor (*68*). Large yield increases of certain American types (*Vitis labrusca*) of grapes were obtained (*176*) when they were grafted on vigorous rootstocks, in comparison with own-rooted plants. Over a six-year period, yield of 'Concord' vines was increased from 30 to 150 percent, depending upon the rootstock used. On the other hand, with European type (*V. vinifera*) scions, use of 'Dog Ridge' (*V. champini*), an extremely vigorous rootstock cultivar, on fertile soils can lead to such strong growing vines that they become unproductive.

SIZE, QUALITY, AND MATURITY OF FRUIT

There is considerable variation among plant species in regard to the effect of the rootstock on fruit characteristics on the scion cultivar. In a grafted tree no transmission of characteristics of the fruit which the stock would produce is encountered in fruit of the scion cultivar. For example, quince, commonly used as a pear rootstock, has fruits with a pronounced tart and astringent flavor, yet this flavor does not appear in the pear fruits. The peach is often used as a rootstock for apricot, yet there is no indication that the apricot fruits have taken on any characteristics of peach fruits.

Although there is no intermingling of fruit characteristics between the stock and the scion, certain rootstocks can affect fruit quality of the scion cultivar. A good example of this is the "black-end" defect of pears. 'Bartlett', 'Anjou', and some other pear cultivars on several different rootstocks often produce fruits that are abnormal at the calyx end. The injury consists of blackened flesh, which in severe cases cracks open. Sometimes the calyx end of the fruit is hard and protruding. Such fruit is worthless commercially. It has been shown (40, 74) that this trouble develops when the trees are propagated on certain rootstocks, such as *Pyrus pyrifolia,* but only rarely when the French pear, *P. communis,* is used. This trouble affects only the fruit; no symptoms of adverse tree growth appear.

The development of black-end fruits disappears if trees on *P. pyrifolia* rootstock are inarched with *P. communis* seedlings, and—after the inarches are able to support the tree—the original *P. pyrifolia* roots are cut away. Black-end fruits continue to appear unless the original connection of the 'Bartlett' top with the *P. pyrifolia* roots is broken. Whether the *P. pyrifolia* roots are producing substances toxic to the fruit or whether there is some other interaction has not been established.

In citrus, striking effects of the rootstock appear in fruit characteristics of the scion cultivar (11). If sour orange (*Citrus aurantium*) is used as the rootstock, fruits of sweet orange, tangerine, and grapefruit are smooth, thin-skinned, and juicy, with excellent quality, and they store well without deterioration. Sweet orange (*C. sinensis*) rootstocks also result in thin-skinned, juicy, high-quality fruits. Citrus fruits on grapefruit (*C. paradisi*) stocks are usually excellent in size, grade, and quality if heavy fertilization is provided. But when rough lemon (*C. limon*) is used as the rootstock, the fruits are often thick-skinned, somewhat large and coarse, inferior in quality, and low in both sugar and acid.

Fruit size of both 'Washington Navel' and 'Valencia' orange is strongly influenced by the rootstock. The largest navel orange fruits are produced on sour orange stocks and the smallest on the Palestine sweet lime. The largest 'Valencia' oranges are associated with the dwarfing trifoliate orange stock, whereas sweet orange rootstocks produce the smallest fruits.

While many such tests have shown that the various characteristics of citrus fruit are affected by the rootstock used, the underlying physiological mechanisms remain unknown.

Tomato (*Lycopersicon esculentum*) grafted on Jimson weed (*Datura stramonium*) roots had been used at one time in the southern part of the United States, owing to the resistance of these roots to nematodes. As it was known that Jimson weed contains poisonous alkaloids, concern was felt about whether such alkaloids might not be translocated from the rootstock to the tomato fruits. Tests showed that this was the case. The alkaloid content of tomato fruits from grafted plants ranged from 3.9 to 28.6 mg per kg of fresh fruit, whereas ungrafted control plants had zero alkaloid content (98).

There are other similar examples of translocation of compounds between stock and scion in intergeneric grafts in Solanaceae. In a series of reciprocal grafts between tomato and tobacco, nicotine was found in tomato scions when they were grown on tobacco rootstocks. But when tobacco scions were grown on tomato rootstocks, tobacco alkaloid production was greatly reduced, most of the alkaloid synthesis being at the graft union and in the tobacco stem immediately above the union. Such localization was not found in tomato scions on tobacco roots (*41, 160*).

MISCELLANEOUS EFFECTS OF THE ROOTSTOCK
ON THE SCION CULTIVAR—
WINTER-HARDINESS, DISEASE RESISTANCE,
AND TIME OF FRUIT MATURITY

In citrus the rootstock used can affect the cold-hardiness of the scion cultivar. During killing freezes in the winter of 1950–51 in the Texas Rio-Grande Valley, young grapefruit trees on 'Rangpur' lime roots survived much better than those on rough lemon or sour orange, whereas trees on 'Cleopatra' mandarin were the most severely damaged (*33*). Survival following a severe winter freeze in Florida in 1962 showed that a wide range in tree hardiness in oranges and grapefruit could be attributed to the rootstock used (*53*).

Different rootstocks respond differently to soil conditions, thus resulting in an altered effect on the behavior of the scion cultivar (Figure 11–23). Almond and myrobalan plum roots tolerate excess boron in the soil better than 'Marianna' plum roots. Thus, in this case, there was fairly good growth of 'French' prune trees when they were grown on almond and myrobalan plum roots under conditions in which the trees were severely injured when grafted on 'Marianna' plum or apricot roots (*66*).

**Figure 11–23**  The tolerance of fruit trees to toxic amounts of certain elements may be influenced by the rootstock used. These 'French' prune trees on four different rootstocks were irrigated with water containing different levels of boron. Rootstocks were: (*top left*) almond seedlings; (*top right*) myrobalan plum seedlings; (*below left*) apricot seedlings; (*below right*) 'Marianna' plum cuttings. Irrigation water containing the following five concentrations of boron were used for each rootstock (left to right): ½ ppm (tap water); 2 ppm; 3 ppm; 5 ppm; 10 ppm. Courtesy C. J. Hansen.

The four rootstocks (plum, peach, apricot, and almond) commonly used for stone fruits differ markedly in their response to adverse soil conditions (43) and consequently can affect growth of the scion cultivar. For example, trees with myrobalan plum as the rootstock are the most tolerant of excessive soil moisture, followed by peach or apricot roots, almond being the most susceptible to injury from such conditions.

In citrus, considerable variability exists in the tolerance of the various rootstocks to adverse soil conditions. The choice of rootstock upon which to work the scion cultivar is very important. In Texas, for example, the severity of lime-induced chlorosis symptoms in trees grown on calcareous soils is greatly influenced by the rootstock used. Tests of grapefruit on 36 different rootstocks showed that with four of the stocks no chlorosis appeared, but severe chlorosis developed with 13 of the rootstocks (34).

It is known that some rootstocks are more tolerant than others to adverse soil pests, such as nematodes (*Meloidogyne* spp.) or oak root fungus (*Armillaria mellea*). The growth of the scion cultivar is subsequently influenced by the rootstock through the latter's relative ability to withstand such adverse conditions.

The above examples illustrate cases in which the behavior of the scion cultivar is affected by the rootstock used, which in turn can be traced to reactions of the rootstock to certain adverse soil situations.

## Effects of the Scion Cultivar on the Rootstock

Although there is a tendency to attribute all cases of dwarfing or invigoration of a grafted plant to the rootstock, the effect of the scion on the behavior of the composite plant may be as important as that of the rootstock. Unquestionably, however, the scion, the interstock, the rootstock, and the graft union itself all interact to influence each other and determine the over-all behavior of the plant. In certain combinations, however, a particular member of the combination could have a marked influence no matter what part of the plant it becomes. For instance, a dwarfing stock will exert a dwarfing influence on the entire plant whether used as rootstock, intermediate stock, or scion.

### EFFECT ON THE VIGOR OF THE ROOTSTOCK

Vigor is the major effect of the scion on the stock, just as it was in the case of rootstock effect on scion cultivar. If a strongly growing scion cultivar is grafted on a weak rootstock, the growth of the rootstock will be stimulated so as to become larger than it would have been if left ungrafted. Conversely if a weakly growing scion cultivar is grafted on a vigorous rootstock, the growth of the rootstock will be lessened from what it might have been if left ungrafted. In citrus, for example, when the scion cultivar is less vigorous than the rootstock cultivar, it is the scion cultivar rather than the rootstock that determines the rate of growth and ultimate size of the tree (82).

That the scion influences the growth of the rootstock was recognized at least as early as the middle of the nineteenth century (57). It has long been known, particularly with apples, that the size, nature, and form of the root system that develops from the seedling rootstocks of grafted trees can be affected by the cultivar of the scion (156, 173). Different scion cultivars may cause a characteristic root growth pattern to develop in the rootstock. For example, if apple seedlings are budded with the 'Red Astrachan' apple, a very fibrous root system with few taproots develops. If other similar seedlings are budded with 'Olden-burg' or 'Fameuse', the subsequent root system is not fibrous but has a two- or three-pronged deep taproot system (*73*). In fact, nurserymen propagating apples often can identify many of the scion cultivars by the appearance of the root system of the grafted nursery trees.

EFFECT ON COLD-HARDINESS OF THE ROOTSTOCK

In some species at least, the cold-hardiness of a particular rootstock can be influenced by the particular scion cultivar grafted on it. This effect is not due necessarily to a winter-hardy scion imparting hardiness to the rootstock. Rather, it is probably related to the degree of maturity attained by the rootstock, certain scion cultivars tending to prolong growth of the roots long into the fall so that insufficient maturity of the root tissues is reached by the time killing low winter temperatures occur. The rootstock, if left ungrafted, or grafted to a scion cultivar that stops growth in early fall, may mature its tissues sufficiently early so as to develop adequate winter-hardiness.

## Effects of an Intermediate Stock on Scion and Rootstock

The ability of certain dwarfing clones, inserted as an interstock between a vigorous top and vigorous root, to produce dwarfed and early-bearing fruit trees has been known for centuries and is used commercially to propagate dwarfed trees. It is reported (*131*) that one of the earliest records (*93*) of such procedures was given in 1681 in England, advocating the use of the Paradise apple as an interstock to induce precocity in apple trees grown on crabapple rootstocks. Figure 11–24 shows, for example, the degree of dwarfing induced in apples by various interstocks.

Tests (*166*) made to determine whether intermediate stocks of various apple cultivars inserted between the apple rootstock 'Malling 2' and the scion cultivar 'Jonathan' or 'Delicious' had any effect on the behavior of the scion showed that in every case a depression of growth occurred in comparison with cases in which the scion cultivar itself was used as the interstock.

An intermediate dwarfing stem piece seems to have a built-in mechanism that causes reduced growth in the rootstock as well as in the scion top (*131*). Repeated comparisons of the influence exerted on the scion cultivar by the rootstock and an interstock show that although both have an influence, the rootstock's effect is greater (*171, 178*).

Dwarfing of apple trees by the use of a dwarfing interstock, such as 'Malling 9', has been widely used commercially for many years. This method has the advantage of allowing the use of well-anchored, vigorous seedlings as the rootstock rather than a brittle, poorly anchored dwarfing clone. However, excessive suckering from the roots may occur due to the dwarfing interstock, even in rootstock types that normally do not sucker freely.

This interstock effect could, in some cases, be due to the introduction of an additional graft union with the possibility of translocation restrictions. Imperfect graft unions are indicated as a cause of the dwarfing exerted on orange trees by a lemon interstock. In contrast to the dwarfing situation in apples with a 'Malling 9' interstock, the lemon itself is strong-growing.

**Figure 11-24**    Effect of interstock on size of six-year-old Cox's Orange Pippin apple grafted on the strong-growing 'MM 104' roots. *Top left:* Cox/'M.9'/'MM.104'. *Top right:* Cox/'M.27'/'MM.104'. *Lower left:* Cox/'MM.104'/'MM.104'. *Lower right:* Cox/'M.20'/'MM.104'. Courtesy M. S. Parry, W. S. Rogers, and the Editors, *Journal of Horticultural Science* (*131*).

On the other hand, there is evidence that the observed effects of the interstock are due directly to an influence of the interstock piece rather than to abnormalities at the graft union (*150*). The dwarfing effect of 'Malling 9' seems to be due to something more than restrictions at the graft union, since 'Malling 9' is an early-bearing, dwarf tree itself.

Studies (*141*) have shown quantitatively that the initial response obtained from a 'Malling 9' interstock in composite 'Starking Delicious'/'M. 9'/'M. 16' trees was early and heavy flowering. This was followed—as a result of the heavy cropping—by a reduction in tree size. The degree of the response obtained was proportional to the length of the 'Malling 9' interstock. This finding supports earlier evidence that the interstock effect is due to a direct influence of the stock, since increasing the length of the interstock intensifies its effect (*39, 59*). However, studies with 'Old Home' pear as an interstock between 'Bartlett' and quince showed no effects resulting from different interstock lengths (*185*).

## Possible Mechanisms for the Effects of Stock
## on Scion and Scion on Stock

The nature of the rootstock-scion relationship is very complex and probably differs among genetically different combinations. The fundamental mechanisms by which stock and scion influence each other are not well understood. Some of the explanations offered for the observed effects are speculative, often conflicting, and sometimes not well substantiated. Several theories have been advanced as possible explanations for the interaction between stock and scion.

One suggested mechanism is that the rootstock influences are the result of *translocation effects* rather than the *absorbing ability* of the root system. That the stem portion of the tree has, to some extent at least, a definite influence is shown by experiments (*13*) in which commonly used rootstock materials were used as intermediate stocks between a vigorous root system and the scion cultivar. The expected effects were still present, although to a lesser degree, even though the materials were used as interstocks rather than as the entire root-absorbing system. This same influence on the tree was noted if the intermediate stock tissue was reduced to just a ring of bark (*142*).

By this reasoning, the uniformity of trees produced on vegetatively propagated rootstocks is due to the uniformity of the *stems* of such stocks, whereas the variable stems of seedlings are responsible for the variation encountered in the growth of the scion cultivar. This is especially so if they are grafted or budded high on the stem so that there is more opportunity for stem influence to be exerted. But if the grafting is done on the roots of seedlings (those with relatively passive influence)—as is done in root grafting—this variability does not appear, owing to the absence of influence contributed by the stem section. Therefore a fairly uniform group of trees could be produced, even on seedling rootstocks. This situation has been noted with cherries. Orchards consisting of nursery trees that were low-worked on 'Mazzard' seedlings were quite uniform, whereas orchards from nursery trees budded high on such seedlings were quite variable.

However, some workers in England concluded from their experiments (*71, 178*) that the *root system* itself, rather than the *stem* of the rootstock, plays the major role in rootstock effects on the scion. Beakbane and Rogers (*9*), after experiments conducted with apple trees for 19 years, concluded that the characteristic rootstock influence was due to the roots themselves. This effect did not depend upon the presence of a piece of rootstock stem, although it did tend to increase the rootstock influence.

Ramirez and Tabuenca (*138*) may have resolved such contradictions in stem and root influence by showing that if the scion or interstock is to influence root morphology, it must be a type that has dominating characteristics.

Chandler (*29*) and Gardner, Bradford, and Hooker (*54*), discussing the subject in general terms, asserted that the effects of stock on scion and scion on stock can be explained by *physiological factors,* chiefly the influences due to *changes in vigor.* Chandler pointed out that when the scion is the more vigorous part of the combination, the carbohydrate supply to the roots should be greater. And since certain roots supply and are supplied by certain branches, it would be expected that the branching habit of the top would influence the branching habit of the roots, thus explaining the different root types obtained in using different scion cultivars. If trees of different cultivars were pruned to exactly the same number and distribution of branches, then the difference in rootstock growth associated with the different scion cultivars might not occur.

Rootstock effects are not invariably related to vigor. Such things as flowering, fruit setting, fruit size, and fruit color or quality may be affected by the rootstock used even on trees showing an equal amount of vigor. For example, the marked effect of the rootstock on the development of black-end in pears is certainly not due to an alteration in vigor.

Tukey and Brase (*173*) concluded from their studies that no one part of a grafted tree could be considered to have complete control, but that all—rootstock, interstock, and scion—influenced the growth of the whole, although generally the rootstock had the dominant role.

Although no completely satisfactory explanation exists of how the three genetically different components of a grafted plant—rootstock, interstock, and scion—interact to influence the growth, flowering, and fruiting responses of the composite plant, three approaches can be considered: (a) *nutritional uptake and utilization,* (b) *translocation of nutrients and water,* and (c) *alterations in endogenous growth factors.*

### NUTRITION

It could possibly be that dwarfing rootstocks cause small trees by starvation effects. This has not been the case, however. Dwarf trees often contain higher concentrations of organic and mineral nutrients than vigorous ones (*30, 140*). The fruitful condition existing in young apple trees worked on the dwarfing 'Malling 9' roots was found to be associated with accumulation of starch in the shoots early in the season (*30*). Such an early starch storage would be expected to be favorable for the initiation of flower bud primordia. Nonfruitful trees on the vigorous 'Malling 12' roots failed to show such a starch accumulation. The increased supply of water and nutrients from the vigorous roots would stimulate production of new growth rather than retard growth and would not allow for the accumulation of carbohydrates, as would be the case with weaker, dwarfing rootstocks.

Mineral absorption by the various rootstocks, which is made available for use by the scion cultivar, can explain certain rootstock influences. For example, the very vigorous 'Shalil' peach root system grown at low nutrient levels was able to pick up and furnish the scion cultivar with a greater supply of nutrients and water than the less vigorous 'Lovell' rootstock. Under these conditions, the scion cultivar showed better growth on 'Shalil' than on 'Lovell' roots. But when the salt level was higher and included toxic chloride ions, the greater accumulation of such salts, subsequently translocated to the scion cultivar, caused injury and growth depression. In this case, the difference in vigor of two rootstocks caused opposite effects under two different soil conditions—a low-salt and a high-salt condition (*72*).

Studies (*4, 5*) in England with the dwarfing 'Malling 9' and 'Malling 27' have provided evidence that trees on these stocks are smaller because they are restricted to fewer growing points, which continue growth for a shorter period. The even slower growth rate of the root system so limits the trees that they are unable to make use of photosynthates completely in continued growth,

### TRANSLOCATION

The fact that interstocks of such dwarfing apple clones as 'Malling 9' will cause a certain amount of dwarfing could indicate that translocation is involved, owing either to partial blockage at the graft unions or to a reduction in movement of water or nutrient materials (or both) through the interstock piece itself.

One explanation advanced (181) for the dwarfing effect of certain rootstocks is supported by studies concerning the efficiency of water conductivity of the graft union. There are indications that the graft union does introduce an additional resistance to the flow of water. This resistance was greater in unions of which the 'Malling 9' stock was one of the components. Certain of the growth characteristics of trees on 'Malling 9' roots, such as small leaves, short internodes, and the early cessation of seasonal shoot growth, are those generally associated with a slight water deficit in the tree. As mentioned before, however, although a restricted graft union may contribute to the dwarfing influence of certain rootstocks, the primary influence probably lies more in the nature of the growth characteristics of such stocks.

In a study of the translocation of radioactive phosphorus ($P^{32}$) and calcium ($Ca^{45}$) from the roots to the tops of one-year 'McIntosh' apple trees grown in solution culture, it was shown that over three times as much of both elements was found in the scion top when the vigorous 'Malling 16' root was used in comparison with the dwarfing 'Malling 9' (22). This may indicate a superior ability of the vigorous stock to absorb and translocate mineral nutrients to the scion in comparison with the dwarfing stock. Or it may only mean that the 'Malling 9' roots, with their high percentage of living tissue, formed a greater "sink" for these materials, retaining them in the roots.


ENDOGENOUS GROWTH FACTORS

The idea was proposed by Sax (147, 150) that some of the growth alterations noted when certain interstocks are used, or when a ring of bark in the trunk of young trees is inverted, may be due to interference with the normal translocation of natural growth substances and nutrients from the leaves to the roots. It is known that certain factors do exist (e.g., vitamin $B_1$) that are necessary for root growth, and anything that would stop or reduce the flow of such substances would limit growth of the rootstock and subsequently dwarf the entire tree.

Dwarfing rootstocks may exhibit their characteristic effects because of their own low production of endogenous growth promoters (auxins and gibberellins) or their inability to conduct or utilize such substances produced by the scion. Young trees on dwarfing stocks can make vigorous growth for a year or two in the nursery (172) when growth-regulating materials may still be present in sufficient amounts, but several years later dwarfing develops, possibly because growth promoters decrease.

There is evidence (65) showing that the amount of indoleacetic acid (a growth-promoting auxin) that is destroyed by enzymes in root and shoot bark tissue of various apple rootstocks is correlated inversely with the scion vigor induced by the rootstock. In a study (111) of leaf extracts from various size-controlling apple rootstocks, it was found that those giving the greatest dwarfing contained materials that stimulated oxidative breakdown of indoleacetic acid.

In a comparison of endogenous growth-regulating factors in the bark of the dwarfing 'Malling 9' apple rootstock with those in the bark of the invigorating 'Malling 16', it was noted (100) that 'Malling 9' contained lower amounts of growth-promoting materials—but more growth inhibitors—than did 'Malling 16'.

Tissue extracts from four size-controlling apple rootstocks 'Malling 16' (invigorating), 'Malling 1' (semi-invigorating), 'Malling 7' (semidwarfing), and 'Malling 9' (dwarfing), showed increasing levels of an endogenous inhibitor—in the same order—which apparently was abscisic acid (*189*).

There is the possibility, too, that different levels of endogenous gibberellin, as a growth promoter, may account, in part, for the size-controlling characteristics of various rootstocks. It is known that roots do produce gibberellin and that the amounts of gibberellin in the transpiration stream are sufficient to have a decided growth-controlling influence (*88*). Extracts of the dwarfing 'Malling 9' apple rootstocks were found (*86*) to contain lower amounts of gibberellic acid-like substances than extracts of the more invigorating 'Malling 1' and 'Malling 25'. Such low gibberellin levels in the more dwarfing stock could be due either to lower production or to more rapid destruction. The differential dwarfing due to different rootstocks thus could possibly be accounted for, in part, by different levels of gibberellin translocated from the roots to the shoot system (*27*). Injections of gibberellic acid into grafted apple trees have given increasing stimulation to the top growth as rootstock vigor decreased, while injections of the growth inhibitor, abscisic acid, caused shoot growth to cease (*144*).

Grafting studies (*97*), where complete rings of the dwarfing 'Malling 26' apple bark were grafted into 'Gravenstein' stems grafted on the invigorating 'M111' roots resulted in dwarfed trees, gave rise to the following postulation of the dwarfing mechanism in apples: Auxin produced by the shoot tip moves downward through the phloem. The amount arriving in the root influences root metabolism, including the amount and kind of cytokinins synthesized and translocated upward from the roots to the shoot system via the xylem. These cytokinins influence the amount of shoot growth.

In these studies the grafted bark ring of the dwarfing 'Malling 26' apple caused reduction in the downward movement of auxins, subsequently reducing cytokinin production in the roots; this lack of cytokinins available for upward movement reduced top growth, eventually giving a dwarfed tree.

The various dwarfing and invigorating types of rootstocks apparently either contain different amounts of naturally occurring growth-promoting and growth-inhibiting materials or can affect the amounts passing through their tissues. They could thus not only control their own growth, but similarly affect growth of their graft partners—which seems to be the best explanation advanced so far for the mechanisms involved in size-controlling rootstocks.

## REFERENCES

*1.* Altman, A., and R. Goren. 1971. Promotion of callus formation by abscisic acid in citrus bud cultures. *Plant Phys.* 47:844–46.

*2.* Amos, J., T. N. Hoblyn, R. J. Garner, and A. Witt. 1936. Studies in incompatibility of stock and scion. I. Information accumulated during twenty years of testing fruit tree rootstocks with various scion varieties at East Malling. *Ann. Rpt. E. Malling Res. Sta. for 1935,* pp. 81–99.

*3.* Argles, G. K. 1937. A review of the literature on stock-scion incompatibility in fruit trees, with particular reference to pome and stone fruits. *Imp. Bur. of Fruit Prod. Tech. Comm. No. 9.*

4. Avery, D. J. 1969. Comparisons of fruiting and deblossomed maiden apple trees and of non-fruiting trees on a dwarfing and an invigorating rootstock. *New Phytol.* 68:323–36.

5. ———. 1970. Effects of fruiting on the growth of apple trees on four rootstock varieties. *New Phytol.* 69:19–30.

6. Bailey, L. H. 1891. *The nursery book.* New York: Rural Publishing Company.

7. Batjer, L. P., and H. Schneider. 1960. Relation of pear decline to rootstocks and sieve tube necrosis. *Proc. Amer. Soc. Hort. Sci.* 76:85–97.

8. Beakbane, A. B. 1953. Anatomical structure in relation to rootstock behavior. *Rpt. 13th Inter. Hort. Cong.* Vol. 1, pp. 152–58.

9. Beakbane, A. B., and W. S. Rogers. 1956. The relative importance of stem and root in determining rootstock influence in apples. *Jour. Hort. Sci.* 31:99–110.

10. Bitters, W. P. 1950. Citrus rootstocks for dwarfing. *Calif. Agr.* 4(2):5–14.

11. ———. 1961. Physical characters and chemical composition as affected by scions and rootstocks. Chapter 3 in *The orange: Its biochemistry and physiology,* W. B. Sinclair, ed. Berkeley: Univ. of Calif. Div. of Agr. Sci.

12. Bitters, W. P., and E. R. Parker. 1953. Quick decline of citrus as influenced by top-root relationships. *Calif. Agr. Exp. Sta. Bul. 733.*

13. Blair, D. S. 1938. Rootstock and scion relationship in apple trees. *Sci. Agr.* 19:85–94.

14. Bloch, R. 1952. Wound healing in higher plants. *Bot. Rev.* 18:655–79.

15. Blumenfield, A., and S. Gazit. 1969. Interaction of kinetin and abscisic acid in the growth of soybean callus. *Plant Phys.* 45:535–36.

16. Bradford, F. C., and B. G. Sitton. 1929. Defective graft unions in the apple and pear. *Mich. Agr. Exp. Sta. Tech. Bul. 99.*

17. Brase, K. D., and R. D. Way. 1959. Rootstocks and methods used for dwarfing fruit trees. *N.Y. State Agr. Exp. Sta. Bul. 783.*

18. Buchloh, G. 1960. The lignification in stock-scion junctions and its relation to compatibility. In *Phenolics in plants in health and disease,* J. B. Pridham, ed. Long Island City, N.Y.: Pergamon Press.

19. ———. 1962. Verwachsung und Verwachsungsstorungen als Ausdruck des Affinitatsgrades bei Propfungen von Birnenvarietaten auf *Cydonia oblonga. Beit Biol. Pfl.* 37:183–240.

20. Buck, G. J. 1953. The histological development of the bud graft union in roses. *Proc. Amer. Soc. Hort. Sci.* 62:497–502.

21. Buck, G. J., and B. J. Heppel. 1970. A bud-graft incompatibility in *Rosa. Jour. Amer. Soc. Hort. Sci.* 95(4):442–46.

22. Bukovac, M. J., S. H. Wittwer, and H. B. Tukey. 1958. Effect of stock-scion inter-relationships on the transport of $P^{32}$ and $Ca^{45}$ in the apple. *Jour. Hort. Sci.* 33:145–52.

23. Bunyard, E. A. 1920. The history of the Paradise stocks. *Jour. Pom.* 2:166–76.

24. Camus, G. 1949. Recherches sur le rôle des bourgeons dans les phénomènes de morphogénèse. *Revue Cytol. et Biol. Vég.* 11:1–199.

25. Caponetti, J. D., G. C. Hall, and R. E. Farmer, Jr. 1971. In vitro growth of black cherry callus: Effects of medium, environment, and clone. *Bot. Gaz.* 132(4):313–18.

26. Carlson, R. F. 1967. The incidence of scion-rooting of apple cultivars planted at different soil depths. *Hort. Res.* 7(2):113–15.

27. Carr, D. J., D. M. Reid, and K. G. M. Skene. 1964. The supply of gibberellins from the root to the shoot. *Planta* 63:382–92.

28. Cation, D., and R. F. Carlson. 1962. Determination of virus entities in an apple scion/rootstock test orchard. *Quart. Bul. Mich. Agr. Exp. Sta., Rpt.* I, 43(2):435–43, 1960. Rpt. II, 45(17):159–66.

29. Chandler, W. H. 1925. *Fruit growing*. Boston: Houghton Mifflin.

30. Colby, H. L. 1935. Stock-scion chemistry and the fruiting relationships in apple trees. *Plant Phys.* 10:483–98.

31. Collins, R. P., and S. Waxman. 1958. Dogwood graft failures. *Amer. Nurs.* 108(8):12.

32. Colquhoun, T. T. 1929. Polarity in *Casuarina paludosa. Trans. and Proc. Roy. Soc. South Australia* 53:353–58.

33. Cooper, W. C. 1952. Influence of rootstock on injury and recovery of young citrus trees exposed to the freezes of 1950–51 in the Rio Grande Valley. *Proc. 6th Ann. Rio Grande Valley Hort. Inst.,* pp. 16–24.

34. Cooper, W. C., and E. O. Olson. 1951. Influence of rootstock on chlorosis of young Red Blush grapefruit trees. *Proc. Amer. Soc. Hort. Sci.* 57:125–32.

35. Copes, D. A. 1969. Graft union formation in Douglas-fir. *Amer. Jour. Bot.* 56(3): 285–89.

36. ———. 1970. Initiation and development of graft incompatibility symptoms in Douglas-fir. *Silvae Genet.* 19:101–7.

37. Crafts, A. S. 1934. Phloem anatomy in two species of *Nicotiana,* with notes on the interspecific graft union. *Bot. Gaz.* 95:592–608.

38. Crane, M. B. and E. Marks. 1952. Pear-apple hybrids. *Nature* 170:1017.

39. Dana, M. N., H. L. Lantz, and W. E. Loomis. 1962. Effects of interstock grafts on growth of Golden Delicious apple trees. *Proc. Amer. Soc. Hort. Sci.* 81:1–11.

40. Davis, L. D., and W. P. Tufts. 1936. Black end of pears III. *Proc. Amer. Soc. Hort. Sci.* 33:304–15.

41. Dawson, R. F. 1942. Accumulation of nicotine in reciprocal grafts of tomato and tobacco. *Amer. Jour. Bot.* 29:66–71.

42. Day, L. H. 1951. Cherry rootstocks in California. *Calif. Agr. Exp. Sta. Bul. 725.*

43. ———. 1953. Rootstocks for stone fruits. *Calif. Agr. Exp. Sta. Bul. 736.*

44. Dimalla, G. G., and J. A. Milbrath. 1965. The prevalence of latent viruses in Oregon apple trees. *Plant Dis. Rpt.* 49(1):15–17.

45. Doesburg, J. van. 1962. Use of fungicides with vegetative propagation. *Rpt. XVI Inter. Hort. Cong.,* pp. 365–72.

46. Doley, D., and L. Leyton. 1970. Effects of growth regulating substances and water potential on the development of wound callus in *Fraxinus. New Phytol.* 69:87–102.

47. Dorsman, C. 1966. Grafting of woody plants in the glasshouse. *Proc. XVII Inter. Hort. Cong.* 1:366.

48. Eames, A. J., and L. G. Cox. 1945. A remarkable tree-fall and an unusual type of graft union failure. *Amer. Jour. Bot.* 32:331–35.

49. Epstein, A. H. 1978. Root graft transmission in tree pathogens. *Ann. Rev. Phytopathol.* 16:181–92.

50. Evans, G. E., and H. P. Rasmussen. 1972. Anatomical changes in developing graft unions of *Juniperus. Jour. Amer. Soc. Hort. Sci.* 97(2):228–32.

51. Evans, W. D., and R. J. Hilton. 1957. Methods of evaluating stock/scion compatibility in apple trees. *Can. Jour. Plant Sci.* 37:327–36.

52. Fujii, T., and N. Nito. 1972. Studies on the compatibility of grafting of fruit trees. I. Callus fusion between rootstock and scion. *Jour. Jap. Soc. Hort. Sci.* 41(1):1–10.

53. Gardner, F. E., and G. H. Horanic. 1963. Cold tolerance and vigor of young citrus trees on various rootstocks. *Proc. Fla. State Hort. Soc.* 76:105–10.

54. Gardner, V. R., F. C. Bradford, and H. D. Hooker, Jr. 1939. *Fundamentals of fruit production* (2nd ed.). New York: McGraw-Hill.

55. Garner, R. J., and D. H. Hammond. 1939. Studies in nursery technique. Shield budding. Treatment of inserted buds with petroleum jelly. *Ann. Rpt. E. Malling Res. Sta. for 1938,* pp. 115–17.

56. Gautheret, R. J. 1947. La culture des tissus végétaux. *Proc. 6th Inter. Cong. Exp. Cytol.* pp. 437–49.

57. Goodale, S. L. 1846. Influence of the scion upon the stock. *Hort.* 1:290.

58. Graham, B. F., Jr., and F. H. Bornmann. 1966. Natural root grafts. *Bot. Rev.* 32(3):255–92.

59. Grubb, N. H. 1939. The influence of intermediate stem pieces in double-worked apple and pear trees. *Scient. Hort.* 7:17–23.

60. Guengerich, H. W., and D. F. Milliken. 1966. Bud transmission of dwarfing in sweet cherry. *Plant Dis. Rpt.* 50:367–68.

61. ————. 1965. Root grafting, a potential source of error in apple indexing. *Plant Dis. Rpt.* 49:39–41.

62. Gur, A. 1957. The compatibility of the pear with quince rootstock. *Spec. Bul. No. 10,* pp. 1–9º, Agr. Res. Sta., Rehovot (Israel).

63. Gur, A., and R. M. Samish. 1965. The relation between growth curves, carbohydrate distribution, and compatibility of pear trees grafted on quince rootstocks. *Hort. Res.* 5:81–100.

64. Gur, A., and E. Lifshitz. 1968. The role of the cyanogenic glycoside of the quince in the incompatibility between pear cultivars and quince rootstocks. *Hort. Res.* 8:113–34.

65. Gur, A., and R. M. Samish. 1968. The role of auxins and auxin destruction in the vigor effect induced by various apple rootstocks. *Beitr. Biol. Pflanz.* 45:91–111.

66. Hansen, C. J. 1948. Influence of the rootstock on injury from excess boron in French (Agen) prune and President plum. *Proc. Amer. Soc. Hort. Sci.* 51:239–44.

67. Hansen, C. J., and H. T. Hartmann. 1951. Influence of various treatments given to walnut grafts on the percentage of scions growing. *Proc. Amer. Soc. Hort. Sci.* 57:193–97.

68. Harmon, F. N. 1949. Comparative value of thirteen rootstocks for ten vinifera grape varieties in the Napa Valley in California. *Proc. Amer. Soc. Hort. Sci.* 54:157–62.

69. Hartmann, H. T. 1958. Rootstock effects in the olive. *Proc. Amer. Soc. Hort. Sci.* 72:242–51.

70. Hatton, R. G. 1927. The influence of different rootstocks upon the vigor and productivity of the variety budded or grafted thereon. *Jour. Pom. and Hort. Sci.* 6:1–28.

71. ——. 1931. The influence of vegetatively raised rootstocks upon the apple, with special reference to the parts played by the stem and root portions in affecting the scion. *Jour. Pom. and Hort. Sci.* 9:265–77.

72. Hayward, H. E., and E. M. Long. 1942. Vegetative responses of the Elberta peach on Lovell and Shalil rootstocks to high chloride and sulfate solutions. *Proc. Amer. Soc. Hort. Sci.* 41:149–55.

73. Hedrick, U. P. 1915. Stocks for fruits. *Rpt. N. Y. State Fruit Growers Assn.*, pp. 84–94.

74. Heppner, M. 1927. Pear black-end and its relation to different rootstocks. *Proc. Amer. Soc. Hort. Sci.* 24:139.

75. Heppner, M., and R. D. McCallum. 1927. Grafting affinities with special reference to plums. *Calif. Agr. Exp. Sta. Bul. 438.*

76. Herrero, J. 1955. Incompatibilidad entre patrón e injerto. II, Effecto de un intermediario en la incompatibilidad entre melocotonero y mirobalán. *An. Aula Dei* 4:167–72.

77. ——. 1951. Studies of compatible and incompatible graft combinations with special reference to hardy fruit trees. *Jour. Hort. Sci.* 26:186–237.

78. Herrero, J., and M. V. Tabuenca. 1969. Incompatibilidad entre patron e injerto. X. Comportamiento de la combinacion melocotonero/mirobalan injertado en estado cotiledonor. *An. Estac, exp. Aula Dei* 10:937–45.

79. Hesse, H. 1710. *Teutscher Gartner.* Leipsig.

80. Hibino, H., and H. Schneider. 1970. Mycoplasma-like bodies in sieve tubes of pear trees affected with pear decline. *Phytopath.* 60:499–501.

81. Hibino, H., G. H. Kaloostian, and H. Schneider. 1971. Mycoplasma-like bodies in the pear psylla vector of pear decline. *Virology* 43:34–40.

82. Hodgson, R. W. 1943. Some instances of scion domination in citrus. *Proc. Amer. Soc. Hort. Sci.* 43:131–38.

83. Hodgson, R. W., and S. H. Cameron. 1935. On bud union effect in citrus. *Calif. Citrog.* 20(12):370.

84. Homes, J. 1965. Histogenesis in plant grafts. In *Proc. Inter. Conf. Plant Tissue Cult.*, P. R. White and A. R. Groves, eds. Amer. Inst. Biol. Sci., pp. 553.

85. Honma, S. 1977. Grafting eggplants. *Scient. Hort.* 7:207–11.

86. Ibrahim, I. M., and M. N. Dana. 1971. Gibberellin-like activity in apple rootstocks. *HortScience* 6(6):541–42.

87. Jensen, D. D., W. H. Griggs, C. Q. Gonzales, and H. Schneider. 1964. Pear decline virus transmission by pear psylla. *Phytopath.* 54:1346–51.

88. Jones, O. P., and H. J. Lacey. 1968. Gibberellin-like substances in the transpiration stream of apple and pear trees. *Jour. Exp. Bot.* 19:526–31.

89. Juliano, J. B. 1941. Callus development in graft union. *Philippine Jour. Sci.* 75:245–51.

90. Keane, F. W. L., and J. May. 1963. Natural root grafting in cherry and spread of cherry twisted-leaf virus. *Can. Plant Dis. Sur.* 43(2):54–60.

91. Kester, D. E., C. J. Hansen, and C. Panetsos. 1965. Effect of scion and interstock variety on incompatibility of almond on Marianna 2624 rootstock. *Proc. Amer. Soc. Hort. Sci.* 86:169–77.

92. Kunkel, L. O. 1938. Contact periods in graft transmission of peach viruses. *Phytopath.* 28:491–97.

93. Langford, G. T. 1681. *Plain and full instructions to raise all sorts of fruit trees that prosper in England.* London: J. M. at the Rose and Crown.

94. Langford, M. H., J. B. Carpenter, W. E. Manis, A. M. Gorenz, and E. P. Imle. 1954. *Hevea* diseases of the Western Hemisphere. *Plant Dis. Rpt. Suppl. 225.*

95. Langford, M. H., and C. H. T. Townsend, Jr. 1954. Control of South American leaf blight of *Hevea* rubber trees. *Plant Dis. Rpt. Suppl. 225.*

96. Lapins, K. 1959. Some symptoms of stock-scion incompatibility of apricot varieties on peach seedling rootstock. *Can. Jour. Plant Sci.* 39:194–203.

97. Lockard, R. G., and G. W. Schneider. 1981. Stock and scion growth relationships and the dwarfing mechanism in apple. *Hort. Rev.* 3:315–75.

98. Lowman, M. S., and J. W. Kelley. 1946. The presence of mydriatic alkaloids in tomato fruit from scions grown on *Datura stramonium* rootstock. *Proc. Amer. Soc. Hort. Sci.* 48:249–59.

99. Luckwill, L. C. 1962. New developments in the study of graft incompatibility in fruit trees. *Adv. in Hort. Sci. and their Appl.*, Vol. 2. Long Island City, N.Y.: Pergamon Press.

100. Martin, G. C., and E. A. Stahly. 1967. Endogenous growth regulating factors in bark of EM IX and XVI apple trees. *Proc. Amer. Soc. Hort. Sci.* 91:31–38.

101. McClintock, J. A. 1948. A study of uncongeniality between peaches as scions and the Marianna plum as a stock. *Jour. Agr. Res.* 77:253–60.

102. McKenzie, D. W. 1961. Rootstock-scion interaction in apples with special reference to root anatomy. *Jour. Hort. Sci.* 36:40–47.

103. McQuilkin, W. E. 1950. Effects of some growth regulators and dressings on the healing of tree wounds. *Jour. For.* 48(9):423–28.

104. Mendel, K. 1936. The anatomy and histology of the bud-union in citrus. *Palest. Jour. Bot.* (R), 1(2):13–46.

105. Mircetich, S. M. J., J. Refsguard, and M. E. Matheron. 1980. Blackline of English walnut trees traced to graft-transmitted virus. *Calif. Agr.* 34(11–12):8–10.

106. Mergen, F. 1954. Anatomical study of slash pine graft unions. *Quart. Jour. Fla. Acad. Sci.* 17:237–45.

107. Milbraith, J. A., and S. M. Zeller. 1945. Latent viruses in stone fruits. *Science* 101:114–15.

108. Miller, L., and F. W. Woods. 1965. Root grafting in Loblolly pine. *Bot. Gaz.* 126:252–55.

109. Miller, P. W. 1965. The etiology of blackline in grafted Persian walnuts. *Plant Dis. Rpt.* 49:954.

110. ———. 1952. Technique for indexing strawberries for viruses by grafting to *Fragaria vesca. Plant Dis. Rpt.* 36.

111. Miller, S. R. 1965. Growth inhibition produced by leaf extracts from size controlling apple rootstocks. *Can. Jour. Plant Sci.* 45(6):519–24.

112. Millner, M. E. 1932. Natural grafting in *Hedera helix. New Phytol.* 31:2–25.

113. Montgomery, H. B. S. 1963. Fruit tree raising. *Bul. 135, Minist. Agr., Fish. and Foods,* London.

114. Moore, R. 1981. Graft compatibility/incompatibility in higher plants. *What's new in plant phys.* 12(4):13–16.

115. ———. 1982. Graft formation in *Kalanchoe blossfeldiana. Jour. Exp. Bot.* 33:533–40.

116. Moore, R., and D. B. Walker. 1981. Graft compatibility-incompatibility in plants. *BioScience* 31(5):389–91.

117. ———. 1981. Studies of vegetative compatibility-incompatibility in higher plants. I. A structural study of a compatible autograft in *Sedum telephoides* (Crassulaceae). II. A structural study of an incompatible heterograft between *Sedum telephoides* (Crassulaceae) and *Solanum penellii* (Solanaceae). *Amer. Jour. Bot.* 68(6):820–42.

118. ———. 1981. Studies of vegetative compatibility-incompatibility in higher plants. III. The involvement of acid phosphatase in the lethal cellular sensescence associated with an incompatible heterograft. *Protoplasma* 109:317–34.

119. ———. 1982. Studies of vegetative compatibility-incompatibility in higher plants. IV. The development of tensile strength in a compatible and an incompatible graft. *Amer. Jour. Bot.* (in press).

120. Mosse, B. 1958. Further observations on growth and union structure of double-grafted pear on quince. *Jour. Hort. Sci.* 33:186–93.

121. ———. 1962. Graft-incompatibility in fruit trees. *Tech. Comm. No. 28, Comm. Bur. Hort. and Plant. Crops,* East Malling, England.

122. Mosse, B. and M. V. Labern. 1960. The structure and development of vascular nodules in apple bud unions. *Ann. Bot.* 24:500–507.

123. Murashige, T., and F. Skoog. 1962. A revised medium for rapid growth and bioassays with tobacco tissue cultures. *Phys. Plant.* 15:473–97.

124. Muzik, T. J. 1958. Role of parenchyma cells in graft union in *Vanilla* orchid. *Science* 127:82.

125. Muzik, T. J., and C. D. LaRue. 1954. Further studies on the grafting of monocotyledonous plants. *Amer. Jour. Bot.* 41:448–55.

126. Nelson, R., S. Goldweber, and F. J. Fuchs. 1955. Top-working for mangos. *Fla. Grower and Rancher,* p. 45, Jan.

127. Nichols, C. W., H. Schneider, H. J. O'Reilly, T. A. Shalla, and W. H. Griggs. 1960. Pear decline in California. *Bul. Calif. State Dept. Agr.* 49:186–92.

128. Nickell, L. G. 1946. Heteroplastic grafts. *Science* 108:389.

129. Nyland, G., and W. J. Moller. 1973. Control of pear decline with a tetracycline. *Plant Dis. Rpt.* 57:634–37.

130. Overbeek, J. van. 1966. Plant hormones and regulators. *Science* 152:721–31.

131. Parry, M. S., and W. S. Rogers. 1968. Dwarfing interstocks: Their effect on the field performance and anchorage of apple trees. *Jour. Hort. Sci.* 43:133–46.

132. Posnette, A. F. 1966. Virus diseases of fruit plants. *Proc. XVII Inter. Hort. Cong.* 3:89–93.

133. Posnette, A. F., and R. Cropley. 1962. Further studies on a selection of Williams Bon Chrétien pear compatible with Quince A rootstocks. *Jour. Hort. Sci.* 37:291–94.

134. Preston, A. P. 1958. Apple rootstock studies: Thirty-five years' results with Cox's Orange Pippin on clonal rootstocks. *Jour. Hort. Sci.* 33:194–201.

135. ———. 1958. Apple rootstock studies: Thirty-five years' results with Lane's Prince Albert on clonal rootstocks. *Jour. Hort. Sci.* 33:29–38.

136. Proebsting, E. L. 1928. Further observations on structural defects of the graft union. *Bot. Gaz.* 86:82–92.

137. ———. 1926. Structural weaknesses in interspecific grafts of *Pyrus*. *Bot Gaz*. 82: 336–38.

138. Ramirez, D., and M. C. Tabuenca. 1964. The reciprocal effects of M. IX and M. XVI apples (English summary). *Ann. Estac. Exp. Aula Dei* 7:164–74.

139. Rao, A. N. 1966. Developmental anatomy of natural root grafts in *Ficus globosa*. *Austral. Jour. Bot.* 14:269–76.

140. Rao, Y. V., and W. E. Berry. 1940. The carbohydrate relations of a single scion grafted on Malling rootstocks IX and XIII. A contribution to the physiology of dwarfing. *Jour. Pom.* 18:193–225.

141. Roberts, A. N., and L. T. Blaney. 1967. Qualitative, quantitative, and positional aspects of interstock influence on growth and flowering of the apple. *Proc. Amer. Soc. Hort. Sci.* 91:39–50.

142. Roberts, R. H. 1949. Theoretical aspects of graftage. *Bot. Rev.* 15:423–63.

143. Robitaille, R. H., and R. F. Carlson. 1970. Graft union behavior of certain species of *Malus* and *Prunus*. *Jour. Amer. Soc. Hort. Sci.* 95(2):131–34.

144. Robitaille, H., and R. F. Carlson. 1971. Response of dwarfed apple trees to stem injections of gibberellic and abscisic acids. *HortScience* 6(6):539–40.

145. Samish, R. M. 1962. Physiological approaches to rootstock selection. *Adv. in Hort. Sci. and their Appl.,* Vol. 2. Long Island City, N.Y.: Pergamon Press.

146. Sass, J. E. 1932. Formation of callus knots on apple grafts as related to the histology of the graft union. *Bot. Gaz.* 94:364–80.

147. Sax, K. 1954. The control of tree growth by phloem blocks. *Jour. Arn. Arb.* 35:251–58.

148. ———. 1950. Dwarf trees. *Arnoldia* 10:73–79.

149. ———. 1950. The effect of the rootstock on the growth of seedling trees and shrubs. *Proc. Amer. Soc. Hort. Sci.* 56:166–68.

150. ———. 1953. Interstock effects in dwarfing fruit trees. *Proc. Amer. Soc. Hort. Sci.* 62:201–4.

151. Sax, K., and A. Q. Dickson. 1956. Phloem polarity in bark regeneration. *Jour. Arn. Arb.* 37:173–79.

152. Schroeder, C. A. 1947. Rootstock influence on fruit set in the Hachiya persimmon. *Proc. Amer. Soc. Hort. Sci.* 50:149–50.

153. Serr, E. F., and H. I. Forde. 1959. Blackline, a delayed failure at the union of *Juglans regia* trees propagated on other *Juglans* species. *Proc. Amer. Soc. Hort. Sci.* 74:220–31.

154. Shalla, T. A., and L. Chiarappa. 1961. Pear decline in Italy. *Bul. Calif. State Dept. Agr.* 50:213–17.

155. Sharples, A., and H. Gunnery. 1933. Callus formation in *Hibiscus rosasinensis L.* and *Hevea brasiliensis* Mull. Arg. *Ann. Bot.* 47:827–39.

156. Shaw, J. K. 1915. The root systems of nursery apple trees. *Proc. Amer. Soc. Hort. Sci.* 12:68–72.

157. Shippy, W. B. 1930. Influence of environment on the callusing of apple cuttings and grafts. *Amer. Jour. Bot.* 17:290–327.

158. Sitton, B. G. 1931. Vegetative propagation of the black walnut. *Mich. Agr. Exp. Sta. Tech. Bul. 119.*

159. Smith, J. W. M., and P. Proctor. 1965. Use of disease resistant rootstocks for tomato crops. *Exp. Hort.* 12:6–20.

160. Solt, M. L., and R. V. Dawson. 1958. Production, translocation, and accumulation of alkaloids in tobacco scions grafted on tomato rootstocks. *Plant Phys.* 33:375–81.

161. Soule, J. 1971. Anatomy of the bud union in mango (*Mangifera indica* L.). *Jour. Amer. Soc. Hort. Sci.* 96(3):380–83.

162. Stoddard, F. L. and M. E. McCully. 1980. Effects of excision of stock and scion organs on the formation of the graft union in coleus: a histological study. *Bot. Gaz.* 141:401–2.

163. Stone, E. L., J. E. Stone, and R. C. McKittrick. 1973. Root grafting in pine trees. *Food and Life Sci. Quart.* 6(2):19–21.

164. Sussex, I. M., and Mary E. Clutter. 1959. Seasonal growth periodicity of tissue explants from woody perennial plants *in vitro. Science* 129:836–37.

165. Thiel, K. 1954. Untersuchungen zur Frage des Unvertraglichkeit bei Birnenedelsorten auf Quitte A (*Cydonia* E. M. A). *Gartenbauwiss* 1(19):127–59.

166. Thomas, L. A. 1954. Stock and scion investigations. X. Influence of an intermediate stem-piece upon the scion in apple trees. *Jour. Hort. Sci.* 29:150–52.

167. Thouin, A. 1821. *Monographie des greffes, ou description technique* (in Royal Hort. Soc. Library, London).

168. Tomaselli, R., and E. Refatti. 1960. The nonexistence of constant relationship between root anatomy and vigor in grafted apple and pear trees. *Atti Ist. bot., Univ. Pavia,* Ser. 5, 18:130–40.

169. Torrey, J. G., D. E. Fosket, and P. K. Hepler. 1971. Xylem formation: A paradigm of cytodifferentiation in higher plants. *Amer. Sci.* 59:338–52.

170. Toxopeus, H. J. 1936. Stock-scion incompatibility in citrus and its cause. *Jour. Pom. and Hort. Sci.* 14:360–64.

171. Tukey, H. B. 1943. The dwarfing effect of an intermediate stem-piece of Malling IX apple. *Proc. Amer. Soc. Hort. Sci.* 42:357–64.

172. ———. 1941. Similarity in the nursery of several Malling apple stock and scion combinations which differ widely in the orchard. *Proc. Amer. Soc. Hort. Sci.* 39:245–46.

173. Tukey, H. B., and K. D. Brase. 1933. Influence of the scion and of an intermediate stem-piece upon the character and development of roots of young apple trees. *N. Y. (Geneva) Agr. Exp. Sta. Tech. Bul. 218.*

174. Tydeman, H. M. 1937. Experiments on hastening the fruiting of seedling apples. *Ann. Rpt. E. Malling Res. Sta. for 1936,* pp. 92–99.

175. Tydeman, H. M., and F. H. Alston. 1965. The influence of dwarfing rootstocks in shortening the juvenile phase of apple seedlings. *Ann. Rpt. E. Malling Res. Sta. for 1964,* pp. 97–98.

176. Vaile, J. E. 1938. The influence of rootstocks on the yield and vigor of American grapes. *Proc. Amer. Soc. Hort. Sci.* 35:471–74.

177. Vöchting, H. 1892. Veber transplantation am pflanzenköper.

178. Vyvyan, M. C. 1938. The relative influence of rootstock and of an intermediate piece of stock stem in some double-grafted apple trees. *Jour. Pom. and Hort. Sci.* 16:251–73.

179. Wagner, D. F. 1969. Ultrastructure of the bud graft union in *Malus.* Ph.D. Diss., Iowa State Univ., Ames.

180. Wareing, P. F., C. E. A. Hanney, and J. Digby. 1964. The role of endogenous hormones in cambial activity and xylem differentiation. In *The formation of wood in forest trees,* M. H. Zimmerman, ed. New York: Academic Press.

181. Warne, L. G. G., and Joan Raby. 1939. The water conductivity of the graft union in apple trees, with special reference to Malling rootstock No. IX. *Jour. Pom. and Hort. Sci.* 16:389–99.

182. Waugh, F. A. 1904. The graft union. *Mass. Agr. Exp. Sta. Tech. Bul. 2.*

183. ———. 1943. The "tristeza" disease of sour orange rootstock. *Proc. Amer. Soc. Hort. Sci.,* 43:160–68.

184. Wellensiek, S. J. 1949. The prevention of graft-incompatibility by own foliage on the stock. *Meded. Landbouwhoogesch. Wageningen* 49:255–72.

185. Westwood, M. N., and H. O. Bjornstad. 1972. Length of Old Home interstem makes little growth difference. *Ore. Orn. Nurs. Dig.* 16(1):3–4.

186. Wetmore, R. H., and J. P. Rier. 1963. Experimental induction of vascular tissues in callus of angiosperms. *Amer. Jour. Bot.* 50:418–30.

187. Williams, R. R., and A. I. Campbell. 1956. Rosetting and incompatibility of pears on Quince A. *Ann. Rpt. Long Ashton Res. Sta.* pp. 51–56.

188. Winter, H. F. 1963. Prevalence of latent viruses in Ohio apple trees. *Ohio Farm and Home Res.* 48:58–59, 63.

189. Yadava, U. L., and D. F. Dayton. 1972. The relation of endogenous abscisic acid to the dwarfing capability of East Malling apple rootstocks. *Jour. Amer. Soc. Hort. Sci.* 97(6):701–5.

190. Yeager, A. F. 1944. Xylem formation from ring grafts. *Proc. Amer. Soc. Hort. Sci.* 44:221–22.

191. Zebrak, A. R. 1937. Intergeneric and interfamily grafting of herbaceous plants. *Timirjazey Seljskohoz Akad.* 2:115–33. *Herb. Abst.* 9:675. 1939.

## SUPPLEMENTARY READING

ARGLES, G. K. 1937. A review of the literature on stock-scion incompatibility in fruit trees with particular reference to pome and stone fruits. *Imp. Bur. Fruit Prod., Tech. Comm. No. 9.*

BEAKBANE, A. B. 1956. Possible mechanisms of rootstock effect. *Ann. of App. Bio.* 44:517–21.

BRASE, K. D., and R. D. WAY. 1959. Rootstocks and methods used for dwarfing fruit trees. *N.Y. Agr. Exp. Sta. Bul. 783.*

CHANG, W. T. 1938. Studies in incompatibility between stock and scion with special reference to certain deciduous fruit trees. *Jour. Pom. and Hort. Sci.* 15:267–325.

DANIEL, L. 1927, 1929. *Etudes sur la greffe,* Vols. 1 and 2. Rennes, Paris: Imprimerie Oberthur.

ESAU, K. 1977. *Anatomy of seed plants* (2nd ed.). New York: John Wiley.

GARDNER, V. R., F. C. BRADFORD, and H. D. HOOKER. 1939. Propagation. Section 6 in *Fundamentals of fruit production* (2nd ed.). New York: McGraw-Hill.

HATTON, R. G. 1930. The relationship between scion and rootstock with special reference to the tree fruits. *Jour. Roy. Hort. Soc.* 55:169–211.

LOCKARD, R. G. and G. W. SCHNEIDER. 1981. Stock and scion growth relationships and the dwarfing mechanism in apple. *Hort. Rev.* 3:315–75.

MOORE, R., and D. B. WALKER. 1981. Graft compatibility-incompatibility in plants. *BioScience* 31(5):389–91.

MOSSE, B. 1962. Graft incompatibility in fruit trees. *Commonwealth Agr. Bur. Tech. Comm. No. 28.*

NELSON, S. H. 1968. Incompatibility survey among horticultural plants. *Proc. Inter. Plant Prop. Soc.* 18:343–93.

ROBERTS, R. H. 1949. Theoretical aspects of graftage. *Bot. Rev.* 15:423–63.

ROGERS, W. S., and A. B. BEAKBANE. 1957. Stock and scion relations. *Ann. Rev. Plant Phys.* 8:217–36.

TUBBS, F. R. 1973. Research fields in the interaction of rootstocks and scions in woody perennials. I and II. *Hort. Abst.* 43:247–53 and 43:325–35.

**Figure 12-1**
Cultivars of the Persian (English) walnut (*Juglans regia*) grafted on *J. hindsii* rootstocks. The characteristics of these two species remain distinctly different after grafting, exactly to the junction of the graft union.

For any successful grafting operation, producing a plant as shown in Figure 12-1, there are five important requirements:

    **1.** *The stock and scion must be compatible.* They must be capable of uniting. Usually, but not always, plants closely related, such as two apple cultivars, can be grafted together. Distantly related plants, such as an oak tree and an apple tree, cannot be used to make a successful graft combination. (See chapter 11 for a discussion of this factor.)

    **2.** *The cambial region of the scion must be placed in intimate contact with that of the stock.* The cut surfaces should be held together tightly by wrapping, nailing, wedging, or some similar method. Rapid healing of the graft union is necessary so that the scion may be supplied with water and nutrients from the stock by the time the buds start to open.

# 12

# Techniques of Grafting

**3.** *The grafting operation must be done at a time when the stock and scion are in the proper physiological stage.* Usually this means that the scion buds are dormant while, at the same time, the cut tissues at the graft union are capable of producing the callus tissue necessary for healing of the graft. For deciduous plants, dormant scionwood is collected during the winter and kept inactive by storing at low temperatures. The rootstock plant may be dormant or in active growth, depending upon the grafting method used.

**4.** *Immediately after the grafting operation is completed, all cut surfaces must be protected from desiccation.* This is done by covering the graft union with tape or grafting wax, or by placing the grafts in moist material or in a covered grafting frame.

**5.** *Proper care must be given the grafts for a period of time after grafting.* Shoots coming from the stock below the graft will often choke out the desired growth from the scion. Or, in some cases, shoots from the scion will grow so vigorously that they break off unless staked and tied or cut back.

## METHODS OF GRAFTING

### Whip Grafting

Whip grafting, as shown in Figures 12–2 and 12–3, is particularly useful for grafting relatively small material, 6 to 13 mm ($\frac{1}{4}$ to $\frac{1}{2}$ in. in diameter). It is highly successful if properly done because there is considerable cambial contact. It heals quickly and makes a strong union. Preferably, the scion and stock should be of equal diameter. The scion should contain two or three buds with the graft made in the smooth internode area below the lower bud.

The cuts made at the top of the stock should be exactly the same as those made at the bottom of the scion. First, a smooth, sloping cut is made, 2.5 to 6 cm (1 to $2\frac{1}{2}$ in.) long. The longer cuts are made when working with large material. This first cut should be made preferably with one single stroke of the knife, so as to leave a smooth, flat surface. To do this, the knife must be razor sharp. Wavy, uneven cuts made with a dull knife will not result in a satisfactory union.

On each of these cut surfaces, a reverse cut is made. It is started downward at a point about one-third of the distance from the tip and should be about one-half the length of the first cut. To obtain a smooth-fitting graft, this second cut should not just split the grain of the wood but should follow along under the first cut, tending to parallel it.

The stock and scion are then inserted into each other, with the tongues interlocking. It is extremely important that the cambium layers match along at least one side, preferably along both sides. The lower tip of the scion should not overhang the stock, as there is a likelihood of the formation of large callus knots. In some species, such callus overgrowths are often mistaken for crown gall knots, caused by bacteria. The use of scions larger than the stock should be avoided for the same reason. If the scion is smaller than the stock, it should be set at one side of the stock so that the cambium layers will be certain to match along that side. If the scion is much smaller than the stock, the first cut on the stock consists only of a slice taken off one corner.

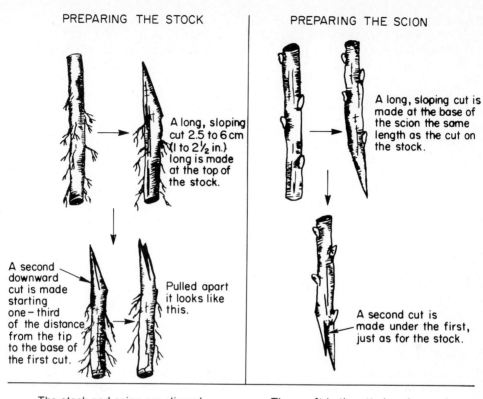

PREPARING THE STOCK

A long, sloping cut 2.5 to 6 cm (1 to 2½ in.) long is made at the top of the stock.

A second downward cut is made starting one-third of the distance from the tip to the base of the first cut.

Pulled apart it looks like this.

PREPARING THE SCION

A long, sloping cut is made at the base of the scion the same length as the cut on the stock.

A second cut is made under the first, just as for the stock.

The stock and scion are slipped together, the tongues interlocking.

The graft is then tied and waxed.

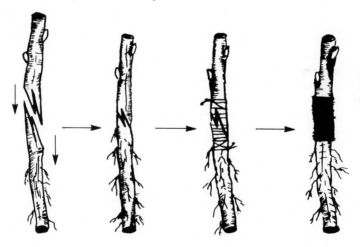

**Figure 12–2**    The whip graft. This method is widely used in grafting small plant material and is especially valuable in making root grafts as illustrated here.

**Figure 12–3** Method of making a whip graft when the scion is considerably smaller than the stock.

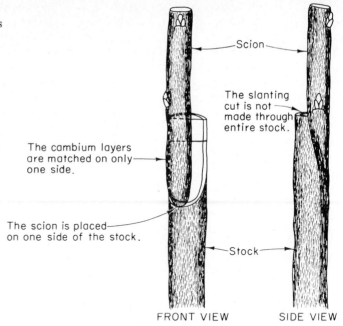

Scion

The slanting cut is not made through entire stock.

The cambium layers are matched on only one side.

The scion is placed on one side of the stock.

Stock

FRONT VIEW        SIDE VIEW

After the scion and stock are fitted together, they should be held securely in some manner until the pieces have united. There are a number of possible ways of doing this.

**1.** If the unions are very well made with a tight, snug fit, it is possible that no additional wrapping or tying is needed, but it is safer to provide some type of wrapping. If not wrapped, the grafts must be protected from drying by burying in moist sand, peat moss, or sawdust until the union has healed. Or they may be planted directly in the nursery with the union below soil level. If the whip graft is used in topworking, the exposed union must be protected in some manner.

**2.** With a secure fit it may be sufficient to omit tying and merely cover the union with hot grafting wax, which will secure the pieces to some extent and give good protection against drying. This method is not recommended for inexperienced grafters.

**3.** A common method is to wrap the union with budding rubbers or possibly raffia or waxed string, such as No. 18 knitting cotton. After wrapping, the whole union can be covered with grafting wax. Waxing may be omitted if the grafts are to be protected from drying by burying in moist sand or peat moss, or if the grafts are planted immediately with the union below the soil surface.

Grafts wrapped with budding rubbers and covered with soil should be inspected later, since the rubber decomposes very slowly below ground and may cause a constriction at the graft union.

**4.** A practice widely used is to wrap the grafts with some type of adhesive tape. A special nurseryman's tape is available. The tape is drawn tightly around the graft union with the edges slightly overlapping. This holds the parts together very well and prevents drying, thus eliminating the need for waxing. If just one thickness of tape is used, it will decompose sufficiently fast (if the union is below ground) that no constriction of growth will develop. If used above ground, the tape should be cut in three to four weeks, after the graft has healed. The use of tight wrapping material such as this is especially recommended when difficulty is encountered with the formation of excessive callus.

**5.** Plastic tapes are available for wrapping grafts. They are used just as adhesive tape, although they are not adhesive. The final turn of the tape is secured by slipping it under the previous turn. This tape has some elasticity. Also, it deteriorates more slowly below ground than above.

## Splice Grafting

This method is the same as the whip graft except that the second, or "tongue," cut is not made in either the stock or scion. A simple slanting cut of the same length and angle is made in both the stock and the scion. These are placed together and wrapped or tied as described for the whip graft. The splice graft is simple and easy to make. It is particularly useful in grafting plants that have a very pithy stem or that have wood that is not flexible enough to permit a tight fit when a tongue is made as in the whip graft.

## Side Grafting

There are numerous variations of the side graft. As the name suggests, the scion is inserted into the side of the stock, which is generally larger in diameter than the scion.

### STUB GRAFT

The stub graft is useful in grafting branches of trees that are too large for the whip graft yet not large enough for other methods such as the cleft or bark graft. For this type of side graft, the best stocks are branches about 2.5 cm (1 in.) in diameter. An oblique cut is made into the stock branch with a chisel or heavy knife at an angle of 20 to 30 degrees. The cut should be about 2.5 cm (1 in.) deep, and at such an angle and depth that when the branch is pulled back the cut will open slightly but will close when the pull is released.

The scion should contain two or three buds and be about 7.5 cm (3 in.) long and relatively thin. At the basal end of the scion, a wedge about 2.5 cm (1 in.) long is made. The cuts on both sides of the scion should be very smooth, each made by one single cut with a sharp knife. The scion must be inserted into the stock at an angle as shown in Figure 12–4 so as to obtain maximum contact of the cambium layers. The grafter inserts the scion into the cut while the upper part of the stock is pulled backward, using care to obtain the best cambium contact. Then the stock is released. The pressure of the stock should grip the scion tightly, making tying unnecessary, but if desired, the scion can be further secured by driving two small flat-headed wire nails (20 gauge, 1.5 cm [⅝ in.] long) into the stock through the scion. Wrapping the stock and scion at the point of union with nurseryman's tape also may be helpful. After the graft is completed the stock may be cut off just above the union. This must be done very carefully or the scion may become dislodged. The entire graft union must be thoroughly covered with grafting wax, sealing all openings. The tip of the scion also should be covered with wax (*35*).

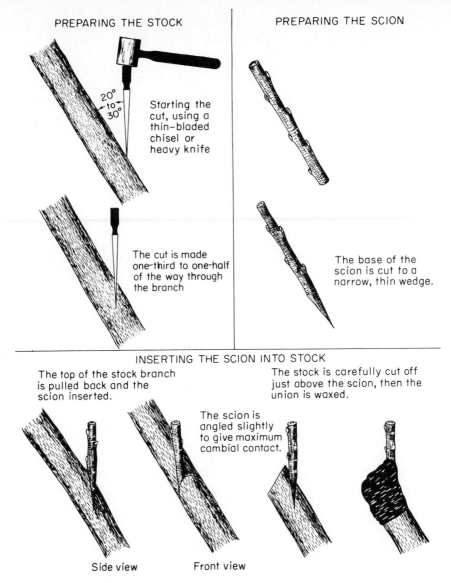

PREPARING THE STOCK

20° to 30°

Starting the cut, using a thin-bladed chisel or heavy knife

The cut is made one-third to one-half of the way through the branch

PREPARING THE SCION

The base of the scion is cut to a narrow, thin wedge.

INSERTING THE SCION INTO STOCK

The top of the stock branch is pulled back and the scion inserted.

The stock is carefully cut off just above the scion, then the union is waxed.

The scion is angled slightly to give maximum cambial contact.

Side view          Front view

**Figure 12–4**    Steps in preparing the side, or stub, graft. A thin-bladed chisel as illustrated here is ideal for making the cut, but a heavy butcher knife could be used satisfactorily.

SIDE-TONGUE GRAFT

The side-tongue graft, shown in Figure 12–5, is useful for small plants, especially some of the broad- and narrow-leaved evergreen species. The stock plant should have a smooth section in the stem just above the crown of the plant. The diameter of the scion should be slightly smaller than that of the stock. The cuts at the base of the scion are made just as for the whip graft. Along a smooth portion of the stem of the stock, a thin piece of bark and wood, the same length as

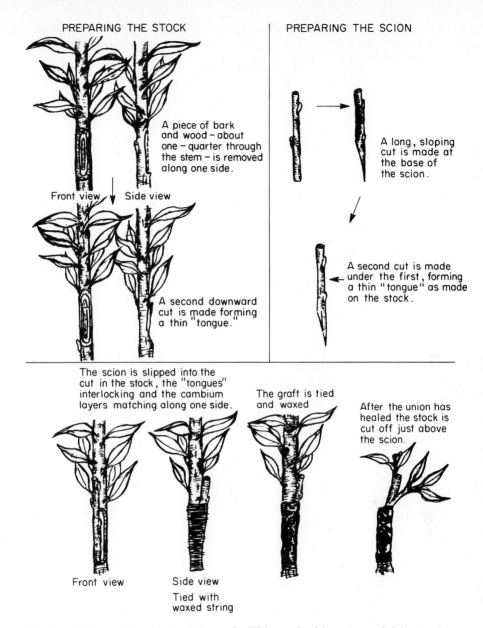

PREPARING THE STOCK

A piece of bark and wood – about one – quarter through the stem – is removed along one side.

Front view    Side view

A second downward cut is made forming a thin "tongue."

PREPARING THE SCION

A long, sloping cut is made at the base of the scion.

A second cut is made under the first, forming a thin "tongue" as made on the stock.

The scion is slipped into the cut in the stock, the "tongues" interlocking and the cambium layers matching along one side.

The graft is tied and waxed

After the union has healed the stock is cut off just above the scion.

Front view     Side view

Tied with waxed string

**Figure 12–5**    The side-tongue graft. This method is very useful for grafting broad-leaved evergreen plants.

the cut surface of the scion, is completely removed. Then a reverse cut is made downward in the cut on the stock, starting one-third of the distance from the top of the cut. This second cut in the stock should be the same length as the reverse cut in the scion. The scion is then inserted into the cut in the stock, the two tongues interlocking, and the cambium layer(s) matching. The graft is wrapped tightly, using one of the methods described for the whip graft.

The top of the stock is left intact for several weeks until the graft union has healed. Then it may be cut back above the scion gradually or all at once. This forces the buds on the scion into active growth.

**Figure 12–6**     Steps in making the side-veneer graft. This method is widely used in propagating narrow-leaved evergreen species that are difficult to start by cuttings.

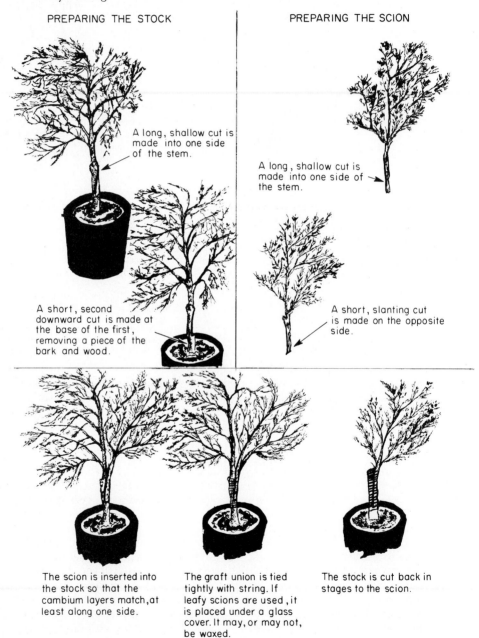

PREPARING THE STOCK

A long, shallow cut is made into one side of the stem.

A short, second downward cut is made at the base of the first, removing a piece of the bark and wood.

PREPARING THE SCION

A long, shallow cut is made into one side of the stem.

A short, slanting cut is made on the opposite side.

The scion is inserted into the stock so that the cambium layers match, at least along one side.

The graft union is tied tightly with string. If leafy scions are used, it is placed under a glass cover. It may, or may not, be waxed.

The stock is cut back in stages to the scion.

## SIDE-VENEER GRAFT (SPLICED SIDE GRAFT)

The side-veneer variation of side grafting, shown in Figure 12–6, is widely used, especially for grafting small potted plants, such as seedling evergreens, to named cultivars. A shallow downward and inward cut from 25 to 38 mm (1 to 1½ in.) long is made in a smooth area just above the crown of the stock plant. At the base of this cut, a second short inward and downward cut is made, intersecting the first cut, so as to remove the piece of wood and bark. The scion is prepared with a long cut along one side and a very short one at the base of the scion on the opposite side. These scion cuts should be the same length and width as those made in the stock so that the cambium layers can be matched as closely as possible.

After inserting the scion, the graft is tightly wrapped with waxed or paraffined string, with budding rubbers, or with nurserymen's adhesive tape. The graft may or may not be covered with wax, depending upon the species. A common practice in side grafting small potted plants of some of the woody ornamental species is to plunge the grafted plants into a damp medium, such as peat moss, so that it just covers the graft union. The newly grafted plants may be placed for healing in a mist propagating house or set in grafting cases. The latter are closed boxes with a transparent cover, which permits retention of a high humidity around the grafted plant until the union has healed. The grafting cases are kept closed for a week or so after the grafts are put in, then gradually opened over a period of several weeks; finally, the cover is taken off completely.

After the union has healed, the stock can be cut back above the scion either in gradual steps or all at once.

## Cleft Grafting

Cleft grafting is one of the oldest and most widely used methods of grafting, being especially adapted to topworking trees, either in the trunk of a small tree or in the scaffold branches of a larger tree (Figures 12–7 and 12–8). Cleft grafting is useful also for smaller plants, as in crown grafting established grapevines or camellias. In topworking trees, this method should be limited to stock branches about 2.5 to 10 cm (1 to 4 in.) in diameter and to species with fairly straight-grained wood that will split evenly. Although cleft grafting can be done any time during the dormant season, the chances for successful healing of the graft union are best if the work is done in early spring just when the buds of the stock are beginning to swell but before active growth has started. If cleft grafting is done after the tree is in active growth, it is likely that the bark of the stock will separate from the wood, causing difficulties in obtaining a good union. When this separation occurs, the loosened bark must be firmly nailed back in place. The scions should be made from dormant, one-year-old wood. Unless the grafting is done early in the season (when the dormant scions can be collected and used immediately), the scionwood should be collected in advance and held under refrigeration until it is used.

In sawing off the branch for this and other topworking methods, the cut should be made at right angles to the main axis of the branch.

In making the cleft graft, a heavy knife, such as a butcher knife, or one of several special cleft grafting tools, is used to make a vertical split for a distance of 5 to 8 cm (2 to 3 in.) down the center of the stub to be grafted. This split is made by pounding the knife in with a hammer or mallet. It is very important to saw the branch off in such a position that the end of the stub which is left is smooth,

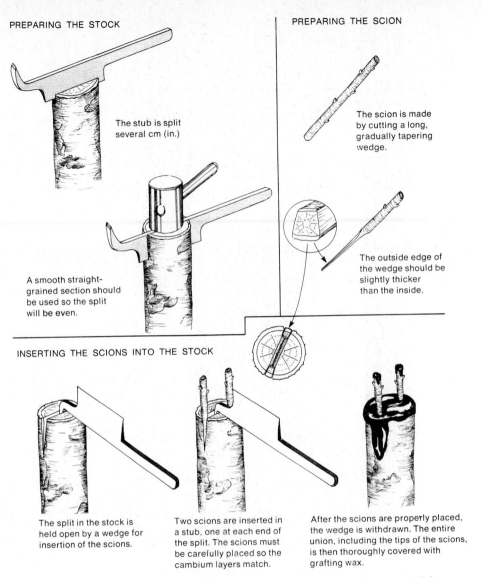

**PREPARING THE STOCK**

The stub is split several cm (in.)

A smooth straight-grained section should be used so the split will be even.

**PREPARING THE SCION**

The scion is made by cutting a long, gradually tapering wedge.

The outside edge of the wedge should be slightly thicker than the inside.

**INSERTING THE SCIONS INTO THE STOCK**

The split in the stock is held open by a wedge for insertion of the scions.

Two scions are inserted in a stub, one at each end of the split. The scions must be carefully placed so the cambium layers match.

After the scions are properly placed, the wedge is withdrawn. The entire union, including the tips of the scions, is then thoroughly covered with grafting wax.

**Figure 12–7**    Steps in making the cleft graft. This method is very widely used and is quite successful if the scions are inserted so that the cambium layers of stock and scion match properly.

straight-grained, and free of knots for at least 15 cm (6 in.). Otherwise, the split may not be straight, or the wood may split one way and the bark another. The split should be in a tangential rather than radial direction in relation to the center of the tree. This permits better placement of the scions for their subsequent growth. Sometimes the cleft is made by a longitudinal saw cut rather than by splitting. After a good, straight split is made, a screwdriver, chisel, or the

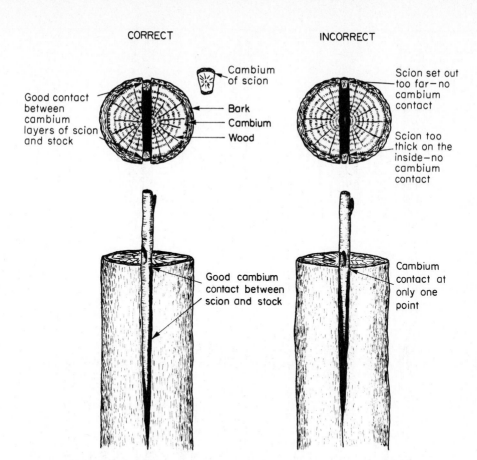

CORRECT                                    INCORRECT

Cambium
of scion

Good contact
between
cambium
layers of scion
and stock

Bark
Cambium
Wood

Scion set out
too far—no
cambium
contact

Scion too
thick on the
inside—no
cambium
contact

Good cambium
contact between
scion and stock

Cambium
contact at
only one
point

**Figure 12–8**     In making the cleft graft, the proper placement of the
scions is very important. The correct way of doing this is shown on the left.
Scions inserted as shown on the right probably would not grow.

wedge part of the cleft-grafting tool is driven into the top of the split to hold it
open.

Two scions are inserted, one at each side of the stock where the cambium
layer is located. The scions should be 8 to 10 cm (3 to 4 in.) long, about 10 to 13
cm (⅜ to ½ in.) in thickness, and should have two or three buds. The basal end
of each scion should be cut into a gently sloping wedge about 5 cm (2 in.) long. It
is not necessary that the end of the wedge come to a point. The side of the wedge
which is to go to the outer side of the stock should be slightly wider than the inside edge. Thus, when the scion is inserted and the tool is removed, the full
pressure of the split stock will come to bear on the scions at the position where
the cambium of the stock touches the cambium layer on the outer edge of the
scion. Since the bark of the stock is almost always thicker than the bark of the
scion, it is usually necessary for the outer surface of the scion to set slightly in
from the outer surface of the stock in order to match the cambium layers.

In all types of grafting the scion must be inserted right side up. That is, the points of the buds on the scion should be pointing upward and away from the stock. Unless this rule is observed, the graft will not be successful.

The long, sloping wedge cuts at the base of the scion should be smooth, made by a single cut on each side with a very sharp knife. Both sides of the scion wedge should press firmly against the stock for their entire length. A common mistake in cutting scions for this type of graft is to make the cut on the scion too short and the slope too abrupt, so that the only point of contact is at the top. Slightly shaving the sides of the split in the stock will often permit a smoother contact.

After the scions are properly made and inserted, the tool is withdrawn, using care not to disturb the scions. They should be held so tightly by the pressure of the stock that they cannot be pulled loose by hand. No further tying or nailing is needed unless very small stock branches have been used, in which case the top of the stock can be wrapped tightly with string or adhesive tape to hold the scions in place more securely.

Thorough waxing of the completed graft is essential. The top surface of the stub should be entirely covered, permitting the wax to work into the split in the stock. The sides of the grafted stub should be well covered with wax as far down the stub as the length of the split. The tops of the scions should be waxed but not necessarily the bark or buds of the scion. Two or three days later all the grafts should be inspected and rewaxed where openings appear. Lack of thorough and complete waxing in this type of graft is almost certain to result in failure.

**Wedge Grafting**

Wedge grafting, illustrated in Figure 12–9, is similar to the saw-kerf graft but simpler and easier. If properly done the wedge graft gives excellent results. Like the cleft and saw-kerf grafts it can be made in late winter (in mild climates) or early spring before the bark begins to slip (separates easily from the wood).

The diameter of the stock to be grafted is the same as for the cleft and saw-kerf grafts—5 to 10 cm (2 to 4 in.), and the scions also are the same size—10 to 13 cm (4 to 5 in.) long and 10 to 13 mm (⅜ to ½ in.) in thickness.

A sharp, heavy, short-bladed knife is used for making a **V**-wedge in the side of the stub, about 5 cm (2 in.) long. Two cuts are made coming together at the bottom and as far apart at the top as the width of the scion. These cuts extend about 2 cm (¾ in.) deep into the side of the stub. After these cuts are made, a screwdriver is pounded downward behind the wedge chip from the top of the stub to knock out the chip, leaving a **V** opening for insertion of the scion. The base of the scion is trimmed to a wedge shape exactly the same size and shape as the opening. With the two cambium layers matching, the scion is tapped downward firmly into place and slanting outward slightly at the top so that the cambium layers cross. If the cut is long enough and gently tapering, the scion should be so tightly held in place that it would be difficult to dislodge.

## PREPARING THE STOCK

A heavy sharp knife is pounded into the side of the stub to make two cuts to form a V.

A screwdriver is used to flip out the V-shaped chip, leaving a space for the insertion of the scion.

## PREPARING THE SCION

The scion should be about 10 to 13 cm (4 to 5 in.) long, 10 to 12 mm (3/8 to 1/2 in.) thick, and with 2 or 3 healthy vegetative buds. The basal ends should be cut to a V-shaped wedge, matching the opening in the stock.

Front view      Side view

## INSERTING THE SCIONS INTO THE STOCK

The scion is gently tapped into the V-shaped opening in the stock, matching the cambium layers at a slight angle so that the cambium of stock and scion cross.

Scion should be inserted at an angle so that the cambium layers of stock and scion are closely matched, barely crossing each other.

After scions are in place all cut surfaces are thoroughly covered with grafting wax.

**Figure 12–9**    Wedge grafting.

In a 5-cm (2-in.) wide stub, two scions should be inserted 180° apart; in a 10-cm (4 in.) stub three scions should be used, 120° apart. After all scions are firmly tapped into place, all cut surfaces, including the tips of the scion, should be thoroughly waxed.

## Bark Grafting

Bark grafting is rapid, simple, readily performed by amateurs, and if properly done, gives a high percentage of "takes." It requires no special equipment and can be performed on branches ranging from 2.5 cm (1 in.) up to 30 cm (1 ft) or more in diameter. The latter size is not recommended as it is difficult to heal over such large stubs before decay-producing organisms attack. The bark graft, since it depends on the bark separating readily from the wood, can only be done after active growth of the stock has started in the spring. As dormant scions must be used, it is necessary to gather the scionwood for deciduous species during the dormant season and hold it under refrigeration until the grafting operation is done. For evergreen species, freshly collected scion wood can be used. In the bark graft scions are not as securely attached to the stock as in some of the other methods and are more susceptible to wind breakage during the first year even though healing has been satisfactory. Therefore the new shoots arising from the scions probably should be staked during the first year, or cut back to about half their length, especially in windy areas. After a few years' growth, the bark graft union is as strong as the unions formed by other methods.

Two modifications of the bark graft are described as follows:

### BARK GRAFT (METHOD NO. 1)

Several scions are inserted into each stub. For each scion, a vertical knife cut 2.5 to 5 cm (1 to 2 in.) long is made at the top end of the stub through the bark to the wood. The bark is then lifted slightly along both sides of this cut, in preparation for the insertion of the scion. The scion should be of dormant wood, 10 to 13 cm (4 to 5 in.) long, containing two or three buds, and be 6 to 13 mm (¼ to ½ in.) in thickness. One cut about 5 cm (2 in.) long is made along one side at the base of the scion. With large scions, this cut extends about one-third of the way into the scion, leaving a "shoulder" at the top. The purpose of this shoulder is to reduce the thickness of the scion to minimize the separation of bark and wood after insertion in the stock. The scion should not be cut too thin, however, or it will be mechanically weak and break off at the point of attachment to the stock. If small scions are used, no shoulder is necessary. On the side of the scion opposite the first long cut, a second, shorter cut is made, as shown in Figure 12–10, thereby bringing the basal end of the scion to a wedge shape. The scion is then inserted between the bark and the wood of the stock, centered directly under the vertical cut through the bark. The longer cut on the scion is placed against the wood, and the shoulder on the scion is brought down until it rests on top of the stub. The scion is then ready to be fastened in place. Nail the scion into the wood, using two nails per scion. Flat-headed nails 15 to 25 mm (⅝ to 1 in.) long, of 19 or 20 gauge wire, depending on the size of the scions, are satisfactory. The bark on both sides of the scion also should be nailed down securely or it will tend to peel back from the wood.

Another method commonly used with soft-barked trees, such as the avocado, is to insert all the scions in the stub and then hold them in place by wrapping string, adhesive tape, or waxed cloth around the stub. This is more effective than nailing in preventing the scions from blowing out but probably does not give as tight a fit. Both nailing and wrapping are advisable for maximum strength. If a wrapping material is used, it must be cut later to prevent constriction.

## PREPARING THE STOCK

A vertical cut 2.5 to 5 cm (1 to 2 in.) long is made through the bark to the wood.

The bark on both sides of the cut is slightly separated from the wood.

## PREPARING THE SCION

The scion is cut as shown below, a long cut with a shoulder on one side, and a shorter cut on the opposite side.

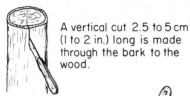

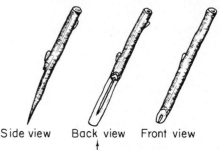

Side view    Back view   Front view

( This side is placed next to the wood of the stock.)

## INSERTING THE SCIONS INTO THE STOCK

The scions are pushed downward between the bark and the wood just under each cut. They are nailed in place, as is the bark on each side of the scion.

The grafted stub is then thoroughly waxed.

**Figure 12–10**     Steps in preparing the bark graft (Method No. 1). In grafting some thick-barked plants the vertical cut in the bark is unnecessary, the scion being inserted between the bark and wood of the stock.

After the stub has been grafted and the scions fastened by nailing or tying, all cut surfaces, including the end of the scions, should be thoroughly covered with grafting wax.

BARK GRAFT (METHOD NO. 2)

In this method, as shown in Figure 12–11, *two* knife cuts about 5 cm (2 in.) long are made through the bark of the stock down to the wood, rather than just one. The distance between these two cuts should be exactly the same as the

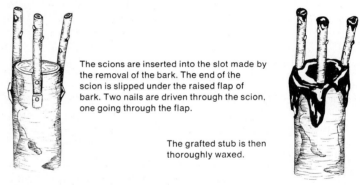

PREPARING THE STOCK

Two parallel, vertical cuts 2.5 to 5cm (1 to 2 in.) long are made through the bark to the wood. The distance between the cuts should equal the width of the scion.

A horizontal cut is made between the two vertical cuts and most of the piece of bark is removed. A small flap is left at the bottom.

PREPARING THE SCION

The scions are made with a long sloping cut on one side and a shorter cut on the opposite side.

Side View    Back View    Front View

(This side is placed next to the wood of the stock.)

INSERTING THE SCION INTO THE STOCK

The scions are inserted into the slot made by the removal of the bark. The end of the scion is slipped under the raised flap of bark. Two nails are driven through the scion, one going through the flap.

The grafted stub is then thoroughly waxed.

**Figure 12–11**    Bark graft, Method No. 2.

width of the scion. The piece of bark between the cuts should be lifted and the terminal two-thirds cut off. The scion is prepared with a smooth slanting cut along one side at the basal end completely through the scion. This cut should be about 5 cm (2 in.) long but *without* the shoulder, in contrast to the other methods. On the opposite side of the scion, a cut about 13 mm ($\frac{1}{2}$ in.) long is made, forming a wedge at the base of the scion. The scion should fit snugly into the opening in the bark with the longer cut inward and with the wedge at the base slipped under the flap of remaining bark. Rapid healing can be expected because both sides of the scion are touching undisturbed bark and cambium cells, which is not the case in the other types of the bark graft.

The scion should be nailed into place with two nails, the lower nail going through the flap of bark covering the short cut on the back of the scion. If the bark along the sides of the scion should accidentally become disturbed, it must be nailed back into place.

Method No. 2 is well adapted for use with thick-barked trees, such as walnuts, on which it is not feasible to insert the scion under the bark.

## Approach Grafting

The distinguishing feature of approach grafting is that two independent, self-sustaining plants are grafted together. After a union has occurred, the top of the stock plant is removed above the graft and the base of the scion plant is removed below the graft. Sometimes it is necessary to sever these parts gradually rather than all at once. Approach grafting provides a means of establishing a graft union between certain plants in which successful graft unions are difficult to obtain. Approach grafting is usually performed with one or both of the plants to be grafted growing in a container. Rootstock plants in containers may be placed adjoining an established plant which is to furnish the scion part of the new, grafted plant (see Figure 12–12).

**Figure 12–12**     Approach grafting used in obtaining a desirable scion cultivar on a seedling camellia. *Top:* seedling plant in container is set close to a large plant of the desired cultivar. The graft union is made and tightly wrapped with adhesive tape. *Below:* after the graft union has healed, which may take several months, the stock plant is cut off above the graft union, and the scion is severed from the parent plant just below the graft union. Approach grafting is sometimes necessary for plants very difficult to graft by other methods.

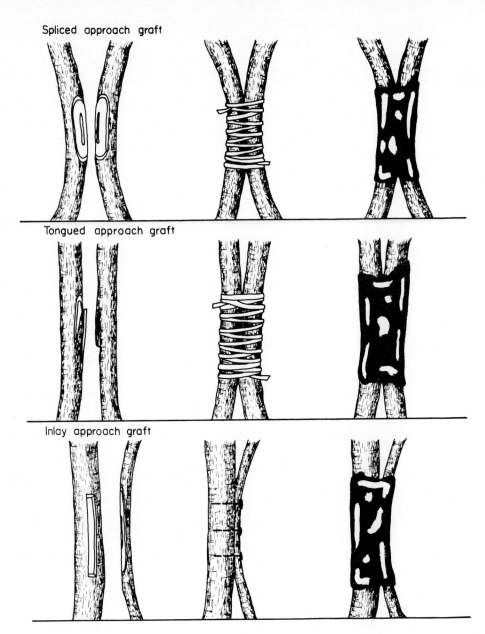

**Figure 12-13**    Three methods of making an approach graft.

This type of grafting should be done at times of the year when growth is active and rapid healing of the graft union will take place. As in other methods of grafting, the cut surfaces should be securely fastened together, then covered with grafting wax to prevent drying of the tissues.

Three useful methods of making approach grafts are described below and illustrated in Figure 12-13.

In the spliced approach graft the two stems should be approximately the same size. At the point where the union is to occur, a slice of bark and wood 2.5 to 5 cm (1 to 2 in.) long is cut from both stems. This cut should be the same size on each so that identical cambium patterns will be made. The cuts must be perfectly smooth and as nearly flat as possible so that when they are pressed together there will be close contact of the cambium layers. The two cut surfaces are then bound tightly together with string, raffia, or nurseryman's tape. The whole union should then be covered with grafting wax. After the parts are well united, which may require considerable time in some cases, the stock above the union and the scion below the union are cut, and the graft is then completed. It may be necessary to reduce the leaf area of the scion if it is more than the root system of the stock can sustain. Figure 12–12 shows the use of this method with camellias.

### TONGUED APPROACH GRAFT

The tongued approach graft is the same as the spliced approach graft except that after the first cut is made in each stem to be joined, a second cut—downward on the stock and upward on the scion—is made, thus providing a thin tongue on each piece. By interlocking these tongues a very tight, closely fitting graft union can be obtained.

### INLAY APPROACH GRAFT

The inlay approach graft may be used if the bark of the stock plant is considerably thicker than that of the scion plant. A narrow slot, 7.5 to 10 cm (3 to 4 in.) long, is made in the bark of the stock plant by making two parallel knife cuts and removing the strip of bark between. This can be done only when the stock plant is actively growing and the bark "slipping." The slot should be exactly as wide as the scion to be inserted. The stem of the scion plant, at the point of union, should be given a long, shallow cut along one side, of the same length as the slot in the stock plant and deep enough to go through the bark and slightly into the wood. This cut surface of the scion branch should be laid into the slot cut in the stock plant and held there by nailing with two or more small, flat-headed wire nails. The entire union must then be thoroughly covered with grafting wax. After the union has healed, the stock can be cut off above the graft and the scion below the graft.

## Inarching

Inarching is similar to approach grafting in that both stock and scion plants are on their own roots at the time of grafting; it differs in that the top of the new rootstock plant usually does not extend above the point of the graft union as it does in approach grafting. Inarching is generally considered to be a form of "repair grafting," used to repair roots damaged by cultivation implements, rodents, or disease. It can be used to very good advantage in saving a valuable tree or improving its root system (Figure 12–14.).

Seedlings (or rooted cuttings) planted beside the older, damaged tree, or suckers arising near its base, are grafted into the trunk of the tree to provide a new root system to supplant the damaged roots. The seedlings to be inarched into the tree should be spaced about 13 to 15 cm (5 to 6 in.) apart around the cir-

**Figure 12–14** *Left:* inarches that have just been inserted. The one on the left has been waxed. The one on the right has been nailed into place and is ready for waxing. *Right:* inarching can be used for invigorating established trees by replacing a weak rootstock with a more vigorous one. Here a Persian walnut tree has been inarched with vigorous Paradox hybrid seedlings (*Juglans hindsii* × *J. regia*) seedlings.

cumference of the tree if the damage is extensive. A damaged tree will usually stay alive for some time unless the injury is very severe. A satisfactory procedure for inarching is to plant seedlings of a compatible species around the tree during the dormant season. Then the grafting operation can be done as active growth commences in early spring.

Inarching old, weakly growing trees with strong, vigorous seedling rootstocks has on some occasions (*17*) proved beneficial in promoting renewed active growth of the old trees.

The seedling plants to provide the new root system are usually considerably smaller than the tree to be repaired. As illustrated in Figure 12–15, the graft union is made in a manner similar to that described for method No. 2 of the bark graft. The upper end of the seedling, which should be 6 to 12 mm (¼ to ½ in.) thick, is given a long shallow cut along the side for 10 to 15 cm (4 to 6 in.) This cut should be on the side next to the trunk of the tree and made deep enough to remove some of the wood, thus exposing two strips of cambium tissue. At the end of the seedling another shorter cut, about 13 mm (½ in.) long, is made on the side opposite the long cut; this makes a sharp, wedge-shaped end on the seedling stem.

A long slot is made in the trunk of the older tree by removing a piece of bark the exact width of the seedling and just as long as the cut surface made on the seedling. A small flap of bark is left at the upper end of the slot, under which the wedge end of the seedling is inserted. Then the seedling is nailed into the slot with four or five small, flat-headed wire nails. The nail at the top of the slot should go through the flap of bark and through the end of the seedling. If any of the bark of the tree along the sides of the seedling should accidentally be pulled loose, it is necessary to nail it back in place. After nailing, the entire area of the graft union should be thoroughly waxed.

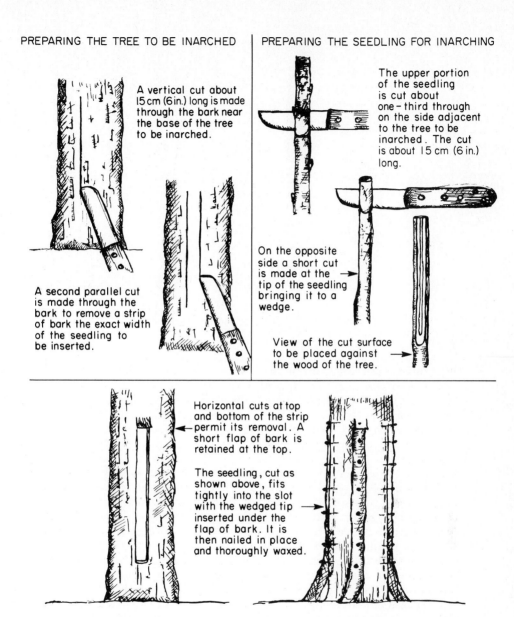

PREPARING THE TREE TO BE INARCHED | PREPARING THE SEEDLING FOR INARCHING

A vertical cut about 15 cm (6 in.) long is made through the bark near the base of the tree to be inarched.

A second parallel cut is made through the bark to remove a strip of bark the exact width of the seedling to be inserted.

The upper portion of the seedling is cut about one-third through on the side adjacent to the tree to be inarched. The cut is about 15 cm (6 in.) long.

On the opposite side a short cut is made at the tip of the seedling bringing it to a wedge.

View of the cut surface to be placed against the wood of the tree.

Horizontal cuts at top and bottom of the strip permit its removal. A short flap of bark is retained at the top.

The seedling, cut as shown above, fits tightly into the slot with the wedged tip inserted under the flap of bark. It is then nailed in place and thoroughly waxed.

**Figure 12–15**    Steps in inarching a large plant with smaller ones planted around its base.

## Bridge Grafting

Bridge grafting is a form of repair grafting, and is used when the root system of the tree has not been damaged but there is injury to the trunk. Sometimes cultivation implements, rodents, disease, or winter injury damage a con-

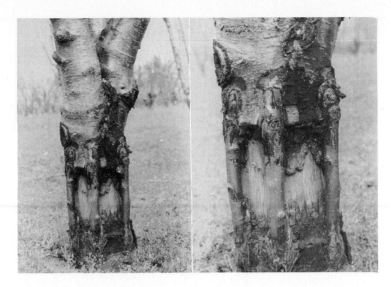

**Figure 12-16**     Injured trunk of a cherry tree successfully bridge grafted by a modification of the bark graft.

siderable trunk area, often girdling the tree completely. If the damage to the bark is extensive, the tree is almost certain to die, because the roots will be deprived of their food supply from the top of the tree. Trees of some species, such as the elm, cherry, and pecan, can heal over extensively injured areas by the development of callus tissue. But trees of most species with severely damaged bark should be bridge grafted if they are to be saved, as illustrated in Figure 12-16.

The bridge grafting operation is best performed in early spring just as active growth of the tree is beginning and the bark is slipping easily. The scions to be used should be taken when dormant from one-year-old growth, 6 to 12 mm (¼ to ½ in.) in diameter, of the same or compatible species, and held under refrigeration until the grafting work is to be done. In an emergency, one may successfully perform bridge grafting late in the spring, using scionwood whose buds have already started to grow. Remove the developing buds or new shoots.

The first step in bridge grafting is to trim the wounded area back to healthy, undamaged tissue by removing dead or torn bark. Then every 5 to 7.5 cm (2 to 3 in.) around the injured section a scion is inserted, attached at both the upper and lower ends into live, undamaged bark. It is important that the scions be inserted right side up. If they are put in reversed, they may make a union and stay alive for a year or two, but the scions will not grow and enlarge in diameter as they would if inserted correctly.

Figure 12-17 shows the details of making a satisfactory type of bridge graft.

After all the scions have been inserted, the cut surfaces must be thoroughly covered with grafting wax, particular care being taken to work the wax around the scions, especially at the graft unions. The exposed wood of the injured section may also be covered with grafting wax to prevent the entrance of decay organisms and excessive drying of the wood, which is important, being the path for upward movement of water and nutrients in the tree.

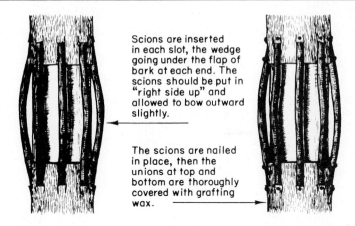

All dead and damaged bark around the wound is trimmed back to live healthy tissue.

One long, slanting cut is made at each end of the scion, with both cuts on the same side.

Cuts are made in the bark at top and bottom of the wound, just as for Method 2 of the bark graft. The slots in the bark should be the same width as the scions to be inserted.

A second, short, slanting cut is made on the back side of the scion, bringing the ends to a sharp wedge. Buds can be trimmed off the scions if desired.

Scions are inserted in each slot, the wedge going under the flap of bark at each end. The scions should be put in "right side up" and allowed to bow outward slightly.

The scions are nailed in place, then the unions at top and bottom are thoroughly covered with grafting wax.

**Figure 12–17**  A satisfactory method of making a bridge graft, using a modification of Method No. 2 of the bark graft.

## TOOLS AND ACCESSORIES FOR GRAFTING

Special equipment needed for any particular method of grafting has been illustrated along with the description of the method. There are some pieces of equipment, however, that are used in all types of grafting.

### How to Sharpen a Knife Properly

To sharpen the grafting knife, the initial grinding may be done with a medium-grit stone, but a hard, fine-grit stone should be used for the final sharpening. The stone should be wet with water or oil during sharpening. Do not use a carborundum stone, because it is too abrasive and will grind off too much metal. Most prefer to use knives beveled only on one side, the back side being flat, whereas others prefer a knife beveled on both sides. In sharpening the knife, hold it so that only the edge of the blade touches the stone in order that a stiff edge for cutting can be obtained. Use the whole width of the stone so that its surface will remain flat. A correctly sharpened knife of high-quality steel should retain a good edge for several days' work, with only occasional stropping on a piece of leather.

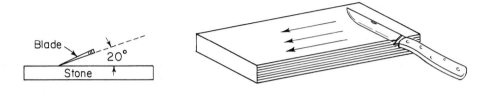

## Knives

For propagation work, the two general types of knives used are the budding knife and the grafting knife (Figure 12–18). Where a limited amount of either budding or grafting is done, the budding knife can be used satisfactorily for both operations. The knives have either a folding or a fixed blade. The fixed-blade type is stronger, and if a holder of some kind is used to protect the cutting edge, it is probably the most desirable. A well-built, sturdy knife of high carbon steel is essential if much grafting work is to be done. The knife must be kept razor sharp in order to do good work.

**Figure 12–18**    Types of folding knives used in plant propagation. *Top:* grafting knife. *Below:* budding knife. The blunt part on the right of this knife is used in T-budding to open the bark flap for insertion of the shield bud piece.

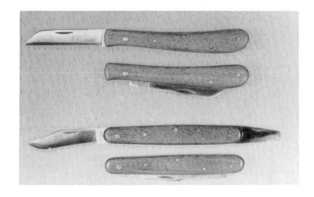

## Grafting Waxes

Grafting wax has two chief purposes: (a) It seals over the graft union, thereby preventing the loss of moisture and death of the tender, exposed cells of the cut surfaces of the scion and stock. These cells are essential for callus production and healing of the graft union. (b) It prevents the entrance of various decay-producing organisms that may lead to wood rotting.

An ideal grafting wax should adhere well to the plant surfaces, not be washed off by rains, not be so brittle as to crack and chip during cold weather, or so soft that it will melt and run off during hot days, but still be pliable enough to allow for swelling of the scion and growth enlargement of the stock without cracking.

### HOT WAXES

The ingredients used in hot waxes are resin, beeswax, either raw linseed oil or tallow, and lampblack or powdered charcoal. The proportions of each ingredient may vary somewhat without affecting the results appreciably. The purpose of the lampblack is to give some color to the otherwise colorless wax so that it will show more readily when the grafts have been well covered. Lampblack also imparts a more workable consistency to the wax, eliminating some of the stickiness and stringiness. After application, the dark-colored wax tends to absorb heat from the sun and thus remain soft and pliable. During hot weather, however, the wax may become too soft.

**Directions for making hot grafting wax (hard type)**

*Ingredients:*

Resin.................................................................2.25 kg (5 lb)
Beeswax.............................................................. 337 g (¾ lb)
Raw linseed oil.....................................................235 ml (½ pt)
Lampblack..............................................................28 g (1 oz)
Fish glue.............................................................42 g (1½ oz)

Heat the glue in a double boiler with just enough water to dissolve it. Melt the other ingredients in another container and allow the mixture to cool but still remain fluid. Add the glue slowly to the partly cooled mixture, stirring continually. Pour out into shallow greased pans or wooden boxes lined with greased paper. Allow to harden. To use, chip or break into small lumps. Reheat in a grafting wax melter and apply with a small paint brush.

The hard type of hot grafting wax solidifies upon cooling and must be reheated just before being applied to the graft union. It is important that this wax be at the right temperature when it is used. If the wax is boiling, it will injure the plant tissues. At the other extreme, if the wax is too cool, it will not flow easily into all the crevices in the bark, thus leaving openings for the entrance of air. *The wax should be hot enough to flow easily, yet not be boiling.*

For heating the wax, any small burner is satisfactory. A brush is used to apply the wax, but provision should be made to suspend it from the side of the container for if the brush rests at the bottom, the heat will burn the bristles. Homemade grafting wax melters can easily be constructed (*32*).

COLD WAXES

A commercially prepared type of grafting wax consisting of an emulsion of asphalt and water is available. It has proved quite satisfactory and is widely used (*34*). This material is about 50 percent water, which evaporates after application, leaving a coat of asphalt over the graft. For the remaining wax to be thick enough to protect the graft adequately, the original application should be fairly heavy. Since this material is water soluble until it dries, rains occurring within 24 hours after application are likely to wash it off. It would then, of course, be necessary to rewax the grafts immediately. If grafting is done during rainy weather, it is advisable to use the hot type of wax, which is not affected by rains.

The emulsions of this type of cold wax are broken down by freezing, so it is very important that the containers be stored in a warm place during cold weather. *Do not confuse these materials with roofing compounds, which have a similar appearance but are entirely unsuitable for use as grafting wax.*

For small-scale operations, grafting waxes are available in aerosol applicator cans. Several repeated applications of the wax are generally necessary to give sufficient coverage for adequate protection.

## Tying and Wrapping Materials

Some of the grafting methods, particularly the whip graft, require that the graft union be held together by tying until the parts unite. Tying can be done in several ways—the simplest would be merely tying with ordinary string and covering with grafting wax. For large-scale operations, waxed string is convenient because it will adhere to itself and to the plant parts without tying. It should be strong enough to hold the grafted parts together yet weak enough to be broken by hand.

A special nurseryman's adhesive tape is manufactured that is similar to surgical adhesive tape but lighter in weight and not sterilized. It is more convenient to use than waxed cloth tape. Adhesive tape is useful for tying and sealing whip grafts. When using any kind of tape or string for wrapping grafts, it is important not to use too many layers or the material may eventually girdle the plant unless it is cut. When this type of wrapping is covered with soil, it usually rots and breaks before damage can occur, unless it has been wrapped in too many layers. It is best to observe such wrappings carefully and cut them after the graft has healed to prevent constriction.

Plastic polyethylene or polyvinyl chloride (PVC) grafting and budding tapes also are available. Since they are not adhesive, they must be secured by folding the end of the tape under the last turn. These plastic tapes are slightly elastic and will allow for some diameter growth of the grafts, but must eventually be removed. Masking tape is a satisfactory wrapping material also.

Parafilm tape has been used with successful results to wrap graft unions rapidly (*4*). This material is a waterproof, flexible, stretchable, thermoplastic film with a paper backing. The film is removed from the paper and two layers are placed over the graft union and pressed into place with the fingers.

## Grafting Machines

Several machines or devices have been developed to prepare graft and bud unions and a few have been widely used, especially in propagating grapevines. (*1, 2,*).

**Figure 12–19**   A type of wedge grafting made by a French grafting device. *Top left:* appearance of graft as made by the cutting blades (before wrapping with grafting tape). *Top right:* apple graft union after one year's growth in the nursery. *Below:* grafting device in use. Grafts can be made with this device much faster than by the whip graft method.

One of the most successful machines is the hand-operated device shown in Figure 12–19, originally developed in France and called the L. B. Grafting Tool.[1] This device makes a type of wedge graft, cutting out a long V-notch in the rootstock and a corresponding long, tapered cut at the base of the scion. Although the cuts fit together very well, the operation is slow because the graft union must be either tied with a budding rubber or stapled together. This machine has been used successfully in propagating both grapes and fruit trees.

Another machine[2] widely used in California is shown in Figure 12–20. It is powered by an electric motor and has two sets of saw blades on a single shaft for cutting notches, one in the base of the scion and one at the top end of the rootstock, as shown in Figure 12–21. Three persons work together, one cutting out the scions, one cutting the rootstocks, and one assembling the two. After the stock and scion have been pushed together, the graft fit is so tight that no wrapping or stapling is necessary.

[1] Available in the United States from Heitz Wine Cellars, 500 Taplin Road, St. Helena, Calif. 94574.

[2] Available from Spink Refrigeration Company, 677 St. Helena Highway South, St. Helena, Calif. 94574.

**Figure 12–20** Machine in operation for cutting notches in ends of grape scion and stocks for making the type of graft illustrated in Figure 12–21. The blades in front are making cuts in the stocks; those in the back are making cuts for the scions.

A third device, widely used for grapes, is the Pfropf-Star grafting machine[3] manufactured in Germany. It cuts through both stock and scion, one laid on top the other, making an omega-shaped cut and leaving the two parts interlocked. The device is available with either foot-operated or electromagnetic-operated mechanism. It is the fastest of the three machines but the graft union is not held together as tightly as the other two and it is not as suitable for tree fruits as the other two devices.

[3] Manufactured by E. & H. Wahler, 7056 Weinstadt-Schnait, BuchhaldenstraBe 21, Stuttgart, Germany. Available in the United States from Tree & Vine Management Co., P. O. Box 311, Reedley, Calif. 93654.

**Figure 12–21** Grapes propagated by machine grafting. Small, one-budded scions are grafted on rooted cuttings during the dormant season. The graft union is wrapped with a budding rubber, then allowed to callus before planting.

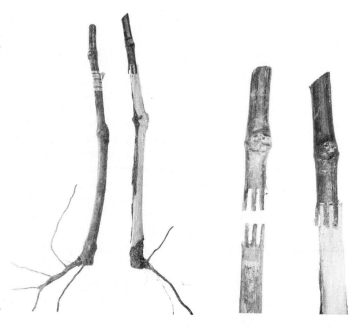

# SELECTION AND HANDLING
# OF SCIONWOOD

## Kind of Wood

Since grafting of deciduous species takes place in late winter or early spring, the use of scionwood that grew the previous summer is necessary.

In selecting such scion material the following points should be observed:

**1.** For most species, the wood should be one year old or less (current season's growth). Avoid including older growth although with certain species, such as the fig or olive, two-year-old wood is satisfactory, or even preferable, if it is of the proper size.

**2.** Healthy, well-developed vegetative buds should be present. Avoid wood with flower buds. Usually, vegetative buds are narrow and pointed, whereas flower buds are round and plump (see Figure 13-1).

**3.** The best type of scion material is vigorous (but not overly succulent), well-matured, hardened shoots from the upper part of the tree, which have made 60 to 90 cm (2 to 3 ft) of growth the previous summer. Such growth develops on relatively young, well-grown, vigorous plants; high production of scion material can be promoted by pruning the plant back heavily the previous winter. Water sprouts from older trees sometimes make satisfactory scionwood, but suckers arising from the base of grafted trees should not be used since they may consist of rootstock material. A satisfactory size is from 0.6 to 1.2 cm (¼ to ½ in.) in diameter.

**4.** The best scions are obtained from the center portion or from the basal two-thirds of the shoots. The terminal sections, which are likely to be too succulent, pithy, and low in stored carbohydrates, should be discarded. Well matured wood with short internodes should be selected.

## Source of Material

Scionwood should be taken from source plants of the correct cultivar known to be pathogen-tested and genetically true-to-type (see chapter 8). Virus diseased, undesirable sports, chimeras, and viruslike genetic disorders must be avoided. Source plants may be of two basic types:

**1.** Plants in an orchard, vineyard, or landscape, growing under conditions where the flowering, fruiting, and growth habits can be observed. For fruit-bearing species, it is best to take propagation material from bearing plants where production history is known. Visual inspection, however, may not reveal the true condition of the proposed source plant and, to be sure, appropriate indexing and progeny tests would be required (see chapter 8).

**2.** In commercial nurseries special scion blocks, where plants are grown particularly for propagation may be maintained. Such plants are handled differently than they would be for producing a crop. For example, fruit trees may be pruned back each year to produce a large annual supply of long, vigorous shoots well-suited for scionwood. Such special blocks would usually be handled to conform to registration and certification programs and would be subject to isolation, indexing, and inspection requirements. In addition, it is important to maintain source identity of scion material through the entire propagation sequence, so that over a period of time proper sources of the various cultivars can be identified and maintained.

## Collection and Handling

For deciduous plants to be grafted in early spring, the scionwood can be collected almost any time during the winter season when the plants are fully dormant (5). In climates with severe winters, the wood should not be gathered when it is frozen. Any wood that shows freezing injury should not be used. Where considerable winter injury is likely, it is best to collect the scionwood and put it in cold storage after leaf fall but before the onset of winter.

## Storage

Scionwood collected prior to grafting must be properly stored. It should be kept slightly moist and at a low enough temperature to prevent development of the buds. A common method is to wrap the wood, in bundles of 25 to 100 sticks, in heavy, waterproof paper or in polyethylene sheets or bags. A small amount of some clean, slightly moist material, such as sawdust, wood shavings, or peat moss, should be sprinkled through the bundle. Sand should not be used, because it will adhere to the scionwood and dull the edge of the knife during the grafting operation. If the packing material is wet, various fungi may develop and damage the buds, even at low storage temperatures. The packing should be barely moist; scionwood is more apt to be damaged by being too wet than by being too dry. If it is to be stored for a prolonged period, the bundle should be examined every few weeks to see that the wood is not becoming either dried out or too wet. When the buds show signs of swelling, the wood should either be used for grafting immediately or moved to a lower storage temperature.

Polyethylene plastic sheeting is a good material for wrapping scionwood. It allows the passage of oxygen and carbon dioxide, which are exchanged during the respiration process of the stored wood, but retards the passage of water vapor. Therefore, if this type of wrapping is sealed, no moist packing material is needed; the natural moisture in the wood is sufficient since little will be lost. Polyethylene bags are useful for storing small quantities of scionwood. All bundles should be labeled accurately.

The temperature at which the wood is stored is important. If it is to be kept only two or three weeks before grafting, the temperature of the home refrigerator—about 5°C (40°F)—is satisfactory. If stored for a period of one to three months, scionwood should be held at about 0°C (32°F) (5) to keep the buds dormant. However, buds of some species, such as the almond and sweet cherry, will start growth after about three months even at this temperature. Do not store scionwood in a home freezer because the very low temperatures—about −18°C (0°F)—may injure the buds.

If cold storage facilities are not available, the scionwood can be kept for a period of time in cold winter regions by burying the bundles in the ground, below frost level, on the north side of a building or tall hedge. Drainage should be provided so that excess water does not remain around the scions and cause the buds to deteriorate.

Storage of scions should not be attempted if succulent, herbaceous plants are being grafted; such scions should be obtained at the time of grafting and used immediately. Certain broad-leaved evergreen species, such as avocados, olives, and citrus, can be grafted in the spring before much active growth starts without previous collection and storage of the scionwood. It is taken directly from the tree as needed, using the basal part of the shoots containing dormant, axillary buds. The leaves are removed at the time of collection.

*Attempting to use scionwood in which the buds are starting active growth is almost certain to result in failure.* In such cases, the buds quickly leaf out before the graft union has healed; consequently the leaves withdraw water from the scions by transpiration, and cause the scions to die.

In topworking pecans (*3*) it was found that good results could be obtained by using precut scions; that is, scions were cut in advance by skilled persons at a convenient time, then held in cold storage in polyethylene bags for periods up to nine days before inserting in the graft unions. Grafting success was reduced only slightly by the use of precut scions.

## GRAFTING CLASSIFIED BY PLACEMENT

Grafting may be classified according to the part of the plant on which the scion is placed—a root, the crown (the junction of the stem and root at the ground level), or various places in the top of the plant.

### Root Grafting

In root grafting the rootstock seedling, rooted cutting, or layered plant is dug up, and the roots are used as the stock for the graft. The entire root system may be used (**whole-root graft**—Figures 12–22 and 12–23), or the roots may be

**Figure 12–22** Bundles of one-year-old apple seedlings of a size suitable for whole root bench grafting.

**Figure 12–23** Whole root apple grafts made by the whip or tongue method and wrapped with adhesive nursery tape.

cut up into small pieces and each piece used as a stock (**piece-root graft**). Both methods give satisfactory results. As the roots used are relatively small (0.6 to 1.3 cm [¼ to ½ in.] in diameter), the whip graft is generally used. Root grafting is usually performed indoors during the late winter or early spring. The scion-wood collected previously is held in storage, while the rootstock plants are also dug in the late fall and stored under cool (1.5° to 4.5°C; 35° to 40°F) and moist conditions until the grafting is done.

The term **bench grafting** is sometimes given to this process, because it is often performed at benches by skilled grafters as a large-scale operation. A number of plants are propagated commercially by root grafting—apples, pears, grapes, and such ornamentals as the wisteria and rhododendron.

In making root grafts, the root pieces should be 7.5 to 15.0 cm (3 to 6 in.) long and the scions about the same length, containing two to four buds. After the grafts are made and properly tied, they are bundled together in groups of 50 to 100 and stored for callusing in damp sand, peat moss, or other packing material. They may be placed in a cool cellar or under refrigeration at approximately 7°C (45°F) for about two months. The callusing period for apples can be shortened to around 30 days if the grafts are stored at a temperature of about 21°C (70°F) and at a high humidity. To use this higher callusing temperature the material should be collected in the fall and the grafts made before any cold weather has overcome the rest period of the scion buds. After the unions are well healed, the grafts must be stored at cool temperatures—2° to 4°C (35° to 40°F)—to overcome the ''rest period'' of the buds and to hold them dormant until planting (*19*). The grafts are lined-out in early spring in the nursery row directly from the low temperature storage conditions. For general callusing purposes, temperatures from 7°C (45°F) to 21°C (70°F) are the most satisfactory. By the proper regulation of temperature, callusing processes may be accelerated by increased temperature or retarded by decreased temperature so that, within reasonable limits, a desired degree of callus formation may be had within a given length of time.

With some plants the graft union should be kept warm, 24° to 27°C (75° to 82°F), but the roots and the buds on the scion should be kept cool, about 7°C (45°F) to prevent premature growth before the graft union has callused and healed together. An ingenious system (Figure 12–24) has been developed for whip grafting filberts, which are notoriously difficult to graft, that overcomes this problem. The graft union itself is kept warm by a plastic pipe to which electric heating cables are taped or which contains hot water. The graft union is laid against the pipe, but the scion and roots protrude into areas having lower temperatures. This ''hot callusing'' system, when used outdoors in late winter or early spring, has increased the grafting ''takes'' in filberts from 10 percent to more than 90 percent. It has also been used successfully in root grafting apples, pears, peaches, and plums (*25*).

Some aeration of the callusing grafts is required, so that airtight containers should not be used (*31*).

As soon as the ground can be prepared in the spring, the grafts are lined out in the nursery row 10 to 15 cm (4 to 6 in.) apart. They should be planted before growth of the buds or roots begins. If growth starts before the grafts can be planted, they should be moved to lower temperatures (−1° to 2°C; 30° to 35°F). The grafts are usually planted deep enough so that the graft union is just below the ground level, but if the roots are to arise only from the rootstock, the graft should be planted with the union well above the soil level. It is very impor-

**Figure 12-24** ''Hot grafting'' system for root grafting difficult plants. The graft union is placed in a slot in a large plastic pipe. Inside the large pipe is a smaller pipe through which thermostatically-controlled hot water flows or to which electric heating cables are taped. Insulating material laid over this pipe retains the heat. The roots and scions protrude into areas having lower temperatures, which retards their development. Courtesy H. B. Lagerstedt and USDA (*25*).

tant to prevent **scion rooting** where certain definite influences, such as dwarfing or disease resistance, are expected from the rootstock.

After one summer's growth, the grafts should be large enough to transplant to their permanent location. If not, the scion may be cut back to one or two buds, or headed-back somewhat to force out scaffold branches, and then allowed to grow a second year. With the older root system a strong, vigorous top is obtained the second year.

NURSE-ROOT GRAFTING

Under certain conditions, it is desired to have a stem cutting of a difficult-to-root species on its own roots. One way this can be done is by making a root graft, using the plant to be grown on its own roots as the scion and a root of a compatible species as the stock. The scion may be made longer than usual and the graft planted deeply with the major portion of the scion below ground. **Scion rooting** can be promoted by rubbing a rooting stimulant, such as indolebutyric acid, into several vertical cuts made through the bark at the base of the scion, just above the graft union. This is done just before planting, and the grafts are set deeply so that most of the scion is covered with soil (*23, 24*). After one or two seasons of growth many of the scions have roots. The temporary nurse rootstock is then cut off and the top reduced in proportion to the root system. The rooted scion is replanted to grow on its own roots.

Scion rooting is often better if the nurse root is buried deeply enough so that a new shoot of the current season's wood, developing from a bud on the scion, will be rooted, rather than trying to root the older, lignified tissue of the original scion. This is essentially a form of mound layering. As the new scion shoot grows, soil is gradually mounded up around it to a height of 13 to 15 cm (5 to 6 in.), although the terminal leaves are at no time covered (*26*).

Several methods of handling eliminate the necessity of digging up the graft and cutting off the rootstock. (a) The rootstock piece will eventually die if it is

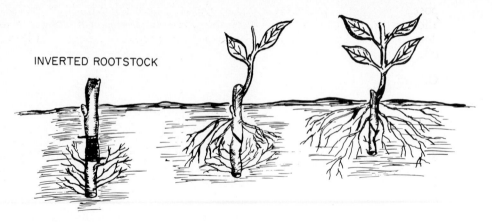

INVERTED ROOTSTOCK

**Figure 12-25** The ''nurse-root'' graft is a temporary graft used to induce the scion to develop its own roots. The nurse root sustains the plant until the scion roots form, then it dies. An easy method is shown here for preparing a nurse-root graft by inverting the rootstock piece.

grafted onto the scion in an inverted position (Figure 12–25) (*26*). The union heals and the inverted stock piece sustains the scion until scion roots are formed, but the stock fails to receive food from the scion and eventually dies, thus leaving the scion on its own roots. (b) In a second method an incompatible rootstock is used. Hence, if the graft is planted deeply, scion roots will gradually become more important in sustaining the plant, and the incompatible rootstock will finally cease to function. An example of this is apple scion on pear rootstock. (c) In another method the base of the scion, just above the graft union, is bound with some type of wrapping material to girdle and cut off the rootstock (Figure 12–26). Excellent results have been obtained with ordinary budding rubber strips (0.016 gauge) (*6*). Budding rubbers disintegrate within a month when ex-

**Figure 12-26** Nurse-root graft. A 'Malling 9' apple scion was root-grafted to an apple seedling nurse root. Just above the graft union the scion was wrapped with a budding rubber strip (see arrow). After two years in the nursery, vigorous scion roots were produced. The budding rubber has effectively constricted development of the seedling nurse root, which can now be broken off and discarded. Courtesy D. S. Brown.

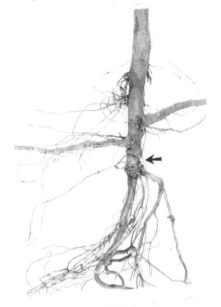

posed to sun and air; when buried in the soil, they will last as long as two years, allowing sufficient time for the scion to become rooted. Yet, owing to the slow deterioration of the rubber below ground, the rootstock is finally girdled and cut off.

## Crown Grafting

A graft union made at the root-stem transition region—the ''crown'' of the plant—on an established rootstock is termed a **crown graft.** Methods commonly used in crown grafting include the whip, side, cleft, saw-kerf, and bark graft, the choice depending upon the species and size of the rootstock.

Crown grafting of deciduous plants is done from late winter to late spring. In each species grafting should take place shortly before new growth starts. The scions should be prepared from well-matured, dormant wood of the previous season's growth.

If the graft is above the soil level, the union must be well wrapped to hold it solid and prevent desiccation. However, when the operation is performed just below, at, or just above the soil level, it is possible to cover the graft union, or even the entire scion, with soil and thus eliminate the necessity for waxing. In all cases the union should be tied securely with string or tape to hold the grafted parts together until healing takes place.

## Double-Working

A double-worked plant has three parts, all different genetically: the rootstock, the interstock, and the scion, or fruiting, top (*13*). (See Figure 12–27.) Such a plant has two graft unions, one between the rootstock and in-

**Figure 12–27**    There are three distinct parts and two graft unions in a double-worked plant, illustrated here by a 'Bartlett' pear grafted on an 'Old Home' pear which in turn is grafted on quince as a rootstock. The 'Old Home' pear, in this case, is the intermediate stock, or interstock.

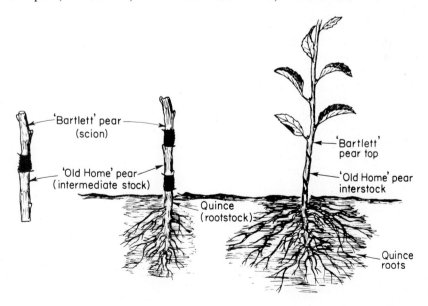

terstock, and one between the interstock and the scion. The interstock may be less than 25 mm (1 in.) in length or extensive enough to include the trunk and secondary scaffold branches of a tree.

Double-working is used for various purposes, such as:

1. Overcoming graft incompatibility between a desired top cultivar and the rootstock
2. Providing a cold- or disease-resistant trunk
3. Obtaining a dwarfing effect from the use of certain intermediate stocks
4. Obtaining the strong trunk or crotch systems characteristic of certain cultivars.

Examples of double-working are (a) the propagation of 'Bartlett' pears on quince as a dwarfing rootstock by using a mutually compatible interstock, such as 'Old Home' or 'Hardy' pear (Figure 12–27), and (b) the propagation of dwarfed apple trees consisting of the scion cultivar grafted onto a 'M 9' interstem piece that is grafted onto a more vigorous rootstock, such as 'MM 106', 'MM 111', or apple seedlings (8).

Several methods are used for developing double-worked nursery trees. The grafting in these techniques can be done by machine grafting devices as illustrated in Figure 12–19 and 12–20, or by use of the whip graft.

1. Rootstock "liners," either seedlings, clonal rooted cuttings, or rooted layers, are set out in the nursery row in early spring. These are then fall-budded with the interstock buds, growth from which, a year later, is fall-budded with the scion cultivar buds. Three years are required to produce a nursery tree by this method.
2. The interstock piece is bench-grafted onto the rooted rootstock—either a seedling or a clonal stock—in late winter. After callusing, the grafts are lined-out in the nursery row in the spring. These are then fall-budded to the scion cultivar. By this method the nursery tree is propagated in two years.
3. A variation of method (2) is to prepare, by bench grafting, two graft unions—the scion grafted to the interstock, and the interstock grafted to the rooted rootstock. After callusing, the completed graft, with two unions, is lined-out in the nursery row. Depending upon growth rate, a nursery tree can be obtained in one or two years.
4. The scion piece is bench-grafted onto the interstem piece in late winter or early spring; then, after callusing, this component is grafted—as shown in Figure 12–27—onto rootstocks that have been grown in place in the nursery row. A nursery tree is obtained in one or two years by this method.
5. Double-shield budding, in which the double-working is done in one operation by budding, is illustrated in Figure 13–18. A nursery tree is produced in one year, or if growth is slow, two years after budding.
6. The interstock shoots still on the plant can be T–budded in late summer with the scion buds inserted about 15 cm (6 in.) apart. These heal into place and during late winter the budded interstock shoots are cut apart with a scion bud at the terminal end of each piece. The budded (shoot) pieces are then whip grafted onto the rooted seedling stocks. After they have callused, the completed graft, consisting of rootstock and interstem piece with scion bud in place, is ready for planting in the nursery row (12).

**Figure 12-28**  An extensive top-working operation. Young apple trees have been topgrafted to a more profitable cultivar. Each tree has a small "nurse" branch.

## Topworking (Topgrafting or Topbudding)

Topworking is used primarily to change the cultivar of an established plant—tree, shrub, or vine—either by grafting (Figures 12-28 and 12-29) using one of the methods described earlier in this chapter, or by budding (see chapter 13). This procedure may be preferred to removal and replacement of the entire plant, since return to flowering and fruiting is faster with topworking than with a new nursery plant, particularly if the top-worked plant is young, healthy, and well cared for. Plants that are old, diseased, or of a short-lived species are not satisfactory candidates for topworking.

**Figure 12-29**  Proper method of topworking trees. *Left:* scions, which have been inserted into fairly small branches, are starting to grow. Tree has been whitewashed to prevent sunburn injury. *Right:* same tree about six weeks later. Stakes have been nailed to branches; the shoots developing from scions are tied to these stakes to prevent their breaking off in winds.

Sometimes virus diseases may be introduced into the new topworked plant, from either the stock or the scion. Because such diseases interfere with growth, flowering, and fruiting, one should know the virus status of both components before considering topworking.

PREPARATION FOR TOPWORKING

Topgrafting is usuallly done in the spring, shortly before new growth starts. The exact time depends upon the method to be used. The cleft, side, whip, and notch grafts can be done before the bark is slipping, but the bark graft must be done when the bark is slipping, preferably just as the buds of the stock tree are starting to grow. Topbudding can be done with the T, patch, or chip bud methods.

It is usually advisable to obtain an ample amount of good-quality scionwood prior to grafting and store it under the proper conditions, although for broad-leaved evergreens, such as avocado or citrus, scionwood can be collected at the time of the grafting operation.

In preparing for topworking, one must decide, for each individual stock tree, how many scaffold branches should be used (usually three to five). If many grafts are made high in the tree in the small, secondary scaffold branches (frameworking), the tree will return to bearing earlier than it will if fewer and larger limbs are grafted lower in the tree. Frameworking, however, is expensive to do and necessitates pruning out all new shoots from the stock below the grafts as they develop.

The branches to be grafted should be well distributed around the tree and up and down the main trunk, avoiding branches with weak, narrow crotches. All others can be removed unless one or more nurse branches are used. Figure 12–30 shows a worker preparing the branches for grafting.

Topworking is an extremely severe pruning operation for the tree and results in a considerable inbalance between the root system and the top. However, deciduous trees soon recover and the new growth from the scions and from latent buds on the stock restores a balance without damage, providing the tree is healthy and vigorous and the grafting is done when the tree is dormant or shortly after growth starts in the spring. However, under some situations, not always well understood, the tree will be adversely affected by the heavy cutting back, and will show intense leaf burning and may even fail to survive. This situation is most likely to occur when the grafting is delayed until after new spring growth is well underway, where all foliage and new shoots are stripped off the stock plants, or where high temperatures prevail shortly after grafting.

Also when all branches of the tree are cut back and grafted, considerable energy goes into the scion growth, producing vigorous, succulent shoots that may break out if not properly pruned or supported. In addition, such rank, succulent growth is very susceptible to winter freezing damage in cold climates.

To avoid the problems cited above it is advisable to retain some of the foliage on the stock plant as well as small lower branches and even large limbs to be held temporarily as **nurse branches.** Such nurse limbs should be pruned back fairly severely, otherwise the grafted scions may not make adequate growth.

It is essential that nurse limbs be left when topworking broad-leaved evergreen trees, such as citrus or olive. If they are retained on the south and west parts of the tree they will shade the grafted portions and reduce the chances of sunburn injury to the exposed branches.

**Figure 12–30** Sawing off a branch in preparation for top-grafting so that no bark-tearing occurs. *Top left:* the first cut is made in a smooth area starting on the under side of the branch and continuing about one-third of the distance through. *Top right:* the second cut is made starting from the upper part of the branch and cutting downward. It should be back 2.5 or 5 cm (1 or 2 in.) on the branch from the first cut. *Below left:* after the branch breaks off and falls, the second cut is continued. *Below right:* final smooth cut, ready for grafting.

A practice often recommended, especially for older trees, is to topwork them during a two-year period, grafting perhaps two main branches on the northeast side of the tree the first year and retaining two branches on the southwest as nurse branches. The second year these branches are topworked.

Topworking is most successful when done on relatively young trees where the branches to be grafted are no larger than 7.5 to 10 cm (3 to 4 in.) in diameter and are relatively close to the ground. When attempting to topwork large, old trees, it is often necessary to go high up in the trees to find branches with a diameter as small as 10 cm. If the grafting is done on such branches, the new top is inconveniently high for the various orchard operations, such as thinning and

**Figure 12–31**    An improper method of top-working trees. This walnut tree was cut off close to the ground, and a number of scions were inserted around its circumference by the bark-graft method. Two of the scions grew successfully, but the remainder failed. Most of the original tree is dead; healing of such a large cut is almost impossible. Better methods of top-grafting trees are shown in Figures 12–28 and 12–29.

harvesting. The other alternative is cutting off the branches or main trunk close to the ground and inserting the scions into wood 30 cm (1 ft) or more in diameter. Although many scions can be inserted around the tree between the bark and the wood by the bark-graft method and may grow well for several years, it is quite likely, as shown in Figure 12–31, that wood rot will develop in the center of the stub before the growth of the scions can heal it over. Also, some scions are not mechanically held in place securely and may be blown out by strong winds after they reach considerable size.

It is important that the branch to be topworked be cut off in such a location that the region just below the cut is smooth and free from knots or small branches, so that there will be a satisfactory place for inserting the scions. The branches are best cut off about 23 to 30 cm (9 to 12 in.) from the main trunk to keep the tree headed low. The branch should not be cut off more than a few hours before the grafting is to be done.

#### In preparation for topworking (topgrafting)

**1.** Do the work in the spring when the trees are dormant or shortly after growth starts.

**2.** Select for grafting three to five well-placed scaffold branches no larger than about 10 cm (4 in.) in diameter and conveniently close to the ground.

**3.** Retain nurse branches for broad-leaved evergreen trees and for deciduous trees where the winters are severe.

**4.** Cut off the branches properly so that the bark is not torn down the trunk.

Grafting on a cool, overcast day with no wind blowing offers the most protection from drying of the cut surfaces of the scion and stock until they can be covered with grafting wax. Grafting on hot, sunny, and windy days should be avoided. During grafting, the scion wood must not dry out by being exposed to the sun. It should be kept moist and cool in some container or be wrapped in moist burlap.

Immediately after each stub is grafted it should be thoroughly covered with grafting wax. The need for prompt and thorough coverage of all cuts, including the tip end of the scions, cannot be stressed too strongly. The wax should be worked into the bark of the stock, sealing all small cuts or cracks where air could penetrate in and around the cut surfaces where healing tissue is expected to develop. Waxes containing beeswax will attract bees, which may remove the wax, necessitating rewaxing.

SUBSEQUENT CARE OF TOPWORKED TREES

After the actual top-grafting (or topbudding) operation is finished, much important work needs to be done before the topworking is successfully completed. A good grafting job can be ruined by improper care of the grafted trees.

If the grafting has been done in late spring when growth is active, trees of some species, such as the walnut, will ''bleed'' to a considerable extent from the grafted stub, even though it has been covered with grafting wax. This flow of sap around the scions can be so heavy as to interfere with the normal healing processes at the graft union. If this condition appears, it can often be corrected by making several slanting cuts around the base of the tree with a knife or saw through the bark, into the water conducting tissues, in the trunk of the tree several feet below the grafted stubs. The bleeding will then take place at these cuts rather than around the graft union. This extensive sap flow is not particularly harmful to the tree and will usually stop within a few days. Boring a series of random holes into the trunk around the tree will accomplish the same purpose.

If certain bacterial diseases, such as crown gall, are present, this practice may spread the bacteria. In such cases, dipping the knife, saw, or bit into a disinfectant, such as 10 percent Clorox, is advisable.

In three to five days after grafting, the trees should be carefully inspected and the graft unions rewaxed if cracks or holes appear in the wax.

It is essential to prevent sunburn on the trunk and large branches exposed to the sun by the removal of the protecting top foliage. Protection is especially important if the grafting has been done late in the season when hot weather can be expected, and when no protecting nurse branches have been retained. The energy from the sun absorbed by the dark-colored bark can raise the temperature of the living cells below the bark to a lethal level.

It is advisable to whitewash the trunk, branches, and scions of the grafted trees. Various cold-water paints, some made especially for this purpose, are available. The white color reflects a considerable portion of the sun's radiant energy, thus keeping the temperature of the living tissues within safe limits. Interior, water-base white house paints (both latex and acrylic) mixed with water, 1:1, prevent sunburn for one season. Exterior paints give longer protection but are more likely to cause injury to the tree (*15, 27*).

Another help in preventing sunburn is to retain some of the watersprouts which soon start growth along the trunk and branches of the grafted tree. However, they must be kept under control or they will quickly shade out the developing scions. Rather than removing the watersprouts completely, they can be headed back to 20 to 25 cm (8 to 10 in.) in length, and they will shade the bark underneath. An additional benefit is that the food manufactured by this leaf area will help sustain the tree until the new scions develop sufficient foliage to do so.

Trees just grafted should be amply supplied with water so that the tissues are in a high state of turgidity. This is necessary in order to obtain good callus production, which is essential for healing of the graft union.

During the first summer after topworking, new shoots arising from below the grafted branches must be kept pruned back so as not to interfere with the growth of the scions. Nitrogen fertilizers should be withheld for a year or two after grafting because no stimulation of growth is usually needed. The water requirement of the trees will be less, owing to the removal of a considerable amount of leaf area when the tops were cut back for grafting.

Two to four scions are generally inserted in each stub. If all the scions grow, they should all be retained during the first year, because they will help heal over the stub. However, just one branch, from the best placed and strongest growing scion, should be retained permanently. Growth from the remaining scions should be retarded by rather severe pruning, keeping them alive to help heal the branch, but allowing the permanent scion to become the dominant one. To keep two or more scions for permanent branches at one point will undoubtedly result in a weak crotch, which will eventually break. The first year, then, the best practice is to retain all scions to help heal the stub but, by pruning, to retard the growth of all but the best one. Such a procedure is illustrated in Figure 12–32. If two shoots arise from the scion to be retained permanently, select the best of these and remove the other. After the stub has healed over, in the second or third year, the temporary scions can be removed completely.

If the permanent scion grows rather vigorously, the danger arises of its becoming topheavy and breaking off during winds. This may be handled in two ways—either by retarding the growth of the permanent branch by pruning it back or by nailing a lath or other type of stick onto the tree and tying the new branch to this stick. *When tying a cord around a branch, always make a loop so that there is no chance of the branch being girdled as it grows.*

Should only one scion grow at each stub, the problem of securing adequate healing of the stub on the side opposite the living scion may prove to be serious. Healing may be helped by sawing off the stub at an angle away from the surviving scion. The cut surface of the stub can be covered with grafting wax to retard wood decay. It may take a number of years for the single scion to heal over the stub, and it is possible that wood decay may start before it can do so.

When none of the scions grow in a stub, there are still some possibilities for getting it topworked. One is by allowing several well-placed watersprouts arising just below the cut surface to grow, then top-budding them during the summer. Or the watersprouts may be allowed to grow so as to keep the branch alive and healthy, then the grafting operation repeated the following year, making a fresh cut a foot or so below the original cut.

If nurse branches have been used, they should be cut back and away from the scions at intervals throughout the summer so that scion shoots are always fully exposed to the sun and have sufficient space to grow. Nurse branches

**Figure 12-32**    Problems encountered in handling limbs that have been top-grafted. *Left:* where both scions "take," one is cut back heavily to allow the other to become predominant so as not to form a weak crotch. It is retained for two or three years, however, to assist in healing the grafting wound, but will eventually be removed. *Right:* where only one scion in a fairly large branch "takes," difficulty may occur in rapid healing of the wound. The wood under the dead scion should be cut off at an angle, then waxed; this will give a minimum cut surface for healing.

should be removed entirely or grafted also by the beginning of the second or third year.

When the top of the tree has been finally worked over to the new cultivar, it will grow vigorously for a few years. Good pruning practices are needed to prevent badly placed branches from developing. Usually by the third year the tree will again be in production.

## HERBACEOUS GRAFTING

Grafting herbaceous types of plants is used for various purposes, such as studying virus transmission, stock-scion physiology, and grafting compatibility, as well as for the commercial greenhouse production of certain cucurbitaceous crops, particularly in Europe. Usually such grafts are made while the plants are quite small, the stock being grafted shortly after seed germination. Such material is generally very soft, succulent, and susceptible to injury. In one technique (see Figure 12-33), a simple splice graft is used with a diagonal cut made through the seedling stock, just above the cotyledons (*18*). A piece of thin-walled polyethylene tubing of the proper size to give a snug fit is slipped over the cut end of the stock. The basal end of the scion receives a diagonal cut similar in length and angle to that given the stock. The scion is then slipped down into the plastic tubing so that the two cut surfaces make intimate contact. The tubing holds the graft in place until healing occurs—about 12 days after grafting. Then the tubing may be slipped off over the scion if there are no leaves and if the bud has not expanded; otherwise, it may be cut off with a razor blade.

**Figure 12-33** Grafting young, herbaceous plant material together by using a splice graft. A single, slanting cut is made in the upper part of the stock. A piece of transparent plastic tubing is slipped over this to give a snug fit. A similar slanting cut is made in the lower portion of the scion, which is then slipped into the plastic tubing so that the cut surfaces match. After the graft has healed, the tubing is cut away with a razor blade.

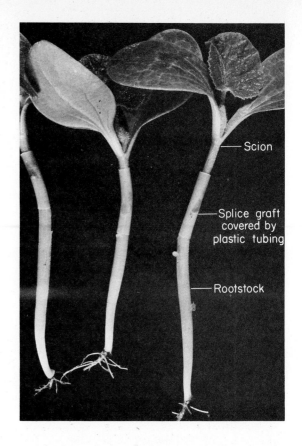

In another procedure (9) used in grafting older herbaceous stock plants, several leaves are retained below the graft union. The cleft graft is used (but with only one scion), and the graft union is bound with raffia, budding rubbers, or adhesive latex tape. To prevent drying out, the entire plant—following grafting—is covered with a supported polyethylene bag. The grafted plant is then set in the shade until the graft has healed; then the plastic cover can be removed.

A method of establishing *Vitis vinifera* grapes on phylloxera or nematode-resistant rootstocks is by ''greenwood'' grafting, in which the grafting is done in the spring, placing the scion—taken from new growth—on a new green shoot developed from the rootstock plant. A simple splice graft is made with the sloping cuts 2.5 to 4 cm (1 to 1½ in.) long. The stock and scion pieces must be the same diameter. The scion has only one bud. The cuts are matched as closely as possible, and the graft union is completely covered by wrapping with a budding rubber. The graft union should be healed and the bud growing by two weeks after grafting. Since these grafts are quite fragile, they must be tied to a stake ( 7, 16).

## NURSE–SEED GRAFTING

Germination of some large-seeded woody species, such as the chestnut, is hypogeal (that is, the cotyledons remain below ground in the seed coats with the

**Figure 12-34**    Rooted chestnut grafts 19 days after grafting by the nurse-seed grafting method (*21*). Courtesy Connecticut Agricultural Experiment Station.

shoot tip appearing above ground). In a grafting procedure using certain species with seeds of this type, when the seeds have just germinated, the petioles of the cotyledons are cut off transversely just at the seed, leaving the cotyledons inside. A knife point is inserted into the seed between the cut petioles, making an opening for the scion. The scion, which is prepared from dormant wood of the previous season's growth, is cut to a wedge shape at the base, as for a cleft graft. The scion is inserted into the cut between the cotyledons so that the exposed cambium surfaces of the scion are in close contact with the cut surfaces of the cotyledons. The ''seed grafts'' are then lined-out in the rooting medium with the union about 4 cm (1½ in.) below the surface. A graft union takes place, and roots arise from the cut cotyledon petioles. This type of grafting is done in early spring, using properly stratified seed and dormant scionwood (*21, 22, 28, 30*). Chestnuts (see Figure 12-34), avocados, and camellias have been grafted successfully by this method; in fact, grafted camellia plants of the desired cultivar can be obtained in the same length of time it would take for the seedlings to become large enough to graft by the conventional cleft graft method.

In a modification of the nurse-seed graft (*28*) which has worked well in propagating chestnuts, pecans, walnuts, and oaks, the actual graft union is made into the seedling hypocotyl at the point of attachment of the cotyledons (see Figure 12-35). The scion is cut to a thin wedge and inserted into a split made in the hypocotyl. After insertion of the scion, the graft is wrapped firmly with rubber strips or waxed string.

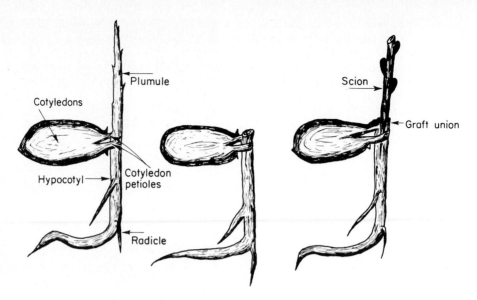

**Figure 12–35**    Modified nurse-seed graft. *Left:* seedling plant, ready for grafting. *Center:* hypocotyl cut open longitudinally, ready for insertion of the scion. *Right:* scion inserted; the graft is now ready for tying (*28*). Redrawn from Moore.

## CUTTING–GRAFTS

In the cutting-graft a leafy scion is grafted onto a leafy, unrooted stem piece (which is to become the rootstock), and the combination is then placed in a rooting medium under intermittent mist for simultaneous healing of the graft union and rooting of the stock. Leaves must be retained on the rootstock piece in order for it to root. This procedure was utilized many years ago in studying stock-scion physiology in citrus (*14*). More recently it has been used in commercial propagation of various types of citrus on clonal dwarfing rootstocks (*10*). It is also of value in propagating certain difficult-to-root conifers (*33*) and rhododendrons (*11*).

For citrus a simple splice graft is used. The slope of the cut is at a 30-degree angle 1⅓ to 2 cm (½ to ¾ in.) long; the union is tied with a rubber band. The base of the stock is dipped into a root-promoting material, such as indolebutyric acid, and then the grafts are placed under mist, or in a closed case, in flats of the rooting medium over bottom heat. After healing of the union and rooting of the stock, the grafts are allowed to harden by discontinuing the mist and bottom heat for about two weeks. Then the grafts are ready for planting in 4-liter (1–gal.) cans or other containers.

## MICRO–GRAFTING

Grafting of tiny plant parts can be done aseptically using the techniques described in chapters 16 and 17, in which the small grafts are grown in closed containers until they are large enough to be transferred to open conditions.

Micro-grafting has been used mostly with citrus, apple, and some *Prunus* species, particularly in developing virus-free plants, where a virus-free shoot tip can be obtained but cannot be rooted. The shoot tip is grafted aseptically onto a virus-free seedling thus providing a complete virus-free plant from which other "clean" plants can then be propagated (*20, 29*). The procedures of micro-grafting are described in chapter 16.

The non-aseptic propagation of very small mini nursery trees by grafting tiny seedlings with match-like scions, then growing the minute grafts long enough to have a viable plant, is a promising procedure (*36*). This is useful particularly in the tropics, where the nursery plants sometimes must be shipped long distances, often by air, into inaccessible regions. Quantities of such tiny plants can be transported much more readily than full-sized nursery trees.

# REFERENCES

1. Alley, C. J. 1957. Mechanized grape grafting. *Calif. Agr.* 11(6):3, 12.

2. ———. 1970. Can grafting be mechanized? *Proc. Inter. Plant Prop. Soc.* 20:244–48.

3. Anonymous. 1968. Pre-cut scions. *Agr. Res.* 17(6):11

4. Beineke, W. F. 1978. Parafilm: A new way to wrap grafts. *HortScience* 13(3):284.

5. Bhar, D. S., R. J. Hilton, and G. C. Ashton. 1966. Effect of time of cutting and storage treatment on growth and vigor of scions of *Malus pumila* cv. McIntosh. *Can. Jour. Plant Sci.* 46:69–72.

6. Brase, K. D. 1951. The nurse-root graft, an aid in rootstock research. *Farm Res.* 17(1):16.

7. Carlson, V. 1963. How to green graft grapes. *Calif. Agr. Ext. Publ. AXT. 115.*

8. Cummins, J. N. 1973. Systems for producing multiple-stock fruit trees in the nursery. *Plant Prop.* 19(4):7–9.

9. Denna, D. W. 1962. A simple grafting technique for cucurbits. *Proc. Amer. Soc. Hort. Sci.* 81:369–70.

10. Dillon, D. 1967. Simultaneous grafting and rooting of citrus under mist. *Proc. Inter. Plant Prop. Soc.* 17:114–17.

11. Eichelser, J. 1967. Simultaneous grafting and rooting techniques as applied to rhododendrons. *Proc. Inter. Plant Prop. Soc.* 17:112.

12. Fisher, E. 1977. The pre-budded interstem: A new technique. *Fruit Var. Jour.* 31(1):14–15.

13. Garner, R. J. 1940. Studies in nursery technique: The production of double worked pear trees. *Ann. Rpt. E. Malling Res. Sta. for 1939,* pp. 84–86.

14. Halma, F. F., and E. R. Eggers. 1936. Propagating citrus by twig-grafting. *Proc. Amer. Soc. Hort. Sci.* 34:289–90.

15. Hansen, C. J., and H. T. Hartmann. 1951. Influence of various treatments given to walnut grafts on the percentage of scions growing. *Proc. Amer. Soc. Hort. Sci.* 57:193–97.

16. Harmon, F. N., and E. Snyder. 1948. Some factors affecting the success of greenwood grafting of grapes. *Proc. Amer. Soc. Hort. Sci.* 52:294–98.

17. Hearman, J., A. B. Beakbane, R. G. Hatton, and W. A. Roach. 1936. The reinvigoration of apple trees by the inarching of vigorous rootstocks. *Jour. Pom. and Hort. Sci.* 14:376–90.

18. Holt, J. 1958. A simple way of grafting herbaceous plants. *Gard. Chron.* 143:332.

19. Howard, G. S., and A. C. Hildreth. 1963. Introduction of callus tissue on apple grafts prior to field planting and its growth effects. *Proc. Amer. Soc. Hort. Sci.* 82:11-15.

20. Huang, S., and D. F. Millikan. 1980. *In vitro* micrografting of apple shoot tips. *HortScience* 15(6):741-43.

21. Jaynes, R. A. 1965. Nurse seed grafts of chestnut species and hybrids. *Proc. Amer. Soc. Hort. Sci.* 86:178-82.

22. Jaynes, R. A., and G. A. Messner. 1967. Four years of nut grafting chestnut. *Proc. Inter. Plant Prop. Soc.* 17:305-11.

23. Jones, F. D. 1950. Hormone on root graft. *Amer. Nurs.* 72(11):6-7.

24. Kerr, W. L. 1936. A simple method of obtaining fruit trees on their own roots. *Proc. Amer. Soc. Hort. Sci.* 33:355-57.

25. Lagerstedt, H. B. 1981. The hot callusing pipe, a grafting aid. *Ann. Rpt. Northern Nut Growers Assn.* 72:27-33.

26. Lincoln, F. B. 1938. Layering of root grafts—a ready method for obtaining self-rooted apple trees. *Proc. Amer. Soc. Hort. Sci.* 35:419-22.

27. Micke, W. C., J. A. Beutel, and J. A. Yeager. 1966. Water base paints for sunburn protection of young fruit trees. *Calif. Agr.* 20(7):7.

28. Moore, J. C. 1963. Propagation of chestnuts and camellias by nurse seed grafts. *Proc. Inter. Plant Prop. Soc.* 13:141-43.

29. Navarro, L., C. N. Roistacher, and T. Murashige. 1975. Improvement of shoot tip grafting *in vitro* for virus-free citrus. *Jour. Amer. Soc. Hort. Sci.* 100:471-79.

30. Park, Kyo S. 1968. Studies on juvenile tissue grafting of some special use trees, II. *Korean Jour. Bot.* 11(3): 88-97.

31. Shippy, W. B. 1930. Influence of environment on the callusing of apple cuttings and grafts. *Amer. Jour. Bot.* 17:290-327.

32. Sitton, B. G., and E. P. Akin. 1940. Grafting wax melter. *USDA Leaflet 202.*

33. Teuscher, H. 1962. Speeding production of hard-to-root conifers. *Amer. Nurs.* 116(7):16.

34. Thompson, L.A., and C. O. Hesse. 1950. Some factors which may affect the choice of grafting compounds for top-working trees. *Proc. Amer. Soc. Hort. Sci.* 56:213-16.

35. Upshall, W. H. 1946. The stub graft as a supplement to budding in nursery practice. *Proc. Amer. Soc. Hort. Sci.* 47:187-89.

36. Verhey, E. W. M. 1982. Minute nursery trees, a breakthrough for the tropics? *Chronica Hort.* 22(1):1-2.

## SUPPLEMENTARY READING

BALTET, C. 1910. *The art of grafting and budding* (6th ed.). London: Crosby, Lockwood.

BANTA, E. S. 1967. *Fruit tree propagation.* Ohio Agr. Ext. Bul. 481.

CHANDLER, W. H. 1957. Chapter 13 in *Deciduous orchards* (3rd. ed). Philadephia: Lea & Febiger.

———. 1958. *Evergreen orchards* (2nd ed.). Philadelphia: Lea & Febiger.

GARNER, R. J. 1979. *The grafter's handbook* (4th ed.). New York: Oxford Univ. Press.

HARTMANN, H. T., and J. A. BEUTEL. 1979. *Propagation of temperate zone fruit plants.* Calif. Agr. Exp. Sta. Leaflet 21103.

INTERNATIONAL PLANT PROPAGATORS' SOCIETY. Proceedings of annual meetings.

MINISTRY OF AGRICULTURE, FISHERIES, AND FOOD. 1965. *Grafting fruit trees.* London: Advisory Leaflet 326.

SNYDER, J. C., and R. D. BARTRAM. 1965. *Grafting fruit trees.* Pacific Northwest Ext. Publ. (Idaho, Oregon, and Washington).

SPANGELO, L. P. S., R. WATKINS, and E. J. DAVIES. 1968. *Fruit tree propagation.* Ottawa: Can. Dept. Agr. Publ. 1289.

WAY, R. D., F. G. DENNIS, and R. M. GILMER. 1967. *Propagating fruit trees in New York.* N.Y. Agr. Exp. Sta. Bul. 817.

In contrast to grafting, in which the scion consists of a short detached piece of stem tissue with several buds, budding utilizes only one bud and a small section of bark, with or without wood. Budding is often termed ''bud grafting,'' since the physiological processes involved are the same as in grafting.

The commonly used budding methods depend upon the bark's ''slipping.'' This term indicates the condition in which the bark can be easily separated from the wood. It denotes the period of year when the plant is in active growth, when the cambium cells are actively dividing, and newly formed tissues are easily torn as the bark is lifted from the wood. Beginning with new growth in the spring, this period should last until the plant ceases growth in the fall. However, adverse growing conditions, such as lack of water, insect or disease problems, defoliation, or low temperatures, may reduce growth and lead to a tightening of the bark and can seriously interfere with the budding operation. Getting the stock plants in the proper condition for budding is an important consideration. Of the methods described here, only one—the chip bud—can be done when the bark is not slipping.

The budding operation, particularly T-budding, can be performed more rapidly than the simplest method of grafting. Some rose budders insert as many as 2000 to 3000 or more T-buds a day with the tying done by helpers. If performed under the proper conditions, the percentage of successful unions in T-budding is very high—90 to 100 percent. Budding is widely used in producing nursery stock of rose and fruit tree cultivars, where hundreds of thousands of individual plants are propagated each year. Therefore, for propagation operations involving large numbers of plants, where speed and low mortality are essential, budding upon selected rootstocks is the method likely to be chosen.

The use of budding is confined generally to young plants or the smaller branches of large plants where the buds can be inserted into shoots from 6 to 26 mm ($1/4$ to 1 in.) in diameter. Topworking young trees by topbudding is quite successful. Here the buds are inserted in small, vigorously growing branches in the upper portion of the tree.

Budding may result in a stronger union, particularly during the first few years, than is obtained by some of the grafting methods, and thus the shoots are not as likely to blow out in strong winds. Budding makes more economical use of

# 13

# Techniques of Budding

propagating wood than grafting, each bud potentially being capable of producing a new plant. This may be quite important if propagating wood is scarce. In addition, the techniques involved in budding are simple and can be performed easily by the amateur.

## ROOTSTOCKS FOR BUDDING

In propagating nursery stock of the various fruit and ornamental species by budding, a rootstock plant is used. It should have the desired characteristics of vigor, growth habit, and resistance to soil-borne pests, as well as being easily propagated. This rootstock plant may be a rooted cutting, a rooted layer, or, more commonly, a seedling (see chapter 11). Usually, one year's growth in the nursery row before the budding is to be done is sufficient to produce a rootstock plant large enough to be budded, but seedlings of slow-growing species, and those grown under unfavorable conditions, may require two seasons.

To produce nursery trees free of harmful pathogens (such as viruses, fungi, or bacteria) it is essential that the rootstock plant, as well as the budwood, be free of such organisms.

## TIME OF BUDDING— FALL, SPRING, OR JUNE

Most budding methods are used at seasons of the year when the stock plant is in active growth and the cambial cells are actively dividing so that the bark separates readily from the wood (although chip budding can be done when the bark is not slipping). It is also necessary that well-developed buds of the desired cultivar be available at the same time. These conditions exist for most plant species at three different times during the year. In the Northern Hemisphere, these periods are late July to early September (**fall budding**), March and April (**spring budding**), and late May and early June (**June budding**). In the Southern Hemisphere, similar periods would be late January to early March (*fall budding*), September and October (*spring budding*), and late November and early December (*''June'' budding*).

### Fall Budding

Late summer and early fall is the most important time of budding in the propagation of fruit tree nursery stock. Most of the budding is done in late summer. The rootstock plants are usually large enough by late summer to accommodate the bud, and the plants are still actively growing, with the bark slipping easily. Once growth has stopped and the bark adheres tightly to the wood, T- or patch budding can no longer be done.

In fall budding, the budsticks, consisting of the current season's shoots, are obtained at or near the time of budding. They should be vigorous and should contain healthy vegetative, or leaf, buds (Figure 13–1). Do not select shoots whose buds have been broken off. Short, slowly growing shoots on the outer portion of the tree should be avoided, because they may have chiefly flower buds rather than vegetative buds. Flower buds are usually round and plump, whereas leaf buds are smaller and pointed. Some species have mixed buds, the node con-

**Figure 13-1** In budding it is important to use vegetative rather than flower buds. Vegetative buds are usually small and pointed, while flower buds are larger and more plump. Differences between vegetative and flower buds in three fruit species are illustrated here. *Left:* almond. The shoot on the left has primarily flower buds and should not be selected for budding. The shoot on the right has vegetative buds, which are more suitable. *Center:* peach. The shoot on the right has excellent vegetative buds while those on the left shoot are mostly flower buds. *Right:* pear. All the buds on the shoot at the left are flower buds. Buds on the shoot at the right are good vegetative buds, suitable for budding.

taining both vegetative and flower buds. These are satisfactory for use in budding.

Every effort should be made to make sure that the trees from which the budsticks are obtained are free of any bacterial, fungus, or virus diseases. *Using infected budsticks can infect every budded nursery tree with the disease.*

As the budsticks are selected, the leaves should be removed immediately, leaving only a short piece of the leaf stalk or petiole attached to the bud; this will aid in handling the bud later on. The budsticks should be kept from drying by wrapping in some material such as clean, moist burlap and keeping them in a cool, shady location until they are needed. The budsticks should be used promptly after cutting, although they can be stored for a short time if kept cool and moist. It is best, if possible, when a considerable amount of budding is being done, to collect the budsticks as they are being used, a day's supply at a time.

The best buds to use on the stick are usually those in the middle and basal portions. Buds on the succulent terminal portion of the shoot should be discarded. However, in certain species, such as the sweet cherry, buds on the basal portion of the shoots are flower buds which, of course, should not be used.

In fall budding, after the buds have been inserted and tied, there is nothing more to be done until the following spring. *Although eventually the rootstock is to be cut off above the bud, in no case should this be done immediately after the bud has been inserted.* Healing of the bud piece to the stock is greatly facilitated by the normal movement of water and nutrients up and down the stem of the rootstock. This would, of course, be stopped if the top of the rootstock were cut off above the bud.

If the budding operation is done properly, the bud piece should unite with the stock in two to three weeks, depending upon the growing conditions. If the leaf stalk or petiole drops off cleanly next to the bud, this is a good indication that the bud has united, especially if the bark piece retains its normal light brown or green color and the bud stays plump. On the other hand, if the leaf petiole does not drop off cleanly but adheres tightly and starts to shrivel and darken, while the bark piece also commences to turn black, it is likely that the operation has failed. If the bark of the rootstock is still slipping easily and budwood is still available, there may be time for the budding to be repeated.

Even though the bud union has healed, in most deciduous species the bud usually does not grow or "push out" in the fall, since it is in a physiological rest period or is inhibited by "apical dominance" (presence of other buds terminal to it). It remains just as it is until spring, at which time the chilling winter temperatures have overcome the rest influence and the bud is ready to grow. There are some exceptions to this; for example, in fall budding of maples, roses, honey locust, and certain other plants, some of the buds may start growth in the fall. In northern areas, if such fall-forced buds do not start early enough for the shoots to mature before cold weather starts, they are likely to be winter-killed.

In the spring, just before new growth begins, the rootstock is cut off immediately above the bud. It is desirable to make a sloping cut, slanting away from the bud. Although this cut may be waxed, it is usually not essential unless the stock is large in diameter. Cutting back the rootstock forces the inserted bud into growth. In citrus budding, it is a common practice to cut the stock partially above the bud and to lop or bend it over away from the bud. (See Figure 18–1.) The leaves of the stock plant still supply the roots with some nutrients, but the partial cutting forces the bud into growth. After the new shoot from the bud has started growth, the top is completely removed.

In colder northern regions, fall-inserted buds are sometimes covered with soil during the winter until danger of frost has passed and are then uncovered and topped-back in late spring.

Where strong winds occur, and in species in which the new shoots grow vigorously, support for the newly developing shoot may be necessary. One practice sometimes followed is to cut off the rootstock several inches above the bud, using this projecting stub as a support on which to tie the tender young shoot arising from the bud. This stub is removed after the shoot has become well established. In another procedure, stakes may be driven into the ground next to the stock to which the developing shoot is tied at intervals during its growth.

Cutting back to force the main bud to grow also forces many latent buds on the rootstock into growth. These must be rubbed off as soon as they appear, or they will soon choke out the inserted bud. It may be necessary to go over the budded plants several times before these "sprouts" stop appearing. Nurserymen refer to this procedure as "suckering."

As shown in Figure 13–2, the shoot arising from the inserted bud becomes the top portion of the plant. After one season's growth in the nursery with favorable conditions of soil, water and nutrients, temperature, and insect and disease control, this shoot will have developed sufficiently to enable the plant to be dug and moved to its permanent location during the following dormant season. Such a tree would have a one-year-old top and a two- or perhaps three-year-old root, but it is still considered a "yearling" tree. If the top makes insufficient growth the first year, it can be allowed to grow a second year, and is then known as a two-year-old tree. However, abnormal, slow-growing, stunted trees should be discarded.

**Figure 13-2**    Row of fall-budded nursery trees about one year after budding. The inserted buds have grown through the spring and summer. The roots have grown through two seasons. The tops of the rootstocks were cut off above the inserted buds the spring following budding.

## Spring Budding

Spring budding is similar to fall budding except that insertion of the bud takes place the following spring as soon as active growth of the rootstock begins and the bark separates easily from the wood. The period for successful spring budding is limited, and budding should be completed before the rootstocks have made much new growth.

Budsticks are chosen from the same type of shoots—in regard to vigor of growth and type of buds (Figure 13-1)—that would have been used in fall budding, except that they are not collected until the dormant season the following winter. The leaves would, of course, have fallen by this time, and the buds would have experienced sufficient chilling to overcome their rest period. Budwood must be collected while it is still dormant—before there is any evidence of the buds swelling. Since the buds must be dormant when they are inserted, and since the rootstocks must be in active growth, it is necessary in spring budding that the budsticks be gathered some time in advance of the time of budding and stored at about 0° to 4° C (32° to 40° F) to hold the buds dormant. The budsticks should be wrapped in bundles with damp peat moss or some similar material to prevent drying out.

In spring budding, the actual budding operation should be done just as soon as the bark on the rootstock slips easily. Then, about two weeks after budding, when the bud unions have healed, the top of the stock must be cut off above the bud to force the inserted bud into active growth. At the same time, latent buds on the rootstock begin to grow and should be removed. Sometimes it is helpful to permit such shoots from the rootstock to develop to some extent to prevent sunburn and help nourish the plant. They must be held in check, however, and eventually removed.

Although the new shoot from the inserted bud gets a later start in spring budding than in fall budding, spring buds will usually develop rapidly enough, if growing conditions are favorable, to make a satisfactory top by fall. Fall budding, however, is to be preferred for several reasons: the higher temperatures at that time promote more certain healing of the union, the budding season is longer, there is no necessity to store the budsticks, the inserted buds start growth earlier in the spring, and the pressure of other work is usually not so great for propagators in the late summer as in the spring. Spring budding is used sometimes on rootstocks that were fall-budded but on which the buds failed to take.

## June Budding

June budding is used to obtain a "one-year-old" budded tree in a single growing season. Both roots and tops of the budded tree develop during only one growing season. Budding is done in the early part of the growing season and the inserted bud forced into growth immediately. As a method of nursery propagation, June budding is confined to regions that have a relatively long growing season—in the United States this region includes California and the southern states. In the propagation of fruit trees, June budding is used mostly in producing such stone fruits as peaches, nectarines, apricots, almonds, and plums. Peach seedlings are generally used as the rootstock, but almond seedlings and plum hardwood cuttings also can be used. Budding is done by the T-bud method. If seeds are planted in the fall, or stratified seeds as early as possible in the spring, the seedlings usually attain sufficient size (30 cm [12 in.] high and at least 3 mm [⅛ in.] in diameter) to be budded by mid-May or early June (in the Northern Hemisphere). Preferably, June budding should not be done much after mid-June, or a nursery tree of satisfactory size will not be obtained by fall. June-budded trees are not as large by the end of the growing season as those propagated by fall or spring budding, but they are of sufficient size—10 mm (⅜ in.) to 16 mm (⅝ in.) caliper and 90 cm (3 ft) to 150 cm (5 ft) tall—to produce entirely satisfactory trees (9).

Budwood used in June budding consists of current season's growth, that is, of new shoots which have developed since growth started in the spring. By late May or early June, these shoots will usually have grown sufficiently to have a well-developed bud in the axil of each leaf. At this time of year these buds will not have entered the rest period, so when they are used in budding they continue their growth on through the summer, producing the top portion of the budded seedling.

For June-budded trees, handling subsequent to the actual operation of budding is somewhat more exacting than for fall- or spring-budded trees. The rootstocks are smaller and have less stored food than those used in fall or spring budding. The object behind the following procedures—shown in Figure 13–3—is to keep the rootstock (and later the budded top) actively and continuously growing so as to allow no check in growth, while at the same time changing the seedling shoot to a budded top. The bud should be inserted high enough (about 14 cm; 5½ in.) on the stem so that a number of leaves—at least three or four—can be retained below the bud. The method of T-budding with the "wood out" should be used. Healing of the inserted bud should be very

**Figure 13–3**    It is important in June budding that the stock be cut back to the bud properly. *Far left:* the bud is inserted high enough on the stock so that there are several leaves below the bud. *Center left:* three or four days after budding, the stock is partially cut back about 9 cm (3½ in.) above the bud. *Center:* ten days to two weeks after budding, the stock is completely removed just above the bud. *Center right:* this forces the bud and other buds on the stock into growth; the latter must subsequently be removed. *Far right:* appearance of the budded tree after the new shoot has made considerable growth.

rapid at this time of year, since temperatures are relatively high, and rapidly growing, succulent plant parts are used. By four days after budding, healing should have started, and the top of the rootstock can be cut back somewhat—about 9 cm (3½ in.) above the bud—leaving at least one leaf above the bud and several below it. This operation will force the inserted bud into growth and will check terminal growth of the rootstock. It will also stimulate shoot growth from basal buds of the rootstock, which will produce additional leaf area. This continuous leaf area is necessary so that there always will be enough leaves to keep manufacturing food for the small plant. Ten days to two weeks after budding, the rootstock can be cut back to the bud, which should be starting to grow. If the budding rubber has not broken, it should be cut at this time. Other shoots arising from the rootstock should be headed back to retard their growth. After the inserted bud grows and develops a substantial leaf area, it can supply the plant with the necessary nutrients. By the time the shoot from the inserted bud has grown about 25 cm (10 in.) high, it should have enough leaves so that all other shoots and leaves can be removed. Later inspections should be made to remove any shoots arising from the rootstock below the budded shoot.

The steps in fall, spring, and June budding are compared in Figure 13–4.

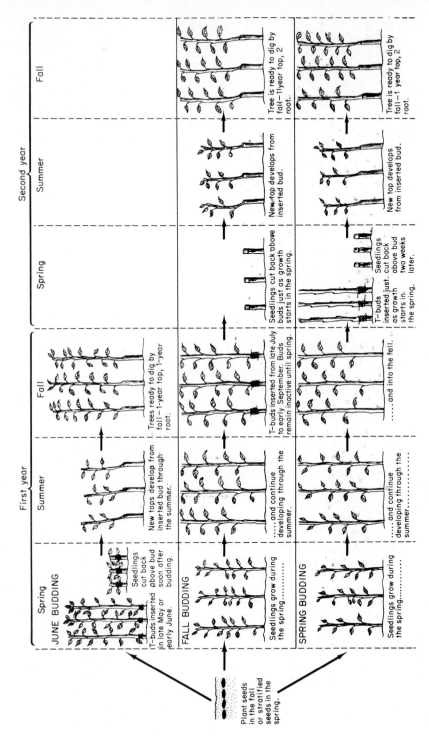

**Figure 13-4** Comparison of the steps in June, fall, and spring budding. The actual techniques in budding are not difficult, but it is very important that the various operations be done at the proper time.

# METHODS OF BUDDING

## T-Budding (Shield Budding)

This method of budding is known by both names, the "T-bud" designation arising from the T-like appearance of the cut in the stock, whereas the "shield bud" name is derived from the shield-like appearance of the bud piece when it is ready for insertion in the stock.

T-budding is by far the most common method of budding and is widely used by nurserymen in propagating nursery stock of most fruit tree species, roses, and some ornamental shrubs. Its use is generally limited to stocks that are about 6 to 25 mm ($\frac{1}{4}$ to 1 in.) in diameter, and are actively growing so that the bark will separate readily from the wood. If the bark is so tight on the wood that it has to be pried loose forcibly, the chances of the bud healing successfully are poor. The operation should then be delayed until the bark is slipping easily.

The bud is inserted into the stock 5 to 25 cm (2 to 10 in.) above the soil level in a smooth bark surface. There are different opinions as to which is the proper side of the stock in which to insert the bud. If extreme weather conditions are likely to occur during the critical healing period just following budding, it may be desirable to place the bud on the side of the stock on which as much protection as possible may be obtained. Some believe that if the bud is placed on the windward side, there is less chance of the young shoot breaking off. Otherwise, it probably makes little difference where the bud is inserted, the convenience of the operator and the location of the smoothest bark being the controlling factors. When rows of closely planted rootstocks are budded, it is more convenient to have all the buds on the same side for later inspection and manipulations.

The cuts to be made in the stock plant are illustrated in Figure 13–5. There are various modifications of this technique; most budders prefer to make the vertical cut first, then the horizontal crosscut at the top of the T. As the horizontal cut is made, the knife is given a twist to throw open the flaps of bark for insertion of the bud. It is important that neither the vertical nor horizontal cut be made longer than necessary, because this requires additional tying later to close the cuts.

After the proper cuts are made in the stock and the incision is ready to receive the bud, the shield piece is cut out of the budstick.

To remove the shield of bark containing the bud, a slicing cut is started at a point on the stem about 13 mm ($\frac{1}{2}$ in.) below the bud, continuing under and about 25 mm (1 in.) above the bud. The shield piece should be as thin as possible but still thick enough to have some rigidity. A second horizontal cut is then made 13 to 19 mm ($\frac{1}{2}$ to $\frac{3}{4}$ in.) above the bud, thus permitting the removal of the shield piece.

There are two methods of preparing the shield—with the "wood in" or with the "wood out." This refers to the little sliver of wood just under the bark of the shield piece and which will remain attached to it if the second, or horizontal, cut is deep and goes through the bark and wood, joining the first slicing cut. Some professional budders believe it is best to remove this sliver of wood, but others retain it. In budding certain species, however, such as maples and walnuts, much better success is usually obtained with "de-wooded" buds. If it is desired to prepare the shield with the wood out, the second horizontal cut should be just deep enough to go through the bark and not through the wood. Then if the bark is slipping easily, the bark shield can be snapped loose from the wood (which still

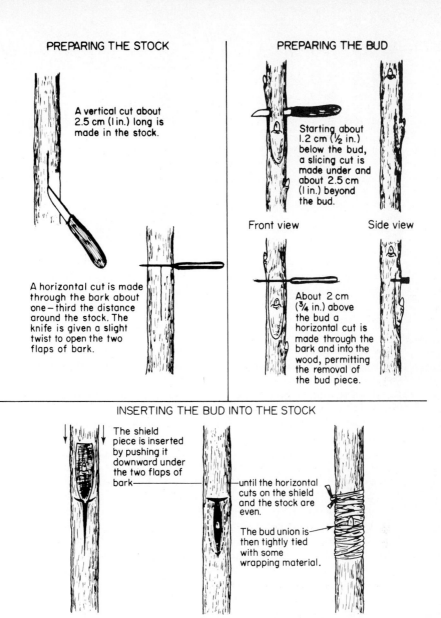

**PREPARING THE STOCK**

A vertical cut about 2.5 cm (1 in.) long is made in the stock.

A horizontal cut is made through the bark about one-third the distance around the stock. The knife is given a slight twist to open the two flaps of bark.

**PREPARING THE BUD**

Starting about 1.2 cm (½ in.) below the bud, a slicing cut is made under and about 2.5 cm (1 in.) beyond the bud.

Front view          Side view

About 2 cm (¾ in.) above the bud a horizontal cut is made through the bark and into the wood, permitting the removal of the bud piece.

**INSERTING THE BUD INTO THE STOCK**

The shield piece is inserted by pushing it downward under the two flaps of bark

until the horizontal cuts on the shield and the stock are even.

The bud union is then tightly tied with some wrapping material.

**Figure 13-5**    Basic steps in making the T-bud (shield bud).

remains attached to the budstick) by pressing it against the budstick and sliding it sideways. A small core of wood comprising the vascular tissues supplying the bud is present, and this should remain in the bud, rather than adhering to the wood and leaving a hole in the bud. If the shield is pulled outward rather than being slid sideways from the wood, this core usually pulls out of the bud, eliminating the chances of success. In June budding of fruit trees, the shield piece is usually prepared with the wood out. In most other instances, however, the wood is left in. In spring budding, using dormant budwood, this sliver of wood is tightly attached to the bark and cannot be removed.

**Figure 13–6**  Steps in the development of a T-bud. *Left:* bud after being inserted and wrapped. *Center:* bud has healed in place, the budding rubber has dropped off, and the stock has been cut back above the bud. *Right:* shoot development from the inserted bud. All buds arising from the stock have been rubbed off.

The next step is the insertion of the shield piece containing the bud into the incision in the stock plant. The shield is pushed under the two raised flaps of bark until its upper, or horizontal, cut matches the same cut on the stock. The shield should fit snugly in place, well covered by the two flaps of bark, but with the bud itself exposed. (See Figure 11–11.)

No waxing is necessary, but the bud union must be wrapped, using tape, budding rubbers, or raffia to hold the two components firmly together until healing is completed. Rubber budding strips, especially made for wrapping, are widely used for this purpose (Figure 13–6). Their elasticity provides sufficient pressure to hold the bud securely in place. The rubber, being exposed to the sun and air, usually deteriorates, breaks, and drops off after several weeks, at which time the bud should be healed in place. If the budding rubber is covered with soil, the rate of deterioration will be much slower. This material has the advantage of eliminating cutting the wrapping ties, which can be a costly operation if many thousands of plants have been budded. The rubber will expand as the rootstock grows, and thus there is little danger of constriction.

In tying the bud, the ends of the budding rubbers are held in place by inserting them under the adjacent turn. The bud itself should not be covered. The amount of tension given the budding rubber is quite important. It should not be too loose, or there will be too little pressure holding the bud in place. On the other hand, if the rubber is stretched extremely tight, it may be so thin that it will deteriorate rapidly and break too soon—before the bud union has taken place. Often the tying is done from the top down to avoid forcing the bud out through the horizontal cut. Proper wrapping of the buds is very important.

Parafilm tape, which is a waterproof, flexible, stretchable, thermoplastic film with a paper backing, has been successfully used (*3*) for covering the bud in bench chip budding of roses.

Raffia (fiberlike leaf segments of certain *Raphia* species) is used in some countries for wrapping buds. This material is soaked in water overnight before it is used so that it will be flexible. Raffia must be cut later—about ten days after budding—to prevent constriction at the bud union as the plant grows. If such nonelastic ties are not promptly cut, the resultant constriction can have a very adverse affect on subsequent growth of the bud (*11*).

Plastic ties of polyvinyl chloride (PVC) film, 10 mm (⅜ in.) wide, are quite useful for budding. Such material is moisture-proof and elastic, and since it is transparent, it permits inspection of the buds after covering (*1*).

### Inverted T-Budding

In rainy localities, water running down the stem of the rootstock may enter the T-cut, soak under the bark, and prevent the shield piece from healing into place. Under such conditions an inverted T-bud may give better results, since it is more likely to shed excess water. In citrus budding, the inverted T method is widely used, even though the conventional method also gives good results. In species that bleed badly during budding, such as chestnuts, the inverted T-bud allows better drainage and better healing. Proponents of both conventional and inverted T-budding can be found, and in a given locality the usage of either with a given species tends to become traditional.

The techniques of the inverted T-bud method are the same as those already described, except that the incision in the stock has the transverse cut at the bottom rather than at the top of the vertical cut, and in removing the shield piece from the budstick the knife starts above the bud and cuts downward below it. The shield is removed by making the transverse cut 13 to 19 mm (½ to ¾ in.) below the bud. The shield piece containing the bud is inserted into the lower part of the incision and pushed upward until the transverse cut of the shield meets that made in the stock.

It is important in using this inverted T-bud method that a normally oriented shield bud piece should not be inserted into an inverted incision in the stock. The bud would then have a reversed polarity. Although such upside down buds do live and grow, at least in some species, their use may not result in the expected shoot development.

### Patch Budding

The distinguishing feature of patch budding and related methods is that a rectangular patch of bark is removed completely from the stock and replaced with a patch of bark of the same size containing a bud of the cultivar to be propagated.

Patch budding is somewhat slower and more difficult to perform than T-budding, but it is widely and successfully used on thick-barked species, such as walnuts and pecans, in which T-budding sometimes gives poor results, presumably owing to the poor fit around the margins of the bud. Patch budding, or one of its modifications, is also extensively used in propagating various tropical species, such as the rubber tree (*Hevea brasiliensis*).

Patch budding requires that the bark of both the stock and budstick be slipping easily. It is usually done in late summer or early fall, but can be done in the spring also. In propagating nursery stock, the diameter of the rootstock and the budstick should preferably be about the same, about 13 to 26 mm (½ to 1 in.).

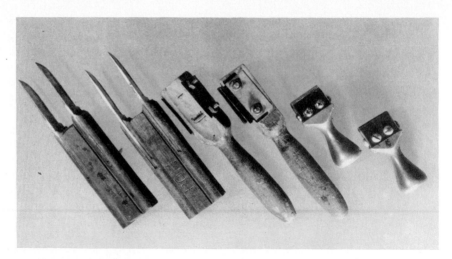

**Figure 13–7**    Tools used in patch budding and related methods. *Left:* manufactured double-bladed knives. *Center:* double-bladed knives made with two razor blades. *Right:* manufactured patch-budding tools consisting of four rectangular cutting blades.

Although the budstick should not be much larger than about 26 mm (1 in.) in diameter, the patch can be inserted successfully into stocks as large as 10 cm (4 in.) in diameter, although healing of such large stubs may be inadequate (*12*).

Special knives (see Figure 13–7) have been devised to remove the bark pieces from the stock and the budstick. Some type of double-bladed knife that will make two transverse parallel cuts 25 to 35 mm (1 to 1⅜ in.) apart is necessary. These cuts, about 25 mm (1 in.) in length, are made through the bark to the wood in a smooth area of the rootstock about 10 cm (4 in.) above the ground. Then the two transverse cuts are connected at each side by vertical cuts made with a single-bladed knife.

The patch of bark containing the bud is cut from the budstick in the same manner in which the bark patch is removed from the stock. Using the same two-bladed knife, the budder makes two transverse cuts through the bark, one above and one below the bud. Then two vertical cuts are made on each side of the bud so that the bark piece will be about 25 mm (1 in.) wide. The bark piece containing the bud is now ready to be removed. It is important that it be slid off sideways rather than being lifted or pulled off. There is a small core of wood, the bud trace, which must remain inside the bud if a successful "take" is to be obtained. By sliding the bark patch to one side, this core is broken off, and it stays in the bud. If the bud patch is lifted off, this core of wood is likely to remain attached to the wood of the budstick, leaving a hole in the bud, as shown in Figure 13–8.

After the bud patch is removed from the budstick, it must be inserted immediately on the stock, which should already be prepared, needing only to have the bark piece removed. The patch from the budstick should fit snugly at the top and bottom into the opening in the stock, since both transverse cuts were made

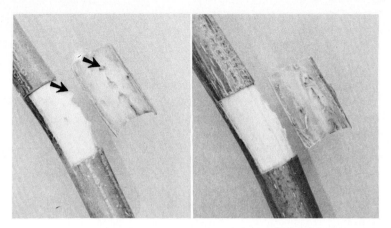

**Figure 13-8**     Removing the bud patch from the budstick in patch budding. *Left:* incorrect. The core of wood in the bud, comprising the vascular tissues, has broken off, leaving a hole in the bud. Such a bud is not likely to grow. *Right:* correct. The patch was pushed off sideways, and the core of wood has remained inside the bud.

with the same knife. It is more important that the bark piece fit tightly at top and bottom than that it fit along the sides. These procedures are illustrated in Figures 13-9, 13-10, and 13-11.

The inserted patch is now ready to be wrapped. Often the bark of the stock will be thicker than the bark of the inserted bud patch so that, upon wrapping, it is impossible for the wrapping material to hold the bud patch tightly against the stock. In this case it is necessary that the bark of the stock be pared down around the bud patch so that it will be of the same thickness, or preferably slightly thinner, than the bark of the bud patch. Then the wrapping material will hold the bud patch tightly in place.

In cases in which difficulty is experienced in obtaining successful unions with the patch bud method, it may help to make the four cuts of the rectangle in the stock one to three weeks ahead of the time when the actual budding is to be done. This bark patch is not removed from the stock, however, until the patch containing the bud is ready to be inserted. When the cuts are made ahead of time, the wounding causes the callusing process to start, so that when the new bark patch is inserted, it heals very rapidly.

In wrapping the patch bud, a material should be used that not only will hold the bark tightly in place but will cover all the cut surfaces to prevent the entrance of air under the patch, with subsequent drying and death of the tissues. The bud itself must not be covered during wrapping. The most satisfactory material is nurserymen's adhesive tape. Another method is to tie the bud patch with heavy cotton string and cover the cut edges with grafting wax.

It is important in patch budding, especially with walnuts, that the wrapping not be allowed to cause a constriction at the bud union. When the stock is growing rapidly, it is necessary to cut the tape about ten days after budding. A single vertical knife cut on the side opposite the bud is sufficient, but care should be

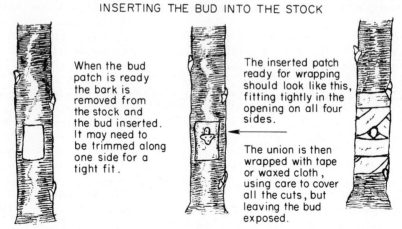

**PREPARING THE STOCK**

A double-bladed knife is used to make two parallel horizontal cuts about one-third the distance around the stock.

The two horizontal cuts are connected at each side by vertical cuts

**PREPARING THE BUD**

The patch containing the bud is cut from the bud stick by two horizontal cuts with the double-bladed knife —

— followed by two vertical cuts on each side of the bud. The bud patch is removed by sliding it off to one side.

**INSERTING THE BUD INTO THE STOCK**

When the bud patch is ready the bark is removed from the stock and the bud inserted. It may need to be trimmed along one side for a tight fit.

The inserted patch ready for wrapping should look like this, fitting tightly in the opening on all four sides.

The union is then wrapped with tape or waxed cloth, using care to cover all the cuts, but leaving the bud exposed.

**Figure 13-9**     Steps in making the patch bud. This method is widely used for propagating thick-barked plants.

taken not to cut into the bark. Sometimes an extra strip of tape is placed vertically before wrapping to protect the bark when the tape is cut. The cut tape should not be pulled off.

**Figure 13–10**    Steps in the development of a patch bud. *Top left:* patch bud after being inserted and wrapped with tape. *Top right:* after about ten days, the tape is slit along the back side to release any constricting pressure. *Below left:* the tape is completely removed after about three weeks, at which time the patch should be healed in place. *Below right:* cutting back the stock above the bud forces it into growth.

Patch budding is best performed in late summer when both the seedling stock and the source of budwood are growing rapidly and their bark slipping easily. The budsticks for patch budding done at this time should have the leaf blades cut off two to three weeks before the budsticks are taken from the tree. The petiole or leaf stalk is left attached to the base of the bud, but by the time the budstick is removed, this petiole has dropped off or is easily pulled off.

Patch budding can also be done in the spring after new growth has appeared on the stocks and it has been determined that the bark is slipping. There is a problem, however, in obtaining satisfactory buds to use at this time of year, since it is necessary that the bark of the budstick separate readily from the wood. At the same time, the buds should not be starting to swell. There are two methods by which satisfactory buds can be obtained for patch budding in the spring. In one, the budsticks are selected during the dormant winter period and stored at low temperatures (about 2° C; 36° F) and wrapped in moist sphagnum or peat moss to prevent their drying out. Then, about two or three weeks before the spring budding is to be done, they are brought into a warm room. The budsticks may be left in the damp peat moss or set with their bases in a container of water. The increased temperature will cause the cambium layer to become active, and soon the bark will slip sufficiently for the buds to be used. Although a few of the more terminal buds on each stick may start swelling in this time and cannot be used, there should be a number of buds in a satisfactory condition.

The second method of obtaining buds for spring patch budding is to take them directly from the tree that is the source of the budwood, but at the time the budding is to be done. If the trees are inspected carefully, it will be seen that not all of the buds start pushing at once. The terminal buds are usually more advanced than the basal ones. There is a period when the bark is slipping easily throughout the shoot containing the desired buds, but when only a few of the buds have developed so far that they cannot be used. The remaining buds, which are still dormant but upon bark that can be removed readily, may be taken and used immediately for budding. It is easier to obtain suitable buds from young trees that made vigorous shoot growth the previous year than it is from old trees. When the budding is done will be governed then by the stage of development of the buds. This will vary considerably with the species and cultivar being used. Stage of development of the rootstock is not as critical. Its bark must be slipping well, and the budding should be done before the stock plant has made much new growth.

## I-Budding

In I-budding the bud patch is cut just as for patch budding, that is, in the form of a rectangle or square. Then, with the same parallel-bladed knife, two transverse cuts are made through the bark of the stock. These are joined at their centers by a single vertical cut to produce the figure I. The two flaps of bark can then be raised for insertion of the bud patch beneath them. It may make a better fit to slant the side edges of the bud patch. In tying the I-bud, care should be taken to see that the bud patch does not buckle upward and fail to touch the stock. (See Figure 13–12.)

**Figure 13–11** *Left:* shoot development from patch budding operation shown in Figure 13–11 after two years' growth. *Right:* strong, smooth union developed after 14 years. *Juglans regia* 'Hartley' on Paradox rootstock (*Juglans regia* × *J. hindsii*).

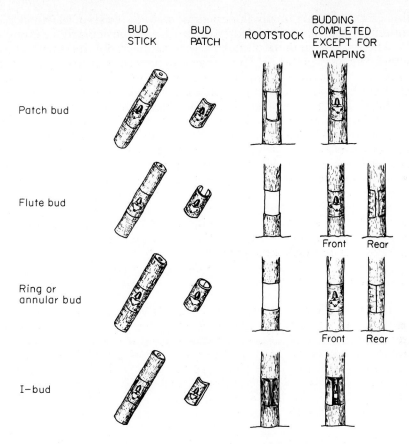

|                  | BUD STICK | BUD PATCH | ROOTSTOCK | BUDDING COMPLETED EXCEPT FOR WRAPPING |
|------------------|-----------|-----------|-----------|---------------------------------------|

Patch bud

Flute bud

Ring or annular bud

I-bud

Front    Rear

Front    Rear

**Figure 13–12**    There are many variations of the patch bud, some of which are shown here. The naming of these types is somewhat confused; the most generally accepted names are given here.

I-budding should be considered for use when the bark of the stock is much thicker than that of the budstick. In such cases, if the patch bud were used, considerable paring down of the bark of the stock around the patch would be necessary. This operation is not necessary in the I-bud method (Figure 13–12).

### Flute Budding and Ring Budding

See details in Figure 13–12.

### Chip Budding

Chip budding can be used at times when the bark is not slipping, that is, early in the spring before growth starts or during the summer when active growth has stopped prematurely owing to lack of water or some other cause. Chip budding is generally used with small material, 13 to 25 mm (½ to 1 in.) in diameter. Chip budding ordinarily is not as fast or as simple as T-budding. For many years, however, chip budding in the fall has given excellent results in budding grape cultivars on phylloxera or nematode-resistant rootstocks (*5, 8*). Chip budding is also used on deciduous fruit trees, particularly in Great Britain (*6, 7*), where more vigorous initial growth has been obtained with chip buds than with T-buds (Figure 13–13). Studies (*3*) have shown that chip budding is also

**Figure 13–13**    *Above:* a block of clonal apple rootstocks that have been budded to the desired cultivar by chip budding in England. Arrows point to insertion of bud. *Below:* close-up of two stems in each of which two chip buds have been inserted. Top buds, in both cases, have been covered by transparent plastic tape. Lower buds have been covered by opaque white tape; buds are completely covered by tape, which will be removed after healing takes place.

quite satisfactory for propagating field roses by indoor bench grafting in winter on dormant, unrooted rootstock cuttings.

As illustrated in Figure 13–14, a chip of bark is removed from a smooth

**Figure 13–14** The chip bud is really a form of grafting, a variation of the side-veneer graft, with the scion reduced to a small piece of wood containing only a single bud. Chip budding is widely used in propagating vinifera grapes on nematode and phylloxera resistant rootstocks. It is also used successfully for propagating fruit trees, with the bud piece cut thinner than shown here and covered completely with wide plastic or thin rubber tape.

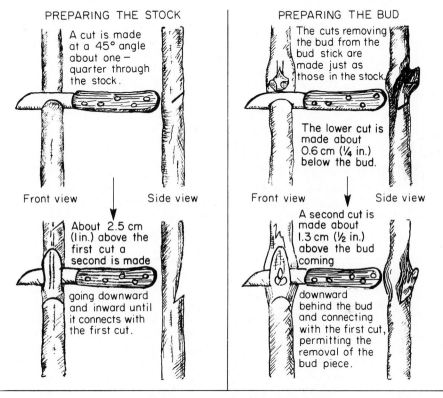

PREPARING THE STOCK

A cut is made at a 45° angle about one–quarter through the stock.

Front view     Side view

About 2.5 cm (1 in.) above the first cut a second is made

going downward and inward until it connects with the first cut.

PREPARING THE BUD

The cuts removing the bud from the bud stick are made just as those in the stock.

The lower cut is made about 0.6 cm (¼ in.) below the bud.

Front view     Side view

A second cut is made about 1.3 cm (½ in.) above the bud coming

downward behind the bud and connecting with the first cut, permitting the removal of the bud piece.

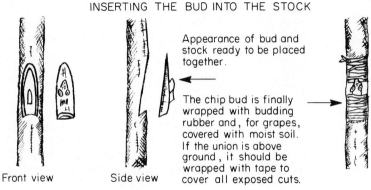

INSERTING THE BUD INTO THE STOCK

Appearance of bud and stock ready to be placed together.

The chip bud is finally wrapped with budding rubber and, for grapes, covered with moist soil. If the union is above ground, it should be wrapped with tape to cover all exposed cuts.

Front view     Side view

place between nodes near the base of the stock and replaced by another chip of the same size and shape from the budstick which contains a bud of the desired cultivar. The chips in both stock and budstick are cut out in the same manner. In the budstick the first cut is made just below the bud and down into the wood at an angle of 30 to 45 degrees. The second cut is started about 25 mm (1 in.) above the bud and goes inward and downward behind the bud until it intersects the first cut. (The order of making these two cuts may be reversed.) The chip is removed from the stock and replaced by the one from the budstick. To obtain a good fit, both should be cut to the same size and shape. The cambium layer of the bud piece must be placed to coincide with that of the stock, preferably on both sides of the stem, but at least on one side.

In chip budding there are no protective flaps of bark to prevent the bud piece from drying out, as there are in T-budding. It is very important, then, that the chip bud be wrapped to seal the cut edges as well as to hold the bud piece tightly into the stock. Nurserymen's adhesive tape works well for this purpose, although white or transparent plastic tape is more often used, covering the bud. When the bud starts growth, the tape is cut. After the bud has been inserted, it must be wrapped immediately so that it will not dry out (1).

In grape propagation, if the bud is inserted into the stock close to the ground level, drying can be prevented by wrapping the bud with budding rubber and covering the whole bud union immediately with several inches of finely pulverized moist soil, which can be removed after the bud has united. Since budding rubbers are slow to disintegrate in the soil, they should be cut or removed before constriction occurs.

In chip budding, as in the other methods, the stock is not cut back above the bud until the union is complete. If the chip bud is inserted in the fall, the stock is cut back just as growth starts the next spring. If the budding is done in the spring, the stock is cut back about ten days after the bud has been inserted.


## TOPBUDDING

In young trees with an ample supply of vigorous shoots at a height of 1.2 to 1.8 meters (4 to 6 ft), topbudding is a fast and certain method of topworking (Figure 13–15). It can be used in older trees, too, if they are cut back rather severely the year before to provide a quantity of vigorous watersprout shoots fairly close to the ground.

Depending upon the size of the tree, 10 to 15 buds are placed in vigorously growing branches 6 to 19 mm ($\frac{1}{4}$ to $\frac{3}{4}$ in.) in diameter in the upper portion of the tree—about shoulder height. Although a number of buds could be placed in a single branch, usually only one will be saved to develop into secondary branches, which will then form the permanent new top of the tree. The T-bud method is used on thin-barked species, and the patch bud on those with thick bark.

Topbudding is usually done in midsummer, as soon as well-matured budwood can be obtained and while the stock tree is still in active growth with the bark slipping easily. Orchard trees generally stop growth earlier in the season than young nursery trees; therefore the budding must be done earlier. When topbudding is done at this time of year, the buds usually remain inactive until the following spring. At that time, just as vegetative growth starts, the stock branches are cut back just above the buds. This forces the buds into active

**Figure 13–15**    Topworking a young tree by topbudding. T-buds were inserted in the positions shown by arrows and have grown for one season. All other shoots have been removed. From L. H. Day, "Apple, Quince, and Pear Rootstocks in California," *Calif. Agr. Exp. Sta. Bul. 700.*

growth, and they should develop into good-sized branches by the end of the summer. At the time the shoots are cut back to the buds, all other unbudded branches should be removed at the trunk. It is important that the trees be inspected carefully through the summer and all shoots removed that are arising from any but the inserted buds.

Topbudding can also be done in the spring just as the tree to be topworked is starting active growth and the bark is slipping easily. The techniques for spring budding of nursery stock are used. Although not commonly done, June budding could be used for topbudding.

## DOUBLE-WORKING BY BUDDING

In propagating nursery trees, some budding methods can be used in developing double-worked trees. The intermediate stock can be budded on the rootstocks; then the following year the desired cultivar is budded on the interstock. Although quite effective, this is a rather lengthy process, taking three years. As illustrated in Figure 13–16, it is possible to develop a double-worked tree in one operation in one year by the double-shield bud method (*2, 4, 6*). A T-bud is used, but just under and below it a thin budless shield piece of the desired interstock is inserted.

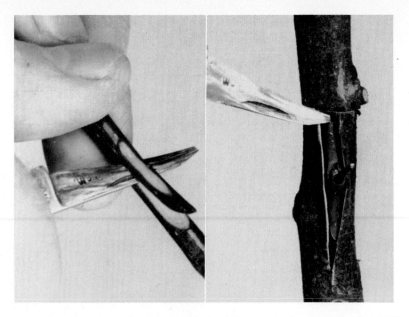

**Figure 13–16** Steps in the double-budding technique of Nicolin (*10*). *Left:* cutting out the oval bark piece of the mutually compatible interstock variety. *Right:* placing the cultivar bud shield on top of the previously inserted oval bark piece. The budding operation is now ready for tying with a budding rubber. From K. D. Brase and R. D. Way, ''Rootstocks and Methods Used for Dwarfing Fruit Trees,'' *N. Y. State Agr. Exp. Sta. Bul. 783.* Courtesy Prof. Karl Brase.

## MICRO–BUDDING

This type of budding is used successfully in propagating citrus trees and probably could be utilized also for other tree and shrub species. It has been of commercial importance in the citrus districts of southeastern Australia (*13*). As described here, it is not done under aseptic conditions. Micro-budding is similar to ordinary T-budding, except that the bud piece is reduced to a very small size. The leaf petiole is cut off just above the bud, and then the bud is removed from the budstick by a flat cut just underneath the bud, with a razor-sharp knife. Only the bud itself and a small piece of wood under it are used. In the stock an inverted T-cut is made, and the micro-bud is slipped into this, right side up. The entire T-cut, including the bud, is covered with thin plastic budding tape. The tape is allowed to remain for 10 to 14 days for spring budding and three weeks for fall budding, after which it is removed by cutting with a knife. By this time the buds should have healed in place; subsequent handling is the same as for conventional T-budding.

## REFERENCES

*1.* Bremer, A. H. 1977. Chip budding on a commercial scale. *Proc. Inter. Plant Prop. Soc.* 27:366–67.

2. Bryden, J. D. 1957. Use of plastic ties in fruit tree budding. *Agr. Gaz. New S. Wales (Austral.)* 68:87–88.

3. Davies, F. T., Jr., Y. Fann, and J. E. Lazante. 1980. Bench chip budding of field roses. *HortScience.* 15(6):817–18.

4. Garner, R. J. 1953. Double-working pears at budding time. *Ann. Rpt. E. Malling Res. Sta. for 1952,* pp. 174–75.

5. Harmon, F. N., and J. H. Weinberger. 1969. The chip-bud method of propagating vinifera grape varieties on rootstocks. *U. S. Dept. Agr. Leaflet 513.*

6. Howard, B. H., D. S. Skene, and J. S. Coles. 1974. The effect of different grafting methods upon the development of one-year-old nursery apple trees. *Jour. Hort. Sci.* 49:287–95.

7. Howard, B. H. 1977. Chip budding fruit and ornamental trees. *Proc. Inter. Plant Prop. Soc.* 27:357–64.

8. Lider, L. A. 1963. Field budding and the care of the budded grapevine. *Calif. Agr. Ext. Ser. Leaflet 153.*

9. Mertz, W. 1964. Deciduous June-bud fruit trees. *Proc. Inter. Plant Prop. Soc.* 14:255–59.

10. Nicolin, P. 1953. Nicolieren: A new method of grafting. *Deutsche Baumschule* 5:186–87.

11. Smith, N. G., R. J. Garner, and W. S. Rogers. 1962. Delayed growth of apple scions in relation to early budding, bud constriction, and some other factors. *Ann. Rpt. E. Malling Res. Sta. for 1961,* pp. 51–56.

12. Taylor, R. M. 1972. Influence of gibberellic acid on early patch budding of pecan seedlings. *Jour. Amer. Soc. Hort. Sci.* 97(5):677–79.

13. Wishart, R. D. A. 1961. Microbudding of citrus. *S. Austral. Dept. Agr. Leaflet 3660.*

## SUPPLEMENTARY READING

BALTET, C. 1910. *The art of grafting and budding* (6th ed.). London: Crosby, Lockwood.

BRASE, K. D. 1952. Propagation of fruit trees by budding. *Farm Res.* (N.Y. Agr. Exp. Sta.) Vol. 18. No. 3.

CARLSON, R. F., and A. E. MITCHELL. 1971. Budding and grafting fruit trees. *Mich. Agr. Ext. Bul. 508.*

GARNER, R. J. 1979. *The grafter's handbook* (4th ed.). New York: Oxford University Press.

HARTMANN, H. T., and J. A. BEUTEL. 1979. Propagation of temperate zone fruit plants. *Calif. Agr. Exp. Sta. Leaflet 21103.*

INTERNATIONAL PLANT PROPAGATORS' SOCIETY. Proceedings of annual meetings.

PLATT, R. G., and K. W. OPITZ. 1973. Propagation of citrus. In *The citrus industry,* vol. 3, W. Reuther, ed. Berkeley: Univ. of Calif. Press.

Layering is a propagation method by which adventitious roots are caused to form on a stem while it is still attached to the parent plant. The rooted, or layered, stem is detached to become a new plant growing on its own roots. Layering may be a natural means of reproduction, as in black raspberries and trailing blackberries, or it may be induced artificially in many kinds of plants.

## FACTORS AFFECTING REGENERATION OF PLANTS BY LAYERING

*Nutrition*    The stem remains attached to the plant during rooting and is continually supplied with water and minerals through the intact xylem. Induction of adventitious roots on the intact stem is affected by several significant factors:

*Stem treatments*    Adventitious roots are induced to form on the attached stems by various manipulations of the stems (as shown in Figure 14–1) that cause an interruption in the downward translocation of organic materials—carbohydrates, auxin, and other growth factors—from the leaves and growing shoot tips. These materials accumulate near the point of treatment, and rooting occurs as it would on a stem cutting (see chapter 9).

*Light exclusion*    Elimination of light from the part of the stem where roots are to develop is a feature common to all methods of layering. A distinction should be made between **blanching,** the covering of an intact stem after it has already formed, and **etiolation,** the effect produced as the stem elongates in the absence of light (*16*). Intact stems of some plants are able to produce roots after blanching only, but in others phloem interruption also may be required. However, the greatest stimulus to root induction results when the initially developing shoots are continuously covered by the rooting medium—as in trench layer-

## 14

# Layering and Its Natural Modifications

BEFORE ROOTING     AFTER ROOTING

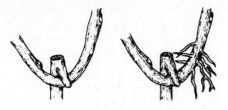

Shoot bent to a sharp v, or it
may be cut, notched, girdled,
or wired.

Shoot cut or broken on lower side

**Figure 14-1**     Treatments used to stimulate rooting during layering.

ing—so that approximately 2.5 cm (1 in.) of the base of the layered shoots is
never exposed to light (*16*). A large measure of the success with which shy-
rooting plants are rooted by layering apparently results from etiolation and
blanching.

*Physiological conditioning*     Root induction during layering may be associated with particular physiological conditions in the stem associated with the time
of year. For many types of layering, the timing is associated with the movement
of carbohydrates and other substances toward the roots at the end of a seasonal
cycle of growth.

*Rejuvenation*     Cutting back shoots in mound and trench layering and
regenerating new shoots from the base annually has a parallel in the hedging
methods used to rejuvenate stock plants for improved rooting of the cuttings that
are taken from them.

Procedures that improve rooting of cuttings also apply to layering. For example, the use of root-promoting substances, such as indolebutyric acid, during
layering is often beneficial, as it is with cuttings, although the methods of application may be somewhat different (*9, 18, 38, 39, 44*). Applying the material
to girdling cuts as a powder, in lanolin, or as a solution in 50 percent alcohol can
be utilized effectively.

Root formation on layers depends upon supplying continuous moisture,
good aeration, and moderate temperatures to the rooting zone (*32, 41, 43*).

## Uses of Layering

There are four principal uses of layering:

**1.** Propagation of plants of species such as black raspberries and trailing blackberries that reproduce naturally by this method.

**2.** Propagation of plants of clones whose cuttings do not root easily, yet are sufficiently valuable to justify the cost and labor required in layering. Filberts are commercially propagated by simple layering, for example, and Muscadine grapes ( *Vitis rotundifolia* ) by compound layering. Specific size-controlling clonal rootstocks of apple, pear, and some other fruit trees are produced by mound or trench layering. Certain tropical fruits, for example, mango and litchi, are propagated by air layering.

**3.** Layering is useful for producing a large-sized plant in a short time ( *24* ). Air layering is used in greenhouses to propagate relatively large specimen plants of rubber plant ( *Ficus elastica* ) (Figure 14–6) and croton. Dieffenbachia and similar plants may be propagated by simple layering (Figure 14–4).

**4.** Layering, particularly simple or air, is valuable for producing relatively small numbers of plants of good size with minimum propagation facilities.

As a commercial propagation method, layering requires considerable labor and is cumbersome and expensive. Special layering beds, or *stool beds,* are established but are somewhat difficult to manage. Nevertheless, for those plants requiring layering, nurseries have developed effective management techniques ( *8, 10, 48* ). Trends in propagation have tended, however, to cause shifts from layering to other methods as they are developed. Improved procedures for hardwood cutting propagation can replace apple layering procedures. More significantly, micropropagation techniques (see chapter 17) can replace many of the layering procedures now in use.

## PROCEDURES IN LAYERING

### Tip Layering

Tip layering is a natural method of reproduction characteristic of trailing blackberries, dewberries, and black and purple raspberries.

Stems of these plants are biennial in that the canes are vegetative during the first year, fruitful the second, and pruned out after fruiting. In the nursery it is advisable to set aside stock plants solely for propagation. Healthy young plants are set 36 m (12 ft) apart to give room for subsequent layering. The plants are cut down to within 9 in. of the ground as soon as they are planted. Vigorous new canes are ''summer topped'' by pinching off 7.6 to 9.2 cm (3 to 4 in.) of the tip after growth of 45 to 76 cm (18 to 30 in.). This encourages lateral shoot production, and will increase the number of potential tip layers, and also next year's fruit crop. By late summer, the canes begin to arch over, and their tips assume a characteristic appearance in that the terminal ends become elongated and the leaves small and curled to give a ''rat-tail'' appearance. The best time for layering is when only part of the lateral tips have attained this appearance. If the operation is done too soon, the shoots may continue to grow instead of forming a terminal bud. If it is done too late, the root system will be small.

Rooting takes place near the tip of the current season's shoot. The shoot tip recurves upward to produce a sharp bend in the stem from which roots develop.

Tips are layered by hand, a spade or trowel being used to make a hole with one side vertical and one sloping slightly toward the parent plant. The tip is

**Figure 14–2**    Tip layer of boysenberry.

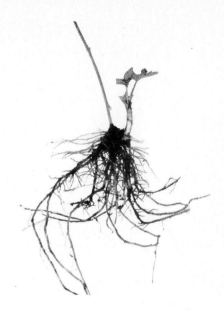

placed in the hole with the shoot lying along the sloping side and the returned soil is pressed firmly against it. Placed thus, the tip cannot continue to grow in length and becomes "telescoped," soon forming an abundant root system and developing a vigorous young vertical shoot.

The plants are ready for digging by the end of the same season. The rooted tip consists of a terminal bud, a large mass of roots, and 15 to 20 cm (6 to 8 in.) of the old cane to serve as a "handle" and to mark the location of the new plant (Figure 14–2). Since the tip layers are tender, easily injured, and subject to drying out, digging should be done preferably just before replanting. The remainder of the layered shoots attached to the parent plant are cut back to 23 cm (9 in.) as in the first year. Economical quantities of shoots are produced annually for as long as ten years. Rooted tip layers are planted in the late fall or early spring. New canes develop rapidly during the first season.

## Simple Layering

Simple layering is illustrated in Figures 14–3, 14–4, and 14–5.

The usual time for layering is in the early spring, and dormant, one-year-old shoots are used. Low, flexible branches of the plant, which can be bent easily to the ground, are chosen.

Shoots layered in the spring will usually be rooted adequately by the end of the first growing season and can be removed either in the fall or in the next spring before growth starts. Mature shoots layered in summer should be left through the winter and either removed the next spring before growth begins or left until the end of the second growing season. When the rooted layer is removed from the parent plant, it is treated essentially as a rooted cutting of the same plant (*11*).

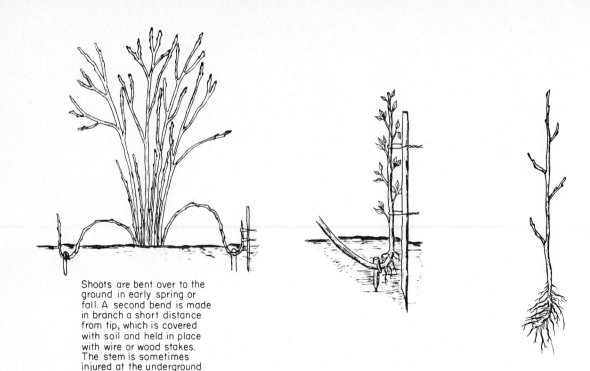

Shoots are bent over to the ground in early spring or fall. A second bend is made in branch a short distance from tip, which is covered with soil and held in place with wire or wood stakes. The stem is sometimes injured at the underground section which stimulates rooting. Includes notching, bending, wiring, or girdling.

Roots form on the buried part of the shoot near the bend.

The rooted layer is removed from the parent plant.

**Figure 14-3**    Steps in propagation by simple layering.

A supply of rooted layers can be produced over a period of years by establishing a stool bed composed of stock plants far enough apart to allow room

**Figure 14-4**    Propagation of two Dieffenbachia plants by simple layering. Leggy stems were curved and placed into containers of soil. After several months, strong root systems formed at the curved portion of stems (right); new plants are then severed from the mother plant for independent growth.

**Figure 14–5** Propagation of filberts in Oregon by simple layering. Center row consists of mother plants. Two outer rows are layered plants, which will be dug at end of growing season. Arrows point to shoots from mother plants, which have been bent over and placed under the soil for rooting in the manner shown in Figure 14–3.

for all shoots to be layered. This procedure has been used commercially to propagate certain hard-to-root shrubs (*10, 25*) as well as filberts (Figure 14–5).

### Compound or Serpentine Layering

Compound layering is essentially the same as simple layering, except that the branch is alternately covered and exposed along its length. Several new plants are thus possible from a single branch.

This method is used for propagating plants that have long, flexible shoots, such as the Muscadine grape. Ornamental vines, such as *Wisteria* and *Clematis* also can be propagated this way, although this method is used more often by amateurs than by commercial propagators.

### Air Layering (Chinese Layering, Pot Layerage, Circumposition, Marcottage, Gootee)

Air layering is used to propagate a number of tropical and subtropical trees and shrubs (*31, 46*), including the litchi (*17*) and the Persian lime (*Citrus aurantifolia*) (*40*). It is used in the greenhouse for *Ficus* species, croton, *Monstera,* and philodendron to produce large plants quickly (*24, 34*). Air layering is effective also for rooting mature pines (*2, 6, 19, 29*). With polyethylene film for wrapping the layers, it is possible to air layer plants out-of-doors (Figure 14–6) (*15, 17, 47*).

Air layers are made in the spring on stems of the previous season's growth or, in some cases, in the late summer with partially hardened shoots. Stems older than one year can be used in some cases, but rooting is less satisfactory and the larger plants produced are somewhat more difficult to handle after rooting. The presence of numerous active leaves on the layered shoot speeds root formation. With tropical greenhouse plants, layering should be done after several leaves have developed during a period of growth.

The first step in air layering is to girdle or cut the bark of the stem (see Figure 14–6). A strip of bark 1.8 to 2.5 cm (½ to 1 in.) wide, depending upon the kind of plant, is completely removed from around the stem. Scraping the exposed surface to insure complete removal of the phloem and cambium is desirable to retard healing. Another procedure is to make slanting upward cuts on one or

**Figure 14–6** Steps in making an air layer on a *Ficus elastica* plant using polyethylene film. *Left:* roots on a *Ficus elastica* air layer showing through the clear plastic covering. At this stage the layer is ready to be removed from the parent plant and potted. *Right (top):* the stem should be girdled for a distance of about 2.5 cm (1 in.) to induce adventitious root formation above the cut. A ball of slightly damp sphagnum moss is placed around the girdled section (*middle*). A wrapping of polyethylene film is placed around the sphagnum moss and tied at each end (*below*).

both sides of the stem about 3 cm (2 in.) long, keeping the two surfaces apart by sphagnum or a piece of wood. In *Ficus* (5), girdling has been shown to cause water deficits and leaf spotting, whereas cutting did not. Growing plants in 50 percent shade to reduce water stress is also effective. Application of a root-promoting material, such as indolebutyric acid, to the exposed wound is beneficial. Increasing concentration, up to 4 percent, of IBA in talc has increased rooting and survival in pecan air layers (39). About two handfuls of moistened sphagnum moss with excess moisture squeezed out are placed around the stem to enclose the cut surfaces. If the moisture content of the sphagnum moss is too high, the stem will decay.

A piece of polyethylene film 20 to 25 cm (8 to 10 in.) square is wrapped carefully about the branch so that the sphagnum moss is completely covered. The ends of the sheet should be folded (as in wrapping meat) with the fold placed on the lower side. The two ends must be twisted to make sure that no water can seep inside. Aluminum foil also is useful for this purpose (6, 45). Adhesive tape, such as electricians' waterproof tape, serves well to wrap the ends; the winding should be started well above the end of the cover to enclose the ends, particularly the upper one, securely. Budding rubbers and florist's ties are other materials that can be used for this purpose.

The layer is removed from the parent plant when roots are observed through the transparent film (Figure 14–6). In some plants, rooting occurs in two to three months or less. Layers made in spring or early summer are best left until the shoots become dormant in the fall, and are removed at that time. Holly, lilac, azalea, and magnolia should be left for two seasons (11). In general, it is desirable to remove the layer for transplanting when it is not actively growing.

Pruning is usually advisable to reduce the top in proportion to the roots. Pot the rooted layer into a suitable container and place it in mist or under cool, humid conditions, such as in an enclosed frame, where the plants can be syringed frequently. If potted in the fall, a sufficiently large root system usually develops by spring to permit successful growth in the open. Placing the rooted layers under mist for several weeks, followed by gradual hardening-off, is probably the most satisfactory procedure (35).

## Mound (Stool) Layering

Mound or stool layering is primarily used commercially to produce various apple and pear clonal rootstocks, quince, currants, and gooseberries (see Figures 14–7 and 14–8) (4, 7, 14, 16).

A stool bed is established by planting healthy mother plants of suitable size (8 to 10 mm diameter) in loose, fertile, well-drained soil one year before the propagation is to begin. The mother plants should be set 30 to 38 cm (12 to 15 in.) apart in the row, but the spacing between rows varies with different conditions and types of nursery equipment used. Width between rows should be sufficient to allow for cultivation and hilling operations during spring and summer. In England 1 m (3½ ft) rows are recommended (16), but in New York a minimum of 2.4 m (8 ft) was found desirable (4).

Before new growth starts the following spring, all plants are cut back to an inch above the ground level. Two to five new shoots usually develop from the crown the second year, more in later years. When these shoots have grown 7.6 to 12.7 cm (3 to 5 in.), loose soil, sawdust, or a soil-sawdust mixture is drawn up around each shoot to one-half its height. When the shoots have grown to a total height of 19 to 25 cm (8 to 10 in.), a second hilling operation takes place. Additional rooting medium is added, to be mounded around the bases of the shoots but not to more than half their total height. A third and final hilling operation is made in midsummer when the shoots have developed to a total length of approx-

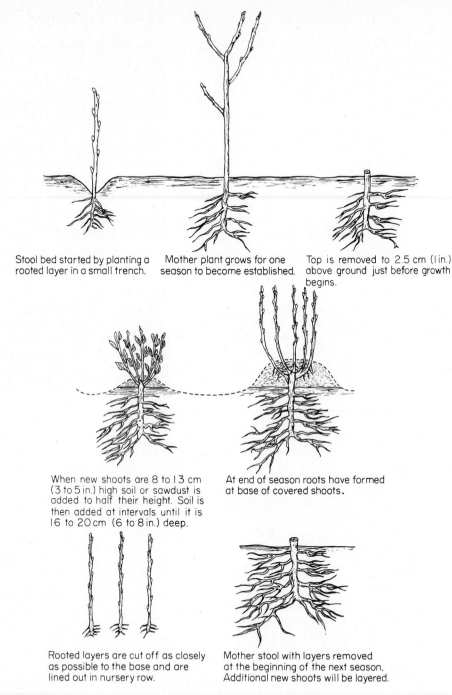

Stool bed started by planting a rooted layer in a small trench.

Mother plant grows for one season to become established.

Top is removed to 2.5 cm (1 in.) above ground just before growth begins.

When new shoots are 8 to 13 cm (3 to 5 in.) high soil or sawdust is added to half their height. Soil is then added at intervals until it is 16 to 20 cm (6 to 8 in.) deep.

At end of season roots have formed at base of covered shoots.

Rooted layers are cut off as closely as possible to the base and are lined out in nursery row.

Mother stool with layers removed at the beginning of the next season. Additional new shoots will be layered.

**Figure 14-7**     Steps in propagation by mound (stool) layering.

imately 45 cm (18 in.). The base of the shoots will then have been covered with soil to a depth of 15 to 20 cm (6 to 8 in.).

**Figure 14–8**  *Above:* rooting medium of wood shavings and soil pulled away showing root production by end of summer. *Below:* stool bed used in propagating clonal apple rootstocks.

Layered shoots of easily propagated plants should have rooted sufficiently by the end of the growing season to be separated from the parent stool for lining out in the nursery row. The rooted layers are cut close to their base to keep the height of the stool plant low. These rooted layers are handled as liners and transplanted directly into the nursery row.

After the shoots have been cut away, the mother stool remains exposed until new shoots have grown 7.6 to 12.7 cm (3 to 5 in.), when the hilling-up is begun for the next year.

Cutting back whole plants, then mound layering the vigorous juvenile shoots, has been described as a method of rooting six- to seven-year-old seedling cashew (*33*), one- to two-year-old seedling pecan (*28*), and other difficult-to-root plants.

A stool bed can be used for 15 to 20 years with proper handling, providing it is maintained in a vigorous condition, with disease, insect, and weed control.

Girdling the bases of the shoots by wiring about six weeks after they begin to grow will stimulate rooting in many plants (*15, 23, 26, 31, 37*). The size of the root system in apple layers has been increased on shoots growing through the spaces of a galvanized screen 0.5 cm (3/16 in.) square laid in a 45 cm (18 in.) strip down the row over the top of the cut-back stumps. New shoots growing through this screen gradually become girdled as the season progresses (*22*). Roots form on the stem above the screen.

Budded plants of apple and citrus (*13, 27*) have been produced by budding the layer in place in the stool bed. This may be done in the middle of the growing season, whereupon the budded layer is transplanted to the nursery in the fall for an additional season's growth.

## Trench Layering

Trench layering (**etiolation** method) consists of growing a plant or a branch of a plant in a horizontal position in the base of a trench and filling in soil around the new shoots as they develop, so that the shoot bases are etiolated. Roots develop from the base of these new shoots (see Figure 14–9). Trench layering is used primarily for woody species difficult to propagate by mound layering.

**Figure 14–9**     Steps in propagation by trench layering. *Upper left:* Mother plant after one year's growth in nursery. The trees were planted in the row at an angle of 30° to 45°. The trees are 46 to 78 cm (18 to 30 in.) apart down the row. *Upper right:* Just before growth begins, the plant is laid flat on the bottom of a trench about 5 cm (2 in.) deep. Shoots are cut back slightly and weak branches removed. Tree must be kept completely flat with wooden pegs or wire fasteners. *Lower left:* Rooting medium, as peat moss or sawdust, is added at intervals to produce etiolation on 5 to 7.5 cm (2 to 3 in.) of the base of the developing shoots. Apply first 2.5 to 5 cm (1 to 2 in.) layer before buds swell. Repeat as shoots emerge and before they expand. Later coverings are less frequent and only cover half of shoot. Final media depth is 15 to 19 cm (6 to 8 in.). *Lower right:* at the end of the season the media is removed and the rooted layers are cut off close to the parent plant. Layering may be repeated annually.

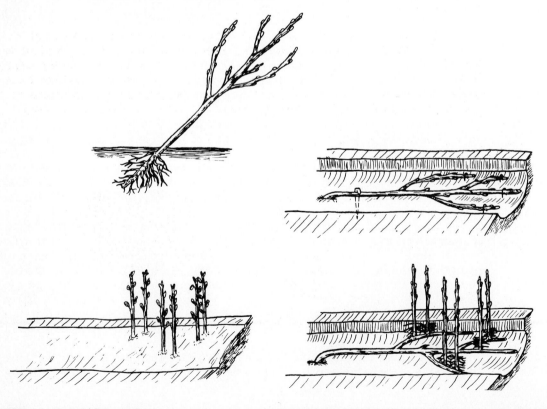

The first step in this procedure involves the establishment of the mother bed which, as in mound layering, can be used over a period of years. Rooted layers or one-year-old nursery-budded or grafted trees are planted 45 to 76 cm (18 to 30 in.) apart at an angle of 30 to 45 degrees down a row. The rows should be 1.2 to 1.5 m (4 to 5 ft) apart—wide enough to allow for cultivation and to draw soil up around the plants to a height of 15 cm (6 in.). The plants are then cut back to a uniform length—45 to 60 cm (18 to 24 in.)—and left to grow one season. In some cases (in layering walnuts, for instance) plants can be placed horizontally in the trench and the developing shoots layered the first year. Further details are given in Figure 14–9.

Trench layering could be practiced on established shrubs or trees by bending long, flexible shoots or vines to the ground as is done in simple layering, but laying them flat in trenches. The shoot is covered along its entire length, but the tip is left exposed. New shoots which develop from the buds along the stem grow upward through the soil, with roots forming at their base. This latter procedure is sometimes known as **continuous layering.**

## PLANT MODIFICATIONS REPRESENTING NATURAL LAYERING

Some plants exhibit modifications of their vegetative structure or method of growth that lead to their natural vegetative increase. Those listed below could be considered natural forms of layering and are often utilized for propagation.

### Runners

A **runner** is a specialized stem that develops from the axil of a leaf at the crown of a plant, grows horizontally along the ground, and forms a new plant at one of the nodes. The strawberry is a typical plant propagated in this way (Figure 14–10). Other plants propagated by runners include bugle (*Ajuga*), the strawberry geranium (*Saxifraga sarmentosa*), and the ground cover, *Duchesnea indica*. Plants of these species grow as a rosette or crown. Some ferns, such as

**Figure 14–10**    Runners arising from the crown of a strawberry plant. New plants are produced at every second node. The daughter plants, in turn, produce additional runners and runner plants.

**Figure 14–11** *Left:* masses of runners developing from spider plants (*Chlorophytum comosum*). *Right:* rooted plantlets at end of runners can be cut off and planted.

Boston fern (*Nephrolepsis*), produce runner-like branches. Runners of the spider plant (*Chlorophytum comosum*) are shown in Figure 14–11.

In most strawberry cultivars, runner formation is related to the length of day and temperature. Runners are produced in long days of 12 to 14 hours or more with high midsummer temperatures. New plants are produced at alternate nodes. These take root but remain attached to the mother plant. New runners are in turn produced by the daughter plants. The connecting stems die in the late fall and winter, and each daughter plant becomes separate from the others.

In propagating by runners, the rooted daughter plants are dug when they have become well rooted, and then transplanted to the desired locations. Strawberry plants are also propagated extensively by micropropagatio (*3, 12*).

### Stolons

**Stolons,** produced by some plants, are modified stems that grow horizontal to the ground. These may be prostrate or sprawling stems growing aboveground as found in some woody species, such as *Cornus stolonifera*. The term also describes the horizontal stem structure occurring in Bermuda grass (*Cynodon dactylis*), *Ajuga,* mint (*Mentha*), and *Stachys* (*36*). Stolonlike underground stems are involved in tuberization; they are the stems that develop into a potato tuber.

The stolon can be treated as a naturally occurring rooted layer and can be cut from the parent plant and planted.

### Offsets

An **offset** is a characteristic type of lateral shoot or branch that develops from the base of the main stem in certain plants. This term is applied generally to a shortened, thickened stem of rosettelike appearance. Many bulbs reproduce by producing typical offset bulblets from their base (see chapter 15 for details).

**Figure 14–12**     Removing a date offshoot with chisel and sledge hammer.
From R. W. Nixon, Date culture in the United States, *USDA Cir. 728.*

The term offset (or **offshoot**) also applies to lateral branches arising on stems of monocotyledons as in date palm, pineapple, or banana (see chapter 18) (Figure 14–12).

Offsets are removed by cutting them close to the main stem with a sharp knife. If it is well rooted, the offset can be potted and rooted like a cutting. If insufficient roots are present, the shoot is placed in a favorable rooting medium and treated as a leafy stem cutting.

In cases in which offset development is meager, cutting back the main rosette may stimulate the development of more offsets from the old stem just as removing the terminal bud stimulates lateral shoots in any other type of plant.

Natural increase by offsets tends to be slow and not suited for commercial propagation. However, the natural rate in many of these monocotyledonous plants can be stimulated greatly by the multiplicative techniques used in micropropagation in aseptic culture (see chapters 16 and 17). With the development of such procedures, commercial propagation of the desirable plants of this group has been greatly enhanced. Shoot increase takes place either by stimulation of axillary branching or by induction of adventitious shoots.

Offshoots of the date palm do not root readily if separated from the parent plant. They are usually layered for a year prior to removing. Alternate methods are available utilizing micropropagation of shoot tips or lateral buds to multiply specific clones vegetatively on aseptic culture. Plantlets can be produced by somatic embryos from callus or from proliferation and rooting of shoot tips and lateral buds, although rates of production may be relatively low (*42*).

### Suckers

A **sucker** is a shoot that arises on a plant from below ground, as shown in Figure 14–13. The most precise use of this term is to designate a shoot that arises

**Figure 14–13**    Suckers arising as adventitious shoots from the roots of a red raspberry plant. After they are well rooted the suckers may be cut from the parent plant and transplanted to their permanent location.

from an adventitious bud on a root. However, in practice, shoots that arise from the vicinity of the crown are called *suckers* even though originating from stem tissue. Nurserymen generally designate any shoot produced from the rootstock below the bud union of a budded tree as a *sucker* and refer to the operation of removing them as *"suckering."* In contrast, a shoot arising from a latent bud of a stem several years old, as, for instance, on the trunk or main branches, should be termed a **watersprout.**

Suckers are dug out and cut from the parent plant. In some cases part of the old root may be retained, although most new roots arise from the base of the sucker. It is important to dig the sucker out rather than pull it, to avoid injury to its base. Suckers are treated essentially as a rooted layer or as a cutting, in case few or no roots have formed. They are usually dug during the dormant season.

### Crown Division

The term **crown** as generally used in horticulture designates that part of a plant at the surface of the ground from which new shoots are produced. In trees or shrubs with a single trunk, the crown is principally a point of location near the ground surface marking the general transition zone between stem and root. In herbaceous perennials, the crown is the part of the plant from which new shoots arise annually. The crown of herbaceous perennials consists of many branches, each being the base of the current season's stem, which originated from the base of the preceding year's branch. These lateral shoots are stimulated to grow from the base of the old stem as it dies back after blooming. Adventitious roots develop along the base of the new shoots. These new shoots eventually flower either the same year they are produced or the following year. As a result of the annual production of new shoots and the dying back of old shoots, the crown may become extensive within a period of a relatively few years and may need to be divided every few years to prevent overcrowding.

Multibranched woody shrubs may develop extensive crowns. Although an individual woody stem may persist for a number of years, new, vigorous shoots are continuously produced from the crown, and eventually crowd out the older shoots. Crowns of such shrubs can be divided in the dormant season and treated as a large rooted cutting.

**Crown division** is an important method of propagation for herbaceous perennials, and to some extent for woody shrubs, because of its simplicity and reliability. Such characteristics make this method particularly useful to the amateur or professional gardener who is generally interested in only a modest increase of a particular plant.

Crowns of outdoor herbaceous perennials are usually divided in the spring just before growth begins or in late summer or autumn at the end of the growing season. As a general rule, plants that bloom in the spring and summer and produce new growth after blooming should be divided in the fall. Those that bloom in summer and fall and make little or no new growth until spring should be divided in early spring. Potted plants are divided when they become too large for the particular container in which they are growing. Division is necessary to maintain the variegated form in some plants usually propagated by leaf cuttings, such as *Sanseviera* (*20*).

In crown division, plants are dug and cut into sections with a knife. In herbaceous perennials, as the Shasta daisy (*Chrysanthemum superbum*) or day lily (*Hemerocallis*) (see Figure 14–14), where an abundance of new rooted offshoots are produced from the crown, each may be broken from the old crown and planted separately, the older part of the plant clump being discarded.

For commercial production of improved cultivars, multiplication is very slow. Increased production of shoots for rooting can be produced by cutting back to the crown in the early spring after new growth starts and treating with a cytokinin (*1*). More rapid multiplication is achieved in tissue culture propagation, regenerating plants from callus produced on flower petals and sepals (*21, 30*) (see chapters 16 and 17).

Herbaceous perennials generally propagated by division can be readily propagated also by micropropagation in aseptic culture systems (*49*). Axillary branching is greatly enhanced and multiplication can be made at very high rates with minimum stock plants.

**Figure 14–14**    Propagation by crown division illustrated by division of day lily *(Hemerocallis)* clump.

# REFERENCES

1. Apps, D. A., and C. W. Heuser. 1975. Vegetative propagation of *Hemerocallis*—including tissue culture. *Proc. Inter. Plant Prop. Soc.* 25:362–67.

2. Barnes, R. D. 1974. Air-layering of grafts to overcome incompatibility problems in propagating old pine trees. *New Zealand Jour. For. Sci.* 4(2):120–26.

3. Boxus, P., M. Quoirin, and J. M. Laine. 1977. Large scale propagation of strawberry plants from tissue culture. In *Applied and fundamental aspects of plant cell, tissue and organ culture,* J. Reinert and Y. P. S. Bajaj, eds. New York: Springer-Verlag, pp. 130–43.

4. Brase, K. D., and R. D. Way. 1959. Rootstocks and methods used for dwarfing fruit trees. *N.Y. Agr. Exp. Sta. Bul. 783.*

5. Broschat, T. K., and H. M. Donselman. 1981. Effects of light intensity, air layering, and water stress on leaf diffusive resistance and incidence of leaf spotting in *Ficus elastica. HortScience* 16(2):211–12.

6. Cameron, R. J. 1968. The leaching of auxin from air layers. *New Zealand Jour. Bot.* 6(2):237–39.

7. Carlson, R. F., and H. B. Tukey. 1955. Cultural practices in propagating dwarfing rootstocks in Michigan. *Mich. Agr. Exp. Sta. Quart. Bul.* 37:492–97.

8. Chase, H. H. 1964. Propagation of oriental magnolias by layering. *Proc. Inter. Plant Prop. Soc.* 14:67–69.

9. Ching, F., C. L. Hamner, and F. Widmoyer. 1956. Air-layering with polyethylene film. *Mich. Agr. Exp. Sta. Quart. Bul. 39*:3–9.

10. Congdon, M. L. 1954. Mass production of deciduous shrubs by layering. *Proc. Plant Prop. Soc.* 4:39–45.

11. Creech, J. L. 1954. Layering. *Nat. Hort. Mag.* 33:37–43.

12. Damiano, C. 1980. Strawberry micropropagation. In *Proc. conf. on nursery production of fruit plants through tissue culture; Applications and feasibility,* R. H. Zimmerman, ed. U.S. Dept. of Agr. Sci. and Education Administration ARR–NE–11, pp. 11–22.

13. Duarte, O., and C. Medina. 1971. Propagation of citrus by improved mound layering. *HortScience* 6:567.

14. Dunn, N. D. 1979. Commercial propagation of fruit tree rootstocks. *Proc. Inter. Plant Prop. Soc.* 29:187–90.

15. Du Preez, D. 1954. Propagation of guavas. *Farming in South Africa* 29:297–99.

16. Garner, R. J. 1979. *The grafter's handbook* (4th ed.). New York: Oxford University Press.

17. Grove, W. R. 1947. Wrapping air layers with rubber plastic. *Proc. Fla. State Hort. Soc.* 60:184–89.

18. Hanger, F. E. W., and A. Ravenscroft. 1954. Air layering experiments at Wisley. *Jour. Roy. Hort. Soc.* 79:111–16.

19. Hare, R. C. 1979. Modular air-layering and chemical treatments improve rooting of loblolly pine. *Proc. Inter. Plant Prop. Soc.* 29:446–54.

20. Henley, R. W. 1979. Tropical foliage plants for propagation. *Proc. Inter. Plant Prop. Soc.* 29:454–67.

21. Heuser, C. W., and J. Harker. 1976. Tissue culture propagation of daylilies. *Proc. Inter. Plant Prop. Soc.* 25:269–72.

22. Hogue, E. J., and R. L. Granger. 1969. A new method of stool bed layering. *Hort-Science* 4:29–30.

23. Hostermann, G. 1930. Versuche zur vegetativen Vermehrung von Gehozen nach dem Dahlemer Drahtungsverfahren. *Ber. Deutsch. Bot. Ges.* 48:66–70.

24. Joiner, J. N., ed. 1981. *Foliage plant production.* Englewood Cliffs, N.J.: Prentice-Hall.

25. Knight, F. P. 1945. The vegetative propagation of flowering trees and shrubs. *Jour. Roy. Hort. Soc.* 70:319–30.

26. Maurer, K. J. 1950. Möglichkeiten der vegetativen Vermehrung der Walnuss. *Schweiz. Z. Obst. V. Weinb.* 59:136–37.

27. Medina, C., and O. Duarte. 1971. Propagating apples in Peru by an improved mound layering method. *Jour. Amer. Soc. Hort. Sci.* 96:150–51.

28. Medina, J. P. 1981. Studies of clonal propagation on pecans at Ica, Peru. *The Plant Propagator* 27(2):10–11.

29. Mergen, F. 1955. Air layering of slash pine. *Jour. For.* 53:265–70.

30. Meyer, M. M. 1976. Propagation of daylilies by tissue culture. *HortScience* 11:485–87.

31. Mowry, H., L. R. Toy, and H. S. Wolfe. 1953. Miscellaneous tropical and subtropical Florida fruits. *Fla. Agr. Ext. Ser. Bul. 156,* pp. 1–110.

32. Modlibowska, I., and C. P. Field. 1942. Winter injury to fruit trees by frost in England (1939–1940). *Jour. Pom. Hort. Sci.* 19:197–207.

33. Nagabhushanam, S., and M. A. Menon. 1980. Propagation of cashew (*Anacardium occidentale* L.) by etiolation, girdling and stooling. *The Plant Propagator* 26:11–13.

34. Neel, P. L. 1979. Macropropagation of tropical plants as practiced in Florida. *Proc. Inter. Plant Prop. Soc.* 29:468–80.

35. Nelson, R. 1953. High humidity treatment for air layers of lychee. *Proc. Fla. State Hort. Soc.* 66:198–99.

36. Nitsch, J. P. 1971. Perennation through seeds and other structures. In *Plant physiology,* Vol. 6A, F. C. Steward, ed. New York: Academic Press, pp. 413–79.

37. Oppenheim, J. E. 1932. A new system of citrus layers. *Hadar* 5:2–4.

38. Singh, L. B. 1960. *The mango.* London: Leonard Hill.

39. Sparks, D., and J. W. Chapman. 1970. The effect of indole-3-butyric acid on rooting and survival of air-layered branches of the pecan, *Carya illinoensis* Koch, cv. 'Stuart'. *HortScience* 5(5):445–46.

40. Sutton, N. E. 1954. Marcotting of Persian limes. *Proc. Fla. State Hort. Soc.* 67:219–20.

41. Thomas, L. A. 1938. Stock and scion investigations. II. The propagation of own-rooted apple trees. *Jour. Counc. Sci. Industr. Res. Org., Austral.* 11:175–79.

42. Tisserat, B. 1981. Date palm tissue culture. *USDA, Agri. Res. Serv., Adv. in Agr. Tech., Western Series No. 17,* pp. 1–50.

43. Tukey, H. B., and K. Brase. 1930. Granulated peat moss in field propagation of apple and quince stocks. *Proc. Amer. Soc. Hort. Sci.* 27:106–13.

44. Vieitez, E. 1974. Vegetative propagation of chestnut. *New Zealand Jour. For. Sci.* 4(2):242–52.

45. Vinzant, K. L. 1978. Propagation of *Schefflera arboricola* by air layerage and factors affecting long term storage. *The Plant Propagator* 24(1):5–6.

46. Watkins, J. V. 1952. Propagation of ornamental plants. *Fla. Agr. Exp. Sta. Bul. 150,* pp. 1–15.

47. Wyman, D. 1952. Air layering with polyethylene films. *Jour. Roy. Hort. Soc.* 77:135–40.

48. ———. 1953. Layering plants in Holland. *Arnoldia* 13:25–28.

49. Zilis, M., D. Zwagerman, D. Lamberts, and L. Kurtz. 1979. Commercial propagation of herbaceous perennials by tissue culture. *Proc. Inter. Plant Prop. Soc.* 29:404–13.

## SUPPLEMENTARY READING

GARNER, R. J. 1979. *The grafter's handbook* (4th ed.). New York: Oxford University Press.

GARNER, R. J., and S. A. CHAUDHRI. 1976. *The propagation of tropical fruit trees.* Hort. Rev. No. 4. Commonwealth Bureau of Hort. and Plant. Crops. East Malling, Maidstone, Kent.

McMILLAN-BROWSE, P. D. A. 1979. *Plant propagation.* New York: Simon and Schuster.

MINISTRY OF AGRICULTURE, FISHERIES AND FOOD (London). 1969. Fruit tree raising—rootstocks and propagation (5th ed.). *Bul. 135.*

TUKEY, H. B. 1964. *Dwarfed fruit trees.* New York: Macmillan.

Bulbs, corms, tubers, tuberous roots and stems, rhizomes, and pseudobulbs are specialized vegetative structures that function primarily in the storage of food for the plant's survival during adversity. Plants possessing these modified plant parts are generally herbaceous perennials in which the shoots die down at the end of a growing season, and the plant survives in the ground as a dormant, fleshy organ that bears buds to produce new shoots the next season. Such plants are well suited to withstand periods of adverse growing conditions in their yearly growth cycle. The two principal climatic cycles for which such performance is adapted are the warm-cold cycle of the temperate zones and the wet-dry cycle of tropical and subtropical regions.

These specialized organs also function in vegetative reproduction. The propagation procedure that utilizes the production of naturally detachable structures, such as the bulb and corm, is generally spoken of as **separation.** In cases in which the plant is cut into sections, as is done with the rhizome, stem tuber, and tuberous root, the process is spoken of as **division.**

## BULBS

### Definition and Structure

A **bulb** is a specialized underground organ consisting of a short, fleshy, usually vertical stem axis (**basal plate**) bearing at its apex a growing point or a flower primordium enclosed by thick, fleshy scales. (See Figure 15-1.) Bulbs are produced by monocotyledonous plants in which the usual plant structure is modified for storage and reproduction.

Most of the bulb consists of **bulb scales,** which morphologically are the continuous, sheathing leaf bases (see Figure 15-2). The outer bulb scales are

# 15

# Propagation
# by Specialized Stems
# and Roots

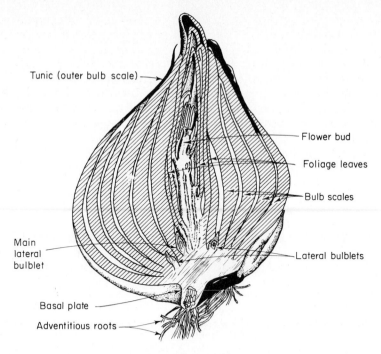

Tunic (outer bulb scale)

Flower bud

Foliage leaves

Bulb scales

Main lateral bulblet

Lateral bulblets

Basal plate

Adventitious roots

**Figure 15–1**   The structure of a tulip bulb—an example of a **tunicated laminate** bulb. Longitudinal section representing stage of development shortly after the bulb is planted in the fall. Redrawn from Mulder and Luyten (*41*).

**Figure 15–2**   Cross section of a daffodil bulb. The continuous, concentric leaf scales found in the laminate bulb are shown. Also shown is the perennial nature of the daffodil bulb, which continues to grow by producing a new bulb annually at the main meristem. Lateral **offset** bulbs are also produced; parts of five individual, differently aged bulbs are shown here. Redrawn from Huisman and Hartsema (*27*).

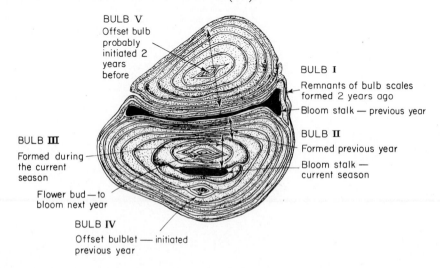

BULB **V**
Offset bulb probably initiated 2 years before

BULB **I**
Remnants of bulb scales formed 2 years ago
Bloom stalk — previous year

BULB **III**
Formed during the current season

BULB **II**
Formed previous year
Bloom stalk — current season

Flower bud — to bloom next year

BULB **IV**
Offset bulblet — initiated previous year

generally fleshy and contain reserve food materials, whereas the bulb scales toward the center function less as storage organs and are more leaflike. In the center of the bulb, there is either a vegetative meristem or an unexpanded flowering shoot. Meristems develop in the axil of these scales to produce miniature bulbs, known as **bulblets,** which when grown to full size are known as **offsets.** In various species of lilies, bulblets may form in the leaf axils either on the underground portion or on the aerial portion of the stem. The aerial bulblets are called **bulbils,** while the underground organs are called **stem bulblets.** There are two types of bulbs:

TUNICATE BULBS

**Tunicate (laminate)** bulbs are represented by the onion, daffodil, and tulip. These bulbs have outer bulb scales that are dry and membranous. This covering, or **tunic,** provides protection from drying and mechanical injury to the bulb. The fleshy scales are in continuous, concentric layers, or **lamina,** so that the structure is more or less solid.

There are three basic bulb structures, defined by type of scales and growth pattern. The amaryllis (*Hippeastrum*) is an example of one type, in which the scales are expanded bases of leaves (Figure 15–3). A second type, which includes the *Narcissus,* has both expanded leaf bases and true scales. Tulip is an example of the third type of bulb, which has only true scales, with leaves produced on the flowering or vegetative shoot (*45*).

Adventitious root primordia are present on the dormant stored bulb. They do not elongate until planted under proper conditions and at the proper time. They occur in a narrow band around the outside edge on the bottom of the basal plate.

**Figure 15–3**     Diagram showing morphology and growth cycle of *Hippeastrum* bulb. New bulbs continually develop from the center in a cycle of four leaves and an inflorescence. Bases of these leaves enlarge to become the scales that contain stored food. Oldest scales disintegrate. From A. R. Rees, *The growth of bulbs,* Academic Press, 1972.

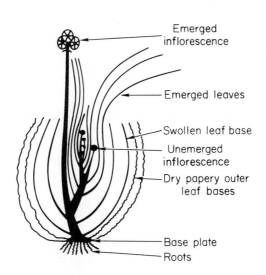

Emerged inflorescence

Emerged leaves

Swollen leaf base

Unemerged inflorescence

Dry papery outer leaf bases

Base plate

Roots

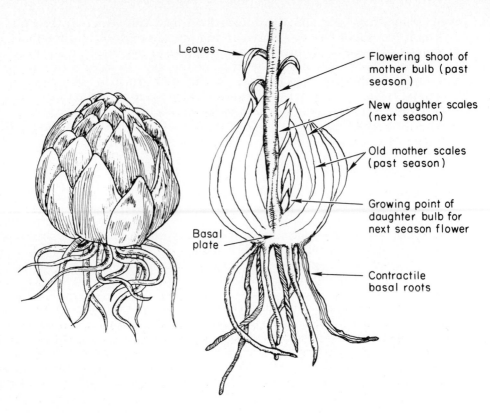

Leaves

Flowering shoot of
mother bulb (past
season)

New daughter scales
(next season)

Old mother scales
(past season)

Growing point of
daughter bulb for
next season flower

Basal
plate

Contractile
basal roots

**Figure 15-4**     *Left:* outer appearance of a **scaly** bulb of lily (*Lilium hollandicum*). *Right:* longitudinal section of bulb of *L. longiflorum* 'Ace', after flowering stage, showing old mother bulb scales and new daughter bulb scales. Bulb obtained in fall near digging time (*11*).

NONTUNICATE BULBS

**Nontunicate (scaly)** bulbs are represented by the lily (Figure 15-4). These bulbs do not possess the enveloping dry covering. The scales are separate and attached to the basal plate. In general, the nontunicate bulbs are easily damaged and must be handled more carefully than the tunicate bulbs; they must be kept continuously moist because they are injured by drying. In the nontunicate lily bulb new roots are produced in midsummer or later, and persist through the following year (*49*). In most lily species, roots also form on the stem above the bulb.

In many species thickened **contractile** roots shorten and pull the bulb to a given level in the ground (*45, 58*). Tulips do not produce contractile roots but produce **droppers** (*45*), stolonlike structures that grow from the bulb and produce a bulb at the tip.

**Growth Pattern**

An individual bulb goes through a characteristic cycle of development, beginning with its initiation as a meristem and terminating in flowering and seed production. This general developmental cycle is composed of two stages: (a) vegetative and (b) reproductive. In the vegetative stage the bulblet grows to flowering size and attains its maximum weight. The subsequent reproductive stage includes the induction of flowering, differentiation of the floral parts, elongation of the flowering shoot, and finally flowering and (sometimes) seed production. Various bulb species have specific environmental requirements for the individual phases of this cycle that determine their seasonal behavior, environmental adaptations, and methods of handling. They can be grouped into classes according to their time of bloom and method of handling.

SPRING-FLOWERING BULBS

Important commercial crops included in the spring-flowering group are the tulip, daffodil, hyacinth, and bulbous iris, although other kinds are grown in gardens.

*Bulb formation*    The vegetative stage begins with the initiation of the bulblet on the basal plate in the axil of a bulb scale. In this initial period, which usually occupies a single growing season, the bulblet is insignificant in size, since it is present within another growing bulb and can be observed only if the bulb is dissected. Its subsequent pattern of development and the time required for the bulblet to attain flowering size differ for different species. The bulbs of the tulip and the bulbous iris, for instance, disintegrate upon flowering and are replaced by a cluster of new bulbs and bulblets initiated the previous season. The largest of these may have attained flowering size at this time, but smaller ones require additional years of growth (see Figures 15–1 and 15–5). The flowering bulb of the daffodil, on the other hand, continues to grow from the center year by year, producing new offsets, which may remain attached for several years (see Figures 15–2 and 15–5). The hyacinth bulb also continues to grow

**Figure 15–5**    Propagation by offset bulbs. *Top left:* bulbous iris. The old bulb disintegrates, leaving a cluster of bulbs. *Top right:* daffodil. Bulb continues to grow from the inside each year, but continuously produces lateral bulbs, which eventually split away. *Below left:* daffodil. Three types of bulbs: "split," or "slab," bulb; "round" bulb; and "double-nose" bulb. *Below right:* Hyacinth. Lateral bulblets produced, but old bulb continues to develop from inside.

year by year, but because the number of offsets produced is limited, artificial methods of propagation are usually used.

The size and quality of the flower are directly related to the size of the bulb. A bulb must reach a certain minimum size to be capable of initiating flower primordia. Commercial value is largely based on bulb size (3), although the condition of the bulb and freedom from disease are also important quality factors.

Increase in size and weight of the developing bulb takes place in the period during and (mostly) after flowering, as long as the foliage remains in good condition (10). Cultural operations that include irrigation; weed, disease, and insect control; and fertilization encourage vegetative growth. The benefit, however, is to the next year's flower, because larger bulbs are produced. Conversely, adverse situations, such as poor growing conditions, removal of foliage, and premature digging of the bulb, result in smaller bulbs and reduced flower production.

Moderately low temperatures tend to prolong the vegetative period, whereas higher temperatures may cause the vegetative stage to cease and the reproductive stage to begin. Thus a shift from cool to warm conditions early in the spring, as occurs in mild climates, will shorten the vegetative period, result in smaller bulbs, and consequently produce inferior blooms the following year (45). Commercial bulb-producing areas for hardy spring-flowering bulbs are largely in regions of cool springs and summers, such as The Netherlands and the Pacific Northwest of the United States.

The relative length of photoperiod apparently is not an important factor affecting bulb formation in most species. It has been shown, however, to be significant in some *Allium* species, such as onion and garlic (26, 36).

*Flower bud formation and flowering*     The beginning of the reproductive stage and the end of the vegetative stage is indicated by the drying of the foliage and the maturation of the bulb. From then on, no additional increase in size or weight of the bulb takes place. The roots disintegrate, and the bulb enters a seemingly "dormant" period. However, important internal changes take place, and in some species the vegetative growing point undergoes transition to a flowering shoot. In nature, all bulb activity takes place underground during this period; in horticultural practice, the bulbs are dug, stored, and distributed during this three- to four-month period.

Temperature controls the progression from the vegetative stage to flowering (25, 26). Differentiation of flower primordia for the spring flowering group occurs at moderately warm temperatures in late summer or early fall either in the ground or in storage. Subsequent exposure to lower but above-freezing temperatures is required to promote flower stalk elongation. As temperatures increase in the spring, the flower stalk elongates and the bulb plant subsequently flowers. Bulbous iris is an exception in that flower induction is induced by low temperatures in fall and early winter. For species in groups II and III, induction occurs in spring after the storage period.

The optimum temperature—determined by the shortest time in which the bulb would flower—has been established for these developmental phases for the important bulb species, such as tulip (57), hyacinth (9), or daffodil (57), depending also upon cultivar. Such information is important in establishing forcing schedules for retail florist sales (5). Holding the bulbs continuously at high temperatures (30° to 32° C; 86° to 90° F or more), or at temperatures near freezing, will inhibit or retard floral development and can be used to lengthen the period required for flowering. With a shift to favorable temperatures, flower bud development will continue. This treatment (below or above optimum) can be used when shipping bulbs from Northern- to Southern-Hemisphere countries.

### SUMMER FLOWERING BULBS

Lilies are important plants with a growth cycle geared to the seasonal pattern of the summer-winter cycles of the temperate zone. Their nontunicate bulbs do not go "dormant" in late summer and fall as does the tunicate type of bulb, but have unique characteristics that must be understood for proper handling. Although different lily species have somewhat different methods of reproduction (*49, 59*) the pattern, as determined for the Easter lily (*Lilium longiflorum*), can serve as a model (*6, 11*).

Lilies flower in the late spring or early summer at the apex of the leaf-bearing stem axis. The flower-producing bulb is known as the **mother bulb** made up of the basal plate, fleshy scales, and the flowering axis (Figure 15–3). Prior to flowering, a new daughter bulb(s) is developing within the mother bulb. It had been initiated the previous fall and winter from a growing point in the axil of a scale at the base of the stem axis. During spring the daughter bulb initiates new scales and leaf primordia at the growing point. Natural chemical inhibitors in the daughter scales prevent elongation of the daughter axis, which remains dormant—but it can be promoted to grow by exposure to high temperature (37.5° C; 100° F), to low temperature (4.5° C; 40° F), or by treatment with gibberellic acid (*54*).

After flowering of the mother bulb, no more scales are produced by the daughter bulb, but it increases in size (circumference) and weight until it equals the weight of the mother bulb surrounding it. Inhibitory effects of the daughter scales decrease, as does the response to dormancy-breaking treatments (*54*). Fleshy basal roots persist on the mother bulb through the fall and winter, and new adventitious roots develop in late summer or fall from the basal plate. Warm temperatures promote root formation (*12*). Bulbs should be dug for transplanting after they "mature" in the fall. The top may or may not have died down. Bulbs should be handled carefully to avoid injury and to prevent drying. The commercial value of the bulb depends on size (transverse circumference) and weight at the time of digging (*34*); and on the condition of the fleshy roots and the freedom of the bulb from disease.

Transition of the meristem to a flowering shoot does not take place until first, the stem axis has protruded through the "nose" of the bulb (*6*), and then the bulb has been subjected to chilling temperatures (*32*). The critical temperatures are 15.5° to 18.5° C (60° to 65° F) or less, and the chilling effect becomes more and more effective down to 2° to 4.5° C (35° to 40° F). Storing bulbs at warm (21° C; 70° F or more), or low (−0.5° C; 31° F) temperatures will keep bulbs dormant and delay blooming (*53*). Moisture content of the storage medium is important; if too dry the bulb will deteriorate and if too wet, it will decay.

Following the flower induction stage, and with the onset of higher temperatures, the stem elongates, initiating first leaves and then flowers. The outer scales of the "old" mother bulb rapidly disintegrate early in the spring as the new mother bulb produces the flower.

Stem bulblets may develop in the axils of the leaves underground or, in some species, bulbils may develop above ground. These appear about the time of flowering.

Flowering time and the size and quality of the Easter lily bloom can be closely regulated by manipulating the temperature at various stages following the digging of bulbs in the fall (*11*). Day length also influences development, but to a lesser extent (*55*).

TENDER, WINTER-FLOWERING BULBS

There are a number of flowering bulbs from tropical areas whose growth cycle is related to a wet-dry, rather than a cold-warm climatic cycle. The amaryllis (*Hippeastrum vittata*) is an example (Figure 15–4) (*25, 57*). This bulb is a perennial, growing continuously from the center with the outer scales disintegrating. New leaves are produced continuously from the center during the vegetative period extending from late winter to the following summer. In the axil of every fourth leaf (or scale) that develops, a meristem is initiated. Thus throughout the vegetative period a series of vegetative offsets is produced. By fall the leaves mature and the bulb becomes dormant, during which time the bulb should be dry. In this period, the *fourth* growing point from the center and any external to it differentiate into flower buds, and the shoot begins to elongate slowly. After two or three months of dry storage, the bulbs can be watered, causing the flowering shoots to elongate rapidly, with flowering taking place in midwinter. Maximum foliage development and bulb growth are essential to produce a bulb large enough to form a flowering shoot.

## Propagation

### OFFSETS

Offsets are used to propagate many kinds of bulbs. This method is sufficiently rapid for the commercial production of tulip, daffodil, bulbous iris, and grape hyacinth but, in general, is too slow for the lily, hyacinth, and amaryllis.

If undisturbed, the offsets may remain attached to the mother bulb for several years. They can also be removed at the time the bulbs are dug and replanted into beds or nursery rows to grow into flowering-sized bulbs. This may require several growing seasons, depending upon the kind of bulb and size of the offset.

*Tulip*    Tulip bulb planting takes place in the fall (*9, 17*). Two systems of planting are used: the bed system, used extensively in Holland, and the row or field system, used mostly in the United States and England. Beds are usually 1 m (3 ft.) wide and separated by 31 to 45 cm (12- to 18-in.) paths. The soil is removed to a depth of 9 cm (4 in.), the bulbs set in rows 15 cm (6 in.) apart, and the soil replaced. In the other system, single or double rows are placed wide enough apart to permit the use of machines. To improve drainage, two or three adjoining rows may be planted on a ridge. Bulbs are spaced one to two diameters apart, with small bulbs scattered along the row. A mulch may be applied after planting but removed the following spring before growth.

Planting stock consists principally of those bulbs of the minimum size for flowering, 9 to 10 cm (3.6 to 4 in.) or smaller in circumference. Since the time required to produce flowering size varies with the size of the bulb, the planting stock is graded so that all those of one size can be planted together. For instance, an 8-cm (3.2 in.) or larger bulb normally requires a single season to become flowering size; a 5- to 7-cm (2 to 3 in.) bulb, two seasons; and those 5 cm or less, three years (*17*).

During the flowering and subsequent bulb growing period of the next spring, good growing conditions should be provided so that the size and weight of the new bulbs will be at a maximum. Foliage should not be removed until it dries or matures. Important cultural operations include removal of competing weed growth, irrigation, fungicidal sprays to control *Botrytis blight* (*15*), and fertilization. Beds should be inspected for disease early in the season and for trueness-to-cultivar at the time of blossoming. All diseased or off-type plants should be rogued out (*23*). It is desirable to remove the flower heads at blooming time, because they may serve as a source of *Botrytis* infection and can lower bulb weight (*1*).

Bulbs are dug in early to midsummer when the leaves have turned yellow or the outer tunic of the bulb has become dark brown in color. In the Pacific Northwest of the United States, where summer temperatures are cool and the leaves remain green for a longer period, digging may take place before the leaves dry. If the bulbs are dug too early or if warm weather causes early maturation, the bulbs may be small in size. The bulbs are dug by machine or by hand with a short-handled spade. After the loose soil is shaken from the bulbs, they are placed in trays in well-ventilated storage houses for drying, cleaning, sorting, and grading. General storage temperatures are 18° to 20°C (65° to 68° F). To force early flowering, the bulbs should be held at 20°C for three to five weeks and then placed at 9° C (48° F) for eight weeks. Later flowering can be produced by holding bulbs at 22° C (72° F) for ten weeks. For shipment from Northern- to Southern-Hemisphere countries, the bulbs can be held at −1° C (31° F) until late December, when they are shifted to a higher temperature (25.5° C; 78° F) (*9*).

*Daffodil*    Daffodil bulbs are perennial and produce a new meristem growing point at the center every year (*18*). Offsets are produced that grow in size for several years until they break away from the original bulb, although they are still attached at the basal plate. An offset bulb, when it first separates from the mother bulb, is known as a "split," "spoon," or "slab," and can be separated from the mother bulb and planted. Within a year it becomes a "round," or "single-nose," bulb containing a single flower bud. One year later a new offset should be visible, enclosed within the scales of the original bulb, indicating the presence of two flower buds. At this stage the bulb is known as a "double-nose." By the next year the offsets split away, then the bulb is known as a "mother bulb." Grading of daffodil bulbs is principally by age, that is, as splits, round, double-nose, and mother bulbs. The grades marketed commercially are the round and the double-nose bulb. The mother bulbs are used as planting stock to produce additional offsets, and only the surplus is marketed. Offsets, or splits, are replanted for additional growth.

Storage should be at 13° to 16° C (55° to 60° F) with a relative humidity of 75 percent. To force earlier flowering, they can be stored at 9° C (48° F) for eight weeks. To delay flowering, store at 22° C (72° F) for 13 to 15 weeks. For shipment from the Northern to the Southern Hemisphere, the bulbs can be held at 30° C (86° F) until October, then stored at −1° C (31° F) until late December, and then at 25° C (77° F) (9).

Hot water treatment plus a fungicide for stem and bulb nematode control is important (23). A three- or four-hour treatment at 43° C (110° F) is used, but that temperature must be carefully maintained or the bulbs may be damaged.

*Lilies*    Lilies increase naturally, but except for a few species this increase is slow and of limited propagation value except in home gardens (20, 49, 59). Several methods of bulb increase are found among the different species. For instance, *Lilium concolor, L. hansonii, L. henryi,* and *L. regale* increase by bulb splitting. Two to four lateral bulblets are initiated about the base of the mother bulb, which disintegrates during the process, leaving a tight cluster of new bulbs. *Lilium bulbiferum, L. canadense, L. pardalinum, L. parryi, L. superbum,* and *L. tigrinum* multiply from lateral bulblets produced from the rhizome-like bulb. This process is sometimes called "budding-off."

### BULBLET FORMATION ON STEMS

*Underground stem bulblets* are used to propagate the Easter lily (*Lilium longiflorum*) and some other lily species (Figure 15–6). In the field, flowering of the Easter lily occurs in early summer. Bulblets form and increase in size from spring throughout summer (6, 48). Between mid-August and mid-September in the Northern Hemisphere the stems are pulled from the bulbs and stacked upright in the field. Periodic sprinkling keeps the stems and bulblets from drying out. Similarly, the base of the stem can be "heeled in" the ground at an angle of 30 to 45 degrees or laid horizontally in trays at high humidity.

About mid-October the bulblets are planted in the field 10 cm (4 in.) deep and an inch apart in double rows spaced 36 in. apart. Here they remain for the following season. They are dug in September as yearling bulbs and again replanted, this time 12.5 cm (6 in.) deep and 10 to 12.5 cm (4 to 6 in.) apart in single rows. At the end of the second year, they are dug and sold as commercial bulbs.

Digging is done in September after the stem is pulled. The bulbs are graded, packed in peat moss, and shipped. Commercial bulbs range in size from 17.5 to 20 cm (7 to 10 in.) in circumference. Lily bulbs must be handled carefully so they will not be injured and must be kept from drying out. The fleshy roots should also be kept in good condition. For long-term storage that will prevent

**Figure 15–6**    Two methods of lily propagation. *Left:* bulblets are produced at the base of individual bulb scales; this method of propagation is possible for nearly all lily species. *Right:* underground stem bulblets; these are produced only by some lily species.

flowering, the bulbs should be packed in polyethylene-lined cases with peat moss at 30 to 50 percent moisture and stored at $-1°$ C ($31°$ F) (53).

Control of viruses, fungus diseases, and nematodes during propagation is important in bulb production. Methods of control include using pathogen-free stocks for propagation (4, 48), growing plants in pathogen-free locations with good sanitary procedures, and treating the bulbs with fungicides.

*Aerial stem bulblets,* commonly known as *bulbils,* are formed in the axil of the leaves of some species, such as *Lilium bulbiferum, L. sargentiae, L. sulphureum,* and *L. tigrinum.* Bulbils develop in the early part of the season and fall to the ground several weeks after the plant flowers. They are harvested shortly before they fall naturally and are then handled in essentially the same manner as underground stem bulblets. Increased bulbil production can be induced by disbudding as soon as the flower buds form. Likewise, some lily species that do not form bulbils naturally can be induced to do so by pinching out the flower buds and a week later cutting off the upper half of the stems. Species that respond to the later procedure include *Lilium candidum, L. chalcedonicum, L. hollandicum, L. maculatum,* and *L. testaceum* (49).

### STEM CUTTINGS

Lilies may be propagated by stem cuttings. The cutting is made shortly after flowering. Instead of roots and shoots forming on the cutting, as would occur in other plants, bulblets form at the axils of the leaves and then produce roots and small shoots while still on the cutting.

Leaf-bud cuttings, made with a single leaf and a small heel of the old stem, may be used to propagate a number of lily species. A small bulblet will develop in the axil of the leaf. It is handled in the same manner as for the other methods described here.

### BULBLET FORMATION ON SCALES (SCALING)

In **scaling,** individual bulb scales are separated from the mother bulb and placed in growing conditions so that adventitious bulblets form at the base of each scale (Figure 15-6). Three to five bulblets will develop from each scale. This method is particularly useful for rapidly building up stocks of a new cultivar or to establish pathogen-free stocks. Almost any lily species can be propagated by scaling (20, 21).

Scaling is done soon after flowering in midsummer, although it might be done in late fall or even in midwinter. The bulbs are dug, the outer two layers of scales are removed and the mother bulb is replanted for continued growth. It is possible to remove the scales down to the core, but this will reduce subsequent growth of the mother bulb. The scales should be kept from drying and handled so as to avoid injury. Scales with evidence of decay should be discarded and the remaining ones dusted with or dipped into a fungicide. Naphthaleneacetic acid (1 ppm) will stimulate bulblet formation.

Scales may be handled by several methods: (a) They may be field planted in beds or frames no more than 6.25 cm ($2\frac{1}{2}$ in.) deep. Bulblets form on the scale during the first year to produce **yearlings.** These are replanted for a third year to produce **commercials.** (b) Scales may be placed in trays or flats of moist sand, peat moss, sphagnum moss, or vermiculite for six weeks at $18°$ to $21°C$ ($65°$ to $70°F$). The scales are inserted vertically to about half their length. Small bulblets and roots should form at the base within three to six weeks. The scales are transplanted either into the open ground or into pots or flats of soil, and then planted in the field the following spring. Subsequent treatment is the same as described for underground bulblets. (c) Scales may be packed in layers in moist vermiculite in plastic lined boxes. These are incubated for 6 to 12 weeks at 15 to 26°C (60° to 80°F). The lower temperature will encourage rooting. The boxes are then placed in cold temperature over winter and planted in rows in the spring. Two years is required to reach planting size (*39*).

A simple method of propagating lilies by scales, after removing them from the bulb, is to dust them with a fungicide then place the scales so they are not touching in damp vermiculite in a polyethylene bag. The bag is closed and tied, then set for six to eight weeks where the temperature is fairly constant at about 21° C (70° F). After bulblets are well developed at the base of the scales, the bag with the scales still inside should be refrigerated at 2° to 4.5° C (35° to 40°F) for at least eight weeks to overcome dormancy. The small bulblets can then be potted and placed in the greenhouse or out-of-doors for further growth.

Tissue culture utilizing aseptic techniques has been used as a means of scaling, particularly to obtain, maintain, and multiply virus-free stock (see chapters 16 and 17).

### BASAL CUTTAGE

The hyacinth is the principal plant propagated by this method, although others such as *Scilla* can be handled in this way. Specific methods include "scooping" and "scoring" (*9, 19*). Mature bulbs which have been dug after the foliage has died down and are 17 to 18 cm or more in circumference are used. In **scooping,** the entire basal plate is scooped out with a special curve-bladed scalpel, a round-bowled spoon, or a small-bladed knife. Adventitious bulblets develop from the base of the exposed bulb scales. Depth of cutting should be enough to destroy the main shoot. In **scoring,** three straight knife cuts are made across the base of the bulb, as shown in Figure 15–7, each deep enough to go through the basal plate and the growing point. Growing points in the axils of the bulb scales grow into bulblets.

To combat decay that may develop during the later incubation period, any infected bulbs should be discarded; the tools disinfected frequently with an alcohol, formalin, or mild carbolic acid solution; and the cut bulbs dusted with a fungicide. Most important is to callus the bulbs at about 21° C (70° F) for a few days to a few weeks in dry sand or soil or in open trays, cut side down. After

**Figure 15–7**    Basal cuttage. Hyacinth bulb which has been scored. Note the bulblets starting to appear.

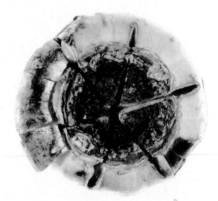

callusing, the bulbs are incubated in trays or flats, in dark or diffuse light, at 21°C (70°F), which is increased to 29.5° to 32° C (85° to 90° F) over a two-week period and held at high humidity (85 percent) for 2½ to 3 months.

The mother bulbs are planted about 10 cm (4 in.) deep in nursery beds in the fall. The next spring bulblets produce leaves profusely. Normally the mother bulb disintegrates during the first summer. Annual digging and replanting of the graded bulblets is required until they reach flowering sizes. Bulbs for greenhouse forcing should be 17 cm (6¾ in.) or more in circumference; bulbs for bedding should be 14 to 17 cm (5½ to 6¾ in.) in circumference (3). On the average, a scooped bulb will produce 60 bulblets, but four to five years will be required to produce flowering sizes; and a scored bulb will produce 24, requiring three to four years (9).

Hot water treatment of hyacinth bulbs used for controlling *Xanthomonas hyacinthi* has been reported to induce bulblet formation and could substitute for basal cuttage (2). Bulbs must be at least 1.5 cm in circumference and treated from mid-July to early September. Treatment is 43°C (110°F) for four days or 38°C (100°F) for 30 days at relative humidity of 60 to 70 percent.

### LEAF CUTTINGS

This method is successful for blood lily (*Haemanthus*), grape hyacinth (*Muscari*), hyacinth, and Cape cowslip (*Lachenalia*) (13), although the range of species is probably wider.

Leaves are taken at a time when they are well developed and green. An entire leaf is cut from the top of the bulb and may in turn be cut into two or three pieces. Each section is placed in a rooting medium with the basal end several inches below the surface, as described for rooting cuttings. The leaves should not be allowed to dry out, and bottom heat is desirable. Within two to four weeks small bulblets form on the base of the leaf and roots develop. At this stage the bulblets are planted in soil.

### BULB CUTTINGS

Among plants that respond to the bulb-cutting method of propagation are the *Albuca, Chasmanthe, Cooperia, Haemanthus, Hippeastrum, Hymenocallis, Lycoris, Narcissus, Nerine, Pancratium, Scilla, Sprekelia,* and *Urceolina* (13).

A mature bulb is cut into a series of eight to ten vertical sections, each containing a part of the basal plate. These sections are further divided by sliding a knife down between each third or fourth pair of concentric scale rings and cutting through the basal plate. Each of these fractions makes up a bulb cutting consisting of a piece of basal plate and segments of three or four scales.

The bulb cuttings are planted vertically in a rooting medium, such as peat moss and sand, with just their tips showing above the surface. The subsequent technique of handling is the same as for ordinary leaf cuttings. A moderately warm temperature, slightly higher than for mature bulbs of that kind, is required. New bulblets develop from the basal plate between the bulb scales within a few weeks, along with new roots. At this time they are transferred to flats of soil to continue development.

A variation of this method is called **twin scaling** (46, 51) and involves dividing bulbs into portions, each containing a pair of bulb scales and piece of basal plate. These are kept in plastic bags with damp vermiculite (12:1 vermiculite:water) for three to four weeks at 21°C (70° F).

**Figure 15–8** Twin scale propagation of narcissus cut from whole bulbs and kept at 23°C in moist vermiculite for two months. From A. R. Rees, *The growth of bulbs,* Academic Press, 1972.

### Micropropagation by Tissue Culture

Many bulb species are highly adapted to micropropagation techniques, utilizing enhanced axillary shoot formation, adventitious shoot formation, bulblet induction on scales, or, very often, flower scapes. These methods are especially valuable to multiply new cultivars rapidly and to develop, maintain, and produce the initial stock of specific virus-tested propagation material. These methods are described in chapters 16 and 17.

## CORMS

### Definition and Structure

A **corm** is the swollen base of a stem axis enclosed by the dry, scalelike leaves. In contrast to the bulb, which is predominantly leaf scales, a corm is a solid stem structure with distinct nodes and internodes. The bulk of the corm consists of storage tissue composed of parenchyma cells. In the mature corm, the dry leaf bases persist at each of these nodes and enclose the corm. This covering, known as the **tunic,** protects it against injury and water loss. At the apex of the corm is a terminal shoot that will develop into the leaves and the flowering shoot. Axillary buds are produced at each of the nodes. In a large corm, several of the upper buds may develop into flowering shoots, but those nearer the base of the corm are generally inhibited from growing. However, should something pre-

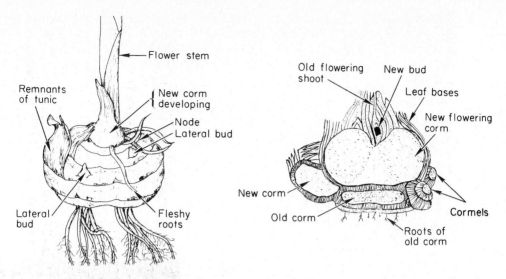

**Figure 15–9** Gladiolus corm. *Left:* external appearance. *Right:* longitudinal section showing solid stem structure.

vent the main buds from growing, these lateral buds would be capable of producing a shoot. (See Figures 15–9 and 15–10.)

Two types of roots are produced from the corm: a *fibrous* root system developing from the base of the mother corm and enlarged, fleshy *contractile* roots

**Figure 15–10** Stage of gladiolus corm development during the latter part of the growing season. The remnants of the originally planted form (see arrow) are evident just below the newly-formed corm. Many small white cormels also have been produced.

developing from the base of the new corm. The latter roots apparently develop in response to the fluctuating temperatures near the soil surface and with exposure of the leaves to light. At lower soil depths temperature fluctuations decrease (*29*), and contraction ceases once the corm is at a given depth.

## Growth Pattern

Gladiolus and crocus are typical cormous plants. The gladiolus is semihardy to tender and, in areas with severe winters, the corm must be stored over winter and replanted in the spring. At the time of planting, the corm is a vegetative structure (*24, 44*). New roots develop from its base, and one or more of the buds begin to develop leaves. Floral initiation takes place within a few weeks after the shoot begins to grow. At the same time the base of the shoot axis thickens, and a new corm for the succeeding year begins to form above the old corm. Stolonlike structures bearing miniature corms or **cormels** on their tip develop from the base of the new corm. The new corm continues to enlarge, and the old corm begins to shrivel and disintegrate as its contents are utilized in flower production. After flowering, the foliage continues to manufacture food materials, which are stored in the new corm. At the end of the summer, when the foliage dries, there are one or more new corms and perhaps a great number of little cormels. The corms are dug and stored over winter until they are planted the following spring.

## Propagation

### NEW CORMS

Propagation of cormous plants is principally by the natural increase of new corms. Flower production in corms, as in bulbs, depends upon food materials stored in the corm the previous season, particularly during the period following bloom. In gladiolus, cool nights and long growing periods are favorable for production of very large corms. Fertilization and other good management practices during bloom have their greatest effect on the next year's flowers. Plants are left in the ground for two months following blooming, or until frost kills the tops. After digging, the plants are placed in trays with a screen or slat bottom arranged to allow air to circulate between them, and cured at about 32°C (90°F) at 80 to 85 percent relative humidity. A few hours at 35° C (95° F) may be helpful. Then the new corms, old corms, cormels, and tops can be easily separated. The corms are graded according to size, sorted to remove the diseased ones, treated with a fungicide, and returned to a 35°C (95° F) temperature for an additional week. This curing process suberizes the wounds and helps combat *Fusarium* infection. The corms are then stored at 5° C (40° F) with a relative humidity of 70 to 80 percent in well-aerated rooms to prevent excessive drying. It may also be desirable to treat them with a suitable fungicide (*35*) immediately before planting.

### CORMELS

Cormels are miniature corms that develop between the old and the new corms. One or two years' growth is required for them to reach flowering size. Shallow planting of the corms, only a few inches deep, results in greater production of cormels; increasing the depth of planting reduces cormel production.

Cormels are separated from the mother corms and stored over winter for planting in the spring. Dry cormels become very hard and may be slow to start growth the following spring, but if they are stored at about 5° C (40° F) in slightly moist peat moss, they will stay plump and in good condition. Soaking dry cormels in cool running water for one to two days and holding them moist until planting at first sign of root development will hasten the onset of growth.

Disease-free cormels can be obtained by hot-water treatments, which should be done between two and four months after digging. Holding cormels at room temperatures to keep them dormant will increase tolerance to this treatment. The cormels are soaked in water at air temperature for two days, then placed in a 1:200 dilution of commerical 37 percent formaldehyde for four hours, and then immersed in a water bath at 57° C (135° F) for 30 minutes. At the end of the treatment, the cormels are cooled quickly, dried immediately, and stored at 5° C (40° F) in a clean area with good air circulation.

The cormels are planted in the field in furrows about 5 cm (2 in.) deep in the manner of planting large seeds. Only grass-like foliage is produced the first season. The cormel does not increase in size but produces a new corm from the base of the stem axis, in the manner described for full-sized corms. At the end of the first growing season, the beds are dug and the corms separated by size. A few of the corms may attain flowering size, but most require an additional year of growth.

Size grades in gladiolus are determined by diameter. There are seven grades, the smallest .9 to 1.2 cm (⅜ to ½ in.) in diameter, the largest 5 cm (2 in.) or more (3).

DIVISION OF THE CORM

Large corms can be cut into sections, retaining a bud with each section. Each of these should then develop a new corm. Segments should be dusted with a fungicide because of the great likelihood of decay of the exposed surfaces.

# TUBERS

## Definition and Structure

A **tuber** is a special kind of swollen, modified stem structure that functions as an underground storage organ (Figure 15–11). The potato (*Solanum tuberosum*) is a notable example of a tuber-producing plant, as is the *Caladium*, grown for its striking foliage, and the Jerusalem artichoke (*Helianthus tuberosus*).

A tuber has all the parts of a typical stem but is very much swollen. Externally the **eyes,** present in regular order over the surface, represent nodes, each consisting of one or more small buds subtended by a leaf scar. The arrangement of the nodes is a spiral, beginning with the terminal bud on the end opposite the scar resulting from the attachment to the stolon. The terminal bud is at the apical end of the tuber, oriented farthest (distally) from the crown of the plant. Consequently tubers show the same apical dominance as any stem.

Internally a potato tuber is composed of enlarged parenchyma-type cells containing large amounts of starch. It has the same internal structure as any stem with pith, vascular areas, and cortex.

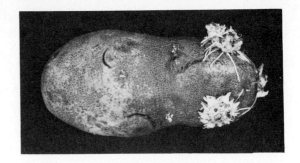

**Figure 15-11**    Potato tuber. Note "eyes" (axillary buds). Those on terminal (distal) end of tuber are starting to grow into new shoots, which are already producing adventitious roots. Basal (proximal) end of tuber, which was attached to a stolon, is to the left.

### Growth Patterns

The tuber is a storage structure that is produced in one growing season, remains dormant during the winter, and then functions to regenerate new shoots the following spring. After a new seasonal cycle begins, the shoots utilize the stored food in the old tuber, which then disintegrates (*7, 16, 52*). As the main shoot develops, adventitious roots are initiated at the base, and lateral buds grow out horizontally into the soil to produce elongated, etiolated stems (*stolons*) as shown in Figure 15-12. Continued elongation of the stolon takes place during long photoperiods and is associated with the presence of auxin and a high gibberellin level. Tuberization begins with inhibition of terminal growth and the initiation of cell enlargement and division in the subapical region of the stolons. This process is associated with short or intermediate day lengths, reduced

**Figure 15-12**    Tubers of white (Irish) potato showing their development from stolons arising from stem tissue. Note adventitious root system originating from main plant stem. Tuber is attached to stolon at the tuber's morphological basal (proximal) end.

temperatures (particularly at night), high light intensity, low mineral content, and a reduction in gibberellin levels in the plant.

Tuberization is caused by the production of a tuber-inducing substance that is produced in the leaves and the mother tuber (*43*). It seems to be necessary for the stolon tip to have attained a particular physiological age. Continued tuber enlargement is dependent on a continuing adequate supply of photosynthate. Conditions that favor rapid and luxurious plant growth above ground, such as an abundance of nitrogen, or high temperatures, are not conducive to tuber production (*37*). In the fall, the tops of the plants die down and the tubers are dug. At this time, the buds of potato tubers are dormant for six to eight weeks. This condition must disappear before sprouting will take place.

## Propagation

### DIVISION

Propagation by tubers can be done either by planting the tubers whole or by cutting them into sections, each containing one or more buds or "eyes." These small pieces of tuber to be used for propagation of the potato are commonly referred to as "seed." The weight of the tuber piece should be 28 to 56 g (1 to 2 oz) to provide sufficient stored food for the new plant to become well established.

Division of tubers is done with a sharp knife shortly before planting. The cut pieces should be stored at warm (20° C; 68° F) temperatures and relatively high humidities (90 percent) for two to three days prior to planting. During this time the cut surfaces heal (**suberization**) and the "seed" piece is effectively protected against drying and decay. Treatment of potato tubers prior to cutting for the control of *Rhizoctonia* and scab may be desirable (*37*). Caladium tubers are produced commercially in Florida (*50*). The tubers are cut into sections, usually two buds per piece. These are planted 7.6 to 9 cm (3 to 4 in.) deep, 9 to 15 cm (4 to 6 in.) apart in rows 45 to 60 cm (18 to 24 in.) apart. Harvest begins in November. After harvest, the tubers are dried in open sheds for six weeks or artificially dried for 48 hours. Further storage should be at temperatures above 16° C (60° F).

## TUBERCLES

*Begonia evansiana* and the cinnamon vine (*Dioscorea batatas*) produce small aerial tubers, known as **tubercles,** in the axils of the leaves. These tubercles may be removed in the fall, stored over winter, and planted in the spring (*13*). Short days induce tuberization (*43*).

## TUBEROUS ROOTS AND STEMS

### Definition and Structure

The tuberous roots and stems class includes several types of structures with thickened tuberous growth that functions as storage organs. Botanically these differ from true tubers, although common horticultural usage sometimes utilizes the term "tuber" for all of them.

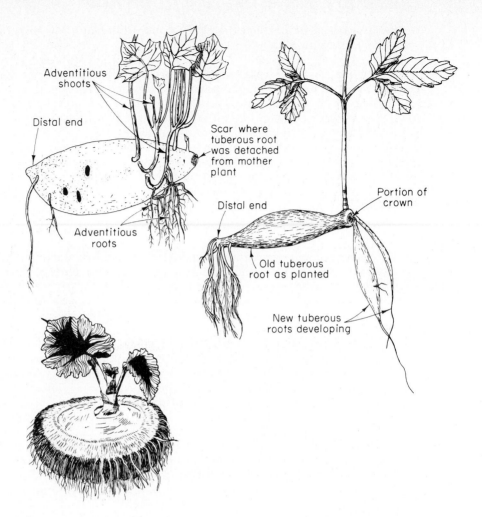

**Figure 15–13** Types of tuberous roots and stems. *Top left:* sweet potato showing adventitious shoots. *Top right:* dahlia during early stages of growth. The old root piece will disintegrate in the production of the new plant; the new roots can be used for propagation. *Below left:* tuberous begonia, showing its vertical orientation. This type continues to enlarge each year.

*Tuberous roots*    Various herbaceous perennial species show massive enlargement of secondary roots. Typical examples are sweet potato (*Ipomoea batatus*) (Figures 15–13 and 15–14) and *Dahlia* (Figures 15–13 and 15–15). Internal and external structures are those of roots. No nodes or internodes are present and the buds are produced only on the crown or stem (*proximal*) end; fibrous roots are commonly produced only on the opposite (*distal*) end. Polarity is the reverse of that of the true tuber.

*Tuberous stems*    **Tuberous stems** are produced by the enlargement of the hypocotyl section of the seedling plant, but may include the first nodes of the epicotyl and the upper section of the primary root (*22, 28, 47*). Typical

**Figure 15–14**    Propagation of sweet potato. Adventitious shoots ("slips") develop when the mother root is placed under warm, moist conditions. *Left:* root on left has been subjected to 34.5°C (100°F) for 26 hours to overcome the proximal dominance shown on the unheated root on the right. Photo courtesy Welch and Little (*56*). *Right:* after slips are well rooted, they are removed and planted.

**Figure 15–15**    Propagation of dahlia. To produce a new plant each separate tuberous root must have a section of the crown bearing a shoot bud, as shown by the detached root on the left.

plants with this structure are the tuberous begonia (*Begonia* × *tuberhybrida*) and cyclamen (*Cyclamen persicum*). These structures have a vertical orientation with one or more vegetative buds produced on the upper end or crown. Fibrous roots are produced on the basal part of the structure.

## Growth Pattern

Tuberous roots are biennial. They are produced in one season, after which they go dormant as the herbaceous shoots die. These function as storage organs to allow the plant to survive the dormant period. In the following spring, buds from the crown produce new shoots, which utilize the food materials from the old root during their initial growth. The old root then disintegrates, and new tuberous roots are produced, which in turn maintain the plant through the following dormant period (*16, 33, 40*).

Tuberous stems of tuberous begonia and cyclamen, on the other hand, are perennial and continue to enlarge laterally every year (*22*). Normally these species have been propagated by seed but the "tuber" can be dug, stored, and used for annual propagation over a period of years.

## Propagation

### DIVISION

The usual method for propagating tuberous roots is by dividing the crown so that each section bears a shoot bud. Dahlia, for example, is dug with its cluster of roots intact, dried for a few days, and stored at 4° to 10°C (40° to 50°F) in sawdust or vermiculite. Open storage may result in shriveling. The root cluster is divided in the late winter or spring shortly before planting. In warm, moist conditions the buds begin to grow, and the tubers can be divided with assurance that each section will have a bud.

The tuberous stem of the tuberous begonia can be divided shortly after growth starts in the spring as long as each section has a bud. To combat decay, the cut surface should be dusted with a fungicide and each section dried for several days after cutting and before placing in a moist medium.

### ADVENTITIOUS SHOOTS

The fleshy roots of a few species of plants such as sweet potato have the capacity to produce adventitious shoots if subjected to the proper conditions. The roots are laid in sand so that they do not touch one another and are covered to a depth of about 5 cm (2 in.). The bed is kept moist. The temperature should be about 27° C (80° F) at the beginning and about 21° to 24° C (70° to 75°F) after sprouting has started. As the new shoots, or **slips,** come through the covering, more sand is added so that eventually the stems will be covered for 10 to 12.5 cm (4 to 5 in.). Adventitious roots develop from the base of these adventitious shoots. After the slips are well rooted, they are pulled from the parent plant and transplanted into the field (*37*). If sweet potato roots are cut in half and the pieces are subjected to 43° C (110° F) for about 26 hours, slip production increases. This treatment overcomes the apical dominance and also controls nematodes and fungus diseases (*56*). (See Figure 15–14.) This procedure can be modified in certain cultivars of the sweet potato by dividing the tuberous root into 20- to 25-g pieces, treating with a fungicide, then giving a presprouting treatment for four weeks of 26.5°C (80°F) and 90 percent relative humidity before planting (*8*).

Cyclamen can be multiplied vegetatively by cutting off the upper one-third of the tuberous stem and notching the surface into 1-cm squares. Adventitious shoots develop (12 to 13 per tuber) and can be used for propagation (*42*).

### LEAFY CUTTINGS

Vegetative propagation in plants of this group, such as dahlia or tuberous begonia, is often more satisfactory with stem, leaf, or leaf-bud cuttings. The cuttings will develop tuberous roots at their base. This process can be stimulated if the stem cutting initially includes a small piece of the fleshy root or stem. Vine cuttings from established beds also can be used in sweet potato propagation.

# RHIZOMES

## Structure

A **rhizome** is a specialized stem structure in which the main axis of the plant grows horizontally at or just below the ground surface. A number of economically important plants, such as bamboo, sugar cane, banana, and many grasses, as well as a number of ornamentals, such as rhizomatous *Iris* and lily-of-the-valley, have rhizome structures. Most are monocotyledons, although a few dicotyledons—for example, low bush blueberry (*Vaccinium angustifolium*)—have analagous underground stems classed as rhizomes. Many ferns and lower plant groups have rhizomes or rhizome-like structures.

Figure 15–16 shows structural features of a rhizome (*60*). The stem appears segmented because it is composed of nodes and internodes. A leaf-like *sheath* is attached at each node; it encloses the stem and, in an expanded form, becomes the foliage leaves. When the leaves and sheaths disintegrate, a scar is left at the point of attachment identifying the node and giving a segmented appearance. Adventitious roots and lateral growing points develop in the vicinity of the node. Upright-growing, above-ground shoots and flowering stems (*culms*) are produced either terminally from the rhizome tip or from lateral branches.

Two general types of rhizomes are found (*38*). The first (the **pachymorph**) is illustrated by rhizomatous *Iris* in Figure 15–17 and by ginger in Figure 15–18. The rhizome is thick, fleshy, and shortened in relation to length. It appears as a many-branched clump made up of short, individual sections. It is determinate; that is, each clump terminates in a flowering stalk, growth continuing only from lateral branches. The rhizome tends to be oriented horizontally with roots arising from the lower side.

The second type (the **leptomorph**) is illustrated by the lily-of-the-valley in Figure 15–16. The rhizome is slender with long internodes. It is indeterminate; that is, it grows continuously in length from the terminal apex and from lateral branch rhizomes. The stem is symmetrical and has lateral buds at most nodes, nearly all remaining dormant. This type does not produce a clump but spreads extensively over an area.

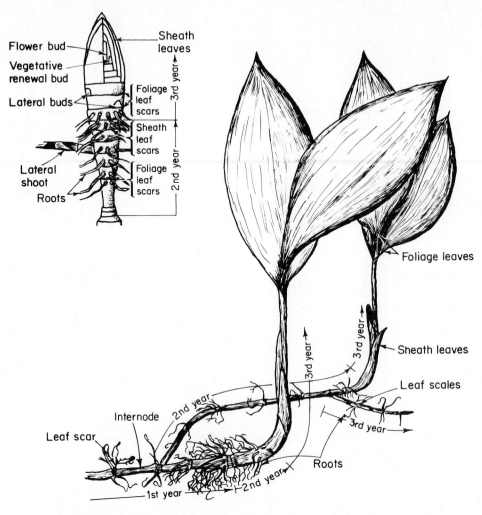

**Figure 15–16**    Structure and growth cycle of lily-of-the-valley (*Convallaria majalis*). *Right:* section of rhizome as it appears in late spring or early summer with one-, two-, or three-year-old branches. A new rhizome branch begins to elongate in early spring and terminates in a vegetative shoot bud by the fall. The following spring the leaves of the bud unfold; food materials manufactured in the leaves by photosynthesis are accumulated in the rhizome. Growth the second season is again vegetative. Early in the third season a flower bud begins to form, and at the same time a vegetative growing point forms in the axil of the last leaf. *Top left:* section of the three-year-old branch showing terminal flower bud and lateral shoot bud enclosed in leaf sheaths. Such a section is sometimes known as a **pip** or **crown** and is forced for spring bloom. In the early spring the flowering shoot expands, blooms, and then dies down, the shoot bud beginning a new cycle of development. Redrawn from Zweede (*60*).

**Figure 15–17**    Structure of an iris (rhizomous type) plant as it appears about the time of flowering. A two-year-old section that had flowered the previous year and is now dying back is shown at D. The lateral branch arising from it consists of the one-year-old vegetative section (B); the current-season, lateral vegetative branches (A); and the current-season, terminal flowering shoot (C). The vegetative shoot (A) will flower the following year.

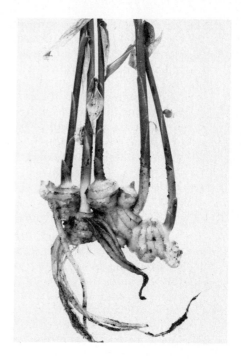

**Figure 15–18**    Rhizome of the tropical ginger plant (*Zingiber officinale*). This is easily propagated by division of the thickened rhizome, which is the source of commercial ginger.

Intermediate forms between these two types also exist. These are called **mesomorphs** (*38*).

## Growth Pattern

Rhizomes grow by elongation of the growing points produced at the terminal end and on lateral branches. Length also increases by growth in the intercalary meristems in the lower part of the internodes. As the plant continues to grow and the older part dies, the several branches arising from one plant may eventually become separated to form individual plants of a single clone.

Rhizomes exhibit consecutive vegetative and reproductive stages, but the growth cycle differs somewhat in the two types described. In the pachymorph rhizome of *Iris* (Figure 15–17), a growth cycle begins with the initiation and growth of a lateral branch on a flowering section. The flowering stalk dies, but these new lateral branches produce leaves and grow vegetatively during the remainder of that season. Continued growth of the underground stem, storage of food, and the production of a flower bud at the conclusion of the vegetative period are dependent upon photosynthesis. Consequently foliage should not be removed during this period. A flowering stalk is produced the following spring and no further terminal growth can take place. In general, plants with this structure flower in the spring and grow vegetatively during the summer and fall.

Plants with a leptomorph habit as a general rule (with exceptions) grow vegetatively during the beginning of the growth period and flower later in the same period. The length of time during which an individual rhizome section remains vegetative varies with different kinds of plants. An individual branch in the lily-of-the-valley in Figure 15–16, for instance, is vegetative three years before a flower bud forms. Some bamboo species remain vegetative for many years, but then they change abruptly and the entire plant produces flowers.

In some rhizomatous plants, such as blueberry, rhizome development is increased by higher temperatures and long photoperiod, and is correlated with vigorous above-ground growth (*30*).

## Propagation

### DIVISION OF CLUMPS AND RHIZOMES

Division is the usual procedure for propagating plants with a rhizome structure, but the procedure may vary somewhat with the two types. In pachymorph rhizomes, individual sections (or culms) are cut off at the point of attachment to the rhizome, the top is cut back, and the piece is transplanted to the new location. Leptomorph rhizomes can be handled in essentially the same way by removing a single lateral "offshoot" from the rhizome and transplanting it. The tip of the lily-of-the-valley rhizome bearing a flower bud (see Figure 15–16) called a "pip" is removed along with the rooted section below and transplanted.

Division is usually carried out at the beginning of a growth period (as in early spring) or at or near the end of a growth period (i.e., in late summer or fall).

Propagation is carried out by cutting the rhizome into sections, being sure that each piece has at least one lateral bud, or "eye"; it is essentially a stem cutting. Bananas, for instance, are propagated in this way. This general method works well for the leptomorph rhizomes, in which a dormant lateral growing point is present at most nodes. The rhizomes are cut or broken into pieces, and adventitious roots and new shoots develop from the nodes. Rhizome-producing turf grasses, for instance, are cut up into sections and the individual "sprigs" transplanted. New plants can be established readily by this method.

### CULM CUTTINGS

In large rhizome-bearing plants, such as bamboos, the aerial shoot, or culm, may be used as a cutting. These may be whole culm cuttings, in which the entire aerial shoot is laid horizontally in a trench. New branches arise at the nodes. A stem cutting of three- or four-node sections may be planted vertically in the ground.

# PSEUDOBULBS

## Definition and Structure

A **pseudobulb** (literally "false bulb") is a specialized storage structure, produced by many orchid species, consisting of an enlarged, fleshy section of the stem made up of one to several nodes (see Figure 15–19). In general, the ap-

**Figure 15-19** *Cattleya* orchid showing rhizome structure and upright elongated pseudobulbs as basal part of shoots.

pearance of the pseudobulb varies with the orchid species. The differences can be used to identify species.

## Growth Pattern

These pseudobulbs arise during the growing season on upright growths that develop laterally or terminally from the horizontal rhizome. Leaves and flowers form either at the terminal end or at the base of the pseudobulb, depending upon the species. During the growth period, they accumulate stored food materials and water and assist the plants in surviving the subsequent dormant period.

## Propagation

### OFFSHOOTS

In a few orchids, such as the *Dendrobium* species, the pseudobulb is long and jointed, being made up of many nodes. Offshoots develop at these nodes. From the base of these offshoots roots develop. The rooted offshoots are then cut from the parent plant and potted.

### DIVISION

Most important commercial species of orchids, including the *Cattleya, Laelia, Miltonia,* and *Odontoglossum,* may be propagated by dividing the rhizome into sections, the exact procedure used depending upon the particular kind of orchid. Division is done during the dormant season and preferably just before the beginning of a new period of growth. The rhizome is cut with a sharp knife back far enough from the terminal end to include four to five pseudobulbs in the new section, leaving the old rhizome section with a number of old pseudobulbs, or "back bulbs," from which the leaves have dehisced. The section is then potted,

**Figure 15–20**  A "back bulb" of a *Cymbidium* orchid was removed from the parent plant and placed in a rooting medium; the offshoot shown above then developed. This offshoot is now ready for removal and potting. When this is done, a second offshoot, or "break," should appear. Courtesy A. Kofranek.

whereupon growth begins from the bases of the pseudobulbs and at the nodes. The removal of the new section of the rhizome from the old part stimulates new growth, or "back breaks," to occur from the old parts of the rhizome. These new growths grow for a season and can be removed the following year.

An alternate procedure is to cut partly through the rhizome and leave it for one year. New back breaks will develop, which can be removed and potted.

BACK BULBS AND GREEN BULBS

**Back bulbs** (i.e., those without foliage) are commonly used to propagate clones of *Cymbidium*. These are removed from the plant, the cut surface is painted with a grafting compound, and they are placed in a rooting medium for new shoots to develop. When the stage shown in Figure 15–20 is reached, the shoot can be removed from the bulb and potted. This back bulb can be repropagated and a second shoot developed from it.

**Green bulbs** (i.e., those with leaves) also can be used in *Cymbidium* propagation. Treatment with indolebutyric acid, either by soaking or by painting with a paste, has been shown to be beneficial (*31*).

# REFERENCES

*1.* Allen, R. C. 1937. Factors affecting the growth of tulips and narcissi in relation to garden practice. *Proc. Amer. Soc. Hort. Sci.* 35:825–29.

*2.* Amano, N., and K. Tsutsui. 1980. Propagation of hyacinth by hot water treatment. *Acta Hort.* 109:279–87.

*3.* Amer. Assoc. Nurserymen, Inc., Comm. on Hort. Stand. 1980. *American standard for nursery stock.* Washington, D.C.: Amer. Assoc. Nurs., Inc.

*4.* Baker, K. F., and P. A. Chandler. 1957. Development and maintenance of healthy planting stock. Sect. 13 in *Calif. Agr. Exp. Sta. Man. 23.*

5. Ball, V., ed. 1972. *The Ball red book.* (12th ed.) Chicago: Geo. J. Ball, Inc.

6. Blaney, L. T., and A. N. Roberts. 1966. Growth and development of the Easter lily bulb *Lilium longiflorum* Thunb. 'Croft'. *Proc. Amer. Soc. Hort. Sci.* 89:643-50.

7. Booth, A. 1963. The role of growth substances in the development of stolons. In *The growth of the potato,* J. D. Ivins and F. L. Milthorpe, eds. London: Butterworth, pp. 99-113.

8. Bouwkamp, J. C., and L. D. Scott. 1972. Production of sweet potatoes from root pieces. *HortScience* 7(3):271-72.

9. Crossley, J. H. 1957. Hyacinth culture; narcissus culture; tulip culture. *Handbook on bulb growing and forcing.* Northwest Bulb Growers Assoc., pp. 79-84, 99-104, 139-44.

10. Curtis, A. H. 1938. Growth studies of King Alfred narcissus bulbs. *Proc. Amer. Soc. Hort. Sci.* 36:781-82.

11. De Hertogh, A. A., A. N. Roberts, N. W. Stuart, R. W. Langhans, P. G. Linderman, R. H. Lawson, H. F. Wilkins, and D. C. Kiplinger. 1971. A guide to terminology for the Easter lily. *HortScience* 6:121-23.

12. De Hertogh, A. A., and N. Blakely. 1972. The influence of temperature and storage time on growth of basal roots of nonprecooled and precooled bulbs of *Lilium longiflorum* Thunb. cv. 'Ace'. *HortScience* 74:409-10.

13. Everett, T. H. 1954. *The American gardener's book of bulbs.* New York: Random House.

14. Genders, R. 1973. *Bulbs.* New York: Bobbs-Merrill.

15. Gould, C. J. 1953. Blights of lilies and tulips. *Plant Diseases: USDA yearbook of agriculture.* Washington, D.C.: U.S. Govt. Printing Office, pp. 611-16.

16. Gregory, L. E. 1965. Physiology of tuberization in plants (tubers and tuberous roots). In *Encyclopedia of Plant Physiology,* vol. 15 Berlin: Springer-Verlag, pp. 1328-54.

17. Griffiths, D. 1922. The production of tulip bulbs. *USDA Bul. 1082.*

18. ———. 1930. Daffodils. *USDA Cir. 122.*

19. ———. 1930. The production of hyacinth bulbs. *USDA Cir. 112.*

20. ———. 1930. The production of lily bulbs. *USDA Cir. 102.*

21. ———. 1932. Artificial propagation of the lily. *Proc. Amer. Soc. Hort. Sci.* 29:519-21.

22. Haegeman, J. 1979. *Tuberous begonias.* A. R. Gantner Verlag K.-G., FL-9490 Vaduz.

23. Harrison, A. D. 1964. Bulb and corm production. London: *Bul. 62, Minist. Agr., Fish. and Foods,* pp. 1-84.

24. Hartsema, A. M. 1937. Periodieke ontwikkeling van *Gladiolus hybridum* var. Vesuvius. *Verh. Koninkl. Ned. Akad. van Wet.* 36(3):1-34.

25. ———. 1961. Influence of temperatures on flower formation and flowering of bulbous and tuberous plants. In *Encyclopedia of Plant Physiology,* vol. 16. Berlin: Springer-Verlag, pp. 123-67.

26. Heath, O. V., and M. Holdsworth. 1948. Morphologenic factors as exemplified by the onion plant. *Symposia for the Soc. Exp. Biol.* II:326-50.

27. Huisman, E., and A. M. Hartsema. 1933. De periodieke ontwikkeling van *Narcissus pseudonarcissus* L. *Meded. Landbouwhoogesch., Wageningen,* DL. 37 (Meded. No. 38, Lab. v. Plantenphys. onderz., Wageningen).

28. Jacobi, E. F. 1950. *Plantkunde voor tuinbouwscholen.* Zwolle, The Netherlands: W. E. J. Tjeenk Willink.

29. Jacoby, B., and A. H. Halevy. 1970. Participation of light and temperature fluctuations in the induction of contractile roots of gladiolus. *Bot. Gaz.* 131(1):74–77.

30. Kender, W. J. 1967. Rhizome development in the lowbush blueberry as influenced by temperature and photoperiod. *Proc. Amer. Soc. Hort. Sci.* 90:144–48.

31. Kofranek, A. M., and G. Barstow. 1955. The use of rooting substances in *Cymbidium* green bulb propagation. *Amer. Orch. Soc. Bul.* 24(11):751–53.

32. Langhans, R. W., and T. C. Weiler. 1968. Vernalization in Easter lilies. *HortScience* 3:280–82.

33. Lewis, C. A. 1951. Some effects of daylength on tuberization, flowering, and vegetative growth of tuberous-rooted begonias. *Proc. Amer. Soc. Hort. Sci.* 57:376–78.

34. Lin, P. C., and A. N. Roberts. 1970. Scale function in growth and flowering of *Lilium longiflorum,* Thunb. 'Nellie White'. *Jour. Amer. Soc. Hort. Sci.* 95(5):559–61.

35. Magie, R. O. 1953. Some fungi that attack gladioli. *Plant Diseases: USDA yearbook of agriculture.* Washington, D.C.: U.S. Govt. Printing Office, pp. 601–7.

36. Mann, L. K. 1952. Anatomy of the garlic bulb and factors affecting bulb development. *Hilgardia* 21:195–251.

37. MacGillivray, J. H. 1953. *Vegetable production.* New York: Blakiston.

38. McClure, F. A. 1966. *The bamboos: A fresh perspective.* Cambridge, Mass.: Harvard University Press.

39. McRae, E. A. 1978. Commercial propagation of lilies. *Proc. Inter. Plant Prop. Soc.* 28:166–69.

40. Moser, B. C., and C. E. Hess. 1968. The physiology of tuberous root development in dahlia. *Proc. Amer. Soc. Hort. Sci.* 93:595–603.

41. Mulder, R., and I. Luyten. 1928. De periodieke ontwikkeling van der Darwin tulip. *Verh. Koninkl. Ned. Akad. van Wet.* 26:1–64.

42. Nakayama, M. 1980. Vegetative propagation of cyclamen by notching of tuber. II. Effect of scooping size and notching size on the regeneration of cyclamen tuber. *Jour. Japan. Soc. Hort. Sci.* 49(2):228–34.

43. Nitsch, J. P. 1971. Perennation through seeds and other structures. In *Plant physiology,* vol. 6A, F. C. Steward, ed. New York: Academic Press, pp. 413–79.

44. Pfieffer, N. E. 1931. A morphological study of *Gladiolus. Contrib. Boyce Thomp. Inst.* 3:173–95.

45. Rees, A. R. 1972. *The growth of bulbs.* New York: Academic Press.

46. Rees, A. R., and G. R. Hanks. 1980. The twin-scaling technique for narcissus propagation. *Acta Hort.* 109:211–16.

47. Reinders, E., and R. Prakken. 1964. *Leerboek der Plantkunde.* Amsterdam: Scheltema & Holkema N. V.

48. Roberts, A. N., and L. T. Blaney. 1957. Easter lilies: Culture. *Handbook on bulb growing and forcing.* Northwest Bulb Growers Assoc., pp. 35–43.

49. Rockwell, F. F., E. C. Grayson, and J. de Graaf. 1961. *The complete book of lilies.* Garden City, N.Y.: Doubleday.

50. Sheehan, T. J. 1955. Caladium production in Florida. *Fla. Agr. Ext. Circ. 128.*

51. Skelmersdale, L. 1978. Propagation of bulbous and bulbous-like plants. *Proc. Inter. Plant Prop. Soc.* 28:209–15.

52. Slater, J. W. 1963. Mechanisms of tuber initiation. In *The growth of the potato*, J. D. Ivins and F. L. Milthorpe, eds. London: Butterworth, pp. 114–20.

53. Stuart, N. W. 1954. Moisture content of packing medium, temperature and duration of storage as factors in forcing lily bulbs. *Proc. Amer. Soc. Hort. Sci.* 63:488–94.

54. Wang, S. Y., and A. N. Roberts. 1970. Physiology of dormancy in *Lilium longiflorum* Thunb. 'Ace'. *Jour. Amer. Soc. Hort. Sci.* 95(5):554–58.

55. Weiler, T. C., and R. W. Langhans. 1972. Growth and flowering responses of *Lilium longiflorum* Thunb. 'Ace' to different day lengths. *Jour. Amer. Soc. Hort. Sci.* 97(2):176–77.

56. Welch, N. C., and T. M. Little. 1967. Heat treatment and cutting for increased sweet potato slip production. *Calif. Agr.* 21(5):4–5.

57. Went, F. W. 1948. Thermoperiodicity. In *Vernalization and photoperiodism*, A. E. Murneck and R. O. Whyte, eds. Waltham, Mass.: Chronica Botanica.

58. Wilson, K., and J. N. Honey. 1966. Root contraction in *Hyacinthus orientalis*. *Ann. Bot.* 30:47–61.

59. Woodcock, H. B. D., and H. T. Stearn. 1950. *Lilies of the world*. New York: Scribner's.

60. Zweede, A. K. 1930. De periodieke ontwikkeling van *Convallaria majalis*. *Verh. Koninkl. Ned. Akad. van Wet.* 27:1–72.

## SUPPLEMENTARY READING

CROCKETT, J. V. 1971. *Bulbs*. New York: Time-Life Books.

Daffodil handbook. 1966. *American Horticultural Magazine* 45 (1):227.

EVERETT, T. H. 1954. *The American gardener's book of bulbs*. New York: Random House.

GENDERS, R. 1973. *Bulbs, a complete handbook*. New York: The Bobbs-Merrill Company.

GOULD, C. J. 1959. The flower bulb industry. *Wash. Agr. Exp. Sta. Cir. 318*, pp. 1–59.

HARRISON, A. D. 1964. Bulb and corm production. London: *Ministry of Agriculture, Fisheries, and Food Bul. No. 62*, pp. 1–84.

HARTSEMA, A. M. 1961. Influence of temperatures on flower formation and flowering of bulbous and tuberous plants. In *Encyclopedia of plant physiology*, Vol. 16, W. Ruhland, ed. Berlin: Springer-Verlag, pp. 123–67.

IVINS, J. D., and F. L. MILTHORPE, eds. 1963. *The growth of the potato*. London: Butterworth & Co., Ltd.

KIPLINGER, D. C., and R. W. LANGHANS, eds. 1967. *Easter lilies, the culture, diseases, insects and economics of Easter lilies*. Columbus: Ohio State Univ.; and Ithaca, N.Y.: Cornell Univ.

N. A. GLADIOLUS COUNCIL. 1972. *The world of the gladiolus*. Edgewood, Md.: Edgewood Press.

REES, A. R. 1972. *The growth of bulbs*. London: Academic Press.

———. 1966. The physiology of ornamental bulbous plants. *Bot. Rev.* 32:1–22.

Third international symposium on flower bulbs. 1980. *Acta Hort.* 109:1–533.

**Micropropagation** involves the production of plants from very small plant parts, tissues, or cells grown aseptically in a test tube or other container where the environment and nutrition can be rigidly controlled. The ability to grow plant tissue, such as callus and cell suspensions, and various plant organs, such as stems, flowers, roots, and embryos, more or less indefinitely, has been utilized in scientific laboratories for many decades as a research tool for geneticists, botanists, and plant pathologists. The methods used have been known collectively as **tissue culture,** a term that is sometimes used synonymously with micropropagation. Tissue culture procedures utilize an **in vitro system** of production that requires a laboratory-type facility and aseptic techniques similar to those used in culturing fungi, bacteria, and other microorganisms.

## HISTORY

### Tissue Culture

The establishment of the biological principles of tissue and organ culture has been credited to the German plant physiologist Haberlandt, who first enunciated them in 1902 (*45*). By 1934 P. R. White (*164*) was able to grow tomato roots continuously in vitro by supplying them with yeast extract. The essential ingredients turned out to be certain B vitamins, notably $B_1$ (thiamine). In 1939, three investigators—Nobecourt and Gautheret in France and White in the United States—reported independently the indefinite culture of plant callus tissue in a synthetic medium (*164*). The discovery of cytokinins and the hormonal control of shoot and root regeneration from tobacco callus by Skoog (*138*) and his co-workers in 1948 at the University of Wisconsin established the basis for manipulating organ initiation and provided the principle on which all micropropagation depends. Another major development was the regeneration of embryolike structures **(somatic embryos** or **embryoids)** from callus cell

# 16

# Principles
# of Tissue Culture
# for Micropropagation

suspensions (*46, 128, 144*). Other important developments included the discovery of haploid plant formation from developing pollen grains (*15, 44, 114*), the isolation of plant protoplasts (*30, 39*), and the artificial (parasexual) hybridization of plant protoplasts in aseptic culture (*24*).

### Micropropagation

The application of tissue culture techniques to the regeneration and commercial propagation of whole plants is a more recent development. It has become an important alternative for more conventional propagation procedures for a wide range of plant species.

The discovery of micropropagation through shoot-tip culture resulted from attempts to obtain virus-free plants through the isolation of noninfected meristems (*55, 123, 147*). Using this procedure for *Cymbidium* orchids, Morel (*100*) discovered that the shoot-tip proliferated into masses of protocorms, which could be divided and recultured to produce new plants. Since the traditional multiplication of orchid clones by division is very slow, the potential for vegetative propagation was greatly enhanced (*101, 165*). Morel's procedures have since been applied to many other plant species (*49, 104, 106, 120*).

## USES

For propagation, tissue culture systems have two primary uses: (1) rapid mass propagation of clones, and (2) development, maintenance, and distribution of specific pathogen-tested (SPT) clones.

Secondarily, in vitro culture systems have the potential for long distance shipment of propagation material, and long-term storage of clonal material (*6, 72, 162*).

Parallel to these propagation uses, tissue culture systems have potential for production of various secondary products, such as pharmaceuticals in cell suspension systems (*37*) and many applications in plant breeding (*17, 43, 160*).

### Mass-Propagation

Mass-propagation is characterized by potentially very high rates of clonal multiplication that can be achieved in a relatively short time. Theoretical rates are very large. Starting with a single plantlet in culture and multiplying geometrically at monthly intervals at the rate of ten per culture, one could predict 1 million plants in six months. Although such rates may not be achievable in practice, the high rates that have been attained are impressive. Commercial laboratories in operation can produce 1 to 3 million plants per year (see chapter 17). Examples of plants in this category include Boston fern (*Nephrolepis*), strawberry (*Fragaria*), *Gerbera*, lily (*Lilium*), *Cymbidium* orchids, *Anthurium,* and *Philodendron* (*145*).

In one case, starting with an initial number of 30 cultures, 20,000 *Hemerocallis* plants were produced in a 20 ft$^2$ space in eight months. One additional year was required to produce salable plants. Conventional methods would have produced perhaps 1500 plants in this same period with much larger space required for maintaining stock plants (*116, 117*). Such rates of nursery operations require careful scheduling to accommodate the year-round flow of material in various stages.

Such rapid increases have particular application to species that multiply slowly by such conventional propagation methods as division, separation, or offsets (145). Examples include orchids, Boston fern, many foliage plants, date palm, and others. Another important use is the rapid multiplication of: (a) new or improved cultivars being introduced to the trade (53) from breeding programs, (b) pathogen-tested source material in disease control programs (21, 34), and (c) parental stock for hybrid seed production (3, 166). Sufficient stock plants can be produced in a single year, instead of four or five years.

The continuous year-round scheduling of propagated material is a potential advantage. Conventionally, most herbaceous and woody perennials are propagated in relation to season. Micropropagation enables propagators to carry out a continuous year-round operation with production scheduled closer to marketing. This advantage applies even for plants that propagate easily by conventional methods.

### Pathogen Control

Control of pathogens in the stock plants is facilitated by in vitro systems but tests to detect known pathogens, such as viruses and bacteria, should be included as a part of the procedure. A major application of in vitro systems is the combination of rapid multiplication plus pathogen-control programs.

## DISADVANTAGES OF MICROPROPAGATION

Micropropagation, however, has its problems. The facilities required are costly, and economic considerations may not justify their use in commercially propagating many kinds of plants. Particular skills are required to carry out the procedures. Errors in maintenance of identity, introduction of an unknown pathogen, or appearance of an unobserved mutant may be multiplied to very high levels in a short time. Some system of cultivar verification is needed.

Specific kinds of genetic and epigenetic modifications of the plant potentially can develop with some cultivars and some systems of culture. These may alter the plants produced. A knowledge of these potential problems and an evaluation of their effect on production of certain cultivars should be considered in a given micropropagation program.

## TYPES OF REGENERATION

There are five fundamental types of vegetative regeneration in tissue culture systems: (a) meristem-tip elongation, (b) axillary shoot production, (c) adventitious shoot initiation, (d) organogenesis, and (e) embryogenesis.

### Meristem-Tip Elongation

A growing point consists of the **apical meristem** on the tip of a telescoped stem made up of nodes and internodes. Rudimentary and undeveloped leaves with **lateral meristems** in their axil occur at each node.

Meristem-tip elongation occurs when the apical meristem on an excised shoot-tip continues to elongate in culture and is rooted to produce a small plantlet. This procedure is used primarily to produce "virus-free" plants ( *123* ), in which only the apical meristem (less than 0.5 mm) is removed with a few subtending leaves. If roots cannot be induced to form on the explant, then a **micrograft** onto a seedling or rooted plantlet rootstock might be performed under aseptic culture. Regeneration occurs by elongation of the apical meristem. Only a single plant per culture is produced by this method.

## Axillary Shoot Proliferation

In axillary shoot formation, lateral growing points on the explant at the nodes below the apical meristem are stimulated to grow and the apical meristem is inhibited. Growth from these axillary shoots provides a rapid multiplication system in which the number of potential plants is increased exponentially by repeated reculturing.

## Adventitious Shoot Initiation

Induction of adventitious shoots directly on roots, leaves, bulb scales, and other organs of intact plants is a common method of propagation (see chapters 9 and 10). On the other hand, regeneration of adventitious shoots on stems of intact dicotyledonous plants is relatively rare, although shoots sometimes occur as natural outgrowths, known as **sphaeroblasts** ( *148* ) on some plants, or as regeneration from callus masses after wounding ( *130* ). In culture, however, excised plant parts can be induced to form adventitious shoots at high rates in many species. The procedure is useful not only for the micropropagation of plants traditionally propagated by adventitious shoots, such as African violet ( *Saintpaulia* ), but also for a wide range of other species.

Adventitious shoots may develop either *directly* on the explant itself or *indirectly* in unorganized masses of callus tissue ( *52* ).

## Organogenesis in Callus Cultures

**Organogenesis** refers to the initiation of both adventitious shoots and roots from within masses of callus cells ( *41, 154* ). These highly vacuolated and largely parenchymatous callus cell masses can develop **meristemoids** which initiate organs under particular cultural conditions. The process is similar to the initiation of adventitious shoots on explants except that an intervening period of independent callus growth has occurred.

## Embryogenesis

In the normal seedling cycle, embryogenesis proceeds from the single-celled zygote to the initiation and development of an embryo. An important discovery was that carrot ( *Daucus* ) cells grown in a suspension culture with unautoclaved coconut milk and auxin and subsequently placed in a hormone-free culture medium, could develop into millions of individual embryos.

These structures have been called **embryoids** or **somatic embryos** to distinguish them from either the *sexual* or *apomictic* embryos produced naturally.

Considerable progress has been made in the techniques of **somatic embryogenesis** as a potential propagation procedure (*133, 150, 157, 163*).

## MICROPROPAGATION AND TISSUE CULTURE SYSTEMS

All micropropagation and tissue culture systems begin with the excision of a small piece of plant, freeing it from contaminating microorganisms, and placing it in culture. The plant part used to start the process is known as the **explant** and is the basic unit for tissue culture propagation, corresponding to such terms as cutting, layer, scion, or seed.

The new shoots or callus that this explant produces by proliferation are divided into **propagules,** which are recultured for further multiplication. Eventually new roots or new shoots and roots are developed so that new **plantlets** are produced.

A general classification of micropropagation and tissue culture systems utilizing aseptic in vitro culture is as follows:

**Class I.** Regeneration of new plants from vegetative structures or tissues

1. Meristem-tip culture
2. Micrografting
3. Shoot-tip culture
4. Adventitious shoot culture
5. Tissue and cell cultures
   a. Callus culture
   b. Cell suspensions
   c. Protoplast culture

The interrelations among callus, cells, and protoplasts and their ability to regenerate whole plants is shown as follows:

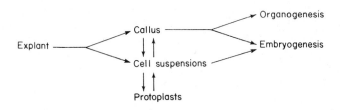

**Class II.** Reproduction of seedling plants by excision of existing reproductive structures

1. Anther and pollen culture
2. Ovule culture
3. Embryo culture
4. Seed culture
5. Spore culture

**Figure 16–1** Shoot-tip of carnation stem with outer leaves removed, showing the apical and lateral meristems (growing points). Part of shoot tip to be excised for culturing is indicated by lines. Courtesy W. P. Hackett.

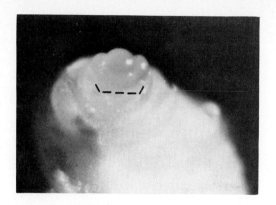

**Figure 16–2** Meristem-tip culture of carnation. Following the arrows: the carnation cutting is obtained, the larger leaves are stripped away, and then the small, enclosing leaves at the extreme tip are removed to expose the growing point. The tip and the next subtending leaf primordia are removed with a scalpel and placed on the surface of a paper wick in a test tube with nutrient media. The shoot tip grows in the tube until large enough to be transplanted to a container. A full-size plant is shown at bottom left. Redrawn from Holley and Baker *(54)*.

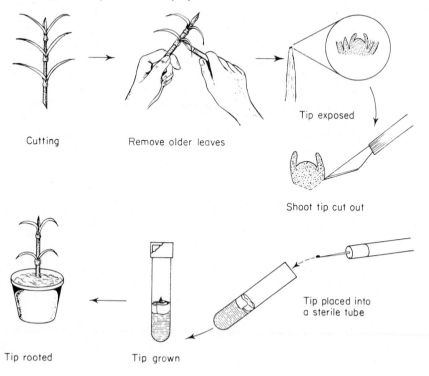

Cutting        Remove older leaves

Tip exposed

Shoot tip cut out

Tip placed into a sterile tube

Tip rooted      Tip grown

## Class I. Vegetative Explants

### MERISTEM-TIP CULTURE

In this procedure, the very smallest part of the shoot tip is excised as the explant (see Figure 16–1). Only the meristem dome and a few subtending leaf primordia are included. The purpose is to produce a small, elongated, rooted plantlet that is free of fungi, bacteria, viruses, and virus-like diseases, which may not be present in the meristem. The size may be 0.25 to 1.0 mm long, or even smaller. The smaller the size, the more effective is the procedure for eliminating pathogens (*105*). On the other hand, the smaller the piece, the more difficult the procedure and the lower the survival rate. In strawberry (*89*) the optimum size to accomplish both plant regeneration and pathogen elimination was found to be 0.5 to 0.9 mm. The culture conditions and medium used are similar to those described for other types of culture (see chapter 17). A low concentration of cytokinin and a moderate level of auxin are usually added, although transfer to an auxin-free medium may be necessary to improve root development. Addition of low levels of gibberellic acid (0.1 mg/l) is helpful in some cases but too much may inhibit rooting.

This procedure has been most successful with herbaceous plants such as carnations (Figure 16–2), potatoes, chrysanthemums, and orchids, which root relatively easily in culture (*55, 69, 92, 123, 139, 140, 147*). Because reproduction by such small apices has been more difficult with woody plants, micrografting has been substituted (*58, 109*).

### SHOOT-TIP GRAFTING

**Shoot-tip grafting** very small shoot-tips consisting of the meristem and three leaf primordia (0.14 to 0.18 mm) to prepared seedling rootstocks in vitro has been used to propagate citrus (*110*), apple (*59*), and *Prunus* plants (*111*). The primary use of this technique has been to produce "virus-free" plants of woody plant species that do not regenerate shoots and roots easily from shoot-tips in culture. For citrus, the method has been used to produce plants in the adult phase and thus bypass the strong juvenile growth characteristics that result when nucellar seedlings are used to produce "virus-free" source plants of established cultivars.

### SHOOT-TIP CULTURE

**Shoot-tip culture** parallels standard cutting propagation (see chapters 9 and 10) but uses a miniaturized stem cutting and includes high rates of multiplication (Figure 16–3). Success varies with the explant used and depends on the application of proper hormones. The explant may be all or part of an apical or lateral growing point on a stem or it may be a stem section of several nodes (*49, 104, 105, 168*). The size of explants varies. The smallest sizes (0.1 to 0.5 mm) described for meristem-tip culture are not normally utilized for propagation because of difficulty in excision.

A more common type is the unexpanded shoot-tip, ranging from about 0.5 to 2.0 mm in length. Although this size of explant is more convenient for propagation, it may not be as free of viruses and other systemic pathogens as the smaller sizes.

**Figure 16–3**    Shoot tip propagation by axillary shoot formation on poplar *(Populus)*. *Upper left:* vegetative bud explant after two weeks in culture (3 mm in diameter). *Upper right:* axillary buds on explant after four to six weeks (15 mm in diameter). *Lower left:* advanced shoot proliferation on propagule (5 cm in diameter). *Lower right:* rooted plantlet. Courtesy C. B. Christie *(29)*.

A third type utilizes a 1.0 to 2.0 cm or larger section of the immature expanding shoot-tip, including immature leaves. This type is convenient for propagation but is even more likely to be contaminated with pathogens. Shoot-tip explants may be taken also from lateral bud segments, which are essentially single node cuttings.

Shoot-tip explants become established through elongation of the terminal meristem accompanied by limited development of the axillary meristems. The multiplication phases result from the suppression of the terminal meristem and the stimulation of axillary buds to grow and elongate. Repeated division and transfer of propagules into fresh media results in extensive increase, but the amount is still limited to the number of axillary buds on the explant or propagule.

Production of axillary shoots may be accompanied by proliferation of adventitious buds from callus on the base of the propagule *(1)*. In such cases,

## Shoot-Tip Grafting in Vitro (Figure 16–4)

*Citrus* embryos are excised from rootstock seeds, surface sterilized, and planted in standard inorganic salt medium with 1 percent agar (*109, 110*). Embryos germinate in the dark for two weeks. Seedlings are then removed and decapitated to a 1 to 1.5 cm length; cotyledons and lateral buds are excised with a mounted razor blade. A 0.14 to 0.18 mm tip with three leaf primordia is used as the scion. This gives reasonable success and the short tip eliminates viruses.

An inverted T-bud cut is made in the seedling rootstock, cutting 1 mm down the stem, followed by a horizontal cut on the bottom. The excised shoot-tip is placed inside the flap next to the cambium. Grafted plants are placed in a liquid medium. A filter paper bridge with a center hole supports the stem. Cultures are kept in the light for three to five weeks to heal. When two expanded leaves appear on the scion, the grafted plant is transplanted.

A similar procedure has been used for apple (*59*) and plum (*111*). Rootstocks are either seedling plants or rooted stems. Shoot-tip scions taken from cultured plants reduce contamination problems.

**Figure 16–4**     Micrografting of apple meristems. *Left:* seedling hypocotyl sections before and after decapitation. *Left center:* grafted plant one week after grafting. *Right center:* grafted plant six weeks after grafting with scion growing. *Right:* eight weeks after grafting, ready for transplanting. Courtesy D. F. Millikan and S. C. Huang (*59*).

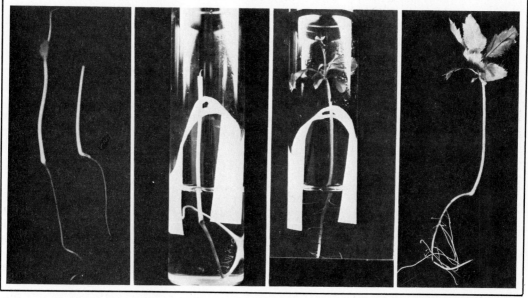

higher rates of multiplication can occur since the increase is not dependent on the number of nodes (Figure 16–5).

Shoot-tip propagation is adapted to in vitro propagation of almost any kind of plant if the proper sequences and cultural requirements are worked out. This

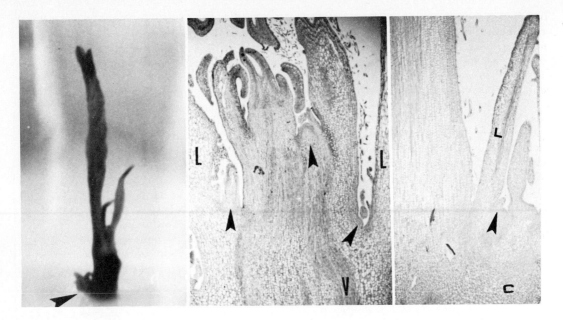

**Figure 16-5** Shoot multiplication on shoot-tip culture of EMLA 26 apple. *Left:* after 10 days propagule consists of apical, axillary, and basal adventitious shoots. *Center:* apical portion of the shoot-tip culture on left after six weeks. *Axillary* shoots shown by arrows. *Right:* basal lateral shoot shown on left after six weeks. Shows that the shoot is *adventitious* and arises from a mass of basal callus (C). Leaf primordia shown by (L). Vascular area identified by (V). Courtesy Neil Miles. From Nasir, F. R., and N. Miles. 1981. Histological origin of EMLA 26 apple shoots generated during micropropagation. *HortScience* 16:53.

system is often preferred to some of the other methods because it most closely parallels conventional cutting propagation, and is the least likely to produce genetic changes.

ADVENTITIOUS SHOOT INITIATION

New shoot apices can develop either (a) directly on the explant, or (b) indirectly from the callus that develops on the cut surfaces of the explant (*52*). The two most important factors that affect the initiation of adventitious shoots are: (a) the choice of explant, and (b) the hormone regime to which the plants are subjected.

Some of the kinds of explants are as follows:

> **Pieces of leaves** are used in such plants as *Saintpaulia* (*13*), *Salpiglossis* (*83*), and horseradish (*96*) that naturally regenerate in this manner. Experimentally, whole plants have been regenerated from epidermis strips or other, deeper layers (*159*). An interesting variation is the "fragmented shoot apex culture" with grape (*9, 10*), in which shoot apices 0.1 mm in length have been cut into two to four segments and cultured in drops of the medium. These small segments proliferate into small leafy structures that can develop into whole plants.

**Figure 16–6** Adventitious bud formation on Douglas fir *(Pseudotsuga menzesii)* cotyledons. (a): cotyledons are excised from germinated seed, cut into segments, and placed on medium with 1 ppm BAP plus 0.001 ppm NAA to initiate adventitious buds. (b): medium is changed by omitting hormones; mass of shoots develop. (c): individual shoots are excised and rooted in culture with 0.001 ppm NAA, 0.5 percent sucrose, and temperature reduced to 19°C. (d): Plantlets grown in greenhouse. Courtesy Dr. Tsai-Ying Cheng *(26)*.

**Shoot-tip explants** are used in a number of species, such as orchids or ferns, in which proliferations of callus-like masses develop and regenerate large numbers of shoots. Plants started from shoot-tips initially for axillary shoot formation may revert to higher percentages of adventitious shoots with time (*1*).

**Cotyledons, hypocotyls, and other seedling structures** have been particularly useful as starting points (Figure 16–6) for conifers in which excised cotyledon segments can be used to regenerate new shoots in the presence of a cytokinin (*23, 28, 31*).

**Young needle fascicles** have been used to regenerate shoots from older conifer trees (*31*). Explants from plant tissue that have been rejuvenated to a juvenile condition are particularly responsive (*14*).

**Segments of immature inflorescences of flower scapes** are highly regenerative in some species, particularly some monocots, such as *Gladiolus* (*169*), *Hemerocallis* (*51*), *Iris* (*97*), *Hosta* (*95*), and *Freesia* (*121*), as well as some dicotyledons such as *Gerbera* (*122*) or *Chrysanthemum* (*129*).

**Bulb scales** of various monocotyledonous plants (*60, 61, 62*) characteristically have rings of meristematic tissue at the base near the basal plate (see chapter 15). When scale explants are excised and put into culture, adventitious shoots develop directly from the explants.

**Tissue discs** of potato tubers (a modified stem), excised from the cortex just inside or outside the vascular ring have the ability to regenerate adventitious shoots (*66*), but those from the pith do not.

**Direct initiation** begins with parenchyma cells that are located either in the epidermis or just below the surface of the stem; some of these cells become

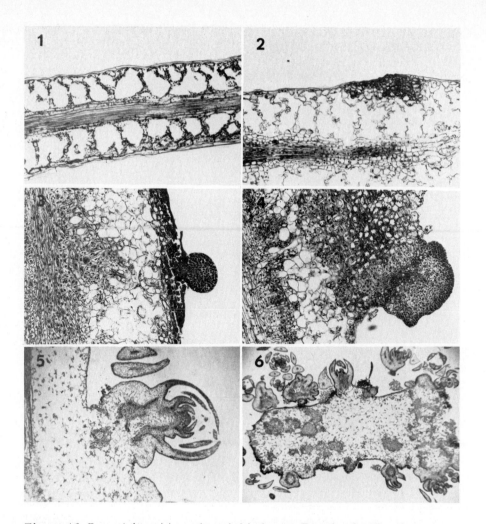

**Figure 16-7**     Adventitious shoot initiation on Douglas fir (*Pseudotsuga menzesii*) cotyledons. *Top left:* cross-section of cotyledon before culture. *Top right:* initiation of a *meristemoid* on surface of cotyledon. *Center left:* development of meristemoid into a tiny globular mass of tissue. *Center right:* differentiation into a shoot primordium. *Lower left:* a completely developed shoot tip. *Lower right:* cotyledon showing masses of adventitious shoot tips. Courtesy Dr. Tsai-Ying Cheng (*25*).

meristematic and pockets of small, densely staining cells termed **meristemoids** develop (*52*). These apparently originate from single cells.

Response of explants, however, depends on hormone levels (Figure 16-7). In a study of shoot initiation on cotyledons of Douglas fir (*25*), cytokinin (BAP 5 mM) was necessary to induce adventitious buds, but three different response patterns resulted depending on the level of auxin also supplied. With low auxin (NAA 5 mM) shoots only developed. With a higher auxin concentration (NAA 5 mM), the cotyledon produced both callus and many shoots. When only auxin (NAA 5 mM) was added, only callus resulted.

**Indirect** development of adventitious shoots first involves the initiation of basal callus from excised shoots in culture (*1*). Shoots arise from the periphery of the callus and are not initially directly connected to the vascular tissue of the explant. Similarly, in intact plants adventitious shoots can arise from callus on the apex of decapitated epicotyls if cytokinin is applied in cylindrical agar blocks (*130*).

Adventitious shoot formation can result in very high rates of multiplication, in general, higher than rates resulting from axillary shoots. On the other hand, adventitious shoots can increase rates of aberrant plant production, resulting from the breakdown of chimeras with resultant loss of variegation in some cultivars and reversions to a more juvenile condition (*85*). When this system is used, the plants that are produced must be evaluated carefully for variation.

### TISSUE AND CELL CULTURE SYSTEMS

The culture of plant cells either as callus tissue (Figure 16–8) or as liquid suspensions (Figure 16–9) provides an important technique that can be preliminary to the regeneration of whole plants. Because of the potential genetic variability associated with these systems, however, cell and tissue culture is less significant in the propagation of cultivars than for genetic, plant breeding, and genetic engineering activities. On the other hand, the mass propagation of somatic embryos, incorporating such seed propagation techniques as fluid drilling, described in chapter 7, might become a feasible procedure for some plants.

**Figure 16–8**    Tissue cultures. *Left:* undifferentiated callus of tomato (*Lycopersicon esculentum*) growing in vitro. *Right:* callus of corn (*Zea mays*) showing dark spots that are green meristemoid-producing areas that will initiate shoots. Courtesy Dr. Carole Meredith.

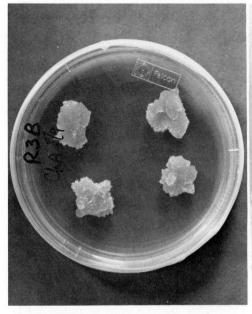

**Figure 16–9** Suspension culture of cells of celery (*Apium graveolens* L.). Courtesy Dr. Lawrence Rappaport.

Mass culture of plant cells for industrial production of secondary compounds, such as pharmaceuticals, might be feasible (*11, 37*).

*Callus culture* **Callus** is produced on explants in vitro as a result of wounding and in response to hormones, either endogenous or supplied in the medium. Explants from almost any plant structure or part, such as seeds, stems, roots, leaves, storage organs, or fruits, can be excised, disinfested, and placed on the surface of a culture medium to produce callus (*151, 155*). Continued subculture of these callus masses can continue for long periods separate from the original explant (Figure 16–8).

A number of different culture media have been used by various research workers to grow callus, but the most common is the **Murashige-Skoog (MS) medium** developed especially for tobacco callus (*107*). This medium is rich in macroelements, particularly nitrogen, including both nitrate ($NO_3$) and ammonium ions ($NH_4$), sucrose, and certain vitamins. Initiation of cell division and subsequent callus production requires that both a cytokinin and an auxin be supplied in the proper proportion (*138*). Auxin at a moderate to high concentration is the primary hormone used to produce callus. The principal auxins include indoleacetic acid (IAA), naphthaleneacetic acid (NAA), and 2,4-dichlorophenoxyacetic acid (2, 4-D), in increasing order of effectiveness. Cytokinin, as kinetin, or benzyladenine (BA) is supplied in a lesser amount if not adequate within the explant.

Although callus tissue cultures may appear outwardly to be uniform masses of cells, in reality their structure is relatively complex with considerable morphological, physiological, and genetic variation within the callus (*151*). Growth follows a typical logarithmic pattern. There is: (a) a slow initial cell division *induction* period requiring auxin, (b) a rapid cell *division* phase involving active synthesis of DNA, RNA, and protein, followed by (c) a gradual *cessation* of cell division along with (d) *differentiation* into larger parenchyma and vascular-type cells. Cell division does not take place throughout the culture mass but is located primarily in a meristematic layer on the outer periphery of cells. The inner parts of the callus remain as an undividing mass of older tissue and, in time, may differ physiologically and genetically from cells of the outer layer. Division in the exterior layer decreases and the appearance of the callus may become "knobby" as cell division becomes restricted to specific islands of cells. Thus, variations in cell age and type may occur within the tissue culture mass (*158*). The inner cells are older and the exterior cells are younger as a meristematic region persists around the periphery of the callus mass.

*Cell suspensions*    A **cell suspension** culture is an extension of tissue or callus culture consisting of cells and/or groups of cells dispersed and growing in an aerated liquid culture medium (*151, 155*).

Cell suspensions have an advantage over callus cultures in that the cells develop more or less individually, have more nutritional access to the culture medium, and some of the gradients and physiological variations of callus cultures are avoided.

A suspension culture is started by placing a piece of friable callus or homogenized tissue in a liquid medium so that the cells disassociate from each other. These cells are grown in various types of devices (Figure 16-9). In one **(batch culture),** cells are grown in a flask placed on a shaking device that allows air and liquid to mix. Rotating devices that continuously bathe the tissue are available. Another device, called a **chemostat** or **turbidostat,** continuously cycles the media through the cell culture essentially the same as in microorganism culture. In a third method, cells on a **filter paper layer** are placed on a shallow liquid medium in a petri dish (*56*) with no agitation.

Growth of cells follows a typical pattern based on changes in rates of cell division. Cells first divide slowly (*lag phase*), then more rapidly (*exponential*), increasing to a steady state (*linear*), followed by a declining rate (*deceleration*) until a *stationary* state is reached. This pattern is referred to as a **logarithmic growth curve.** If a small inocula of liquid with new cells is transferred to new liquid medium, the process will be repeated. Under proper environmental conditions with media control, the process can go on indefinitely.

Culture media are usually similar to those for callus tissue culture and include a complete range of ingredients: inorganic salts, sucrose, vitamins, and a proper balance of hormones.

*Protoplast culture*    **Protoplasts** are the living parts of plant cells containing the nucleus, cytoplasm, vacuole, and various cellular structures surrounded by a semipermeable membrane. The plant cell, in contrast to the animal cell, is surrounded by a firm nonliving cell wall composed of cellulose and hemicellulose and held together by pectin materials.

The major advance that permitted protoplast cultures to be made (*30, 151*) was the discovery that the walls of plant cells could be removed by enzymes that digest the pectin and allow the protoplast surrounded by the cellular membrane to survive (Figure 16-10). Commercial enzyme preparation involving cellulase complexes are available for this purpose. Maintaining an adequate osmotic pressure to prevent disruption of membranes is necessary, and mannitol (0.45 to 0.8 M) has been used for this purpose.

Protoplasts are cultured in media similar to those for cells except for the presence of the enzymes and osmoticum. Once the enzyme is removed, regeneration of walls takes place rapidly within several days. When new cell

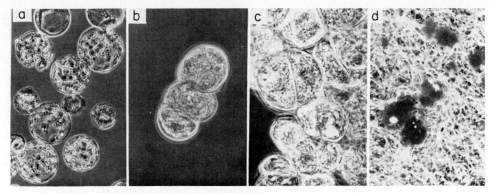

**Figure 16-10**  Protoplasts of Douglas fir *(Pseudotsuga menziesii)* cotyledons. (a): freshly isolated protoplasts in which cell walls have been removed. (b): four-cell stage after protoplast had resynthesized a cell wall and divided twice. (c): cells from a colony. (d): callus formation. Courtesy Dr. Tsai-Ying Cheng *(26)*.

walls are produced, the regenerated cells can be used to start tissue cultures or cell suspensions. It is then necessary to regenerate new plants.

Protoplast culture is significant in plant biology because many manipulations with the cells have been found possible *(17, 134, 151, 155)* once they are freed of the enclosing cell walls. Plant pathology research is facilitated because viruses can be more easily incorporated into protoplasts. The fusion of protoplasts of two different genotypes, such as two species, which combine two nuclei and two cytoplasms, has been accomplished in a process called **somatic** (or **parasexual) hybridization** *(24)*. Protoplasts can absorb DNA, proteins, and other large macromolecules such that new genetic material can be directly incorporated into cells of an organism. This procedure is called **transformation.** Isolated protoplasts are also capable of taking up nuclei and chloroplasts **(organelle transfer).** All of these procedures are important techniques of **genetic engineering** which, however, requires the subsequent regeneration of whole plants followed by their mass propagation in a stable form.

*Differentiation*  The induction of roots, shoots, or embryoids begins with the dedifferentiation of groups of parenchyma cells to produce centers of meristematic activity. These have been termed **meristemoids** *(86, 154, 159)* in the case of organogenesis and **preembryonic masses** *(PEM)* *(41, 133, 163)* in the case of embryogenesis. Whether organogenesis or embryogenesis occurs depends largely on the source of the explant, but the processes may also be directed to some degree by manipulations of ingredients of the culture system. For any one species or cultivar, experiments should be conducted to determine the best source of explant and the optimum culture conditions. In early studies, tobacco callus produced shoots *(138)* if a relatively high level of cytokinin was supplied, with a correspondingly low concentration or complete lack of auxin (Figure 16-11). Adenine has been synergistic with cytokinin and increased inorganic phosphate ($PO_4$) has been useful. Although the same pattern tends to follow with most other plants, an exact formula for optimizing conditions is needed for each species or cultivar.

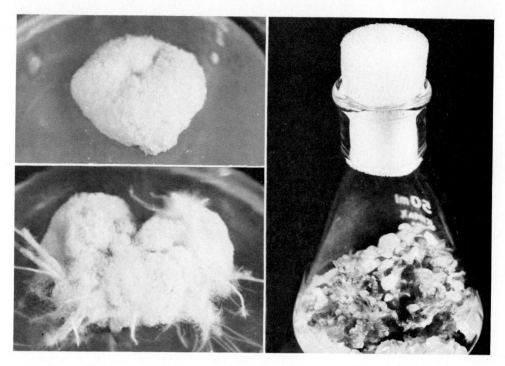

**Figure 16-11**     Organogenesis in tobacco callus. *Left top:* undifferentiated callus tissue developed from a single pith cell produced by nurse-culture technique. *Left below:* adventitious roots develop on the callus culture when it is grown on a medium high in auxin (IAA) and low in kinin (kinetin). *Right:* stems and leaves are produced when the callus is grown on a medium low in IAA and high in kinetin. With proper balance both roots and stems develop and new plants are produced. Courtesy T. Murashige.

**Somatic embryos** have been produced by two general processes (Figure 16-12).

*Direct embryogenesis* utilizes, as explants, tissues that are immediately able to produce somatic embryos. Examples of these are nucellar tissue of polyembryonic citrus cultivars (*20*), grape (*Vitis*) ovules (*102*), and immature cacao (*Theobroma*) embryos (*119*), although this type of embryogenesis may also involve some callus formation. These explants are said to have **preembryonically determined cells (PEDC).**

*Indirect embryogenesis* utilizes an intermediate stage of callus, or a suspension culture during which a change in the potentiality of the cells is induced. These somatic embryos are described as arising from **induced embryonically determined cells (IEDC)** (*133*). Other views favor the idea that the two processes are actually the same and somatic embryos arise from the continuation of special cells in the original explant (*150*). In either case, embryo-producing (**embryogenic**) callus, once developed, continues to produce embryos (*78, 79*). Somatic embryos arise from single cells located within clusters of meristematic

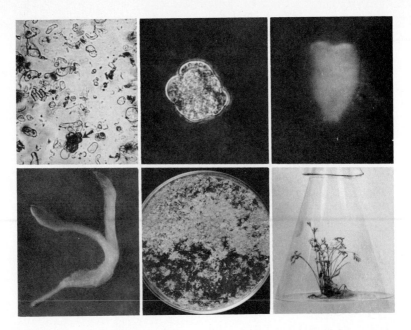

**Figure 16–12**     Embryogenesis in carrot *(144). Top left:* highly magnified view of suspended cells and cell clumps derived from tissue cultures of carrot growing in a liquid medium. Arrow points to a clump of cells—the beginning stage for an embryoid. *Top center:* single cell clump (higher magnification) illustrating the globular stage of development. *Top right:* a more advanced heart-shaped stage of embryoid development. *Below left:* mature embryoid that developed continuously from the globular stage, through the heart and torpedo (not shown) stages. *Below center:* thousands of carrot plantlets that developed as embryoids. *Below right:* single carrot plant that grew from a single embryoid transplanted from previous stage. Courtesy F. C. Steward and M. O. Mapes.

cells either in the callus mass or in suspension. Such cells develop into proembryos with polarity following a pattern that tends to mimic the general pattern associated with the development of embryos in the ovule (see chapter 3) *(163)*.

In both cases embryogenesis follows a two-step procedure in coffee *(131)*, date palm *(156)*, and grape *(78, 79)*. The first step **(conditioning)** involves a callus forming period in which relatively high levels of auxin are supplied; 2,4–D has been successful in this operation. The second step **(induction)** involves a shift to a medium free of auxin and other hormones but with a high nitrogen supply.

## Reproductive Explants

ANTHER AND POLLEN CULTURE

Whole plants have been produced in some plant species by the differentiation of the immature pollen grain to produce either a small embryo by **embryogenesis** *(32, 112, 114)* or a mass of callus that might, in turn, undergo organogenesis *(135, 152)*. Anther culture is significant in plant breeding to pro-

## Embryogenesis in Coffee

Explants (7 mm$^2$) cut from mature coffee leaves on lateral branches were placed lower side up on a simple medium (minerals and sugar) for 36 hours in the dark; then the healthy explants were transferred to a conditioning medium: MS medium with thiamine (30 uM), L-cysteine (210 uM), meso-inositol (117 uM), sucrose, 0.8 to 1.0 percent agar, kinetin (20 uM), and 2,4-D (5 uM). Material was incubated in the dark for 45 to 50 days during which time typical massive white calluses developed (*131*). These calluses were then transferred to the inducing medium and grown in light. This medium was ½ strength MS except that KNO$_3$ was increased 2 times, and included sucrose, kinetin (2.5 uM), and NAA (0.5 uM). In the inducing medium the callus slowly turned brown. In 13 to 15 weeks, somatic embryos began to appear but in low frequency (1 to 20 per culture).

In 16 to 19 weeks, small white globular *proembryonic masses,* or PEMs, began to develop from the brown callus mass. These new tissues eventually developed into about 100 somatic embryos. In an additional stage, the PEMs were transferred into a liquid medium of the same composition, but without kinetin, and grown in light for four to six weeks. The embryoids and plantlets could then be plated out onto a medium of inorganic salts and sugar (0.5 to 1 percent) and grown in light, later being transferred to containers in a high-humidity chamber.

duce haploid plants in which the chromosomes subsequently may be doubled, resulting in an isogenic, homozygous line (Figure 16–13) (*18*).

Success in this technique has varied greatly among different genera and cultivars (*136, 160*) so that considerable prior research may be needed for it to be used with any given kind of plant. Success in producing entire plantlets has been obtained with *Datura, Nicotiana, Hyoscyamus, Solanum, Brassica, Pelargonium,* and some members of the Gramineae family. In other cases, only partially developed proembryos or callus has been produced (Gramineae, *Anemone, Paeonia,* and most woody plants tested). In grape, success has been achieved with a single cultivar (*102, 125*) illustrating the strong influence of genetic control.

Nevertheless, certain principles can be established to aid in anther culture of other plants. The stage of microspore development in pollen development is apparently critical since the normal programmed sequence of development can be diverted from pollen to embryo only if the process has not proceeded too far. Normal pollen development in the plant can be described in three general phases:

**a.** Meiosis resulting in the formation of four haploid cells in a *tetrad*

**b.** Release of the four individual microspores from each other within the anther, each with a single nucleus (*uninucleate*)

**c.** Division of the nucleus to a *binucleate* stage followed by change in cytoplasm and cell wall to develop the mature pollen grain

For diversion to embryo production, the microspore should not have proceeded beyond the unincleate stage.

**Figure 16–13**    Anther culture. Procedure for obtaining haploid plants from *Nicotiana tabacum*, then diploid plants from the haploid ones. A flower bud at the right stage is excised from a flowering plant *(upper left)*. The immature stamen is removed and planted aseptically on the proper nutrient medium. Pollen grains at the uni-nucleated, microspore stage develop into haploid embryos, which germinate and form plantlets. These plantlets are transplanted into pots in the greenhouse. The haploid plants flower abundantly but do not set seed as they contain only $1n$ number of chromosomes per cell. In a second step, stem sections of haploid plants are surface sterilized then planted aseptically on a nutrient medium that favors proliferation of a callus *(lower right)*. This callus can be transferred to the same medium in order to eliminate the initial explant and to let the process of endomitosis produce diploid cells. The callus is then transferred to a new medium that favors the formation of adventitious shoots *(lower left)* from which whole plants can be raised, the majority of which are diploid and capable of setting seed. Courtesy J. P. Nitsch *(114)*.

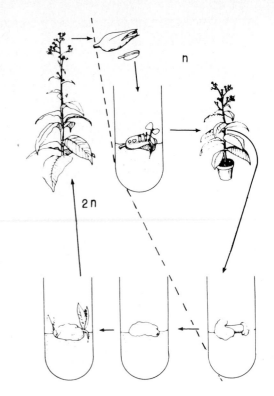

However, callus may develop from the older microspore, which, if haploid, could be used in later embryoid initiation. Some of the conditions that have been helpful in producing embryogenesis in anther culture are the use of liquid media, separation from surrounding anther walls, using activated charcoal in the medium, prechilling (e.g., 4° to 8°C for four days), and use of amino acids such as glutamine (*112*).

OVARY (FRUIT)
AND OVULE (SEED) CULTURE

In the flower, the ovary is the immature fruit and the fertilized ovule is the immature seed (see chapter 3). In some species these organs have been excised at flowering or shortly after and grown to maturity in aseptic culture. Embryos have been produced that subsequently germinated and grew in experiments with tomato, gherkin (*113*), strawberry (*7*), onion (*67*), dill (*68*), cotton (*12*), poppy (*87*), and others.

In other experiments, unfertilized ovules have been excised, grown in culture, then supplied with pollen, with subsequent fertilization taking place in vitro (*127*). Such a technique has potential applications to produce hybrid seeds not possible by conventional means (*167*), including recovery from wide genetic crosses (*63, 64*) as well as self-fertilization of self-incompatible species (*36*). Such procedures have been used most successfully with plants having multiple fruits. Pollen can be placed directly on the placenta inside the ovule where pollen tubes can develop and grow immediately into the ovules without passing down the style.

The technique of growing ovaries and ovules in sterile culture is not difficult, but careful timing in relation to flowering is required for success.

Flowers are sterilized using the same procedures as for any explant. An agar medium is generally used, but in cotton, ovules were floated on a liquid medium. Some modifications of standard basic culture medium are used. Nitrate ($NO_3$) nitrogen is preferable to ammonium ($NH_4$), which tends to be toxic. Amino acids, as supplied by casein hydrolysate, glutathione, glutamine, and asparagine are useful. A solution of tomato juice (red or green), 25 percent v/v, has stimulated ovary growth, as has coconut milk. Both gibberellin ($GA_3$ at 10 to 40 ppm) and auxin appear to be essential for fruit development.

Treatment of the pedicel of the developing fruitlet with an auxin has caused initiation of roots. Presence of such roots in culture was particularly stimulatory to fruit growth (113).

EMBRYO CULTURE

Embryos can be excised at various stages of development and germinated aseptically on a sterile medium. This technique can be used to rescue embryos that would have aborted in their normal sequence on the plant (48), but such embryos must be excised early enough to save them for germination (108, 115, 167). Abortion may result from interspecific hybrid crosses (17) or from early fruit ripening before the embryo is fully formed as occurs in many *Prunus* cultivars (73).

Developmental stages of the embryo are described in chapter 3. Attempts to culture and germinate embryos in Stage I have been largely unsuccessful for many, if not most, species. Early in Stage II, when one-third to one-half size, embryos begin to acquire the ability to germinate and survive in aseptic culture.

Culture of embryos at immature stages appears to require conditions that promote continued embryogenesis rather than precocious germination. Such conditions may include high sucrose concentrations (8 to 18 percent), casein hydrolysate (5 to 30 percent), coconut milk (1 to 15 percent), and glutamine in the medium. High osmotic pressure promotes development but may (22) or may not (143) be essential. Reduced (ammonia or amino) nitrogen appears necessary. The role of regulatory hormones in seed development and germination is well recognized (see chapters 3 and 6), but their use in aseptic culture is unclear. Combinations of GA, ABA, and cytokinin (115) may need to be tested. ABA may play a particularly important role in regularizing embryo development in vitro. Chilling the partially developed embryo at 2°C (36°F) for one to three months has promoted normal germination in *Prunus* species (50).

At more advanced stages of embryo development, approximately half- to full size (or in Stage III), a simple medium of only sucrose and minerals may be adequate. Decreased concentrations of sucrose are desirable with increased development.

Embryo culture is useful also to induce prompt germination of mature dormant seeds and thus shorten breeding cycles (80). Using excised embryos can eliminate dormancy-inducing restraints due to the surrounding seed covering or the endosperm, as is done with *Iris* (126, 146), *Maranta* (48), peach (73, 80), rose (5), and olive. Special handling may be required, however, to provide the most appropriate environmental sequences needed for embryos with internal epicotyl dormancy, as in *Paeonia* (94) (Figure 16–14).

**Figure 16–14**    Embryo culture of peony *(Paeonia)* seed with epicotyl dormancy. *Left top:* seed (left). Seed coat removed (right) to show massive endosperm. *Left center:* very small rudimentary embryo is located inside base on endosperm. Shows cutting away sections of endosperm to extract the embryo on probe. *Right top:* first stage of embryo germination after two months in test tube on agar media of mineral salts (MS medium; see chapter 17) and 4 percent sucrose. Only radicle grows at warm temperature. *Lower right:* a subsequent cold treatment stimulates the epicotyl to grow. Shows seedling after seven weeks at 27°C (80°F), five weeks at 3°C (37°F), and four weeks at 27°C (80°F) in light. *Lower left:* seedling transplanted to growing medium. Courtesy M. M. Meyer, Jr. Reprinted from *Amer. Peony Soc. Bul. 217,* 1976 *(94).*

The use of in vitro culture for germinating whole seeds may be practical with very small seeds. An example is the orchid, in which aseptic culture systems have been used commercially for many years as a standard propagation procedure. Orchid seeds are very minute and dustlike and have embryos that are not fully developed (*4*). In nature these seeds depend on symbiotic relationships with certain microorganisms in the barks of trees to provide needed nutrition. In 1922 Knudson (*76, 77*) reported that these organisms could be replaced by an in vitro system utilizing a simple medium of inorganic salts and sucrose and revolutionized orchid production. The procedure is described in chapter 17.

Bromeliad seed has been grown in the manner described for orchids except the seeds are placed in liquid medium on a shaker, with the medium changed each seven days until germination (*99*).

SPORE CULTURE

Fern spores are the reproductive structures of the sporophyte generation produced on the undersides of leaves. These are haploid and produce the gametophyte generation, which grows as a separate structure known as the prothallus. Spores are discharged and "germinate" on moist media. In vitro culture can be used to improve the normal spore production system. The procedures are described in chapter 17.

# FACTORS AFFECTING SUCCESS IN PRODUCING PLANTS BY MICROPROPAGATION

Four sequential stages are recognized in micropropagation systems: (I) establishment, (II) multiplication, (III) pretransplant, and (IV) transplant (*2, 104, 105, 106*). Strict adherence to these stages is essential for some species. For others, the sequence may be varied to adapt to the requirements of the species and to the needs of the propagator.

The medium has two major functions. The first is to supply the basic nutritional ingredients for continued growth of the isolated explant and subsequent propagules (see chapter 17).

The second function is to direct growth and development through *hormonal control*. The primary classes of hormones are *auxins* and *cytokinin*, but *gibberellins* and *abscisic acid* are utilized in specific situations. Hormonal control is exerted by

**a.** The kind of hormone or growth regulator
**b.** The concentration
**c.** The sequence in which they are supplied

Different plants may respond differently to the various cytokinins and auxins, in part because of their natural hormone content.

A proper sequence of hormones is particularly important in that a hormone may have an inducing effect but must be absent or reduced for the organ to grow. For example, high cytokinin concentrations stimulated shoot bud initiation on Douglas fir cotyledons but had to be removed for subsequent growth of the shoot (*28*). Likewise, auxin stimulates root initiation on shoots, but may inhibit or reduce subsequent root growth. Embryogenesis requires auxin for induction, but embryo development occurs in its absence. These sequences must be established empirically for various kinds of plants.

## Stage I: Establishment

The function of this stage is to establish a sterile explant in culture. Factors that affect the success of this stage include choice of explant, elimination of contaminants from the explant, and culture conditions including ingredients, light, temperature, and choice of explant support. The choice of the explant is, perhaps, the most critical since it must be physiologically competent to survive the initial culture and to elicit the appropriate response. Although the kinds of explants and the kinds of regeneration that one can utilize are described in this chapter, and information on appropriate explant use can come from knowledge of the plant species and their cultivars, the propagator may have to obtain additional data by systematically testing different explant and handling procedures.

In general, younger tissues, such as terminal or axillary shoot-tips or tips of adventitious shoots, will regenerate better than older, more mature tissues of the same stem. Immature flower buds and inflorescences are often quite regenerative. Seasonal dormancy patterns must be considered in relation to buds as well as to storage organs.

Since the ability of various tissues to respond in particular ways seems to be due to its internal inherent capabilities, choice of explant is obviously most important in determining the success of the operation.

Proper handling of the stock plants may be important to obtain successful regeneration and reduce contamination. Growing stock plants under controlled light and temperature to produce flushes of growth of the proper physiological stage might be useful. Inducing adventitious shoots, as on root cuttings, also may be useful. For woody plants, rejuvenation of the source trees to induce juvenile growth or selection of juvenile material may be required (see chapter 8). A method is essential to disinfect the tissue from fungi, bacteria, and other contaminants, without harming the regenerative capacity of the explant.

In general, the ingredients of the culture medium in the first stage are determined by kind of response needed. For example, axillary shoot formation usually requires relatively low levels of both auxin and cytokinin. Adventitious shoot initiation needs a higher cytokinin concentration plus an approximately equal amount of auxin. For callus production, higher levels of auxin are combined with low levels of cytokinin.

Explants of some species contain endogenous substances that exude from the cut surfaces into the medium and inhibit development. These exudates may be leached or washed away with frequent changes of water. A liquid medium or frequent reculturing can overcome this inhibition. Antioxidants, such as ascorbic acid or citric acid, also have been used, either in a preliminary washing solution or in the medium itself. Likewise, an absorbent material, such as activated charcoal, may be placed in the medium.

In some cases, particularly with explants of woody species, an initial preselection or ''conditioning'' substage is utilized for one to two weeks (*26, 27, 28*). The explants are first grown on a simple medium without hormones. The use of this treatment (a) allows the detection of obviously contaminated explants; (b) makes possible the selection of the most actively growing and healthy shoots for further multiplication; and (c) permits some "conditioning" of the plantlets, although the physiological basis of this conditioning is not clear.

Light and temperature are not usually critical for the establishment stage. Usually a light intensity of about 1000 lux is used, with either continuous light or a photoperiod required for optimum growth of the species involved. Temperatures of 20 to 25°C (68 to 77°F) are commonly used.

The establishment stage usually lasts four to six weeks. During this period the explant grows into a small, leafy structure perhaps a centimeter long with several axillary branches.

## Stage II: Multiplication

The function of the multiplication stage is to increase the number of propagules for later rooting to the plantlet stage. The expanded explants from stage I are cut apart and propagules are recultured onto a new medium. Multiplication of vegetative shoots depends on either the continued production of axillary shoots or the initiation of adventitious shoots from the expanded calluslike mass of tissue at the base of the shoots.

Multiplication is repeated at regular intervals. In consecutive multiplication stages, rates of multiplication may vary from about 5 to 50, depending on the species and the method of reproduction. Under ideal conditions of culture, rates of increase may become very high. Success in multiplication requires that uniform unrooted plantlets of the proper size be produced that recover quickly after transfer and begin to grow immediately. The minimum critical mass of tissue, the method of dividing the proliferating mass, and the proper positioning of the propagule on the medium must be established. Often a proliferating mass of shoots will have one elongated shoot that tends to inhibit the larger number of shorter shoots. This elongated shoot can be removed and laid horizontally on new medium for further multiplication or divided into its separate nodes. The basal mass of propagules can then be divided and transferred to a new medium.

Frequency of transfer is important. If transfer is delayed, deterioration and slow recovery often follow. Transfer may need to be two to four weeks apart and should ideally take place when shoots begin to increase in length (106). An analysis and recording of the time sequence of shoot production, along with observations on the health of the cultures, would be helpful to determine the optimum conditions (2).

Size of the culture vessels depends on the kind of plantlets being produced and the space required for proliferation. A choice between agar or liquid media is needed, although a dilute agar solution is usually used. Agar provides support and allows aeration but may reduce the contact of the propagule for nutrient absorption. Modifications to overcome the latter problem include:

**a.** The use of a very dilute agar solution (0.3 to 0.4%) in which the growing plantlet gradually sinks

**b.** The use of a low level of liquid in the container, with or without agitation

**c.** Rotating liquid cultures in a wheel at about 1 rpm

The latter has been used, for instance, with orchids, to bathe the tissue continually with nutrient solution.

Selecting the optimum hormone levels in the culture medium is essential for maximum production of uniform plantlets. For many plant species this information can be obtained from research or commercial experience. For some species, however, optimizing the various media conditions and other factors may require systematic experimentation. As a general guide, if axillary shoots are desired with minimum callus, then a high cytokinin:auxin ratio such as 100:1 is required. If adventitious shoots are needed, use nearly equal amounts at reduced concentrations.

Various procedures for testing the medium requirements have been proposed. De Fossard has described a "broad spectrum experiment" to use in initiating studies for a given species (*35*).

Anderson (*2*) has described a factorial approach to establish medium requirements. First, a single basal medium that produces adequate growth is adopted and enough material is produced to provide 100 propagules for testing auxin and cytokinin requirements. In the first test auxin is kept at a constant level (for example 1 ppm) while the concentration of available cytokinins (kinetin, BAP, and 2iP) is varied (e.g., 1, 5, and 10 ppm) (Figure 16–15). Using "no cytokinin" as the control, plus 3 cytokinins and 3 concentrations, 10 treatments are produced, each with 10 replicates. Using the optimum treatment from this test, a second experiment is then prepared to test auxin requirements. First, all cultures should be recultured through one transfer to equalize growth before proceeding. In the test, auxins (IAA, NAA, and IBA) at 1, 2.5, and 5 ppm are compared but keeping the best cytokinin treatment constant. A subsequent test involves a further experiment in which the best auxin and the best cytokinin sources are tested over the range of concentrations in all combinations.

Tests need to be made for different cultivars (*81*). Usually the appropriate auxin and cytokinin will be the same, but the optimum range of concentrations may be different.

Other factors of culture also should be tested using the procedures described above.

**Figure 16–15**   Cytokinin (6-benzyl-adenine) concentration changes the kind of shoot development in multiplication stage. At 0.1 mg/l the terminal shoot elongates; at 1.0 mg/l lateral shoots develop and the terminal shoot is inhibited. Propagules are almond-peach hybrid.

## Stage III:
## Pretransplant

The function of the pretransplant stage is to prepare the plantlet for transplanting and establishment outside the artificial, closed environment of the culture vessel. In the multiplication stage the plant is conditioned in a high cytokinin medium to favor shoot proliferation rather than elongation. Root initiation is inhibited.

Thus a major change in the pretransplant stage is the shift to conditions that favor root initiation and shoot elongation. For example, cytokinin concentration is reduced or completely eliminated and auxin supply is increased. In this modified medium the propagule can be grown for one passage of two to four weeks to allow roots to develop, or the auxin treatment period can be reduced to a short period of a few days to a week and the propagule transferred to an auxin-free medium for roots to grow, or the shoots may be excised as small cuttings, treated with auxin, and placed in a high humidity chamber to root.

Other conditions may favor root initiation in culture, particularly with hard-to-root plants. Reduction of the inorganic salt concentrations may be important for various plants (142). Addition of phloroglucinol and phloridzin has improved rooting and reduced callusing (65, 70). Phloroglucinol may act as a synergist with auxin; it is effective with some apple and plum cultivars and is stimulative over certain ranges of auxin concentrations but inhibitive in others (161). Its use should be established with tests for each cultivar.

In some species shoot growth becomes inhibited in Stage II because of high cytokinin levels. A single passage without cytokinin or at low cytokinin levels combined with gibberellic acid (1 ppm) is helpful in such cases, such as for *Prunus* (47, 170).

Another characteristic of the plant in the multiplication stage is that it depends for its energy source on the sucrose in the medium rather than its own manufactured supply. Although leaves may be green, they may not actually be capable of photosynthesis. This condition is described as **heterotrophic** in contrast to **autotrophic,** in which energy comes from photosynthesis and the plant is independent of an artificial source. Thus, one of the changes that must take place in the establishment of the propagule in Stage IV is the shift from a heterotrophic to an autotrophic (free-living) existence. This process involves the initiation of new leaves by the plantlet *after* it is removed from the culture environment. Thus, it may be important to keep the shoot in an active state of growth. If the plantlet has gone dormant, shown some growth inhibition, or lost leaves, establishment may be difficult. This problem may be intensified by excessively long culture periods.

Other characteristics of plantlets growing in the highly protected, artificial culture environment are sensitivity to moisture stress and susceptibility to pathogen attack. Sensitivity to stress is apparently due in part to lack of cuticle on the leaves (153). Leaves of such plantlets may appear *nonglaucous* (i.e., smooth, shiny) as compared to the *glaucous* (waxy) appearance of leaves with cuticle. In addition, leaves may be thin, their stomates may not function effectively, and they may be highly sensitive to dehydration (16, 42).

An increased agar concentration (1.0 to 1.2 percent) has been used in Stage III to improve survival of herbaceous perennial plants after transplanting (168). The effect is associated with decreased growth rate and improved hardiness.

## Stage IV: Transplant

The transplant stage involves the transfer of the plantlet from the aseptic cultural environment to the free-living environment of the greenhouse and ultimately to the final location. At the beginning of this stage the plantlet may be rooted or unrooted. In either case, the plantlet or propagule must undergo a period of **acclimation** to enable it to survive. Hence, it must become autotrophic, develop functional roots and shoots, and increase its resistance to desiccation and pathogen attack.

Several environmental conditions are essential in the initial period after transplanting (*141*). One is maintenance of high relative humidity (50 to 100 percent) for two to three weeks to protect the plant from desiccation and enable it to initiate new roots and new shoots. The second requirement is a loose, aerated, well-drained rooting medium, which allows new roots to develop quickly. Another is protection from various pathogens until some resistance has developed. Another is the control of growth after transplanting to overcome or prevent dormancy and resulting lack of growth. As an example of handling, growth of plum plantlets produced in aseptic culture was promoted by chilling for two months at 0°C (32°F) to overcome a rest period before potting or by treating with gibberellic acid (200 ppm) after potting (*57*).

## CONTROL OF PATHOGENS THROUGH MICROPROPAGATION

A principal aspect of micropropagation is the control of pathogens. External pathogens that are present on the explant must be eliminated in order to produce a completely sterile culture.

Internal pathogens that are endogenous to the particular plant source also must be eliminated, but these may not be detected while the explant and subsequent propagules are being grown in culture.

### External Pathogens

External contaminants include fungi, molds, bacteria, yeasts, and other microorganisms. These are present literally everywhere—in the air and on the surface of plants, tables, hands, and so on. Spores of these organisms move by air currents particularly on dust particles. To control these organisms one must disinfest the explants, the tools, and the working area to remove such contaminants from the surface. All of this work must be done in special transfer areas in which such contaminants have been eliminated and precautions have been made to prevent recontamination.

Reduction in surface contaminants begins with control of the stock plants used as the source of explants. Usually contaminants are present only on the surface of plant parts although they may be lodged in cracks, between bud scales, and elsewhere, and sometimes are quite difficult to eliminate. Internal structures, such as growing points of buds, inside seeds and fruits, tend to be relatively free of pathogens. However, if the plant is growing in a humid atmosphere, mycelia may invade plant interiors and become a persistent problem.

Reduction in the surface contaminants on stock plants aids in later disinfection attempts (*48*). In general, stock plants growing in containers in protected environments in a greenhouse are better sources of explants than those growing out-of-doors. Overhead watering, sprinkling, or any activity that increases humidity around the plant should be avoided. Likewise, keep plant parts off the ground and avoid use of roots or underground portions as explant sources, if possible. Insect and mite populations should be controlled by spraying. In one case, withholding water and keeping plants cool and dry for three weeks was necessary to reduce contaminants on *Dieffenbachia,* even inside a greenhouse (*74*). If plants are growing out-of-doors, covering shoots with plastic bags, or spraying with fungicides may be useful (*35*).

Disinfestation requires use of chemicals that are toxic to the microorganism but relatively nontoxic to plant material. Tissue culture became possible with the discovery of convenient and effective disinfestants, such as calcium hypochlorite and sodium hypochlorite (now widely available as commercial household bleaches under various trade names). Both effectiveness of treatment and damage to living tissue increase as time-dosage responses so that an effective compromise must be worked out.

Alcohol (ethyl, methyl, or isopropyl) at 70 to 75 percent is used widely as a disinfestant but is highly toxic to plant material, so its use is restricted to short rinses and sterilization of external surfaces where the explant is extracted from the interior of the organ.

Placing such antibiotics as streptomycin in the culture medium has been tried (*88*) for inhibiting microorganisms but with varying degrees of success.

## Internal Pathogens

Organisms that are present inside the tissue of the explant at the time of culture present a most serious problem. One of the primary reasons for using micropropagation procedures is the ability to produce pathogen-free materials. Nevertheless, various workers in the field have repeatedly warned that one should not take for granted that the materials are free of pathogens unless specific indexing or other detection procedures are utilized as part of the procedure.

### VIRUSES AND VIRUSLIKE ORGANISMS

Viruses and viruslike organisms may be present in the tissues of the plant without manifesting specific symptoms (*82, 105*). Some viruses, such as dasheen mosaic virus in *Dieffenbachia,* have seriously inhibited growth in culture (*49*). Thus, one should select explants where possible from virus-tested source plants, such as those in a Registration or Certification program (see chapter 8). The extreme meristem-tip of the shoot is utilized to separate the explants from any viruses in the rest of the plant. This procedure may be combined with heat treatments and other decontamination procedures.

### BACTERIAL AND FUNGAL CONTAMINANTS

Certain bacterial contaminants, such as *Bacillus subtilis* (*145*), *Erwinia* (*74, 75*), or *Pseudomonas* (*74*), sometimes are present inside the plantlet but do not grow readily on the medium, either because the medium has become more acid (*49*) or because cytokinin in the medium depresses their growth (*53*). These contaminants can greatly inhibit growth and rooting potential (*49*) or remain suppressed without effect until the explant is transferred to a new culture medium (*53*).

A useful procedure is to culture-index pieces of individual plantlets and eliminate those containers showing positive results for the presence of pathogens. Doing this at the initial explant stage, however, does not always identify all potential contaminants (*49*).

# GENETIC, CHIMERAL, AND EPIGENETIC VARIATION IN PLANTS DURING MICROPROPAGATION

Three general sources of variants in vegetatively propagated clones were described in chapter 8: *genetic, chimeral,* and *epigenetic.* Since these variants originate at the cell level, their impact can be significant in micropropagation and tissue culture when cells are the starting points of new plants. These variants can be both beneficial and detrimental in micropropagation. On the one hand, success in directing development in culture systems appears to be based in part on controlling the epigenetic state of the explant, propagule, and plantlet. On the other hand, because of genetic and epigenetic changes, the final product may be plants that look different or perform somewhat differently from comparable plants produced by traditional propagation. One may find a proportion of aberrant "off-type" plants that need to be eliminated or "rogued out" and monitoring procedures need to be part of the propagation sequence.

## Types of Variation

### GENETIC VARIATION

Genetic changes, as described in chapter 8, are permanent changes in the genome or plastome that give rise to mutant plants. Chlorophyll controlling mutants involving appearance of albino plants (Figure 16–16) or albino leaf sec-

**Figure 16–16** Albino plantlet appearing on an African violet *(Saintpaulia)* after micropropagation. Origin is likely to have been from a single cell; hence a solid mutant resulted.

**Figure 16-17**    Genetic variation shown between two tobacco plants originating from single cells. Callus was developed and organogenesis occurred as described in Figure 16-11. *Left:* normal plant. *Right:* variant plant. Courtesy T. Murashige.

tors are fairly common but usually can be identified easily (*34, 84*). Other kinds of mutants affecting growth habit (Figure 16-17), vigor, or productivity may not be identified easily, or not until the plant has been grown in its ultimate location. Usually such off-types appear as individual plants at a relatively low rate.

Gross changes that tend to occur in the genotype of cells, such as polyploidy or aneuploidy, characteristically are associated with unorganized cell systems, such as tissue or suspension cultures, and their rate of appearance may increase with time and number of transfers (*106*). In some cases, the variants may result from cellular changes that had already occurred in the source plant. In some intact plants, cells with increased chromosome numbers have been found to develop during the transition from meristematic cells to mature parenchyma cells (*33*).

Although induced or natural genetic variability in cell populations in aseptic culture can be a hazard in propagation, this potentiality is valuable in plant breeding. In one study (*137*) geranium plants, referred to as **calliclones,** were regenerated from callus produced from various geranium cultivars. The population of new plants showed a wider range of variability as compared to plants propagated from the same cultivars by conventional stem, root, or petiole cuttings.

The variants could have been the result of segregation of cells from different chimeral sectors in the source cultivars, mutation, elimination of viruses, or epigenetic variations. In a somewhat similar study in potato (*134*), a series of intraclonal variants have been obtained, which were referred to as **protoclones.** Separate protoplasts were cultured from leaves of 'Burbank' potato, a vegetatively propagated cultivar grown for over 100 years. Callus was produced from them and new plants regenerated. Some of the new intraclonal variants involved useful economic characteristics, such as size or flowering, and the method has been suggested for plant breeders.

CHIMERAS

Plants that are chimeras are made up of tissues with different genotypes growing adjacent to each other. These exist in various cellular patterns and vary in their stability in vegetative propagation.

The stability of chimeral cultivars grown by micropropagation depends upon the nature of the chimera and the type of culture system used. Since the variant is present in the original explant source as a chimera, regeneration by shoot-tip elongation or by axillary buds would offer the best possibility of maintaining the chimeral nature of the cultivar (21). If adventitious shoots initiate directly on the explant, segregation among the genotypes represented in the chimera would likely occur, depending on which genotype was present in the cells where the adventitious shoot originated (19). Adventitious shoots arise primarily from one or a few cells near the surface of the organ from which the explant is taken. If an intermediate period of callusing occurs between explant selection and plantlet regeneration, then the potential for chimeral breakdown is great enough to preclude use of methods where callusing is part of the procedure.

Spontaneous mutations for propagation of known chimeral cultivars occur at regular rates in somatic tissue. Sometimes two "new" genotypes result from *somatic crossing-over* (40) to give rise to "twin" sectors differing from the surrounding tissue. The appearance of aberrant plants in culture systems may not actually represent an abnormal increase over expectations. Rather, the large numbers of plants involved in micropropagation and the stimulation of many axillary shoots may provide an efficient system for their detection.

New mutants often appear early as visible streaks or sectors in leaves.

With adventitious shoot formation, where single cells or a limited number participate, one might expect recovery of whole plant mutants. Similarly, regeneration by organogenesis in callus or embryogenesis from callus is likely to give rise to whole plant, solid mutants (20). Variants that affect growth or productivity might not be detected until much later stages.

With callus systems, rates of spontaneous mutations and chromosomal multiplications have been shown to increase above those of cells in organized tissues and the number to increase with more transfer passages (105). Thus, for propagation purposes the use of callus intermediary phases is generally avoided. If needed, the period of time in terms of numbers of transfers should be held to a minimum.

EPIGENETIC VARIATION

Epigenetic variants are those that appear to be more or less constant, persist with continued vegetative propagation (90, 91), but do not involve a permanent change in the genotype. Such a precise distinction between genetic and epigenetic variants is hard to prove (93), and is often assumed to exist by indirect evidence. Direct evidence for a genetic variant would be transmission of the trait to seedling offspring or demonstration of segregation by classical genetic tests. Epigenetic variants are not transmitted during the formation of an embryo, whereas genetic traits should be.

Epigenetic differences may involve variation among different explants taken from intact plants. The control of the juvenile-to-mature phase changes has been described as epigenetic (see chapter 8). Explants taken from different parts and organs of the plant may represent different epigenetic states. Callus cultures started from adult and juvenile phases of ivy (*Hedera*), for example, have been shown to have different growth characteristics, which persisted over many transfers (*149*). When new plants were regenerated, "adult" callus underwent embryogenesis, whereas "juvenile" callus underwent organogenesis. However, all plantlets produced were juvenile, indicating that reversion and loss of the epigenetic state occurred (*8*).

In annual seedling plants, e.g., tobacco and others (*132, 159*), cell layer explants obtained consecutively from base to the apex of stems showed a gradient in which the basal explants regenerated vegetative shoots and the apical nodes flowering shoots. Thus, different parts of the plant were "programmed" toward a specific morphological direction.

Shoot-tip cultures (*85, 103*) often produce plantlets that tend to show juvenile growth. In part, this tendency may be associated with adventitious shoot production, which is known to result in reversion (see chapter 8). In other cases, as in a 1000-year-old grape clone, reversion to a juvenile state has been produced by consecutive subculturing of shoot-tips in aseptic culture (*103*). It is reported that woody plants produced after several consecutive passages in culture, then planted in the field, have produced easy-to-root cuttings (*14*).

Embryogenesis is dependent on achieving the proper epigenetic state in specific *embryogenic callus*. This potentiality depends on the plant part used for the original explant as well as the sequence of handling in culture. However, once such callus is produced, it remains embryogenic and continues to produce somatic embryos (*78*).

Cells growing in culture either in suspensions or as callus commonly change their growth characteristics with consecutive passages. Usually there is a gradual improvement in ability of cells to grow. This tendency represents the "sorting out" of the most adapted cells to grow in culture. Whether this phenomenon represents selection of genetic variants or epigenetic shifts in cell characteristics is difficult to establish.

With continued culture, cells very often lose their capacity to undergo organogenesis or embryogenesis. Such loss has been associated with change in their ploidy level, indicating genetic change (*33, 158*). In other cases, the change has been reversed, suggesting an epigenetic change (*150*).

In some cases, tissue cultures have been found to lose their requirement for specific growth hormones, as cytokinin or auxin (*90*). This process has been called **habituation**.

Similarly, the changes in plant tumor cells have been shown to result from epigenetic effects rather than from genetic effects (*151*).

### Monitoring Variation in Micropropagation

Certain guidelines are suggested to handle micropropagated material to minimize unwanted variation and the production of aberrant plants or those not performing properly. Such a program should be part of the process to adapt micropropagation for particular species or cultivars and should continue into the evaluation of the propagated material. It may be necessary to include the program as part of a continuing monitoring process. Control of variation might occur at any of these four steps: (a) selection of cultivar, (b) selection of specific explant and source, (c) selection of appropriate propagation procedure, (d) observations of the propagated plants.

In case of micropropagated plants one should

1. Know the nature of the cultivar in terms of its chimerism, genetic characteristics, and the like. This knowledge may determine which propagation system will be selected.

2. Be sure the source plant utilized is the correct cultivar and does not itself represent a variant. If possible, use a source plant previously indexed for virus or other internal pathogens, or carry out indexing. Using a very small meristem explant helps to eliminate viruses but does not guarantee a pathogen-free state without confirmation through appropriate tests.

3. Use an appropriate propagation procedure. In general, explants with intact growing points, terminal or axillary, are preferred over adventitious shoot formation. An intermediate callusing step should be avoided except in special situations. If callusing is part of the procedure, limit the number of passages to three and then start with new explants. In each case, of course, the method to be used should be researched through step 5.

4. At the end of the multiplication stage, check for obvious aberrant plants and off-types and "rogue" them out. Such monitoring can be done at any time, of course. At this time, plantlets can also be sorted for uniformity to go into the pretransplant stage.

5. Carry out *progeny tests* on the propagated material to the final growing area to establish that the plants produced are appropriate for their purpose.

6. Develop a system to maintain *source identity* of plants going through the propagation procedure. This system makes it possible to trace a problem back to its origin.

# REFERENCES

1. Abbott, A. J., and E. Whiteley. 1976. Culture of *Malus* tissues in vitro. 1. Multiplication of apple plants from isolated shoot apices. *Scient. Hort.* 4:183–89.

2. Anderson, W. C. 1980. Mass propagation by tissue cultures: Principles and techniques. In *Proc. conf. on nursery production of fruit plants: Applications and feasibility,* R. H. Zimmerman, ed. USDA ARR–NE–11, pp. 1–14.

3. Anderson, W. C., and J. B. Carstens. 1977. Tissue culture propagation of broccoli, *Brassica oleracea* (Italian group) for use in $F_1$ hybrid seed production. *Jour. Amer. Soc. Hort. Sci.* 102(1):69–73.

4. Arditti, J. 1967. Factors affecting the germination of orchid seeds. *Bot. Rev.* 33(1):1–97.

5. Asen, S., and R. E. Larson. 1951. Artificial culturing of rose embryos. Pennsylvania State College School of Agriculture, Agr. Exp. Sta. *Progress Report* 40:1–4.

6. Bajaj, Y. P. S. 1979. Establishment of germ plasma banks through freeze storage of plant tissue culture and their application to agriculture. In *Plant cell tissue culture: Principles and application,* W. R. Sharp et al., eds. Columbus: Ohio State Univ., pp. 745–74.

7. Bajaj, Y. P. S., and W. B. Collins. 1968. Some factors affecting the in vitro development of strawberry fruits. *Proc. Amer. Soc. Hort. Sci.* 93:326–33.

8. Banks, M. S. 1979. Plant regeneration from callus from two growth phases of English ivy, *Hedera helix* L. *Zeits. fur Pflanzenphys.* 92:349–53.

9. Barlass, M., and K. G. M. Skene. 1980. Studies on the fragmented shoot apex of grapevine. I. The regenerative capacity of leaf primordial fragments in vitro. *Jour. Exp. Bot.* 31(121):483–88.

10. ———. 1980. Studies on the fragmented shoot apex of grapevine. II. Factors affecting growth and differentiation *in vitro. Jour. Exp. Bot.* 31(121):489–95.

11. Barz, W., E. Reinhard, and M. H. Zenk, eds. 1977. *Plant tissue culture and its biotechnological applications.* Berlin: Springer-Verlag.

12. Beasley, C. A. 1977. Ovule culture: Fundamental and pragmatic research for the cotton industry. In *Plant cell, tissue and organ culture,* J. Reinert and Y. P. S. Bajaj, eds. Berlin: Springer-Verlag, pp. 160–78.

13. Bilkey, P. C., B. H. McCown, and A. C. Hildebrandt. 1978. Micropropagation of African violet from petiole cross-sections. *HortScience* 13(1):37–38.

14. Boulay, M., H. Chaperon, and A. David. 1979. *Micropropagation d'arbres forestiers: Etudes et recherches.* Nangis, France: Association foret-cellulose domain de'etaneon.

15. Bourgin, J. P., and J. P. Nitsch. 1967. Obtention de Nicotiana haploides a'partir d'elamines cultivees *in vitro. Ann. Phys. Veg.* 9:377–82.

16. Brainerd, K. E., L. H. Fuchigami, S. Kwiatkowski, and C. S. Clark. 1981. Leaf anatomy and water stress of aseptically cultured 'Pixy' plum grown under different environments. *HortScience* 16(2):173–75.

17. Brettell, R. I. S., and D. S. Ingram. 1979. Tissue culture in the production of novel disease-resistant crop plants. *Bio. Rev.* 54:329–45.

18. Burk, L. G., G. R. Gwynn, and J. R. Chaplin. 1972. Diploidized haploids from aseptically cultured anthers of *Nicotiana tabacum. Jour. Hered.* 63(6):355–60.

19. Bush, S. R., E. D. Earle, and R. W. Langhans. 1976. Plantlets from petal segments, petal epidermis, and shoot tips of the periclinal chimera, *Chrysanthemum morifolium* 'Indianapolis'. *Amer. Jour. Bot.* 63(6):729–37.

20. Button, J., and J. Kochba. 1977. Tissue culture in the citrus industry. In *Plant cell, tissue and organ culture,* J. Reinert and Y. P. S. Bajaj, eds. Berlin: Springer-Verlag, pp. 70–92.

21. Boxus, P. H., M. Quoirin, and J. M. Laine. 1977. Large scale propagation of strawberry plants from tissue culture. In *Plant cell, tissue and organ culture,* J. Reinert and Y. P. S. Bajaj, eds. Berlin: Springer-Verlag, pp. 131–43.

22. Cameron-Mills, V., and C. M. Duffus. 1977. The *in vitro* culture of immature barley embryos on different culture media. *Ann. Bot.* 41:1117–27.

23. Campbell, R. A., and D. S. Durzan. 1975. Induction of multiple buds and needles in tissue cultures of *Picea glauca. Can. Jour. Bot.* 53:1652–57.

24. Carlson, P. S., H. H. Smith, and R. D. Dearing. 1972. Parasexual interspecific plant hybridization. *Proc. Nat. Acad. Sci. USA* 69(8):2292–94.

25. Cheah, Kheng-Tuan, and Tsai-Ying Cheng. 1978. Histological analysis of adventitious bud formation in cultured Douglas fir cotyledon. *Amer. Jour. Bot.* 65(8):845–49.

26. Cheng, Tsai-Ying. 1978. Clonal propagation of woody plant species through tissue culture techniques. *Proc. Inter. Plant Prop. Soc.* 28:139–55.

27. ———. 1978. Propagating woody plants through tissue culture. *Amer. Nurs.* 147(10):7–8, 94–102.

28. Cheng, Tsai-Ying, and Thanh H. Voqui. 1977. Regeneration of Douglas fir plantlets through tissue culture. *Science* 198:306–7.

29. Christie, C. B. 1978. Rapid propagation of aspens and silver poplars using tissue culture techniques. *Proc. Inter. Plant Prop. Soc.* 28:255–60.

30. Cocking, E. C. 1960. A method for the isolation of plant protoplasts and vacuoles. *Nature* 187:927–29.

31. Coleman, W. K., and T. A. Thorpe. 1977. *In vitro* culture of western red-cedar (*Thuja plicata* Donn). I. Plantlet formation. *Bot. Gaz.* 138(3):298–304.

32. Collins, G. B. 1978. Pollen and anther culture: Progress and application. In *Propagation of higher plants through tissue culture,* K. W. Hughes, R. Henke, and M. Constantin, eds. U.S. Dept. of Energy Tech. Information Center, pp. 179–88.

33. D'Amato, F. 1977. Cytogenetics of differentiation in tissue and cell cultures. In *Plant cell, tissue and organ cultures,* J. Reinert and Y. P. S. Bajaj, eds. Berlin: Springer-Verlag, pp. 343–57.

34. Damiano, C. 1980. Strawberry micropropagation. In *Proc. of the conf. on nursery production of fruit plants through tissue culture: Applications and feasibility,* R. H. Zimmerman, ed. Beltsville, Md.: USDA ARR–NE–11, pp. 11–22.

35. de Fossard, R. A. 1976. *Tissue culture for plant propagators.* Armidale, N.S.W., Aust.: Univ. of New England Printery.

36. de Nettencourt, D., and M. Deuvreaux. 1977. Incompatibility and *in vitro* cultures. In *Plant cell, tissue and organ cultures,* J. Reinert and Y. P. S. Bajaj, eds. Berlin: Springer-Verlag, pp. 426–35.

37. Dougall, D. K. 1979. Factors affecting yields of secondary products in plant tissue cultures. In *Plant cell and tissue cultures: Principles and applications,* W. R. Sharp et al., eds. Columbus: Ohio State Univ. Press, pp. 727–44.

38. Dunstan, D. I. 1981. Transplantation and post-transplantation of micropropagated tree-fruit rootstocks. *Proc. Inter. Plant Prop. Soc.* 31:39–45.

39. Eriksson, T., and K. Jonasson. 1969. Nuclear division in isolated protoplasts from cells of higher plants grown *in vitro*. *Planta* 89:85–89.

40. Evans, D. A., and E. F. Haddock. 1978. Mitotic crossing-over in higher plants. In *Plant cell and tissue culture: Principles and applications,* R. W. Sharp et al., eds. Columbus: Ohio State Univ. Press, pp. 315–52.

41. Evans, D. A., W. R. Sharp, and C. E. Flick. 1981. Plant regeneration from cell cultures. *Hort. Rev.* 3:214–314.

42. Fuchigami, L. H., T. Y. Cheng, and A. Soeldner. 1981. Abaxial transpiration and water loss in aseptically cultured plum. *Jour. Amer. Soc. Hort. Sci.* 106(4):519–22.

43. Gamborg, O. L., K. Ohyama, L. E. Pletcher, L. C. Fowke, K. Kartha, F. Constabel, and K. Kao. 1979. Genetic modification of plants. In *Plant cell and tissue culture: Principles and applications,* W. R. Sharp et al., eds. Columbus: Ohio State Univ. Press, pp. 371–98.

44. Guha, S., and S. C. Maheshwari. 1964. *In vitro* production of embryos from anthers of *Datura*. *Nature* 204:497.

45. Haberlandt, G. 1902. Kulturversuche mit isolierten Pflanzenzellsn. *Sitzungsber. Akad. der Wiss. Wien, Math.-Naturwiss.* K1. 111:69–92.

46. Halperin, W., and D. F. Wetherell. 1964. Adventive embryony in tissue cultures of the wild carrot, *Daucus carota. Amer. Jour. Bot.* 51:274–83.

47. Hammerschlag, F. 1980. Peach micropropagation. In *Proc. of conf. on nursery production of fruit plants through tissue culture.* USDA. Arr–NE–11, pp. 48–52.

48. Henny, R. J. 1980. *In vitro* germination of *Maranta leuconeura* embryos. *HortScience* 15(2):198–200.

49. Henny, R. J., J. F. Knauss, and A. Donnan, Jr. 1981. Foliage plant tissue culture. In *Foliage plant production,* J. Joiner, ed. Englewood Cliffs, N.J.: Prentice-Hall.

50. Hesse, C., and D. E. Kester. 1955. Germination of embryos of *Prunus* related to degree of embryo development and method of handling. *Proc. Amer. Soc. Hort. Sci.* 65:251–64.

51. Heuser, C. W., and J. Horker. 1976. Tissue culture propagation of daylilies. *Proc. Inter. Plant Prop. Soc.* 26:269–72.

52. Hicks, G. S. 1980. Patterns of organ development in plant tissue culture and the problem of organ determination. *Bot. Rev.* 46:1–23.

53. Holdgate, D. P., and J. S. Aynsley. 1977. The development and establishment of a commercial tissue culture laboratory. *Acta Hort.* 78:31–36.

54. Holley, W. D., and R. Baker. 1963. *Carnation production.* Dubuque, Iowa: William C. Brown.

55. Hollings, M. 1965. Disease control through virus-free stock. *Ann. Rev. Phytopath.* 3:367–96.

56. Horsch, R. B., J. King, and G. E. Jones. 1980. Measurement of cultured plant cell growth on filter paper discs. *Can. Jour. Bot.* 58:2402–6.

57. Howard, B. H., and V. H. Oehl. 1981. Improved establishment of *in vitro* propagated plum micropropagules following treatment with $GA_3$ or prior chilling. *Jour. Hort. Sci.* 56(1):1–7.

58. Huang, Shu-Ching, and D. F. Millikan. 1979. Disease-free plants through micropropagation. *Proc. Inter. Plant Prop. Soc.* 29:393–98.

59. ———. 1980. *In vitro* micrografting of apple shoot-tips. *HortScience* 15:741–43.

60. Hussey, G. 1976. *In vitro* release of axillary shoots from apical dominance in monocotyledonous plantlets. *Ann. Bot.* 40:1323–25.

61. ———. 1977. *In vitro* propagation of some members of the Liliaceae, Iridaceae and Amaryllidaceae. *Acta Hort.* 78:303–9.

62. Hussey, G., and A. Falavigna. 1980. Origin and production of *in vitro* adventitious shoots in the onion, *Allium cepa* L. *Jour. Exp. Bot.* 31(125):1675–86.

63. Inomata, N. 1978. Production of interspecific hybrids in *Brassica campestris* × *B. oleracea* by culture *in vitro* of excised ovaries. I. Development of excised ovaries in the crosses of various cultivars. *Japan. Jour. Genetics* 53(3):161–73.

64. ———. 1978. Production of interspecific hybrids between *Brassica campestris* × *Brassica oleracea* by culture *in vitro* of excised ovaries. II. Effects of cocoanut milk and casein hydrolysate on the development of excised ovaries. *Japan. Jour. Genetics* 53:1–11.

65. James, D. J., and Isobel J. Thurbon. 1979. Rapid *in vitro* rooting of the apple rootstock M.9. *Jour. Hort. Sci.* 54(4):309–11.

66. Jarret, R. L., P. M. Hasegawa, and H. T. Erickson. 1980. Factors affecting shoot initiation from tuber discs of potato (*Solanum tuberosum*). *Phys. Plant.* 49:177–84.

67. Johri, B. M., and S. Guha. 1963. *In vitro* development of onion plants from flowers. In *Plant tissue and organ culture,* P. Maheshwari and N. S. Rangaswamy, eds. Delhi: Inter. Soc. of Plant Morph., Univ. of Delhi, pp. 215–23.

68. Johri, B. M., and C. B. Sehgal. 1963. Growth of ovaries of *Anethum graveolens* L. In *Plant tissue and organ culture,* P. Maheshwari and N. S. Rangaswamy, eds. Delhi: Int. Soc. of Plant Morph., Univ. of Delhi.

69. Jones, J. B. 1979. Commercial use of tissue culture for the production of disease-free plants. In *Plant cell and tissue culture: Principles and applications,* W. R. Sharp et al., eds. Columbus: Ohio State Univ. Press, pp. 441–52.

70. Jones, O. P. 1976. Effect of phloridzin and phloroglucinol on apple shoots. *Nature* 262(5567):392–93.

71. ———. 1979. Propagation in vitro of apple trees and other woody fruit plants: Methods and applications. *Sci. Hort.* 30:44–48.

72. Kartha, K. K., N. L. Leung, and K. Pahl. 1980. Cryopreservation of strawberry meristem and mass propagation of plantlets. *Amer. Soc. Hort. Sci.* 105:481–84.

73. Kester, D. E., and C. O. Hesse. 1955. Embryo culture of peach varieties in relation to season of ripening. *Proc. Amer. Soc. Hort. Sci.* 65:265–73.

74. Knauss, J. F. 1976. A tissue culture method of producing *Dieffenbachia picta* cv. Perfection free of fungi and bacteria. *Proc. Fla. State Hort. Soc.* 89:293–96.

75. Knauss, J. F., and J. W. Miller. 1978. A contaminant, *Erwinia carotovera,* affecting commercial plant tissue culture. *In Vitro* 14:754–56.

76. Knudson, L. 1922. Non-symbiotic germination of orchid seeds. *Bot. Gaz.* 73:1–25.

77. ———. 1951. Nutrient solutions for orchids. *Bot. Gaz.* 112:528–32.

78. Krul, W. R., and J. Myerson. 1980. *In vitro* propagation of grape. In *Proc. conf. on nursery production of fruit plants through tissue culture: Applications and feasibility,* R. H. Zimmerman, ed. USDA ARR-NE-11, pp. 35–43.

79. Krul, W. R., and J. F. Worley. 1977. Formation of adventitious embryos in callus cultures of 'Seyval', a French hybrid grape. *Jour. Amer. Soc. Hort. Sci.* 102:360–63.

80. Lammerts, W. E. 1942. Embryo culture, an effective technique for shortening the breeding cycle of deciduous trees and increasing germination of hybrid seed. *Amer. Jour. Bot.* 29:166–171.

81. Lane, W. D. 1982. Plant manipulation *in vitro* with hormones. *Proc. Inter. Plant Prop. Soc.* 31:101–8.

82. Langhans, R. W., R. K. Horst, and E. D. Earle. 1977. Disease-free plants via tissue culture propagation. *HortScience* 12:149–50.

83. Lee, C. W., R. M. Skirvin, A. I. Soltero, and J. Janick. 1977. Tissue culture of *Salpiglossis sinuata* L. from leaf discs. *HortScience* 12(6):547–49.

84. Lo, P. F., C. H. Chen, and J. G. Ross. 1980. Vegetative propagation of temperate forage grasses through callus culture. *Crop. Sci.* 20:363–67.

85. Lyrene, P. M. 1981. Juvenility and production of fast-rooting cuttings from blueberry shoot cultures. *Jour. Amer. Soc. Hort. Sci.* 106:396–98.

86. Maeda, E., and T. A. Thorpe. 1979. Shoot histogenesis in tobacco callus cultures. *In vitro* 15(6):415–24.

87. Maheshwari, P., and N. S. Rangaswamy. 1963. Plant tissue and organ culture from the viewpoint of an embryologist. In *Plant tissue and organ culture,* P. Maheshwari and N. S. Rangaswamy, eds. Delhi: Inter. Soc. of Plant Morph., Univ. of Delhi, pp. 390–420.

88. McGarrity, G. L. 1975. Control of microbiological contamination. *Tissue Culture Assn. Man.* 1:181–85.

89. McGrew, J. R. 1980. Meristem culture for production of virus-free strawberries. In *Proc. conf. on nursery production of fruit plants through tissue culture: Applications and feasibility,* R. H. Zimmerman, ed. USDA. ARR–NE–11, pp. 80–85.

90. Meins, F., Jr., and A. N. Binns. 1978. Epigenetic clonal variation in the requirements of plant cells for cytokinins. In *The clonal basis of development,* S. Subtelny and I. M. Sussex, eds. New York: Academic Press, pp. 185–201.

91. ———. 1979. Cell determination in plant development. *BioScience* 29:221–25.

92. Mellor, F. C., and R. Stace-Smith. 1977. Virus-free potatoes by tissue culture. In *Plant cell, tissue and organ culture,* J. Reinert and Y. P. S. Bajaj, eds. Berlin: Springer-Verlag, pp. 616–35.

93. Meredith, C., and P. S. Carlson. 1978. Genetic variation in cultured plant cells. In *Propagation of higher plants through tissue culture,* K. W. Hughes, R. Henke, and M. Constantin, eds. U.S. Dept. of Energy, Tech. Information Center, pp. 166–76.

94. Meyer, M. M., Jr. 1976. Culture of Paeonia embryos by *in vitro* techniques. *Amer. Peony Soc. Bul.* 217:32–35.

95. ———. 1980. *In vitro* propagation of *Hosta sieboldiana. HortScience* 15:737–38.

96. Meyer, M. M., and G. M. Milbrath. 1977. *In vitro* propagation of horseradish with leaf pieces. *HortScience* 12(6):544–45.

97. Meyer, M. M., Jr., L. H. Fuchigami, and A. N. Roberts. 1975. Propagation of tall bearded irises by tissue culture. *HortScience* 10:479–80.

98. Meyer, M. M., Jr. 1976. Propagation of daylilies by tissue culture. *HortScience* 11(5): 485–87.

99. Miyata, N. S. 1978. Embryo culture of bromeliad seed. *The Plant Propagator* 24(1):5.

100. Morel, G. M. 1960. Producing virus-free cymbidiums. *Amer. Orch. Soc. Bul.* 29:495–97.

101. ———. 1964. Tissue culture: A new means of clonal propagation of orchids. *Amer. Orch. Soc. Bul.* 33:473–78.

102. Mullins, M. G., and C. Srinivasan. 1976. Somatic embryos and plantlets from an ancient clone of the grapevine (cv. Cabernet-Sauvignon) by apomixis *in vitro. Jour. Exp. Bot.* 27(100):1022–30.

103. Mullins, M. G., G. Y. Nair, and P. Sampet. 1979. Rejuvenation *in vitro:* Induction of juvenile characters in an adult clone of *Vitis vinifera* L. *Ann. Bot.* 44:623–27.

104. Murashige, T. 1974. Plant propagation through tissue cultures. *Ann. Rev. Plant Phys.* 25:135–66.

105. ———. 1977. Plant cell and organ cultures as horticultural practices. *Acta Hort.* 78:17–30.

106. ———. 1978. Principles of rapid propagation. In *Propagation of higher plants through tissue culture,* K. M. Hughes, R. Henke, and M. Constantin, eds. U.S. Dept. of Energy, Tech. Information Center, pp. 14–24.

107. Murashige, T., and F. Skoog. 1962. A revised medium for rapid growth and bioassays with tobacco tissue cultures. *Phys. Plant.* 15:473–97.

108. Narayanaswami, S., and K. Norstog. 1964. Plant embryo culture. *Bot. Rev.* 30(4): 587–625.

109. Navarro, L., and J. Juarez. 1977. Tissue culture techniques used in Spain to recover virus-free citrus plants. *Acta Hort.* 78:425–35.

110. Navarro, L., C. N. Roistacher, and T. Murashige. 1975. Improvement of shoot-tip grafting *in vitro* for virus-free citrus. *Jour. Amer. Soc. Hort. Sci.* 100:471–79.

111. Negueroles, J., and O. P. Jones. 1979. Production *in vitro* of rootstock/scion combinations of *Prunus* cultivars. *Jour. Hort. Sci.* 54(4):279–81.

112. Nitsch, C. 1981. Production of isogenic lines: Basic technical aspects of androgenesis. In *Plant tissue culture: Methods and applications in agriculture,* T. Thorp, ed. New York: Academic Press, pp. 241–52.

113. Nitsch, J. P. 1963. *In vitro* culture of flowers and fruit. In *Plant tissue and cell culture,* P. Maheshwari and N. S. Rangaswamy, eds. Delhi: Inter. Soc. of Plant Morph., Univ. of Delhi, pp. 198–214.

114. Nitsch, J. P., and C. Nitsch. 1969. Haploid plants from pollen grains. *Science* 163:85–87.

115. Norstog, K. 1977. Embryo culture as a tool in the study of comparative and developmental morphology. In *Plant cell, tissue and organ culture,* J. Reinert and Y. P. S. Bajaj, eds. Berlin: Springer-Verlag, pp. 190–202.

116. Oglesby, R. P. 1978. Tissue culture of ornamentals and flowers: Problems and perspectives. In *Propagation of higher plants through tissue culture,* K. M. Hughes, R. Henke, and M. Constantin, eds. U.S. Dept. of Energy, Tech. Information Center, pp. 59–61.

117. Oglesby, R. P., and R. E. Strode. 1979. Commercial tissue culturing at Oglesby Nursery. *Proc. Inter. Plant Prop. Soc.* 29:341–44.

118. Papachatzi, M., P. A. Hammer, and P. M. Hasegawa. 1980. *In vitro* propagation of *Hosta plantaginea.* *HortScience* 15(4):506–7.

119. Pence, V. C., P. M. Hasegawa, and J. Janick. 1979. Asexual embryogenesis in *Theobroma cacao* L. *Jour. Amer. Soc. Hort. Sci.* 104(2):145–48.

120. Pierik, R. L. M. 1979. *In vitro culture of higher plants.* Amsterdam: Posen en Looyen.

121. Pierik, R. L. M., and H. H. M. Steegmans. 1976. Vegetative propagation of Freesia through the isolation of shoots *in vitro. Neth. Jour. Agr. Sci.* 24:274–77.

122. Pierik, R. L. M., J. L. M. Jansen, A. Maasdam, and C. M. Binnendijk. 1975. Optimalization of *Gerbera* plantlet production from excised capitulum explants. *Sci. Hort.* 3:351–57.

123. Quak, F. 1977. Meristem culture and virus-free plants. In *Plant cell, tissue and organ culture,* J. Reinert and Y. P. S. Bajaj, eds. Berlin: Springer-Verlag, pp. 598–615.

124. Quoirin, M., P. Boxus, and T. Gaspar. 1974. Root initiation and isoperoxidases of stem tip cuttings from mature *Prunus* plants. *Phys. Veg.* 12(2):165–74.

125. Rajasekaran, K., and M. G. Mullins. 1979. Embryos and plantlets from cultured anthers of hybrid grapevines. *Jour. Exp. Bot.* 30(116):399–407.

126. Randolph, L. F., and L. G. Cox. 1943. Factors influencing the germination of *Iris* seed and the relation of inhibiting substances to embryo dormancy. *Proc. Amer. Soc. Hort. Sci.* 43:284–300.

127. Rangaswamy, N. S. 1977. Application of *in vitro* pollination and *in vitro* fertilization. In

*Plant cell, tissue and organ culture,* J. Reinert and Y. P. S. Bajaj, eds. Berlin: Springer-Verlag, pp. 412–25.

128. Reinert, J. 1959. Über die kontrolle de morphogenese und die induktion von adventiveembryonen an gewebekulturen aus karotten. *Planta* 53:318–33.

129. Roest, S., and G. S. Bokelmann. 1973. Vegetative propagation of *Chrysanthemum cinerariaefolium in vitro. Sci. Hort.* 1:120–22.

130. Salomon, E. 1976. Formation of adventitious buds in decapitated citrus seedlings and the effect of some growth regulators. *Jour. Exp. Bot.* 27:69–75.

131. Sandahl, M. R., and W. R. Sharp. 1979. Research in coffee sp. and applications of tissue culture methods. In *Plant cell and tissue culture: Principles and applications,* W. R. Sharp et al., eds. Columbus: Ohio State Univ. Press, pp. 527–84.

132. Scorza, R., and J. Janick. 1980. *In vitro* flowering of *Passiflora suberosa* L. *Jour. Amer. Soc. Hort. Sci.* 105:892–97.

133. Sharp, W. R., M. R. Sandahl, L. S. Caldas, and S. B. Maraffa. 1980. The physiology of in vitro asexual embryogenesis. *Hort. Rev.* 2:268–309.

134. Shepard, J. F., D. Bidney, and E. Shakin. 1980. Potato protoplasts in crop improvement. *Science* 208:17–24.

135. Sinha, S., R. P. Roy, and K. K. Jha. 1979. Callus formation and shoot bud differentiation in anther culture of *Solanum surattense. Can. Jour. Bot.* 57:2524–27.

136. Sink, K. C., Jr., and V. Padmanabhan. 1977. Anther and pollen culture to produce haploids: Progress and application for the plant breeder. *HortScience* 12:19–23.

137. Skirvin, R. M., and J. Janick. 1976. Tissue culture-induced variation in scented *Pelargonium* spp. *Jour. Amer. Soc. Hort. Sci.* 101(3):281–90.

138. Skoog, F., and C. O. Miller. 1957. Chemical regulation of growth and organ formation in plant tissues cultured *in vitro. Symp. Soc. for Exp. Biol.* 11:118–31.

139. Slack, S. A. 1980. Pathogen-free plants: Meristem-tip culture. *Plant Disease* 64(1): 15–17.

140. Smith, S. H., and W. A. Oglevee-O'Donovan. 1979. Meristem-tip culture from virus-infected plant material and commercial implications. In *Plant cell and tissue culture: Principles and applications,* W. R. Sharp et al., eds. Columbus; Ohio State Univ. Press, pp. 453–60.

141. Smith, W. A. 1981. The aftermath of the test tube. *Proc. Inter. Plant Prop. Soc.* 31:47–49.

142. Sriskandarajal, C., and M. G. Mullins. 1981. Micropropagation of Granny Smith apple: Factors affecting root formation *in vitro. Jour. Hort. Sci.* 56:71–76.

143. Stafford, A., and D. R. Davies. 1979. The culture of immature pea embryos. *Ann. Bot.* 44:315–21.

144. Steward, F. C., M. O. Mapes, A. E. Kent, and R. D. Holstein. 1964. Growth and development of cultured plant cells. *Science* 143:20–27.

145. Stokes, M. J. 1980. Current aspects of commercial micropropagation. *Proc. Inter. Plant Prop. Soc.* 30:255–67.

146. Stoltz, L. P. 1971. Agar restriction of the growth of excised mature iris embryos. *Jour. Amer. Soc. Hort. Sci.* 96:611–84.

147. Stone, O. M. 1978. The production and propagation of disease-free plants. In *Propagation of higher plants through tissue culture,* K. M. Hughes, R. Henke, and M. Constantin, eds. U.S. Dept. of Energy, Tech. Information Center, pp. 25–34.

148. Stoutemyer, V. T. 1957. Regeneration in various types of apple wood. *Iowa Agr. Exp. Sta. Res. Bul.* 220:308-52.

149. Stoutemyer, V. T., and O. K. Britt. 1969. Growth and habituation in tissue cultures of English ivy, *Hedera helix. Amer. Jour. Bot.* 56(2):222-26.

150. Street, H. E. 1977. Embryogenesis and chemically induced organogenesis. In *Plant cell and tissue culture: Principles and applications,* W. R. Sharp et al., eds. Columbus: Ohio State Univ. Press, pp. 123-54.

151. ———, ed. 1977. *Plant tissue and cell culture.* Berkeley, Calif.: Univ. of Calif. Press.

152. Sunderland, N., and J. M. Dunwell. 1977. Anther and pollen culture. In *Plant tissue and cell culture,* H. E. Street, ed. Berkeley, Calif.: Univ. of Calif. Press, pp. 223-66.

153. Sutter, E., and R. W. Langhans. 1979. Epicuticular wax formation on carnation plantlets regenerated from shoot-tip culture. *Jour. Amer. Soc. Hort. Sci.* 104(4):493-96.

154. Thorpe, T. A. 1979. Regulation of organogenesis *in vitro.* In *Propagation of higher plants through tissue culture,* K. W. Hughes, R. Henke, and M. Constantin, eds. Springfield, Va.: Natl. Tech. Inf. Ser. U.S. Dept. of Commerce, pp. 87-101.

155. ———, ed. 1981. *Plant tissue culture: Methods and applications in agriculture.* New York: Academic Press.

156. Tisserat, B. 1981. *Date palm tissue culture.* USDA Agr. Res. Ser., Adv. in Agr. Tech., Western Series No. 17, pp. 1-50.

157. Tisserat, B., E. B. Esan, and T. Murashige. 1979. Somatic embryogenesis in angiosperms. *Hort. Rev.* 1:1-78.

158. Torrey, J. G. 1977. Cytodifferentiation in cultured cells and tissues. *HortScience* 12(2):14-15.

159. Tran Thank Van, K., and H. Trinh. 1978. Plant propagation: Non-identical and identical copies. In *Propagation of higher plants through tissue culture,* K. W. Hughes, R. Henke, and M. Constantin, eds. Springfield, Va.: Natl. Tech. Inf. Ser., U.S. Dept. of Commerce, pp. 134-58.

160. Vasil, I. K., M. R. Ahuja, and V. Vasil. 1979. Plant tissue cultures in genetics and plant breeding. *Advances in Genetics* 20:127-215.

161. Welander, M., and I. Huntrieser. 1981. The rooting ability of shoots raised *in vitro* from the apple rootstock A2 in juvenile and in adult growth phase. *Phys. Plant.* 81:36-41.

162. Westcott, R. J., G. G. Henshaw, B. W. W. Grout, and W. M. Roca. 1977. Tissue culture methods and germ plasm storage in potato. *Acta Hort.* 78:45-49.

163. Wetherell, D. F. 1978. *In vitro* embryoid formation of cells derived from somatic plant tissues. In *Propagation of higher plants through tissue culture,* K. M. Hughes, R. Henke, and H. Constantin, eds. Springfield, Va.: Natl. Tech. Inf. Ser., U.S. Dept. of Commerce, pp. 102-24.

164. White, P. R. 1963. *The cultivation of animal and plant cells* (2nd ed.). New York: Ronald Press.

165. Wimber, D. E. 1963. Clonal multiplication of cymbidiums through tissue culture of the shoot meristem. *Amer. Orch. Soc. Bul.,* pp. 105-7.

166. Yang, H. 1977. Tissue culture technique developed for asparagus propagation. *HortScience* 12(2):16-17.

167. Yeung, E. C., T. A. Thorpe, and C. J. Jensen. 1981. *In vitro* fertilization and embryo culture. In *Plant tissue culture: Methods and applications in agriculture,* T. A. Thorpe, ed. New York: Academic Press, pp. 253-71.

168. Zilis, M., D. Zwagerman, D. Lamberts, and L. Kurtz. 1979. Commercial propagation of herbaceous perennials by tissue culture. *Proc. Inter. Plant Prop. Soc.* 29:404–13.

169. Ziv, M., A. H. Halevy, and R. Shilo. 1970. Organs and plantlets regeneration of *Gladiolus* through tissue culture. *Ann. Bot.* 34:671–76.

170. Zuccherelli, G. 1979. Moltiplicazione *in vitro* dei Portainnesti clonali del pesco. *Frutticoltura* XLI(2):15–20.

## SUPPLEMENTARY READING

BARZ, W., E. REINHARD, and M. H. ZENK, eds. 1977. *Plant tissue culture and its biotechnological applications.* Berlin: Springer-Verlag.

BONGA, J. M., and D. J. DURZAN. 1982. *Tissue culture of forest trees.* Amsterdam: Elsevier.

CHALEFF, R. S. 1981. *Genetics of higher plants: Applications of cell culture.* Cambridge: Cambridge Univ. Press.

CONNOR, A. J., and M.B. THOMAS. 1981. Re-establishing plantlets from tissue culture: a review. *Proc. Inter. Plant Prop. Soc.* 31:342–57.

CONSTANTIN, M. J., R. P. HENCKE, K. W. HUGHES, and B. V. CONGER, eds. 1982. Propagation of higher plants through tissue culture. *Env. and Exp. Bot.* 21 (3/4):1–452.

HUGHES, K. W., R. RENKE, and M. CONSTANTIN, eds. 1977. *Propagation of higher plants through tissue culture: A bridge between research and application.* Washington D.C.: U.S. Dept. of Energy, Tech. Information Center.

KOSUGE, T., C. MEREDITH, and A. HOLLAENDER, eds. 1983. *Genetic engineering in plants.* New York: Plenum Publishing Corporation.

SALA, F., B. PARISI, R. CELLA, and O. CIFERRI, eds. 1980. *Plant cell cultures: Results and perspectives.* Amsterdam: Elsevier North Holland Bio-Medical Press.

SHARP, W. R., P. O. LARSEN, E. F. PADDOCK, and V. RAGHAVAN, eds. 1979. *Plant cell and tissue culture: Principles and applications.* Columbus: Ohio State Univ. Press.

THORPE, T. A., ed. 1978. *Frontiers of plant tissue culture.* Calgary: Univ. of Calgary Press.

TISSUE CULTURE ASSOCIATION. *In Vitro.* (Series)

UNIVERSITY OF CALIFORNIA, DIVISION OF AGRICULTURAL SCIENCES. 1982. Genetic engineering in plants. *Calif. Agr.* (special issue) 36(8):1–35.

VASIL, I. K. 1980. *Perspectives in plant cell and tissue culture. Inter. Rev. Cytol. Suppl. 11B.* New York: Academic Press.

———. 1979. Plant tissue cultures in genetics and plant breeding. *Adv. in Gen.* 20:127–215.

Micropropagation has significant uses in vegetative propagation of horticulturally important species and cultivars. The list of plants propagated by this method can be expected to increase as technological advances are made. This chapter concentrates primarily on procedures used for in vitro reproduction of specific cultivars for propagation. These procedures have a wide range of other applications (*52, 65*) involving manipulation and regeneration using callus, cells, or protoplasts. These are significant tools in plant breeding, genetics, and physiological studies (see chapter 16).

Considerable information is available on in vitro tissue culture propagation requirements of specific plants (see general references). Nevertheless, new ventures into micropropagation systems and the extension to new species and cultivars require considerable empirical research effort to establish feasibility and optimum conditions of culture and handling. This testing phase should include an evaluation of the genetic quality of the resulting plants in the production system where they are to be grown.

Successful in vitro propagation has several specific requirements:

*Facilities*    Special laboratory type facilities and equipment must be provided in order to prepare the cultures, produce and maintain sterile conditions, and provide proper growing conditions.

*Selection*    Selection must be made of the plant parts to be used as explants and also the required culture conditions, such as sterilization, media, and environmental conditions, which vary with different plant cultivars and species. Systematic empirical trials are needed to adapt any cultivar to the system.

*Contaminant elimination*    Contaminants invariably present on the surface of plant parts must be eliminated by disinfestation. All culture media must be sterilized, and the rooms where the cultures are to be prepared and maintained must be kept as free of contaminants as possible. Special precautions are used to

# 17

# Techniques of In Vitro Micropropagation

prevent the cultures from being recontaminated by microorganisms that are present everywhere—on plant material, on surfaces of equipment and benches, on the hands and clothing of operators, and on dust particles blown about by air currents. Even with these precautions, some undetected contaminants may invade, and indexing tests for them may need to be incorporated into the procedures.

*Staging*    Once established in sterile culture, most species are carried through a series of developmental stages. The general sequence is characterized as (a) explant establishment, (b) multiplication, (c) pretransplant (including root initiation and acclimation), and finally (d) transplanting. Considerable variation may exist among different propagators for different species in the actual handling procedures in this sequence.

*Transplanting*    Proper procedures must be utilized to enable the plant to be transferred successfully from the sterile, artificial environment of the culture vessel to the more rigorous conditions of the open greenhouse or nursery.

## FACILITIES AND EQUIPMENT

Tissue culture facilities may be placed into three categories as determined by their scope, size, sophistication, and cost (*6*). These categories are: (a) research facilities where precise work requiring highly sophisticated equipment is needed (*21, 36, 61*), (b) large commercial facilities where several millions of plants can be mass produced annually (*9, 15, 23, 27, 48, 69, 70*), and (c) limited facilities for small research laboratories, individual nurseries, or hobbyists where a relatively small volume of material is handled (*14, 42, 60*). There is much variation in kinds of facilities and opportunities for cost cutting but the basic principles of in vitro culture must be followed (*16, 17, 25, 60*). In any case, for economic reasons careful consideration must be given to costs and labor required (*4, 8, 42, 47, 49, 52, 59*).

No matter what the size, the facility should include three basic components: (a) preparation area, (b) transfer area, and (c) growing area (*14, 24, 55*). It is advisable to separate this facility from the regular nursery and greenhouse production area and maintain restricted entry to avoid introducing contaminants into the culture rooms (*15, 26, 62*). Floors, benches, and table tops should be kept hospital clean. Technicians should use clean lab coats, foot coverings, hair nets, and the like.

### Preparation Area

The preparation area is essentially a kitchen with three basic functions: (a) cleaning glassware, (b) preparation and sterilization of media, and (c) storage of glassware and supplies.

An efficient method of washing is required, either by hand or by machine. Normal washing is followed by rinsing in dilute acid, and then in distilled or deionized water. A sink, running water, and a gas outlet for heating are necessary, and air or vacuum outlets are often useful. Table surfaces should be made of a material that can be cleaned easily.

The following equipment items are needed in the preparation area:

1. Refrigerator to store chemicals, stock solutions, and small batches of media.
2. Scales or analytical balance, preferably top loading.
3. Autoclave capable of reaching 120°C (250°F). A household pressure cooker can be used to sterilize small batches of media.
4. A pH meter. Indicator papers can be used as a substitute in less precise work.
5. Gas or heating plate.
6. Filters for sterilizing nonautoclavable ingredients. These are special funnellike units fitted with a bacteria-proof membrane through which a solution is pulled by vacuum or passed under positive pressure. Various commercial types are available.
7. Equipment to purify water, either a glass still or an ion exchanger. In smaller operations, distilled water may be purchased and stored in plastic containers.

Storage facilities should be provided for flasks, bottles, and other supplies.

### Transfer Area

The transfer area is a sterile place where explants are inserted into culture flasks and the resulting plantlets are subcultured. This procedure is most conveniently carried on in an open-sided **laminar air-flow hood** (Figure 17–1) where filtered air is passed from the rear of the hood outward on a positive pressure gradient. With such equipment the transfer operation may be carried out in the same room as other activities.

Less expensive alternatives, as an enclosed walk-in room or various enclosed transfer boxes, can be used if they are carefully sterilized before use.

Ultraviolet (UV) germicidal lamps are often used to sterilize the interior of the chamber. These are turned on about two hours prior to using the chamber but must be turned off during operations. The light should be directed inward, not toward the worker's face, because of danger to the eyes. UV light will not penetrate glass.

**Figure 17–1**   Laminar flow transfer hood within which sterile operations are carried out. Operator is sterilizing end of test tube by flaming.

**Figure 17–2** Growing area where racks of tubes are placed for growing plants in vitro. Equipped with lights and controlled temperatures.

### Growing Area

Cultures should be grown in a separate, lighted facility where both day length and intensity can be controlled and where specific temperature regimes can be provided (Figure 17–2). If various kinds of plants are to be propagated commercially, it is useful to have several rooms, each programmed to meet the temperature and light needs of specific plants. Light requirements vary from about 1000 to 10,000 lux. Cool-white or Gro-Lux fluorescent lights are generally used. A 16-hour photoperiod is most common. Temperatures of 21° to 30°C (70° to 86°F) are generally adequate, although some kinds of plants may need lower temperatures. The relative humidity at the temperatures given is about 30 to 50 percent. Dehydration of the medium and increase in salt concentration might occur during long culture periods if the relative humidity is too low. If too high, contamination may occur.

Various kinds of rolling drum culture devices or shakers are sometimes used. These provide aeration in liquid culture systems.

## MEDIA PREPARATION

### Equipment and Supplies

The following supplies are needed:

1. Beakers and Erlenmeyer flasks of various sizes in which to mix media
2. Graduated cylinders, flasks, and pipettes of various sizes to measure and dispense solutions
3. Bottles to store reagents
4. Magnetic stirring devices to mix media
5. Dispensing pipettes and large funnels or burettes mounted on rings for distributing media

## Containers for Growing Cultures

Test tubes, Erlenmeyer flasks, petri dishes, and shell vials of various sizes are used routinely in research laboratories. These are usually Pyrex glass, which can be sterilized and flamed during transfer operations. Various kinds of plastic containers also are available and may be less expensive and less likely to break.

Petri dishes are glass or plastic, reusable or disposable. Less expensive, larger glass containers commonly used for various home canning or commercial canning operations are widely used for the later stages of development in large-scale propagation. Some kinds of soft glass give off impurities and may need to be discarded after about a year's use. Most of these supplies are available from commercial chemical supply houses.

Various kinds of closures are useful for the containers. Nonabsorbent cotton plugs have been used in research laboratories for a long time but are inconvenient in large operations. Various kinds of metal or plastic covers or polyurethane or polypropylene plugs are more convenient. Covering wide-mouthed jars with petri dish halves works well. Parafilm (paraffined paper in a sterile roll) is widely used to seal petri dishes and other containers.

## Ingredients

Ingredients of the culture medium vary with the kind of plant and the propagation stage at which one is working (*16, 20, 22, 28*). Some leeway is allowed in the proportions of ingredients but in general certain standard mixtures are used. These can be made from the pure chemicals or purchased as commercial premixed culture media. Empirical trials are necessary to test available combinations of ingredients when dealing with a new kind of plant. These ingredients can be grouped into the specific categories of (a) inorganic salts, (b) organic compounds, (c) complex natural ingredients, and (d) inert supports.

### INORGANIC SALTS
### (GROUPS A TO E, TABLE 17-1)

Inorganic salts provide the macroelements (nitrogen, phosphorus, potassium, calcium, and magnesium) and microelements (boron, cobalt, copper, manganese, iodine, iron, and zinc). These salts can be made up in stock solutions by group and stored in a refrigerator. The Murashige-Skoog (MS) medium (*46*) has been used widely for a range of culture types and species, particularly herbaceous plants, and for tissue cultures in general. For other types, such as woody plants, a dilution of 3 to 10× in the inorganic salts or a shift to other media is often desirable. Thus a Woody Plant Medium (WPM) (*40*) was developed for woody plants and the Anderson (And) medium was developed for rhododendrons. It also has a low salt (*2, 37*) concentration. The Gamborg B5 medium has been used widely for cell and tissue cultures (*21*).

### ORGANIC COMPOUNDS
### (GROUP F TO H, TABLE 17-1)

*Carbohydrates*    Sucrose at 2 to 4 percent is used for most kinds of cultures, but concentrations as high as 12 percent might be used in some cases, as for young embryos. Glucose has sometimes been used for monocots. Fructose and

**Table 17-1**   Stock solutions used in preparing various media for micropropagation. *

| Group | Compound | Murashige and Skoog (MS) | | Woody Plant Medium (WPM) | | Anderson (AND) | | Gamborg B5 | |
|-------|----------|-----------|---|-----------|---|-----------|---|-----------|---|
| A | $NH_4NO_3$ | 165.00 | g/l | 40.00 | g/l | 40.00 | g/l | — | |
|   | $KNO_3$ | 190.00 | g/l | — | | 48.00 | g/l | 250.00 | g/l |
|   | $Ca(NO_3 \cdot 4H_2O)$ | — | | 55.6 | g/l | — | | — | |
| B | $K_2SO_4$ | — | | 99.00 | g/l | — | | — | |
|   | $MgSO_4 \cdot 7H_2O$ | 37.00 | g/l | 37.00 | g/l | 37.00 | g/l | 25.00 | g/l |
|   | $MnSO_4 \cdot H_2O$ | 1.69 | g/l | 2.23 | g/l | 1.69 | g/l | 1.00 | g/l |
|   | $ZnSO_4 \cdot 7H_2O$ | 0.86 | g/l | 0.86 | g/l | 0.86 | g/l | 0.2 | gl/ |
|   | $CuSO_4 \cdot 5H_2O$ | 0.0025 | g/l | 0.0025 | g/l | — | | 0.0025 | g/l |
|   | $NH_4SO_4$ | — | | — | | — | | 13.4 | g/l |
| C | $CaCl_2 \cdot 2H_2O$ | 44.00 | g/l | 9.6 | g/l | 44.00 | g/l | 15.00 | g/l |
|   | KI | 0.083 | g/l | — | | 0.083 | g/l | 0.075 | g/l |
|   | $CoCl_2 \cdot 6H_2O$ | 0.0025 | g/l | — | | 0.083 | g/l | 0.0025 | g/l |
| D | $KH_2PO_4$ | 17.00 | g/l | 17.00 | g/l | — | | — | |
|   | $H_3BO_3$ | 0.62 | g/l | 0.62 | g/l | 0.62 | g/l | 0.30 | g/l |
|   | $Na_2MoO_4 \cdot 2H_2O$ | 0.025 | g/l | 0.025 | g/l | 0.025 | g/l | 0.025 | g/l |
|   | $NaH_2PO_4 \cdot H_2O$ | † | | — | | 38.00 | g/l | 15.00 | g/l |
| E | $FeSO_4 \cdot 7H_2O$ | 2.784 | g/l | 2.78 | g/l | 5.57 | g/l | 2.78 | g/l |
|   | $Na_2 \cdot EDTA$ | 3.724 | g/l | 3.73 | g/l | 7.45 | g/l | 3.725 | g/l |
| F | Thiamin·HCl | 0.10 | g/l | 0.10 | g/l | 0.04 | g/l | 1.00 | g/l |
|   | Nicotinic acid | 0.05 | g/l | 0.05 | g/l | — | | 0.10 | g/l |
|   | Pyridoxine·HCl | 0.05 | g/l | 0.05 | g/l | — | | 0.10 | g/l |
|   | Glycine | 0.20 | g/l | 0.20 | g/l | — | | — | |
| G | Myo-inositol | 10.00 | g/l | 10.00 | g/l | 10.00 | g/l | 10.00 | g/l |
|   | | | | | | † | | | |

*These are 100× the final solution. Use 10 ml of each stock solution for preparing one liter of the culture medium.

†Commonly added as a supplement directly to the culture medium at 85 to 255 mg/l.

‡Adenine sulfate added, 80 mg/l.

even starch have been used occasionally. These materials are added when the cultures are made.

*Vitamins*    Thiamin (0.1 to 0.5 mg/l) is almost always essential and nicotinic acid (0.5 mg/l) and pyridoxine (0.5 mg/l), which are required for some plant tissues, are usually added. Inositol at 100 mg/l is beneficial in many cultures and also is added routinely. Other materials sometimes beneficial include pantothenic acid (0.1 mg/l) and biotin (0.1 mg/l). All of these materials are soluble in water and should be prepared as stock solutions, ready for dilution, at 100× the final concentration. These stock solutions should be stored in a refrigerator.

*Hormones and growth regulators*    The two most important hormone classes are the **auxins** and **cytokinins** that control root, shoot, and callus formation. **Gibberellins** have sometimes been used to induce shoot elongation. Synthetic auxins include naphthaleneacetic acid (NAA) used at 0.1 to 10 mg/l, 3-indolebutyric acid (IBA) used at 0.1 to 10 mg/l, 4-chlorophenoxyacetic acid (CPA) and 2,4 dichlorophenoxyacetic acid (2,4-D), both used at 0.05 to 0.5 mg/l, and indoleacetic acid (IAA) used at 1 to 50 mg/l.

Cytokinins include $N^6$-benzyladenine (BA), kinetin, $N^6$-isopentenyladenine (2iP), and zeatin. These are used at a range of 0.01 to 10 mg/l. Adenine sulfate benefits growth and development in many cultures and is often added at 40 to 120 mg/l. All of these materials should be prepared in advance and maintained as stock solutions in a refrigerator at near freezing.

*Miscellaneous*    Citric acid (150 mg/l) or ascorbic acid (10 mg/l) is often used as an antibrowning factor. Malic acid (100 mg/l) is added in some embryo culture media. These should be filter sterilized and not autoclaved as autoclave temperatures cause decomposition. These organic acids may be incorporated in the media or used only in the prewashing steps. Activated charcoal of fine grade, preferably prewashed, is sometimes added (0.1 to 1 percent) to counteract certain inhibiting substances released by some tissues.

COMPLEX NATURAL INGREDIENTS

Various materials of unknown composition have been used to establish cultures when known substances fail to do so; sometimes these materials promote extra growth. Protein hydrolysates from casein or other proteins (30 to 3000 mg/l) are often helpful, primarily to provide organic nitrogen and amino acids. Coconut milk (endosperm of green or ripe coconuts, 10 to 20 percent by volume) is useful but must be filter sterilized. Malt (500 mg/l), yeast extract (50 to 5000 mg/l), and such substances as tomato juice (30 percent v/v) and orange juice (30 percent v/v) also have been used successfully. These substances are dissolved in water and added during preparation of the media.

INERT SUPPORTS

*Agar*    Agar is a powdered product obtained from certain species of red algae (*28, 44*). It owes its value in culture systems to two properties: (a) it melts when heated but cools to a semisolid gel at room temperature; (b) it is essentially inert biologically. However, it may contain certain impurities in its natural state, notably salt. Therefore, it should be prewashed or only the purified product (USP grade) should be used. Quality may vary, however, among commercial products.

Two factors that affect agar usage are concentration and pH. A pH of 5.0 to 6.0 is usually adequate, but for acid-requiring plants a pH of 4.5 is better (*1*). At very low pH (acid), the agar does not solidify well and tends to deteriorate with heating.

The lowest concentration of agar that will support the explants (0.5 to 0.6 percent) is recommended. There should be good medium-plant contact to allow adequate uptake of nutrients. Lower concentrations allow more nutrient uptake and contact, but if the explant sinks into the agar, aeration is impaired. The minimum concentration used is about 6 g/l. However, in some systems, an even lower concentration is used in which the small explant is supported initially. The small explant will gradually sink as it grows. Improved growth has resulted from better medium contact with the explant. Higher concentrations (1.0 to 1.2 gm/l) tend to depress growth but in some cases have been used to harden plants and to improve their ability to survive transplantation.

Substitutions for part of the agar (2 g/l) with selected kinds of commercial pectin (8 to 10 g/l) have been used successfully (*70*).

*Liquid*     Nutrient solutions without agar have an advantage for some plants by improving nutrient uptake and avoiding the impurities in agar. However, some type of support is required to keep the explants and cultures from sinking, or a shaker or rotating drum must be used to provide aeration. Filter paper bridges are usually inserted into the liquid. Plastic fabric supports (*12*) made of 100 percent polyester fleece are available commercially. Liquid media are preferred for some plant species whose explants exude toxic substances from cut surfaces. The liquid medium can be changed without reculturing (*12*).

## Preparation of Stock Solutions

Inorganic substances, and many of the organic materials, should be maintained in stock solutions of $100\times$ the required concentration. An aliquot of 10 ml of the stock solution can then be added to each liter of the medium to provide the required concentration. From the list in Table 17–1, the following materials can be combined without forming precipitates: (a) nitrates, (b) sulfates, (c) halides, i.e., chlorides and iodides, (d) phosphates, borates, and molybdates. Iron is prepared separately by combining the substances in group E. It may also be supplied as inorganic iron—$FeCl_3$-$6H_2O$ (1 mg/l), $FeSO_4$ (2.5 mg/l), or iron tartrate (10 mg/l). Iron solutions should be stored in a dark bottle.

Some organic compounds are relatively insoluble in water. Addition of dimethylsulfoxide (DMSO) in a low amount (not more than 0.5 percent in the final medium) is effective in dissolving the organic compounds. Cytokinins are weak bases and are best dissolved in a dilute acid, then diluted to final volume with water. Auxins are weak acids; they are best dissolved in a dilute base or ethanol. Use 0.3 ml of 1N HCl (for cytokinin) or NaOH (for auxin) for each 10 mg of the compound. Stock solutions and organic substances should always be refrigerated.

## Preparation of Media

Ordinary distilled water is used routinely in commercial micropropagation, but research studies usually require the purer grades produced by deionization or redistilling. To prepare media, use a large Pyrex flask or beaker, add a portion (½ to ⅔) of the final volume of water. Add sugar, dissolve, and stir. Add to this by pipettes or a graduated cylinder proper aliquots of each ingredient from stock bottles. Adjust to the desired pH (usually about 5.7) with drop-by-drop additions of 1N HCl or 1N NaOH using a pH meter (or pH indicator strips) to measure. Agar is added next and then the solution is made up to the prescribed volume by adding water.

The solution is then heated to melt the agar, such as in an autoclave at 121°C (218°F) for 3 to 7 minutes or on a hot plate while stirring continuously to prevent the agar from settling and burning. The hot solution is then dispensed into containers, which are closed with stoppers immediately and sterilized in an autoclave for 10 to 20 minutes at 121°C. The greater the volume of media to be sterilized, the longer the sterilization time.

Substances unstable during autoclaving must be filter-sterilized; an aliquot of such material is added to the medium after heat sterilization when the medium has cooled to 35° to 40°C (95° to 104°F). The medium should be used within about two weeks and refrigerated if kept longer.

# GENERAL METHODS OF MICROPROPAGATION

## Preparation for Establishment of the Culture

### HANDLING OF THE STOCK PLANTS

Two principal considerations are involved in handling stock plants. One is reducing the potential for contamination by fungi, bacteria, viruses, and other pathogens. The other is control of the physiological condition of the explant.

In general, stock plants should be healthy and actively growing, not in or entering dormancy. Pruning or use of supplementary light control (increasing day length) is sometimes useful in keeping new flushes of growth active (3).

### CHOICE OF EXPLANT

The kinds of explant and their location on the stock plant vary tremendously with the purpose for which the culture is being made, as well as with the species, and sometimes the cultivar. In the absence of definite information on a given species or cultivar, one can use available information on a closely related species or cultivar or carry out experimentation with the particular plant. For shoot tips, plants in relatively active growth are usually most desirable, utilizing either quiescent axillary growing points or actively growing terminal shoots. The size of the explant may vary from the 1 to 5 mm meristem tip itself to a piece of shoot several centimeters long. A nodal section bearing a lateral bud also can be used. For woody plants, a shoot-tip from a dormant (but not resting) bud may be utilized but is often difficult to disinfest. Removal of the bud scales, plus cutting away the leaf scar, may give sterile tissue.

Pieces of leaves with veins present, bulb scales, flower scapes, and cotyledons also may be used for obtaining explants.

---

**Pathogen Elimination**

Systemic pathogens, such as viruses and some bacteria, are not automatically eliminated by in vitro techniques unless this goal is built into the system (7, 26, 30, 39, 45, 54, 59). If available, preselected "virus-free" or pathogen-indexed stock plants may be used (see chapter 8). Otherwise, special techniques such as meristem-tip culture, heat treatment, and various pathogen indexing tests may be incorporated into the procedure (50). Maintaining "clean" stock plants under aseptic conditions from previous propagations has been a useful procedure in some situations (26).

---

**Table 17–2**    Common surface sterilants for treating explants prior to aseptic culture (*61*)

| Agent | Concentration | Treatment Time (min) | Removal from Explants | Remarks |
|-------|---------------|----------------------|-----------------------|---------|
| Calcium hypochlorite | 9 to 10% | 5 to 30 | Easy | Very effective |
| Sodium hypochlorite* | 2% | 5 to 30 | Easy | Very effective |
| Bromine water | 1 to 2% | 2 to 10 | Easy | Very effective but may be toxic to plant tissue |
| Hydrogen peroxide | 10 to 12% | 5 to 15 | Very easy | Moderately effective |
| Mercuric chloride | 0.1 to 1% | 2 to 10 | Difficult | Effective but may be toxic to plant tissue |

\* Available as commercial bleach usually with 5.25% active ingredient. Common usage rate is 20% v/v (1 part bleach to 4 parts water) but a range of concentrations also should be tested.

DISINFESTATION

The primary disinfestants include alcohol (ethyl, methyl, or isopropyl) and calcium or sodium hypochlorite, sold as commercial household bleach, usually with 5.25 percent active ingredient (see Table 17–2). To the latter, a few drops of a surfactant or detergent should be added to improve surface coverage. The effectiveness of these materials is essentially a time-dosage response in which the effectiveness to disinfest increases with increase in both factors, but the ability to damage tissue also increases. Consequently, an effective compromise must be worked out for the kind of explant used.

A typical disinfestation procedure during shoot-tip propagation is to cut the shoot into short pieces several centimeters long and dip them into alcohol (70 to 95 percent) for a preliminary surface sterilization. Wrapping these with small pieces of sterile gauze will hold the material intact during sterilization.

Plant parts are then placed in the disinfesting solution (with surfactant or detergent added) using a solution of 1 to 10 percent of the prepared material for 5 to 15 minutes. This solution is then poured off and the material rinsed two or three times with sterile water.

Modifications of this basic procedure that can be used to improve efficiency include prewashing, mechanical agitation, vacuum, and consecutive treatments.

The contaminants, consisting of yeasts and various species of fungi and bacteria, usually appear on the agar surface within a few days or a week as white or opaque slime or as variously colored colonies, sometimes with black spores and mycelium. When these contaminants can be identified visually, the cultures containing them are quickly discarded. Contaminated culture vessels should be sterilized in the autoclave without removing the covers to avoid contaminating the laboratory. Some bacterial contaminants, such as *Bacillus subtilis, Erwinia* (*26*), or *Pseudomonas* (*3*), sometimes remain inside the plantlets without being detected until late in the culture period. Procedures to identify these contaminants may be included. Antibiotics are sometimes used in the medium to improve control of bacterial growth (*19*).

ASEPTIC PROCEDURE

Microorganisms occur primarily as spores resting on surfaces of tables, hands, arms, clothing, and various objects and settling down from the air or blown in the dust on air currents.

Aseptic transfer procedures are done in some kind of transfer hood or box. These should have a high-efficiency particulate air (HEPA) filter with a positive air pressure blowing outward from the rear of the hood to prevent movement of any air-borne spores into the chamber. The sides and surfaces of the chamber should be washed with a disinfestant, such as sodium or calcium hypochlorite or 95 percent alcohol. Hands and arms should be washed thoroughly, and gloves and a clean lab coat should be worn. Equipment placed in the transfer hood includes an automatic gas burner or alcohol lamp, a dissecting microscope (if needed), and (sometimes) a top-loading balance. These devices should be wiped with disinfectant and covered. The dissecting microscope should have a plastic or glass tunnel over the viewing area to protect against contamination.

Wash hands and arms thoroughly with detergent. Rinse with 75 percent alcohol to which a few crystals of iodine are added before dilution (*25*).

Supplies needed for a transfer operation are as follows:

**1.** Burner for sterilizing dissecting tools and containers. This can be an automatic gas burner or an inexpensive glass alcohol lamp with wick. Certain types of burner-incinerators available commercially avoid the presence of alcohol or gas fumes.

**2.** Tweezers to hold and manipulate tissue. Three useful sizes (*60*) are 300 mm, 200 mm straight-tipped, and 115 mm curve-tipped.

**3.** Dissecting needles with wood handles, or a metal needle holder with replaceable tips.

**4.** Dissecting scalpels. Metal scalpel holders with replaceable knife blades are most convenient. A piece of broken razor blade glued to a holder can be used for fine work, or various kinds of microtools can be made.

**5.** Sterile petri dishes for holding explants and sectioning them. Small squares of sterile moist paper toweling inside each petri dish are convenient to assist cleaning later (*37*). Also, the moist surface keeps small explants from drying out.

Scalpels, needles, and other dissecting tools are moved into position, usually on holders. Sterile petri dishes are placed in the chamber to use as working trays. Sterile water containers are included, as are containers of media, to be used as needed. These containers are generally kept to the front of the working area so that reaching beyond a particular "sterile" area to the rear of the hood can be avoided. Outside surfaces of containers should be wiped with a disinfectant. Occasionally containers of media may be placed on shelves or on mobile carts next to the transfer chamber.

In preparing explants for transferring, the tips of implements are dipped into alcohol and flamed before use. This step is repeated after a certain amount of material has been handled, or before changing to a new explant material. However, simply dipping and flaming the scalpels may not be sufficient to kill some of the more resistant pathogens or spores. Longer exposure to the hot flame or use of a chemical sterilant may be required (59). It is very important to allow needles, scalpels, and tweezers to cool after flaming and before use to prevent damage to tender tissue. When working in an open area, the propagator flames the neck of the container before opening it and after inserting the material, to avoid entrance of contaminants. When working in a sterile chamber, the procedure may be omitted.

### PREPARING THE EXPLANT

The exact method for preparing the explant varies with the kind of explant. In general, all procedures described above should be followed. If shoot-tips are used, remove the disinfested shoot pieces from the rinsing solution, and cut away the enclosing leaves with sharp needles or scalpels to expose the immature tip. Removal of only the meristem dome and a few primordial leaves requires very fine microtools or scalpels and use of a dissecting microscope. If a larger piece of the meristem or the entire shoot-tip is used as the explant, the operation is less exacting. In either case the work must be done rapidly and the piece inserted on the agar medium quickly to minimize drying of the tender tissues.

## Growing the Cultures

### PRETREATMENT

Micropropagation operations may include a preliminary pretreatment culture step in which the explants are placed initially in a petri dish on the surface of an agar medium of basic salts and sugar, but without hormones.

If the explants contain phenolic or other substances that exude from the cut surface and inhibit development, charcoal, ascorbic acid, citric acid, or other substances can be added to the medium to prevent the effects. Using the antioxidants—ascorbic acid and citric acid—in the preliminary washing also may be useful. Use of liquid media in the initial stages and frequent changes for several days may leach out the material.

### STAGE I: ESTABLISHMENT

The function of this stage is to establish the explant in culture and to induce multiple shoot development for further multiplication, which may involve

(a) the stimulation of axillary shoots, (b) the initiation of adventitious shoots on excised shoots, leaves, bulb scales, flower scapes, cotyledons, and so on, or (c) the initiation of callus from the cut surfaces. Thus the medium selected varies with the species, cultivar, and kind of explant to be used. The basic medium (BM) includes the ingredients listed in Table 17–1, plus sucrose and sometimes adenine and other supplements. Control of development is achieved by manipulating the levels of auxin and cytokinin. If axillary shoots are desired, a moderate level of cytokinin (BA, kinetin, or 2iP) is used (0.5 to 1 mg/l). The auxin level must be kept very low (0.01 to 0.1 mg/l) or omitted altogether.

To develop adventitious shoots on the explants (excised shoots, leaves, bulb scales, cotyledons, flower scapes, and so on) a somewhat higher cytokinin level is needed. For callus formation, increased auxin levels also are needed but must be adjusted to an appropriate level of cytokinin. Usually four to six weeks is required to complete Stage I and produce explants ready to be transplanted to Stage II.

## How to Determine Concentration in Media

Establishing the optimum concentration and range of individual chemicals in culture media is as important as establishing the proper choice of chemical, whether inorganic salt, sugar, hormone, or other substance. Concentration is established by two main systems. Concentrations are often given as **parts per million** (ppm) or **parts per billion** (ppb), or expressed as milligrams per liter (mg/l) or grams per liter (g/l) of water or solution weight/weight (w/w) or weight/volume (w/v). In the case of liquids, relative volumes (v/v) are used. In either case a dilution range can be established as 1:1, 1:5, 1:10, 1:100, and so on. A progression is chosen to discover the range of concentrations within which a particular substance is effective, i.e., its minimum threshold value and the maximum value where injury may occur or further increase in response does not occur. With hormones the kind of response changes with the concentration.

Research laboratories and considerable literature on tissue culture utilize methods of **chemical equivalents** as a better way to express effective concentrations and compare one chemical substance with another more realistically.

Chemical equivalents are calculated from the grammolecular weight (*mole*) of a substance by adding together the atomic weights of all the atoms in the molecule plus any attached water. Such weights are available in any chemistry textbook. One molar (1M) concentration means the weight of one mole in grams dissolved in enough water to make one liter of solution. Usually physiological concentrations are in much smaller amounts, such as:

$$1 \text{ molar} = 1 \text{ mole/l} = 1 \text{ M} = 1 \text{ mol.l}^{-1}$$
$$1 \text{ millimolar} = .001 \text{ M/liter} = 1 \text{ mM} = \text{mmol.l}^{-1}$$
$$1 \text{ micromolar} = 0.000001 \text{ M/liter} = 1 \mu\text{M} = \mu\text{mol.l}^{-1}$$

The significance of this system is that 1 mole of any substance has the same *number* of molecules per volume as any other substance at that molar concentration.

For other types of regeneration, such as embryogenesis or organogenesis from callus, the basic techniques are similar to those described above. The differences are in the choice of explant, ingredients in the medium, and hormone levels.

STAGE II: MULTIPLICATION

In the multiplication stage each explant has expanded into a cluster of shoots arising from a common expanded mass of basal, calluslike tissue. This structure is now divided into separate propagules, which are transplanted into a fresh culture medium.

The kind of medium depends on the species, cultivar, and type of culture. Usually it is essentially the same as that used in Stage I, but often the cytokinin level is increased. In the absence of specific directions, experimentation is necessary to adjust the concentrations to optimize the rate of multiplication and production of uniform plantlets (3, 16, 45). Propagules are removed from Stage I containers and placed onto fresh medium in other containers.

Different species vary in the optimum size of the propagule and method of cutting them apart. There is generally a critical mass of tissue for the explant to develop at which size cuts can be made in order to produce uniform rapid multiplication in the next transfer. Division may be made vertically by dividing the tissue mass into sections, keeping some of the basal section on each piece. In some cases, one dominant shoot may develop and inhibit elongation of others. Reculturing is done by cutting off the shoots and transferring the base to the new medium. Elongated shoots laid horizontally on the agar surface, acting as a type of layering, often stimulate lateral shoots to develop.

The number of propagules produced for transfer varies from 5 to 25 or more from each explant. Multiplication stages may be repeated several times to build up the supply of material to a predetermined level for subsequent rooting and transplanting.

During these multiplication stages, off-type propagules sometimes appear, depending on the kind of plant and method of regeneration. In some plants, such as Boston fern, restricting the multiplication phase to only three propagations is recommended to avoid development of off-type propagules.

STAGE III: PRETRANSPLANT

The purpose of the third stage is to prepare the plantlets for transplanting from the artificial heterotrophic environment of the test tube to an autotrophic free-living existence in the greenhouse and on to their ultimate location. This preparation may involve rooting, but it also involves a change in the physiology of the plantlet so as to stimulate photosynthesis, nutrient and water absorption through roots, and resistance to desiccation and to pathogens.

At this stage, propagules can be handled in three basic ways:

1. Individual cuttings can be made and planted directly into a rooting medium in mist or under high humidity with or without rooting hormones. All remnants of the agar medium should be washed off the individual cuttings to remove a potential source of contamination.

2. Individual propagules may be recultured into new containers in a sterile medium with reduced or omitted cytokinin, an increased auxin concentration, and often a reduced inorganic salt content. For some kinds of plants, adventitious roots

then develop readily on these propagules; better root hair development may occur in an aerated liquid medium than in an agar medium. With other plants, rooting is best if the propagule is kept in the auxin medium for only one or two days and then transferred to an auxin-free medium. Or the propagule may simply be dipped into a rooting (auxin) solution immediately and inserted directly into an auxin-free medium.

**3.** The third method is to include an "elongation" phase between Stage II and Stage III by placing the propagules into an agar medium for two to four weeks without cytokinins (or at very low levels) and, in some cases, adding (or increasing) gibberellic acid. The propagules are then handled as described in methods 1 or 2 above.

Selection for uniform propagules, and roguing out of obviously abnormal, aberrant, or diseased plantlets should be made at the start of Stage III (*3*). Sometimes plant cultures deteriorate with time, lose leaves, fail to grow, develop tip-burn, go dormant, or lose potentiality to regenerate (*3*). Plants of species having an inherent dormancy or rest requirement may need to be chilled to stimulate new growth and elongation (*3*).

STAGE IV: TRANSPLANTING

Rooted or unrooted plantlets are removed from the culture vessel, agar is washed away completely to remove a potential source of contamination, and the plantlets are transplanted into a standard pasteurized soil mix in small pots in a more or less conventional manner. Initially, plantlets should be protected from desiccation in a shaded, high-humidity tent or under mist. Several days may be required for new functional roots to form (*55*).

Once the plantlets are established in the rooting medium, they should be gradually exposed to a lower humidity and a higher light intensity. Any dormancy or resting condition that develops in the plant may need to be overcome as part of the establishment process.

# EXAMPLES OF METHODS OF HANDLING REPRESENTATIVE SPECIES

## Orchids

SEED CULTURES

Orchid seeds are extremely tiny; it is estimated that some 30,000 seeds are produced in a single Cattleya pod. In one method of seed culture, the seeds may be extracted from the green pod when it is about 60 percent mature and not split open. Dip the pod in alcohol or other disinfestant. Open under sterile conditions and excise seeds aseptically.

The usual method, however, is to remove the seeds from the mature fruit when it dehisces naturally and treat them with a disinfestant such as calcium or sodium hypochlorite (*25, 34, 35, 51, 53*). A small amount of seed is placed in a small vial or flask and covered with five to ten times their volume with sodium or calcium hypochlorite solution plus a drop or two of a wetting agent. Seeds are left in the solution for five minutes (sometimes longer) and shaken periodically.

**Figure 17–3** Orchid seedlings growing aseptically on nutrient agar medium. Seedlings are ready for transplanting.

At the end of this time, seeds will have sunk to the bottom of the tube. Gently pour off the disinfestant, then add sterile water to fill the vial half-full. After shaking a few times, pour off the water containing the seeds into the germination flask so that it spreads over the surface of the sterile agar medium. An alternate procedure is to dip out the seeds with a platinum wire loop, as used for bacteriological studies, or with a spatula.

The culture medium contains 1.0 to 1.5 percent agar. Germination is apparent within a few weeks. By the fourth to sixth month the small plants must be transplanted to a new flask with sterile medium (Figure 17–3). About a year after sowing, the young plants are moved to larger containers for growing in the greenhouse. A flowering plant may not be produced for seven years or longer.

PROTOCORM CULTURES

In protocorm culture, an actively growing vegetative shoot is excised, the outer leaves are stripped away, the bulblike section (pseudocorm) is dipped into alcohol, soaked in a calcium hypochlorite solution for 15 to 20 minutes, then washed in sterile water (*25, 51, 53, 68*). The outer leaves are removed to expose the shoot apex. A section of the tip is then excised and placed on the sterile nutrient medium, which may be agar, filter paper, or liquid. Within four to six weeks the shoot proliferates to produce a small, round protocorm similar to that produced when an orchid seed germinates. (See Figure 17–4.)

For Stage II the protocorm is removed from the flask and cut into eight to ten sections, which are replanted separately onto new media. At monthly intervals the process is repeated as the additional protocorm pieces continue to enlarge. By monthly transfers and consecutive sectioning this multiplication process can build up a very large culture stock of any one clone within a relatively short time.

**Figure 17-4**    Steps in the asexual production of *Cymbidium* orchids. This procedure is known often as "mericloning." *Top row* (left to right): growing shoot—source of meristem; excised shoot apical meristem; protocorm-like bodies, after enlarging in nutrient medium and before dividing. *Second row* (left to right): dividing each enlarged protocorm-like structure into several pieces; before and after dividing; rotating wheels with divided mericlones in nutrient solutions for enlargement and prevention of root-shoot polarity.

**Figure 17-4** *(cont.)* *Third row* (left to right): protocorm-like material placed on nutrient agar for polarizing growth and development of roots and shoots; greenhouse filled with flasks for growing on into plants; well developed plants large enough for flatting. *Bottom row* (left to right): well developed orchid plants, ready for potting; potted *Cymbidium* orchid, originating from a divided protocorm-like body; orchid house with plants coming into flowering. Courtesy Richard Smith, Rod McLellan Co., South San Francisco.

**Table 17–3** Culture media for orchids

*(A) Knudson's medium C\* for seed culture*

| | | | |
|---|---|---|---|
| $Ca(NO_3)_2 \cdot 4H_2O$ | 1000 mg/l | $MnSO_4 \cdot 4H_2O$ | 7.5 mg/l |
| $(NH_4)_2SO_4$ | 500 mg/l | $FeSO_4 \cdot 7H_2O$ | 25.0 mg/l |
| $KH_2PO_4$† | 250 mg/l | Sucrose | 2% |
| $MgSO_4 \cdot 7H_2O$ | 250 mg/l | Agar | 1.5% |

*(B) Wimber's medium for protocorms (Cymbidium)*

| | | | |
|---|---|---|---|
| $KNO_3$ | 0.525 g/l | Ferric tartrate | 0.03 g/l |
| $CaHPO_4$ | 0.200 g/l | Tryptone | 2.0 g/l |
| $KH_2PO_4$ | 0.250 g/l | Sucrose | 2.0 g/l |
| $(NH_4)_2SO_4$ | 0.500 g/l | Agar ‡ | 12.0 g/l |
| $MgSO_4$ | 0.250 g/l | | |

\*Add minor elements; also, iron can be supplied by other means (see Table 17–1).

†May be modified by replacing with 18 ml of potassium phosphate buffer prepared by combining 97.5 mg of 0.1M $KH_2PO_4$ solution (13.6 g/l) and 2.5 ml of 0.1M $K_2HPO_4$ solution (17.4 g/l). Maintains pH at 5.3 (*64*).

‡May be omitted and liquid used in rotating wheel.

In an alternate procedure for multiplications, the shoot-tip is placed into a liquid medium where it is continuously agitated on a shaking device or rotated on a circulating wheel. Under these conditions, new lateral protuberances continue to develop and are recultured every three to four weeks after slicing. Production of new protocorms will continue as long as the shaking or rotating continues, or as long as they are resliced and recultured. When new plants with roots and shoots are desired for Stage III, individual pieces are placed onto an agar medium, the shaking discontinued, and plants allowed to develop without reculturing. Within eight months leaves and roots are well developed and the young plants are removed from the flasks and planted. A flowering plant should be produced in two years since these plants are in the adult phase.

## Ferns

SPORE CULTURE

Spores of various species can be sterilized and sown on nutrient agar. Spore germination itself may be favored by using a nutrient-free medium, but growth of the prothallus is improved by the addition of inorganic salts and sucrose (*33*).

A complete aseptic culture system is described as follows (*26*): Fern spores are placed in 2 percent Clorox® solution and centrifuged. The pellet is retrieved, resuspended, and recentrifuged several times. Resuspended spores are streaked on a culture medium (MS salts, 3 percent sucrose, thiamine, and 0.8 percent agar). After 20 days' incubation in light at 27°C (81°F), gametophyte tissue is divided, recultured, and grown for 30 days. Then one gram of gametophyte tissue in 200 ml ½ MS salts may be ground in a kitchen blender for 5 seconds. Approximately 50 ml of solution is deposited over sterile soil. Subsequent handling is by conventional methods.

A modified method also has been described (*38, 63*). One part spores to 5 parts of 5 percent Clorox® solution is shaken for 1½ minutes and then poured through a Buchner funnel filter with vacuum, followed by rinsing, followed by an additional 5 minutes' air vacuum. Spores are allowed to air dry in a sterile hood. The medium (2 percent agar and inorganic nutrients) is melted, cooled to approximately 49°C (120°F), and poured into a sterile pyrex oven dish or petri dishes. Spores are sown over the surface using a fine mesh screen (270 mesh). As the medium cools, the surface tension facilitates uniform distribution of spores. These are placed in light at room temperature. Germination occurs in two to three weeks and transplanting to soil is done in two to three months.

### BOSTON FERN
#### (*NEPHROLEPIS EXALTATA* 'BOSTONIENSIS')

Boston fern is a hybrid plant that produces no viable spores and propagates slowly by conventional crown division, but is one of the plants most adaptable to micropropagation and can be multiplied at high rates. Greenhouse stock plants are best grown as hanging baskets with drip irrigation to avoid overhead watering. Actively growing one-inch-long stolon tips are excised from the plant and placed in running water for one hour. These are then treated with five percent sodium or calcium hypochlorite solution (plus surfactant) for 20 minutes, followed by three rinses in sterile water. The enclosing leafy coverings are removed and the 3 to 6 mm apical tip explant is inserted upright in sterile agar medium.

In Stage I an MS medium containing sucrose (3 percent) and agar (0.8 percent) is used. The pH is 5.7, temperature 27°C (81°F), and light at 1000 lux for 16 hours per day. After two to three months the new plants are divided.

In Stage II the MS medium is modified to include kinetin (2.0 mg/l), NAA (0.1 mg/l), and increased phosphorus ($NaH_2PO_4$ 255 mg/l). An alternate modification is to use IAA (2 mg/l), kinetin (0.04 mg/l), and BAP (1.126 mg/l) (*10*). Light is increased to 3000 lux. Reculturing can be done every six weeks with five or more new plantlets per culture. No more than three subcultures are recommended from any one sequence of operation (*14, 44*) to avoid producing aberrant plants (Figure 17–5).

**Figure 17–5**  Plantlet of Boston fern (*Nephrolepis exaltata* 'Bostoniensis') in culture. Ready for transplanting into Stage III.

In Stage III the same medium used in Stage I is utilized, but light intensity is increased. At the end of six weeks cultures are divided, and the small plantlets, perhaps less than 13 mm tall, are planted into a pasteurized soil mixture in cell packs, and protected from drying for about two weeks in high-humidity tents. They can then be moved to the open greenhouse. In six to eight weeks they can be transplanted to small pots in which they grow to a commercially salable size.

## African Violet (*Saintpaulia ionantha*)

African violet is traditionally propagated by leaf cuttings. However, it can be easily micropropagated by various tissue culture techniques. Under proper hormonal conditions, regeneration can occur from such explants as petioles (*5, 24*), leaf sections (*13, 58*), or even epidermal cells (*66*).

The following procedure is used commercially. Mature leaves are detached and washed with a detergent, then sterilized for about 15 minutes in a 1 to 2 percent sodium hypochlorite (plus surfactant) solution, followed by rinsing four to five times in sterile water. Leaf margins are discarded, and the remaining leaf is cut into 10 mm squares. Each piece is placed with the underside down on a sterile nutrient agar medium.

In Stages I and II, a basal MS medium is supplemented by $NaH_2PO_4 \cdot H_2O$ (170 mg/l), IAA (2 mg/l), BA (0.8 mg/l), sucrose (3 percent), and agar (0.9 percent). Growing conditions include a temperature of 27°C (81°F) and light at 1000 lux for 16 hours daily. Shoots develop in 30 to 60 days. Added adenine sulfate at 125 mg/l will promote shoot formation (*58*).

In Stage III, rooting occurs readily with the same medium used in Stages I and II but with the hormones omitted; some (*58*) recommend adding IAA at 10 mg/l. Transplanting to a pasteurized soil mix is done along with protection by shading. Flowering can occur in three to six months.

In an alternate procedure (*5*), petiole cross-sections are placed on a basal MS medium using the following hormone level: NAA (0.1 mg/l) plus BA (0.01 mg/l) in Stages I and II. Rooting is induced by placing the cultures in low light intensity (500 to 10,000 lux) for two weeks.

## Rhododendrons (*Rhododendron* spp.)

Rhododendron has been used successfully in micropropagation (*1, 2, 37, 41*) and provides a model for other woody species (Figure 17–6). Short, soft shoots completing a flush of growth but with no terminal bud set are used for preparing explants. All but the smallest leaves are removed and the apical shoot trimmed to 1 to 1.5 cm. These are washed in soapy water and then placed in 1 to 2 percent sodium hypochlorite solution for 10 to 15 minutes with agitation, followed by rinsing. The base is trimmed and the small microcutting explants are placed in 25 × 150 mm test tubes. The following sequence is used:

*Stage I: Establishment*    The Anderson medium is used in a liquid form (5 ml/tube) with the addition of inorganic salt at ½ strength, sucrose (3 percent), 2iP (2 mg/l), and IAA (0.5 mg/l). Axillary shoots 1 to 2 cm long develop in two to eight weeks and are cut off and transferred as they develop.

*Stage II: First Subculture*    In the first subculture the same medium is continued but with agar added, and includes 2iP (1.5 mg/l) and IAA (4 mg/l). As

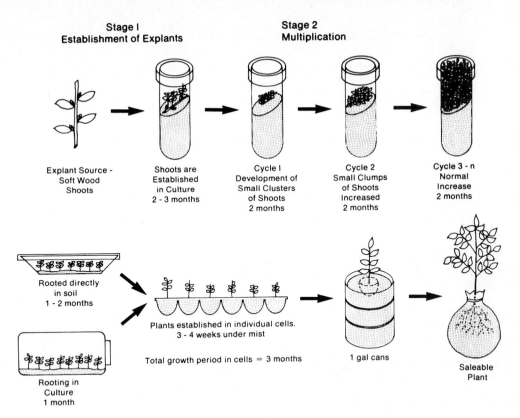

**Stage I**
**Establishment of Explants**

Explant Source -
Soft Wood
Shoots

Shoots are
Established
in Culture
2 - 3 months

**Stage 2**
**Multiplication**

Cycle I
Development of
Small Clusters
of Shoots
2 months

Cycle 2
Small Clumps
of Shoots
Increased
2 months

Cycle 3 - n
Normal
Increase
2 months

Rooted directly
in soil
1 - 2 months

Plants established in individual cells.
3 - 4 weeks under mist

Total growth period in cells = 3 months

1 gal cans

Rooting in
Culture
1 month

Saleable
Plant

**Figure 17–6**    Sequence of operations used in micropropagating rhododendrons. Courtesy W. C. Anderson *(2)*.

new shoots arise, apparently (in part at least) adventitiously, 2.5 cm sections are cut off, gradually eliminating the original explant stem. These are transferred to fresh medium. At first, transfers can be made every two weeks, but later they are made six to eight weeks apart.

*Stage II: Second Subculture*    For the second subculture, the same medium is used, but 2iP is varied from 1 to 15 mg/l and IAA 0.5 to 4 mg/l, depending on cultivar. Rates of 20 to 40 shoots per culture can be achieved every two months.

*Stage III: Pretransplant*    Either of two procedures can be followed. In one the shoots are removed and rooted directly into a pasteurized soil mix in a covered or misted chamber. In the other, shoot propagules are transferred to a sterile medium but 2iP, IAA, and KI are omitted, inorganic salts are reduced to one-third strength, and activated charcoal (600 mg/l) is added. Addition of IBA (0.5 mg/l) may induce rooting in one month. The pH should be 4.5 in all cases.

*Stage IV: Transplanting*    Plants are rooted and grown in a porous soil mix (for example, sawdust, peat moss, perlite, pumice, vermiculite, and fertilizers) in a high-humidity area *(55)*. The tender plants are subject to both pathogens and fungicide toxicity so that fungicide treatment must be carefully adjusted.

Space plants to avoid crowding and thus reduce disease problems and slow growth. Plants are salable in six to twelve months.

*Flower Buds*  Excised florets from rhododendron flower buds also have been used successfully as explants (*43*). Florets are more easily sterilized than shoots and proliferate shoots readily on AND medium with 2iP (5 to 15 mg/l) and IAA (1 to 4 mg/l). These root in subsequent subcultures in medium with 2iP omitted but with 1 g/l activated charcoal added.

## Apple (*Malus pumila*)

Micropropagation of both rootstocks (*11, 18, 29, 56*) (Figure 17–7) and fruiting cultivars (*71*) is possible with careful attention to the requirements of the various stages. Explants can be taken either from actively growing shoots or from meristems of dormant but nonresting buds. Three procedures have been described:

**a.** Shoots are rinsed in water with detergent for 5 minutes with agitation, rinsed two to three times in sterile water, treated with half-strength calcium hypochlorite plus detergent for 20 minutes, and rinsed again in sterile distilled water (*71*).

**b.** In another procedure (*32*) a two-step sodium hypochlorite treatment is used for sterilization. Shoots are dipped first into detergent and then into sodium hypochlorite (0.14 percent) for 1 minute. They are then washed in three changes of sterile water and placed on a culture medium. The next day, the shoots are treated again with 0.4 percent sodium hypochlorite solution for 40 minutes.

**c.** Dormant buds are agitated in 95 percent alcohol plus detergent for 10 minutes, in full strength calcium hypochlorite for 20 minutes, and then rinsed in sterile water. Buds are dissected by removing bud scales and leaf primordia sequentially to

**Figure 17–7**  Apple rootstock propagation by shoot-tip proliferation. (A): proliferated shoots in Stage II. (B): rooted plant ready for transplanting. (C): apple plant after transplanting. Courtesy Dr. Iona Snir and Dr. A. Erez (*56*).

expose the growing point. The meristem tip is excised and planted using special care to prevent desiccation.

*Stage I: Establishment*    Several procedures have been successfully used. (a) In one (*12*), a preconditioning step is included by growing the shoot-tip on a simplified medium in a petri dish for one week before transferring. (b) A second procedure (*71*) is to grow the shoot-tips to a length of 10 to 15 mm in 15 ml of a rotating liquid MS medium in 125 ml Erlenmeyer for two to four days and then transfer them to an agar medium in which they are planted horizontally to half their thickness. (c) Use agar medium for four to eight weeks (*67*). (d) Polyvinyl-pyrrolidine (PVP) can be used in the agar medium for the first four to eight weeks (*56*). The medium used has included BA (1 mg/l), IBA (0.1 to 1 mg/l), and GA (0.1 to 0.5 mg/l).

*Stage II: Multiplication*    The same medium is used as described for Stage I with multiplications necessary every three to four weeks. Much improved proliferation has been achieved by placing small shoots into liquid, 10 ml/100 ml flask, and agitating for four days before placing them on a solid medium (*56*).

*Stage III: Pretransplant*    (Figure 17–8) Rooting is achieved by reducing the inorganic salt content by one-half, increasing the auxin concentration, and omitting both the cytokinin and the GA. Adding phloroglucinol (*32*) has been helpful in some cases to reduce callusing, which can be a problem at some concentrations of auxin and with some cultivars (*29, 31, 67*). A continuously agitated liquid medium was necessary with the hard-to-root Granny Smith cultivar (*57*). Wounding also has been useful (*57*).

**Figure 17–8**    Culture of apple rootstocks and cultivars by shoot-tips in vitro. Shoots are rooted in agar in glass jars under lights in a controlled growth chamber.

**Figure 17–9**    Culture of apple rootstocks and cultivars by shoot-tips in vitro. *Left:* rooted shoots are transplanted into a porous medium and placed under mist. *Right:* rooted plantlets are hardened off in greenhouse for transplanting into nursery.

*Stage IV: Transplanting*    Plantlets need to develop new shoots with good cuticle development after removal from culture before they can withstand the drier greenhouse environment (Figure 17–9) (*71*). Thus, the plantlets should be kept in mist or under high-humidity conditions in a well-drained rooting medium for about 10 days for roots to form.

# REFERENCES

*1.* Anderson, W. C. 1975. Propagation of rhododendrons by tissue culture: Part 1. Development of a culture medium for multiplication of shoots. *Proc. Inter. Plant Prop. Soc.* 25:129–35.

*2.* ———. 1978. Rooting of tissue cultured rhododendrons. *Proc. Inter. Plant Prop. Soc.* 28:135–39.

*3.* ———. 1980. Mass propagation by tissue culture: Principles and practice. In *Proc. of the conf. on nursery production of fruit plants through tissue culture: Applications and feasibility,* R.H. Zimmerman, ed. U.S. Dept. of Agr. Sci. and Education Adminstration ARR–NE–11, pp. 1–10.

*4.* Anderson, W. C., and J. B. Carstens. 1977. Tissue culture propagation of broccoli, *Brassica oleracea* (italica group), for use in $F_1$ hybrid seed production. *Jour. Amer. Soc. Hort. Sci.* 102(1): 69–73.

5. Bilkey, P. C., B. H. McCown, and A. C. Hildebrandt. 1978. Micropropagation of African violet from petiole cross-sections. *HortScience* 13(1):37–38.

6. Biondi, S., and T. A. Thorpe. 1981. Requirements for a tissue culture facility. In *Plant tissue culture: Methods and applications in agriculture,* T. A. Thorpe, ed. New York: Academic Press, pp. 1–20.

7. Boxus, P. 1978. The production of fruit and vegetable plants by *in vitro* culture: Actual possibilities and perspectives. In *Propagation of higher plants through tissue culture: A bridge between research and application,* K.W. Hughes, R. Henke, and M. Constantin, eds. U.S. Dept. of Energy, Tech. Information Center, pp. 44–58.

8. Boxus, P., and P. Druart. 1980. Micropropagation, an industrial propagation method of quality plants true to type and at a reasonable price. In *Plant cell cultures: Results and perspectives,* F. Sala et al., eds. Amsterdam: Elsevier/North Holland Biomedical Press.

9. Boxus, P., M. Quoirin, and J. M. Laine. 1977. Large-scale propagation of strawberry plants from tissue culture. In *Applied and fundamental aspects of plant cell, tissue and organ culture,* J. Reinert and Y. P. S. Bajaj, eds. Berlin: Springer-Verlag, pp. 130–43.

10. Burr, R. W. 1975. Mass production of Boston fern through tissue culture. *Proc. Inter. Plant Prop. Soc.* 25:122–24.

11. Cheng, Tsai-Ying. 1978. Clonal propagation of woody plant species through tissue culture techniques. *Proc. Inter. Plant Prop. Soc.* 28:139–55.

12. ———. 1978. Propagating woody plants through tissue culture. *Amer. Nurs.* 147(10): 7–8, 94–102.

13. Cooke, R. C. 1977. Tissue culture propagation of African violets. *HortScience* 12(6):549.

14. Crehan, M. J. 1980. Profitable tissue culture. *Proc. Inter. Plant Prop. Soc.* 30:38–40.

15. Damiano, C. 1980. Strawberry micropropagation. In *Proc. conf. on nursery production of fruit plants through tissue culture: Applications and feasibility,* R. H. Zimmerman, ed. U.S. Dept. of Agr. Sci. and Education Administration ARR–NE–11, pp. 11–22.

16. de Fossard, R. A. 1976. *Tissue culture for plant propagators.* Armidale, Australia: University of New England Printery.

17. de Fossard, R. A., and R. A. Bourne. 1977. Reducing tissue culture costs for commercial propagation. *Acta Hort.* 78:37–44.

18. Dunstan, D. I. 1981. Transplantation and post-transplantation of micropropagated tree-fruit rootstocks. *Proc. Inter. Plant Prop. Soc.* 31: 39–45.

19. Fogh, J., ed. 1973. *Contamination in tissue culture.* New York: Academic Press.

20. Gamborg, O. L., and J. P. Shyluk. 1981. Nutrition, media, and characteristics of plant cell and tissue cultures. In *Plant tissue culture: Methods and applications in agriculture,* T. A. Thorpe, ed. New York: Academic Press, pp. 21–44.

21. Gamborg, O. L., and L. R. Wetter, eds. 1975. *Plant tissue culture methods.* Saskatoon, Canada: Prairie Regional Laboratory.

22. Gamborg, O. L., T. Murashige, T. A. Thorpe, and I. K. Vasil. 1976. Plant tissue culture media. *In Vitro* 12:473–78.

23. Ganzer, J. 1979. Commercial application of tissue culture in fruit production. *Proc. Inter. Plant Prop. Soc.* 29:401–3.

24. Harney, P. M., and A. Knap. 1979. A technique for the *in vitro* propagation of African violets using petioles. *Can. Jour. Plant Sci.* 59:263–66.

25. Hartmann, H. T., and J. Whisler. 1977. Micropropagation exercises in teaching plant propagation. *Proc. Inter. Plant Prop. Soc.* 27:407–13.

26. Henny, R. J., J. F. Knauss, and A. Donnan, Jr. 1981. Foliage plant tissue culture. In *Foliage plant production,* J. Joiner, ed. Englewood Cliffs, N.J.: Prentice-Hall.

27. Holdgate, D. P., and J. S. Aynsley. 1977. The development and establishment of a commercial tissue culture laboratory. *Acta Hort.* 78:31–36.

28. Huang, Li-Chun, and T. Murashige. 1977. Plant tissue culture media: Major constituents, their preparation and some applications. *TCA Man.* 3(1):539–48.

29. James, D. J., and I. J. Thurbon. 1979. Rapid *in vitro* rooting of the apple rootstock M.9. *Jour. Hort. Sci.* 54(4):309–11.

30. Jones, J. B. 1979. Commercial use of tissue culture for the production of disease-free plants. In *Plant cell and tissue culture: Principles and applications,* W. R. Sharp et al., eds. Columbus: Ohio State Univ. Press, pp. 441–52.

31. Jones, O. P. 1979. Propagation *in vitro* of apple trees and other woody fruit plants: Methods and applications. *Sci. Hort.* 30:44–48.

32. Jones, O. P., C. A. Pontikis, and M. E. Hopgood. 1979. Propagation *in vitro* of five apple scion cultivars. *Jour. Hort. Sci.* 54:155–58.

33. Khoo, S. I., and M. B. Thomas. 1980. Studies on the germination of fern spores. *The Plant Propagator* 26:11–15.

34. Knudson, L. 1922. Nonsymbiotic germination of orchid seeds. *Bot. Gaz.* 73:1–15.

35. ———. 1951. Nutrient solution for orchids. *Bot. Gaz.* 112:528–32.

36. Kruse, P. F., Jr., and M. K. Patterson, Jr., eds. 1973. *Tissue culture: Methods and application.* New York: Academic Press.

37. Kyte, L., and B. Briggs. 1979. A simplified entry into tissue culture production of rhododendrons. *Proc. Inter. Plant Prop. Soc.* 29:90–95.

38. Lane, B. C. 1980. A procedure for propagating ferns from spores using a nutrient-agar solution. *Proc. Inter. Plant Prop. Soc.* 30:94–97.

39. Langhans, R. W., P. K. Horst, and E. D. Earle. 1977. Disease-free plants via tissue culture propagation. *HortScience* 12:149–50.

40. Lloyd, G., and B. McCown. 1980. Commercially feasible micropropagation of mountain laurel, *Kalmia latifolia,* by use of shoot-tip culture. *Proc. Inter. Plant Prop. Soc.* 30:421–27.

41. Ma, S. S., and S. O. Wang. 1977. Clonal multiplication of azaleas through tissue culture. *Acta Hort.* 78:209–15.

42. Matsuyama, J. 1980. Overview of tissue culture at K. M. Nursery. *Proc. Inter. Plant Prop. Soc.* 30:40–42.

43. Meyer, M. M., Jr. 1981. *In vitro* propagation of rhododendron from flower buds (abstract). *HortScience* 16(3):452.

44. Murashige, T. 1974. Plant propagation through tissue cultures. *Ann. Rev. Plant Phys.* 25:135–66.

45. ———. 1977. Plant cell and organ cultures as horticultural practices. *Acta Hort.* 78:17–30.

46. Murashige, T., and F. Skoog. 1962. A revised medium for rapid growth and bioassays with tobacco tissue cultures. *Phys. Plant.* 15:473–97.

47. Oglesby, R. P. 1978. Tissue culture of ornamentals and flowers: Problems and perspectives. In *Propagation of higher plants through tissue culture: A bridge between research and application,* K. M. Hughes, R. Henke, and M. Constantin, eds. U.S. Dept. of Energy, Tech. Information Center, pp. 59–61.

48. Oglesby, R. P. and R. E. Strode. 1979. Commercial tissue culturing at Oglesby Nursery. *Proc. Inter. Plant Prop. Soc.* 29:341–44.

49. Oki, G. 1978. Setting up a tissue culture system. *Proc. Inter. Plant Prop. Soc.* 28:344–48.

50. Quak, F. 1977. Meristem culture and virus-free plants. In *Applied and fundamental aspects of plant cell, tissue, and organ culture,* J. Reinert and Y. P. S. Bajaj, eds. Berlin: Springer-Verlag, pp. 598–615.

51. Rao, A. N. 1977. Tissue culture in the orchid industry. In *Applied and fundamental aspects of plant cell, tissue, and organ culture,* J. Reinert and Y. P. S. Bajaj, eds. Berlin: Springer-Verlag, pp. 44–69.

52. Reinert, J., and Y. P. S. Bajaj, eds. 1977. *Applied and fundamental aspects of plant cell, tissue, and organ culture.* Berlin: Springer-Verlag.

53. Smith, R. J. 1972. Orchid propagation by *in vitro* culture techniques. *Proc. Inter. Plant Prop. Soc.* 22:174–77.

54. Smith, S. H., and W. A. Oglevee-O'Donovan. 1979. Meristem-tip culture from virus-infected plant material and commercial implications. In *Plant cell and tissue culture: Principles and applications,* W. R. Sharp et al., eds. Columbus: Ohio State Univ. Press, pp. 453–60.

55. Smith, W. A. 1981. The aftermath of the test tube. *Proc. Inter. Plant Prop. Soc.* 31:47–49.

56. Snir, I., and A. Erez. 1980. *In vitro* propagation of Malling Merton apple rootstocks. *HortScience* 15(5):597–98.

57. Sriskandarajah, S., and M. G. Mullins. 1981. Micropropagation of Granny Smith apple: Factors affecting root formation in vitro. *Jour. Hort. Sci.* 56(1):71–76.

58. Start, N. D., and B. G. Cumming. 1976. *In vitro* propagation of *Saintpaulia ionantha*: *HortScience* 11:204–6.

59. Stokes, M. J. 1980. Current aspects of commercial micropropagation. *Proc. Inter. Plant Prop. Soc.* 30:249–54.

60. Stoltz, L. P. 1979. Getting started in tissue culture: Equipment and costs. *Proc. Inter. Plant Prop. Soc.* 29:375–81.

61. Street, H. E., ed. 1977. *Plant tissue and cell culture* (2nd ed.). Botanical Monographs, vol. 11. Berkeley, Calif.: Univ. of Calif. Press.

62. Tisserat, B. 1981. Date palm tissue culture. U.S. Dept. Agr. Res. Ser., Adv. in Agr. Tech., Western Series No. 17, pp. 1–50.

63. Tjosvold, S., and A. Teasdale. 1980. Uniform fern spore dispersal on warm nutrient-agar solution. *The Plant Propagator* 26:11.

64. Vacin, E., and F. Went. 1949. Some pH changes in nutrient solutions. *Bot. Gaz.* 110:605.

65. Vasil, I. K. 1979. Plant tissue cultures in genetics and plant breeding. *Adv. in Gen.* 20:127–215.

66. Vasquez, A. M., M. R. Davey, and K. C. Short. 1977. Organogenesis in cultures of *Saintpaulia ionantha. Acta Hort.* 78:249–58.

67. Welander, M., and I. Huntrieser. 1981. The rooting ability of shoots raised *in vitro* from the apple rootstock A2 in juvenile and in adult growth phase. *Phys. Plant.* 81:36–41.

68. Wimber, D. E. 1963. Clonal multiplication of Cymbidium through tissue culture. *Amer. Orch. Soc. Bul.* 32:105–7.

69. Zilis, M., D. Zwagerman, D. Lamberts, and L. Kurtz. 1979. Commercial propagation of herbaceous perennials by tissue culture. *Proc. Inter. Plant Prop. Soc.* 29:404–13.

70. Zimmerman, R. H. 1979. The laboratory of micropropagation at Cesena, Italy. *Proc. Inter. Plant Prop. Soc.* 29:398–400.

71. Zimmerman, R. H., O. C. Broome. 1980. Apple cultivar micropropagation. In *Proc. of the conference on nursery production of fruit plants through tissue culture: Applications and feasibility*, R. H. Zimmerman, ed. USDA. ARR–NE–11, pp. 54–58.

## SUPPLEMENTARY READING

Boxus, P. 1978. *"In vitro" multiplication of woody species; round-table conference, Gembloux (Belgium), June 6–8, 1978.* Gembloux: Service des Relations publiques.

Conger, B. V., ed. 1981. *Cloning agricultural plants via in vitro techniques.* Boca Raton, Fla.: CRC Press.

Debergh, I. P., ed. 1977. Symposium on tissue culture for horticultural purposes, Ghent, Belgium, 6–9 September, 1977. *Acta Hort.* 78:1–459.

de Fossard, R. A. 1976. *Tissue culture for plant propagators.* Armidale, Australia: Univ. of New England Printery.

Gamborg, O. L., and L. R. Wetter, eds. *Plant tissue culture methods.* Saskatoon, Saskatchewan: National Research Council of Canada Prairie Regional Laboratory.

Ingram, D. S., and J. P. Helgeson. 1981. *Tissue culture methods for plant pathologists.* Oxford: Blackwell Scientific Publ.

International Plant Propagators Society. Proceedings of annual meetings.

International Plant Tissue Culture Society. Newsletters.

Reinert, J., and Y. P. S. Bajaj, eds. 1977. *Plant cell, tissue and organ culture.* Berlin: Springer-Verlag.

Street, H. E., ed. 1977. *Plant tissue and cell culture* (2nd ed.). Berkeley: Univ. of Calif. Press.

Thorpe, T. A., ed. 1981. *Plant tissue culture: Methods and applications in agriculture.* New York: Academic Press.

Tisserat, B. 1981. Date palm tissue culture. *USDA Agr. Res. Ser., Adv. in Agr. Tech., Western Series* no. 17, pp. 1–50.

Zimmerman, R. H., ed. 1980. *Proceedings of conference on nursery production of fruit plants through tissue culture: Applications and feasibility.* USDA Sci. and Education Administration ARR–NE–11.

Few fruit or nut crop cultivars reproduce true to type when propagated by seed. It is necessary, therefore, that they be propagated by some asexual method. Most tree fruit and nut species are propagated by budding or grafting on rootstocks—seedlings, rooted cuttings, or layered plants. In some cases, cuttings would be the simplest and easiest method to use, but for other than a few species such as the grape, fig, olive, quince, currant, gooseberry, and pomegranate, fruits and nuts are difficult to propagate by cuttings, so other methods are used. Some, such as the filbert, are propagated by layering.

**Almond (** *Prunus dulcis* [Mill.] D. A. Webb. Syn. *P. amygdalus* Batsch) The almond is propagated by T-budding on seedling rootstocks by fall, spring, or June budding ( *79* ). Stem cuttings have given but slight success.

    **Rootstocks for almond** Traditionally, almond seedlings have been used. Peach seedlings are favored for irrigated orchards or where nematodes are a problem. The clonal rootstock, 'Marianna 2624' plum is of value in certain situations. Almond × peach hybrids is a class of rootstocks for almond developed in recent years.

    *Almond (P. dulcis)* Almond seedlings are quite satisfactory in deep and well-drained soils. Seeds of the bitter types, or certain commercial cultivars such as 'Texas' ('Mission'), are commonly used. Almond seeds require stratification for three or four weeks before planting. In poorly drained soils almond roots are often unsatisfactory, owing to their susceptibility to infection by crown rot ( *Phytophthora* spp. ) and crown gall ( *Bacterium tumefasciens* ). Their deep-rooting tendency is an advantage in orchards grown on unirrigated soils or where drought conditions occur. Almond seedlings are tolerant of high-lime soils and, of the rootstocks available for almonds, are the least affected by excess boron salts. Almond seedlings are susceptible to root-knot and root-lesion

# 18

# Propagation Methods and Rootstocks for the Important Fruit and Nut Species

nematodes as well as to oak root fungus (*Armillarea mellea*). Nematode-resistant almond selections have been developed in Israel.

*Peach (P. persica)*    Peach seedlings are widely used as rootstocks for almond where irrigation is practiced and where nematodes are present. Any of the peach stocks are satisfactory, but 'Lovell' and 'Nemaguard' (for nematode resistance) are most commonly used. Peach roots are not as susceptible to crown gall or crown rot as almond roots. Almond on peach roots in irrigated soils will grow faster for the first several years and bear heavier crops during the first 15 to 20 years than those on almond roots, but trees on almond roots tend to live longer and may eventually outgrow those on peach.

*'Marianna 2624' plum*    This clonal rootstock is used in heavy, wet soils or where oak root fungus or root-knot nematodes (to which it is immune) are present. Almond trees on this stock are about one-third smaller than those on the other available rootstocks. Not all almond cultivars are compatible with this stock (*119*); those that cannot be used include 'Nonpareil', 'Milow', and 'Kapareil'.

*Almond-peach hybrids (120)*    These are first-generation offspring of peach and almond. Seedling populations can be produced by natural crossing between adjoining trees of the two species. Hybrid plants are identified in the nursery row by their high vigor and intermediate appearance between the parents. Combinations produced include 'Titan' almond × 'Nemaguard' peach. Rooting of selected clones is possible by hardwood cuttings treated with a hormone plus a fungicide, then planted directly in the nursery; micropropagation also can be used. These stocks are noted for their vigor and excellent compatibility with scion cultivars (*253*).

**Apple** (*Malus pumila* Mill., *M. sylvestris* Mill., *M. domestica* Borkh.)    T-budding is used, either as fall budding or spring budding on seedling or clonal rootstocks. Chip budding also is successful. Root grafting is used in some places, usually as whole-root grafts. Propagation of cultivars by hardwood cuttings, with the exception of certain clonal rootstocks, is seldom used. Softwood cuttings can be rooted under mist, but this method is not used commercially. Micropropagation methods have been used to produce apple nursery trees (*116, 215, 252*).

All apple rootstocks, either seedling or clonal, are in the genus *Malus,* although the apple will grow for a time and even come into bearing grafted on pear (*Pyrus communis*) roots.

Apple seedling rootstocks are widely used in the U.S. western and southeastern states. The clonal dwarfing rootstocks are more favored in the U.S. midwestern and northeastern states and in Europe.

### Rootstocks for apple (*23, 40, 157, 216*)

**Seedling stocks**    Apples on seedling roots produce a larger tree than those grown on most of the clonal stocks.

Seedlings of 'Delicious', 'Golden Delicious,' 'McIntosh', 'Winesap', 'Yellow Newtown', 'Rome Beauty' (particularly 'Delicious') are widely and successfully used as rootstocks. Such seedlings are quite uniform, and no incompatibility problems have arisen. However, in purchasing seed it is often difficult to determine the seed source. In the colder portions of the United States—the Dakotas and Minnesota—the hardier Siberian crabapple (*Malus baccata*) and seedlings of such cultivars as Antonovka, containing some *M. baccata* parentage, are used. In Poland 'Antonovka' seedlings are the chief apple rootstock. Some nurseries in British Columbia, Canada, use 'McIntosh' seedlings for their winter-hardiness, upright growth in the nursery, and early fall shedding of leaves, although they tend to be somewhat susceptible to *hairy root,* a form of *crown gall,* and to *powdery mildew.* The resulting trees tend to be variable in size and performance.

Apples with the triploid number of chromosomes, such as 'Gravenstein', 'Baldwin', 'Stayman Winesap', 'Arkansas', 'Rhode Island Greening', 'Bramley's Seedling', and 'Tompkins King', produce seeds that are of low viability and are not recommended as a seed source. Seeds of 'Wealthy', 'Jonathan', or 'Hibernal' have given unsatisfactory results.

The principal source of seeds is the pomace from processed apples. For spring sowing, seeds require stratification for 60 to 90 days at 2° to 7°C (35° to 45°F) to germinate. Some nurseries fall-plant seeds to receive the natural winter chilling. To avoid seedlings that become crooked in breaking through the crust, soil must be raked over in the spring. To obtain a branched root system, the seedlings may be undercut while small to prevent the development of a taproot; a straight root may be preferred, however, when bench grafting is to be done. Seedlings that do not grow to a satisfactory size in one year should be culled out.

Various Asiatic species of *Malus* have been tried as dwarfing or semidwarfing rootstocks or interstocks for apple cultivars (*192, 193*). Some of these, *Malus hupehensis, M. toringoides, M. sargentii,* and *M. sikkimensis,* are apomictic. These stocks are moderately hardy and resistant to crown gall. *Malus sikkimensis* seedlings are uniform, and scion cultivars worked on this stock are restricted in growth and start bearing early. However, such trees are relatively unproductive compared to those on the dwarfing Malling stocks.

Apple roots are resistant to root-knot and root-lesion nematodes, moderately resistant to oak root fungus, and highly resistant to verticillium wilt.

**Clonal stocks**    Numerous clonal, asexually propagated apple rootstocks have been developed. These are all in the species, *Malus pumila.* It is important in using these clonal stocks to obtain only virus-tested material. All of the clonal apple rootstocks listed below have deficiencies. In various apple producing countries rootstock breeding and testing programs are underway to develop new dwarfing and semidwarfing rootstocks to replace them (*41*).

*'Alnarp 2'*    This rootstock was developed in Sweden and is widely used there for its winter-hardy properties. The roots are well anchored and give about 20 percent size reduction as compared to trees on seedling roots. It is susceptible to woolly apple aphid and to fire blight.

*'Robusta No. 5' (M. robusta [M. baccata × M. prunifolia])*    This vigorous, very hardy clonal apple rootstock, propagated by stooling or stem cuttings, was developed at the Central Experimental Farm, Ottawa, Canada. It is apparently resistant to fire blight and crown rot and seems to be compatible with most apple cultivars. It is extensively used as an apple rootstock in eastern Ontario and Quebec, as well as in the New England states. It is the best stock for use where extreme winter-hardiness is required, but it has a low chilling requirement and has the unfavorable habit of starting growth too early in the spring following three or four warm days in late winter. It is not a dwarfing stock.

*Malling series*    Beginning in 1912 the East Malling Research Station in England selected and classified a series of vegetatively propagated apple rootstocks that ranged from very dwarfing to very invigorating in their effect on the scion cultivar (*248*). This size-controlling influence is modified by the scion cultivar used. The dwarfing influence of these various rootstocks does not extend to the fruit, however; fruit size, especially on young dwarfed trees, is often larger than on standard-size trees.

The Malling stocks apparently give completely compatible graft unions with most apple cultivars. Trees on this series of rootstocks have been planted in varying degrees in many parts of the world. These stocks have proved to be hardy except in regions with extremely severe winters, such as the northern United States and Canada. They have done well on both heavy and light soils, but none are resistant to woolly aphid.

*Malling-Merton series (178)*     The John Innes Horticultural Insititution in England and the East Malling Research Station began work jointly in 1928 on breeding a new series of apple rootstocks to provide resistance to woolly aphids and to give a range in tree vigor. Of this group only 'MM 111' and 'MM 106' are now widely used. Both show resistance to woolly aphids. However, these stocks have been severely attacked by this insect in South Africa, presumably because of the presence there of a different strain of woolly aphid (*66*).

Other improved cultivar characteristics associated with these stocks include high yield, induced precocity of flowering (with some stocks), well-anchored trees (with some stocks), freedom from suckering, and good propagation qualities.

All the Malling and Malling-Merton stocks are readily propagated, mostly by stool-bed layering (*24*). A number of rootstock clones, notably 'Malling 26', 'MM 106', and 'MM 111', respond very well to propagation by the hardwood cutting method. Certain of these clonal apple rootstocks have been produced by micropropagation methods (*112, 212*). Virus-tested material of both of these groups of rootstocks has been developed by a joint effort of the *E*ast *M*alling and *L*ong *A*shton Research Stations in England and is distributed under the designations EMLA 7, EMLA 9, EMLA 26, EMLA 27, EMLA 106, EMLA 111, and so on. These stocks give trees with 10 to 15 percent more vigor than the older virus-infected stocks of the same clone.

A summary of the characteristics of the most useful of these two groups of clonal rootstocks follows, with the rootstocks grouped according to their effect on the vigor of the scion cultivar. However, the particular scion cultivar used has a definite influence on the size of the composite tree. The most popular of these rootstocks in the United States are Malling 7, 9, 26 and Malling-Merton 106 and 111.

**Dwarfing stocks**

*'Malling 27' (179)*     This is, by far, the most dwarfing of all the Malling stocks, producing trees only about four feet tall, about half the size of those on 'Malling 9'. It is a cross between 'Malling 13' and 'Malling 9'. It may have use in high-density plantings. Virus-tested propagating material was released by the East Malling Research Station in 1970. It can also be used as an interstock to give a dwarfing effect.

*'Malling 9' ('Jaune de Metz')*     This originated as a chance seedling in France in 1879 and has been used widely in Europe for many years as an apple rootstock. It is a dwarfed tree itself and a valuable dwarfing rootstock much in demand for producing small trees for the home garden or for commercial high-density plantings. Recommended tree spacing is 1.8 × 3.6 meters (6 × 12 feet). Such trees are seldom over nine feet tall when mature and usually start bearing in the first year or two after planting. Virus-tested propagating material is available.

'Malling 9' has numerous thick, fleshy, brittle roots and requires a fertile soil; the trees require staking or trellising for support. It is resistant to collar rot (*Phytophthora cactorum*) but susceptible to crown gall, fire blight, and woolly apple aphid. Roots are sensitive to low winter temperatures. It is propagated by stooling.

When used as an intermediate stock in double-working it will cause dwarfing of the scion cultivar but less than when it is used as the rootstock (*193, 229*).

*'Malling 26'*     This stock was introduced in 1959 at the East Malling Research Station, originating from a cross between 'Malling 16' and 'Malling 9'. It is better an-

chored than 'M. 9' and it produces a tree somewhat larger and sturdier than 'Malling 9'—less so than 'Malling 7' or 'MM 106', but still a tree that requires staking. Suggested tree spacing is 3 × 4.2 meters (10 × 14 feet). It is propagated by softwood cuttings under mist or by hardwood cuttings (*99*), but is a poor producer in stool beds. It is quite winter-hardy but does not tolerate heavy or poorly drained soils and is very susceptible to fire blight.

### Semidwarfing stocks

*'Malling 7a'*    This stock produces trees somewhat larger than those on 'Malling 26' roots. 'Malling 7a' has a stronger, deeper, root system than 'Malling 9' and produces an early-bearing, semidwarf tree. It is tolerant of excessive soil moisture but susceptible to crown gall and not very well anchored, requiring staking for the first few years. It seems to make good growth over a wide soil temperature range (*145*). Suggested tree spacing is 4.2 × 4.8 meters (14 × 16 feet). This stock has the undesirable characteristic of suckering badly, and the trees are not very winter-hardy. 'Malling 7a' is easily propagated by stooling or by leafy cuttings under mist. (The ''a'' designation indicates a clone free of certain viruses present in the original 'Malling 7' but removed by heat therapy.

*'Malling-Merton 106'*    This stock produces trees about half the size of those on seedling rootstocks and about the same as those on 'Malling 7a', but is more productive, with earlier cropping. It is the most extensively used clonal apple rootstock in the United States and, in England, it is the most popular apple rootstock for bush trees and hedgerows. On good soils, with some scion cultivars, it can produce a standard-size tree. The roots are well anchored and do not sucker. Suggested planting distances are 4.2 × 5.4 meters (14 × 18 feet). In some areas 'MM 106' is susceptible to collar rot, which may be its chief weakness. It has not been affected by fire blight. It grows well in the nursery but drops its leaves later in the fall than most other understocks and is not resistant to early fall freezes. Hardwood and softwood cuttings root easily and stool beds are quite productive.

### Vigorous stocks

*'Malling 2' ('Doucin')*    This was the most commonly used apple understock in England but has been replaced by 'MM 111'. Trees on 'Malling 2' tend to be vigorous and fruitful, come into bearing early, and are smaller than trees on seedling roots. It is moderately susceptible to crown gall but resistant to collar rot. It is somewhat difficult to propagate by cuttings and the layers on young stool plants root rather sparsely. It is no longer planted in the United States.

*'Malling-Merton 111'*    Trees on this stock are comparable in vigor to those on 'Malling 2', but grow better when young, start bearing earlier, and survive periods of drought better. It does well on a wide range of soil types. Stool beds are highly productive with heavy root systems developing. Hardwood cuttings root well with proper treatment (*99*) and softwood cuttings under mist root well. 'MM 111' is more winter-hardy than 'Malling 7' or 'MM 106'. Suggested planting distances are 4.8 × 6 meters (16 × 20 feet). It shows excessive vigor in some situations.

*'Malling-Merton 104'*    Trees on this stock are drought-resistant and well anchored (depending upon the scion cultivar), but its lack of tolerance for poorly drained soils and its susceptibility to collar rot has limited its usefulness. The trees bear heavily and are more vigorous than those on 'Malling 2'. Trees on 'MM 104' are almost as large as those on seedling roots. It is difficult to propagate by hardwood cuttings (*99*) and is no longer planted in the United States.

#### Very vigorous stocks

*'Malling 16' ('Ketziner Ideal')*    Trees on this rootstock are large, well anchored, and come into bearing about the same time as those on seedling roots (or later). 'Malling 16' seems to do well over a wide range of soil temperatures (*159*). It is usually propagated by stooling, but can also be started by root cuttings. It is very susceptible to woolly aphid and is no longer used in the United States.

*'Malling-Merton 109'*    Trees on this stock are about the same size as those on seedling stocks. Its chief advantage is its resistance to woolly aphid and its stimulation of early bearing. Trees on this stock sometimes tend to lean badly and are not tolerant of waterlogged soils. It is not used in the United States.

*'Malling 25'*    Although not as resistant to woolly aphid as the Malling-Merton stocks, it is more resistant than 'Malling 16'. It produces large, well-anchored, vigorous trees, induces early fruit-bud formation, and gives excellent fruit set and yields but has not gained acceptance by U.S. apple growers.

**Apricot** (*Prunus armeniaca L.*)    Apricot cultivars are propagated commercially by T-budding or chip budding on various seedling rootstocks in the genus *Prunus*. Fall budding is the usual practice, but spring and June budding may be used. Bench grafting also has been successful (*44*).

**Rootstocks for apricot** (*163, 214*)    Three stocks are commercially suitable—apricot seedlings, peach seedlings, and in some cases, myrobalan plum seedlings. Seeds of all these species require low-temperature 5°C (41°F) stratification before planting in the spring—three to four weeks for the apricot and about three months for the peach and myrobalan plum. On good, well-drained soils, apricot seedlings are the recommended rootstock for apricot cultivars.

*Apricot (P. armeniaca)*    Apricot seeds can be obtained from drying yards and canneries. Seeds of 'Royal' or 'Blenheim' produce excellent rootstock seedlings in California. In the eastern United States seedlings of 'Manchurian', 'Goldcot', and 'Curtis' are recommended. Since the apricot root is almost immune to the root-knot nematode (*Meloidogyne* spp.), it should be used where this pest is present. In addition, it is somewhat resistant to the root-lesion nematode. It is susceptible to crown rot (*Phytophthora* spp.) and intolerant of poor soil-drainage conditions. Apricot roots are not as susceptible to crown gall (*Agrobacterium tumefaciens*) as are peach and plum roots. Apricot seedling roots are susceptible to oak root fungus and highly susceptible to verticillium wilt. Apricot trees on apricot roots live longer and produce heavier crops than trees on either peach or plum roots, if grown on well-drained loam soils.

*Peach (P. persica)*    In California peach seedlings are satisfactory as a rootstock for apricot cultivars. Although the peach itself is short-lived, apricot trees 85 years old growing satisfactorily on peach are known. In unirrigated orchards or where drought conditions prevail, apricots on peach seedling roots make better growth than those on apricot roots. Peach roots are not tolerant of wet soils, growing better on light or well-drained soils. For trees to be planted in a location formerly occupied by peach roots, some stock other than peach should be used, because new peach roots often grow poorly on these soils. In the eastern United States and Canada apricot cultivars show definite incompatibility on peach seedlings, so it cannot be assumed that all cultivars will do well on peach roots (*22, 126*).

*Myrobalan plum (P. cerasifera)*    Although there are successful high-yielding apricot orchards grown on this rootstock, it cannot be recommended without qualification. In a few instances the trees have broken off at the graft union in heavy winds, and die-back conditions have been noted. Nurserymen often have trouble starting apricots on myrobalan roots, some of the trees failing to grow rapidly and upright or else having weak or rough unions. After these weaker trees are culled out, the remaining ones seem to grow satisfactorily. This stock is useful for apricot when the trees are to be planted in heavy soils or under excessive soil moisture conditions, which the myrobalan root will tolerate. An alternative to myrobalan plum seedlings is the related vegetatively propagated 'Marianna 2624' plum, on which apricots seem to do well.

**Avocado** (*Persea americana* Mill.) (*16, 173*)    Nursery trees of the avocado are propagated commercially in California by T-budding, tip grafting (whip or splice graft) (*220*), or wedge or cleft grafting on avocado seedlings. In Florida and some of the Caribbean countries T-budding is used occasionally, but the usual nursery practice is to graft mature tip scions, either as side grafts—or sometimes the side-veneer—or cleft grafts on young succulent seedlings. To avoid sunblotch virus, the seeds (and budwood) must be taken from source trees that have been registered by state certifying agencies as free of the disease.

Seeds are generally planted shortly after removal from fruit taken from the tree (not picked from the ground), care being taken not to allow them to dry out, although they can be stored six to eight months if packed in dry peat moss and held at 5°C (41°F) with 90 percent relative humidity. To eliminate infection from *Phytophthora cinnamomi* (avocado root rot), seeds should be dipped in hot water at 49° to 52°C (120° to 125°F) for 30 minutes before planting (*249*). Germination of the seed is hastened by removing the brown seed coats and cutting a thin slice from the apical and basal end of each seed before planting. The seed coats can be removed by wetting the seeds and allowing them to dry in the sun. Seeds of the Mexican race, which ripens its fruit in the fall, may be planted in beds in late fall or early winter. They should be placed with the large, basal end down, just deep enough to cover the tips. If grown in a warm area, the sprouted seeds are ready to line out in the nursery row the following spring. By summer or fall the seedlings are usually large enough to permit T-budding; if not, they can be budded the following spring.

It is important to select the budwood properly. The best buds are usually near the terminal ends of completed growth cycles with fully matured, leathery leaves. To prevent drying, the leaves should be removed when the budwood is taken.

Four to six weeks after budding, the seedling rootstocks should be cut off 20 to 25 cm (8 to 10 in.) above the bud, or bent over a few inches above the bud. The remaining portion of the seedling above the bud is not cut off until the bud shoots have completed a cycle of growth. The new shoots are usually staked and tied. In digging from the nursery, the trees are ''balled and burlapped'' for removal to their permanent location following the first or second growth cycle or just as the first flush starts.

In California, avocado propagation has changed in recent years to a container operation. Seeds are planted in polyethylene bags and the resulting seedlings are cleft or wedge grafted about 9 cm above the seed two to four weeks after germination. Scions are taken from freshly cut terminal shoots showing strong plump buds. The plastic containers, with the grafts, are placed on raised benches in plastic-covered houses, which facilitates sanitation procedures to avoid *P. cinnamomi*. After four to six weeks the grafts are moved to a 50 percent shade house to harden, then transplanted into large containers for moving to an oudoor growing area (*174*).

In Florida, side grafting on young, succulent West Indian seedlings is used (*136*). The seedlings, grown in gallon containers, are grafted when they are 15 to 25 cm (6 to 10 in.) high and 6 to 10 mm (¼ to ⅜ in.) in diameter. The scions are shoot terminals, 5 to 7.5 cm (2 to 3 in.) long, with a plump terminal bud, taken just as it resumes growth.

Semihardwood stem cuttings from mature trees of the Mexican race can be rooted under intermittent mist using bottom heat at 25°C (77°F), provided seven or more leaves are retained on a 20 cm (8 in.) cutting. If the leaves drop, rooting ceases (*184*). It is difficult to root cuttings taken from mature trees of the Guatemalan or West Indian race, but they can be rooted if the basal portion of the leafy shoot to be made into the cutting is etiolated, that is, allowed to grow only in complete darkness. The terminal portion of such a shoot develops in the light until three to five leaves have formed. Those shoots with etiolated bases can then be detached and rooted in a propagating case (*61, 62*). Use of rooting hormones has not been beneficial. Avocados started as rooted cuttings eventually make satisfactory trees but, in general, grow poorly in the initial stages.

### Rootstocks for avocado

*Mexican race (P. americana var. drymifolia)*   Seedlings of this race are preferred in California for their cold-hardiness and their partial resistance to *Phytophthora cinnamomi,* lime-induced chlorosis, and *Dothiorella* and *Verticillium.* They are, however, susceptible to injury from high salinity. In Florida, where seedlings of large diameter are preferred for grafting, the Mexican types are little used, owing to their thin shoots.

*Guatemalan race (P. americana)*   These are occasionally used in California when there is a scarcity of Mexican seeds. Guatemalan seedlings are often initially more vigorous than Mexican seedlings but are more susceptible to diseases and to injury from cold.

*West Indian race (P. americana)*   These seedlings are too liable to frost injury to be used commercially under California conditions but are widely used in Florida. The large seed produces a pencil-size shoot suitable for side grafting in two to four weeks after germination.

**Banana** (*Musa* spp.) (*190, 199, 205*)   The banana "tree" is a large perennial herb. The "stem" consists of compressed, curved leaf stalk bases arranged spirally in strips. The bases of the leaf stalks are attached to the true stem, a rhizome (a horizontal, underground stem that develops into a so-called "corm" structure). New "suckers" grow from buds on the corm and soon develop their own roots and a base as large as the parent plant, which dies and deteriorates shortly after the fruit bunch is harvested. A banana plant may live for a considerable time but it is really a succession of new plants, each arising as a sucker from a rhizomatous bud; any given sucker fruits only once, then dies.

Since the edible types rarely produce seeds, commercial propagation of the banana is asexual, consisting of division of the rhizome and replanting of the pieces or the suckers. A large rhizome is cut into pieces weighing 3 to 4.5 kg (7 to 10 lb), depending on the cultivar, which are termed "heads." Each head should contain at least two buds capable of growing into suckers. Each sucker produces two branches in the first crop. Fairly large "sword" suckers 0.9 to 1.8 meters (3 to 6 ft) high with well-developed roots also are used, but the leaves must be shortened considerably to reduce water loss after the sucker is cut from the parent plant. These suckers are removed with a sharp cutting tool inserted vertically about half way between the parent stalk and stem of the sucker. These sword suckers produce only one bunch of fruit in the first crop, but they are often preferred, because of the large size of the bunch.

**Blackberry** (*Rubus* spp.)   The upright type of blackberry produces suckers readily. Its propagation consists of digging up suckers with attached root pieces during early spring and replanting them in a new location. The suckers are often grown an additional year in the nursery to develop stronger plants before being set out in a new planting.

The trailing type of blackberry, such as the youngberry, boysenberry, loganberry, or dewberry, does not produce many suckers. This type is usually propagated by tip layering.

All blackberries can be propagated by root cuttings, but some thornless forms, such as the thornless youngberry and 'Evergreen Thornless', revert to the thorny type if propagated in this manner.

Both the upright and the trailing types of blackberries may be started easily by conventional stem cuttings, by one-node stem cuttings (251), or by leaf-bud cuttings taken from leafy shoots and rooted under high humidity, particularily in mist-propagating beds. Treatment of such cuttings with root-promoting chemicals is beneficial.

Micropropagation methods also can be used for rapid, large-scale production of new plants (18).

**Blueberry, Highbush** (*Vaccinium corymbosum* L. and *V. australe* Small) (*36, 43, 49, 197*)    Dormant hardwood stem cuttings are used. The blueberry can also be started by leaf-bud cuttings. Most plants are grown in the nursery for one year after the year of rooting and then sold as two-year plants for setting in their permanent location. Micropropagation of some highbush blueberry cultivars has been successful (31).

Blueberry cultivars can be propagated by T-budding in midsummer to seedling plants or rooted cuttings. The highbush blueberry is successfully worked onto the rabbiteye blueberry (*V. ashei*) as a rootstock to take advantage of the latter's wide soil adaptability and vigor (64).

**Hardwood cuttings**    The blueberry is not easy to propagate by hardwood cuttings, but good results can be obtained. Special stock blocks with plants free of viruses and mycoplasma-like diseases should be used for the cutting material rather than production fields, which may be highly diseased (52). Cutting material—consisting of vigorous, firm, pencil-size, unbranched shoots of the previous season's growth—should be taken from dormant plants in winter and placed in cold storage until early spring when the cuttings are made and planted. Only vegetative wood, without fruit buds, should be used. The cuttings should contain three or four buds and be 10 to 13 cm (4 to 5 in.) long. A polyethylene-covered frame in a lathhouse is a suitable rooting structure. Bottom heat is helpful, but root-promoting chemicals have generally failed to improve rooting. A mixture of half sand and half ground sphagnum peat moss is a satisfactory rooting medium.

Cuttings should be spaced about 5 cm (2 in.) apart in the rooting bed and set with the top bud just showing. When leaves appear, the frame should be raised slightly to allow for ventilation. Either mist or frequent watering to maintain a high humidity is required. Roots start to form in about two months.

**Blueberry, Lowbush** (*Vaccinium angustifolium* Ait.) (*78*)    This is probably best propagated by leafy softwood cuttings under intermittent mist with bottom heat, using sand and peat moss (1:1) as a rooting medium (118). Cuttings taken in late spring and early summer from actively growing shoots root well, some clones giving almost 100 percent rooting. Transfer rooted cuttings to peat pots for further growth and overwintering. Cuttings made from rhizomes also can be rooted. These are best taken in early spring or late summer and fall, avoiding the midsummer rest period of the rhizome buds.

Seed propagation is sometimes used. Seeds are removed from ripe berries, then spread over a well-drained, acid type soil mix containing one-third peat moss. Cover with a layer of finely ground sphagnum moss and keep it moist until the seeds germinate, usually in three to four weeks. Seedlings 2 cm (¾ in.) tall are transferred to peat pots.

**Blueberry, Rabbiteye** (*Vaccinium ashei* Reade)    This can be propagated by hardwood cuttings as well as by leafy softwood cuttings taken in midsummer and rooted under mist with a 1 percent indolebutyric acid in talc treatment. Added lights to give a

16-hour daylength may improve root production (*37*). Rabbiteye blueberry plants also can be produced by micropropagation (*138*).

**Butternut** (*Juglans cinerea* L.)   There are several butternut cultivars and these are propagated by grafting onto *Juglans nigra* seedlings, using a form of the bark graft or any method successful with the other nut species. Excessive sap "bleeding" is a problem in graft union healing and necessary steps must be taken to overcome it.

**Cacao** (*Theobroma cacao* L.) (*230*)   Almost all commercial cacao plantings consist of seedling trees, which are highly variable. Cacao can be propagated vegetatively, however, and with the development of superior clones and the use of modern techniques for rooting softwood cuttings, it is probable that more vegetatively propagated plantings will be made (*55*).

In the large cacao-producing areas of West Africa and South America, emphasis is on seedling propagation with seed taken from selected clones. Vegetative propagation is used to produce the seed parents. Cross-pollination between high-yielding clones and use of the resultant seeds to produce bearing trees is an important propagation method.

**Seedling propagation**   Freshly harvested mature seeds should be planted immediately. Cacao seeds quickly deteriorate after harvesting, normally losing all capacity to germinate within a week after removal from the pod. Prevention of drying plus storage at 24° to 29°C (75° to 85°F) prolongs seed life (*8*).

A common practice is to plant three or four seeds in a hole. If they all germinate they are allowed to grow, the plants being treated as branches of one tree. Another method is to start the seedlings in a nursery bed and transplant the small trees to their permanent location. Alternatively, the seedlings may be started in polyethylene bags, baskets, bamboo or paper cylinders, or clay pots, from which they are removed later and planted. A germination temperature of about 26° C (80° F) should be used.

**Asexual propagation**   After young seedling trees have attained sufficient size, they can be used as rootstocks on which superior clones are budded. The patch bud is most successful, but T-budding may be used (*54*). In areas where unfavorable soil conditions occur, superior clones are sometimes grafted on resistant clonal rootstocks. In the world industry, however, budding and grafting are done on a negligible scale. Air layering is quite successful.

**Carob** (*Ceratonia siliqua* L.)   This subtropical evergreen tree is usually propagated by seeds, which germinate without difficulty when freshly harvested. If seeds dry out and the seed coats are hard, they should be softened by hot water or sulfuric acid treatment (see chapter 7). Transplanting of bare root seedlings gives poor results so the seeds are planted in their permanent location or started in containers for later transplanting. Seedlings are best budded to selected cultivars; this is most successful in late spring. Cuttings can be rooted if taken in mid-spring and treated with indolebutyric acid at about 7500 ppm (see Figure 9–20). Air layering in late summer is successful.

**Cashew** (*Anacardium occidentale* L.) (*5, 152*)   This tender tropical evergreen tree, grown principally in India, is usually propagated by seed; two or three are planted directly in place in the orchard, since seedlings transplant with difficulty. They are later thinned to one tree per location. There is no seed dormancy but seeds should be tested for the presence of embryos by placing in water; those that float should be discarded. Germination takes place in 15 to 20 days. For transplanting, the seeds are started in some type of container which will disintegrate, and the container—with the seedling—is set in the ground. Cashew seedlings show great variability in growth habit, yield, and nut quality. There are few named cultivars but efforts are being made to select superior, high-yielding types and propagate them by asexual methods, particularly by stooling

(*158*), approach-grafting (*181*), and air layering (*162*). Other methods, such as T-budding, patch budding, and rooting leafy cuttings, also have been successfully used in limited trials.

**Cherimoya** (*Annona cherimola* Mill.)     This is propagated by cleft grafting or T-budding selected cultivars on seedlings of cherimoya or the related sugar apple (*Annona squamosa*), which gives a dwarf plant, or on custard apple (*Annona reticulata*). The latter two species should not be used as rootstocks in cold areas or in poorly drained soils where they are subject to root rot.

Some seedling forms developed in Mexico and South America come nearly true from seed, and in many regions seed propagation is used exclusively. The seeds retain their viability for many years if kept dry, and germinate in a few weeks after planting. After the young seedlings are 7.5 to 10 cm (3 to 4 in.) high they should be transferred from flats to small pots, and when about 20 cm (8 in.) high, to larger pots or to open ground.

**Cherry** (*Prunus avium* L., *P. cerasus* L.)     Cherry trees are propagated by T-budding or chip budding the desired cultivar on a seedling rootstock. Rootstock seedlings are often grown closely planted in a seed bed for one year, then are lined-out about 10 cm (4 in.) apart in the nursery and grown a second year before budding. If growing conditions are good, seeds may be planted directly in the nursery in early spring, and the seedlings will be large enough for budding by late summer or early fall.

**Rootstocks for cherry** (*45, 164, 187*)     The two most commmon stocks are Mazzard (*Prunus avium*) and Mahaleb (*P. mahaleb*) seedlings. The vegetatively propagated 'Stockton Morello' (*P. cerasus*) is occasionally used in California (*93, 108*), and the 'Colt' stock in England. These rootstocks are used for sweet cherry (*P. avium*) cultivars. Sour cherry (*P. cerasus*) and Duke cherry (hybrids between sweet and sour cherries) cultivars are propagated on either Mahaleb or Mazzard roots. All these cherry rootstocks are susceptible to verticillium wilt. *P. avium* is moderately resistant to oak root fungus where *P. mahaleb* and 'Stockton Morello' are susceptible.

*Mazzard (P. avium)*     Mazzard seedlings used in the United States are available from sources in Oregon and Washington. There is considerable variation among the sources of Mazzard seeds, some undoubtedly better than others. Seeds from trees indexed for freedom from ring spot virus should be used if possible. In germinating Mazzard seeds it is beneficial to soak them in water changed daily for about eight days prior to stratification. A warm, moist stratification period at 21° C (70° F) for four to six weeks prior to cold stratification should improve germination. Finally a cold (4° C; 40° F) stratification period of 150 days is used. When a fair percentage of the seeds show cracking of the endocarp with root tips emerging, they should be removed and planted (*127*).

The clonal Mazzard stock, 'Malling 12/1', developed at the East Malling Research Station, produces trees that are vigorous, uniform, and resistant to bacterial canker. Propagation is by trench or mound layering. In Oregon this stock has worked well for the root, trunk, and primary scaffold system onto which the scion cultivar is worked.

Sweet cherry cultivars make an excellent graft union with Mazzard roots, giving trees that are vigorous and long-lived, but may be too large for economical harvest. Mazzard roots are not particularly suitable for heavy, poorly aerated, wet soils, but will tolerate such conditions better than Mahaleb. Under dry, unirrigated, drought conditions Mahaleb is more likely to survive than Mazzard, presumably because of the deep, vertical rooting habit of Mahaleb in contrast to Mazzard's shallow, horizontal root system.

Mazzard roots are immune to one species of root-knot nematode (*Meloidogyne in-*

*cognita*) and resistant to *M. javanica,* but susceptible to the root-lesion nematode (*Pratylenchus vulnus*).

*Mahaleb (P. mahaleb)*    The seeds should be soaked in water for 24 hours, then stratified for about 100 days at 4°C (40°F). Leafy Mahaleb cuttings are easily rooted under intermittent mist if treated with indolebutyric acid, thus permitting the establishment of clonal Mahaleb stocks (*93*). This is the principal rootstock in the United States for 'Montmorency', the leading sour cherry cultivar. Mahaleb roots may produce a somewhat dwarfed tree, particularly if the budding is done high, e.g., 38 to 50 cm (15 to 20 in.) on the rootstock; there is evidence, however, that in good soils the trees will become as large as those on Mazzard roots. Trees on Mahaleb roots are resistant to the buckskin virus. Mahaleb-rooted trees do not grow satisfactorily in heavy, wet soils with high water tables, being susceptible to *Phytophthora* root rot. This rootstock should be used for nonirrigation or drought conditions. Trees on Mahaleb roots are more cold-hardy than those on Mazzard. Sweet cherries often grow faster for the first few years on Mahaleb than on Mazzard roots, and some cultivars start heavy bearing rather early, which may result in some dwarfing of the tree. There is evidence, especially in England, that trees on Mahaleb roots are relatively short-lived, but in the United States good, productive trees known to be more than 50 years old are grown on this stock. Mahaleb roots are more resistant to root-lesion nematode (*Pratylenchus vulnus*) than Mazzard (*45*). They are also resistant to one species of the root-knot nematode (*Meloidogyne incognita*) but susceptible to *M. javanica.* Mahaleb roots are more resistant to bacterial canker than Mazzard roots.

'Colt' (*Prunus avium* × *P. pseudocerasus*) is a dwarfing clonal cherry rootstock developed at the East Malling Research Station. It can be used for both sweet and sour cherries and is easily propagated by cuttings.

**Chestnut, Chinese** (*Castanea mollissima* Blume.) (*113, 114*)    This species is resistant to the blight fungus, *Endothia parasitica,* which killed most American chestnut (*C. dentata*) trees in eastern United States upon its introduction from China in the late 1800s. The well known edible chestnut is *C. sativa,* the European sweet chestnut, imported in large quantities into the United States from Europe.

*C. mollissima* is usually propagated by seed, many plantings consisting of seedling trees. Clonal selections have been made, however, and named cultivars are available. Seeds should be prevented from drying. The nuts are gathered as soon as they drop and either planted in the fall or kept in moist storage one or two months at 0° to 2° C (32° to 40° F) for spring planting. Seed nuts are satisfactorily stored in tight tin cans, with one or two very small holes for ventilation, at 0° C (32° F) or slightly higher; this storage temperature also aids in overcoming embryo dormancy (*144*). Weevils in the nuts, which will destroy the embryo, can be killed by hot-water treatment at 49° C (120° F) for 30 minutes.

After one year's growth, the seedlings should be large enough to transplant to their permanent location or be grafted to the desired cultivar (*171*). Although the chestnut is difficult to graft or bud, splice or bark grafting and inverted T-budding have given good results. In regular T-budding, the buds tend to ''drown,'' owing to excessive bleeding. Only *C. mollissima* seedlings should be used as rootstocks.

**Chinese Gooseberry**    *See* Kiwifruit.

**Citrus** (*Citrus* spp.) (*175, 182, 185, 250*)    Propagation methods are the same for all species of citrus. Members of this genus are readily intergrafted and can also be grafted to other closely related genera such as *Fortunella* (kumquat) and *Poncirus* (trifoliate orange). Citrus cultivars are propagated commercially by T-budding on seedling rootstocks. In all types of citrus propagation it is very important to use true-to-type material free of transmissible pathogens (*121*). Citrus propagation has changed but little in the past century. There are some new practices, however, such as container production, more fumigation of nursery soils, use of plastic tape in budding, and more attention paid to controlling viruses during propagation (*247*).

Many citrus species can be propagated by rooting leafy cuttings, or by leaf-bud cuttings, although nursery trees are not commonly propagated in this manner (*42*). Except for the psorosis virus, transmissible diseases do not appear in seedlings unless they become infected from insect vectors or by budding.

In Florida the Persian lime (*C. aurantifolia*) is propagated to some extent by air layering, as is the pummelo (*C. grande*) in southeast Asian countries.

Probably the most rapid method of obtaining a citrus cultivar worked on a given rootstock is the use of "cutting-grafts" to produce dwarf plants (*47*).

**Growing citrus nursery stock—field production**     It is important to avoid using soils infested with citrus nematodes (*Tylenchulus semipenetrans*) or burrowing nematodes (*Radopholus similis*), or soilborne diseases, although citrus is resistant to verticillium wilt. For nurseries it is preferable to use virgin soil or at least a soil that has not been formerly planted to citrus. For small operations, raised seedbeds enclosed by 12-in. boards can be used. A soil mixture of ¾ sandy loam and ¼ peat moss is satisfactory. Treating the soil with a fumigant such as DD (dichloropropane-dichloropropene) at the rate of 700 to 1000 lb per acre will minimize the chances of nematode infestations. To reduce the hazard of fungus infection, methyl bromide is often used to fumigate the seedbed and nursery site. Following treatment, planting should be delayed for six to ten weeks to allow the fumigant to dissipate. Some soils in California, however, have remained toxic to citrus for a year following such treatment. The seedbed should be in a lathhouse, or some other provision should be made for screening the young seedlings from the full sun.

Since there is considerable variation in the performance of seedlings taken from different trees, it is best to select the seeds from healthy, virus-tested old trees, known to produce vigorous, uniform seedlings that develop into satisfactory orchard trees after being budded to the desired cultivar.

Citrus seeds generally have no dormancy but are injured by being allowed to dry; they may be planted immediately after being extracted from the ripe fruit. Certain species, such as the trifoliate orange or its hybrids, mature their fruits in the fall. If the seeds are to be planted at that time, they should be held in moist storage at $-1°$ to $4°C$ ($30°$ to $40°F$) for at least four weeks before planting. Trifoliate orange seedlings, if grown during times of the year having short days, will respond strongly with increased growth when supplementary light is provided to lengthen the day. The same is true for seedlings of certain citrange and sweet orange cultivars (*233*).

Seeds may be stored in polyethylene bags at a low temperature ($4°C$; $40°F$); before storage they should be soaked for ten minutes in water at $49°C$ ($120°F$) to aid in eliminating seed-borne diseases. Treatment of the seed with a fungicide, such as thiram, also is beneficial.

The best time to plant the seeds is in the spring after the soil has warmed (above $15°C$; $60°F$). Seeds should be planted in rows 5 to 7.5 cm (2 to 3 in.) apart, and 2.5 cm (1 in.) apart in the row. They are pressed lightly in the soil and covered with a 2 cm (¾ in.) layer of clean, sharp river sand. The sand prevents crusting and aids in the control of "damping-off" fungi. The soil should be kept moist at all times until the seedlings emerge. Either extreme, allowing the soil to become dry and baked or overly wet, should be avoided. Electric soil-heating cables placed below the seedbed to maintain a temperature of $27°$ to $29°C$ ($80°$ to $85°F$) will hasten germination. By this method, seeds may be planted in the winter months, and the seedlings will be large enough to line-out in the nursery in the spring. Many can be budded by fall or the following spring. This often shortens the propagation time by six to twelve months.

After the seedlings are 20 to 30 cm (8 to 12 in.) tall, they are ready to be transplanted from the seedbed to the nursery row, preferably in the spring after danger of frost has passed. The seedlings are dug with a spading fork after the soil has been wet thoroughly to a depth of 45 cm (18 in.). They can then be loosened and removed with little danger of root injury. All stunted or off-type seedlings or those with crooked, misshapen roots should be discarded.

The nursery site should be in a frost-free, weed-free location on a medium-textured, well-drained soil at least 61 cm (24 in.) deep and with irrigation water available. Old citrus soils should be avoided unless heavily fumigated with DD or methyl bromide before planting. The seedlings should be planted at the same depth as in the seedbed and spaced 25 to 30 cm (10 to 12 in.) apart in 0.9 to 1.2 meters (3 to 4 ft) rows.

Citrus seedlings are usually budded in the fall in Florida and California, starting mid-September, early enough so that warm weather will insure a good bud union, yet late enough so that bud growth does not start and the wound callus does not grow over the bud itself.

**Growing citrus nursery stock—container production**    Citrus seeds are germinated and the seedlings grown in plastic tubes in a temperature-controlled structure, such as a greenhouse. After three or four months the rootstock plants are large enough for microbudding; finished budded trees can be developed in about 12 months grown in plastic containers (25). Sometimes cleft grafting is used on the young seedlings after they are growing in their final containers (147). Citrus nursery trees apparently grow well in containers and develop into good orchard trees upon field planting providing they are not held in the container so long that root binding occurs (151, 153).

Most commercial citrus-producing areas of the world have programs to determine the presence of virus and virus-like diseases in trees used as budwood sources (121). There are about 12 known viruses, a viroid causing exocortis, and a mycoplasmalike organism causing "stubborn" disease, which can infect citrus. good indexing procedures have been worked out for most of them (20, 185). Budwood should be taken only from known high-producing, disease-free trees. It is desirable to select the budsticks from a single tree, avoiding any off-type "sporting" branches. The best type of budwood is that next to the last flush of growth, or the last flush after the growth hardens. A round budstick gives more good buds than an angular one. The best buds are those in the axils of large leaves. The budsticks are usually cut at the time of budding, the leaves removed, and the budsticks protected against drying. Budsticks may be stored for several weeks if kept moist and held under refrigeration at 4° to 13° C (40° to 55° F).

The T-bud method is satisfactory for citrus. The bud piece is cut to include a sliver of wood beneath the bud. Fall buds are unwrapped in six to eight weeks after budding, spring buds in about three weeks. In California and Texas the buds are usually inserted at a height of 30 to 45 cm (12 to 18 in.) but in Florida the buds are inserted very low on the stock—2.5 to 5 cm (1 or 2 in.) above the soil. Such low budding is often necessitated by profuse branching in rough lemon and sour orange seedlings, which is caused by partial defoliation due to scab and anthracnose spot.

Micro-budding is used to some extent, particularly in Australia, for citrus.

Buds inserted in the fall are forced into growth in the spring by "lopping" the top of the seedlings 5 to 7.5 cm (2 or 3 in.) above the bud. This is done just before spring growth starts, and consists of partly severing the top, allowing it to fall over on the

**Figure 18–1** Budded citrus seedlings are often ''lopped over'' just above the inserted bud. This forces the bud into growth, even though the top continues to provide nourishment for the plant. Courtesy Department of Pomology, University of California, Davis.

ground. The top thus continues to nourish the seedling roots, but the bud is forced into growth. Lopping of spring buds is done when the bud wraps are removed—about three weeks after budding. If possible, the ''lops'' should be left until late summer, at which time they are cut off just above the bud union (Figure 18–1). Although lopping is satisfactory, it may make irrigation and cultivation difficult. An alternative practice is to first cut the seedling completely off 30.5 to 35.5 cm (12 to 14 in.) above the bud, then later cut it back immediately above the bud. However, this practice does not force the bud as well as lopping or cutting the seedling just above the bud.

Young citrus nursery trees may be dug ''balled and burlapped'' or bare-root. Bare-rooted trees should have the tops pruned back severely before digging. Transplanting of such trees is best done in early spring, but balled trees can be moved any time during spring before hot weather starts.

Polyembryony is high in seeds of most citrus species used as rootstocks. The sexual seedlings are often weak and make a poor rootstock so are usually rogued out. The other seedlings arising from the nucellus are apomictic and have the same characteristics as the seed-bearing plant. Consequently they are uniform, and make good rootstocks if the parent tree is desirable. If used as orchard trees, nucellar seedlings are vigorous, thorny, and upright-growing, but slow to start bearing. These undesirable qualities (thorniness and delayed bearing) are less pronounced in nursery trees propagated from budwood taken from the upper part of old nucellar seedling trees. Nucellar cultivars have been developed for all the commercial citrus cultivars, and are used because of their increased vigor, tree size, and yields, and their freedom from viruses (*21*). (See chapter 8.)

**Rootstocks for Citrus** (*101, 167, 247*)

*Sweet Orange (C. sinensis)*     This is a good rootstock for all citrus cultivars, producing large, vigorous trees resistant to tristeza. Sweet orange is adapted to well-drained, light- to medium-loam soils, but owing to its susceptiblity to gummosis (*Phytophthora* spp.), it is not suited to poorly drained, heavy soils. It produces standard-size fruits that are thin-skinned, juicy, and of fairly high quality. The seeds germinate readily, but the seedlings, which are 70 to 90 percent nucellar, are relatively slow growers and tend to produce low-branched, bushy trunks in the nursery. It is not widely used today principally because of its susceptibility to *Phytophthora*.

*Sour Orange (C. aurantium)* This is an excellent stock for most citrus species, owing to its vigor, hardiness, deep root system, resistance to gummosis diseases, and to the high quality, smooth, thin-skinned, and juicy fruit produced by cultivars worked on it. However, it lacks tolerance to "tristeza" (quick decline), due to a virus disease that is transmitted by an insect vector or from infected budwood (*12, 80*). The scion top itself may be quite tolerant to the virus, but in combination with sour orange root the stock is affected, owing to death of the phloem tissues in the bud union area, which then results in starvation of the root. Grapefruit and mandarin, as well as orange cultivars, on sour orange are subject to tristeza. In California sour orange is no longer recommended as a rootstock for orange or grapefruit cultivars. In Florida, however, over one-third of the orange trees are on sour orange roots, but tristeza has not been serious there, except in certain limited areas. Presumably the strains of tristeza in Florida are less virulent than those in California and other citrus areas. In Texas, Mexico, Cuba, Venezuela, and Israel also, sour orange rootstock is widely used. In Australia strains of sour orange—originating in Israel—have been tested that are tolerant to tristeza (*218*).

*Rough Lemon (C. limon)* This stock is well adapted to sandy soils, and because of this about 60 percent of all citrus trees in Florida are on rough lemon roots. It has an extensive root system, which makes it very drought tolerant. Trees on this rootstock outyield those on all other stocks, although the fruit produced is of lower quality. Its use in California is restricted to the lighter soil areas, particularly in the desert regions. Its susceptibility to *Phytophthora* precludes its use on heavy soils. Both trees and fruit are more susceptible to cold injury on this stock than on most other commercial stocks (*65*). Fruits from trees on rough lemon roots are early maturing, but have a thick rind, are low in both sugar and acid, and are often coarse-textured compared with fruit from trees on other rootstocks.

Rough lemon produces numerous seeds that germinate well. Ninety to 100 percent of the seedlings are nucellar. They grow upright with single, unbranched trunks that are easy to bud and handle in the nursery. Rough lemon also can be easily propagated by cuttings. Sweet orange cultivars are resistant to quick decline when worked on rough lemon roots. The chief advantages of this stock are its vigor and its ability to produce a bearing tree quickly, particularly on light, sandy soils. Some orange cultivars on rough lemon develop an incompatible bud union.

*Trifoliate orange (Poncirus trifoliata)* This dwarfing citrus rootstock has been used to some extent for many years. It is the main citrus rootstock in Japan. In northern Florida and along the Gulf coast to Louisiana, it has long been used as a stock for Satsuma oranges and kumquats, for which it is excellent. *Poncirus trifoliata,* as a rootstock, does best on medium-textured soils. It will tolerate heavy soil but grows quite slowly. On light, sandy soil its growth may be so poor that trees on this stock are worthless. Trifoliate orange is commonly used as a stock for ornamental citrus and in home orchards for dwarfed trees. Trees on this stock yield heavily and produce high-quality fruits. Trifoliate orange is a deciduous species noted for its winter-hardiness and its *Phytophthora* and nematode resistance. Both the tree and its fruits, when worked on this stock, are more resistant to cold than other rootstock-top combinations, making it particularly adaptable to the colder citrus-growing regions. A citrus cultivar worked on trifoliate orange is one of the few examples of an evergreen top on a deciduous rootstock.

Trees on trifoliate roots are often affected by exocortis, or "scaly butt." This may be avoided by using uncontaminated nucellar buds or by taking buds from trees on trifoliate stock that are free of the disease. This stock is quite resistant to gummosis, but it is very susceptible to citrus canker (*Phytomonas citri*) and moderately so to scab. It is susceptible to lime-induced chlorosis but tolerant of excess boron.

Trifoliate orange fruits produce large numbers of plump seeds that germinate easily. The upright-growing, thorny seedlings, about 60 percent of which are nucellar, are easy to bud and handle in the nursery, but their slow growth often necessitates an extra year in the nursery before salable trees are produced.

*'Cleopatra' mandarin (C. reticulata)*     This is a promising rootstock, particularly in Florida, as a stock for other mandarin types, and has come into use in California and Texas as a replacement for sour orange. Its resistance to gummosis, comparative salt tolerance, and resistance to tristeza seemingly justify its greater use. In addition, trees on this stock show good yields of high-quality fruit, although fruit size is somewhat smaller than average. Its chief disadvantages are the slow growth of the seedlings, slowness in coming into bearing, and susceptibility to *Phytophthora parasitica* root rot. About 80 percent of the seedlings are nucellar. Some cases of incompatibility, especially with 'Eureka' lemon, have appeared.

*Citranges (Trifoliate orange × sweet orange hybrids)*     There are many named cultivars—'Savage', 'Morton', 'Troyer', 'Carrizo', and so on—some of which are proving useful as rootstocks both in California and in Florida.

'Savage' is especially suitable as a dwarfing stock for grapefruit and is also satisfactory for the mandarins. It is resistant to gummosis, and trees on this stock are more cold-hardy than those on many of the other rootstocks (*11, 65*).

'Morton' citrange is not particularly dwarfing. Trees worked on it are heavy producers of excellent-quality fruit. It produces so few seeds, however, that to get quantities of nursery trees started is difficult. In addition, orange cultivars on this stock appear to be susceptible to quick decline.

Sweet orange on 'Troyer' citrange is vigorous, cold-hardy, and resistant to gummosis, with high-quality fruit. This stock is more widely used than any other for oranges in California. In replanting citrus on old citrus soils, trees on the 'Troyer' citrange have shown outstanding vigor. 'Troyer' itself is relatively fruitful and produces 15 to 20 plump seeds per fruit, facilitating the propagation of nursery trees.

Seedlings of citrange cultivars are mostly nucellar and develop strong, single trunks, easily handled in the nursery. As with trifoliate orange, only exocortis-free buds should be used on citrange rootstocks; otherwise, dwarfing—and eventual low production—will result.

*Rangpur lime (C. aurantifolia × C. reticulata)*     This is the most widely used citrus rootstock in Brazil. It produces vigorous, fruitful trees, which are resistant to tristeza but highly susceptible to exocortis. In Texas it has been more salt-tolerant than other citrus rootstocks. Some strains of Rangpur lime are susceptible to *Phytophthora*.

*Alemow (Citrus macrophylla)*     This is widely used in California as a rootstock for lemons in high-boron areas, owing to its boron tolerance. It is susceptible to tristeza when sweet orange scion cultivars are used.


**Coconut** (*Cocos nucifera* L.) (*29, 176*)     Trees are propagated only by seed, and there are some named cultivars that maintain their characteristics quite dependably by seed propagation. It is important to select seeds from trees that produce large crops of high-quality nuts.

The nuts are usually germinated in seedbeds. Seedlings that develop rapidly and have strong, vigorous shoots are selected. The nuts, still in the husk, are set at least 12 in. apart in the bed and laid on their sides with the stem end containing the "eyes" slightly raised. The sprout emerges through the eye on the side that has the longest part of the triangular hull. As soon as this occurs (about a month after planting), the sprout sends roots downward through the hull and into the soil. In 6 to 18 months the seedlings are large enough to transplant to their permanent location.

**Coffee** (*Coffea arabica* L.) (*238*)    The most common method of propagation is by seed, preferably obtained from selected superior trees. Coffee seeds lose viability quickly and are subject to drying through the seed coverings. Seeds held at a moisture content of 40 to 50 percent and at 4° to 10°C (40° to 50°F) will keep for several months. There are no dormancy problems. Seeds are usually planted in seedbeds under shade (*219*). Sometimes the seedlings are started in soil in containers formed from leaves, or in polyethylene bags, to facilitate transplanting. Germination takes place in four to six weeks. When the first pair of true leaves develops, the seedlings are transplanted to the nursery and set 12 in. apart. After 12 to 18 months in the nursery, by which time they have formed six to eight pairs of laterals, the young trees are ready to set out in the plantation.

Coffee can be propagated asexually by almost all methods, but leafy cuttings probably hold the most promise for commercial use. Cutting material should be taken only from upright shoot terminals in order to produce the desired upright-growing tree. Leafy cuttings of partially hardened wood can be rooted fairly easily, especially if treated with a root-promoting chemical. High-humidity conditions, such as in intermittent mist, must be maintained, and the cuttings should be partially shaded during rooting (*208*). Coffee plants have been propagated by the formation of adventitious embryos in aseptic callus cultures (*104*).

**Crabapple**    Siberian crabapple (*Malus baccata* Borkh.), Western crabapple (*M. ioensis* Britt.), and other *Malus* species. The usual propagation method is to bud or graft the desired cultivars on seedling rootstocks, either one of the crabapple species or the common apple, *Malus pumila*. In areas where winter-hardiness is important, *M. baccata* seedlings should be used.

Crabapples can be propagated, although with difficulty, by softwood or hardwood cuttings, especially if root-promoting chemicals are used.

**Cranberry** (*Vaccinium macrocarpon* Ait.) (*39, 77*)    This vine type of evergreen plant produces trailing runners upon which are numerous short upright branches. Propagation is by cuttings made from either runners or upright branches. Cutting material is obtained by mowing the vines in early spring before new growth has started. The cuttings are then set directly in place in their permanent location without previous rooting at distances of 6 to 18 in. apart each way. Two to four cuttings are set in sand in each "hill." The cuttings are 13 to 25 cm (5 to 10 in.) long and set deep enough so that only an inch is above ground. A more rapid method of starting a cranberry bog is to scatter the cuttings over the ground and work them into the soil with a special disk-type planter. This is justifiable when there is an abundance of cutting material and a scarcity of labor for setting the cuttings by hand. Water is applied to the bog immediately after planting. The cuttings root during the first year and make some top growth, but the plants do not start bearing until three or four years later.

**Currant** (*Ribes* spp.)    Currants are readily propagated by hardwood cuttings prepared from well-matured wood of the previous summer's growth. Cuttings 20 to 25 cm (8 to 10 in.) long are made in late fall, stored in moist sand, sawdust, or peat moss at about 2° C (35° F), then planted in early spring. The plants can be transplanted to their permanent location in one or two years, depending upon their growth. Currants can also be propagated by mound layering.

**Date** ( *Phoenix dactylifera* L.) ( *1, 2, 161, 191* )    Propagation of the date palm is either by seed or by offshoots. This is a monocotyledonous plant having no continuous cambial cylinder, so it cannot be propagated by budding or grafting. In commercial plantings, most of the trees are female, but a few male trees are necessary for pollination. The date palm is capable of regeneration from various tissues in in vitro culture systems ( *225* ).

In seed propagation about half of the trees produced are males, whereas the seedling female trees produce fruits of variable and generally inferior types. The commercial grower is therefore little interested in seedling trees, preferring the superior named cultivars, even though the latter must be propagated by the vegetative offshoot method. Date seeds germinate readily.

Offshoots arise from axillary buds near the base of the tree. If they are near the ground level, they will develop roots in the soil after three to five years on the parent tree. Large, well-rooted offshoots, weighing 18 to 45 kg (40 to 100 lb), are more likely to grow than smaller ones. Offshoots higher on the trunk can be induced to root if moist rooting medium is held against the base of the offshoot by means of a box or a polyethylene tube. Unrooted offshoots arising higher on the stem can be cut off and rooted in the nursery, but in relatively low percentages.

Considerable skill is required to cut off a date offshoot properly. Soil is dug away from the rooted offshoot, but with a ball of moist earth as thick as possible attached to the roots. The connection with the parent tree should be exposed on each side by removing loose fiber and old leaf bases. A special chisel, having a blade flat on one side and beveled on the other, is used to sever the offshoot. The first cut is made to the side of the base of the offshoot close to the main trunk. The beveled side of the chisel is toward the parent tree, which gives a smooth cut on the offshoot. A single cut may be sufficient, but usually one or more cuts from each side are necessary to remove the offshoot. The offshoot should never be pried loose; it should be cut off cleanly. After removal, it should be handled carefully and replanted as soon as possible, care being taken to prevent the roots from drying out. (See Figure 14–12.)

**Feijoa** ( *Feijoa sellowiana* Berg.) ( *110* )    Propagation can be done by seeds, which germinate without difficulty. They should be started in flats of soil and later transplanted to the nursery row. Cultivars can be grafted on feijoa seedlings, although it is difficult to obtain a high percentage of successful unions. Leafy softwood cuttings treated with root-promoting substances and started under closed frames can be rooted.

**Fig** ( *Ficus carica* L.) ( *34, 122* )    The fig is easily propagated by hardwood cuttings. Two- or three-year-old wood or basal parts of vigorous one-year shoots with a minimum of pith are suitable for cuttings. For best results cuttings should be prepared in early spring, well before bud break, with the bases allowed to callus for about 10 days at about 24°C (75°F) in damp wood shavings or sawdust before planting. The cuttings are grown for one or two seasons in the nursery, then transplanted to their permanent location. A common method in European countries is to plant long cuttings, 0.9 to 1.2 meters (3 to 4 ft), their full length in the ground where the tree is to be located permanently; sometimes two cuttings are set in one location to increase the chance that one will grow.

The fig can be budded, using either T-buds inserted in vigorous one-year-old shoots on heavily pruned trees, or patch buds on older shoots.

Fig roots are resistant to oak root fungus (*Armillaria mellea*) and verticillium wilt but quite susceptible to both root-knot (*Meloidogyne* spp.) and lesion nematodes (*Pratylenchus vulnus*) (*34*).

Seed propagation is used only for breeding new cultivars. The small seeds can be germinated easily in flats of well-prepared soil. Fertile seeds should first be separated from the sterile ones by placing all of them in water; fertile seeds sink, but sterile ones float.

Figs can be air-layered. One-year-old branches, if layered in early spring, are usually well-rooted by midsummer.

**Filbert** (*Corylus avellana* L. and *C. maxima* Mill.) (*124, 146*)     Simple layering is the usual method of commercial filbert propagation. The suckers arising from the base of vigorous young trees four to eight years old are layered in early spring, or special stool mother plants are maintained in a bush form just for layering purposes (see Figure 14–5). After one season's growth, a well-rooted tree 0.6 to 1.8 cm (2 to 6 ft) tall may be obtained, ready to set out in the orchard. Old orchard trees are not suitable to use for layering. Sometimes suckers arising from the roots are dug and grown in the nursery row for a year or, if well-rooted, planted directly in place in the orchard.

Filberts are difficult to propagate by cuttings. Budding or grafting filbert cultivars on seedling stocks is rarely practiced, because of the difficulty in obtaining successful unions, but see page 430.

Filbert seeds are easily germinated but require a stratification period of several months at 0° to 4°C (32° to 40°F).

**Gooseberry** (*Grossularia* or *Ribes* spp.)     Mound layering is used commercially. Layered shoots of American cultivars usually root well after one season. They are then cut off and transferred to the nursery row for a second season's growth before they are set out in their permanent location. The slower-rooting layers of European cultivars may have to remain attached to the parent plant for two seasons before they develop enough roots to be detached. Some cultivars, such as the 'Houghton', 'Poorman', and 'Van Fleet', can be started fairly easily by hardwood cuttings (*180*).

**Grape** (*Vitis* spp.) (*3, 234, 245*)     Grapevines are propagated by seeds, cuttings, layering, budding, or grafting. New plants have been produced by several in vitro techniques, including embryoid formation and fragmented shoot tip cultures (*123*). Seeds are used only in breeding programs to produce new cultivars. Grape propagation methods have been modernized by the use of virus-indexed, "clean" planting stock, mist-propagation techniques for leafy cuttings, and rapid machine-grafting procedures. Most commercial propagation is by dormant hardwood cuttings. For types difficult to root, such as the Muscadine (*Vitis rotundifolia*), layering or the use of leafy cuttings under mist is necessary (*203*). Budding or grafting on rootstocks is used occasionally to increase vine life, plant vigor, and yield. Where noxious soil organisms, such as phylloxera (*Dactylosphaera vitifoliae*) or root-knot nematodes (*Meloidogyne* spp.), are present, and cultivars of susceptible species such as *V. vinifera* are to be grown, it is necessary to graft or bud onto a resistant rootstock.

Root-knot nematodes can be eradicated from grapevine rootings by dipping them in hot water (51.5° to 54.5°C; 125° to 130°F) for five to three minutes, respectively (*128*).

**Seeds**     Grape seeds are not difficult to germinate. Best results with vinifera grape seeds are obtained after a moist stratification period at 0.5° to 4°C (33° to 40°F) for about 12 weeks before planting (*87*).

**Dormant cuttings**    Most grape cultivars are traditionally propagated by dormant hardwood cuttings, which root readily. Cutting material should be collected during the winter from healthy, vigorous, mature vines. Well-developed current season's canes should be used; they should be of medium size and have moderately short internodes. Cuttings 8 to 13 mm (⅓ to ½ in.) in diameter and 36 to 46 cm (14 to 18 in.) long are generally used, planted in the spring deep enough to cover all but one bud. One season's growth in the nursery should produce plants large enough to transplant to the vineyard.

Root-promoting auxin-type chemicals have not been particularly helpful in rooting hardwood grape cuttings (*85*).

**Leafy cuttings**    Leafy greenwood grape cuttings root profusely under mist in about ten days if given relatively high (26.5° to 29.5°C; 80° to 85°F) bottom heat and if treated with indolebutyric acid. Scarce planting stock (such as virus-indexed material) can be increased very rapidly by using one-budded stem cuttings (see Figure 18–2), then consecutively taking additional cuttings from the shoot arising from the bud on the rooted cutting, and so on.

**Layers**    Grape cultivars difficult to start by cuttings can be propagated by simple, trench, or mound layering. (See chapter 14.)

**Grafting**    Bench grafting (*88*) is widely used; scions are grafted on either rooted or unrooted disbudded rootstock cuttings by the whip graft or, better, by machine grafting (see Figures 12–22 to 12–24). The grafts are made in late winter or early spring from completely dormant scion and stock material. The stocks are cut to 30.5 to 35.5 cm (12 to 14 in.) with the lower cut just below a node and the top cut an inch or more above a node. All buds are removed from the stock to prevent subsequent suckering. Scion wood should have the same diameter as the stock.

After grafting, using a one-bud scion, the union is stapled together or wrapped with budding rubber. The grafts should be held for three to four weeks in well-aerated, moist wood shavings or peat moss at about 26.5°C (80°F) for callusing.

The callused grafts are removed from the callusing boxes or plastic bags and any roots and the scion shoot are carefully trimmed back to an 18 mm (½ in.) stub. The scion is dipped quickly into a temperature controlled container of melted (low melting point) paraffin to a depth of 2.5 cm (1 in.) below the graft union and then quickly into cool water. The paraffined bench graft is then planted into a 5.0 × 5.0 × 25 cm

**Figure 18–2**    Leafy, one-node grape stem cutting rooted under mist. In the grape, roots arise readily from the internodes.

**Figure 18-3** Steps in grafting a grape cultivar onto a resistant rootstock. *Left:* a one-budded scion grafted on an unrooted rootstock cutting using a French grafting device. *Center:* after callusing, followed by paraffining scion and graft union, the stock is planted in a tube of felt building paper (or a plastic-coated cardboard cylinder) and then set in a protected place, such as a greenhouse or lathhouse. *Right:* several weeks later the scion bud starts growth. Graft union has healed and the rootstock cutting is well rooted, so that after seven to ten days hardening-off, the plant can be set out (still in the tube) directly in the vineyard.

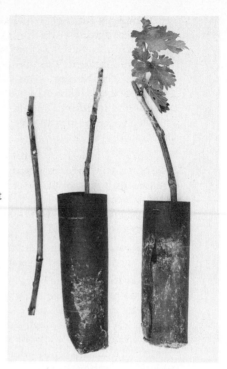

(2 × 2 × 10 in.) milk carton or planting tube, which contains a mixture of perlite and pumice, or perlite and peat moss (see Figure 18-3). The cartons or tubes are placed upright in flats. The flats are set on pallets and moved into a heated greenhouse for six to eight weeks. The growing bench grafts are then transferred to a 50 percent shade screen house for about two weeks for hardening off prior to planting in the nursery or vineyard.

The bench grafts are planted deeply so that the top of the carton or planting tube is at least 2 to 3 inches below the soil level to insure that water will get inside the carton or tube. At no time, however, should bench grafts be planted in the vineyard with the graft union at or below ground level.

**Greenwood grafting** Greenwood grafting is a simple and rapid procedure for propagating vinifera grapes on resistant rootstocks (*86*). A one-budded greenwood scion is splice-grafted during the active growing season on new growth arising either from a one-year-old rooted cutting or from a cutting during midseason of the second year's growth.

**Budding** (*131*) A satisfactory method of establishing grape cultivars on resistant rootstocks is by field budding on rapidly growing, well-rooted cuttings that were planted in their permanent vineyard location the previous winter or spring. T-budding in late spring can be done using dormant budwood held under refrigeration. Shortly after budding, the trunks should be cut with diagonal slashes at the base to allow the "bleeding" to take place there rather than where the bud was inserted (*4*). In another method chip budding is performed in late summer or early fall just as soon as fresh mature buds from wood with light brown bark can be obtained and before the stock begins to go dormant. In areas where mature buds cannot be obtained early in the fall, growers may store under refrigeration budsticks collected in the winter and bud them in late spring or early summer.

The bud is inserted in the stock 5 to 10 cm (2 to 4 in.) above the soil level, preferably on the side adjacent to the supporting stake. It is tied in place with budding rubber but is not waxed. The bud is then covered with 13 to 25 cm (5 to 10 in.) of well-pulverized, *moist* soil to prevent drying. In areas of extremely hot summers, or in soils of low moisture, variable results are likely to be obtained, and bench or nursery grafted vines should be used. If the buds are tied with white 13 mm (½ in.) plastic tape, it is unnecessary to mound them over with soil.

**Top-grafting grapevines**     Two methods can be used for changing cultivars of mature grapevines: By one method the tops of the vines are cut off in early spring 30 to 53 cm (12 to 21 in.) below the lower wire and side whip-grafted, using a two-bud scion of the desired cultivar. The scions are wrapped with one-inch white plastic tape and the cut surfaces are covered with grafting wax.

After the bark "slips" in late spring, the vines may be T-budded with the inverted T wrapped with white plastic tape. No grafting compound is needed. Both methods give highly successful takes.

**Rootstocks for American grapes** ( *135, 160* )

*'Salt Creek' (Ramsay), 'Dog Ridge', 'Champanel', 'Lukfata' (V. champini)*     These stocks have been very effective in the southern coastal region of the United States in increasing yields and prolonging vine life of the "bunch" type of grapes.

*V. rupestris 'Constantia', 'Couderc 3309', 'V. cordifolia × V. riparia 125-1', 'Cynthiana', 'Wine King', 'Lenoir'*     These stocks have been useful in the inland states of the United States in increasing yields, plant vigor, and evenness of fruit ripening.

*'Couderc 3309', 'SO–4' ('Oppenheim #4'), 'Teleki 5–A', 'Couderc 1616', 'Mal gue 44–53',* and *'Castel 188–15'*     These stocks are used in New York State for their resistance to root parasites, especially in replant situations.

**Rootstocks for vinifera grapes** ( *129, 130, 160* )

*'St. George' (V. rupestris)*     This vigorous, phylloxera-resistant stock, noted for its drought tolerance, is especially suitable for shallow, nonirrigated soils; it is not resistant to root-knot nematodes. It is readily propagated by cuttings and easily grafted. Because of high vigor and low vine productiveness with this stock, it is planted less frequently than in the past. Its best use is in replant situations where phylloxera has built up in the soil.

*'Ganzin No. 1' ('A × R. #1')*     This phylloxera-resistant rootstock is highly recommended for fertile, irrigated soils. Vines on this stock under such conditions are generally more productive than those on *'St. George'*. It is susceptible to root-knot nematodes and does not do well on dry, hillside soils. Cuttings root easily but it does not bench graft well, as it calluses poorly.

*'Couderc 1613'*     This stock is the one most widely used in the interior valley grape growing region of California. It is resistant to root-knot nematodes. Although suitable for fertile, irrigated, sandy loam soils, it produces weak, unproductive vines in nonirrigated or sandy soils of low fertility. Cuttings are easily rooted.

*'Harmony'*     This rootstock, developed by the USDA, is a cross between a *'Dog Ridge'* seedling and a *'1613'* seedling. It has more vigor and more phylloxera and nematode resistance than *'1613'*. Cuttings root readily and it buds and grafts easily. It is recommended as a replacement for *'1613'*, especially for table grape vineyards.

*'Freedom'*     This stock originated as a sister seedling of *'Harmony'*. It is very vigorous, equal to *'Salt Creek'*. It has the same resistance to phylloxera and nematodes as *'Harmony'*, but cuttings root more easily and grow more uniformly and vigorously. It

buds and grafts easily. It should be used as a replacement for *'Couderc 1613'* where high vigor is desired.

*'Dog Ridge' and 'Salt Creek' ('Ramsay') (V. champini)*    These closely related stocks are resistant to phylloxera and to root-knot nematodes and are extremely vigorous. They should be used only in low-fertility and sandy soils. In fertile soils, the vines are often so vigorous that they are unproductive. Cuttings of these stocks are difficult to root, especially *'Salt Creek'*. On better soils, *'Salt Creek'* is usually preferred to *'Dog Ridge'* because the latter's extreme vigor causes poor fruit set. These two stocks have performed best for raisin and wine grape vineyards in sandy soils.

Poorly growing vinifera grapevines on infertile soil can be invigorated by inarching with rooted cuttings of one of these cultivars—'Dog Ridge' for very sandy soils or 'Salt Creek' for sandy soils.

**Grapefruit**    *See* Citrus.

**Guava** (*Psidium cattleianum* Sabine—the Cattley or strawberry guava—and *Psidium guajava* L.—the common, tropical, or lemon guava) (*136*)    The Cattley guava has no named cultivars, and most nursery plants are propagated by seed. This species comes nearly true from seed, large-fruited, superior trees being used as the seed source. It is difficult to propagate by vegetative methods.

The common guava has several cultivars, but they are not grown extensively, because of difficulties in asexual propagation. Most trees of this species are propagated by seeds, which germinate easily and in high percentages, but do not come true to type. Seeds should be taken from the best type of tree available, and the flowers to produce the seeds should be self-pollinated, which will reduce seedling variability.

Seedlings are somewhat susceptible to damping-off organisms, and should be started in sterilized soil or otherwise protected by fungicides. When about 4 cm (1½ in.) high, the seedlings should be transplanted into individual containers. In six months the plants should be about 30 cm (12 in.) high and can be transplanted to their permanent location.

For large-scale propagation of guava cultivars, grafting or budding is necessary. Chip budding has been successful, done any time during the summer using greenwood buds from selected cultivars inserted into seedling stocks about 5 mm in thickness. Plastic wrapping tape is used to cover the buds. The stock is cut off above the inserted bud after about three weeks (*111*).

Cuttings are difficult to root, best results being obtained by rooting succulent shoots under mist after treating them with indolebutyric acid (*172*).

Air layering gives good results. Simple and mound layering are effective methods of starting new plants; the layers may be tightly wrapped with wire just below the point where roots are wanted, or can be ringed, with indolebutyric acid (500 ppm) in lanolin rubbed into the cuts to promote rooting (*143*).

**Hickory** (*Carya ovata* [Mill] C. Koch—Shagbark hickory)    Hickory cultivars are propagated by grafting or budding on seedlings of the Shagbark hickory (*C. ovata*), or of the pecan (*C. illinoensis*) (*140*). However, owing to grafting and transplanting difficulties and slow growth of the trees, there is little planting of grafted trees. Most trees grown are seedlings.

To avoid transplanting failure caused by the long taproot, several stratified nuts are planted in the spring where the trees are to be located permanently. The best tree is saved and topworked, usually by bark grafting, to the desired cultivar, although standards for cultivars have not been well established. If grafted nursery trees are to be used, the taproot should be cut a year previous to digging or the tree transplanted in the nursery once or twice to force out lateral roots.

Hickory seeds will germinate when planted in the spring without previous stratification but should be kept in moist, cool storage until planting. Fall planting is successful if the soil in cold climates is well mulched to prevent excessive freezing and thawing.

Patch budding is used by nurserymen in the commercial propagation of hickories, usually in late summer. The seedling stocks are grown for two years or more before they are large enough to bud.

**Jujube** (*Zizyphus jujuba Mill.*)—Chinese Date     The leading cultivars—'Lang' and 'Li'—are propagated by budding or grafting on jujube seedlings. The seeds should be stratified at about 4°C (40°F) for several months before planting. Jujubes may be propagated also by root cuttings or hardwood stem cuttings.

**Kiwifruit** (*Actinidia chinensis* Planch.) (*17, 57, 148, 168*) Chinese Gooseberry     Both male and female vines of this dioecious subtropical fruit must be planted to ensure fruiting. It is sometimes propagated by budding or grafting cultivars on seedling rootstocks, but it is mostly propagated by leafy, semihardwood cuttings under mist. Hardwood cuttings will root. New plants have also been obtained with in vitro culture techniques using root or stem explants (*83*).

**Seed**     Seed propagation can be used but the sex of the vine cannot be determined until fruiting at seven years or more. Select seed from soft, well-ripened fruit; dry and store at 5°C (41°F). After at least two weeks at this temperature, subject seed to fluctuating temperatures—10°C (50°F) night and 20°C (68°F) day—for two or three weeks before planting.

**Cuttings**     Leafy semihardwood cuttings taken from late spring to midsummer may be rooted under mist when treated with an 0.8 percent indolebutyric acid in talc powder preparation. Dipping the bases of the cuttings in captan is helpful. Hardwood cuttings, taken in mid-winter and planted in a greenhouse, can be rooted if treated with a strong concentration of a rooting hormone. Plants can be started also by root cuttings.

**Grafting and budding**     Seedlings can be grafted successfully by the whip graft in late winter using dormant scion wood. T-budding in late summer is also successful. Plants grafted on seedlings are believed by some to be more vigorous than those started as rooted cuttings.

**Kola** (*Cola nitida* Vent.)     Propagation of the kola, the tropical caffein-containing nut used to make refreshing beverages, is largely by seed planted directly in the field or in nurseries for later field planting. Seeds for planting should be harvested only when completely mature. A pronounced juvenility pattern occurs in kola; seven to nine years are required for the seedlings to flower so that asexual propagation using mature growth is essential for early production.

Rooting of terminal leafy cuttings in polyethylene-covered frames can be done but root-promoting hormones have not been successful (*231*). Air layering is quite successful in obtaining early bearing, commercial production taking place 12 to 18 months after planting the rooted layers. IBA treatments aid in rooting the layers. Patch budding is also a successful method of vegetative propagation (*7*).

**Lemon**     *See* Citrus.

**Lime**     *See* Citrus.

**Loganberry**     *See* Blackberry.

**Loquat** (*Eriobotrya japonica* [Thunb.] Lindl.)    This is propagated by T-budding or side grafting superior cultivars on loquat seedlings or by top-grafting older, established trees, using the cleft graft. The quince can be used as a rootstock, producing a dwarfed tree. Air layering is successful; indolebutyric acid in lanolin at 250 ppm rubbed into the layered surface will increase rooting (*206*). For ornamental use, seedling trees are satisfactory.

**Litchi** (*Litchi chinensis* Sonn.)    Seed-produced litchi plants are inferior, so the many cultivars of this species are propagated by asexual methods, principally air layering. This can be done at any time of the year, but with best results in spring or summer (*136*). Large limbs air layer easier than small ones.

Tip cuttings from a flush of growth in the spring have been rooted in fairly high percentages under mist in the full sun. Root-promoting chemicals are beneficial. Hardwood cuttings from an active flush of new growth have rooted more readily than those from dormant hardwood cuttings.

Litchi seeds germinate in two to three weeks if planted immediately upon removal from the fruit, but they lose their viability in a few days if not planted. Seedling trees are rarely grown for their fruit; they require 10 to 15 years to start bearing, and the fruit is likely to be of low quality (*30*).

**Macadamia** (*Macadamia integrifolia* Maiden & Betche, and *M. tetraphylla* L. Johnson) (*81, 189*)    Trees of this subtropical evergreen nut tree may be grown as seedlings, but vegetative propagation from a few superior selections by grafting is usually practiced. Macadamia is highly resistant to *Phytophthora cinnamomi* and will tolerate heavy clay soil.

Fresh seeds should be planted in the fall as soon as they mature, either directly in the nursery or in sand boxes in a lathhouse, then transplanted to the nursery or into polyethylene containers (see Figure 2–16) after the seedlings are 10 to 15 cm (4 to 6 in.) tall. It is important not to crack the seeds, because they are readily attacked by fungi. Only seeds that sink when placed in water should be used (*81*). Seeds retain viability for about 12 months at 4° C (40°F), but at room temperature viability starts decreasing after about four months. Scarifying or soaking the seed in hot water hastens germination.

Selected clones are propagated by one of the side graft methods. Leaves should be retained for a time on the rootstock. Rapid healing of the union is promoted if the rootstock is checked in growth prior to grafting by water or nitrogen deficiency to permit carbohydrate accumulation. Also, ringing the branches that are to be the source of the scions several weeks before they are taken increases their carbohydrate content and promotes healing of the union. Budding has generally been unsuccessful (*50*). *M. tetraphylla* is the preferred rootstock for *M. integrifolia* cultivars (*81, 217*).

Macadamia can be propagated by rooting leafy, semi-hardwood cuttings of mature, current season's growth; 3- to 4-in. tip cuttings are best. Treatment with indolebutyric acid at 8000 to 10,000 ppm is beneficial. Cuttings should be placed in a closed propagating frame or under intermittent mist for rooting. Bottom heat at 24° C (75° F) is beneficial. There are pronounced cultivar differences in ease of rooting. *M. tetraphylla* cuttings root more readily than those of *M. integrifolia*. Rooted cuttings are not widely used in commercial plantings because of their shallow root system.

**Mandarin**    *See* Citrus.

**Mango** (*Mangifera indica* L.) (*27, 207*)    Most plantings of this evergreen tropical fruit are seedlings, although many superior selections are maintained vegetatively. Polyembryonic cultivars commonly occur in mango. The seedlings may be sexual or nucellar in origin, either condition occurring in the seed, or both conditions occurring simultaneously. However, growth of several shoots from one seed does not necessarily indicate the presence of nucellar embryos, since in certain cultivars shoots develop from below ground, arising in the axils of the cotyledons of one embryo, which may or may not be of zygotic origin (6). Monoembryonic cultivars should not be propagated by seed, as they do not come true.

Mango seeds are used either to produce a true-to-type nucellar seedling of some superior clone or a rootstock on which the desired clone is budded or grafted. The seeds should be planted as soon as they are mature, although they can be stored in the fruits or in polyethylene bags at about 21° C (70° F) for at least two months. Low-temperature (below 10° C; 50° F) storage and excessive drying should be avoided. Removing the tough endocarp that surrounds the seed, followed by planting in a sterilized medium, should result in good germination in two to three weeks. Soon after the seedlings start to grow, they should be transplanted to pots or the nursery.

Mangos are commonly propagated in Florida by a type of veneer grafting or by chip budding. A week after budding, the stock is cut off two to three nodes above the bud, with a final removal to the bud when the bud shoot is 7.5 to 10 cm (3 to 4 in.) long. The best budwood is prepared from hardened terminal growth 6 to 10 mm ($\frac{1}{4}$ to $\frac{3}{8}$ in.) in diameter. The leaves are removed, with the exception of two or three terminal ones. The buds swell in two to three weeks, and are then ready to use. If the buds are to be used on stocks older than three weeks, ringing the base of the shoots from which the buds are to be taken about ten days before they are used increases their carbohydrate supply and seems to promote healing. Budding is best done when the rootstock seedlings are two to three weeks old—in the succulent red stage. Four to six weeks after budding, the inserted bud should start growth (136). T-budding also has been successful.

Approach grafting, termed "inarching" in India, has been used since ancient times in propagating the mango. Veneer grafting also is successful (155), as is saddle grafting (223). Rootstocks are mostly from assorted seedlings, although selected monoembryonic rootstocks can be multiplied by air layering or by cuttings, and subsequently clonally propagated by stooling (154).

Air layering is successful, especially when etiolated shoots are treated with indolebutyric acid at 10,000 ppm; such treatments have given 100 percent rooting with 90 percent survival (156). The mango is difficult to propagate by cuttings, but it can be done by using leafy cuttings under mist along with indolebutyric acid treatments.

**Mulberry** (*Morus* spp.)    The mulberry is readily started by hardwood cuttings 20 to 30 cm (8 to 12 in.) long, made from wood of the previous season's growth and planted in early spring.

**Nectarine** (*Prunus persica*)    *See* Peach.

**Olive** (*Olea europaea* L.) (97)    Olives are propagated by budding or grafting on seedling or clonal rootstocks, by hardwood or semihardwood cuttings, or by suckers from old trees.

Seeds of small-fruited cultivars germinate more easily than those of large-fruited ones. Often germination is prolonged over one or two years. The usual practice is to plant many more seeds than will be needed as seedlings to offset the low germination percentage. The seed is enclosed in a hard endocarp or "pit." Removing, clipping, or cracking this endocarp materially hastens germination which, however, is still slow and erratic.

Removing the seed from the pit and germinating the seed (without the seed coats) in petri dishes on moist filter paper will give high germination percentages if the fruits are collected shortly after pit hardening.

The seedlings grow slowly and may take a year or two to become large enough to be grafted or budded. Several methods have been used successfully, including T-budding, patch budding, whip grafting, and side-tongue grafting. In Italy, a widely used method is to bark graft small seedlings in the nursery row in the spring. The stocks are cut off several inches above ground, and one small scion is inserted in each seedling, followed by tying and waxing. After grafting or budding, one or two more years are required before a tree large enough for transplanting to the orchard is produced.

Hardwood cuttings may be made from two- or three-year-old wood about an inch in diameter and 20 to 30 cm (8 to 12 in.) long. All leaves are removed. It is often helpful to soak the basal ends of the cuttings in a solution of indolebutyric acid (15 ppm for 24 hours), followed by storage in moist sawdust at 15° to 21° C (60° to 70° F) for a month preceding spring planting in the nursery (*90*).

Semihardwood cuttings from vigorous, one-year-old wood about 6 mm (¼ in.) in diameter can be successfully rooted. They should be 10 to 15 cm (4 to 6 in.) long with two to six leaves retained on the upper portion of the cutting. It is best to take the cuttings in early summer or midsummer and root them under high humidity (*96*). Olive cuttings root well under intermittent mist and respond markedly to treatment with indolebutyric acid at about 4000 ppm for five seconds. There is considerable variability among cultivars in ease of rooting.

**Rootstocks for olives**    *Olea europaea* seedlings are often used with budding or grafting, although considerable variation in tree vigor and size many result from this practice (*92*). A more suitable rootstock, giving uniform trees, is rooted cuttings of a strong-growing cultivar. Olive trees are very susceptible to *Verticillium* wilt; where this occurs resistant clonal rootstocks should be used (*100*). Other *Olea* species do not make satisfactory rootstocks for the edible olive.

## Orange    *See* Citrus.

**Papaya** (*Carica papaya* L.) (*84, 222*)    Propagation by seed is the most practical method. They can be sown in flats of soil or in seedbeds in the open with germination in two to three weeks. The seeds do equally well if taken from stored or fresh fruit. Seeds can be stored for up to six years at 5°C (41°F) in sealed, moisture-proof bags without losing viability (*9*). The best seeds are from controlled pollinations between superior trees. When the seedlings are about 10 cm (4 in.) tall they are transplanted. This is generally done once or twice before they are put in their permanent location.

A method of starting seedlings without disturbing the roots is to plant four to eight seeds in a container and then thin to two to four of the strongest when they are about 10 cm (4 in.) tall. They can then be set in the field without disturbing the root system. Young papaya seedlings are very susceptible to damping-off organisms, so the soil in which they are started should be pasteurized if possible.

In Florida, the usual practice is to plant seeds in midwinter and set the young plants in the field by early spring. They grow during spring and summer, and usually mature their first fruits by fall, the plants bearing all winter and the following season.

Papaya trees can also be produced by budding (*213*), by in vitro micropropagation (*133*), and by cuttings (*228*). Papaya is sometimes called pawpaw, but it is not the native American pawpaw (*68*).

**Passion Fruit** (*Passiflora edulis* Sims.) (*51*)     This subtropical tender evergreen fruit is propagated chiefly by grafting onto seedling roots. Seeds germinate two to three weeks after planting. To avoid attacks of fusarium wilt, which is a serious problem, it is necessary to cleft-graft the purple-fruited hybrid fruiting cultivars onto seedlings of golden passion fruit, *P. edulis* forma *flavicarpa* (*221*), which are resistant to this disease and to nematodes.

**Peach and Nectarine** (*Prunus persica* Batsch) (*58, 236*)     Nursery trees of these fruits are propagated by T-budding or chip budding on seedling rootstocks. Fall budding is most common, but spring budding or—in regions with long growing seasons—June budding also is done.

Some peach cultivars can be propagated by leafy, succulent softwood cuttings taken in spring or summer, treated with a root-promoting material, and rooted in a mist propagating bed (*38, 95*). Also, in areas with mild winters, some peach cultivars can be started from hardwood cuttings if they are treated with indolebutyric acid (4000 ppm for 5 sec), then set out in the nursery in the fall (*82*). Propagation of peaches by cuttings has not been done commercially but may be in the future where high density tree production systems, requiring a large number of inexpensive nursery trees, are used (*53*).

**Rootstocks for peach** (*125, 165, 204*)     Most peach cultivars are propagated on peach seedlings. Apricot and almond seedlings are sometimes used, as well as (in Europe) 'Brompton' and 'Common Mussel' plum clones. Peach seeds must be stratified at about 4° C (40°F) for three to four months, and almond and apricot for four weeks, before they are planted.

*Peach (P. persica)*     Peach seedlings are the most satisfactory rootstock for peach and should be used unless certain special conditions warrant other stocks. Seeds of 'Elberta', 'Halford', 'Lovell', or 'Rutgers Red Leaf' are usually used, since they germinate well and produce vigorous seedlings. Seedlings of these cultivars are not resistant to root-knot nematodes, however, and where this is a problem, resistant peach stocks should be considered. Seeds from peach cultivars whose fruits mature early in the season should not be used because their germination percentage is usually low. It is best to obtain seeds from the current season's crops, since viability decreases with each year of storage. Both 'Lovell' and 'Nemaguard' seeds from trees free of known viruses are available and should be used where possible.

Peach roots are quite susceptible to the root-knot nematode (*Meloidogyne* spp.), especially in sandy soils. They are also susceptible to the root-lesion nematode, *Pratylenchus vulnus*. 'Nemaguard', a *Prunus persica* × *P. davidiana* hybrid rootstock introduced by the USDA in 1959, produces seedlings that are uniform and resistant to both *M. incognita* and *M. javanica,* and which form strong, well-anchored trees, although certain peach and plum cultivars have not done well on it and it has shown susceptibility to bacterial canker (*204*) and to crown rot. 'Nemaguard' is not hardy in the colder peach growing areas.

A very winter-hardy peach rootstock, 'Siberian C', withstanding −11° C (12°F) soil temperatures, was introduced in 1967 by the Canadian Department of Agriculture Research Station at Harrow, Ontario. It does well on light, sandy soils and will give about 15 percent dwarfing to the scion cultivar. It is, however, susceptible to root-knot and root-lesion nematodes (*125*).

Nursery trees on peach roots often make unsatisfactory growth when planted on soils previously planted to peach trees. Peach roots are susceptible to oak root fungus, crown rot, crown gall, and verticillium wilt.

*Apricot (P. armeniaca)*     Apricot seedlings are occasionally used as a rootstock for the peach. The graft union is not always successful, but numerous trees and commercial orchards of this combination have produced fairly well for many years. Seedlings of the 'Blenheim' apricot seem to make better rootstocks for peaches than those of 'Tilton'. The apricot root is highly resistant to root-knot but not to root-lesion nematodes.

*Almond (P. amygdalus)*    Almond seedlings have been used with limited success as a rootstock for peaches. There are trees of this combination growing well, but in general this is not a satisfactory combination. The trees are often dwarfed and tend to be short-lived.

*Western sand cherry (P. besseyi)*    When this stock was used in limited experiments as a dwarfing rootstock for several peach cultivars, it was found that although the bud unions were excellent, about 40 percent of the nursery trees failed to survive. The remainder grew well, however, and developed into typical dwarf trees with healthy, dark green foliage. The trees bore normal-size fruit in the second or third year after transplanting to the orchard (*15*).

*Nanking cherry (P. tomentosa)*    This may be suitable for some peach cultivars as a dwarfing rootstock.

*'Brompton' and 'St. Julien A' plum*    In England these rootstocks have been compatible with all peach and nectarine cultivars worked on them, and produce medium to large trees. Peach seedlings are rarely used as peach rootstocks in England.

**Pear** (*Pyrus communis* L.)    Pears are propagated by fall budding, using T-budding or chip budding on either seedling pear rootstocks or rooted quince cuttings. Pear trees are also started by whole-root grafting, using the whip or tongue method. Pear seeds must be stratified for 60 to 100 days at about 4° C (40° F). They are then planted thickly about 13 mm (1/2 in.) deep in a seedbed, where  they are allowed to grow during one season. The following spring they are dug, the roots and top are cut back, and then they are transplanted to the nursery row, where they are grown a second season, ready for budding in the fall.

Some pear cultivars, such as 'Old Home' and 'Bartlett', can be propagated by hardwood cuttings or by leafy cuttings under mist if treated with indolebutyric acid (*94, 98, 240*). In pear rootstock test plantings, own-rooted 'Bartlett' trees have shown excellent production with large, well-shaped fruit; they are resistant to pear decline (*74, 98*) and with age become partially dwarfed, a desirable attribute for high-density plantings.

**Rootstocks for pear** (*73, 74, 89, 239, 241*)    In using rootstocks from seedlings of one of the several *Pyrus* species available, the seed source is very important if a rootstock of a certain species is expected. The various *Pyrus* species hybridize freely, some bloom at the same time, and cross-pollination is necessary for seeds to develop. It is best to use as the seed source an isolated group of trees of a given known species and to avoid collecting seeds from a planting containing trees of several different species, seeds that are likely to produce hybrid seedlings.

"Pear decline," caused by mycoplasmalike bodies (*106*) known to be spread by pear psylla (*Psylla pyricola*), devastated pear orchards, first in Italy (*202*) and later in western North America. Pear decline is associated wih the rootstock used (*10, 13, 194*) so it is an important factor to be considered in rootstock selection. For plantings in decline-prevalent areas only rootstocks known to be resistant to decline should be used.

Seedlings from cultivars that may have resulted from pollination by a decline-susceptible species, such as *Pyrus pyrifolia* or *P. ussuriensis,* should not be used. Where pear psylla exist, and where they may be carrying the mycoplasmalike bodies causing pear decline, seedlings of the Oriental pears—*P. pyrifolia* or *P. ussuriensis*—should *not* be used, as trees with these as rootstocks are highly susceptible to decline.

*French pear (P. communis)*    French pear seedlings are generally grown from seeds of 'Winter Nelis' or 'Bartlett'. This rootstock is moderately vigorous and winter-hardy, produces moderately productive, uniform trees with a strong, well-anchored root system, and is resistant to pear decline. It forms an excellent graft union with all pear cultivars and will tolerate relatively wet (but not waterlogged) and heavy soils. French pear roots are resistant to verticillium wilt, oak root fungus (*Armillaria mellea*), root-knot and root-lesion nematodes, and crown gall.

The two serious defects of French pear roots are their susceptiblity to pear root aphid, *Eriosoma pyricola,* and to fire blight, *Erwinia amylovora*. Millions of pear trees on French pear roots have died from fire blight, owing to the high susceptiblity of this stock.

Seedlings of the 'Kieffer' pear—a hybrid between *P. communis* and *P. pyrifolia*— have been satisfactorily used for many years in Australia as a pear rootstock (*32*).

*Blight-resistant French pear rootstocks*    Pear cultivars resistant to blight, such as 'Old Home' (*72, 98*) and 'Farmingdale' (*183*), have been used as an intermediate stock grafted on seedling rootstocks. Topworking with the desired cultivar takes place after the trunk and primary scaffold branches develop. If a blight attack occurs in the top of the tree, it will stop at the resistant body stock, which can subsequently be regrafted after the blight has been cut out.

'Old Home' × 'Farmingdale' crosses have been made in Oregon and clonal blight-resistant pear rootstocks have been developed, some of which give dwarfed trees (*239*).

*Japanese Pear [P. pyrifolia (P. serotina)]*    This was widely used in the United States as a pear rootstock from about 1900 to 1925, but is no longer considered of value, owing to its high susceptibility to pear decline and to the physiological defect, "black-end" or "hard-end," which may occur in 'Bartlett', 'Anjou', 'Winter Nelis', and other cultivars when they are propagated on it (*103*). Fruits affected with black-end are unusable because of hardening and cracking of the flesh at the blossom end, often with the development of blackened areas.

*P. calleryana*    This stock is blight resistant and produces vigorous trees with a strong graft union; fruit quality is good, with no black-end. 'Bartlett' orchards with this rootstock have produced well in California, and it is popular in the southern part of the United States for 'Kieffer' and other hybrid pears. In areas with cold winters it lacks winter-hardiness. Where good pear psylla control has been practiced, trees on *P. calleryana* roots are resistant to pear decline. However, trees with *P. calleryana* roots show less resistance to oak root fungus than those with *P. communis* roots. Seedlings of a selection of this species, known as 'D–6', are widely used as a pear rootstock in Australia (*71*).

*P. ussuriensis*    This Oriental stock has been used in the past to some extent; many pear cultivars on this root develop black-end, although not to the extent found with *P. pyrifolia* roots. It also produces small trees that are susceptible to pear decline, and should not be used in areas where this may occur.

*P. betulaefolia*    This species has vigorous seedlings, resistance to leaf spot and pear root aphid, tolerance to alkali soils, an adaptability to a wide range of climatic conditions, good resistance to pear decline, and produces large high-yielding trees.

*Quince (Cydonia oblonga) (242)*    This species has been used for centuries as a dwarfing stock for pear and is now universally used as a pear rootstock in England. Some cultivars, however, fail to make a strong union directly on the quince, hence double-working, with an intermediate stock such as 'Old Home', or 'Hardy', is necessary. Cultivars that require such a compatible interstock when worked on quince roots include 'Bartlett', 'Bosc', 'Winter Nelis', 'Seckel', 'Easter', 'Clairgeau',

'Guyot', 'Clapp's Favorite', 'Farmingdale', and 'El Dorado'. A selection of 'Bartlett' compatible with 'Quince A' rootstock originated in Switzerland, presumably as a bud mutation from an incompatible form (*177*). The following pears appear to be compatible when worked directly on quince: 'Anjou', 'Old Home', 'Hardy', 'Packham's Triumph', 'Gorham', 'Comice', 'Flemish Beauty', 'Duchess', and 'Maxine'.

Quince roots are resistant to pear root aphids and nematodes, but are susceptible to oak root fungus, fire blight, and excess lime, and are not winter-hardy in areas where extremely low temperatures occur. In some areas trees on quince roots have developed pear decline, but in others they have not, possibly because different quince stocks were used. The black-end trouble has not developed with pears on quince roots. There are a number of quince cultivars, most of which are easily propagated by hardwood cuttings or layering. 'Angers' quince is commonly used as a pear rootstock because its cuttings root readily, it grows vigorously in the nursery, and does well in the orchard. 'Provence' quince roots produce larger pear trees than 'Angers' and are more winter hardy.

The East Malling Research Station has selected several clones of quince suitable as pear rootstocks and designated them as 'Quince A', 'B', and 'C'. 'Quince A' ('Angers') has proved to be the most satisfactory stock. 'Quince B' (Common quince) is somewhat dwarfing, whereas 'Quince C' produces very dwarfed trees. It is important to use only virus-tested quince stock.

Rootstocks recommended for Bartlett pear orchards in California are: *P. betulaefolia* seedlings, own-rooted 'Old Home', *P. communis* 'Winter Nelis' seedlings, as well as own-rooted 'Bartlett' (no rootstock) (*74*).

**Pecan** (*Carya illinoensis* Wangenh. [*K. Koch*]) (*134, 141, 186*)    Pecans are propagated by budding or grafting selected cultivars on pecan seedling rootstocks (*19*). Hardwood cuttings can be rooted. They should be taken in the fall and treated with about 10,000 ppm IBA. Long (50 cm), thick (20 mm) cuttings root best (*210*). Plants have been started also by root cuttings and by trench and air layering, the latter with the aid of indolebutyric acid treatments.

Pecan seeds sometimes start growing in the hulls even before the nuts are harvested. The seeds lose their viability, however, in warm, dry storage. To prevent this loss, they should be stored at 0°C (32°F) immediately after harvest and until planted. Pecan seeds should be given a stratification treatment for 12 to 16 weeks at about 1° to 5°C (34° to 41°F) to ensure good, rapid germination (*142*), or else germination should take place at high temperatures—30° to 35°C (86° to 95°F)—to overcome growth restrictions of the shell (*48, 232*). Midwinter planting, with seedling emergence in the spring, is a successful procedure.

Only deep, well-drained, sandy soil should be used for growing pecan nursery trees. Young seedlings are tender and should be shaded against sunburn. In the summer, toward the end of the second growing season, the seedlings are large enough to bud to the desired cultivar. Patch budding, or sometimes ring budding, is the usual method employed. After the new top grows for one or two seasons, the nursery tree is ready to transplant to its permanent location. Young pecan trees have a long taproot, which must be handled carefully in digging and replanting.

Sometimes the seedlings are changed to the desired cultivar by crown grafting in late winter or early spring, using the whip graft method. Large pecan seedlings, growing in place in the orchard, can be topworked to the desired cultivar by bark grafting limbs 4 to 9 cm (1 ½ to 3 ½ in.) in diameter.

**Rootstocks for pecans**    Commercially, pecan cultivars are propagated on pecan (*C. illinoensis*) seedlings. Seeds from the 'Apache', 'Riverside', 'Elliott', and 'Curtis' cultivars are reported to produce excellent seedlings. As a possible rootstock for wet soils, seedlings of one of the hickory species, *C. aquatica,* have been used experimentally. Although pecan scions will grow on hickory species, the nuts generally do not attain normal size. Pecan roots are susceptible to verticillium wilt.

**Persimmon** (*Diospyros* spp.) (*69*)     **American persimmon** (*D. virginiana* L.) (*56*). The named cultivars of this species are commonly propagated by budding or grafting on *D. virginiana* seedlings. Root cuttings are also successful.

Seed germination is rather slow, owing to the seed's slow rate of water absorption. Fall sowing, or stratification at about 10° C (50° F) for 60 to 90 days, is advisable. In transplanting any but small seedlings, the taproot should be cut 30 cm (12 in.) below ground a year before moving to force out lateral roots.

**Oriental persimmon** (*D. kaki* L.)     The cultivars of this fruit are propagated by grafting on seedling rootstocks. Crown grafting by the whip method in early spring when both scion wood and stock are still dormant is a common practice, although bench grafting also is used. Budding can be done but it is less successful than grafting.

The usual practice in germinating seeds of both *D. lotus* and *D. kaki* is to stratify the seeds for 120 days at about 10°C (50°F). If the seeds have dried out, they should be soaked in warm water for two days before stratification. Excessive drying of the seed is harmful, especially for *D. kaki.* The seeds are planted either in flats or in the nursery row. Young persimmon seedlings require shading.

**Rootstocks for oriental persimmon** (*107*)

*Diospyros lotus*     This stock has been widely used in California; it is very vigorous and drought-resistant, and produces a rather fibrous type of root system, which transplants easily. This stock is quite susceptible to crown gall and verticillium and will not tolerate poorly drained soils but is highly resistant to oak root fungus. 'Hachiya' does not produce well on *D. lotus* stock, owing to the excessive shedding of fruit in all stages (*195*). 'Fuyu' scions usually do not form a good union with *D. lotus,* although 'Fuyu' topworked on a *D. kaki* cultivar established on *D. lotus* roots makes a satisfactory tree.

*D. kaki*     This stock is the one most favored in Japan and is probably best for general use, because it develops a good union with all cultivars and is resistant to crown gall and oak root fungus but susceptible to verticillium. Trees worked on it grow well and yield satisfactory commercial crops. The seedlings have a long taproot with few lateral roots, making transplanting somewhat difficult.

*D. virginiana*     Seedlings of this species are utilized in the southern part of the United States and seem to be adapted to a wide range of soil conditions. It has not proven satisfactory in some localities, however. In California, 'Hachiya' on this stock is distinctly dwarfed and yields poorly, owing to the sparse bloom (*195*). Most oriental cultivars make a good union with this stock, but there are diseases carried by *D. kaki* scions that will move into the *D. virginiana* roots, causing them to die. This stock seems to be quite tolerant either of drought or of excess soil moisture conditions. It produces a fibrous type of root system easy to transplant, but tending to sucker badly.

**Pineapple** (*Ananas comosus* Merr.) (*33*)     All commercial propagation of the pineapple is by asexual methods. As shown in Figure 18–4, there are three main types of planting material—*suckers, slips,* and *crowns.* Suckers coming from below ground are rarely used as planting material. Those above ground are cut from the mother plant about one

Crown

Slips

Peduncle

Suckers

Ground sucker

Main stem of mother plant

**Figure 18–4**     Parts of the pineapple plant showing the three major types of asexual planting material—crowns, suckers, and slips. In Hawaii slips are, by far, the type most commonly used. Courtesy Pineapple Research Institute of Hawaii.

month after the peak fruit harvest. Slips are taken from the plant two to three months following the peak harvest. When used in propagation, crowns are taken either before or at the time of harvest.

Slips are the most popular type of planting material for commercial use. More of these are produced by the plant than either suckers or crowns. After removal from the mother plant, slips can be stored for a relatively long time and still retain sufficient vigor for replanting. If the slips are all about the same size and age when planted, they will flower and fruit at approximately the same time.

Slips will produce fruit 18 to 20 months after planting, suckers at about 15 months, and crowns at about 22 months. Because of these differences, the three types of planting material are never mixed in one field.

Before planting, all types of planting material must be cured or dried for one to several weeks after they are cut from the mother plant. This allows a callus layer to develop over the cut surface, reducing losses from decay organisms after they are planted.

**Pistachio** (*Pistacia vera* L.) (*79, 115*)     The pistachio nut is usually propagated by T-budding on *Pistacia atlantica* or *P. terebinthus* seedlings as rootstocks. However, these stocks are highly susceptible to verticillium. Seedlings of *P. integerrima* show resistance to this soil-borne disease and are being used in areas having verticillium problems. Budding is possible over a considerable period of time, but if done before midspring, when sap flow may be excessive, the percentage of takes is apt to be low. A marked improvement in bud union occurs as budding is extended through the summer and fall. Best results are usually obtained by planting seedling rootstocks in their permanent orchard location, then T-budding them two feet or more above ground after the seedlings are well established. Seedlings are grown in long tubular, biodegradable pots ready for planting in the orchard.

To collect seeds for rootstocks, obtain fruits when the hulls turn blue-green. The hulls should be removed, as they apparently contain a germination inhibitor. *P. terebinthus* seed should be held in moist sand at 5° to 10° C (41° to 50° F) for six weeks before sowing, then germinated at 20° C (68° F) or slightly below. With early planting of seed and adequate irrigation and fertilization through the summer, the more vigorous seedlings may reach budding size by fall. Buds of *P. vera* cultivars are quite large, requiring a fairly large seedling to accommodate them. Because of their long taproot, bare-root nursery trees should be transplanted to their permanent planting site as early as possible.

**Plum** (*Prunus* spp.)     Plums are propagated by T-budding or chip budding in the fall on seedling rootstocks or, with certain stocks, on rooted cuttings or layers. Budding can also be done in the spring. Some plums can be propagated by hardwood cuttings (*94*) and some by leafy, softwood cuttings under intermittent mist (*95*). Bench grafting in winter and planting the grafts in the nursery in spring gives good results using the whip graft (*44*). Japanese plum (*P. salicina*) nursery trees have been produced by micropropagation methods (*188*).

**Rootstocks for plum** (*67, 102, 166*)     *Myrobalan plum (P. cerasifera).* This is the most widely used plum rootstock, being particularly desirable for the European plums, *P. domestica* (which includes the commercially important prune cultivars). It is also a satisfactory stock for the Japanese plums, *P. salicina*. Some kinds of plums—'President', 'Kelsey', 'Stanley', and 'Robe de Sergeant'—are not entirely compatible with this stock, however. Cultivars that are Japanese-American hybrids (*P. salicina* × *P. americana*) are best worked on American plum seedlings.

Myrobalan roots are adapted to a wide range of soil and climatic conditions. They will endure fairly heavy soils and excess moisture, and are resistant to crown rot but susceptible to root-knot nematodes and oak root fungus. They grow well on light sandy soils.

Myrobalan seeds require stratification for about three months at 2° to 4° C (36° to 40° F). They may then be planted thickly in a seedbed for one season, then transplanted to the nursery and grown for a second season before budding, or the seeds may be planted directly in the nursery row and grown there for one season, the seedlings being budded in late summer or fall.

Certain very vigorous myrobalan selections are propagated by hardwood cuttings. One of these, 'Myro C', is immune to root-knot nematodes. In England a selection, 'Myrobalan B', was developed at the East Malling Research Station and is propagated by hardwood cuttings. It is particularly valuable in producing vigorous trees, although there are cultivars not completely compatible with it (*67*). Most plum cultivars in England belong to the *P. domestica* species.

*Marianna plum (P. cerasifera* × *P. munsoniana?)*     This is a clonal rootstock that originated in Texas as an open-pollinated cross between the myrobalan plum and, supposedly, *P. munsoniana*. It is propagated by hardwood cuttings (*91, 94*). Some plums have grown well on it; others have not. An exceptionally vigorous seedling selection of the parent 'Marianna' plum, made by the California Agricultural Experiment Station in 1926, is widely used under the identifying name, 'Marianna 2624'. It is adaptable to heavy, wet soils and is immune to root-knot nematodes, resistant to crown rot, crown gall, oak root fungus, and verticillium wilt, but is susceptible to bacterial canker. Since it is propagated as rooted cuttings, it has a shallow-root system for the first few years, but as the trees grow older they develop a deeper root system.

*Peach (P. persica)* Many plum and prune orchards in California are on peach seedling rootstocks. This stock has proved satisfactory for light, well-drained soils. However, peach rootstocks should be avoided if the trees are to be planted on a site formerly occupied by a peach orchard. In some areas plum on peach roots tends to overbear and develop a die-back condition. Peach is not satisfactory as a stock for some plum cultivars, including 'Sugar' prune and 'Robe de Sergeant'.

*Apricot (P. armeniaca)* Apricot seedlings can be used as a plum stock in nematode-infested sandy soils for those cultivars that are compatible with apricot. Japanese plums tend to do better than European plums on apricot roots.

*Almond (P. dulcis)* Some plum cultivars can be grown successfully on almond seedlings. The 'French' prune does very well on this stock, the trees growing faster and bearing larger fruit than when myrobalan roots are used. Plum cultivars on almond roots tend to overbear, sometimes to the detriment of the tree. Except where plantings are to be made on well-drained, sandy soils, high in lime or boron, this stock probably should not be used for plums.

*'Brompton' and 'Common' plum (P. domestica)* These two clonal plum stocks are used chiefly in England. 'Brompton' seems to be compatible with all plum cultivars and tends to produce medium to large trees. 'Common' plum, which produces small to medium trees, has shown incompatibility with some cultivars (*67, 150*). Propagation is by hardwood cuttings.

*'St. Julien', 'Common Mussel', 'Damas', and 'Damson' (P. insititia)* The first three stocks are used mostly in England. 'Damas C' produces medium to large trees, whereas 'Common Mussel' generally produces small to medium trees. The latter stock seems to be compatible with all plums. 'St. Julien A' produces small to medium trees for all compatible scion cultivars. Results have been variable with these stocks, since they vary widely in type. At the East Malling Research Station, 'St. Julien', 'Mussel', and 'Damas' clones have been selected and listed as A, B, C, D, and so forth. 'St. Julien' has been used to some extent as a plum stock for the *P. domestica* cultivars in the United States. 'Pixy' has been released by the East Malling Research Station as a dwarfing plum rootstock (*235*).

In England the following clonal stocks are recommended for plums: for *vigorous* trees, 'Myrobalan B'; for *semivigorous* trees, 'Brompton'; for *intermediate* trees, 'Marianna', 'Pershore'; for *semidwarf* trees, 'Common' plum, 'St. Julien A'; for *dwarf* trees, 'St. Julien K.'

*Florida sand plum (P. angustifolia)* This species may be useful as a dwarfing stock for compatible plum cultivars. In California, after 16 years, 'Giant', 'Burbank', and 'Beauty' on this stock were healthy and very productive, with a dwarf type of growth.

*Japanese plum (P. salicina)* Seedlings of this species are used as plum stocks in Japan but apparently not elsewhere. European plums (*P. domestica*), when topworked on Japanese plum stocks, result in very short-lived trees; the reverse combination, though, produces compatible unions.

*Western sand cherry (P. besseyi)*    This stock has produced satisfactory dwarfed plum trees of the Japanese and European types, but poor bud unions and shoot growth developed when it was used as a stock for cultivars of *P. insititia* (*15*).

**Pomegranate** (*Punica granatum* L.)    The pomegranate is easily propagated by hardwood cuttings. After one season's growth, the plants are usually large enough to be moved to their permanent location. Softwood cuttings taken during the summer are also easily rooted if maintained under high-humidity conditions. The pomegranate forms suckers readily, and these may be dug up during the dormant season with a piece of root attached.

**Quince** (*Cydonia oblonga* Mill.)    The quince is usually propagated by hardwood cuttings, which root readily, "heel" cuttings more so than those made entirely from one-year-old wood. Cuttings from two- or three-year-old wood also root easily. The burrs or knots found on this older wood are masses of adventitious root initials. Cuttings made from the basal portion of one-year-old wood root better than those made from the terminal end. The cuttings usually make sufficient growth in one season to transplant to their permanent location. Quince can also be propagated by mound layering.

Occasionally, commercial quince cultivars are T-budded on rooted cuttings (such as 'Angers') or sometimes on quince seedlings.

**Raspberry, Black** (*Rubus occidentalis* L.) **and Purple** (*R. occidentalis* × *R. idaeus*)    Tip layering is the usual method of propagating these species, the black rooting more easily than the purple. The black raspberry can be propagated also by leaf-bud cuttings, roots forming in about three weeks. Cuttings should be taken in early summer and rooted under high-humidity conditions or in a mist-propagating frame.

**Raspberry, Red** (*Rubus strigosus* Michx., *R. idaeus* L.)    The red raspberry is easily propagated by removing suckers of one-year-old growth. These are dug in early spring, care being taken to leave a piece of the old root attached. Also, young, green suckers of new wood may be dug in the spring shortly after they appear above ground. When these plants are dug, they should have some new roots starting. A piece of the old root is taken with them. Such suckers should be dug and transplanted during cool, cloudy weather and irrigated well after resetting, otherwise they will not survive. Sucker production can be stimulated by inserting a spade deeply at intervals in the vicinity of old plants to cut off roots, each root piece then sending up a shoot. Production of suckers can also be stimulated by mulching with straw or sawdust. Viruses and crown gall can be problems with red raspberries so suckers should be taken only from clean plants obtained from nurseries specializing in certified plants.

The red raspberry can be propagated by root cuttings (*109, 226*). For thick root pieces 15 cm (6 in.) root lengths are best; for thin roots use 5 cm (2 in.) pieces. Shallow 13 mm (½ in.) planting is advised (*105*). If given good care, strong nursery plants will be produced in one year.

Leafy softwood cuttings made in early spring from young sucker shoots just emerging from the soil, with a 2.5 to 5 cm (1- to 2-in.) etiolated section from below the surface, will give almost 100 percent rooting when placed in a propagating frame (*244*). Shoot cuttings taken directly from the canes are almost impossible to root.

**Strawberry** (*Fragaria* × *ananassa* Duch.) (*70, 169, 196, 227, 243*)    Strawberries are propagated by runners or in certain everbearing types—which produce few runners—by crown division. Plants can be easily mass-produced by micropropagation using in vitro culture of shoot tips (*42, 117*). Seed propagation is used only in breeding programs for the development of new cultivars. Treatment of seed with sulfuric acid for 15 to 20 minutes before planting gives good germination (*198*).

Commercial strawberry nursery plant production involves some form of registration and certification to make available to growers pathogen-free, true-to-name plants (*60*). Viruses have been eliminated from strawberry clones by meristem-tip culture, with or without additional heat treatment (*35, 211*).

Most strawberry cultivars are short-day plants that produce runners in summer, initiate flower buds in autumn, are chilled during the winter to enable them to produce flowers and fruits in the spring.

Nursery fields are planted in the spring and the new runner plants are dug the following fall and winter after they are mature and dormant. Such planting stock comes from special ''mother blocks,'' maintained in isolation to prevent infection with viruses and other pathogens.

**Figure 18–5**    Equipment used for large-scale digging of strawberry nursery plants. The entire row is scooped up and elevated to a revolving trommel, which separates the plants from the soil. The plants are then packed in wet burlap bags and taken to sheds for trimming and packing. Courtesy Wheeler's Nursery, Los Molinos, California.

In California plants are dug from the nursery row mostly in the fall in order to produce optimum plant response at particular berry production areas. The digging should be completed in the spring before the plants start to grow, particularly if they are to be held in cold storage. Special machine diggers handle great quantities of plants (Figure 18-5). Immediately after digging, the plants are hauled to packing sheds, where all leaves and petioles are removed. Plants are packed in vented cardboard or wood boxes for storing or shipping. A polyethylene liner (0.75 to 1.5 mil) is used to keep the plants from drying out (246). They are usually handled in 1000- or 2000-plant units, depending upon plant size. The plants are placed firmly in the boxes with roots to the center. Such plants can be stored and kept in good condition for one year if held between -2° and -1°C (28° and 30°F) to enable growers in certain commercial producing areas to plant in the summer and fall (76).

Micropropagation has become an important commercial alternative for strawberry plant propagation, and laboratories capable of producing several million plants annually have been built (14, 42). However, actual commercial production may involve a combination of mother block laboratory procedures and field multiplication operations (117).

Everbearing cultivars that produce few runners are propagated by crown division. Certain cultivars, such as the 'Rockhill', may produce 10 to 15 strong crowns per plant by the end of the growing season. In the spring, such plants are dug and carefully cut apart; each crown may then be used as a new plant.

**Tangerine**     *See* Citrus.

**Walnut, Persian, English** (*Juglans regia* L.)     Walnut cultivars are propagated by patch budding (or one of its variations), T-budding, or whip grafting one-year-old seedling rootstocks, or by topworking seedling trees, one to four years old or more, planted in place in the orchard. Bark grafting in late spring or patch budding in the spring or summer works well for the latter method.

**Rootstocks for Persian walnut** (59, 201)     Nuts of most of the *Juglans* species used as seedling rootstocks should be either fall-planted or stratified for about three months at 2° to 4°C (36° to 40°F) before they are planted in the spring, to obtain good germination. Although Persian walnut seeds germinate without any cold treatment, a stratification period hastens germination. It is better to plant the seeds before they start sprouting in the stratification boxes but, with care, sprouted seeds can be planted successfully. At the end of one season, the seedlings should be large enough to bud.

*Northern California black walnut (J. hindsii)*     This is the stock most commonly used in California. The seedlings are vigorous and make a strong graft union. They are resistant to verticillium, oak root fungus (*Armillaria mellea*), and root-knot nematode (*Meloidogyne* spp.), but are susceptible to crown rot (*Phytophthora* spp.), crown gall, and the root-lesion nematode (*Pratylenchus vulnus*). Persian walnut trees grafted on this rootstock are subject to a serious defect known as "black-line." The latter has appeared in walnut districts of California, Oregon, and France. It is characterized by a breakdown of the tissues in the cambial region at the graft union, leading to a girdling of the trees. The cause of this problem is a graft-transmissible virus (149). Symptoms are yellow, drooping leaves, premature leaf drop, poor terminal shoot growth, reduction in crop, black lesions at the graft union, and a heavy production of suckers from the black walnut stock below the graft union. The Persian walnut top above the graft union eventually dies from the virus, leaving the *J. hindsii* stock alive.

*Persian walnut (J. regia)*     Seedlings of this species as a rootstock produce good trees with an excellent graft union and are highly resistant to crown rot. The roots are susceptible to crown gall, oak root fungus, and salt accumulation in the soil, and are not as resistant to root-knot nematodes as *J. hindsii*. Nurserymen object to the slow initial growth of the seedlings. Trees on this stock should be quite satisfactory in soils free of oak root fungus and salt accumulation. Its use is recommended in some localities as a means of avoiding the black-line virus (*149*). In Oregon, where black-line is very serious, seedlings of the 'Manregian' clone (of the Manchurian race of *J. regia*), imported by the USDA as P.I. No. 18256, are used as rootstocks. They are vigorous, cold-hardy, and—in Oregon—no more susceptible to oak root fungus than *J. hindsii* seedlings (*170*).

*Paradox walnut (J. hindsii × J. regia)*    This is a first-generation (F₁) hybrid stock; the seedlings are obtained from seed taken from *J. hindsii* trees, whose pistillate (female) flowers have been wind-pollinated with pollen from nearby *J. regia* trees. When seeds from such a *J. hindsii* tree are planted, some of the seedlings may be the hybrid progeny. The amount varies widely, from none to almost 100 percent, depending upon the individual mother tree. The hybrids are easily distinguished by their large leaves in comparison with the smaller-leaved, self-pollinated *J. hindsii* seedlings (see Figure 4-4). Seedlings from Paradox trees themselves should not be used for producing rootstocks, because of their great variability in all characteristics. Although first-generation (F₁) seedlings are variable in some characteristics, most of them exhibit hybrid vigor and make excellent vigorous rootstocks for the Persian walnut. They are resistant to root-lesion nematodes and crown rot, and are tolerant of saline and heavy, wet soils. Paradox seedlings are very susceptible to oak root fungus and crown gall. Persian walnut or Paradox roots are just as susceptible to the black-line virus as *J. hindsii* seedlings. Trees on Paradox rootstocks grow and yield as well as, or better than, those on *J. hindsii* roots, and may produce large-size nuts with better kernel color (*201*). In very heavy or low-fertility soils, trees on Paradox roots grow faster than those on *J. hindsii*.

Since it is difficult to secure Paradox seeds in quantity, vegetative propagation methods would be very desirable in order to establish superior Paradox clones from which large numbers of rootstock plants could be obtained. Propagation by leafy cuttings under mist, hardwood cuttings, and trench layering can be done but with difficulty (*137, 200*).

**Walnut, Carpathian** (*Juglans regia* L.) (*75*)    The Carpathian strain of *J. regia* originated in cold winter areas of the Carpathian Mountains of Poland and the Kiev and Poltava regions of the Ukraine, whereas the Persian walnuts originated in the area of present-day Iran. Carpathian walnut trees are much more winter hardy than trees of the Persian walnut strain commonly grown in California. Carpathian cultivars have been developed that are winter hardy in the eastern United States. These are propagated by chip budding or bench grafting (using the whip graft) on Persian (*J. regia*) or black (*J. nigra*) walnut seedling rootstocks.

**Walnut, Black** (*Juglans nigra* L.) (*63*)    The several named black walnut cultivars are propagated by patch or ring budding or side grafting. A modified cleft graft (Figure 18–6) has been used with success (*26, 139, 209*).

This method is suitable for seedling stocks up to about 2.5 cm (1 in.) in diameter. The scion is cut to a wedge, slightly thicker on one side than on the other. The stock is cut diagonally rather than split. The scion is inserted into this cleft, the thicker part out, with the cambium of stock and scion matching.

In growing Eastern black walnut for timber production, seedlings are started in nurseries and only the strongest, most vigorous trees are set out in the plantation. Careful planting is necessary to obtain the essential rapid, early growth. Cuttings are very difficult to root.

**Figure 18-6**    Steps in the modified cleft graft used in propagating the Eastern black walnut. Left to right: scion cut to a wedge; stock cut diagonally with the scion inserted; waxed cloth placed over cut surfaces and the union tied with raffia; scion and stub after waxing *(139)*.

# REFERENCES

1.  Albert, D. W  1926. Propagation of date palms from offshoots. *Ariz. Agr. Exp. Sta. Bul. 119.*

2.  Aldrich, W. W., G. H. Leach, and W. A. Dollins. 1945. Some factors influencing the growth of date offshoots in the nursery row. *Proc. Amer. Soc. Hort. Sci.* 46:215–21.

3.  Alley, C. J. 1980. Propagation of grapevines. *Calif. Agr.* 34(7):29–30.

4.  Alley, C. J., and A. T. Koyama. 1978. Vine bleeding delays growth of T-budded grapevines. *Calif. Agr.* 32(8):6.

5.  Argles, G. K. 1976. *Anacardium occidentale* (cashew). In *The propagation of tropical fruit trees,* R. J. Garner and S. A. Chaudri, eds. Hort. Rev. No. 4. East Malling, England; FAO and Commonwealth Agricultural Bureaux.

6.  Arndt, C. H. 1935. Mango polyembryony and other multiple shoots. *Amer. Jour. Bot.* 22:26.

7.  Ashiru, G. A., and T. Quarcoo. 1971. Vegetative propagation of kola (*Cola nitida*). *Trop. Agr.* 48(1):85–92.

8.  Barton, L. V. 1965. Viability of seeds of *Theobroma cacao* L. *Contr. Boyce Thomp. Inst.* 23(4):109–22.

9.  Bass, L. N. 1975. Seed storage of *Carica papaya* L. *HortScience* 10(3):232–33.

10. Batjer, L. P., and H. Schneider. 1960. Relation of pear decline to rootstocks and sieve-tube necrosis. *Proc. Amer. Soc. Hort. Sci.* 76:85–97.

11. Bitters, W. P. 1950. Rootstocks with dwarfing effect. *Calif. Agr.* 4(2):5, 14.

12. Bitters, W. P., and E. R. Parker. 1953. Quick decline of citrus as influenced by top-root relationships. *Calif. Agr. Exp. Sta. Bul. 733.*

13. Blodgett, E. C., H. Schneider, and M. D. Aichele. 1962. Behavior of pear decline disease on different stock-scion combinations. *Phytopathology* 52:679–84.

14. Boxus, P. 1974. The production of strawberry plants by *in vitro* micropropagation. *Jour. Hort. Sci.* 49:209-10.

15. Brase, K., and R. D. Way. 1959. Rootstocks and methods used for dwarfing fruit trees. *N. Y. Agr. Exp. Sta. Bul. 783.*

16. Brokaw, W. H. 1977. Subtropical fruit tree production; avocado as a case study. *Proc. Inter. Plant Prop. Soc.* 27:113-21.

17. ———. 1981. Kiwifruit production. *Proc. Inter. Plant Prop. Soc.* 30:48-54.

18. Broome, O. C., and R. H. Zimmerman. 1978. *In vitro* propagation of blackberry. *Hort-Science* 13(2):151-53.

19. Bryant, M. D. 1969. Propagating the pecan tree. *New Mex. State Univ. Pl. Sci. Guide* 400. H-604.

20. Calavan, E. C., E. F. Frolich, J. B. Carpenter, C. N. Roistacher, and D. W. Christiansen. 1964. Rapid indexing for exocortis of citrus. *Phytopathology* 54:1359-62.

21. Cameron, J. W., and R. K. Soost. 1953. Nucellar lines of citrus. *Calif. Agr.* 7(1):8, 15, 16.

22. Carlson, R. F. 1965. Growth and incompatibility factors associated with apricot scion/rootstock in Michigan. *Mich. Quart. Bul.* 48(1):23-29.

23. ———. 1970. Rootstocks in relation to apple cultivars. In *North American apples: Varieties, rootstocks, outlook,* W. H. Upshall, ed. East Lansing: Michigan State Univ. Press.

24. Carlson, R. F., and H. B. Tukey. 1955. Cultural practices in propagating dwarfing rootstocks in Michigan. *Mich. Agr. Exp. Sta. Quart. Bul.* 37(4):492-97.

25. Castle, W. S., W. G. Adams, and R. L. Dilley. 1979. An indoor container system for producing citrus nursery trees in one year from seed. *Proc. Fla. State Hort. Soc.* 92:3-7.

26. Chase, S. B. 1947. Budding and grafting eastern black walnut. *Proc. Amer. Soc. Hort. Sci.* 49:175-80.

27. Chaudri, S. A. 1976. *Mangifera indica—Mango.* In *The propagation of tropical fruit trees,* R. J. Garner and S. A. Chaudri, eds. Hort. Rev. No. 4., Comm. Bureau Hort. and Plant Crops. East Malling, England; FAO and Commonwealth Agricultural Bureaux.

28. Cheng, T. 1978. Clonal propagation of woody plant species through tissue culture techniques. *Proc. Inter. Plant Prop. Soc.* 28:139-55.

29. Child, R. 1964. *Coconuts.* London: Longmans, Green.

30. Cobin, M. 1954. The lychee in Florida. *Fla. Agr. Exp. Sta. Bul. 546.*

31. Cohen, D., and D. Elliott. 1979. Micropropagation methods for blueberries and tamarillos. *Proc. Inter. Plant Prop. Soc.* 29:177-79.

32. Cole, C. E. 1966. The fruit industry of Australia and New Zealand. *Proc. XVII Inter. Hort. Cong.* 4:321-68.

33. Collins, J. L. 1960. *The pineapple—history, cultivation, utilization.* London: Leonard-Hill.

34. Condit, I. J. 1969. *Ficus, the exotic species.* Berkeley, Calif.: Univ. of Calif. Div. Agr. Sci.

35. Converse, R. A. 1972. The propagation of virus-tested strawberry stocks. *Proc. Inter. Plant Prop. Soc.* 22:73-76.

36. Coorts, G. D., and J. W. Hull. 1972. Propagation of highbush blueberry (*Vaccinium Australe* Small) by hard and softwood cuttings. *The Plant Propagator* 18(2):9-11.

37. Couvillon, G. A., and F. A. Pokorny. 1968. Photoperiod, indolebutyric acid, and type of cutting wood as factors in rooting of rabbiteye blueberry (*Vaccinium ashei Reade*), cv. Woodward. *HortScience* 3(2):74-75.

38. Couvillon, G. A., and A. Erez. 1980. Rooting, survival, and development of several peach cultivars propagated from semi-hardwood cuttings. *HortScience* 15(1):41-43.

39. Cross, C. E., I. E. Demoranville, K. H. Deubert, R. M. Devlin, J. S. Norton, W. E. Tomlinson, and B. M. Zuckerman. 1969. Modern cultural practice in cranberry growing. *Mass. Agr. Exp. Sta. Ext. Ser. Bul. No. 39.*

40. Cummins, J. N., and R. L. Norton. 1974. Apple rootstock problems and potentials. *N. Y. Food Life Sci. Bul. 41.*

41. Cummins, J. N., and H. S. Aldwinckle. 1974. Breeding apple rootstocks. *HortScience* 9(4):367-72.

42. Damiano, C. 1980. Strawberry micropropagation. In *Proc. Conf. Nurs. Prod. Fruit Plants through Tissue Cult.,* R. H. Zimmerman, ed. USDA, SEA, ARR–NE–11.

43. Darrow, G. M., and J. N. Moore. 1966. Blueberry growing. *USDA Farmers' Bul.* 1951 (rev.).

44. Davis B., II. 1976. Bench grafting plum and apricot as compared to T-budding. *Proc. Inter. Plant Prop. Soc.* 26:253-55.

45. Day, L. H. 1951. Cherry rootstocks in California. *Calif. Agr. Exp. Sta. Bul. 725.*

46. Day, L. H., and E. R. Serr. 1951. Comparative resistance of rootstocks of fruit and nut trees to attack by root lesion or meadow nematode. *Proc. Amer. Soc. Hort. Sci.* 57:150-54.

47. Dillon, D. 1967. Simultaneous grafting and rooting of citrus under mist. *Proc. Inter. Plant Prop. Soc.* 17:114-18.

48. Dimalla, G. G., and J. Van Staden. 1978. Pecan nut germination: A review for the nursery industry. *Scient. Hort.* 8:1-9.

49. Doehlert, C. A. 1953. Propagating blueberries from hardwood cuttings. *N.J. Agr. Exp. Sta. Cir. 551.*

50. Doggrell, L. M. 1976. Commercial propagation of macadamias. *Proc. Inter. Plant Prop. Soc.* 26:391-93.

51. Doncaster, T. 1981. The whys and hows of passion fruit growing in Queensland. *Proc. Inter. Plant Prop. Soc.* 30:616-17.

52. Eck, P., and A. W. Stretch. 1979. Cutting wood production from highbush blueberry mother block plants as affected by nitrogen nutrition and fungicide treatment. *HortScience* 14(5):599-600.

53. Erez, A., and Z. Yablowitz. 1981. Rooting of peach hardwood cuttings for the meadow orchard. *Scient. Hort.* 15:137-44.

54. Evans, H. 1951. Investigations on the propagation of cacao. *Trop. Agr.* 28:147-203.

55. ———. 1953. Physiological aspects of the propagation of cacao from cuttings. *Rpt. 13th Inter. Hort. Cong.* 2:1179-90.

56. Fletcher, W. F. 1942. The native persimmon. *USDA Farmers' Bul. 685.*

57. Fletcher, W. A. 1971. Growing Chinese gooseberries. *New Zealand Dept. Agr. Bul. 349.*

58. Fogle, H. W., H. L. Keil, W. L. Smith, S. M. Mircetich, L. C. Cochran, and H. Baker. 1974. Peach production. *USDA, ARS Agr. Handbook 463.*

59. Forde, H. I. 1979. Persian walnut in the western United States. In *Nut tree culture in North America,* R. A. Jaynes, ed. Hamden, Conn.: Northern Nut Growers Assn.

60. Frazier, N. F., and A. F. Posnette. 1958. Relationships of the strawberry viruses of England and California. *Hilgardia* 27(17):455-513.

61. Frolich, E. F. 1966. Rooting citrus and avocado cuttings. *Proc. Inter. Plant Prop. Soc.* 16:51–54.

62. Frolich, E. F., and R. G. Platt. 1972. Use of the etiolation technique in rooting avocado cuttings. *Calif. Avoc. Soc. Yearbook,* 1971–1972, pp. 97–109.

63. Funk, D. T. 1979. Black walnuts for nuts and timber. In *Nut tree culture in North America,* R. A. Jaynes, ed. Hamden, Conn.: North American Nut Growers Assn.

64. Galletta, G. J., and A. S. Fish, Jr. 1971. Interspecific blueberry grafting, a way to extend *Vaccinium* culture to different soils. *Jour. Amer. Soc. Hort. Sci.* 96(3):294–98.

65. Gardner, F. E., and G. E. Horanic. 1963. Cold tolerance and vigor of young citrus trees. *Proc. Fla. State Hort. Soc.* 76:105–10.

66. Giliomee, J. H., D. K. Strydom, and H. J. Van Zyl. 1968. Northern Spy, Merton, and Malling-Merton rootstocks susceptible to woolly aphid, *Eriosoma lanigerum,* in the Western Cape. *S. Afr. Jour. Agr. Sci.* 11:183–86.

67. Glenn, E. M. 1961. Plum rootstock trials at East Malling. *Jour. Hort. Sci.* 36:28–39.

68. Gould, H. P. 1939. The native papaw. *USDA Leaflet 179.*

69. ———. 1940. The Oriental persimmon. *USDA Leaflet 194.*

70. Greathead, A. S., N. Welch, W. S. Seyman, N. F. McCalley, V. Voth, and R. Bringhurst. 1977. Strawberry production in California. *Div. Agr. Sci., Univ. Calif. Leaflet 2959.*

71. Greenhalgh, W. J. 1963. A pear rootstock trial at Bathurst Experiment Farm. *Agr. Gaz. N.S.W. (Aust.)* 74:350–53.

72. Griggs, W. H., and H. T. Hartmann. 1960. Old Home pears show resistance to decline when on own roots. *Calif. Agr.* 14:8, 10.

73. Griggs, W. H., D. D. Jensen, and B. T. Iwakiri. 1968. Development of young pear trees with different rootstocks in relation to psylla infestation, pear decline, and leaf curl. *Hilgardia* 39:153–204.

74. Griggs, W. H., J. A. Beutel, W. O. Reil, and B. T. Iwakiri. 1980. Combination of pear rootstocks recommended for new Bartlett plantings. *Calif. Agr.* 34(10):20–24.

75. Grimo, E. 1979. Carpathian (Persian) walnuts. In *Nut tree culture in North America,* R. A. Jaynes, ed. Hamden, Conn.: North American Nut Growers Assn.

76. Guttridge, C. G., D. T. Mason, and E. G. Ing. 1965. Cold storage of strawberry runner plants at different temperatures. *Exp. Hort.* 12:38–41.

77. Hall, I. V. 1969. Growing cranberries. *Can. Dept. Agr. Publ.* 1282 (rev.).

78. Hall, I. V., L. E. Aalders, L. P. Jackson, G. W. Wood, and C. L. Lockhart. 1975. Lowbush blueberry production. *Can. Dept. Agr. Publ. 1477.*

79. Hall, W. J. 1975. Propagation of walnuts, almonds, and pistachios in California. *Proc. Inter. Plant Prop. Soc.* 25:53–57.

80. Halma, F. F., K. M. Smoyer, and H. W. Schwalm. 1944. Quick decline associated with sour rootstocks. *Calif. Citrogr.* 29:245.

81. Hamilton, R. Z., and W. Yee. 1974. Macadamia: Hawaii's dessert nut. *Univ. Hawaii Cir.* 485.

82. Hansen, C. J., and H. T. Hartmann. 1968. The use of indolebutyric acid and captan in the propagation of clonal peach and peach-almond hybrid rootstock. *Proc. Amer. Soc. Hort. Sci.* 92:135–40.

83. Harada, H. 1975. *In vitro* organ culture of *Actinidia chinensis* Pl. as a technique for vegetative multiplication. *Jour. Hort. Sci.* 50:81–83.

84. Harkness, R. W. 1967. Papaya growing in Florida. *Fla. Agr. Exp. Sta. Cir.* S–180.

85. Harmon, F. N. 1943. Influence of indolebutyric acid on the rooting of grape cuttings. *Proc. Amer. Soc. Hort. Sci.* 42:383–88.

86. ———. 1954. A modified procedure for greenwood grafting of vinifera grapes. *Proc. Amer. Soc. Hort. Sci.* 64:255–58.

87. Harmon, F. N., and J. H. Weinberger. 1959. Effects of storage and stratification on germination of *Vinifera* grape seeds. *Proc. Amer. Soc. Hort. Sci.* 73:147–50.

88. ———. 1963. Bench grafting trials with Thompson seedless grapes on various rootstocks. *Proc. Amer. Soc. Hort. Sci.* 83:379–83.

89. Hartmann, Henry. 1961. Historical facts pertaining to root and trunkstocks for pear trees. *Ore. Agr. Exp. Sta. Misc. Paper 109.*

90. Hartmann, H. T. 1952. Further studies on the propagation of the olive by cuttings. *Proc. Amer. Soc. Hort. Sci.* 59:155–60.

91. ———. 1955. Auxins for hardwood cuttings. *Calif. Agr.* 9(4):7, 12, 13.

92. ———. 1958. Rootstock effects in the olive. *Proc. Amer. Soc. Hort. Sci.* 72:242–51.

93. Hartmann, H. T., and R. M. Brooks. 1958. Propagation of Stockton Morello cherry rootstocks by softwood cuttings under mist. *Proc. Amer. Soc. Hort. Sci.* 71:127–34.

94. Hartmann, H. T., and C. J. Hansen. 1958. Rooting pear, plum rootstocks. *Calif. Agr.* 12(10):4, 14–15.

95. ———. 1955. Rooting of softwood cuttings of several fruit species under mist. *Proc. Amer. Soc. Hort. Sci.* 66:157–67.

96. Hartmann, H. T., and F. Loreti. 1965. Seasonal variation in rooting leafy olive cuttings under mist. *Proc. Amer. Soc. Hort. Sci.* 87:194–98.

97. Hartmann, H. T., and K. Opitz. 1977. Olive production in California. *Div. Agr. Sci., Univ. Calif. Leaflet 2474.*

98. Hartmann, H. T., W. H. Griggs, and C. J. Hansen. 1963. Propagation of own-rooted Old Home and Bartlett pears to produce trees resistant to pear decline. *Proc. Amer. Soc. Hort. Sci.* 82:92–102.

99. Hartmann, H. T., C. J. Hansen, and F. Loreti. 1965. Propagation of apple rootstocks by hardwood cuttings. *Calif. Agr.* 19(6):4–5.

100. Hartmann, H. T., W. C. Schnathorst, and J. Whisler. 1971. Oblonga, a clonal olive rootstock resistant to verticillium wilt. *Calif. Agr.* 25(6):12–15.

101. Hearn, C. J., D. J. Hutchison, and H. C. Barrett. 1974. Breeding citrus rootstocks. *HortScience* 9(4):357–58.

102. Hedrick, U. P. 1923. Stocks for plums. *N. Y. Agr. Exp. Sta. Bul. 498.*

103. Heppner, M. J. 1927. Pear black-end and its relation to different rootstocks. *Proc. Amer. Soc. Hort. Sci.* 24:139–42.

104. Herman, E. B., and G. J. Haas. 1975. Clonal propagation of *Coffea arabica* L. from callus culture. *HortScience* 10(6):588–89.

105. Heydecker, W., and Margaret Marston. 1968. Quantitative studies on the regeneration of raspberries from root cuttings. *Hort. Res.* 8(2):142–46.

106.  Hibino, H., G. H. Kaloostian, and H. Schneider. 1971. Mycoplasmalike bodies in the pear psylla vector of pear decline. *Virology* 43:34–40.

107.  Hodgson, R. W. 1940. Rootstocks for the Oriental persimmon. *Proc. Amer. Soc. Hort. Sci.* 37:338–39.

108.  Howard, W. L. 1925. The "Stockton" Morello cherry. *Proc. Amer. Soc. Hort. Sci.* 21:320–23.

109.  Hudson, J. P. 1954. Propagation of plants by root cuttings I. Regeneration of raspberry root cuttings. *Jour. Hort. Sci.* 29:27–43.

110.  Ivey, I. D. 1979. Feijoas: Selection and propagation. *Proc. Inter. Plant Prop. Soc.* 29:161–68.

111.  Jaffee, A. 1970. Chip grafting guava cultivars. *The Plant Propagator* 16(2):6.

112.  James, D. J., and I. J. Thurbon. 1981. Shoot and root initiation *in vitro* in the apple rootstock M.9 and the promotive effects of phloroglucinol. *Jour. Hort. Sci.* 56(1):15–20.

113.  Jaynes, R. A. 1961. Buried-inarch technique for rooting chestnut cuttings. *North. Nut Grow. Assoc. 52nd Ann. Rpt.*, pp. 37–39.

114.  ———. 1979. Chestnuts. In *Nut tree culture in North America,* R. A. Jaynes, ed. Hamden, Conn.: Northern Nut Growers Assn.

115.  Joley, L. E. 1979. Pistachios. In *Nut tree culture in North America,* R. A. Jaynes, ed. Hamden, Conn.: Northern Nut Growers Assn.

116.  Jones, O. P., C. A. Pontikis, and M. E. Hopgood. 1979. Propagation *in vitro* of five apple scion cultivars. *Jour. Hort. Sci.* 54(2):155–58.

117.  Kartha, K. K., N. L. Leung, and K. Pahl. 1980. Cryopreservation of strawberry meristems and mass propagation of plantlets. *Jour. Amer. Soc. Hort. Sci.* 105(4):481–84.

118.  Kender, W. J. 1965. Some factors affecting the propagation of lowbush blueberries by softwood cuttings. *Proc. Amer. Soc. Hort. Sci.* 86:301–6.

119.  Kester, D. E., C. J. Hansen, and C. Panetsos. 1965. Effect of scion and interstock variety on incompatibility of almond on Marianna 2624 rootstock. *Proc. Amer. Soc. Hort. Sci.* 86:169–77.

120.  Kester, D. E., and R. A. Asay. 1975. Almond breeding. In *Advances in fruit breeding,* J. M. Moore and J. Janick, eds. Lafayette, Ind.: Purdue Univ. Press.

121.  Klotz, L. J., E. C. Calavan, and L. G. Weathers. 1972. Virus and virus-like diseases of citrus. *Calif. Agr. Exp. Sta. Cir. 559.*

122.  Krezdorn, A. H., and G. W. Adriance. 1961. Fig growing in the South. *USDA Agr. Handbook 196.*

123.  Krul, W. R., and J. Meyerson. 1980. *In vitro* propagation of grape. In *Proc. Conf. Nurs. Prod. Fruit Plants through Tissue Culture,* R. H. Zimmerman, ed. USDA, SEA, ARR–NE–11.

124.  Lagerstedt, H. B. 1979. Filberts. In *Nut tree culture in North America,* R. A. Jaynes, ed. Hamden, Conn.: Northern Nut Growers Assn.

125.  Layne, R. E. C. 1974. Breeding peach rootstocks for Canada and the northern United States. *HortScience* 9(4):364–66.

126.  Lapins, K. 1959. Some symptoms of stock-scion incompatibility of apricot varieties on peach seedling rootstock. *Can. Jour. Plant Sci.* 39:194–203.

127.  Lawyer, E. M. 1978. Seed germination of stone fruits. *Proc. Inter. Plant Prop. Soc.* 28:106–9.

128. Lear, B., and L. Lider. 1959. Eradication of root-knot nematodes from grapevine rootings by hot water. *Plant Dis. Rpt.* 43:314–17.

129. Lider, L. 1958. Phylloxera-resistant grape rootstocks for the coastal valleys of California. *Hilgardia* 27:287–318.

130. ———. 1960. Vineyard trials in California with nematode-resistant grape rootstocks. *Hilgardia* 30:123–52.

131. ———. 1963. Field budding and the care of the budded grapevine. *Calif. Agr. Ext. Ser. Leaflet 153.*

132. Lider, L., and N. Shaulis. 1965. Resistant rootstocks for New York vineyards. *N.Y. Agr. Exp. Sta. Res. Cir. No. 2.*

133. Litz, R. E., and R. A. Conover. 1978. *In vitro* propagation of papaya. *HortScience* 13(3):241–42.

134. Livingston, R. L., and T. F. Crocker. 1975. Pecans in Georgia. *Univ. Ga., Col. Agr. Bul. 609.*

135. Loomis, N. H. 1948. Rootstocks for grapes in the south. *Proc. Amer. Soc. Hort. Sci.* 42:380–82.

136. Lynch, S. J., and R. O. Nelson. 1956. Current methods of vegetative propagation of avocado, mango, lychee, and guava in Florida. *Cieba* (Escuela Agricola Panamericana, Tegucigalpa, Honduras) 4:315–77.

137. Lynn, C., and H. T. Hartmann. 1957. Rooting cuttings under mist—Paradox walnut. *Calif. Agr.* 11(5):11, 15.

138. Lyrene, P. M. 1980. Micropropagation of rabbiteye blueberries. *HortScience* 15(1):80–81.

139. MacDaniels, L. H. 1952. Nut growing. *Cornell Ext. Bul. 701.*

140. ———. 1979. Hickories. In *Nut tree culture in North America,* R. A. Jaynes, ed. Hamden Conn.: Northern Nut Growers Assn.

141. Madden, G. E. 1979. Pecans. In *Nut tree culture in North America,* R. A. Jaynes, ed. Hamden, Conn.: Northern Nut Growers Assn.

142. Madden, G. E., and H. W. Tisdale. 1975. Effects of chilling and stratification on nut germination of northern and southern pecan cultivars. *HortScience* 10(3):259–60.

143. Majumdar, P. K., and S. K. Mukherjee. 1968. Guava: A new vegetative propagation method. *Indian Hort.* Jan.-Mar.

144. McKay, J. W., and H. L. Crane. 1953. Chinese chestnut: A promising new orchard crop. *Econ. Bot.* 7:228–42.

145. McKenzie, D. W. 1964. Apple rootstock trials: Jonathan on East Malling, Merton, and Malling-Merton rootstocks. *Jour. Hort. Sci.* 39:69–77.

146. McMillan-Browse, P. D. A. 1970. Vegetative propagation of *Corylus. Proc. Inter. Plant Prop. Soc.* 20:356–58.

147. Maxwell, N. P., and C. G. Lyons. 1979. A technique for propagating container-grown citrus on sour orange rootstock. *HortScience* 14(1):56–57.

148. Milbocker, D. C. 1980. Kiwifruit. *Amer. Hort.* 59(1):21, 34–35.

149. Mircetich, J. S. M., J. Refsguard, and M. E. Matheron. 1980. Blackline of English walnut trees traced to graft-transmitted virus. *Calif. Agr.* 34(11, 12):8–10.

150. Montgomery, H. B. S. 1963. Fruit tree raising: Rootstocks and propagation. *Minist. Agr. Fish. Foods (London) Bul. 135.*

151. Moore, P. W. 1966. Propagation and growing citrus nursery trees in containers. *Proc. Inter. Plant Prop. Soc.* 16:54–62.

152. Morton, J. F., and F. D. Venning. 1972. Avoid failures and losses in the cultivation of cashew. *Econ. Bot.* 26(3):245–54.

153. Moss, G. I., and R. Dalgleish. 1981. Rapid propagation of citrus in containers. *Proc. Inter. Plant Prop. Soc.* 31:262–74.

154. Mukherjee, S. K., and P. K. Majumdar. 1963. Standardization of rootstock of mango. I. Studies on the propagation of clonal rootstocks by stooling and layering. *Indian Jour. Hort.* 20:204–9.

155. ———. 1964. Effect of different factors on the success of veneer grafting in mango. *Indian Jour. Hort.* 21:46–50.

156. Mukherjee, S. K., and N. N. Bid. 1965. Propagation of mango. II. Effect of etiolation and growth regulator treatment on the success of air layering. *Indian Jour. Agr. Sci.* 35:309–14.

157. Mullins, C. A. 1972. Twenty years' research with size-controlling apple rootstocks on the Cumberland plateau. *Tenn. Agr. Exp. Sta. Bul. 488.*

158. Nagabhushanam, S., and M. A. Menon. 1980. Propagation of cashew (*Anacordium occidentale* L.) by etiolation, girdling, and stooling. *The Plant Propagator* 26(1):11–13.

159. Nelson, S. H., and H. B. Tukey. 1955. Root temperature affects the performance of East Malling apple rootstocks. *Quart. Bul. Mich. Agr. Exp. Sta.* 38:46–51.

160. Nesbitt, W. B. 1974. Breeding resistant grape rootstocks. *HortScience* 9(4):359–61.

161. Nixon, R. W. 1969. Growing dates in the United States. *USDA Agr. Inf. Bul. 207.*

162. Northwood, P. J. 1964. Vegetative propagation of cashew (*Anacardium occidentale*) by the air layering method. *E. Afr. Agr. For. Jour.* 30:35–37.

163. Norton, R. A., C. J. Hansen, H. J. O'Reilly, and W. H. Hart. 1963. Rootstocks for apricots in California. *Calif. Agr. Ext. Ser. Leaflet 156.*

164. ———. 1963. Rootstocks for sweet cherries in California. *Calif. Agr. Ext. Ser. Leaflet 159.*

165. ———. 1963. Rootstocks for peaches and nectarines in California. *Calif. Agr. Ext. Ser. Leaflet 157.*

166. ———. 1963. Rootstocks for plums and prunes in California. *Calif. Agr. Ext. Ser. Leaflet 158.*

167. Opitz, K. W., and R. G. Platt. 1969. Citrus growing in California. *Univ. Calif. Agr. Exp. Sta. Man. 39.*

168. Opitz, K. W., and J. Beutel. 1975. Kiwi propagation. *Proc. Inter. Plant Prop. Soc.* 25:63–65.

169. Otterbacher, A. G., and R. M. Skirvin. 1978. Derivation of the binomial *Fragaria* × *ananassa* for the cultivated strawberry. *HortScience.* 13(6):637–39.

170. Painter, J. H. 1967. Producing walnuts in Oregon. *Ore. Agr. Ext. Bul. 795.*

171. Pease, R. W. 1954. Growing chestnuts from seed. *W. Va. Agr. Exp. Sta. Cir. 90.*

172. Pennock, W., and G. Maldonado. 1963. The propagation of guavas from stem cuttings. *Jour. Agr., Univ. Puerto Rico* 47:280–90.

173. Platt, R. G., and E. F. Frolich. 1965. Propagation of avocados. *Calif. Agr. Exp. Sta. Cir. 531.*

174. Platt, R. G. 1976. Current techniques of avocado propagation. In *Proc. 1st Inter. Trop. Short Course: The avocado,* J. W. Sauls, R. L. Phillips, and L. K. Jackson, eds. Gainesville, Fla.: Fruit Crops Dept., Univ. of Fla.

175. Platt, R. G., and K. Opitz. 1973. The propagation of citrus. In *The citrus industry,* vol. 3, W. Reuther, ed. Berkeley, Calif.: Div. Agr. Sci., Univ. Calif.

176. Popenoe, J. 1970. Coconut varieties and propagation. In *Coconuts: Production, processing, products,* J. G. Woodroof, ed. Westport, Conn.: AVI Publishing.

177. Posnette, A., and R. Cropley. 1962. Further studies on a selection of Williams Bon Chretien pear compatible with Quince A rootstocks. *Jour. Hort. Sci.* 37:291–94.

178. Preston, A. P. 1966. Apple rootstock studies: Fifteen years' results with Malling-Merton clones. *Jour. Hort. Sci.* 41:349–60.

179. ———. 1971. Apple rootstocks 3431 (M. 27). *Ann. Rpt. E. Malling Res. Sta. for 1970,* pp. 143–47.

180. Rake, B. A. 1954. The propagation of gooseberries. I. Some factors influencing the rooting of hardwood cuttings. *Ann. Rpt. Long Ashton Res. Sta. for 1953,* pp. 79–88.

181. Rao, V. N. M., and I. K. S. Rao. 1957. Studies on the vegetative propagation of cashew (*Anacardium occidentale* L.). Approach grafting with and without the aid of plastic film wrappers. *Indian Jour. Agr. Sci.* 27:267–75.

182. Ray, R., and L. Walheim. 1980. *Citrus: How to select, grow and enjoy.* Phoenix, Ariz.: HP Books.

183. Reimer, R. C. 1925. Blight resistance in pears and characteristics of pear species and stocks. *Ore. Agr. Exp. Sta. Bul. 214.*

184. Reuveni, O., and M. Raviv. 1981. Importance of leaf retention to rooting of avocado cuttings. *Jour. Amer. Soc. Hort. Sci.* 106(3):127–30.

185. Reuther, W., ed. *The citrus industry,* vol. 1, 1967; vol. 2, 1968; vol. 3, 1973; vol. 4, 1978. Berkeley, Calif: Univ. Calif. Div. Agr. Sci.

186. Rizzi, A. D., and H. I. Forde. 1974. Pecans. *Calif. Coop. Ext. Ser.* AXT–382.

187. Roberts, A. N. 1962. Cherry rootstocks. *Rpt. Ore. Hort. Soc.* 54:95–98.

188. Rosati, P., G. Marino, and C. Swierczewski. 1980. *In vitro* propagation of Japanese plum (*Prunus salicina* Lindl. cv. Calita.). *Jour. Amer. Soc. Hort. Sci.* 105(1):126–29.

189. Rosedale, D. O. 1963. Growing macadamia nuts in California. *Calif. Agr. Ext. Ser. Publ. AXT–103.*

190. Ruehle, G. D. 1958. Growing bananas in Florida. *Univ. Fla. Agr. Ext. Ser. Cir. 178.*

191. Rygg, G. L. 1975. Date development, handling, and packing in the United States. *USDA, ARS Agr. Handbook 482.*

192. Sax, K. 1949. The use of *Malus* species for apple rootstocks. *Proc. Amer. Soc. Hort. Sci.* 53:219–20.

193. ———. 1953. Interstock effects in dwarfing fruit trees. *Proc. Amer. Soc. Hort. Sci.* 62:201–4.

194. Schneider, H. 1970. Graft transmission and host range of the pear decline causal agent. *Phytopathology* 60:204–7.

195. Schroeder, C. A. 1947. Rootstock influence on fruit set in the Hachiya persimmon. *Proc. Amer. Soc. Hort. Sci.* 50:149–50.

196. Scott, D. H. 1977. Growing strawberries in the Southeastern and Gulf Coast states. *USDA ARS Farmers' Bul. 2246.*

197. Scott, D. H., A. D. Draper, and G. M. Darrow. 1973. Commercial blueberry growing. *USDA Farmers' Bul. 2254.*

198. Scott, D. H., and D. P. Ink. 1948. Germination of strawberry seed as affected by scarification treatment with sulfuric acid. *Proc. Amer. Soc. Hort. Sci.* 51:299–300.

199. Seelig, R. A. 1969. Bananas. In *Fruit & vegetable facts & pointers* (3rd ed.). Washington, D.C.: United Fresh Fruit & Vegetable Assn.

200. Serr, E. F. 1954. Rooting Paradox walnut hybrids. *Calif. Agr.* 8(5):7.

201. Serr, E. F., and A. D. Rizzi. 1964. Walnut rootstocks. *Univ. Calif. Agr. Ext. Publ. AXT-120.*

202. Shalla, T., and L. Chiarappa. 1961. Pear decline in Italy. *Bul. Calif. Dept. Agr.* 50:213–17.

203. Sharpe, R. H. 1954. Rooting of muscadine grapes under mist. *Proc. Amer. Soc. Hort. Sci.* 63:88–90.

204. ———. 1974. Breeding peach rootstock for the southern United States. *HortScience* 9(4):362–63.

205. Simmonds, N. W. 1966. *Bananas* (2nd ed.). London: Longmans, Green.

206. Singh, S., and D. V. Chugh. 1961. Marcotting with some plant regulators in loquat (*Eriobotrya japonica* Lindl.). *Indian Jour. Hort.* 18:123–29.

207. Singh, L. B. 1960. *The mango: Botany, cultivation, and utilization.* New York: Interscience Publishers.

208. Singh-Dhaliwal, T., and A. Torres-Sepulveda. 1961. Recent experiments on rooting coffee stem cuttings in Puerto Rico. *Univ. Puerto Rico Agr. Exp. Sta. Tech. Paper 33.*

209. Sitton, B. G. 1931. Vegetative propagation of the black walnut. *Mich. Agr. Exp. Sta. Tech. Bul. 119.*

210. Smith, I. E., B. N. Wolstenholme, and P. Allan. 1974. Rooting and establishment of pecan (*Carya illinoensis* [Wang] K. Koch.) stem cuttings. *Agroplantae* 6:21–28.

211. Smith, S. H., R. E. Hilton, and N. W. Frazier. 1970. Meristem culture for elimination of strawberry viruses. *Calif. Agr.* 24(8):8–10.

212. Snir, I. and A. Erez. 1980. *In vitro* propagation of Malling-Merton apple rootstocks. *HortScience* 15(5):597–98.

213. Sookmark, S., and E. A. Tai. 1975. Vegetative propagation of papaya by budding. *Acta Hort.* 49:85–90.

214. Spiegel-Roy, P. 1962. Experience with rootstocks for the apricot and stock-scion compatibilities in Israel. *Proc. 16th Inter. Hort. Cong.,* pp. 587–92.

215. Sriskandarajah, S., and M. G. Mullin. 1981. Micropropagation of Granny Smith apple: Factors affecting root formation *in vitro. Jour. Hort. Sci.* 56(1):71–76.

216. Stebbins, R., and F. E. Larsen. 1981. Rootstocks for apple in the Pacific Northwest. *PNW Coop. Ext. Publ. 208.*

217. Storey, W. B. 1976. *Macadamia tetraphylla*—the preferred rootstock. Vista, Calif.: *Calif. Macadamia Soc. Yearbook* 22:101–7.

218. Stubbs, L. L. 1963. Tristeza-tolerant strains of sour orange. *F. A. O. Plant Prot. Bul.* 11:8–10.

219. Suarez de Castro, F. 1961. Semilleros o germinadores de cafe (Coffee seed beds or germinators). *Agri. Trop.* (Bogota) 17:317–24.

220. Teague, C. P. 1966. Avocado tip-grafting. *Proc. Inter. Plant Prop. Soc.* 16:50–51.

221. Teulon, J. 1971. Propagation of passion fruit (*Passiflora edulis*) on a fusarium-resistant rootstock. *The Plant Propagator* 17(3):4–5.

222. Theakston, F. E. 1976. Carica papaya. In *The propagation of tropical fruit trees,* R. J. Garner and S. A. Chaudri, eds. Hort. Rev. No. 4 Comm. Bur. Hort. and Plant. Crops. East Malling, England: Comm. Agr. Bur.

223. Thomas, C. A. 1981. Mango propagation by saddle grafting. *Jour. Hort. Sci.* 56(2):173–75.

224. Tisserat, B. 1980. Propagation of date palm (*Phoenix dactylifera* L.) *in vitro. Jour. Exp. Bot.* 30(119):1275–83.

225. ———. 1981. Date palm tissue culture. *USDA ARS, Adv. in Agr. Tech. AAT-W-17.*

226. Torre, L. C., and B. H. Barritt. 1979. Red raspberry establishment from root cuttings. *Jour. Amer. Soc. Hort. Sci.* 104(1):28–31.

227. Tomkins, J. P., and D. K. Ourecky. 1972. Growing strawberries in New York State. *N.Y. State Col. Agr. Inform. Bul. 15.*

228. Traub, H. P., and L. C. Marshall. 1937. Rooting of papaya cuttings. *Proc. Amer. Soc. Hort. Sci.* 34:291–94.

229. Tukey, H. B., and K. D. Brase. 1943. The dwarfing effect of an intermediate stempiece of Malling IX. *Proc. Amer. Soc. Hort. Sci.* 42:357–64.

230. Urquhart, D. H. 1961. *Cocoa* (2nd ed.). London: Longmans, Green.

231. van Eignatten, C. L. M. 1969. Propagation of kola (*Cola nitida*). Commun. NEDERF (Amsterdam), Sept.

232. van Staden, J., B. N. Wolstenholme, and G. G. Dimala. 1976. Effect of temperature on pecan seed germination. *HortScience* 11(3):261–62.

233. Warner, R. M., Z. Worku, and J. A. Silva. 1979. Effect of photoperiod on growth responses of citrus rootstocks. *Jour. Amer. Soc. Hort. Sci.* 104(2):232–35.

234. Weaver, R. J. 1976. *Grape growing.* New York: John Wiley.

235. Webster, A. D. 1974. Dwarfing plum rootstocks emerging at East Malling. *Grower* 81:382–86.

236. Weinberger, J. H. 1975. Growing nectarines. *USDA, ARS. Agr. Inf. Bul. 379.*

237. Weinberger, J. H., and N. H. Loomis. 1972. A rapid method for propagating grapevines on rootstocks. *USDA Agr. Res. Ser. ARS-W-2.*

238. Wellman, F. L. 1961. *Coffee: Botany, cultivation, utilization.* London: Leonard-Hill.

239. Westwood, M. N. 1978. *Temperate zone pomology: Pear rootstocks.* San Francisco: W. H. Freeman & Co. Publishers.

240. Westwood, M. N., and L. A. Brooks. 1963. Propagation of hardwood pear cuttings. *Proc. Inter. Plant Prop. Soc.* 13:261–68.

241. Westwood, M. N., H. R. Cameron, P. B. Lombard, and C. B. Cordy. 1971. Effects of trunk and rootstock on decline, growth, and performance of pear. *Jour. Amer. Soc. Hort. Sci.* 96(2):147–50.

242. Westwood, M. N., and A. N. Roberts. 1965. Quince root used for dwarf pears. *Ore. Orn. and Nurs. Dig.* 9(1):1–2.

243. Wilhelm, S. 1974. The garden strawberry: A study of its origin. *American Scientist* 62(3):264–71.

244. Williams, M. W., and R. A. Norton. 1959. Propagation of red raspberries from softwood cuttings. *Proc. Amer. Soc. Hort. Sci.* 74:401–6.

245. Winkler, A. J., J. A. Cook, L. A. Lider, and W. M. Kliewer. 1974. Propagation. Chapter 9 in *General viticulture* (2nd ed.). Berkeley: Univ. of Calif. Press.

246. Worthington, T., and D. H. Scott. 1957. Strawberry plant storage using polyethylene liners. *Amer. Nurs.* 105(9):13, 56–57.

247. Wutscher, H. K. 1979. Citrus rootstocks. *Hort. Rev.* 1:237–69.

248. Zeiger, D., and H. B. Tukey. 1960. An historical review of the Malling apple rootstocks in America. *Mich. State Univ. Cir. Bul. 226.*

249. Zentmyer, G. A., A. O. Paulus, and R. M. Burns. 1962. Avocado root rot. *Calif. Agr. Exp. Sta. Cir. 511.*

250. Ziegler, L. W., and H. S. Wolfe. 1975. *Citrus growing in Florida.* Gainesville, Fla.: Univ. Presses of Fla.

251. Zimmerman, R. H., G. J. Galleta, and O. C. Broome. 1980. Propagation of thornless blackberries by one-node cuttings. *Jour. Amer. Soc. Hort. Sci.* 105(3):405–7.

252. Zimmerman, R. H., and D. C. Broome. 1980. Apple cultivar micropropagation. *USDA, Agr. Research Results ARR-NE-11,* pp. 54–63.

253. Zuccherelli, G. 1979. Moltiplicazione *in vitro* dei portainnesti clonali del pesco. *Frutticoltura* 41:15.

## SUPPLEMENTARY READING

CARLSON, R. F. 1971. Fruit trees—dwarfing and propagation. *Mich. State Univ. Hort. Rpt. No. 1* (rev.).

CHANDLER, W. H. 1957. *Deciduous orchards* (3rd ed.). Philadelphia: Lea & Febiger.

———. 1958. *Evergreen orchards* (2nd ed.). Philadelphia: Lea & Febiger.

DAY, L. H. 1953. Rootstocks for stone fruits. *Calif. Agr. Exp. Sta. Bul. 736.*

GARNER, R. J., and S. A. CHAUDRI. 1976. *The propagation of tropical fruit trees.* Hort. Rev. No. 5. Comm. Bureau of Hort. and Plant. Crops. East Malling, England: FAO and Commonwealth Agr. Bur.

HARTMANN, H. T., and J. A. BEUTEL. 1979. Propagation of temperate zone fruit plants. Leaflet 21103. Berkeley, Calif.: Div. Agr. Sci., Univ. Calif.

HUTCHINSON, A. 1969. Rootstocks for fruit trees. *Ontario (Canada) Dept. Agr. and Food. Publ. 334.*

INTERNATIONAL DWARF FRUIT TREE ASSOCIATION. Compact tree fruit. *Proceedings of Annual Meetings.*

JAYNES, R. A., ed. 1979. *Nut tree culture in North America.* Hamden, Conn.: Northern Nut Growers Assn.

ROACH, F. A. 1969. Fruit tree raising: Rootstocks and propagation. *Ministry of Agriculture, Fisheries, and Food (London) Bul. 135* (5th ed.).

TUKEY, H. B. 1964. *Dwarfed fruit trees.* New York: Macmillan.

WARNER, R. M. 1972. Propagation of tropical crop plants. *Proc. Inter. Plant Prop. Soc.* 22:181–90.

WESTWOOD, M. N. 1978. Rootstocks: Their propagation, function, and performance. Chapter 4 in *Temperate zone pomology.* San Francisco: W. H. Freeman & Co. Publishers.

This chapter describes propagation methods for those woody perennial ornamentals in which there seems to be the greatest interest.

**Abelia** (*Abelia* spp.)    Leafy cuttings from partially matured current season's growth can be rooted easily under glass or mist in summer or fall, and respond markedly to treatments with indolebutyric acid. Hardwood cuttings also may be rooted in fall or spring.

**Abies spp.**    *See* Fir.

**Abutilon** (*Abutilon* spp.) Flowering Maple, Chinese Bell Flower    Started by leafy cuttings rooted under glass or mist in the fall. Seeds germinate without difficulty.

**Acacia** (*Acacia* spp.)    Acacia is usually propagated by seeds. The impervious seed coats must be softened before planting by soaking in concentrated sulfuric acid for 20 minutes to two hours or by pouring boiling water over the seeds and allowing them to soak for 12 hours in the gradually cooling water. Leafy cuttings of partially matured wood can be rooted under mist if treated with 8000 ppm IBA. All but the youngest plants are difficult to transplant because of a pronounced taproot.

**Acer spp.**    *See* Maple.

**Aesculus spp.**    *See* Buckeye.

**Ailanthus** (*Ailanthus altissima* [Mill] Swingle) Tree of Heaven    Seed propagation is easy, and self-sowing usually occurs when both male and female trees are grown close together. Embryo dormancy apparently is present in freshly harvested seed. Stratification at about 4°C (40°F) for two months aids germination. Seed propagation produces both types of trees, but planting male trees should be avoided, since the staminate flowers produce an obnoxious odor. The more desirable female trees can be propagated by root cuttings planted in the spring.

**Albizzia** (*Albizia julibrissin* Durazz.) Silk Tree    This species is started by seeds with

# 19

# Propagation of Ornamental Trees, Shrubs, and Woody Vines

some treatment to overcome the impermeable seed coat, such as soaking in sulfuric acid for half an hour plus thorough washing before planting. Stem cuttings do not root, but root cuttings about 7.5 cm (3 in.) long and 12 mm (½ in.) or more in diameter taken and planted in early spring are successful (*56*).

**Alder** (*Alnus* spp.)     Alder is propagated by seed, which should be cleaned thoroughly and planted in late fall, or given a low-temperature stratification period before planting. Leafy softwood cuttings of *Alnus incana* root easily when the cuttings are taken from young, actively growing shoots (*85*).

**Amelanchier spp.**     *See* Service Berry.

**Aralia** (*Aralia* spp.)     Seeds should be treated with sulfuric acid for 30 or 40 minutes, then planted in early fall. Do not allow seeds to become dry.

**Araucaria araucana**     *See* Monkey Puzzle Tree.

**Araucaria heterophylla**     *See* Norfolk Island Pine.

**Arborvitae** (American [*Thuja occidentalis* L.] and oriental [T. *orientalis* L.])

**Seeds**     Germination is relatively easy but stratification of seeds for 60 days at about 4°C (40°F) may be helpful.

**Cuttings**

*Thuja occidentalis*     Cuttings can be rooted in midwinter under mist in the greenhouse. Best rooting is often found with cuttings taken from older plants no longer making rapid growth. The cuttings should be about 20 cm (6 in.) long and may be taken either from succulent, vigorously growing terminals or from more mature side growths several years old. Wounding and treating with root-inducing chemicals are beneficial. No shading should be used.

Cuttings may also be made in midsummer and rooted out-of-doors in a shaded, closed frame. They should be several inches long and of current season's growth with somewhat matured wood at the base. They should be rooted by fall.

*Thuja orientalis*     Cuttings of this species are often more difficult to root than those of *T. occidentalis.* Small, soft cuttings several centimeters long, taken in late spring, can be rooted in mist beds if treated with a root-promoting chemical (*170*).

**Grafting**     The side graft is used in propagating selected clones of *T. orientalis,* with two-year-old potted *T. orientalis* seedlings as the rootstock. Grafting is done in late winter in the greenhouse. After making the grafts, the potted plants are set in open benches filled with moist peat moss just covering the union. The grafted plants should be ready to set out in the field for further growth by midspring.

**Arbutus menziesii**     *See* Madrone.

**Arctostaphylos**     *See* Manzanita.

**Aronia arbutifolia**     *See* Chokeberry.

**Ash** (*Fraxinus* spp.)     Seeds of most species germinate if stratified for two to four months at about 4°C (40°F). Seeds of *F. excelsior, F. nigra,* and *F. quadrangulata* should have one to three months moist storage at room temperature, followed by five to six months at about 4°C (40°F). *F. excelsior* and *F. ornus* seedlings are usually used as rootstocks for grafting or budding ash cultivars.

**Aspen**     *See* Poplar.

**Aucuba japonica 'Variegata'**     *See* Gold Dust Plant.

## Azalea (*Rhododendron* spp.) (*95, 105*)

### Evergreen and semievergreen types

Although evergreen azaleas can be propagated by seeds, grafting, and layering, al. most all nursery plants are started by cuttings.

**Seeds**    The seed capsules should be gathered in the fall after they turn brown, then stored at room temperature in a container that will hold the seed when the capsules open. The seeds have no dormancy problems but will lose viability if held for extended periods in open storage; sealed, low-temperature storage is preferable. The seeds germinate satisfactorily in a thin layer of screened sphagnum moss over an acid soil mixture. Seeds are usually planted in the greenhouse from midwinter to early spring. Optimum germinating temperatures are about 21° C (70° F) during the day and 13° C (55° F) at night. Germination usually occurs within a month. In areas having hard water, the seedlings should be watered with distilled or rain water because the alkaline salts will soon cause injury.

**Cuttings**    Azalea cuttings of the evergreen and semievergreen types are not difficult to propagate, roots forming in three to four weeks under proper conditions. They are best taken in midsummer after the current season's new growth has become somewhat hardened but before the wood has turned red or brown. Root-promoting growth regulators are often beneficial. Azaleas do very well under mist if the medium is well drained. After roots form the mist should be reduced and finally discontinued.

After rooting, the cuttings may be potted in equal parts of composted leaf mold, sand, and peat moss.

### Deciduous types

**Seeds**    Deciduous azaleas are often propagated by seed, since it is inexpensive and rooting cuttings is difficult. The same germination procedures can be used as described for the evergreen types. Growth of the seedlings is sometimes slow, but with early spring germination and early transplanting to a good growing medium, it is possible to have plants large enough to set flower buds in their second year.

In a different procedure, which will bring the seedlings into bloom sooner, seeds are sown in late summer (from seed harvested in midwinter) on top of screened peat moss in flats covered with glass. After the seeds germinate, the day length is extended by added light to keep them growing well. The seedlings are transplanted in midfall into flats to 10 percent sand and 90 percent peat moss, where they are left (in the greenhouse) until midsummer (*78*).

**Cuttings** (*15, 21, 127*)    Deciduous azaleas are difficult to propagate by cuttings, but by taking soft, succulent material in the spring, particularly from stock plants brought into the greenhouse and forced, cuttings can be rooted. Timing is very important; it is much better to take the cuttings in spring than in summer.

Wounding, plus the use of root-promoting chemicals, is important in obtaining good rooting. Treatment with indolebutyric acid (75 ppm for 15 hours, or IBA in talc at 0.8 percent) has been found to be effective. Closed frames or intermittent mist with bottom heat (20° C; 68°F) and a rooting medium of sphagnum peat or a sand-peat-perlite mixture provide good rooting conditions.

**Grafting**    This is the chief propagation method for most Ghent and Mollis hybrids. *Rhododendron luteum (Azalea pontica)* seedlings, two or three years old, are used as the understock. This method has the disadvantages of undesirable suckering from the understock and a general lack of plant vigor, perhaps owing to an unsatisfactory graft union or to a lack of adaptability of the seedling stocks to some climates. The seedlings are potted in early spring, the grafting being done during late summer. A side-veneer graft may be used. The grafted plants are set in closed frames in the greenhouse during healing of the union (*110*). Tree-type azaleas can be produced by grafting high in rooted cuttings of *R. concinnum*.

**Layering** Azaleas are easily layered. It is a method worth using for obtaining a limited number of new plants, especially of the deciduous type difficult to start by cuttings. Simple, mound, trench, and air layering have all been used satisfactorily.

**Bald Cypress** (*Taxodium distichum* [L.] Rich.) Bald cypress is propagated by seeds, which should be fall-planted or stratified at 5°C (41°F) for 90 days.

**Bamboo** (*Arundinaria, Bambusa, Dendrocalamus, Phyllostachys, Sasa* spp.) (*111, 115*) Bamboo has about 1000 species in some 50 genera. It is easily propagated vegetatively by division of the clumps or young rhizomes. Use of one- or two-year-old growth taken from the periphery of the clump gives best results. Prevention of drying during transplanting is important. Best results are obtained from rhizomes taken and planted in late winter or early spring before the buds begin to push (*40*).

**Banksia spp.** These native Australian plants are best propagated by seed although they are difficult to remove from the seed-containing structures. One method is to soak several days in water then dry quickly in the hot sun.

**Barberry** (*Berberis* spp.) Barberries are propagated without difficulty by fall-sowing or by spring-sowing seeds that have been stratified for two to six weeks at about 4°C (40°F). It is important to remove all pulp from seeds. Leafy cuttings taken from spring to fall can be rooted under mist. Indolebutyric acid at 5000 ppm aids rooting (*166*). Greenhouse grafting of some selected types also is practiced, and layering is done occasionally. Division of crowns is useful for small quantities of plants.

**Basswood** *See* Linden.

**Bayberry** (*Myrica* spp.) Wax myrtle Propagated by seed or by cuttings, both with difficulty. *M. pennsylvanicum* best started by seed given a scarification, then cold-stratification treatment, followed by kinetin applications. Gibberellic acid treatments also may be helpful (*68*).

**Beech** (*Fagus* spp.) Seeds germinate readily in the spring from fall planting or after being stratified for three months at about 4°C (40°F). Seeds should not be allowed to dry out. Selected clones are grafted by either the cleft, the whip, or the side-veneer method, on seedlings of *F. sylvatica*, the European beech. Frequent transplanting or root pruning of nursery trees is necessary to prevent development of a single long taproot, which, if present, makes subsequent transplanting difficult (*36*).

**Berberis spp.** *See* Barberry.

**Betula spp.** *See* Birch.

**Birch** (*Betula* spp.) Seeds should be either fall-planted or spring-planted following stratification at about 4°C (40°F) for one to three months. *Betula nigra* matures its seeds in the spring. If planted promptly, the seed will germinate at once without treatment. Some of the selected, weeping forms are grafted on *Betula pubescens* or *B. pendula* seedlings (*39*). *B. papyrifera* seeds germinate best if held below freezing temperature for six weeks before planting. Birch is considered difficult to propagate by cuttings, but leafy, semihardwood cuttings root under mist if taken in midsummer. In one test (*12*) use of rooting hormones was not helpful. Low-growing species can be propagated by layering. Micropropagation of birch has been successful (*116*).

**Bittersweet** (*Celastrus* spp.)    These twining vines have male flowers on one plant and female on another, and the two types must be near each other to produce the attractive berries. To propagate, seeds must be removed from the berries and then fall-planted or stratified for about three months at 4°C (40°F) before planting. Clones of known sex can be propagated by softwood cuttings taken in midsummer, or by hardwood cuttings taken in winter. Treatments with indolebutyric acid aid in rooting both types.

**Bottle Brush** (*Callistemon* spp.)    Although seeds germinate without difficulty, seedlings should be avoided because many of them prove worthless as ornamentals. The preferred method of propagation is by leafy cuttings of partially matured wood taken from selected cultivars. These root under glass quite easily.

**Bougainvillea** (*Bougainvillea* spp.)    This showy tropical evergreen vine, growing outdoors only in mild climates, is propagated by hardwood cuttings taken at any time of the year. Some cultivars, difficult to start, should be rooted as leafy cuttings under mist after treatment with a root-promoting substance. Micropropagation by shoot-tip culture has been successful (*26*).

**Boxwood** (*Buxus sempervirens* L. and *B. microphyllas* Sieb. and Zucc.)    Cuttings are commonly used—either softwood taken in spring or summer, or semihardwood taken in the fall. In the latter method, which is ordinarily used, the cuttings are rooted in a cool greenhouse or cold frame during the winter and spring or under mist at any time. Seeds are rarely used, because of the very slow growth of the seedlings. Young plants should always be grown in a container or transplanted with a ball of soil around their roots.

**Broom** (*Cytisus* spp.)    Seeds of many of the species germinate satisfactorily if gathered as soon as mature and treated with sulfuric acid to soften the hard seed coat before planting. The seeds are best germinated in a warm location and then transferred to a cooler place when the seedlings are several inches high. Since the various *Cytisus* species crossbreed readily, stock plants for seed sources should be isolated.

Cuttings can be rooted rather easily under mist in midsummer if treated with indolebutyric acid and given bottom heat.

**Buckeye** (*Aesculus* spp.) (*120*) Horse Chestnut    This tree may be propagated by seeds, but prompt sowing or stratification after gathering in the fall is necessary. If the seeds lose their waxy appearance and become wrinkled, their viability will be reduced. For best germination, seeds should be stratified for four months at about 4°C (40°F) immediately after collecting. In propagating the low-growing species, simple layering in either spring or fall is often used. T-budding or bench grafting, using the whip graft, can be used for selected cultivars worked on *A. hippocastanum* as the rootstock. The dwarf buckeye, *A. parviflora,* is propagated by underground stem pieces. Seeds of buckeye are poisonous if eaten.

**Buckthorn** (*Rhamnus* spp.)    This can be propagated by planting seed out-of-doors in the fall. Macerate fruits and clean seeds. Those of some species may germinate better if given a 20 minute treatment with sulfuric acid before planting.

**Butterfly Bush** (*Buddleia* spp.)    Softwood cuttings can be taken in the summer or fall and rooted under mist or in a closed frame. Seeds started in the greenhouse in early spring will provide flowering plants by fall, although reproduction is not genetically true by seed.

**Buddleia spp.**    *See* Butterfly Bush.

**Buxus spp.**    *See* Boxwood.

**Callistemon spp.**     *See* Bottle Brush.

**Calluna vulgaris**     *See* Heather.

**Calocedrus decurrens**     *See* Incense Cedar.

**Camellia** (*Camellia* spp.) (*118*)     Camellias can be propagated by seed, cuttings, grafting, and layering. They do not come true from seed. To perpetuate cultivars, cuttings, grafts, or layers must be used. Seedlings are used in breeding new cultivars, as rootstocks for grafting, or in growing hedges where foliage is the only consideration.

**Seeds**     In the fall when the capsules begin to turn reddish-brown and split, seeds should be gathered before the seed coats harden and the seeds become scattered. The seed should not be allowed to dry out and should be planted before the seed coat hardens. If the seeds must be stored for long periods, they will keep satisfactorily mixed with ground charcoal and stored in an airtight container placed in a cool location. After the hard seed coat develops, germination is hastened by pouring boiling water over the seeds and allowing them to remain in the cooling water for 24 hours. For germination, a well-drained acid soil high in organic matter should be used. It takes four to seven years to bring a camellia into flowering from seed.

**Cuttings**     Most *C. japonica* and *C. sasanqua* cultivars are produced commercially from cuttings, which are not difficult to root. *Camellia reticulata* cuttings do not root easily, however, and this species is generally propagated by either cleft or approach grafting.

· Cuttings are best taken in midsummer from the spring flush of growth after the wood has matured somewhat and changed from green to light brown in color. Tip cuttings are used, 3 to 6 in. long, with two or three terminal leaves. Rooting will be much improved if the cuttings are treated with indolebutyric acid used at 20 ppm for 24 hours. Wounding the base of the cuttings before they are treated is also likely to improve rooting. The cuttings root best either in a closed frame or under mist.

Camellias may be started also as leaf-bud cuttings, which are handled as stem cuttings. In this case, excessive concentrations of a root-promoting substance should be avoided because this may inhibit development of the single bud.

**Grafting**     Camellias are frequently grafted, not only to produce nursery plants but to change cultivars of older established plants. Vigorous seedlings or rooted cuttings of either *C. japonica* or *C. sasanqua* can be used as stocks for grafting. Any of the side graft methods is suitable.

Both cleft and bark graft methods are used for grafting larger, established plants. This is best done in the spring, two or three weeks before new vegetative growth starts. The stock and the scion should be dormant at the time of grafting. Scions several inches long from terminal shoots containing one or two leaves and several dormant buds are used, inserted into the stock plant, which is cut off 5 to 7.5 cm (2 to 3 in.) above the soil level.

The "nurse-seed" method of grafting has been used successfully for cultivars hard to root (*122*).

**Layering**     To obtain a few additional plants from a single mother plant, simple layering may be performed in the spring. Branches must be present close to the ground which, preferably, are young and not over 13 mm (½ in.) in diameter. One or two years may be required for enough roots to form before the layer can be removed.

**Camphor Tree** (*Cinnamomum camphora* (L.) J. Pressl)     Best propagated by semihardwood cuttings taken in spring and rooted in a closed frame or under mist. Seed propagation also is satisfactory.

**Campsis spp.**　　*See* Trumpet Creeper.

**Cape Jasmine**　　*See* Gardenia.

**Carpinus spp.**　　*See* Hornbeam.

**Casuarina spp.**　　These Australian native trees are easily propagated by seed.

**Catalpa** (*Catalpa* spp.)　　Seeds germinate readily without any previous treatment. They are stored dry over winter at room temperature and planted in late spring. For ornamental purposes, *C. bignonioides* 'Nana' is often budded or grafted high on stems of *C. speciosa,* giving the "umbrella tree" effect. A strong shoot is forced from a one-year-old seedling rootstock, which is then budded with several buds in the fall at a height of 6 ft or more. Catalpa species can also be propagated in summer by softwood cuttings rooted under glass.

**Ceanothus** (*Ceanothus* spp.)　　Propagation is by seed, cuttings, layering, and sometimes grafting. Seeds must be gathered shortly before the capsules open or they will be lost. Those of *C. arboreus, C. cuneatus, C. jepsoni, C. megacarpus, C. oliganthus, C. rigidis,* and *C. thyrsiflorus* have only seed coat dormancy. Placing the seeds in hot water (82° to 87° C; 180° to 190° F) and allowing them to cool for 12 to 24 hours, or even boiling in water for five minutes, will aid germination. To obtain germination in other *Ceanothus* species, which have both seed coat and embryo dormancy, the seed should be immersed in hot water (as described above), then stratified at 2° to 4° C (35° to 40° F) for two to three months.

　　Leafy cuttings can be rooted under mist at any time from spring to fall, especially if treated with a root-promoting substance. Terminal softwood cuttings from vigorously growing plants in containers give good results.

　　*Ceanothus americanus* seedlings are often used as rootstocks for grafting selected clones.

**Cedar** (*Cedrus* spp.)　　Seeds germinate readily if not permitted to dry out. No dormancy conditions occur, but soaking the seeds in water several hours before planting is helpful. Cuttings do not root easily, but if taken in late summer or fall, treated with a root-promoting hormone and placed over bottom heat under a plastic covered frame, some rooting may be obtained (*55*). Side-veneer grafting of selected forms on one- or two-year-old potted seedling stocks may be done in the spring. Scions should be taken from vigorous terminal growth of current season's wood rather than from lateral shoots. *Cedrus atlantica* selections are grafted on *Cedrus deodara* seedlings as a rootstock (*36*).

**Celastrus spp.**　　*See* Bittersweet.

**Celtis spp.**　　*See* Hackberry.

**Cercis spp.**　　*See* Redbud.

**Chaenomeles spp.**　　*See* Quince, Flowering.

**Chamaecyparis** (*Chamaecyparis* spp.)　　False cypress. After collection in the fall, the seeds should be carefully dried in a warm room or in a kiln at 32° to 43°C (90° to 110°F). Stratification at about 4° C (40° F) for two to three months will aid germination. Cuttings of most species are not difficult to root, particularly if juvenile forms are used. They may be taken in fall or winter and rooted in a closed frame in the greenhouse, using lateral shoots of current season's wood. Indolebutyric acid treatments are helpful (*98*).

**Cherry, Flowering** (*Prunus* spp.)    Cultivars of *P. serrulata, P. sargentii, P. sieboldi, P. yedoensis, P. campanulata,* and *P. subhirtella* comprise the flowering cherries. Seedlings of *P. avium,* the Mazzard cherry, or *P. serrulata* are suitable as rootstocks upon which these ornamental forms may be T-budded, either in the fall or in the spring. *Prunus dropmoreana* is a suitable stock for *P. serrulata* 'Kwanzan' (*53*). If cross-pollination with other species can be avoided, *P. sargentii, P. campanulata,* and *P. yedoensis* will reproduce true from seed. Leafy cuttings of some of the flowering cherry species can be rooted under mist in high percentages if treated with indolebutyric acid, but subsequent survival and overwintering are sometimes difficult.

**Chionanthus spp.**    *See* Fringe Tree.

**Chokeberry** (*Aronia arbutifolia* [L.] Pers.)    Propagated by seed, which should be stratified for about three months at 5°C (41°F) before planting. Can also be started by leafy cuttings under glass or mist, or by suckers or layers.

**Cinnamomum camphora**    *See* Camphor Tree.

**Clematis** (*Clematis* spp.) (*47, 48*)    Clematis can be propagated by seed, cuttings, grafting, division of roots, or layering. Seeds of some clematis species have embryo dormancy, so stratification for one to three months at about 4° C (40° F) is likely to aid germination. Clematis is probably best propagated by cuttings taken from young plants; these root under mist in about five weeks. Young wood with short internodes taken in the spring gives satisfactory results, but partially matured wood taken in late spring to late summer is more commonly used. Leaf-bud cuttings taken in midsummer also will root readily under mist. Treatment with indolebutyric acid is helpful. The large-flowering hybrids are generally root-grafted by the cleft or side-veneer graft or a similar method (*143*) in the spring on *C. flammula, C. vitalba,* or *C. viticella* seedlings. The grafts are planted deeply with scion roots forming, the rootstock acting as a nurse-root. When only a few plants are needed, layering the long canes gives satisfactory results.

**Cornus spp.**    *See* Dogwood.

**Cotinus coggygria**    *See* Smoke Tree.

**Cotoneaster** (*Cotoneaster* spp.) (*59*)    Seeds should be soaked for about 90 minutes in concentrated sulfuric acid and then stratifed for three to four months at about 4°C; 40°F. As a substitute for the acid treatment, a moist, warm (15° to 24°C; 60° to 75°F) stratification treatment for three to four months may be used. This must be followed by the cold stratification treatment. Leafy cuttings of most species taken in spring or summer will root under mist without much difficulty. Treatments with indolebutyric acid are helpful. Cotoneaster can be budded high onto a pear nursery tree to produce a "tree" cotoneaster. A blight-resistant pear, such as 'Old Home', should be used. Simple layering can be done.

**Cottonwood**    *See* Poplar.

**Crabapple, Flowering** (*Malus* spp.) (*180*)    Four species of crabapples—*M. toringoides, M. hupehensis, M. sikkimensis,* and *M. florentina*—will reproduce true from seed. Selected forms of all other crabapple species, such as *M. sargentii, M. floribunda,* and *M.* 'Dolgo', should be propagated by asexual methods. Nursery trees are commonly propagated either by root grafting, using the whip graft or by T-budding seedlings in the nursery row; the latter is done either as spring or as fall budding. Fall budding is considered by most nurserymen to be the fastest and most desirable method of propagating crabapples. In addition, older *Malus* trees may be topworked to the desired flowering crab cultivar. Various seedling rootstocks are used, such as *M. pumila* (common apple), *M. baccata*, *M. ionensis,* and *M. coronaria,* as well as a number of the Malling series of apple rootstocks (*53*). The flowering 'Bechtel's Crab' is said to show delayed incompatibility, which appears in 10 to 15 years, when worked on *M. pumila* seedlings. Dwarf trees of the 'Carmine' crabapple (*M. atrosanguinea*) have been produced by grafting onto *Cotoneaster divaricata.*

**Crape Myrtle** (*Lagerstroemia indica* L.) (*45*)    This species is propagated by subterminal leafy cuttings under glass or mist taken in early summer. Transplanting is somewhat difficult, so all but very small plants should be moved with a ball of soil. It is also easily propagated by hardwood cuttings planted in early spring and dug bare root in the fall. Root cuttings taken in late winter also are successful.

**Crataegus spp.**    *See* Hawthorn.

**Cryptomeria, Japanese** (*Cryptomeria japonica* [L.f] D. Don)    This can be propagated either by seeds or by cuttings. Seeds should not dry out. Cuttings, 5 to 15 cm (2 to 6 in.) long, should be taken from green wood at a stage of maturity at which it breaks with a snap when bent. Root in sand over bottom heat; keep the cuttings shaded and cool. After roots start to form, in about two weeks, give more light; transplant to pots when roots are about 13 mm ($\frac{1}{2}$ in.) long. Indolebutyric acid treatments promote rooting.

**× Cupressocyparis leylandii**    *See* Leyland cypress.

**Cupressus spp.**    *See* Cypress.

**Cypress** (*Cupressus* spp.)    Seeds have embryo dormancy, so stratification for about four weeks at 2° to 4°C (35° to 40°F) will improve germination. Cuttings can be rooted if taken during winter months. Treatments with indolebutyric acid at 60 ppm for 24 hours aid rooting. Side-veneer grafting of selected forms on seedling *Cupressus* rootstocks in the spring is often practiced.

**Cytisus spp.**    *See* Broom.

**Daphne** (*Daphne* spp.) (*20, 24*)    Daphne can be propagated by seed, stem and leaf-bud cuttings, layering, and grafting. Seeds should be sown at once after harvest or, if dried, scarified, then given a moist-chilling period before planting. Daphne is probably best propagated by leafy cuttings in sand and peat moss (2:1) under glass or mist; cuttings are taken in summer from partially matured current season's growth. Root-promoting substances are often helpful. The daphnes do not transplant easily and should be moved only when young. Berries are very poisonous if eaten.

**Dawn Redwood**    *See* Metasequoia.

**Delonix regia**    *See* Royal poinciana.

**Deutzia** (*Deutzia* spp.)    Deutzia is easily propagated either by hardwood cuttings lined-out in the nursery row in spring or by softwood cuttings under glass or mist in summer.

**Diosma** (*Ericoides* L.) Breath of Heaven.    Propagated by leafy cuttings taken in summer and rooted under glass or mist.

**Dogwood** (*Cornus* spp.)    Seeds have various dormancy conditions; those of the popular flowering dogwood (*C. florida*) require either fall planting or a stratification period of about four months at 4°C (40°F). Best germination is obtained if the seeds are gathered as soon as the fruit starts to color, and sown or stratified immediately. If allowed to dry out it is best to remove seeds from the fruit and soak in water. Other species require, in addition, a treatment to soften the seed covering. Two months in moist sand at diurnally fluctuating temperatures (21° to 30°C; 70° to 85°F), followed by four to six months at 0° to 4°C (32° to 40°F), is effective. With some species, the warm stratification period may be replaced by mechanical scarification or soaking in sulfuric acid.

Some dogwoods can be started easily by cuttings. Those of *C. florida* are best taken in late spring or early summer from new growth after flowering, then rooted under mist. Treatments with indolebutyric acid at 20,000 ppm liquid formulation have given good rooting. Hardwood cuttings taken in the spring are successful with certain species, such as *C. alba.*

*C. florida* 'Rubra' can be rooted successfully if cuttings are taken in early summer after the second flush growth and rooted under mist in peat moss-sand 1:3, or in perlite-peat, 3:2 (*145*), after treating with IBA, 3000 ppm. To insure survival through the following winter in cold climates, the potted cuttings should be kept in heated cold frames to hold the temperature between 0° and 7°C (32° and 45°F).

Selected types such as the red flowering dogwood, *C. florida* 'Rubra', and the weeping forms, difficult to start by cuttings, are often propagated by T-budding in late summer, or by whip grafting in the greenhouse in winter on *C. florida* seedlings (*29*).

**Douglas Fir** (*Pseudotsuga menziesii* [Mirb.] Franco) (*5, 11*)     Seeds of this important lumber and Christmas tree species exhibit varying degrees of embryo dormancy. For prompt germination, it is best to sow the seeds in the fall or stratify them for two months at about 4°C (40°F). Adventitious root formation on cotyledons in in vitro culture has been successful in producing plantlets (*27*).

Douglas fir cuttings are rather difficult to root, but by taking them in late winter, treating them with indolebutyric acid, and rooting them in a sand-peat moss mixture, it is possible to obtain fairly good rooting. Cuttings from young trees root much more easily than those from old trees, and cuttings from certain source trees are easier to root than those from others (*7*).

**Elaeagnus** (*Elaeagnus* spp.) Russian olive, Silverberry.     Seeds planted in the spring germinate readily following a stratification period of three months at 4°C (40°F). Removal of the pit (endocarp) for silverberry seeds (*E. commutata*) resulted in about 90 percent germination of unstratified seeds, a germination inhibitor apparently being present in the pit (*32*). Seeds of the Russian olive, *E. angustifolia,* should be treated with sulfuric acid for 30 to 60 minutes before fall planting or stratification. This species can also be started by hardwood cuttings planted in spring. Root cuttings or layering also is successful. Leafy cuttings of the evergreen, *E. pungens,* will root if taken in the fall and started under glass or mist.

**Elderberry** (*Sambucus* spp.)     Seed propagation is difficult because of complex dormancy conditions involving both the seed coat and embryo. Probably the best treatment is a warm (21° to 30°C; 70° to 85°F), moist stratification period for two months, followed by a cold (4°C; 40°F) stratification period for three to five months. These conditions could be obtained naturally by planting the seed in late summer, after which germination should occur the following spring.

Since softwood cuttings root easily if taken in spring or summer, this method is generally practiced.

**Elm** (*Ulmus* spp.)     Seed propagation is commonly used. Elm seed loses viability rapidly if stored at room temperature, but it can be kept for several years in sealed containers at 0° to 4°C (32° to 40°F). Seed ripening in the spring should be sown immediately, and germination usually takes place promptly. For those species that ripen their seed in the fall, either fall planting or stratification for two months at about 4°C (40°F) should be used. To obtain tree uniformity, selected clones are propagated by budding on seedling rootstocks of the same species.

Softwood cuttings of several elm species can be rooted under mist when taken in early summer (4). Treatments with indolebutyric acid at 50 ppm for 24 hours have been beneficial (135). Semilignified cuttings have been rooted without mist or hormone treatments (144). Softwood cuttings taken from new growth arising from cut-off stumps root readily if treated with IBA and placed under mist (146). *U. hollandica* can be propagated by hardwood cuttings taken in late winter, treated with IBA at 1500 ppm, then placed in a bin over bottom heat for six weeks before planting (175).

**English Ivy** ( *Hedera helix* L.)    English ivy is readily propagated by rooting cuttings of the juvenile (nonfruiting, lobed-leaf) form. It is also sometimes grafted onto *Fatshedera lizei (Fatsia japonica | Hedera helix)* as a rootstock. *Acanthopanax sieboldianus* is a suitable rootstock where root rot is likely to result from excessive soil moisture.

**Erica spp.**    *See* Heath.

**Escallonia** ( *Escallonia* spp.)    Escallonia is easily started by leafy cuttings taken after a flush of growth. Cuttings root well under mist and respond markedly to treatment with indolebutyric acid.

**Eucalyptus** ( *Eucalyptus* spp.) (33, 131)    Eucalyptus is almost entirely propagated by seeds planted in the spring. Mature capsules are obtained just before they are ready to open. No dormancy conditions occur in most species, the seed being able to germinate immediately following ripening. Seeds of some species—for example, *E. dives, E. niphophila,* and *E. pauciflora*—however, require stratification for about two months at 4°C (40°F) for best germination. Eucalyptus seedlings are very susceptible to damping-off. Seeds are usually planted in flats of pasteurized soil placed in a shady location. From flats they are transplanted into small pots, from which they are later lined out in a nursery row. The roots of young trees will not tolerate drying, so the young plants should be handled as container-grown stock. Seeds may be sown directly into containers in which the seedlings are grown until planting in their permanent location (67).

Eucalyptus is difficult to start from cuttings, but good rooting can be obtained. For example, leafy cuttings of *E. camaldulensis* taken in early spring from shoots arising from the base of young trees, wounded, treated with a 4000 ppm solution of IBA plus NAA, 1:1, and placed in perlite under mist over bottom heat at 21°C (70°F), gave 65 percent rooting (49).

*Eucalyptus ficifolia* has been successfully grafted by a side-wedge method, using young vigorous *Eucalyptus* seedling rootstocks growing in containers and placed under very high humidity following grafting. Use of scions taken from shoots that had been girdled at least a month previously increased success (141).

**Euonymus** ( *Euonymus* spp.) (61)    Stratification for three to four months at 0° to 10°C (32° to 50°F) is required for satisfactory seed germination. Remove seeds from fruit and prevent drying. Euonymus is easily started by cuttings, hardwood in early spring for the deciduous species and leafy, semihardwood under glass or mist after a flush of growth has partially matured, for the evergreen types. Layering also is successful.

**Euphorbia pulcherrima**    *See* Poinsettia.

**Fagus spp.**    *See* Beech.

**False Cypress**  *See* Chamaecyparis.

**Fir** (*Abies* spp.)    Seed propagation is not difficult, but fresh seed should be used; those of most species lose their viability after one year in ordinary storage. Embryo dormancy is generally present, fall planting or stratification at about 4°C (40°F) for one to three months required for good germination. Fir seedlings are very susceptible to damping-off. They should be given partial shade during the first season, since they are injured by excessive heat and sunlight.

Fir cuttings are considered difficult to root, but *Abies fraseri* cuttings can be rooted in high percentages, especially if taken from young trees, wounded, and treated with IBA (*83*). This species, along with white fir (*A. concolor*) and red fir (*A. magnifica*), the California "silver tip," are important Christmas tree species.

**Firethorn**  *See* Pyracantha.

**Forsythia** (*Forsythia* spp.)    Forsythia is easily propagated by hardwood cuttings set in the nursery row in early spring or by leafy softwood cuttings taken during late spring and rooted under high-humidity conditions.

**Fraxinus spp.**  *See* Ash.

**Fringe Tree** (*Chionanthus virginicus* L. and *C. retusus* L.)    Seed propagation can be used, but is very slow. Embryo dormancy, as well as some endosperm inhibition, seems to be present. Probably the best practice is stratification for 30 days or more at room temperature, followed by one or two months' stratification at about 4°C (40°F). With fall-planted seed, germination may not occur until the second spring.

Cutting propagation of the Chinese fringe tree (*C. retusus*) generally has been considered almost impossible, but by taking softwood cuttings in late spring, rooting under mist, treating with indolebutyric acid, and using a mixture of sand and vermiculite as the rooting medium, excellent rooting percentages can be obtained (*153*).

**Fuchsia** (*Fuchsia* spp.) (*163*)    Fuchsia is easily rooted by leafy cuttings maintained under humid conditions. Roots develop in two or three weeks.

**Gardenia** (*Gardenia jasminoides* Ellis) Cape jasmine    Leafy terminal cuttings can be rooted in the greenhouse under glass or mist from fall to spring. Sand and peat moss, 1:1, is a good rooting medium. Gardenias are difficult to transplant and should be moved only when small.

**Ginkgo** (*Ginkgo biloba* L.) (*107*)    Seed propagation may be used for this "living fossil," shown by records in rocks to have existed on earth 150 million years ago during the dinosaur age. A satisfactory procedure is to collect the "fruits" in midfall, remove the pulp, and pack the cleaned seeds in layers of moist sand for ten weeks at temperatures of 15° to 21°C (60° to 70°F) to permit the embryos to finish developing. After this the seeds require a stratification period of several months at about 4°C (40°F) for good germination. Seedlings produce either male or female trees, but the sex cannot be determined until the trees flower—after about 20 years. The plum-like "fruits" on the female trees have a very disagreeable odor, so only male trees are used for ornamental planting. For this reason, seed propagation should not be used; propagation by cuttings from male trees is advisable. Softwood cuttings taken in early summer can be rooted under glass or mist if treated with indolebutyric acid. Commercial propagation is by T-budding or chip budding using buds from male trees inserted into ginkgo seedlings.

**Gleditsia triacanthos**    *See* Honeylocust.

**Goldenrain Tree** (*Koelreuteria* spp.)    This tree is usually propagated by seed. The seeds have double dormancy, germinating best if the seed coats are softened by soaking for about 60 minutes in concentrated sulfuric acid, or by mechanical scarification followed by stratification for about 90 days at 2° to 4°C (35° to 40°F) to overcome the embryo dormancy. Root cuttings can be used and softwood cuttings of new growth taken in the spring can be rooted under glass or mist.

**Grevillea spp.**    These Australian native shrubs or trees are propagated by seed or by cuttings. Cuttings of the low growing species root readily, but larger growing species, such as *G. robusta,* silk oak, are best propagated by seed. Take cuttings from late summer to early winter.

**Hackberry** (*Celtis* spp.) (*117*)    Seeds are ordinarily used, sown either in the fall or stratified for two or three months at about 4°C (40°F) and planted in the spring. Prior to stratification, treatments to soften the seed coat, such as soaking in concentrated sulfuric acid, may hasten germination. Clones of two species, *C. occidentalis* and *C. laevigata* (sugarberry), can be started by cuttings. Grafting and chip budding also have been used.

**Hamamelis**    *See* Witch Hazel.

**Hawthorn** (*Crataegus* spp.) (*35*)    Hawthorns tend to reproduce true by seed. Pronounced seed dormancy is present because of a combination of an impermeable seed coat and embryo conditions. Probably the best procedure for rapid germination is stratification of freshly collected and cleaned seed in moist peat moss for three or four months at 21° to 27°C (70° to 80°F) (or treatment with sulfuric acid), followed by stratification for five months at about 4°C (40°F). Planting the seed in early summer will provide these conditions naturally, resulting in germination the following spring. Untreated seed may require two or three years to germinate. Seeds of some *Crataegus* species do not have an impermeable seed coat, so the initial high-temperature storage period is unnecessary. Since hawthorn develops a long taproot, transplanting is successful only with very young plants.

Selected clones may be T-budded or root-grafted on seedlings of *C. crus-galli,* or *C. coccinea* for the American (entire leaf) types and seedlings of *C. laevigata* or *C. monogyna* for the European (cut-leaf) types.

**Heath** (*Erica* spp.) and **Heather** (*Calluna vulgaris* Hull)    The propagation of these two closely related genera is about the same. Seeds may be germinated in flats in the greenhouse in winter or in a shaded outdoor cold frame in spring. Leafy, partially matured cuttings taken at almost any time of year, but especially in early summer, root readily under glass or mist. Treatment with indolebutyric acid at about 50 ppm for 24 hours is helpful.

**Hebe** (*Hebe* spp.) Veronica    Propagated by seed, by leafy cuttings in summer under mist, or by layering.

**Hedera helix**    *See* English Ivy.

**Hemlock** (*Tsuga* spp.)    Hemlocks are propagated by seed without difficulty. Seed dormancy is variable; some lots exhibit embryo dormancy whereas others do not. To insure good germination it is advisable to stratify the seeds for two to four months at about 4°C (40°F). Fall planting outdoors generally gives satisfactory germination in the spring. The seedlings should be given partial shade during the first season. Hemlock cuttings are somewhat difficult to root, but success has been reported with cuttings taken at all times of the year. They seem to respond to treatments with root-promoting substances. Layering also is successful.

**Heteromeles arbutifolia** (Ait.) M. J. Roemer, Toyon, Christmas Berry     Usually propagated by seed which requires a long (3 month) stratification period, or by fall planting to obtain outdoor winter chilling of the seed. Softwood tip cuttings taken in midspring, treated with 8 percent IBA in talc, and placed under mist will root (*66*). It can also be started by layering.

**Hibiscus** (*Hibiscus syriacus* L.) Shrub-althea, Rose of Sharon.     This is easily propagated, either by hardwood cuttings in the nursery row in spring, or by softwood cuttings in midsummer under glass. Lateral shoots make good cutting material. Softwood cuttings respond well to treatment with indolebutyric acid.

**Hibiscus, Chinese** (*Hibiscus rosa-sinensis* L.) (*164*)     Seeds, cuttings, budding or grafting, division, and air layering can be used.

   **Cuttings**     Cuttings are not difficult to root; terminal shoots of partially matured wood of most cultivars taken in spring or summer usually form roots in about six weeks. Leaf-bud cuttings also can be used. Rooting should take place in a mist rooting bed or in a glass-covered frame. Rooting becomes poorer as the stock plants age. Reduced light intensity is conducive to better rooting.

   **Grafting**     Strongly growing cultivars which are resistant to soil pests and can be started easily by cuttings, such as 'Single Scarlet', 'Dainty', 'Euterpe', or 'Apple Blossom', are used as rootstocks. Some clones develop into much better plants when grafted on these rootstocks than when they are on their own roots, propagated by cuttings. Whip grafting in the spring or cleft grafting or side grafting in late spring or early summer is successful. Scions of current season's growth, about pencil size, are grafted on rooted cuttings of about the same size (*147*).

   **Budding**     T-budding, using an inverted T, is sometimes practiced, generally in the spring, although it is successful at any time during the year when the bark is slipping.

   **Air layering**     Air layering is practiced during the spring or summer, particularly for cultivars difficult to start by cuttings. Roots will usually form in six to eight weeks.

**Holly** (*Ilex* spp.) (*54, 72, 129, 149*)     Holly can be propagated by seeds, cuttings, grafting, budding, layering, and division.

   Most hollies are dioecious. The female plants produce the very desirable decorative berries if male plants are nearby for pollination. In seed propagation, both male and female plants are produced in ratios of one female to three, or sometimes up to ten, male plants. Sex cannot be determined, however, until the seedlings start blooming, at 4 to 12 years.

   **Seeds**     Germination of holly seed is very erratic; those of some species, *I. crenata*, *I. cassine*, *I. glabra*, *I. vomitoria*, *I. amelanchier*, and *I. myrtifolia*, germinate promptly and should be planted as soon as they are gathered. Seeds of other species, *I. aquifolium* (English holly), *I. cornuta* (Chinese holly), *I. vericilliata*, *I. decidua*, and most *I. opaca* (American holly), do not germinate until a year or more after planting even though stratified, owing probably to rudimentary embryos at time of harvest.

   Seeds of *I. aquifolium*, *I. opaca*, and *I. cornuta* should be collected and cleaned as soon as the fruit is ripe in the fall and then stored at about 4°C (40°F) until spring in a mixture of moist sand and peat moss. Germination in these species generally does not occur until a year later, and then growth is very slow, two seasons being required to bring seedlings of *I. opaca* to a size large enough to be used for grafting.

**Cuttings** (*154*)    This is the method most used commercially, permitting large-scale production of choice clones. Semihardwood tip cuttings from well-matured current season's growth produce the best plants. Cuttings taken from flat, horizontal branches of *I. crenata* tend to produce plants having this type of growth (plagiotropic) and those from upright growth (orthotropic) produce upright plants.

Timing is important; best rooting is usually obtained from mid- to late summer, but cuttings may be successfully taken on into the following spring. Wounding the base of the cuttings helps induce root formation. The wounding induced by stripping off the lower leaves may be sufficient.

The use of a root-promoting chemical, particularly indolebutyric acid at relatively high concentrations (8000 to 20,000 ppm), is essential in obtaining rooting of some cultivars, whereas in others this is not needed. Boron, at 50 to 200 ppm, in combination with IBA, has increased root quality in English holly cuttings (*169*). Bottom heat at 21° to 24°C (70° to 75°F) is beneficial. The maintenance of a high relative humidity is essential. The use of intermittent mist in a greenhouse, where high temperatures can be maintained, provides good rooting conditions. A perlite-peat moss (1:1) rooting medium is satisfactory.

**Grafting and budding**    Hollies are easily grafted, the cleft, whip, and side grafts being used. The operation is best performed during the dormant season for field grafting. Grafting is often done on greenhouse potted stock. T-budding also is suitable, and is done in late summer or early spring. *Ilex opaca* is a satisfactory stock for its own cultivars and those of *I. aquifolium,* but probably the best stocks for English holly are *I. aquifolium* and *I. cornuta* 'Burfordii'.

**Air layering**    This is successful for a number of *Ilex* species. Layers are best started in early summer; after 10 to 14 weeks, plants 30 to 60 cm (1 to 2 ft) high are produced.

**Honeylocust, Common** (*Gleditsia triacanthos* L.)    This is readily propagated either by seeds planted in the spring or by cuttings. The thornless honey locust, *G. triacanthos,* var. *inermis,* and the thornless and fruitless patented 'Moraine' locust are usually propagated by grafting on seedlings of the thorny type. In seed propagation, soaking the seed in sulfuric acid for one hour, followed by stratification at 2°C (36°F) for three months, gives good germination. Hardwood cuttings planted in the spring root successfully.

**Honeysuckle** (*Lonicera* spp.)    Seeds show considerable variation in their dormancy conditions, some species having both seed coat and embryo dormancy, some only embryo dormancy, and some no dormancy. This variability also occurs among different lots of seeds of the same species. In *L. tatarica* some lots have no seed dormancy. In general, however, for prompt germination, stratification for two to three months at about 4°C (40°F) is recommended. Seeds of *L. hirsuta* and *L. oblongifolia* should have two months of warm stratification (21° to 30°C; 70° to 85°F), followed by two to three months' stratification at about 4°C (40°F).

Most honeysuckle species are propagated easily by either hardwood cuttings in the spring or leafy softwood cuttings of summer growth under glass or in mist. Layering of vine types, such as 'Hall's' honeysuckle, is very easy, roots forming wherever the canes become buried under moist soil.

**Hornbeam** (*Carpinus* spp.)    For seed propagation, collect seeds while the wings are still soft and pliable. Do not allow seeds to dry out. Sow outdoors in autumn or stratify over winter and sow in spring. If the seed dries, a hard seed coat germination block develops requiring some type of scarification before stratification. Cultivars may be grafted or budded on seedlings of the same species.

**Horse Chestnut**    *See* Buckeye.

**Hydrangea** (*Hydrangea* spp.) (*114, 134*)    This is easily rooted by leafy, softwood cuttings taken in midspring; they do well under mist and respond markedly to indolebutyric acid treatments. Leaf-bud cuttings can be used if propagating material is scarce. Hardwood cuttings planted in early spring are often used in propagating *H. paniculata* 'Grandiflora'.

**Hypericum spp.**    *See* St. Johnswort.

**Ilex** spp.    *See* Holly.

**Incense Cedar** (*Calocedrus decurrens* [Torr.] Florin.)    This is propagated by seed. Germination is promoted by a stratification period of about eight weeks at 0° to 4°C (32° to 40°F).

**Jacaranda** (*Jacaranda acutifolia* Humb. & Bonpl.) (*167*)    This is easily propagated by seed taken from capsules after blooming. Vegetative propagation is not used.

**Jasmine** (*Jasminum* spp.)    Jasmine is propagated without difficulty by leafy semihardwood cuttings taken in late summer and rooted under glass. Layers and suckers also can be used.

**Juniper** (*Juniperus* spp.) (*81, 94, 151*)    The junipers are generally propagated by cuttings, but in some cases difficult-to-root types are grafted on seedlings. The low-growing, prostrate forms are easily layered.

Seeds    Seedlings of the red cedar, *Juniperus virginiana,* or of *J. chinensis,* are ordinarily used as stocks for grafting ornamental clones.

Seeds should be gathered in the fall as soon as the berrylike cones become ripe. For best germination, seeds should be removed from the fruits, then treated with sulfuric acid for 30 minutes before being stratified for about four months at 4°C (40°F). Rather than the acid treatment, two to three months of warm 21° to 30°C (70° to 85°F) stratification, or summer planting, could be used. As an alternative for cold stratification, the seed may be sown in the fall. Germination is delayed at tempertures above 15°C (60°F). Viability of the seed varies considerably from year to year and among different lots, but it is never much over 50 percent. Treated seed is usually planted in the spring, either in outdoor beds or in flats in the greenhouse. Two or three years are required to produce plants large enough to graft.

Cuttings    The spreading, prostrate types of junipers are more easily rooted than upright kinds. Cuttings are made 5 to 15 cm (2 to 6 in.) long from new lateral growth tips stripped off older branches. A small piece of old wood—a heel—is thus left attached to the base of the cutting. Some propagators believe this to be advantageous. In other cases, good results are obtained when the cuttings are just clipped without the heel from the older wood. Terminal growth of current season's wood also roots well.

Cuttings to be rooted in the greenhouse can be taken at any time during the winter (*102*). Exposing the stock plants to subfreezing temperatures seems to give better rooting. For propagating in an outdoor cold frame, cuttings are taken in late summer or early fall. Lightly wounding the base of the cuttings is sometimes helpful, and the use of root-promoting chemicals, especially indolebutyric acid, is beneficial. A medium-coarse sand or a 1:1 mixture of perlite and peat moss is a satisfactory rooting medium. A greenhouse temperature of about 15°C (60°F) is best for the first four to six weeks. Maintenance of a humid environment without excessive wetting of the cuttings is desirable, as is a relatively high light intensity. A light, intermittent misting can be used. Bottom heat of about 27°C (80°F) will aid rooting.

**Grafting** (*133*)    Vigorous seedling understocks with straight trunks, about pencil size, are dug in the fall from the seedling bed and potted in small pots set in peat moss in a cool, dry greenhouse. Seedlings potted earlier—in the spring—may also be used. After about 30 days, the greenhouse is heated and the plants kept well watered. This procedure stimulates growth so that after one or two weeks the plants resume root activity and are in a suitable condition for grafting.

The scions should be selected from current season's growth taken from vigorous, healthy plants and preferably of the same diameter as the stock to be grafted. Scion material can be stored at -1° to 4°C (30° to 40°F) for several weeks until used if kept in a saturated atmosphere.

Side-veneer or side-graft methods are ordinarily used. The unions are best tied with budding rubber strips. The grafted plants are set in a greenhouse bench filled with peat moss deep enough to cover the union. The temperature around the graft union should be kept as constant as possible at 24°C (75°F) with a relative humidity of 85 percent or more around the tops of the plants. A lightly shaded greenhouse should be used to avoid burning the grafts. Adequate healing will take place in two to eight weeks, after which the temperature and humidity can be lowered. The stock plant is then cut off above the graft union to allow the scion to develop.

**Kalmia spp.**    *See* Laurel.

**Koelreuteria spp.**    *See* Goldenrain Tree.

**Laburnum** (*Laburnum* spp.) Golden Chain    Propagated by seeds or layering. Cultivars are propagated by grafting or budding on laburnum seedling rootstocks. Seeds are poisonous if eaten.

**Lagerstroemia indica**    *See* Crape Myrtle.

**Larch** (*Larix* spp.) (*77*)    Most of these deciduous conifers are propagated easily by fall-planted seeds. Cones should be collected before they dry and open on the tree. Several species have empty or improperly developed seed. Seeds of some species have a slight embryo dormancy, so for spring planting, stratification for one month at about 4°C (40°F) is advisable. Cutting propagation is best done by rooting leafy-tipped softwood cuttings in late summer under mist. Indolebutyric acid promotes rooting. Cutting material should be taken from young trees only.

**Laurel** (*Kalmia* spp.) (*44, 88, 89*)    The laurels can be propagated readily by seed germinated at about 20°C (68°F). Germination of *K. latifolia* seed is enhanced by cold stratification for eight weeks, or by a 12-hour soak in 200 ppm gibberellic acid. *Kalmia* cultivars can be propagated by leafy cuttings under mist or in polyethylene-covered frames (*44, 57*), by cleft or side grafting, or by layering; cuttings of some *Kalmia* selections respond to rooting hormones, while others do not.

**Leptospermum spp.** Tree tea    Some species of this Australian native must be propagated by seed. Cultivars of other species, such as *L. scaparium,* can be propagated readily by cuttings.

**Leyland Cypress** (× *Cupressocyparis leylandii*) (*4, 176*)   This bigeneric hybrid of *Cypressus macrocarpa* and *Chamaecyparis nootkatensis* is propagated by cuttings, rooted under mist with bottom heat. Cuttings can be taken any time from late winter to autumn and should be treated with IBA at 3000 ppm.

**Libocedrus decurrens**   *See* Incense Cedar.

**Ligustrum spp.**   *See* Privet.

**Lilac, French Hybrid** (*Syringa vulgaris* cvs.) (*30, 41, 69, 168*)   Grafting or budding on California privet (*Ligustrum ovalifolium*) or Amur privet (*L. amurense*) cuttings, or on lilac or green ash (*Fraxinus pennsylvanica*) seedlings, is used commercially in lilac propagation. Cuttings can be rooted, however, if attention is given to proper timing. Layering or division of old plants is quite satisfactory when only a few new plants are needed. The common purple lilac is propagated by root suckers.

**Seeds**   Seedlings are used mostly as an understock for grafting or in hybridization. Lilac cultivars will not reproduce true from seed. Seeds require fall planting out-of-doors or a stratification period of 40 to 60 days at about 4°C (40°F) for good germination.

**Cuttings**   Ordinarily, good rooting of lilacs can be obtained only with terminal leafy cuttings taken within a narrow period shortly after growth commences in the spring. When the new, green shoots have reached a length of 10 to 15 cm (4 to 6 in.) they should be cut off and trimmed into cuttings. Since they are very succulent at this stage it is difficult to prevent wilting. In a mist propagating bed rooting occurs in three to six weeks; a well-drained rooting medium must be used. Rooting can also be obtained in a polyethylene-covered bed in the greenhouse with bottom heat. Sprays with captan (2 tsp per gal of water) will help avoid fungus attack. The use of indolebutyric acid at about 80 ppm for 24 hours has given good rooting, as has dipping in an 0.8 percent IBA in talc preparation. Cuttings can be rooted in out-of-doors mist beds (*69, 84*).

**Grafting and budding**   Because of the difficulty in rooting lilac cuttings and the fact that they must be taken at a definite time in the spring, often at the peak of the nurseryman's busy season, many propagators practice grafting.

When privet or green ash is used as the understock, the lilac may show incompatibility symptoms, but if the grafts are planted deeply, scion roots rapidly develop from the lilac and soon become the predominant root system of the plant. *Syringa vulgaris* seedlings are sometimes used as the stock, but if suckers arise from this stock, there is difficulty in distinguishing them from the selected hybrid lilac top. The plant thus becomes a mixture of growth from the scion and the rootstock. Should privet understock produce suckers, however, they can be easily identified and removed. It is best in any case, to plant lilacs "on their own roots," either as rooted cuttings or with the privet or lilac "nurse root" already removed.

Grafting is done during the winter season, using rootstock plants which have been dug and brought inside. Vigorous, one-year-old scion wood is used, taken from plants that have been heavily pruned and well fertilized to induce such growth. Different grafting methods can be used, such as the side or the whip graft. Cleft grafting on pieces of privet root is also practiced and tends to eliminate subsequent suckering from the rootstock.

**T-budding**   T-budding in late summer or early fall is sometimes practiced, the lilac buds being inserted below ground on one-year-old privet cuttings or seedlings. The following spring the privet stock is cut off above the bud and soil mounded around the shoot as it develops so as to encourage subsequent rooting from the lilac. Unless this is done, the plant will be short-lived.

**Layering**    Simple layering of one-year shoots arising from the base of plants on their own roots provides an easy propagation method where only a few plants are needed. Air-layering of one- or two-year-old branches also is successful.

**Linden** ( *Tilia* spp.) ( *52, 159* ) Basswood    The seeds have a dormant embryo plus an impermeable seed coat which, in some species, is surrounded by a hard, tough pericarp. Such seeds are slow and difficult to germinate. Removing the pericarp, either mechanically or by soaking the seeds in concentrated nitric acid for one-half to two hours, rinsing thoroughly and drying, then soaking the seeds for about 15 minutes in concentrated sulfuric acid to etch the seed coat, followed by stratification for four months at 2°C (35°F), may give fairly good germination; otherwise, warm (15° to 27°C; 60° to 80°F) stratification for four to five months, followed by an equal period at 2° to 4°C (35° to 40°F) can be used. Collecting the seed from the tree just as the seed coats turn completely brown (but before the seeds drop and the seed coats become hard and dry), followed by immediate planting has given good germination.

Suckers arising around the base of trees cut back to the ground have been successfully mound layered, and softwood cuttings taken from stump sprouts have been rooted.

Propagation of selected clones by cuttings or by grafting is slow and difficult but T-budding in late summer on seedling stocks of the same species gives good results ( *52* ).

**Liquidambar** ( *Liquidambar styraciflua* L.) American sweet gum.    Propagation is usually by seeds, which are collected in the fall but not allowed to dry out. Stratification for one to three months at about 4°C (40°F) is recommended to overcome seed dormancy. Selected clones are grafted or T-budded in midsummer onto *L. styraciflua* seedlings. They can also be propagated by root cuttings.

Leafy softwood cuttings of partially matured wood can be rooted under mist in midsummer. Treatments with naphthaleneacetic acid have been helpful.

**Liriodendron tulipifera**    *See* Tulip Tree.

**Locust, Black** ( *Robinia pseudoacacia* L.)    Black locust is readily propagated by seeds, which are soaked in concentrated sulfuric acid for one hour, followed by thorough rinsing in water, before planting.

**Lonicera spp.**    *See* Honeysuckle.

**Madrone, Pacific** ( *Arbutus menziesii* Pursh.)    This Pacific Coast evergreen tree is usually propagated by seeds, which are stratified for three months at 2° to 4°C (35° to 40°F). Seedlings are started in flats, then transferred to pots. They are difficult to transplant and should be set in their permanent location when not over 18 in. tall. Propagation can be done by cuttings, layering, and grafting also.

**Magnolia** ( *Magnolia* spp.) ( *2, 9, 25* )    Seeds, cuttings, grafting, and layering are utilized in propagating magnolias. Magnolia nursery trees are difficult to transplant, so they should be set out from containers or balled and burlapped, but only in early spring.

**Seeds**    Magnolia seeds are gathered in the fall as soon as possible after the fruit is ripe, when the red seeds are visible all over the fruit. After cleaning, the seeds should either be sown immediately in the fall or—prior to spring planting—stratified for two to three months at about 4°C (40°F). Allowing the seeds to dry out at any time seems to be harmful. After sowing, the germination medium must not become dry. *M. grandiflora* seeds, and perhaps those of other species, lose their viability if stored through the winter at room temperature. If prolonged storage is necessary, the seeds should be held in sealed containers at 0° to 4°C (32° to 40°F).

Magnolia seedlings grow rapidly, and generally are large enough to graft by the end of the first season. Transplanting should be kept at a minimum, since this retards the plants.

**Cuttings**    Some species, such as *M. soulangeana* and *M. stellata,* are successfully propagated commercially by leafy softwood cuttings. These may be taken from late spring to late summer after terminal growth has stopped and the wood has become partly matured.

Excellent rooting can be obtained if cuttings are taken from very young plants, the bases wounded, treated with a root-promoting substance, then rooted in sand in outdoor mist beds. Under such conditions, rooting is rapid and in high percentages, and there is little trouble from diseases.

Leafy cuttings of *M. grandiflora,* taken from late spring to late summer, wounded, and treated with indolebutyric acid at 5000 to 20,000 ppm have rooted well with bottom heat (24°C; 75°F) and under intermittent mist (*152*).

To obtain survival of the rooted cuttings through the following winter, they should be rooted early enough in the season so that some resumption of growth will occur before fall.

**Grafting**    *Magnolia kobus* is probably the best rootstock for the oriental magnolias, whereas *M. acuminata* can be used as a stock for either oriental or American species. *Magnolia grandiflora* seedlings or rooted cuttings are used for *M. grandiflora* cultivars.

One-year-old seedlings are potted in early spring and then grafted while the understocks are in active growth during mid- to late summer. Side or side-veneer grafts are satisfactory, with the union and scion waxed after grafting. Some propagators pot the seedlings in the fall, then bring them into the greenhouse and do the grafting in midwinter. The newly grafted plants may be set on open benches in the greenhouse or placed in closed propagating frames, where they stay for seven to ten days while the union is healing. Air is gradually given until after six weeks they can be removed from the case and the understock cut off above the union.

**Layering**    Simple or mound layering gives good results. One- or two-year-old shoots arising from the base of stock plants are started in spring, but often two seasons are required to produce well-rooted layers.

**Mahonia** (*Mahonia* spp.)    Seeds of most species should not be allowed to dry out; they are stratified through the winter to obtain satisfactory germination. On the other hand, dry seeds of red mahonia (*M. haematocarpa*) will germinate promptly after planting in the spring. Cleaned seeds of fall-planted *M. aquifolium,* over-wintered outdoors, germinate in spring.

Leafy cuttings taken in late summer and treated with indolebutyric acid, then placed under intermittent mist, have rooted well (*14*).

**Malus spp.**    *See* Crabapple, Flowering.

**Manzanita** (*Arctostaphylos* spp.)    Seed propagation can be done, but with difficulty. Soaking the cleaned fruits in sulfuric acid up to 24 hours before planting will aid germination. Cuttings can be rooted if taken from fall to early spring. Stock material should be sterilized with a weak Clorox solution. Cuttings are rooted under high humidity in polyethylene-covered frames. Vermiculite-perlite (1:1) is a suitable rooting medium, and root-promoting growth regulators are helpful (*80*).

**Maple** (*Acer* spp.) (*8, 162*)    Various methods of propagation are used—seeds, grafting, budding, cutting, layering, and in vitro micropropagation (*174*).

**Seeds** (*128*)     Most maples ripen their seed in the fall, but two species—*A. rubrum* and *A. saccharinum*—produce seed in the spring. Such spring-ripening seeds should be gathered promptly when mature and sowed immediately without drying. For other species, stratification, usually for 90 days at 4°C (40°F), followed by spring planting, gives good germination. Fall planting out-of-doors may be done if the seeds are first soaked for a week, changing the water daily. Seeds of the Japanese maple, *A. palmatum,* germinate satisfactorily if they are placed in warm water (about 43°C; 110°F) and allowed to soak for two days, followed by stratification. Soaking seeds of *A. rubrum* and *A. negundo* in cold running water for five days and two weeks, respectively, before planting may increase germination. *Acer* seeds should not be allowed to dry out.

Cultivars of some maples, such as *A. palmatum* 'Atropurpureum', will reproduce fairly true from seed, especially if the stock plants are isolated. The few off-type plants can be removed from the nursery row.

**Cuttings** (*172*)     Leafy Japanese maple cuttings, as well as those of other Asiatic maples, will root readily in a sand–peat moss medium if they are made from tips of vigorous pencil-size shoots in late spring and placed under mist. Wounding and relatively strong applications of indolebutyric acid are helpful. IBA at 10,000 to 20,000 ppm in talc has given good results in rooting leafy cuttings of various *Acer* species under mist and over bottom heat in the greenhouse. A mixture of 1:1 peat moss and perlite is a good rooting medium. It is often difficult to overwinter rooted maple cuttings. To overcome this problem, new growth should be induced on the cuttings, shortly after rooting, by using supplementary lighting and supplying fertilizers. Hardwood cuttings of *Acer palmatum* taken in midwinter have been successfully rooted in the greenhouse after wounding and treating with indolebutyric acid (*22*). Maple species have been propagated in vitro by shoot-tip explants (*174*).

Sugar maple (*A. saccharum*) cuttings are best taken in late spring after cessation of shoot elongation, then rooted under mist. Long, thick cuttings are preferred. IBA treatments give variable results. Overwintering is a problem but survival is best with large, vigorous, well-rooted cuttings. In cold climates it may be necessary to pot the cuttings in late summer, harden them off, then store until spring at about 1°C (34°F) (*42*).

**Grafting and budding**     *Acer palmatum* seedlings are used as the understock for Japanese maple cultivars, *A. saccharum* for the sugar maples, *A. rubrum* for red maple cultivars, and *A. platanoides* for such Norway maple clones as 'Crimson King', 'Schwedler', and the pyramidal forms. There is some evidence of delayed incompatibility in using *A. saccharinum* as a stock for red maple cultivars.

Seedling rootstock plants are grown for one year in a seedbed, then in the fall or early spring are dug and transplanted into small pots in which they grow, plunged in propagating frames, through the second summer. In late winter, the stock plants are brought into the greenhouse preparatory to grafting. As soon as roots show signs of growth, the stock is ready for grafting. Dormant scions are taken from outdoor plants. The side graft or side-veneer graft is ordinarily used. While the union is healing, the plants are set in a grafting case with peat moss covering the union. Sometimes grafting wax is used, covering both the scion and the graft union (*64*).

Maples are also successfully T-budded on one-year seedlings in the nursery row, the buds being inserted from mid- to late summer. The wood is removed from the bud shield, which then consists only of the actual bud and attached bark. The seedling is cut back to the bud the following spring just as growth is starting.

**Melaleuca spp.**     The seeds of these native Australian species are almost dustlike but germinate easily and can be handled like eucalyptus seed.

**Metasequoia** (*Metasequoia glyptostroboides* Hu and Cheng) Dawn redwood (*4, 28*)    Seeds of this "living fossil," discovered in Western China in 1945, germinate without difficulty, and both softwood and hardwood cuttings root readily. The leafless hardwood cuttings may be lined-out in the nursery row in early spring. Leafy cuttings root easily under mist if taken in summer and treated with indolebutyric acid at 20,000 ppm by the concentrated-dip method.

**Mock Orange** (*Philadelphus* spp.)    The many cultivars of mock orange are best propagated by cuttings, which root easily. Hardwood cuttings can be planted in early spring or leafy softwood cuttings under glass in early summer. Removing rooted suckers arising from the base of old plants is an easy means of obtaining a few new plants.

**Monkey Puzzle Tree** (*Araucaria araucana* [Mol.] C. Koch)    Propagated by seeds, which have no dormancy problems. Do not sow too early in spring as seedlings are sensitive to frost.

**Mountain Ash**    *See* Sorbus.

**Mulberry, Fruitless** (*Morus alba* L.)    Some mulberry trees produce only male flowers and hence do not bear fruits. Propagated vegetatively, these are suitable as ornamental shade trees. Some are very rapid growers. They may be propagated by cuttings or by budding or grafting on mulberry seedling rootstocks. Leafy softwood cuttings taken in midsummer will root under mist.

**Myrica**    *See* Bayberry.

**Myrtle** (*Myrtus* spp.)    Myrtle is usually propagated in summer by leafy softwood cuttings of partially matured wood rooted under glass. Treatments with indolebutyric acid have been helpful in some instances.

**Nandina** (*Nandina domestica* Thun.) (*3*)    Propagation is usually by seed, which should not be allowed to dry out. The embryos in the mature fruits are rudimentary, but will develop in cold storage. Seeds can be collected in late fall, held in dry storage at 4°C (40°F), then planted in late summer, germination starting in about 60 days. Germination tends to take place in autumn regardless of planting date. No low-temperature moist stratification period is necessary. Growth of the seedlings is slow, taking several years to produce salable plants. Suckers arising at the base of old plants may be removed for propagation.

**Nerium oleander**    *See* Oleander.

**Norfolk Island Pine** (*Araucaria heterophylla* [Salisb.] Franco.)    This pine is propagated by seed. Cuttings of side branches will root but produce horizontally growing plants. Cuttings from terminals grow upright.

**Oak** (*Quercus* spp.)    Seed propagation is generally practiced. Wide variations exist in the germination requirements of oak seed, particularly between the black oak (acorns maturing the second year) and white oak (acorns maturing the first year) groups. Seeds of the white oak group have little or no dormancy and, with few exceptions, are ready to germinate as soon as they mature in the fall. Seeds of most species of the black oak group have embryo dormancy, requiring either stratification (0° to 2°C; 32° to 35°F) for one to three months of fall planting. Seeds of the following species will germinate without a low-temperature stratification period: *Quercus agrifolia, Q. alba, Q. arizonica, Q. bicolor, Q. chrysolepis, Q. douglassii, Q. garryana, Q. lobata, Q. macrocarpa, Q. montana, Q. petraea, Q. prinus, Q. robur, Q. stellata, Q. suber, Q. turbinella,* and *Q. virginiana.*

Acorns are often attacked by weevils. Soaking in water held at 49°C (120°F) for 30 minutes will rid the acorns of this damaging pest.

Acorns of many species tend to lose their viability rapidly when stored dry at room temperature. Seeds of some species can be stored for several years without losing viability by holding them at 1° to 3°C (34° to 37°F) in polyethylene bags. The seeds should have a 60 to 70 percent moisture level at the start of storage (*46*).

To obtain lateral root branching—which makes the seedlings more adaptable to transplanting—the acorns can be planted in a box that has a copper wire screen (mesh about 6 in.) below the seeds. The tip of the tap root, upon contacting this mesh, will be killed, forcing development of many lateral roots.

Bench grafting of potted seedling stocks in the greenhouse in late winter or early spring is moderately successful. Side or whip grafting is ordinarily used, with dormant one-year-old wood for scions. Seedlings in place in the nursery row are occasionally crown grafted in the spring, after the stock plants start to leaf out. Scions are taken from wood gathered when dormant and stored under cool, moist conditions until used. Various grafting methods are satisfactory—whip, cleft, or bark. Budding generally has been unsatisfactory. In grafting, only seedlings of the black oak group should be used for scion cultivars of the same group and, in the same manner, only seedlings of the white oak group should be used as stocks for other members of the white oaks. The use of seedlings of the same species is preferable. Although some distantly related species of oak will unite satisfactorily, incompatibility symptoms usually appear later.

Attempts to propagate oaks by cuttings or layering usually have been unsatisfactory, although some success has been obtained in rooting leafy softwood cuttings of *Quercus robur* 'Fastigata' under outdoor mist in midsummer after treatment with IBA at 20,000 ppm (*51*). Cuttings from certain source trees of *Q. ilex* have been rooted in moderate percentages (*38*).

**Oleander** (*Nerium oleander L.*)    Seedlings reproduce fairly true-to-type, although a small percentage of plants with different flower colors will appear. The seeds should be collected in late fall after a frost has caused the seed pods to open. Rubbing the seeds through a coarse mesh wire screen removes most of the of the fuzzy coating. The seeds are then planted immediately in the greenhouse in flats without further treatment. Germination occurs in about two weeks. Leafy cuttings root easily under glass or in mist if taken from rather mature wood during the summer. Simple layering also is successful. Plant parts are very poisonous.

**Olive** (*Olea europaea L.*)    A fruitless cultivar, 'Swan Hill', is available for use as a patio or street tree. Cuttings are difficult to root. They should be placed under mist and treated with IBA at 2000 to 3000 ppm, or it can be grafted on easily rooted cultivars used as a rootstock. Olives will not survive outdoor winter temperatures below about -9°C (15°F) (*73*).

**Osmanthus spp.**    Osmanthus is propagated by leafy cuttings taken in mid- to late summer and rooted under glass or mist. IBA treatments at about 4000 ppm should promote rooting. Seeds are slow and difficult to germinate.

**Pachysandra terminalis** (Siebold & Zucc.) Japanese spurge    This evergreen ground cover for shady areas spreads naturally by rhizomes. Can be propagated easily by division or by leafy cuttings under mist or glass.

**Paeonia suffruticosa**    *See* Peony, Tree.

**Palms** (*17, 23, 93, 112, 181*)    There are several thousand species, in many genera, of ornamental palms. They are propagated by mature seed, which should be planted as soon as possible after harvesting, not being allowed to dry out. Palm seeds remain viable for only a short period of time. Palm seeds are susceptible to surface molds and should be protected by dusting with a fungicide. Any seeds that float in water should be discarded. A mixture of one-half peat moss and one-half perlite is a good germination medium. Seeds of most species germinate in one to three months, especially if bottom heat, 28°C (80°F), is maintained, but some may take as long as one to two years. Germination in seed of some species can be accelerated by scarification, plus soaking in gibberellic acid at 1000 ppm for 72 hours, plus germination over bottom heat at 27°C (81°F) (*126*).

**Parthenocissus** (*P. quinquefolia* Planch.) Virginia Creeper; *(P. tricuspidata* Planch.) Boston Ivy    These two ornamental vines can be propagated by seeds planted in the fall or stratified for two months at about 4°C (40°F) before planting in the spring.

Leafy softwood cuttings taken in late summer root easily under glass or mist, as do hardwood cuttings planted outdoors in early spring. Plants may be started by compound layering.

Grafting of named cultivars on *P. quinquefolia* seedlings or rooted cuttings is done by some nurseries using the whip or cleft graft.

**Passiflora × alatocaerulea** (Lindl.) Passion Vine    This tender, subtropical vine is propagated by leafy cuttings under glass or mist.

**Pear, 'Bradford'** (*Pyrus calleryana* Dcne. 'Bradford')    This ornamental pear introduced by the USDA can be propagated by T-budding or root grafting onto *P. calleryana* seedlings. It is not compatible with *P. communis* roots. Leafy cuttings, if treated with IBA, can be rooted under mist—with peat moss, perlite, vermiculite (1:1:1) giving good results as a rooting medium (*1*). Cuttings should be taken in late summer.

**Pear, Evergreen** (*Pyrus kawakamii* Hayata)    Evergreen pear is propagated by cuttings or, more commonly, by grafting on *P. calleryana* seedlings. Bench grafting is done in midwinter using the cleft graft. The grafts can be planted in containers or in the nursery row.

**Peony, Tree** (*Paeonia suffruticosa* Andr.) Moutan (*79, 178*)    Seed propagation is complicated by "epicotyl dormancy." The seeds should be planted in a moist medium and, after roots have developed, transplanted to pots of soil placed in a cold room (4° to 10°C; 40° to 50°F) or outdoors (in winter) for 2½ months. This overcomes dormancy conditions in the shoot tip, which then grows readily in spring or upon transfer of the plants to warm temperatures. Selected cultivars are propagated by grafting in late summer on herbaceous peony (*P. lactiflora*) roots as the understock. The grafts are callused in a sand-peat medium in a greenhouse until fall, when they are potted.

**Philadelphus spp.**    *See* Mock Orange.

**Photina arbutifolia**    *See* Heteromeles Arbutifolia.

**Pieris spp.** (*58*)    *Pieris floribunda* reproduces readily by seed with no treatment necessary. Some *Pieris* species can be started easily by cuttings, but *P. floribunda* cuttings are difficult to root. However, moderate rooting can be obtained by rooting under mist or polyethylene-covered frames, after treating with a rooting hormone (indolebutyric acid at 8000 ppm), and by using bottom heat.

**Picea spp.**    *See* Spruce.

**Pine** (Pinus spp.) (*121, 157*)    Pines are ordinarily propagated by seed. Considerable variability exists among the species in regard to seed dormancy conditions. Seeds of many species have no dormancy and will germinate immediately upon collection, whereas others have embryo dormancy. With the latter, stratification at 0° to 4°C (32° to 40°F) for one to three months will increase or hasten germination. Seed coat dormancy also seems to be present in *P. cembra* and *P. monticola*. With these species, concentrated sulfuric acid treatment for three to five hours and for 45 minutes, respectively, followed by stratification for three months at 2°C (36°F) will aid germination. Species whose seeds have no dormancy conditions and can be planted without treatment include *Pinus aristata, P. banksiana, P. canariensis, P. caribaea, P. clausa, P. contorta, P. coulteri, P. edulis, P. halepensis, P. jeffreyi, P. latifolia, P. mugo, P. nigra, P. palutris, P. pinaster, P. ponderosa, P. pungens, P. radiata, P. resinosa, P. roxburghi, P. sylvestris, P. thunbergi, P. virginiana,* and *P. wallichiana*. However, if seeds of the above species have been stored for any length of time, it is advisable to give them a cold stratification period before planting. Pine seeds can be stored for considerable periods of time without losing viability if held in sealed containers between -15° and 0°C (5° and 32°F). Seeds should not be allowed to dry out.

Pine cuttings are difficult to root, although those of mugo pine (*Pinus mugo*) root easily if taken in early summer (*92*), and *P. radiata* roots well if cuttings are taken from young trees (*18*). Success is more likely if cutting material is taken in winter from low-growing lateral shoots on young trees. Treatment with indolebutyric acid is beneficial. Considerable study has been given to the rooting of cuttings of Monterey pine (*P. radiata*) because of its importance as a lumber tree in New Zealand and Australia and as a Christmas tree species. Best rooting is from cuttings taken in early winter. Wounding, plus a concentrated-dip treatment of IBA at 4000 ppm, was beneficial. A more symmetrical root system could be induced by clipping the ends of the original roots and allowing the root system to develop from the secondary roots. Outdoor rooting under intermittent mist has been successful using IBA at 5 percent, with peat moss, redwood sawdust, and *P. radiata* litter (1:1:1) as the rooting medium (*82, 108*).

Pines can be propagated asexually be rooting needle fascicles (needle leaves held together by the scale leaves, containing a base and a diminutive shoot apex) (*31*). Rooting is best when the fascicles are taken from trees less than four years old. IBA treatments are helpful (*113*). By selecting certain seedlings whose cuttings root easily and by using critical timing in taking cuttings, it is possible to set up clones in which cutting propagation is commercially feasible. This was shown to be true for the mugo pine (*62, 138*). Cuttings of the Scotch pine have been rooted by a unique method of forcing out interfascicular shoots from young stock plants. These shoots, when made into cuttings and given the proper treatment, rooted in high percentages.

Side-veneer grafting is used for propagating selected clones; well-established two-year-old seedlings of the same or closely related species should be used as rootstocks. Scions should be of new growth, taken from firm, partly matured wood. Air layering of loblolly pine has been successful (*70*).

**Pistache, Chinese** (*Pistacia chinensis* Bunge)   Seeds should be collected from relatively large fruits, blue-green in color, in midfall. Pulp must be removed. Soak fruits in water, then rub over a screen. Seeds that float in water have aborted embryos and should be discarded. Stratification at 4° to 10°C (40° to 50°F) for ten weeks gives good germination. Seedlings exhibit a wide range of variability. T-budding selected clones on seedling *P. chinensis* rootstocks in late summer is used to produce uniform, superior trees. Propagation by cuttings is very difficult. Pistache nursery trees do not transplant well if the roots become exposed, so they are usually handled as container-grown nursery stock (*90*).

**Pittosporum** (*Pittosporum* spp.)    These are started by seeds or cuttings. The seeds are not difficult to germinate; dipping a cloth bag containing the seeds for several seconds in boiling water may hasten germination. Leafy semihardwood cuttings taken after a flush of growth has partially matured will root readily, particularly under mist. Indolebutyric acid treatments also are beneficial.

**Plane Tree** (*Platanus* spp.)    Sycamore. Seeds are ordinarily used in propagation but they should not be allowed to dry out. The best procedure is to allow the seeds to over-winter in the seed balls right on the tree. They may then be collected in late winter or early spring and planted immediately. Germination usually occurs promptly. If the seeds are collected in the fall, then stratification over winter at about 4°C (40°F) should be used. The hybrid London plane tree, *P* × *acerifolia,* can be propagated by hardwood cuttings, taken and planted in the nursery in autumn (*4*) or by leafy softwood cuttings taken in midsummer and rooted under mist. Hormone treatments tend to inhibit rooting (*125*).

**Platanus spp.**    *See* Plane Tree.

**Plumbago** (*Plumbago* spp.)    Seeds sown in late winter usually germinate easily. Leafy cuttings taken from partially matured wood can be rooted without difficulty under glass. Root cuttings also can be used, and old plants can be divided.

**Plumeria** (*Plumeria* spp.)    Leafy cuttings 15 to 20 cm (6 to 8 in.) long of this tender tropical shrub widely grown in Hawaii will root readily under mist if treated with 2500 ppm indolebutyric acid. Terminal, leafless shoots, about the same length, can be cut off before new leaves appear and planted. Root and shoot growth develop readily.

**Podocarpus** (*Podocarpus* spp.)    These evergreen trees and shrubs, having foliage resembling the related yews (*Taxus*), make good container plants. They can be propagated by stem cuttings taken in late summer.

**Poinsettia** (*Euphorbia pulcherrima* Willd.) (*43, 100, 104*)    Propagation by leafy cuttings under mist is the usual procedure. Low-strength root-promoting chemicals are helpful. Stock plants making moderate growth should be used as a source of cuttings. Disease control is very important; sanitary procedures should be used throughout the rooting operation. It is best to root cuttings in containers so roots will not be disturbed. Cuttings can be rooted in the greenhouse from spring to fall. Leaves are very poisonous.

**Poplar** (*Populus* spp.) Poplar, Cottonwood, Aspen (*4, 132*)    These trees can be propagated by seeds; they should be collected as soon as the capsules begin to open and planted at once, because they lose viability rapidly, and should not be allowed to dry out. However, if held in sealed containers near 0°C (32°F), seeds of some species can be stored for as long as three years. There are no dormancy conditions and seeds germinate within a few days after planting. The seedlings are highly susceptible to damping-off fungi and will not tolerate excessive heat or drying. Poplars are difficult to propagate in quantity by seed.

Hardwood cuttings of *Populus* (except the aspens) planted in the spring root easily. Treatment with indolebutyric acid is likely to improve rooting. Leafy softwood cuttings (of some species at least) taken in midsummer also root well. In vitro culture using shoot-tip explants has been successful in poplar propagation (*177*).

**Potentilla** (*Potentilla* spp.) (*60*) Cinquefoil    Propagation is usually by cuttings, but seed and division can be used. Cuttings are taken from early summer through fall. Rooting is best under light mist with bottom heat. Rooting hormones are helpful.

**Populus spp.**     *See* Poplar.

**Privet** (*Ligustrum* spp.)     Seed propagation is easily done. The cleaned seed should be stratified for two to three months at 0° to 10°C (32° to 50°F) before planting. Hardwood cuttings of most species planted in the spring root easily, as do softwood cuttings in summer under glass. Japanese privet (*L. japonicum*) is somewhat difficult to start from cuttings, the best results being obtained with actively growing terminal shoots rather than more mature wood.

**Protea spp.** (*130*)     This South African native, popular for its cut flowers, is propagated by seed, cuttings, and grafting. Cuttings are difficult to root but respond to auxin treatments.

**Prunus** *campanulata, P. sargentii, P. serrulata, P. sieboldi, P. subhirtella, P. yedoensis*     *See* Cherry, Flowering.

**Pseudotsuga menziesii**     *See* Douglas fir.

**Pyracantha** (*Pyracantha* spp.) Firethorn     Propagation is almost always by cuttings. Partially matured, leafy, current-season's growth is taken from late spring to late fall and rooted either in the greenhouse under glass or under mist for good results. Treatments with root-promoting substances are beneficial. Large 46 cm (18 in.) leafy cuttings, taken in early spring, wounded at the base, treated with indolebutyric acid, and rooted under mist produced salable nursery plants, potted in one-gallon containers, in six months (*158*).

**Pyrus calleryana**     *See* Pear, 'Bradford'.

**Pyrus kawakami**     *See* Pear, Evergreen.

**Quaking Aspen** (*Populus tremuloides)*     This can be propagated by removing root pieces, inducing adventitious shoots to form from them in vermiculite, then rooting these shoots as stem cuttings under mist with IBA treatments.

**Quercus spp.**     *See* Oak.

**Quince, Flowering** (*Chaenomeles* spp.)     Flowering quince is easily started by seeds, which should be fall-planted or stratified for two or three months at 4°C (40°F) before sowing. Clean the seeds from the fruit. Root cuttings can be taken in late fall, cut into 5 to 10 cm (2- to 4-in.) lengths, and stored at 2° to 4°C (35° to 40°F) until spring, when they can be lined-out horizontally in the nursery row. Leafy cuttings of partially matured wood may be rooted under mist in late spring. Treatment with indolebutyric acid at about 15 ppm for 24 hours is beneficial. Older plants tend to produce suckers freely at the base; these suckers may be removed and used if they are well rooted.

**Redbud** (*Cercis* spp.)     Seed propagation is successful, but seed treatments are necessary because of dormancy resulting from an impervious seed coat plus a dormant embryo. Probably the most satisfactory treatment is a 60 minute soaking period in concentrated sulfuric acid, followed by stratification for three months at 2° to 4°C (35° to 40°F). Fall sowing outdoors of untreated seeds also may give good germination.

Leafy softwood cuttings of some *Cercis* species root readily under mist if taken in spring or early summer. Simple layering also is used successfully. T-budding in midsummer on *C. canadensis* seedlings is used commercially for *Cercis* cultivars (*165*).

**Redwood** Coast redwood. (*Sequoia sempervirens* [D. Don] Endl.). Giant sequoia or Sierra redwood (*Sequoiadendron giganteum* [Lindl.] Buchh.)     Both species are ordinarily propagated by seed. Seeds of *S. sempervirens* are mature at the end of the first season, but those of *S. giganteum* require two seasons for maturity of the embryos. Cones are collected in the fall and allowed to dry for two to four weeks, after which the seeds can be separated. Seeds may be kept for several years in sealed containers under 4°C (40°F) storage without losing viability. Stratification for ten weeks before planting at about 4°C (40°F) promotes germination of *S. giganteum* seed. Seeds of *S. sempervirens* will germinate without a stratification treatment.

Fall planting also may be done, sowing the seed about 3 mm (⅛ in.) deep in a well-prepared seedbed. A burlap cover on the seedbed, which is removed when germination starts, is helpful. The young seedlings should be given partial shade for the first 60 days.

*S. sempervirens* may also be propagated by leafy cuttings taken from sprouts arising from the burls around the bases of older trees. Young *S. giganteum* trees are grown for Christmas trees. *S. giganteum* cuttings root easily if taken from young trees, treated with indolebutyric acid, and rooted under mist (*50, 179*).

**Redwood, Dawn**     *See* Metasequoia.

**Rhamnus spp.**     *See* Buckthorn.

**Rhododendron** (*Rhododendron* spp.) (*103, 160, 161, 173*)     These can be propagated by seeds, cutting, grafting, and layering. Micropropagation of rhododendron also has been successful using shoot-tip explants (*6*).

**Seeds**     Seedlings may be used as rootstocks for grafting or for the propagation of the ornamental species. *Rhododendron ponticum* is the principal rootstock for grafting. The seed should be collected just when the capsules are beginning to dehisce, and may be stored dry and planted in late winter or early spring in the greenhouse. Seed to be kept for long periods should be put in sealed bottles and held at about 4°C (40°F). A good germination medium is a layer of shredded sphagnum moss or vermiculite over a mixture of sand and acid peat. The very small seeds are sifted on the surface of the medium and watered with a fine spray. The flats should be covered with glass and always kept shaded. Careful attention must be given to provide adequate moisture and ventilation as well as even heat: 15° to 21°C (60° to 70°F). The plants grow slowly, taking about three months to reach transplanting size. After two or three true leaves form, they are moved to another flat and spaced 2.5 to 5 cm (1 to 2 in.) apart, where they remain through the winter in a cool greenhouse or in cold frames. In the spring the plants are set out in the field in an acid soil, and by fall they are ready to be dug and potted preparatory to grafting in the winter.

**Cuttings** (*65, 160*)     Rooting cuttings is the chief method of rhododendron propagation. These are best taken, midsummer into fall, from stock plants in full sun grown especially for this purpose. However, stem cuttings—or leaf-bud cuttings—of some hybrids taken in midwinter will root well. Any flower buds should be removed from the cuttings. Treatments with indolebutyric acid at relatively high concentrations are required; IBA in talc at 1 or 2 percent concentration, plus an added fungicide (benomyl) works well. Wounding the base of the cutting on both sides is a strong stimulus to rooting in rhododendrons. A rooting medium of two-thirds sphagnum peat moss and one-third perlite is suitable. Bottom heat at 24°C (75°F) should be used. Rhododendron cuttings are best rooted under mist in the greenhouse and should be lifted soon after roots are well formed (about three months) or the roots will deteriorate. After rooting and transplanting (into German peat moss, with added fertilizers) the cuttings should be held at 4°C (40°F) for about 20 days, after which the night temperature can be raised to a minimum of 18°C (65°F). Supplementary light at this stage to extend the day length will give good growth response. Plants started from cuttings usually develop rapidly and are free of the disadvantage of suckering from the rootstock, which occurs with grafted plants.

**Grafting**     A side-veneer graft is most successful. The best scionwood is taken from straight, vigorous current season's growth. After grafting, the plants are kept in closed frames under high humidity and a temperature near 21°C (70°F) until the union has healed. The plants should then be moved to cooler conditions—10° to 15°C (50° to 60°F)—and the top of the stock removed above the graft union. After the plant has hardened, it is transplanted to the nursery row in acid soil and grown for two years, after which it is ready to dig as a salable nursery plant.

**Layering**     Rhododendrons are easily reproduced by trench and simple layering. All parts of rhododendron and azalea plants are poisonous if eaten.

**Rhus spp.**     *See* Sumac.

**Robinia pseudoacacia**     *See* Locust, Black.

**Rose** (*Rosa* spp.) (*34, 76, 96, 156*)     All the selected rose cultivars are propagated by asexual methods. T-budding on vigorous rootstocks is most common, although the use of softwood or hardwood cuttings or grafting, layering, or the use of suckers is sometimes practiced. Seed propagation is used in breeding new cultivars, in producing plants in large numbers for conservation projects or mass landscaping, and in growing rootstock plants of certain species, such as *R. canina*. Micropropagation of rose cultivars is successful (*75, 76, 150*).

**Seeds** (*19*)     As soon as the rose fruits ("hips") are ripe but before the flesh starts to soften, they should be collected and the seeds extracted. It is best to stratify them immediately at 2° to 4°C (35° to 40°F). Six weeks is suffcient for *Rosa multiflora*, but others—*R. rugosa* and *R. hugonis*—require four to six months, and *R. blanda* ten months (*148*). *Rosa canina* germinates best if the seeds are held at room temperature for two months in moist vermiculite and then transferred to 0°C (32°F) for an additional two months. Hybrid rose seeds usually respond best to a stratification temperature of 1° to 4°C (34° to 40°F) for 60 to 90 days, although some seeds may germinate with no cold stratification treatment. Germination is probably prevented in rose seeds by inhibitors occurring in the seed coverings, as well as by the mechanical restriction imposed by the massive pericarp (fruit wall) (*87*). The seeds may be planted either in the spring or in the fall in seedbeds or in the nursery row. In areas of severe winters, seedlings are likely to be winter-killed if they are smaller than 10 cm (4 in.) by the onset of cold weather.

**Hardwood cuttings**     Hardwood cuttings are widely used commercially in the propagation of rose rootstocks, and to some extent in propagating the strong growing polyanthas, pillars, climbers, and hybrid perpetuals. The hybrid teas and other similar everblooming roses also can be started by cuttings, but more winter-hardy and nematode-resistant plants are produced if they are budded on selected vigorous rootstocks. In mild climates, the cuttings are taken and planted in the nursery in the fall. In areas with severe winters, cutting may be made in late fall or early winter, tied in bundles, and stored in damp peat moss or sand at about 4°C (40°F) until spring, when they are planted in the nursery row. The rootstocks are ready to bud by the following spring, summer, or fall. The cuttings are made into 15 to 20 cm (6 to 8 in.) lengths from previous season's canes of 6 to 9 mm (1/4 to 3/8 in.) diameter. Commercially, large bundles of canes are run through band saws to cut them to the correct length. Disbudding is usually done in rootstock propagation; all buds except the top one or two are removed so as to prevent subsequent sucker growth in the nursery row.

**Softwood cuttings**     Softwood cuttings are made from current season's growth, from early spring to late summer, depending upon the time the wood becomes partially mature. Rooting is fairly rapid, occurring in 10 to 14 days. At the end of the season the cuttings may be transplanted to their permanent location, potted and overwintered in a cold frame, or transferred to the nursery row for another season's growth or to be budded to the desired cultivar. Cultivars of most miniature roses are easily propagated by softwood or semihardwood cuttings under mist.

**Budding**     T-budding is the method ordinarily used. The buds are inserted into 5 to 10 mm ($\frac{3}{16}$- to $\frac{3}{8}$-in.) diameter rootstock plants. In mild climates budding can be done during a long period, from late winter until fall, but mostly in the spring. Early buds will make some growth during the summer and produce a salable plant by fall. Some propagators break over the top of the rootstock about two weeks after budding to force the bud out. After the bud has reached a length of 10 to 20 cm (4 to 8 in.), the top of the stock is entirely removed. In areas with shorter growing seasons, budding is done during the summer. Buds inserted late in the summer either make little growth or remain dormant until the following spring. In this case, the rootstock is cut off just above the bud in late winter or early spring, forcing the inserted bud into growth. Shoots from buds started in the fall are cut back to 13 mm ($\frac{1}{2}$ in.) in the spring. The shoot then grows through the following entire summer, producing a well-developed plant by fall. After the shoot has grown about 15 cm (6 in.), it is generally cut back to 5 to 7.5 cm (2 or 3 in.) to force out side branches.

Budwood may be obtained during the budding season from current season's growth of the desired cultivar, only a day's supply being taken at a time. It is best collected early in the morning, clipping off the leaves immediately and leaving about 6 mm ($\frac{1}{4}$ in.) of the petiole attached to the bud. Lateral buds from the stems producing the flowers are the best to use. Plump, but dormant, buds three or four nodes below the flower are the most desirable. The wood should be at a stage of maturity in which the thorns are easily removed. It is important to use budwood taken only from plants free of verticillium. Buds from diseased wood can infect each budded plant with this fungus (*137*).

An alternative method for obtaining budwood, which has been used widely and successfully, is to store dormant wood under refrigeration just below freezing (-1° to 0°C; 30° to 32°F) until time for budding. The budwood is collected in late fall after the flowers are shed and the thorns become dark. The leaves are removed by hand, but the thorns are left intact. Sticks 25 to 38 cm (10 to 15 in.) long are put up in bundles of 30 or 40 each. The bundles are wrapped tightly as possible in waterproof paper over which a layer of moist wrapping—such as wet newspapers—is placed. Finally, the bundles are covered with another layer of waterproof paper.

Buds are inserted in the wood of the original cuttings rather than into new growth arising from the rootstock.

**Rootstocks for rose cultivars** (*16*)     Most rose rootstock clones have been in use for many years, propagated by cuttings; many of the clones are virus-infected, thus infecting the cultivar top after budding. However, these clonal rootstocks are available with the viruses eliminated by heat treatments. Holding potted rose plants at a dry heat of 37° to 38°C (98° to 100°F) for four to five weeks will rid infected stocks of the virus.

*Rosa multiflora*     This is a useful rootstock, especially in its thornless forms, for outdoor roses. Several "strains" have been developed, some giving better bud unions and bud development than others. Cuttings of this species root readily, develop a vigorous, nematode-resistant root system, and do not sucker excessively. It is adaptable to a wide range of soil and climatic conditions, but does not seem to do well in the southern United States. Seedlings are used in the eastern part of the United States and cuttings on the Pacific Coast. The bark often becomes so thick late in the season that budding is impossible.

*Rosa canina (Dog rose)*    Although this species has not done well under American conditions, the stock is commonly used in Europe. It is usually propagated by seed, since the cuttings do not root easily; however, the seeds are difficult to germinate. The prominent thorns make it difficult to handle. It also tends to sucker. Young plants on this stock grow slowly, but they are long-lived. *Rosa canina* is adaptable to drought and alkaline soil conditions.

*Rosa chinensis* 'Gloire de Rosomanes', 'Ragged Robin'    This old French stock is popular in California for outdoor roses, resisting heat and dry conditions well. It is also resistant to nematodes and does not sucker if the lower buds on the cuttings are removed. This stock grows steadily through the summer, permitting budding at any time. The fibrous root system is easy to transplant but requires good soil drainage. In some areas, however, it is difficult to propagate and is injured by leafspot. Owing to its susceptibility to verticillium wilt, it should not be planted on land previously planted to tomatoes or cotton.

*Rosa* 'Dr. Huey'    This is the principal rootstock in Arizona and the southern San Joaquin Valley, California, rose districts, replacing 'Ragged Robin' to a large extent. It has also performed well in Australia (*139*). It is useful for late season budding because of its thin bark. It is very vigorous and well adapted to irrigated conditions, and its cuttings root readily. It is very good as a stock for weak-growing cultivars. Defects are its injury from subzero temperatures and susceptibility to blackspot, mildew, and verticillium.

*Rosa* × *noisettiana* 'Manetti'    This is an old stock, very popular for greenhouse forcing roses. It is also of value for dwarf roses and for planting in sandy soils. It is easily propagated by cuttings, produces a plant of moderate vigor, and is resistant to some strains of verticillium (*137*).

*Rosa odorata (Odorata 22449)* Tea rose    This stock is excellent for greenhouse forcing roses. Cuttings root easily under suitable conditions, and produce a large symmetrical root system. It is adapted to both excessively dry and wet soil conditions. Since it is not cold-hardy, it should be used only in areas with mild winters. Some propagation stock of this clone is badly diseased and does not root well. The plants are not adaptable to cold-storage handling. It is more susceptible to verticillium than *R. manetti*.

*Rosa dumetorum* 'Laxa'    This stock is the most widely used rose rootstock in Great Britain. It produces few suckers and starts growth early in the season. It is propagated by seed, which is given a sulfuric acid treatment followed by stratification at 24°C (76°F) for 30 days and then at 5°C (42°F) for 85 days.

*IXL (Tausendschon / Veilchenblau)*    This stock is used primarily as a trunk for tree roses. It is very vigorous and has no thorns. The canes tend to sunburn and are somewhat susceptible to low-temperature injury.

*Multiflore de la Grifferaie*    This stock is useful as a trunk for tree roses, producing desirable straight canes. It is vigorous, extremely hardy, and resistant to borers, but very susceptible to mite injury.

*Rosa rugosa*    This form, which is used as an understock, bears single, purplish-red flowers. For bush roses it is propagated by cuttings, and for tree roses by seed. The root system is shallow and fibrous and tends to sucker badly, but the plants are very long-lived. It is also used as the upright stem in producing standard (tree) roses.

**Propagating tree (standard) roses**    A satisfactory method of producing this popular form of rose is to use *Rosa multiflora* as the rootstock, which is budded in the first summer to IXL or, preferably, the Grifferaie stock. These are trained to an upright form and kept free of suckers. In the second summer, at a height of about 0.9 meter (3 ft) three or four buds of the desired flowering cultivar are inserted into the trunk stock. During the winter, the cane above the inserted buds is removed. The buds develop the following summer, as do buds from the stock, which must be removed. In the fall the plants may be dug and moved to their permanent location. Tree roses are sometimes dug as balled and burlapped plants because an extensive root system is formed during the two years in which the top is being developed. (See Figure 10–34.)

**Propagation of miniature roses** (*123*)    Soft or semihardwood cuttings are taken the year around and rooted under mist, after dipping in an indolebutyric acid rooting powder. Miniature rose cultivars are especially bred for their ease of rooting. A good rooting mix is $\frac{1}{3}$ peat moss, $\frac{1}{3}$ fir bark, $\frac{1}{3}$ perlite. Cuttings root in three to four weeks under warm conditions.

**Rose of Sharon**    *See* Hibiscus syriacus.

**Rosemary** (*Rosmarinus officinalis L.*)    This ground cover with aromatic leaves is easily propagated by leafy cuttings under mist or glass, and also by seeds.

**Royal Poinciana** (*Delonix regia* [Bajer.] Raf.)    This spectacular tropical flowering tree is propagated by seed. Germination is rapid when seeds are treated to soften seed coats, as by pouring boiling water over seeds or by soaking in concentrated sulfuric acid for one hour.

**Russian Olive**    *See* Elaeagnus.

**St. Johnswort** (*Hypericum* spp.)    St. Johnswort is easily started by softwood cuttings taken in late summer from the tips of current growth and rooted under high humidity or mist.

**Salix spp.**    *See* Willow.

**Sambucus spp.**    *See* Elder.

**Senicio cineraria**    *See* Dusty Miller.

**Sequoia sempervirens**    *See* Redwood, Coast.

**Sequoiadendron giganteum**    *See* Redwood, Giant Sequoia.

**Serviceberry** (*Amelanchier* spp.) (*71*)    Seeds show embryo dormancy which can be overcome by stratification at about 2°C (36°F) for three to six months. Seeds should not be allowed to dry out. Seeds of some species also require scarification prior to stratification. This is also readily propagated by leafy softwood cuttings taken when the new growth is several inches long and rooted under mist. Indolebutyric acid at 0.3 percent in talc increases rooting. *Amelanchier* is also easily propagated by root cuttings.

**Shrub althea**    *See Hibiscus syriacus.*

**Silk Tree**    *See Albizzia.*

**Silver Berry**    *See Elaeagnus.*

**Smoke Tree** (*Cotinus coggygria* Scop.) (*91*)    This tree can be propagated by leafy soft-wood cuttings under mist. Tip cuttings taken from spring growth and treated with in-dolebutyric acid should root in about five weeks. After the mist is discontinued, the rooted cuttings are left in place to be transplanted bareroot the following spring. Smoke tree should not be propagated by seeds, since many of the seedlings are male plants, lacking the showy flowering panicles. Only vegetative methods should be used, with propagating wood taken from plants known to produce large quantities of the desirable fruiting clusters.

**Snowberry** (*Symphoricarpos* spp.)     Seed propagation is difficult because of a hard, impermeable endocarp and a partially developed embryo at harvest. Give seeds a three- to four-month warm, moist stratification followed by cold stratification at 5°C (41°F) for six months before planting. Propagation is possible also by suckers, division, and cuttings.

**Sorbus** (*Sorbus* spp.) (*101*)     Seeds should be collected as soon as the fruits mature; fleshy parts are removed to eliminate inhibitors. Seeds require stratification for at least two months at about 5°C (41°F) for good germination. Cuttings do not root, and layering is difficult. Either fall budding or bench grafting (whip graft) is successful. Selected cultivars are best worked on seedlings of their own species, although *S. aria, S. aucuparia,* and *S. cuspidata* (European mountain ash) seedlings seem satisfactory as a rootstock for other species.

**Spiraea** (*Spiraea* spp.)     Spiraea is usually propagated by cuttings although some species, such as *S. thunbergi,* are more easily started by seeds, which should not be allowed to dry out. Leafy softwood cuttings taken in midsummer and rooted under high humidity are generally successful. Treatments with one of the root-promoting substances are often of considerable benefit. Some species, such as *S. vanhouttei,* can be started readily by hardwood cuttings, planted in early spring.

**Spruce** (*Picea* spp.)     Spruces are ordinarily propagated without difficulty by seed, either fall-planted or stratified over winter. Most species have embryo dormancy, requiring one to three months' stratification at about 4°C (40°F) for good germination. Seeds of *P. abies, P. engelmannii,* and *P. glauca* var. *albertiana* are among those giving good germination without stratification. Colorado blue spruce (*Picea pungens* 'Glauca') grown from seed produces trees with a slight bluish cast. Only a small percentage of the seedlings have the very desirable bright blue color. Several exceptionally fine blue seedling specimens have been selected as clones and are perpetuated by grafting. The two best known are the Koster blue spruce (*Picea pungens* 'Koster') developed at the Koster Nursery in Holland many years ago, and the compact 'Moerheim' blue spruce (*Picea pungens* 'Moerheimii'), which originated in Europe around 1930.

Selected clones of spruce are difficult to propagate by cuttings, but there are instances in which good percentages of cuttings have rooted (*63, 86*), especially from certain young source trees. Taking cuttings from vigorous containerized trees gives good results.

Cuttings taken in spring, midsummer, and midautumn have been rooted. As cuttings, it is best to use only shoot terminals, which should be gathered in early morning when the wood is turgid. Wounding and high light intensity during rooting are helpful. Medium, slightly acid sand or coarse perlite is a good rooting medium. Use of a light mist is beneficial. Hormone treatments have given variable results. In making cuttings of upright-growing types, terminal shoots should be selected rather than lateral branches, since the latter, if rooted, tend to produce abnormal, sprawling plants rather than the desired upright form.

The Koster blue spruce is propagated commercially by grafting scions on Norway spruce (*Picea abies*) or Sitka spruce (*P. sitchensis*) seedlings (*36, 109*).

**Spurge, Japanese**     *See* Pachysandra.

**Star Jasmine, Chinese** ( *Trachelospermum jasminoides* [Lindl.] Lem.)    Leafy cuttings of partially matured wood root easily, especially when placed under mist and treated with a root-promoting substance.

**Sumac** ( *Rhus* spp.)    Sumacs are commonly propagated by seeds, which are collected in the fall; they should not be allowed to dry out. For prompt germination the seeds should be soaked in concentrated sulfuric acid for one to six hours, depending upon the species, then either fall-planted out-of-doors or stratified for two months at about 4°C (40°F) before planting. Not all species have embryo dormancy, however, and the latter treatment may sometimes be omitted (e.g., with *R. ovata* and *R. integrifolia).* Some species, such as *R. aromatica,* need no pretreatments and will germinate with fall sowing. Some sumac plants bear only female flowers and others only male flowers, whereas still others have both flower types on the same plant. To insure fruiting, plants of the latter type should be propagated asexually. In seed propagation many of the seedlings produced will not bear fruit.

For those species which sucker freely, such as *R. typhina* and *R. copallina,* root cuttings several inches long planted in the nursery row in early spring may be used. Leafy softwood cuttings, at least of some species, such as *R. aromatica,* taken in midsummer, root well under plastic if treated with 1 percent IBA mixed with 50 percent captan, 1:1.

**Sycamore**    *See* Plane Tree.

**Symphoricarpos**    *See* Snowberry.

**Syringa spp.**    *See* Lilac.

**Tamarisk** ( *Tamarix* spp.)    These are easily rooted by hardwood cuttings, which are usually made about 12 in. long and planted deeply. Softwood cuttings taken in early summer also will root readily under glass or mist.

**Taxodium distichum**    *See* Bald cypress.

**Taxus spp.**    *See* Yew.

**Telopea speciosissima**    *See* Waratah.

**Thuja spp.**    *See* Arborvitae.

**Tilia spp.**    *See* Linden.

**Toyon**    *See* Heteromeles.

**Trachelospermum jasminoides**    *See* Star Jasmine, Chinese.

**Tree Tea**    *See* Leptospermum.

**Trumpet Creeper** ( *Campsis* spp.)    This vine is usually propagated by cuttings, but seeds can also be used. With the latter method, stratification for two months at 4° to 10°C (40° to 50°F) hastens but does not increase germination. Both softwood and hardwood cuttings root readily. *C. radicans* can be started by root cuttings. Layering also is successful.

**Tsuga spp.**    *See* Hemlock.

**Tulip Tree** ( *Liriodendron tulipifera* L.) ( *124* ) Yellow Poplar    Seed propagation is somewhat difficult. Artificial cross-pollination gives a higher percentage of "filled" seeds. Seeds should be stratified for about two months before planting and should not be allowed to dry out. A daily varying stratification temperature between freezing and about 10°C (50°F) has given good results, although a constant temperature around 4°C (40°F) would probably be equally satisfactory. Fall planting, with outdoor stratification through the winter, also has given good germination. Seeds of this species are often devoid of embryos, so cutting tests of each seed lot should be made. Although not often done commercially, leafy stem cuttings taken in the summer have been rooted in fairly good percentages. Root cuttings also have been successful. Propagation by both budding and grafting is successful.

Young tulip trees are very difficult to transplant, so they should always be propagated into containers or dug, balled, and burlapped, for transplanting from the nursery row.

**Ulmus spp.**    *See* Elm.

**Viburnum** ( *Viburnum* spp.) ( *10, 13, 119* )    This large group of desirable shrubs can be propagated by a number of methods, including seeds, cuttings, grafting, and layering. At least one species ( *V. dentatum* ) is readily started by root cuttings.

**Seeds**    The viburnums have rather complicated seed dormancy conditions. Seeds of some species, such as *V. sieboldi,* will germinate after a single ordinary low-temperature (4°C; 40°F) stratification period, but for most species a period of two to nine months at high temperatures (20° to 30°C; 68° to 86°F), followed by a two- to four-month period at low temperatures (4°C; 40°F) is required. The initial warm temperatures cause root formation, and the subsequent low temperature causes shoot development. Cold stratification alone will not result in germination. Such rather exacting treatments may best be given by planting the seeds in summer or early fall (at least 60 days before the onset of winter), thus providing the initial high-temperature requirement; the subsequent winter period fulfills the low-temperature requirement. After this, the seeds should germinate readily in the spring. Often, collecting the seeds early, before a hard seed coat has developed, will hasten germination. Viburnum seed can be kept for one or two years if stored dry in sealed containers and held just above freezing. *V. lantana, V. opulus,* and *V. rhytidophyllum* are commonly propagated by seed.

**Cuttings**    Although some viburnum species ( *V. opulus, V. dentatum,* and *V. trilobum* ) can be propagated by hardwood cuttings, softwood cuttings rooted in sand or perlite under glass or mist are successful for most species. Soft, succulent cuttings taken in late spring root faster than those made from more mature tissue in midsummer, but the latter are more likely to grow on into sturdy plants that will survive through the following winter. Treatments with indolebutyric acid are helpful. One of the chief problems with viburnum cuttings is to keep them growing after rooting. Cuttings made from succulent, rapidly growing material often die in a few weeks after being potted. This trouble may be overcome by not digging the cuttings too soon, allowing a secondary root system to form, which will better stand the transplanting shock. It may help, too, to feed the rooted cuttings with a nutrient solution about ten days before the cuttings are to be removed. Placing the rooted cuttings under artificial lights to increase day length also is helpful. Cuttings of some species root more easily than others. *Viburnum carlesii,* for example, is difficult to root, but *V. burkwoodii, V. / rhytidophylloides, V. lantana, V. sargentii,* and *V. plicatum* forma *tomentosum* root readily.

**Grafting**    Selected types of viburnum are often propagated by grafting on rooted cuttings, layers, or seedlings of *V. dentatum* or *V. lantana.* Often, grafted viburnums will develop into vigorous plants more quickly than those started as cuttings. *V. opulus* 'Roseum' (Snowball) is dwarfed when grafted onto *V. opulus* 'Nanum' cuttings. It is important that all buds be removed from the rootstock so that subsequent suckering from the stock does not occur. The understocks are potted in the fall and brought into the greenhouse, where they are grafted in midwinter by the side graft method, using dormant scionwood. After grafting, the potted plants are placed in a closed, glass-covered frame with the unions buried in damp peat moss.

Grafting can be done in late summer also, using potted understock plants and scion material that has stopped growing and become hardened. The grafted plants are plunged in slightly damp peat moss in closed frames in the greenhouse until the unions heal, after which they are moved to outdoor, glass-covered cold frames for hardening-off for the winter.

**Layering**    Simple layering is widely used, especially in Europe, for propagating most viburnum species. Wood of the previous season's growth will produce roots in 18 to 24 months if layered in the spring. Some species are best layered in midsummer, using current season's wood.

**Waratah** (*Telopea speciosissima*)    Seeds of this Australian native shrub with beautiful chrysanthemumlike flowers germinate easily, but the seedlings are difficult to transplant and grow outside their native environment. It requires soil of extremely low phosphorus content.

**Wax Myrtle**    *See* Bayberry.

**Weigela** (*Weigela* spp.)    This shrub is easily propagated either by hardwood cuttings planted in early spring or by softwood tip cuttings under glass or mist taken any time from late spring into fall. Treatment with indolebutyric acid promotes rooting.

**Willow** (*Salix* spp.)    Willow seeds must be collected as soon as the capsules mature (when they have turned from green to yellow) and planted immediately since they retain their viability for only a few days at room temperature. Even under the most favorable conditions, maximum storage is four to six weeks. No dormancy occurs, germination taking place 12 to 24 hours after planting if the seeds are kept constantly moist. Willows are difficult to propagate in quantity by seed.

Willows root so readily by either stem or root cuttings that there is little need to use other methods. Hardwood cuttings planted in early spring root promptly.

**Wisteria** (*Wisteria* spp.)    Wisterias may be started by softwood cuttings under glass or mist taken in midsummer. Indolebutyric acid treatments often aid rooting. Some species can be started by hardwood cuttings set in the greenhouse in the spring. Simple layering of the long canes is quite successful. Choice types are often grafted on rooted cuttings of less desirable types. Suckers arising from roots of such grafted plants should be removed promptly. Wisterias do not transplant easily, so young nursery plants are best started in containers. Seeds and pods are poisonous if eaten.

**Witch Hazel** (*Hamamelis* spp.) (*106*)    Witch hazel is propagated by seed, budding (*136*), grafting, or layering. Cutting propagation is difficult but possible. Seeds are gathered and planted outdoors in early fall. Prevent seeds from drying. *H. japonica* seeds should be soaked for a week with water changed daily. Cultivars of *H. mollis*, *H. × intermedia*, and *H. japonica* are propagated by budding or grafting on *H. virginiana* seedlings. Leafy cuttings of *H. mollis*, *H. virginiana*, *H. japonica*, and *H. vernalis* can be rooted under mist with indolebutyric acid at 8000 ppm treatments. Survival during winter is a problem, however (*99*).

**Xylosma congestum** (Lour.) Merr. (*97*)    This is propagated by rooting leafy cuttings, taken in late summer or early fall, under closed frames, using the first and second subterminal cuttings on the shoot. Cuttings are dipped in IBA at 5000 ppm and rooted in a 1:3 peat moss-perlite mixture with bottom heat (21°C; 70°F). Results under mist have been contradictory. Rooted cuttings should be hardened in a cool, humid greenhouse.

**Yellow Poplar**    *See* Tulip Tree.

**Yew** (*Taxus* spp.) (*37, 142, 171*)    Most clonal selections of yews are propagated by cuttings, which root without much difficulty. Seedling propagation is little used, because of variation in the progeny, complicated seed dormancy conditions, and the slow growth of seedlings. Side or side-veneer grafting is practiced for those few cultivars that are especially difficult to start by cuttings, with easily rooted cuttings used as the rootstock.

**Seeds**    This method is confined in commercial practice almost entirely to the Japanese yew, *Taxus cuspidata,* which comes fairly true from seed if isolated plants can be located as sources of seed. Seed imported from Japan is believed to produce uniform off-spring.

For good germination, seeds should be given a warm 20°C (68°F) stratification period in moist peat moss or other medium for three months, followed by four months at a lower temperature (5°C; 41°F). Seedling growth is very slow. Two years in the seed-bed, followed by two years in a lining-out bed, then three or four years in the nursery row are required to produce a plant of salable size of the Japanese yew.

**Cuttings**    *Taxus* cuttings can be rooted outdoors in cold frames or in the greenhouse under mist, the latter giving much faster results.

For the cold frame, fairly large cuttings, 20 to 25 cm (8 to 10 in.) long, are made in early fall from new growth with a section of old wood at the base. *Taxus* cuttings seem to respond well to treatment with a root-inducing chemical, indolebutyric acid at relatively high concentrations being particularly effective.

Cuttings may be kept in closed frames through the winter; in climates with severe winters, the frames should be kept covered, especially at times when the ground is frozen but the sun is shining. Rooting takes place slowly during the following spring and summer.

For greenhouse propagation, cuttings should be taken in early winter, after several frosts have occurred, and rooted under mist in sand with bottom heat at about 21°C (70°F) and an air temperature of 10° to 13°C (50° to 55°F). Rooting in the greenhouse takes only about two months, but cuttings should not be dug too soon. Allow time for secondary roots to develop from the first-formed primary roots. There is evidence that cuttings from male plants (at least in *T. cuspidata expansa*) root more readily than cuttings from female plants (those that produce fruits) (*29*). Berries and foliage of yew plants are very poisonous if eaten.

# REFERENCES

*1.* Ackerman, W. L., and Seaton, G. A. 1969. Propagating the Bradford pear from cuttings. *Amer. Nurs.* 130(7):8.

*2.* Afanasiev, M. 1937. A physiological study of dormancy in seed of *Magnolia acuminata. N.Y. (Cornell) Agr. Exp. Sta. Mem. 208.*

*3.* ———. 1943. Germinating *Nandina domestica* seeds. *Amer. Nurs.* 78(9):5–6.

*4.* Aldhous, J. R. 1972. *Nursery practice.* London: Her Majesty's Stationary Office.

*5.* Allen, G. S., and J. N. Ownes. 1972. *The life history of douglas fir.* Ottawa, Canada: Forestry Service, Environment Canada.

6. Anderson, W. C. 1978. Rooting of tissue-cultured rhododendrons. *Proc. Inter. Plant Prop. Soc.* 28:135–39.

7. Andison, A., S. Arrowsmith, and M. Crown. 1974. Rooting cuttings of Douglas fir "plus" trees. *The Plant Propagator* 20(1):4–12.

8. Argles, G. K. 1969. Propagating maples (*Acer* species). *Nurs. and Gard. Cent.* 148(5):129–33; 148(6):199–201.

9. ———. 1969. The propagation of magnolias. *Nurs. and Gard. Cent.* 148(10):361–65; 148(11):399–406.

10. Barton, L. V. 1958. Germination and seedling production in species of *Viburnum. Proc. Plant Prop. Soc.* 8:126–34.

11. Bhella, H. S. 1975. Some factors affecting propagation of Douglas fir. *Proc. Inter. Plant Prop. Soc.* 25:420–24.

12. ———. 1977. Propagation of river birch (*Betula nigra* L.) by stem cuttings. *The Plant Propagator* 23(2):5–7.

13. ———. 1980. Vegetative propagation of viburnum, cvs. Alleghany, Mohican, and Onondaga. *The Plant Propagator* 26(3):5–9.

14. Boddy, R. M. 1975. Propagation of *Mahonia aquifolium. Proc. Inter. Plant Prop. Soc.* 25:68–71.

15. Brydon, P. H. 1964. The propagation of deciduous azaleas from cuttings. *Proc. Inter. Plant Prop. Soc.* 14:272–76.

16. Buck, G. J. 1951. Varieties of rose understocks. *Amer. Rose Ann.* 36:101–16.

17. Bunker, E. J. 1975. Germinating palm seed. *Proc. Inter. Plant Prop. Soc.* 25:377–78.

18. Cameron, R. J. 1968. The propagation of *Pinus radiata* by cuttings. *New Zealand Jour. For.* 13:78–89.

19. Carter, A. R. 1969. Rose rootstocks—performance and propagation from seed. *Proc. Inter. Plant Prop. Soc.* 19:172–80.

20. ———. 1979. Daphne propagation. *Proc. Inter. Plant Prop. Soc.* 29:248–51.

21. Carville, L. 1967. Propagation of Knaphill azaleas from softwoods. *Proc. Inter. Plant Prop. Soc.* 17:255–58.

22. ———. 1975. Propagation of *Acer palmatum* cultivars from hardwood cuttings. *Proc. Inter. Plant Prop. Soc.* 25:39–42.

23. Caulfield, H. W. 1976. Pointers for successful germination of palm seed. *Proc. Inter. Plant Prop. Soc.* 26:402–5.

24. Chandler, G. P. 1969. Rooting daphnes from cuttings. *Proc. Inter. Plant Prop. Soc.* 19:205–6.

25. Chase, H. H. 1964. Propagation of Oriental magnolias by layering. *Proc. Inter. Plant Prop. Soc.* 14:67–69.

26. Chaturvedi, A., K. Sharma, and P. N. Prasad. 1978. Shoot apex culture of *Bougainvillea glabra* 'Magnifica'. *HortScience* 13(1):36.

27. Chang, T. Y., and T. H. Vogue. 1977. Regeneration of Douglas fir plantlets through tissue culture. *Science* 198:306–7.

28. Chu, K., and W. S. Cooper. 1950. An ecological reconnaissance in the native home of *Metasequoia glyptostroboides. Ecology* 31:260–78.

29. Coggeshall, R. C. 1960. Whip and tongue grafts for dogwoods. *Amer. Nurs.* 111(2):9, 56, 59.

30. ———. 1977. Propagating French hybrid lilacs by softwood cuttings. *Proc. Inter. Plant Prop. Soc.* 27:442–44.

31. Cohen, M. A. 1975. Vegetative propagation of *Pinus strobus* by needle fascicles. *Proc. Inter. Plant Prop. Soc.* 25:413–19.

32. Corns, W. G., and R. J. Schraa. 1962. Dormancy and germination of seeds of silverberry (*Elaeagnus commutata* Bernh.). *Can. Jour. Bot.* 40:1051–55.

33. Costin, J. J. 1977. Production of eucalyptus. *Proc. Inter. Plant Prop. Soc.* 27:44–48.

34. Crockett, J. U. 1971. *Roses.* New York: Time-Life Books.

35. Cumming, W. A. 1964. *Crataegus* rootstock studies. *Proc. Inter. Plant Prop. Soc.* 14:146–49.

36. Curtis, W. J. 1962. The grafting of Koster spruce, *Cedrus atlantica* 'Glauca', copper beech, and variegated dogwood. *Proc. Inter. Plant Prop. Soc.* 12:249–53.

37. Davidson, H., and A. Olney. 1964. Clonal and sexual differences in the propagation of *Taxus. Proc. Inter. Plant Prop. Soc.* 14:156–62.

38. Deen, J. L. W. 1974. Propagation of *Quercus ilex* by cuttings. *The Plant Propagator* 20(3):18–20.

39. Deering, T. 1979. Bench grafting of *Betula* species. *The Plant Propagator* 25(3):8–9.

40. deRigo, H. T., and W. O. Hawley. 1966. Effect of time of rhizomatous propagation of a temperate zone bamboo on shoot growth. *Agron. Jour.* 58:401–2.

41. de Wilde, R. 1964. Production and breeding of lilacs. *Proc. Inter. Plant Prop. Soc.* 14:107–13.

42. Donnelly, J. R., and H. W. Yawney. 1972. Some factors associated with vegetatively propagating sugar maple by stem cuttings. *Proc. Inter. Plant Prop. Soc.* 22:413–31.

43. Ecke, P., and O. A. Matkin. 1976. *The poinsettia manual.* Encinitas, Calif.: Paul Ecke Poinsettias.

44. Eichelser, J. E. 1978. Propagation of kalmia. *Proc. Inter. Plant Prop. Soc.* 28:133–35.

45. Einert, A. E. 1974. Propagation of dwarf crape myrtles. *Proc. Inter. Plant Prop. Soc.* 24:370–73.

46. Farmer, R. E., Jr. 1975. Long term storage of northern red and scarlet oak seed. *The Plant Propagator* 20(4) and 21(1):11–14.

47. Evison, R. J. 1977. Propagation of clematis. *Proc. Inter. Plant Prop. Soc.* 27:436–40.

48. ———. 1979. *Making the most of clematis.* Nottingham, England: Floraprint.

49. Fazio, S. 1964. Propagating *Eucalyptus* from cuttings. *Proc. Inter. Plant Prop. Soc.* 14:288–90.

50. Fins, L. 1980. Propagation of giant sequoia by rooting cuttings. *Proc. Inter. Plant Prop. Soc.* 30:127–32.

51. Flemer, W., III. 1962. The vegetative propagation of oaks. *Proc. Inter. Plant Prop. Soc.* 12:168–71.

52. ———. 1980. Linden propagation—a review. *Proc. Inter. Plant Prop. Soc.* 30:333–36.

53. Fleming, R. A. 1962. Rootstocks for ornamental trees. *Rpt. Hort. Exp. Sta. and Prod. Lab.* (Vineland, Ontario), pp. 46–49.

54. ———. 1978. Propagation of holly in Southern Ontario. *Proc. Inter. Plant Prop. Soc.* 28:553–57.

55. Fordham, A. J. 1968. *Cedrus deodara* 'Kashmir' and its propagation by cuttings. *Proc. Inter. Plant Prop. Soc.* 18:319–21.

56. ———. 1972. Vegetative propagation of *Albizia. Amer. Nurs.* 128(4):7, 63.

57. ———. 1977. Propagation of *Kalmia latifolia* by cuttings. *Proc. Inter. Plant Prop. Soc.* 27:479–83.

58. ———. 1977. *Pieris floribunda* and its propagation. *Proc. Inter. Plant Prop. Soc.* 27:495–97.

59. Fox, B. S. 1972. Propagation of cotoneasters. *Proc. Inter. Plant Prop. Soc.* 22:213–18.

60. Freeland, K. S. 1977. Propagation of potentilla. *Proc. Inter. Plant Prop. Soc.* 27:441–42.

61. Fuller, C. W. 1979. Container production of *Euonymus alata* 'Compacta'. *Proc. Inter. Plant Prop. Soc.* 29:360–62.

62. Girouard, R. M. 1971. Vegetative propagation of pines by means of needle fascicles—a literature review. *Inform. Rpt. Dept. Environ., Can. For. Ser., Quebec.*

63. ———. 1973. Rooting, survival, shoot formation, and elongation of Norway spruce stem cuttings as affected by cutting types and auxin treatments. *The Plant Propagator* 19(2):16–17.

64. Goddard, A. N. 1970. Grafting Japanese maples. *The Plant Propagator* 16(4):6.

65. Goreau, T. 1980. Rhododendron propagation. *Proc. Inter. Plant Prop. Soc.* 30:532–37.

66. Greever, P. T. 1979. Propagation of *Heteromeles arbutifolia* by softwood cuttings or by seed. *The Plant Propagator* 25(2):10–11.

67. Grossbechler, F. 1981. Mass production of eucalyptus seedlings by direct sowing method. *Proc. Inter. Plant Prop. Soc.* 31:276–79.

68. Hamilton, D. F., and P. L. Carpenter. 1977. Seed germination of *Myrica pennsylvanicum* L. *HortScience* 12(6):565–66.

69. Hand, N. P. 1978. Propagation of lilacs. *Proc. Inter. Plant Prop. Soc.* 28:348–50.

70. Hare, R. C. 1979. Modular air layering and chemical treatments improve rooting of Loblolly pine. *Proc. Inter. Plant Prop. Soc.* 29:446–54.

71. Harris, R. E. 1961. The vegetative propagation of *Amelanchier alnifolia. Can. Jour. Plant Soc.* 41:728–31.

72. Hartline, J. B. 1970. Holly propagation. *Amer. Hort. Mag.* 49(4):213–18.

73. Hartmann, H. T. 1967. 'Swan Hill': a new fruitless ornamental olive. *Calif. Agr.* 21(1):4–5.

74. Hasegawa, P. M. 1979. *In vitro* propagation of rose. *HortScience* 14:610–12.

75. ———. 1980. Factors affecting shoot and root initiation from cultured rose shoot tips. *Jour. Amer. Soc. Hort. Sci.* 105(2):216–20.

76. Hasek, R. F. 1980. Roses. In *Introduction to floriculture*, R. A. Larson, ed. New York: Academic Press.

77. Heit, C. E. 1972. Propagation from seed: Growing larches. *Amer. Nurs.* 135(8):14–15, 99–110.

78. Henny, J. 1963. Exbury azaleas. *Proc. Inter. Plant Prop. Soc.* 13:231–33.

79. Hicks, H. E. 1956. The propagation of tree peonies. *Proc. Inter. Plant Prop. Soc.* 6:31–33.

80. Hildreth, W. R. 1969. The propagation of manzanitas by cuttings. *Jour. Calif. Hort. Soc.* 30(2):45–47, 53.

81. Hill, J. B. 1962. The propagation of *Juniperus chinensis* in greenhouse and mist bed. *Proc. Plant Prop. Soc.* 12:173-78.

82. Hill, S. R., and W. J. Libby. 1969. Outdoor rooting of *Pinus radiata*. *The Plant Propagator* 15(4):13-16.

83. Hinesley, L. E., and F. A. Blazich. 1980. Vegetative propagation of *Abies fraseri* by stem cuttings. *HortScience* 15(1):96-97.

84. Hume, E. P., and P. Owens. 1970. Rooting of hybrid lilac cuttings in outdoor beds. *The Plant Propagator* 16(2):14-17.

85. Huss-Danell, K., L. Eliasson, and I. Öhberg. 1980. Conditions for rooting of leafy cuttings of *Alnus incana*. *Phys. Plant.* 49:113-16.

86. Iseli, J., and D. Howse. 1981. New cultivars of *Picea pungens glauca*—their attributes and propagation. *The Plant Propagator* 27(1):5-8.

87. Jackson, G. A. D., and J. B. Blundell. 1963. Germination in *Rosa*. *Jour. Hort. Sci.* 38:310-20.

88. Jaynes, R. A. 1975. *The laurel book*. New York: Hafner Press.

89. ———. 1976. Mountain laurel selections and how to propagate them. *Proc. Inter. Plant Prop. Soc.* 26:233-36.

90. Joley, L. 1960. Experiences with propagation of the genus *Pistacia*. *Proc. Plant Prop. Soc.* 10:287-92.

91. Kelley, J. D., and J. E. Foret, Jr. 1977. Effect of timing and wood maturity on rooting of cuttings of *Cotinus coggygria* 'Royal Purple'. *Proc. Inter. Plant Prop. Soc.* 27:445-48.

92. Kiang, Y. T., O. M. Rogers, and R. B. Pike. 1974. Rooting Mugho pine cuttings. *HortScience* 9(4):350.

93. Kiem, S. C. 1961. Propagation of palms. *Amer. Hort. Mag.* 40:133-37.

94. Klapis, A. J., Jr. 1964. Grafting junipers. *Proc. Inter. Plant Prop. Soc.* 15:340-41.

95. Kofranek, A. M., and R. Larson, eds. 1975. *Growing azaleas commercially*. Berkeley, Calif.: Div. Agr. Sci., Univ. Calif.

96. Krussmann, G. 1981. *The complete book of roses*. Forest Grove, Ore.: Timber Press.

97. Kubo, E. 1965. Propagation of *Xylosma congestum*. *Proc. Inter. Plant Prop. Soc.* 15:340-41.

98. Lamb, J. G. D. 1970. Trials on propagation of *Chamaecyparis* at Kinsealy. *Proc. Inter. Plant Prop. Soc.* 20:334-38.

99. ———. 1976. The propagation of understocks for *Hamamelis*. *Proc. Inter. Plant Prop. Soc.* 26:127-30.

100. Larson, R. A., J. W. Love, D. L. Strider, R. K. Jones, J. R. Baker, and K. F. Horn. 1978. *Commercial poinsettia production*. Raleigh, N.C.: Dept. Agr. Inf., N.C. State Univ.

101. Lawyer, D. A. 1968. Propagating *Sorbus*. *Amer. Nurs.* 127(2):7, 54, 56, 58, 60.

102. Lanphear, F. O. 1963. The seasonal response in rooting of evergreen cuttings. *Proc. Inter. Plant Prop. Soc.* 20:334-38.

103. Leach, D. G. 1965. Efficient production of rhododendrons. *Amer. Nurs.* 122(9):7, 36, 46.

104. Lebens, N. A. 1979. Vegetative propagation of poinsettia without mist. *The Plant Propagator* 25(1):10-11.

105. Lee, F. P. 1965. *The azalea book* (2nd ed.). Princeton, N.J.: D. Van Nostrand.

106. Leiss, J. 1969. *Hamamelis* propagation. *Proc. Inter. Plant Prop. Soc.* 19:349-52.

107. Li, Hui-Lin. 1961. Ginkgo—the maidenhair tree. *Amer. Hort. Mag.* 40:239–49.

108. Libby, W. J., and M. T. Conkle. 1966. Effects of auxin treatment, tree age, tree vigor, and cold storage on rooting young Monterey pine. *For. Sci.* 12:484–502.

109. Mahlstede, C. 1962. A new technique in grafting blue spruce. *Proc. Plant.Prop. Soc.* 12:125–26.

110. March, S. G. 1959. Propagating Ghent and Mollis azaleas. *Amer. Nurs.* 110(12):98–101.

111. Marden, L. 1980. Bamboo, the giant grass. *National Geographic* 158(4):502–29.

112. Martens, O. 1971. Palms—propagation, production, and uses. *Proc. Inter. Plant Prop. Soc.* 21:110–18.

113. Mergen, R., and B. A. Simpson. 1964. Asexual propagation of pines by rooting leaf fascicles. *Silvae Genet.* 13(5):125–64.

114. McCahon, W. 1964. Propagation of hydrangeas. *Proc. Inter. Plant Prop. Soc.* 14:253–54.

115. McClure, F. A. 1966. *The bamboos: A fresh perspective.* Cambridge, Mass.: Harvard Univ. Press.

116. McCown, B., and R. Amos. 1979. Initial trials with micropropagation of birch selections. *Proc. Inter. Plant Prop. Soc.* 29:387–93.

117. McDaniel, J. D. 1964. A look at some hackberries. *Proc. Inter. Plant Prop. Soc.* 14:143–46.

118. MacDonald, B. 1974. Camellia propagation. *Proc. Inter. Plant Prop. Soc.* 24:152–54.

119. McMillan-Browse, P. D. A. 1970. Notes on the propagation of viburnums. *Proc. Inter. Plant Prop. Soc.* 20:378–86.

120. ———. 1971. Propagation of *Aesculus. The Plant Propagator* 17(2):4–6.

121. Mirov, N. T. 1967. Morphology and reproduction. Chapter 5 in *The genus* Pinus. New York: Ronald Press.

122. Moore, J. C. 1963. Propagation of chestnuts and camellias by nurse seed grafts. *Proc. Inter. Plant Prop. Soc.* 13:141–43.

123. Moore, R. S. 1981. Miniature rose production. *Proc. Inter. Plant Prop. Soc.* 30:54–60.

124. Morsink, W. A. G. 1974. Propagation of hardy strains of tulip trees (*Liriodendron tulipifera* L.) for southern Ontario. *The Plant Propagator* 20(3):14–15.

125. Myers, J. R., and S. M. Still. 1979. Propagating London plane tree from cuttings. *The Plant Propagator* 25(3):8–9.

126. Nagao, M. A., K. Kanegawa, and W. S. Sakai. 1980. Accelerating palm seed germination with gibberellic acid, scarification, and bottom heat. *HortScience* 15(2):200–201.

127. Nienhuys, H. C. 1981. Propagation of deciduous azaleas. *Proc. Inter. Plant Prop. Soc.* 30:457–59.

128. Nordine, R. M. 1952. Collecting, storage, and germination of maple seed. *Proc. Inter. Plant Prop. Soc.* 2:62–64.

129. Orton, E. R., Jr., S. H. Davis, Jr., and L. M. Vasvary. 1966. Growing American holly in New Jersey. *N.J. Agr. Ext. Bul. 388.*

130. Parvin, P., R. A. Criley, and R. M. Bullock. 1973. Proteas: Developmental research for a new cut flower crop. *HortScience* 8(4):299–303.

131. Penfold, A. R., and I. L. Willis. 1961. *The Eucalypts: Botany, cultivation, chemistry and utilization.* New York: Interscience.

132. Phipps, H. M., D. A. Belton, and D. A. Netzer. 1977. Propagating cuttings of some *Populus* clones for tree plantations. *The Plant Propagator* 23(4):8–11.

133. Pinney, J. J. 1970. A simplified process for grafting junipers. *Amer. Nurs.* 131(10):7, 82–84.

134. Pope, D. R. 1974. Propagation of *Hydrangea petiolaris* by cuttings. *Proc. Inter. Plant Prop. Soc.* 24:194–95.

135. Pridham, A. M. S. 1964. Propagation of American elm from cuttings. *Proc. Inter. Plant Prop. Soc.* 14:86–88.

136. Purcell, G. V. 1973. The budding of *Hamamelis. Proc. Inter. Plant Prop. Soc.* 23:129–32.

137. Raabe, R. D., and S. Wilhelm. 1966. Budwood as a source of verticillium wilt in greenhouse roses. *Calif. Agr.* 20(10):5–6.

138. Roberts, A. N., and F. W. Moeller. 1968. Propagation of Mugho pine successful. *Ore. Orn. & Nurs. Dig.* 12(1):1–2.

139. Ross, D. M. 1977. Rose rootstock, 'Dr. Huey', in South Australia. *Proc. Inter. Plant Prop. Soc.* 27:562–63.

140. Rouland, H., and N. E. Pellett. 1974. Propagation of Norway spruce (*Picea abies* [L.] Karst) by stem cuttings. *The Plant Propagator* 20(1):20–26.

141. Ryan, G. F. 1966. Grafting *Eucalyptus ficifolia. The Plant Propagator* 12(2):4–6.

142. Sabo, J. E. 1976. Propagation of *Taxus* in Northern Ohio. *Proc. Inter. Plant Prop. Soc.* 26:174–76.

143. Salter, C. E. 1970. *Clematis armandii* grafting. *Proc. Inter. Plant Prop. Soc.* 20:330–32.

144. Saul, G. H., and L. Zsuffa. 1978. Vegetative propagation of elms by green cuttings. *Proc. Inter. Plant Prop. Soc.* 28:490–94.

145. Savella, L. 1981. Propagating pink dogwoods from rooted cuttings. *Proc. Inter. Plant Prop. Soc.* 30:405–7

146. Schreiber, L. R., and M. Kawase. 1975. Rooting of cuttings from tops and stumps of American elm. *HortScience* 10(6):615.

147. Scott, A. 1976. A successful technique for grafting hibiscus. *Proc. Inter. Plant Prop. Soc.* 26:389–91.

148. Semeniuk, P., and R. N. Stewart. 1964. Low temperature requirements for after-ripening of seed of *Rosa blanda. Proc. Amer. Soc. Hort. Sci.* 85:639–41.

149. Simpson, R. C. 1981. Propagating deciduous holly. *Proc. Inter. Plant Prop. Soc.* 30:338–42.

150. Skirvin, R. M., and M. C. Chu. 1979. *In vitro* propagation of 'Forever Yours' rose. *HortScience* 14:608–10.

151. Snyder, W. E. 1953. The fundamentals of juniper propagation. *Proc. Plant Prop. Soc.* 3:67–77.

152. Stadtherr, R. J. 1967. *Magnolia grandiflora* by cuttings. *Proc. Inter. Plant Prop. Soc.* 17:260–62.

153. Stoutemyer, V. T. 1942. The propagation of *Chionanthus retusus* by cuttings. *Nat. Hort. Mag.* 21:175–78.

154. Schmidt, C. 1961. Propagation of *Ilex aquifolium* from cuttings. *Proc. Plant Prop. Soc.* 11:318–37.

155. Strode, R. E., P. A. Travers, and R. P. Oglesby. 1979. Commercial micropropagation of rhododendrons. *Proc. Inter. Plant Prop. Soc.* 29:439–43.

156. Stump, D. S. 1980. Roses [*Plants & Gardens* 36 (1)]. Brooklyn, N.Y.: Brooklyn Botanic Garden.

157. Ticknor, R. L. 1969. Review of the rooting of pines. *Proc. Inter. Plant Prop. Soc.* 19:132–37.

158. Tinga, J. H., J. J. McGuire, and R. J. Parvin. 1963. The production of pyracantha plants from large cuttings. *Proc. Amer. Soc. Hort. Sci.* 82:557–61.

159. Vanstone, D. E. 1978. Basswood (*Tilia americana*) seed germination. *Proc. Inter. Plant Prop. Soc.* 28:566–69.

160. Van Veen, T. 1971. The propagation and production of rhododendrons. *Amer. Nurs.* 133(8):15–16, 52–58.

161. ———. 1969. Rhododendrons in America. Portland, Ore.: Sweeney, Krist, Dimm.

162. Vertrees, J. D. 1978. *Japanese maples.* Forest Grove, Ore.: Timber Press.

163. Wallis, J. S. 1976. The propagation and training of standard fuchsias. *Proc. Inter. Plant Prop. Soc.* 26:346–48.

164. Walter, H. 1967. Propagation of Chinese hibiscus. *Proc. Inter. Plant Prop. Soc.* 17:263–64.

165. Warren, P. 1973. Propagation of *Cercis* cultivars by summer budding. *The Plant Propagator* 19(3):16–17.

166. Wasley, R. 1979. The propagation of *Berberis* by cuttings. *Proc. Inter. Plant Prop. Soc.* 29:215–16.

167. Watkins, J. V. 1972. Jacaranda. *Horticulture* 50(5):22–23.

168. Wedge, D. 1977. Propagation of hybrid lilacs. *Proc. Inter. Plant Prop. Soc.* 27:432–36.

169. Weiser, C. J., and L. T. Blaney. 1960. The effects of boron on the rooting of English holly cuttings. *Proc. Amer. Soc. Hort. Sci.* 75:704–10.

170. Wells, J. S. 1955. Chapter 31 in *Plant propagation practices.* New York: Macmillan.

171. ———. 1961. Propagation of *Taxus*—review. *Amer. Nurs.* 114(10):11, 12, 91–98.

172. ———. 1980. How to propagate Japanese maples. *Amer. Nurs.* 151(9):14, 117–20.

173. ———. 1981. A history of rhododendron cutting propagation. *Amer. Nurs.* 154(9):14–15, 114–26.

174. Welsh, K., K. C. Sink, and H. Davidson. 1979. Progress on *in vitro* propagation of red maple. *Proc. Inter. Plant Prop. Soc.* 29:382–86.

175. Whalley, D. N. 1975. Propagation of Commelin elm by hardwood cuttings in heated bins. *The Plant Propagator* 21(3):4–6.

176. ———. 1979. Leyland cypress—rooting and early growth of selected clones. *Proc. Inter. Plant Prop. Soc.* 29:190–202.

177. Whitehead, H. C. M., and K. L. Giles. 1976. Rapid propagation of poplars by tissue culture methods. *Proc. Inter. Plant Prop. Soc.* 26:340–43.

178. Wister, J. C., and G. S. Wister, eds. 1962. *The peonies.* Washington, D.C.: American Horticultural Society.

179. Wolford, J. L., and W. J. Libby. 1976. Rooting giant sequoia cuttings. *The Plant Propagator* 22(2):11–13.

180. Wyman, D. 1955. *Crab apples for America.* Rockford, Ill.: American Association Botanical Gardens and Arboreta.

181. Yocum, H. G. 1964. Factors affecting the germination of palm seeds. *Amer. Hort. Mag.* 43:104–6.

## SUPPLEMENTARY READING

CROCKETT, J. U. 1971. *Evergreens.* New York: Time-Life Books.

———. 1972. *Flowering shrubs.* New York: Time-Life Books.

———. 1972. *Trees.* New York: Time-Life Books.

DUNMIRE, J. R., ed. 1979. *New Western garden book.* Menlo Park, Calif.: Lane.

FOWELLS, H. A. 1965. *Silvics of forest trees of the United States.* USDA For. Ser. Handbook No. 386. Washington, D.C.: U.S. Govt. Printing Office.

HENLEY, R. W. 1980. Growing trees for interior use. *Proc. Inter. Plant Prop. Soc.* 30:505–10.

HEPTING, G. H. 1971. *Diseases of forest and shade trees of the United States.* USDA For. Ser. Handbook No. 386. Washington, D.C.: U.S. Govt. Printing Office.

INTERNATIONAL PLANT PROPAGATORS' SOCIETY. *Proceedings of Annual Meetings.*

McCLINTOCK, E., and A. T. LEISER. 1979. *An annotated checklist of woody ornamental plants of California, Oregon and Washington.* Berkeley, Calif.: Univ. Calif. Div. Agr. Sci. Priced Publ. 4091.

LAMB, J. G. D., J. C. KELLY, and P. BOWBRICK. 1975. *Nursery stock manual.* London: Grower Books.

McMILLAN-BROWSE, P. D. A. 1979. *Hardy woody plants from seed.* London: Grower Books.

SCHOPMEYER, C. S., ed. 1974. Seeds of woody plants in the United States. *U.S. Dept. Agr. For. Ser. Handbook 450.*

TEUSCHER, H. 1969. Handbook on conifers. *Plants and Gardens* (Brooklyn Botanic Garden) 25(2):1–105.

WRIGLEY, J. W., and M. FAGG. 1979. Australian native plants: Propagation, cultivation, and use in landscaping. Sydney: William Collins Publishers.

WYMAN, D. 1965. *Trees for American gardens* (2nd ed.). New York: Macmillan.

———. 1969. *Shrubs and vines for American gardens* (2nd ed.). New York: Macmillan.

Herbaceous plants are classifed as *annuals, biennials,* or *perennials,* although the differences among these types may not be obvious. They may also be classified as *hardy, half-hardy,* or *tender.* In general, the propagation procedures for such plants depend upon their categories and the locality where they are to be grown.

In the following list of plants, seed germination data are given for some species, including suggested approximate temperatures that should give the most rapid and complete germination, along with the expected germination time (*4*). If two temperatures are given, separated by a dash, the first is the minimum night temperature, and the second the maximum day temperature. A single figure indicates a constant temperature.

The propagation methods indicated will serve as a guide, but some variation from these methods may be necessary with individual cultivars (*5, 6, 50*).

**Achillea spp.**     Yarrow. Hardy perennial. Seeds germinate in one to two weeks at 20°C (68°F). Also propagated by dividing clumps.

**Achimenes spp.**     Tender perennial. Seeds germinated in a warm greenhouse can be used for propagating species. Plants grow from small scaly rhizomes, which can be divided for propagation. Softwood cuttings in spring or leaf cuttings in summer can be rooted. Partially dried leaf scales can be planted and rooted.

**Aconitum spp.**     Monkshood. Hardy perennial. Seeds often show dormancy and before planting must be moist-chilled below 5°C (41°F) for six weeks. Plants have tuberous roots that can be divided, but once established they should not be transplanted. All parts of the plants are poisonous.

**Adiantum**     *See* Fern.

20

# Propagation of Selected Annuals and Herbaceous Perennials Used as Ornamentals

**African Violet**    *See* Saintpaulia.

**Agapanthus spp.**    Lily-of-the-Nile. The thick rhizomes can be divided to produce new plants, or seed propagation can be used.

**Agave spp.**    Many species of succulents, including Century plant. Perennial. Seeds should be sown in sandy soil when mature. Reproduces vegetatively by offsets from base of plant; these are removed along with roots and repotted in spring. Some species produce bulbils that can be used for propagation.

**Ageratum houstonianum**    Ageratum. Half-hardy annual. Seeds germinate in one to two weeks at 20–30°C (68–86°F). Ageratum may be propagated by cuttings also.

**Agrostemma githago**    Corncockle. Hardy annual. Seeds germinate in two to three weeks at 20°C (68°F).

**Allium spp.**    Ornamental onion; also onion, chives, and garlic. Propagated by seed. Plants grow from bulbs, which produce offsets. Many species produce bulbils.

**Aloe spp.**    Succulents of the lily family. Propagated by seed in well-drained sandy soil. Germination takes place in three to four weeks at 20 to 24°C (68 to 75°F). Plants produce offshoots that can be detached and rooted. Plants with long stems can be made into cuttings, which should be exposed to air for a few hours to allow cut surfaces to suberize.

**Althaea rosea**    Hollyhock. Half-hardy biennial. Seeds germinate in two to three weeks at 20° (68°F). Where winters are not too severe sow seeds in summer, transplant in fall for bloom the following year, or sow seeds in warm greenhouse in winter and transplant outdoors.

**Alyssum saxatile**    Goldentuft. Hardy perennial. Seeds germinate in three to four weeks at 20–30°C (68–86°F). Sow in summer for bloom the following year. Germination may be stimulated by light or exposure of moist seeds at 15°C (50°F) for five days (*4*). Propagate by division or by softwood cuttings in spring. Double forms must be propagated by cuttings or division.

**Amaranthus caudatus**    Love-Lies-Bleeding. Half-hardy annual. Seeds germinate in two to three weeks at 20–30°C (68–86°F). Light may increase germination (*4*). Sow in warm greenhouse for later transplanting or sow out-of-doors when frost danger is past. *A. gangeticus tricolor,* Joseph's Coat. Same as for *A. caudatus.* Sensitive to excess water.

**Amaryllis belladonna**    Belladonna lily. Perennial. Grows from bulbs outdoors in mild areas or in pots in cold climates. Propagate by bulb cuttings or separation of bulbs.

**Anchusa capensis**    Bugloss. Hardy annual or biennial. Seeds germinate in two to three weeks at 20–30°C (68–86°F). Sow seeds in summer for bloom next year or plant in greenhouse in winter for later transplanting to garden. Seeds may be sensitive to temperatures above 15°C (60°F) (*4*). *A. azurea.* Perennial. Selected clones best propagated by root cuttings or clump division.

**Anemone coronaria**    Poppy Anemone. Tender perennials. Seeds germinate in five to six weeks at 20°C (68°F) and may be sensitive to higher temperatures (*4*). Plants develop clusters of small, clawlike tuberous roots. *A. japonica.* Japanese anemone. Hardy perennial. Since seeds do not come true, cultivars are propagated by division or by root cuttings. Roots are dug in fall and cut into 2-in. pieces, which are laid in flats or in a cold frame, then covered with an inch of soil. After shoots appear, plants are potted.

*A. pulsatilla.* Pasque Flower. Hardy perennial. Seeds germinate in five to six weeks at 20°C (68°F), but may be sensitive to high temperatures (*4*). Plants can be divided.

**Anigozanthos spp.**     Kangaroo Paw. This native perennial Australian plant can be propagated easily by seed. The clumps of rhizomes can also be divided to propagate a few plants.

**Anthemis spp.**     Golden Marguerite, Camomile. Hardy perennial. Seeds germinate in one to three weeks at 20°C (68°F). Plants can be divided or propagated by stem cuttings.

**Anthurium andraeanum**     Anthurium. Remove offshoots with attached roots from the parent plant or root two- or three-leaved terminal cuttings under mist. Anthurium can be propagated by in vitro methods using a vegetative bud explant (*36*). Seed propagation is a lengthy process requiring 1½ to 3 years for flowering, and cultivars do not come true from seed (*26*).

**Antirrhinum majus**     Snapdragon. Tender perennial, treated as an annual. Seeds germinate in one to two weeks at 13°C (55°F) and may respond to light (*4*). Some hybrids are best started at 16 to 18°C (60 to 65°F). Seeds germinate well in mist (*6*). Start indoors for later outdoor planting (in fall in mild climates or in spring in severe winter areas). Softwood cuttings root readily.

**Aquilegia spp.**     Columbine. Hardy perennials. Seeds germinate in three to four weeks at 20–30°C (68–86°F) and may respond to light and three to four weeks moist-chilling at 5°C (41°F) (*4*).

**Arabis spp.**     Rockcress. Hardy perennials. Seeds germinate in three to four weeks at 20°C (68°F) and may respond to light (*4*). Softwood cuttings taken from new growth immediately after bloom root readily. Plants can be divided in spring or fall.

**Arctotis stoechadifolia**     African daisy. Half-hardy annual. Seeds germinate in two to three weeks at 20°C (68°F). Sow indoors for later transplanting.

**Armeria spp.**     Thrift. Hardy evergreen perennials. Seeds germinate in three to four weeks at 20°C (68°F). Best propagated by clump division in spring or fall.

**Asclepias tuberosa**     Butterfly weed. Hardy perennial. Seeds germinate in three to four weeks at 20–30°C (68–86°F). Fresh seed may need chilling. Plants should not be disturbed once established. *A. curassavica.* Bloodflower. Tropical perennial. Propagated by seed or by rooting softwood cuttings. Long taproot makes divison difficult.

**Asparagus asparagoides**     Smilax. Tender perennial. Propagated by seeds, which germinate in three to four weeks at 20–30°C (68–86°F). Sow seeds soon after they ripen, since they are short-lived (*6*). Cuttings can be made of young side shoots taken from old plants in spring; clumps can be divided. *A. plumosus,* Fern asparagus, and *A. sprengeri,* Sprenger asparagus. Seeds germinate in four to six weeks at 20–30°C (68–86°F). Crack the seed coats with a knife. This can also be propagated vegetatively as described for Smilax.

**Aster spp.**     Hardy perennials. Seeds germinate in two to three weeks at 20°C (68°F). Cultivars are propagated by lifting clumps in fall and dividing into rooted sections, discarding the older parts.

**Aubrieta deltoidea**     Aubrieta. Hardy perennials, sometimes treated as annuals. Seeds germinate in two to three weeks at 13°C (55°F). Clumps are difficult to divide; cuttings may be taken immediately after blooming.

**Aucuba japonica** *See* Gold Dust Plant.

**Baptisia spp.** False Indigo. Hardy perennial. Seed germination at 20°C (68°F) is slow and uneven. Gather seeds when ripe and sow outdoors to overwinter. Clumps are difficult to divide because of a long taproot.

**Begonia spp.** Begonia (*54*). Tropical perennials. Seeds, which are very fine and need light, germinate in two to four weeks at 20°C (68°F). Sow on moist, light medium with little or no covering. Begonia species, tuberous begonias, and wax begonias are propagated by seed, but other types are propagated vegetatively.

    **Tuberous begonias.** In addition to seed propagation these can be grown from tuberous stems, which are divided into sections, each bearing at least one growing point. Leaf, leaf-bud, and short stem cuttings (preferably with piece of tuberous stem attached) will root readily.

    **Fibrous-rooted begonias** Wax begonias, Christmas begonias, and others are propagated by leaf cuttings or softwood cuttings taken from young shoots in spring and summer.

    **Rhizomatous types** (Various species and cultivars, including *Rex begonia*). Plants are divided or rhizomes are cut into sections. Propagation is usually by leaf cuttings, but stem cuttings also will root. Treatment of leaf cuttings with a cytokinin increases the number of plantlets produced per leaf (*65*). *B. evansiana* produces small tubercles, which are detached and planted. Begonia can also be propagated by in vitro culture methods using leaf-petiole explants (*43*).

**Bellis perennis** English daisy. Hardy perennial often treated as annual or biennial. Seeds germinate in one to two weeks at 20°C (68°F) and may respond to light (*4*). Clumps should be divided every year to prevent crowding.

**Boltonia spp.** Boltonia. Hardy perennial. Seeds germinate in two to three weeks at 20°C (68°F). Divide plants in spring or fall.

**Bromeliads** About 2000 species of tropical herbs or subshrubs in 45 genera. The pineapple (*Ananas*) is the best known. Propagation is mainly by seeds or by asexual division of lateral shoots, but in vitro propagation by tissue culture methods has been used successfully with some species (*28*).

**Browallia spp.** Amethyst flower. Tender, blue-flowered perennial often treated as annual. Seeds germinate in two to three weeks at 20°C (68°F). Softwood cuttings can be taken in fall or spring. Can be used as flowering pot plant indoors in winter.

**Cactus** (*22, 48*) Large group of many genera, species, and some cultivars. Tender to semihardy perennials. Seed propagation can be used for most species, but seeds often germinate slowly. Sow fungicide-treated seed in well-drained, sterile mixture, and water sparingly, but do not allow medium to dry out. Pieces of stem can be broken off and rooted as cuttings; or small offsets, which root readily, can be removed. Allow offsets to dry for a few days to heal (suberize) cut surface before rooting. Cuttings require two to three weeks to heal. High humidity during rooting is unnecessary, but bottom heat is beneficial. Grafting is used to provide a decay-resistant stock for certain kinds and to produce unusual growth forms. For example, the pendulous *Zygocactus truncatus* is sometimes grafted on tall erect stems of *Pereskia aculeata*. Intergeneric grafts are usually successful. A type of cleft graft is used. The stem of the stock is cut off, and a wedge-shaped piece is removed. The scion is prepared by removing a thin slice from each side of the base; this is fitted into the opening made in the stock. The scion is held in place with a pin or thorn. The completed graft is held in a warm greenhouse until healed (*12, 20*). (See Figure 11-4.)

**Caladium bicolor**     This tropical perennial, grown for its strikingly colorful foliage, produces tubers. Propagation is by removing the tubers from the parent plant at the end of the four- to five-month dormancy period just before planting. Sometimes the tubers are cut into pieces, each containing at least two buds ("eyes"). Caladiums do best out-of-doors when planted after the minimum night temperature is above 18°C (65°F) or as pot plants maintained with night temperatures of 18–21°C (65–70°F) and day temperatures of 24–29.5°C (75–85°F).

**Calceolaria spp.**     Tender perennials often grown as annuals. Seeds germinate in two to three weeks at 20°C (68°F). Propagation is also by softwood cuttings.

**Calendula officinalis**     Pot Marigold. Hardy annual; gives winter bloom in mild climates from seed sown in late summer. Seeds germinate in one to two weeks at 20–30°C (68–86°F). Thin plants 30 cm (12 in.) apart.

**Calla**     *See* Zantedeschia spp.

**Callistephus chinensis**     China Aster. Half-hardy annual. Seeds germinate in two to three weeks at 20°C (68°F). Plant only wilt-resistant types.

**Campanula carpatica**     Tussock, Bellflower. Hardy perennial. Seeds germinate in two to three weeks at 20–30°C (68–86°F) and may respond to light (*4*). *C. lactiflora.* Bellflower. Hardy perennial. Seeds germinate in two to three weeks at 13–32°C (55–90°F).
    *C. medium.* Canterbury Bells. Hardy biennial. Seeds, which germinate in two to three weeks at 20–30°C (68–86°F), are sown in late spring or early summer for bloom the following year.
    *C. persicifolia.* Peach Bells. Hardy perennial. Seeds germinate in two to three weeks at 13–32°C (55–90°F) and may respond to light (*4*). Small offsets can be detached and rooted.
    *C. pyramidalis.* Chimney Bellflower. Hardy perennial, often treated as a biennial. Seeds germinate in two weeks at 20–30°C (68–86°F). Campanulas may also be increased by division and by rooting cuttings.

**Canna spp.**     Canna. Tender perennial. Cultivars do not come true from seed. Seeds, which have hard coats and must be scarified before planting, are germinated in a warm greenhouse. Cultivars are propagated by dividing the rhizome, keeping as much stem tissue as possible for each growing point. In mild climates this is done after the shoots die down in the fall or before growth starts in spring. In cold climates the plants are dug in fall, stored over winter, divided in spring, then started in sand or sandy soil for transplanting outdoors when frost danger is over.

**Carnation**     *See* Dianthus.

**Catananche caerulea**     Cupidsdart. Hardy perennial. Seeds germinate in two to four weeks at 20–30°C (68–86°F). Plants may be divided in fall.

**Celosia argentea**     Cockscomb. Tender annual. Seeds germinate in one to two weeks at 20–30°C (68–86°F) and may respond to light (*4*).

**Centaurea cineraria and others**     Dusty Miller. Tender perennial. Seeds germinate in two to four weeks at 20–30°C (68–86°F). Cuttings can be rooted. *C. cyanus.*

Cornflower. Bachelor Button. *C. moschata.* Sweet-Sultan. These are hardy annuals whose seeds germinate in three to four weeks at 20–30°C (68–86°F).

**Cerastium tomentosum**     Snow-in-summer. Hardy perennial. Seeds germinate in two to four weeks at 20°C (68°F). This is easily propagated by division in the fall or by softwood cuttings in summer.

**Cheiranthus cheiri**     Wallflower. Semihardy perennial often treated as a biennial. Seeds germinate in two to three weeks at 13°C (54°F) and may respond to light (*4*). Choice plants may be increased by cuttings taken in early summer.

**Chlorophytum comosum**     Spider plant. Propagated easily by planting miniature plants developing at ends of stolons. Stolon formation is under day-length control; short days (12 hours or less daily) promote stolon production (*23*).

**Chrysanthemum carinatum, C. coronarium, and C. segetum hybrids**     Many cultivars. Hardy annuals. Seeds germinate in two to four weeks at 20°C (68°F).

*C. parthenium*     Feverfew. Hardly perennial usually grown as an annual. Start from seeds, as described above. Plants easily self-seed and can be divided.

*C. maximum*     Shasta Daisy. Hardy perennial but often treated as a biennial, since it is short-lived. Plants increase from seeds or from the stolonlike shoots, which can be lifted and divided in the fall. Individual pieces have roots and can be transplanted.

*C. / morifolium*     Garden and greenhouse chrysanthemum and *C. frutescens.* Marguerite. Hardy and semihardy perennials. After flowering, lateral shoots develop from the base of the flowering stems, particularly if the tops are cut back. When the new side shoots are 9 to 10 cm (3½ to 4 in.) long and firm but not woody, they are cut off and rooted as softwood cuttings under mist and with a treatment of indolebutyric acid rooting hormone. The best source of new cuttings is a mother block (or increase block) grown in an isolated area away from the producing area. Such plants are grown in programs designed to keep them pathogen- and virus-free and true-to-type (*6*). Unrooted cuttings can be held for as long as 30 days at 0.5°C (33°F). For outdoor planting, cuttings may be taken in the same way. In areas with mild winters, cuttings can be taken in late winter for rooting and later transplanting to the garden. In cold-winter areas, the plants should be dug in the fall and brought into the greenhouse or cold frame; cuttings should be made in winter. The plants may also be left in place and divided in spring or fall. If cuttings are taken from the ends of stems high above ground, they are not likely to be infected with soil-borne insects and diseases.

Chrysanthemum is readily propagated by shoot-tip and petal segment culture in vitro (*7*).

**Clarkia spp.**     Hardy annuals. Seeds germinate in one to two weeks at 13–32°C (54–90°F). Seeds of some strains need light.

**Cleome spinosa**     Spiderflower. Tender annual. Seeds germinate in one to two weeks at 13–32°C (54–90°F). Seeds may respond to light (*4*).

**Codiaeum variegatum** *(51)*     Croton. Tropical perennial. Propagated by leafy terminal cuttings in spring or summer. Tall, "leggy" plants can be propagated by air layering.

**Colchicum autumnale**     Autumn Crocus. Saffron. Hardy perennial that grows as a corm. Seeds are sown as soon as ripe in summer but may require chilling over winter to germinate. Several years are required for plants to reach flowering size.

**Coleus blumei**     Tender perennials. Seeds germinate in two to three weeks at 20–30°C (68–86°F) but seedlings will be variable. Selected individuals are propagated by softwood cuttings, which root easily.

**Convallaria majalis**     Lily-of-the-Valley. Hardy perennial that grows as a rhizome, whose end develops a large underground bud, commonly called a "pip." In fall the plants are dug, and the pip, with attached roots, is removed and used as the planting stock. Digging should take place in early autumn, with replanting completed by late autumn. Single pips may be stored in plastic bags in the refrigerator, then planted in late winter for spring bloom (see Figure 15–16).

**Cordyline terminalis**     Ti. Easily propagated by cuttings, also by stem explants in in vitro culture (*35*).

**Coreopsis spp.**     Hardy annuals and perennials. Seeds, which germinate in two to three weeks at 20°C (68°F), may respond to light (*2*). Perennial clumps can divided in spring or fall.

**Cortaderia selloana**     Pampas Grass. Best feathery plumes are found on female plants. Propagated by clump division.

**Cosmos bipinnatus and C. sulphureus**     Half-hardy annual. Seeds germinate in one to two weeks at 20–30°C (68–86°F) and may respond to light (*4*).

**Crassula argentea**     Jade Plant. Can be propagated at any time by leafbud or stem cuttings.

**Crocus vernus**     Dutch Crocus. Also other *C.* species. Hardy perennial that grows from a corm. Seeds germinate as soon as ripe in summer; several years are required for plants to flower. When leaves die in fall, plants are dug and corms and cormels are separated and replanted.

**Cucurbita pepo var. ovifera**     Ornamental gourds. Tender annuals. Seeds germinate in two to three weeks at 20–30°C (68–86°F).

**Cyclamen spp.** *(39, 44)*     Tender perennials. Plants grow from a large tuberous underground stem. Cyclamen is propagated best by seeds, which germinate in three to four weeks in the dark at temperatures about 20°C (68°F), no higher than 22°C. Seeds are planted from midsummer to midwinter. Germination is best in a medium of peat moss to which pulverized limestone and mineral nutrients have been added (*64*). Seedlings require one to several years to flower. The tubers can be divided for the production of a few plants identical to the parent.

**Cymbalaria muralis**     Kenilworth Ivy. Semi-hardy perennial. Seeds germinate in one to four weeks at 12°C (54°F). Self-seeds readily. Softwood cuttings or clump division may be used.

**Cynoglossom amabile**     Chinese Forget-Me-Not. Hardy biennial grown as an annual. Seeds germinate in two to three weeks at 20°C (68°F) and may respond to light (*4*).

**Dahlia**     Tender perennials consisting of hundreds of cultivars. Seeds germinate in two to three weeks at 20–30°C (68–86°F) when planted indoors for later transplanting outdoors. Cultivars must be propagated vegetatively. Plant grows from large tuberous roots. Clumps are dug in the fall before frost and are stored over winter at 2 to 10°C (30 to 50°F), covered with a material such as soil or vermiculite to prevent shriveling. In spring, when new sprouts begin to appear, divide the clumps so that each root section has at least one sprout. Plant outdoors when danger of frost is over. Dahlias can also be propagated by softwood or leaf-bud cuttings.

**Daylily**     *See* Hemerocallis.

**Delphinium spp.**     Hardy perennials, usually propagated by seeds, which germinate in three to four weeks at 12°C (54°F). Seeds are short-lived and should be used fresh, or stored in containers at low temperature and reduced moisture. Seeds are usually sown outdoors in spring or summer to produce plants that flower the following year. Delphiniums can be propagated by softwood cuttings (*18*). Clumps can be divided in spring or fall, but such plants tend to be short-lived.

*D. ajacis*     Larkspur. Hardy annual. Seeds germinate in three to four weeks at 12°C (54°F). Young plants and seeds can be poisonous if eaten.

**Dianthus caryophyllus**     Carnation (*8*). Tender to semihardy perennial that has many cultivars used in florist's trade. Seeds germinate readily but are used primarily for breeding. Carnations are readily propagated by softwood cuttings (*27*). With mist, and with growth regulator treatment, rooting can be done any time of the year. The best source of cuttings is a mother (or increase) block isolated from the producing area, this block originating from cuttings taken from stock plants maintained under a complex program designed to keep them pathogen- and virus-free and true-to-type. (See Figure 8–10.) Such rooted cuttings are produced to a large extent by specialist growers, but commercial growing benches may be another source, if careful disease control and selection is practiced in the blocks. Lateral shoots ("breaks") that arise after flowering are removed and used as cuttings. Cuttings root in two to four weeks and may be planted directly to a greenhouse bench or transplanted to peat pots or to a nursery bed. Carnations can also be propagated on a large scale using shoot tip explants in in vitro culture (*19*).

*D. plumarius* and related species     Garden pinks. Hardy perennials, although some kinds are grown as annuals or biennials. Seeds germinate easily in two to three weeks at 20°C (68°F), but may not reproduce the cultivar. Softwood cuttings are taken in early summer and rooted to produce next year's plants. Layering also can be used.

*D. barbatus*     Sweet william. Perennial but grown as a biennial. Start by seed planted outdoors in spring. In mild-winter areas, transplant to permanent location in fall. In cold-winter areas, overwinter in a cold frame and transplant in spring.

**Dicentra spp.**     Bleedingheart. Hardy perennials. Seeds are sown in late summer or fall for overwintering at low temperatures; alternatively, seeds should be stratified for six weeks below 5°C (41°F) before planting. Divide clumps in spring or fall. Stem cuttings can be rooted if taken in spring after flowering. Root cuttings about 7.5 cm (3 in.) long can be taken from large roots after flowering.

**Dictamnus albus**     Gasplant. Hardy perennial. Propagation is the same as given for *Dicentra*. Plants should not be disturbed after establishment. Some people are allergic to plants of this species.

**Dieffenbachia spp.**     Dumbcane. Tropical perennial. Cut stem into 5-cm (2-in.) segments, and place horizontally in sand. New shoots and roots will develop from nodes. If plant gets tall and "leggy," the top may be cut off and rooted as a cutting, or the plant may be air layered. Leaves and stem are poisonous (Figure 10–11).

**Digitalis spp.**     Foxglove. Seeds germinate in two to three weeks at 20–30°C (68–86°F) and may respond to light (*4*). Sow seeds outdoors in spring, transplant to a nursery row at 9-in. spacing, then transplant to permanent location in fall. Perennial species increased by clump division.

**Dimorphotheca spp.**     Cape marigold. Half-hardy annual. Seeds germinate in two to three weeks at 20–30°C (68–86°F).

**Doronicum spp.**     Leopard Bane. Hardy perennial. Seeds germinate in two to three weeks at 20°C (68°F). Divide plants in spring or fall.

**Dracaena spp.**     Variable group of tropical perennial foliage plants. Seeds germinate in three to four weeks at 30°C (86°F). Leaf-bud cuttings, with part of the leaf removed, root well under intermittent mist if treated with indolebutyric acid at 25,000 ppm.

**Dusty Miller**     *See* Senecio Cineraria.

**Echeveria**     *See* Succulents.

**Echinops exaltatus**     Globe thistle. Hardy perennials. Seeds germinate in one to four weeks at 20–30°C (68–86°F). Plants may be divided in spring. Root cuttings, 5 to 7.5 cm (2 to 3 in.) long, may be made in the fall and planted in sandy soil in a cold frame.

**Epiphyllum spp.**     Leaf-flowering Cactus. Tender perennial. Seeds do not germinate well when fresh but will after 6 to 12 months' storage if planted in a warm greenhouse. Propagated readily by leaf cuttings or by grafting to *Opuntia. See* Cactus.

**Eschscholzia californica**     California Poppy. Hardy annual. Sow seeds outdoors in fall in mild climates or in early spring in colder areas. Tends to self-sow.

**Fatshedera lizei**     Tree Ivy. Cross between *Hedera helix* and *Fatsia japonica*. Propagated by stem cuttings or by air layering.

**Ferns** *(29, 49, 52)*     Many genera and species. Dust-like spores are collected from the spore cases on lower sides of fronds. Examine these sporangia with a magnifying glass to be sure they are ripe but not empty. Place fronds with the spores in a manila envelope and dry for a week at 21°C (70°F). Screen them to separate spores from the chaff. Transfer to a vacuum-tight bottle and store in a dry, cool place. Sow spores evenly on top of sterilized moist soil mixture, e.g., two-thirds peat moss, one-third perlite in flats, paying particular attention to sanitation. Leave 1 in. space on top and cover with a pane of glass. Use 18 to 24°C (65 to 75°F) air temperature; bottom heat may be helpful. Keep moist, preferably using distilled water to avoid salt injury.

Spores germinate and produce moss-like growth ⅛ in. thick, composed of many prothallia. Fertilization of the archegonium on the underside of the prothallus occurs in three to six months. In a first transplanting, a small piece of prothallus is removed with tweezers and transplanted to wider spacing in a new flat of soil mixture. The prothallia expand to about ½ in. in diameter and produce tiny sporophyte plants with primary leaves and roots. The fern plant will grow from a second transplanting.

Procedures have been developed for propagating ferns from spores in vitro using nutrient agar solutions (*33, 38, 63*).

Several vegetative propagation methods are possible. Ferns grow from thick rhizomes, which can be divided. Certain species *(e.g., Cystopteris bulbifera)* produce small "bulblets," about the size of a pea, on the underside of the leaf. These drop when mature, are planted, and produce a fern plant by the second year. Other species produce small vegetative buds on the upper surface or edge of the leaves; these detach and form new plants.

Cultivars of the sterile (nonpropagatable by spores) Boston fern group *(Nephrolepsis)* are now largely propagated by in vitro culture methods starting with rhizome tips (*10*).

This technique is also applicable to other fern genera, such as *Adiantum* (maidenhair fern), *Alsophila* (Australian tree fern), *Pteris* (brake fern), *Microlepia*, *Platycerium* (staghorn fern), and *Woodwardia* (chain fern).

**Ficus benjamina**     Weeping Fig. Easily started by leafy semihardwood cuttings taken in spring or early summer and rooted under mist. Can be propagated by shoot-tip culture in vitro (*41*).

**Ficus elastica** *(13)*     Rubber plant. Tropical perennial. Propagate as cuttings taken from 6- to 12-in. shoots; single buds or "eyes" can be removed and rooted. These cuttings are made in spring, inserted in sand or a similar medium, and held in a warm greenhouse. Shoots of trees growing outdoors in the tropics are air-layered and the rooted air layers shipped to wholesale nurseries for further development. Indoor plants that become too "leggy" also can be air layered (see Figure 14–6).

**Ficus lyrata**     Fiddleleaf Fig. Can be propagated by cuttings, or by air layering. Can also be propagated by adventitious shoots developing on excised leaf pieces in vitro (*17, 41*).

**Freesia spp.**     Tender perennials. Seeds planted in fall germinate in four to six weeks and will bloom the next spring. A germination temperature of about 18.5°C (65°F) is best. Plants grow from corms, which are planted in spring and dug in fall. Small cormels are removed at this time and replanted to grow larger for future flowering.

**Gaillardia spp.**     Blanketflower. Annual and hardy perennials. Seeds germinate in two to three weeks at 20°C (68°F) and may respond to light (*4*). Perennial kinds are planted in spring to bloom the following year. These may also be started from root cuttings or may be divided in spring or fall but are not long-lived.

**Galanthus spp.**     Snowdrop. Hardly perennial. Bulbs are planted in the fall for bloom the following spring. Offsets are removed when bulbs are dug.

**Gazania spp.**     Tender perennial often grown as an annual. Propagated by seeds sown in spring or by softwood cuttings taken in late summer, rooted in a cold frame, then transplanted in spring. Divide clumps after three or four years.

**Gentiana spp.**     Gentian. Many species, mostly hardy perennials, although some are annuals and biennials. Plant fresh seed in the fall to overwinter outdoors. Seeds germinate in one to four weeks at 20°C (68°F), but it is best to hold them at 0°C (32°F) for ten days before planting.

**Geranium**     *See* Pelargonium.

**Gerbera jameson**     Transvaal daisy. Tender perennial. Seeds germinate in two to three weeks at 20°C (68°F); it is important to use fresh seed. Or remove basal shoots from the rhizome and use as cuttings. In vitro culture from shoot tips can be used for rapid, large-scale multiplication (*47*).

**Geum spp.**     Avens. Hardy perennials. Seeds germinate in three to four weeks at 20–30°C (68–86°F). Propagate also by clump division in spring or fall.

**Gladiolus**     Tender perennial grown from a corm. Seed propagation is used for developing new cultivars. Seeds are planted in spring either indoors for later transplanting or outdoors when danger of frost is over (see chapter 15).

**Gloxinia** *(Sinningia speciosa)*     Grown from seed for commercial production. Seeds germinate rapidly if placed in an artificial medium under intermittent mist and held at about 21°C (70°F) night temperature. Flowering plants develop in six to seven months. Gloxinia can also be propagated in vitro using leaf explants (*32*).

**Godetia spp.**     Hardy annuals. Sow seeds in early spring; these germinate in two to three weeks at 20°C (68°F).

**Gold Dust Plant** (*Aucuba japonica* 'Variegata')     Propagated by leafy stem cuttings under mist, which root easily, or by root cuttings. Grows well in shade.

**Gypsophila spp.** (*G. elegans)*     Baby's Breath. Annual. Seed germinates in two to three weeks at 20°C (68°F).

   *G. paniculata*     Hardy perennial. Started by seed as above. Plants can be divided in spring and fall. Double-flowered cultivars are grafted on seedling *G. paniculata* (single-flowering) roots. Grafting can be done in summer and fall, using outdoor-grown plants for rootstocks and placing them in a cold frame for healing of the graft; grafting is also done in winter and early spring, using greenhouse-grown stock plants. *G. paniculata* can be propagated by in vitro methods using shoot tip explants (*37*).

**Haworthia**     *See* Succulents.

**Helenium autumnale**     Sneezeweed. Hardy perennial. Seeds germinate in one to two weeks at 20°C (68°F). Cultivars are increased by division. Separate rooted shoots in spring, line-out in nursery, then transplant in fall and winter.

**Helianthemum nummularium**     Sunrose. Half-hardy perennial. Seeds germinate in two to three weeks at 20–30°C (68–86°F). Cultivars are propagated by softwood cuttings taken from young shoots in spring. Transplant to pots and place in permanent location the following winter or spring. Division of clumps is also possible, but plants tend to be short-lived.

**Helianthus annuus**     Sunflower. Hardy annual. Seeds germinate in two to three weeks at 20–30°C (68–86°F). *H. decapetalus* and other hardy perennial species are increased by division.

**Heliconia spp.** (*59*)     These tropical ornamental herbaceous perennials are prized for their showy inflorescences. They are easily propagated by division of the rhizomes.

**Heliopsis scabra**     Heliopsis. Hardy perennial. Seeds germinate in one to two weeks at 20°C (68°F). Divide clumps in fall.

**Heliotropium spp.**     Heliotrope. Tender perennial usually grown as an annual. Seed germinates in three to four weeks at 20–30°C (68–86°F) and may respond to light (*4*). Start indoors for later spring planting. Take softwood cuttings of side shoots in fall or spring and root at low temperatures 10°C (50°F) in slightly moist conditions.

**Helleborus spp.**     Hellebore. Christmas Rose and Lenten Rose. Hardy perennials. Gather seeds as soon as ripe, but give six weeks moist-chilling before planting. Plants require several years to produce flowers. Roots of both species are poisonous.

**Hemerocallis spp.**     Daylily (*16*). Hardy perennial. Seeds require about six weeks of moist-chilling for good germination. Seed propagation is used only to develop new cultivars. Divide clumps in fall or spring, separating into rooted sections, each with about three offshoots. Clones can also be reproduced by in vitro methods using flower petals and sepals as explants (*1, 42*).

**Hippeastrum spp.**    Amaryllis. Tender bulbous perennial. Remove and pot bulb offsets which will flower the second year. Make bulb cuttings in late summer. Seeds germinate under warm conditions 20–30°C (68–86°F) and sprouting is slow and uneven. Seedlings take two to four years to produce flowers.

**Hunnemannia fumariaefolia**    Goldencup. Tender perennial often grown as annual. Seeds germinate in two to three weeks at 20°C (68°F). For bloom first year, sow seeds early indoors then transplant outdoors when danger of freezing is over.

**Hyacinthus spp.**    Hyacinth. Hardy, spring-flowering perennial; bulbs are planted in the fall. Removal of offset bulbs gives small increase. For commercial propagation, new bulbs are obtained by scoring or scooping mature bulbs. In vitro propagation, using segments of the bulb, leaf, inflorescence, or stem as explant, is successful (*30*). Seeds may be planted outdoors in fall, but up to six years are required to produce bloom (see chapter 15).

**Iberis spp.**    Candytuft. Hardy annual and perennial species. Seeds germinate in one to two weeks at 20–30°C (68–86°F) but may need light. Root softwood cuttings in summer or divide clumps in fall.

**Impatiens spp.**    Snapweed. Touch-Me-Not. Balsam. Perennials and half-hardy annuals. Seeds germinate in two to four weeks at 20°C (68°F) and may respond to light (*4*). Perennial species can be started by cuttings.

**Incarvillea spp.**    Incarvillea. Hardy perennial. Seeds germinate in one to two weeks at 20°C (68°F). Divide in fall or, preferably, in spring.

**Ipomoea spp.**    Morning Glory. Tender perennial grown as an annual. Seeds germinate in one to three weeks at 20–30°C (68–86°F). Notch seed coats or soak seeds overnight in warm water before planting.

**Iresine spp.**    Bloodleaf. Tender perennial. Softwood cuttings root easily. Keep stock plants over winter in greenhouse and take cuttings in late winter or spring.

**Iris spp.**    Perennials. There are several different groups of hardy or semihardy iris, which grow either from rhizomes or from bulbs. Rhizomes are divided after bloom. Discard the older portion and use only the vigorous side shoots. Leaves are trimmed to about 15 cm (6 in.).

Bulbous species follow a typical spring-flowering, fall-planting sequence. The old bulb completely disintegrates, leaving a cluster of various-size new bulbs. These are separated and graded, the largest size being used to produce flowers, the smaller for further growth.

Seeds, which are used to propagate species and to develop new cultivars, should be planted as soon as ripe after being given a moist-chilling period; germination is often irregular and slow. Removal of embryo from the seed and growing it in artificial culture has given prompt germination in some cases (*57*). Iris can be propagated by in vitro methods, greatly hastening production of new cultivars over the customary division of rhizomes (*30*).

**Ixia spp.**    Corn Lily. Tender, summer- or fall-flowering perennials grown from corms. In cold climates these are dug in fall and stored over winter. Small cormels are removed and planted in the ground or in flats to reach flowering size, as is done with gladiolus.

**Kalanchoe spp.**    Tropical perennials. *See* Succulents.

**Kangaroo Paw.**     *See* Anigozanthos.

**Lantana sellowiana, L. camara**     Lantana. Tender perennials. Seeds germinate in six to seven weeks at 20°C (68°F). Softwood cuttings root easily.

**Lathyrus latifolius**     Perennial pea vine. Hardy perennial. Seeds germinate in two to three weeks at 20–30°C (68–86°F). Clumps may be divided. *L. odoratus.* Sweet pea. Hardy annual. Seed germinates in two weeks at 20°C (68°F). Notching seed or soaking in warm water may hasten germination. Plant outdoors in fall where winters are mild, in spring where winters are severe.

**Lavandula officinalis**     Lavender. Half-hardy perennial. Seeds, which may be planted in winter, germinate in two to three weeks at 11–32°C (52–90°F). Take cuttings from side shoots in later summer or fall; plant in soil and cover or start in cold frame. Divide clumps in the fall.

**Lilium spp.** *(45)*     Lily. Hardy perennials. These are spring- and summer-flowering plants grown from scaly bulbs; most have a vertical axis, but in some species growth is horizontal with a rhizomatous structure. Lilies include many species, hybrids, and named cultivars. Seed propagation is used for species and for new cultivars. Seeds of different lily species have different germination requirements (*53*).

*Immediate seed germinators* include most commercially important species and hybrids (*L. amabile, L. concolor, L. longiflorum, L. regale, L. tigrinum,* Aurelian hybrids, Mid-Century hybrids, and others). Germination is epigeous; a shoot should emerge three to six weeks after planting at moderately high temperatures. Treat seeds with a fungicide to control *Botrytis.* Sow ¾ in. deep in flats during winter or outdoors in a seedbed in early spring. Dig the small bulblets in fall, sort for size, store over winter, and replant with same sizes together. Plants normally grow two years in a seedbed and two years in a nursery row before producing good-size flowering bulbs.

Another group is the *slow seed germinators* of the epigeal type (*L. candidum, L. henryi,* Aurelian hybrids, and others) in which seed germination is slow and erratic; the procedures used are essentially the same as described above.

The most difficult group to propagate are the *slow seed germinators* of the hypogeous type (*L. auratum, L. bolanderi, L. canadense, L. martagon, L. parvum, L. speciosum,* and others). Seeds of this group require three months under warm conditions for the root to grow and produce a small bulblet, then a cold period of about six weeks, followed by another warm period in which the leaves and stem begin to grow. This sequence can be provided by planting the seeds outdoors in summer as soon as they are ripe, or planting seeds in flats and then storing under appropriate conditions to provide the required temperature sequence. Vegetative methods of propagation include natural increase of the bulbs, such as bulblet production on stems (either naturally or artificially), aerial stem bulblets (bulbils), or scaling. These procedures are described in chapter 15.

Lilies can be propagated *in vitro* from bulb scales (*61*).

**Linaria spp.**     Toadflax. Hardy annual and perennials. Seeds germinate in two to three weeks at 12°C (54°F). Perennial species take two years to produce bloom from seed. Clumps can be divided in spring or fall.

**Linum spp.**     Flax. Hardy annual and perennial species. Seeds germinate in three to four weeks at 12°C (54°F). Divide clumps of perennial species in fall or spring.

**Lobelia erinus**     Lobelia. Tender perennial grown as an annual. Seeds germinate in two to three weeks at 20–30°C (68–86°F), but seedling growth is slow. May respond to light (*4*). Start indoors 10 to 12 weeks before transplanting outdoors after last frost. Mature plants, if potted in the fall and kept in greenhouse over winter, can be used to provide new growth for cuttings to be taken in late winter.

**Lobelia spp.**     Hardy perennials. Seeds germinate in three to four weeks at 20–30°C (68–86°F). Divide clumps in fall or spring.

**Lobularia maritima**     Sweet Alyssum. Perennial but grown as hardy annual. Seed germinates in one to two weeks at 20°C (68°F) and blooms appear in six weeks.

**Lunaria annua**     Honesty. Biennial, sometimes grown as an annual. Seeds germinate in two to three weeks at 20°C (68°F). *L. rediviva.* Hardy perennial. Propagated by seed as described above. Also increased by division.

**Lupinus hartwegii, L. nanus, and others**     Hardy annuals. Seeds germinate in two to three weeks at 20°C (68°F). *L. polyphyllus.* Perennial. Seeds germinate in three to four weeks at 20°C (68°F). Seed coats may be hard and should be scarified. *L. arboreus.* Tree lupine. Hardy perennial. Start from seeds indoors and transplant to permanent location. *L.* 'Russell Hybrid'. Sow seeds in spring or summer, or propagate by cuttings taken in early spring with small piece of root or crown attached.

**Lychnis spp.**     Campion. Mostly hardy perennials but some are grown as annuals or biennials. Seeds germinate in three to four weeks at 20°C (68°F). Clumps can be divided in spring or fall.

**Lycoris spp.**     Spider Lily. Tender and semihardy bulbous perennials. Propagation is by bulb offsets, which are removed when the dormant bulbs are dug. These are replanted to grow larger. Bulb cuttings can also be used for increase.

**Marigold**     *See* Tagetes.

**Matthiola incana**     Common Stock. *M. longipetala bicornis.* Evening scented stock. Perennial grown as biennial or annual. Seed germinates in two weeks at 12–32°C (54–90°F) and may respond to light (*4*). Seeds are sown in summer or fall for winter bloom, in late winter indoors for spring bloom, or outdoors in spring for summer bloom.

**Mesembryanthemum spp.**     *See* Succulents.

**Mimulus spp.**     Monkey flower. Includes many species of tender to hardy plants. Mostly perennials, but sometimes grown as annuals. Seeds germinate in one to two weeks at 12°C (54°F). Softwood cuttings taken from young shoots can be rooted.

**Molucella laevis**     Bells of Ireland. Half-hardy annual. Seed germinates in three to five weeks at 10°C (50°F). Difficult to transplant; sow seeds in place.

**Monstera deliciosa**     Often misnamed cut-leaf philodendron. Easily propagated by rooting sections of the main stem, by stem cuttings, or by air layering.

**Muscari spp.**     Grape Hyacinth. Hardy bulbous perennial. Plant blooms in spring; bulbs become dormant in fall when they are lifted and divided. Increase by removing bulb offsets. Seed propagation can also be used.

**Myosotis sylvatica**     Forget-Me-Not. Hardy biennial. Seeds germinate in two to three weeks at 20°C (68°F). Sow in summer and transplant to permanent location the following spring. *M. scorpioides* is a perennial started from seed; division in spring is also used.

**Narcissus, Pseudo-narcissus, and other species**     Daffodil. Hardy perennial spring-flowering bulb. Vegetative propagation procedures are described in chapter 15. Species can be grown from seed, but cultivars do not come true; seedlings require several years to produce flowers. Seeds should be planted in fall so they can be moist-chilled (stratified) over winter.

**Nasturtium**     *See* Tropaeolum.

**Nemesia strumosa**     Half-hardy annual. Seeds germinate in two to three weeks at 13°C (55°F). Sow outdoors in spring in cold climates, in fall in mild-winter areas.

**Nepeta mussinii**     Nepeta, Catmint. Hardy perennial. Seeds germinate in two to three weeks at 20°C (68°F). Softwood cuttings of nonflowering side shoots taken in early summer root readily. Cuttings may be inserted directly into soil if protected. Plants may be divided in spring, using newest parts and discarding older portion of clumps.

**Nicotiana spp.**     Flowering tobacco. Half-hardy annuals. Seeds germinate in one to two weeks at 20–30°C (68–86°F) and may respond to light (*4*). Sow indoors four to six weeks before last frost then transplant out-of-doors.

**Nierembergia spp.**     Cupflower. Tender perennial, sometimes grown as an annual. Seeds germinate in two to three weeks at 20–30°C (68–86°F). Softwood cuttings removed from new growth in spring root readily. Clumps can be divided.

**Nymphaea spp.**     Hardy water lily. Consists of numerous species and many named cultivars. Plants grow as rhizomes. Clumps are divided in spring. Seeds are used to grow natural species and to develop new cultivars. Tropical water lily hybrids and species grow from tubers. Seeds do not reproduce hybrids. Both kinds of seeds are planted 2.5 cm (1 in.) deep in sandy soil, then immersed in water 3 to 4 in. deep. Hardy species should be started at 16°C (60°F), tropical species at 21 to 27°C (70 to 80°F). Vegetative propagation is either from small tubers that can be removed from old tubers in fall or from small plantlets growing from the leaf.

**Oenothera spp.**     Evening-primrose. Hardy perennials, but some kinds are biennial. Seeds germinate in one to three weeks at 20–30°C (68–86°F). Plants can also be increased by division of clumps in the fall.

**Opuntia**     *See* Cactus.

**Orchids** *(24, 46, 58, 60)*     Many genera, hybrids, and cultivars are cultivated, and many more are found in nature. Some, such as *Aerides, Arachnis, Phalaenopsis, Renanthera,* and *Vanda,* exhibit *monopodial* habit of growth. This means they are erect and grow continuously from the shoot apex and can be propagated by tip cuttings. Adventitious roots are produced along the stem and inflorescences are produced laterally from leaf axils. Most others, including *Brassovola, Calanthe, Cattleya, Cymbidium, Laelia, Miltonia, Odontoglossom, Oncidium,* and *Phaius* have a *sympodial* habit of growth, are procumbent, and do not grow continuously from the apex. Their main axis is a rhizome in which new growth arises from offshoots, or "breaks." Pseudobulbs are usually present on plants of this type.

Many orchids are *epiphytes* (i.e., air plants), typically growing on branches of trees. Others *(Cymbidium, Cypripedium, Paphiopedium)* are terrestrial and grow in the ground.

Seed propagation is mainly used for hybridization. Many important cultivars are seedling hybrids, either from species or between genera, resulting from controlled crosses of carefully selected parents. Many such important crosses are between tetraploid and diploid parents to produce triploids. These offspring are sterile and are not in turn usable as parents. Seedling variation occurs since orchids are heterozygous. Five to seven years is required for a seedling plant to bloom. Orchid flowers are hand-pollinated. A seed capsule requires 6 to 12 months to mature. A single capsule will contain many thousands of tiny seeds with relatively undeveloped embryos. In vitro culture is universally used for seed propagation. (The procedure is described in chapter 17.) Knudson's C medium is usually used. Arditti (2) has summarized the many experiences of testing various nutritional and other factors for orchid seed germination. Orchid seed can be stored for many years if held in sealed containers over calcium chloride at about 2°C (36°F).

Vegetative methods for orchids are generally slow, difficult for many genera, and usually too low-yielding for extensive commercial use. Sympodial species are increased by division of the rhizome while it is dormant or just as new growth begins. Four or five pseudobulbs are included in each section. "Backbulbs" and "greenbulbs" can be used for some genera.

Orchids with long canelike stems, as *Dendrobium* and *Epidendrum,* sometimes produce offshoots ("keiki") that produce roots. Offshoots can also be produced if the stem is cut off and laid horizontally in moist sphagnum or some other medium. Flower stems of *Phaius* and *Phalaenopsis* can be cut off after blooming and handled in the same way. A drastic method of inducing offshoots is to cut out or mutilate the growing point of *Phalaenopsis,* remove the small leaves, and treat the injured portion with a fungicide. Offshoots may then be produced.

Monopodial species can be propagated by long (30 to 37 cm) tip cuttings with a few roots already present. Air layering also is possible.

Vegetative propagation by proliferation of shoot-tip (meristem) cultures in vitro has revolutionized orchid propagation, particularly for *Cymbidium, Cattleya,* and some other genera (3, 46). The shoot growing point is dissected from the plant and grown on a special, sterile medium; a proliferated mass of tissue and small protocorms develops, which can be divided periodically. Many thousands of separate protocorms can be developed in this way within a matter of months, each of which will eventually differentiate shoots and roots to produce an orchid plant. The procedure is described in chapter 17 and shown in Figure 17–4.

**Paeonia lactiflora and other hardy perennial species**      Herbaceous peony. Numerous hybrid cultivars. Seeds are used for growing species or for breeding new cultivars. Seed propagation is difficult; germination may require one to two years. Sow seeds in fall to give moist-chilling over winter. Roots develop during the first summer, and shoots develop the second spring. Another method is to collect seeds before they become black and completely ripe. Do not allow them to dry out; sow in pots, which should be buried in the ground for six to seven weeks. Roots will develop; dig up and plant in protected location or under mulch over winter. The best method is to divide clumps in fall; each piece should have at least one bud or "eye," preferably three to five.

**Paeonia suffruticosa**      *See* Tree Peony.

**Pansy**      *See* Viola.

**Papaver nudicaule**      Iceland Poppy. Hardy perennial grown as biennial. Seeds germinate in one to two weeks at 12°C (54°F). Sow in permanent location in summer for bloom next year. *P. orientale.* Oriental poppy. Hardy perennial. Very fine seeds, which may respond to light, should be covered very lightly (4). Seeds germinate in one to two weeks at 12°C (54°F). Cultivars are propagated by root cuttings. Dig when leaves die down in fall, cut into 7.5- to 10-cm (3- to 4-in.) sections, lay horizontally in sandy soil in a flat covered to about an inch. Rooted cuttings are transplanted in spring. Or dig plants in spring, prepare root cuttings, and plant directly in permanent location. *P. rhoeas*. Corn Poppy, Shirley Poppy. Hardy annual. Seed germinates in one to two weeks at 13°C (55°F). Sow in late summer for early spring bloom or in early spring for summer bloom. Do not transplant.

**Pelargonium × hortorum**     Geranium. Started by cuttings and by seed. Traditionally propagated by cuttings, which root easily, but *Pythium* and *Botrytis* infection can be serious problems. Pathogen-free stock, identified by culture indexing, should be used and can be supplied by specialists (*62*).

In the mid-1970s large-scale seed propagation of geraniums began with the introduction of certain cultivars that would grow from seed to flower in 14 to 16 weeks. Seeds germinate best at about 22°C (72°F) in an artificial medium (*4, 15*).

**Pentstemon spp.**     Beardtongue. Semihardy to hardy perennials, sometimes handled as annuals. Seeds germinate in two to three weeks at 20–30°C (68–86°F) but growth is slow and uneven; seeds may respond to light (*4*). Plants started indoors in early spring and transplanted outdoors later may bloom the first year. Plants are usually short-lived. Softwood cuttings taken from nonflowering side shoots of old plants root readily. Make cuttings in fall to obtain plants for next season. Clumps may be divided.

**Peperomia spp.**     Tender perennial. Softwood stem, leaf-bud, or leaf cuttings root readily. Plants can be divided. Peperomia can also be propagated in vitro from excised leaf explants (*25*).

**Periwinkle**     *See* Vinca.

**Petunia Hybrids**     Petunia. Tender perennial often grown as annual. Seeds germinate in one to two weeks at 20°C (68°F). Seeds of double-flowered cultivars and some $F_1$ hybrids may need light and high temperature (27 to 29°C; 80 to 85°F) for good germination. It is best to start plants indoors for later outdoor planting. Softwood cuttings taken in late summer or fall from side shoots root easily.

**Philodendron spp.**     Tropical vines. Seeds germinate readily at about 25°C (77°F) if sown as soon as they are ripe and before they become dry. Best propagation methods are use of stem cuttings, rooting sections of main stem, or by air layering.

**Phlox drummondi**     Annual phlox. Hardy. Seed germinates in two to three weeks at 20°C (68°F). Start indoors for later outdoor planting, or outdoors after frost.

**Phlox divaricata**     Sweet William. Hardy perennial. Expose seeds to cold during winter before planting. Softwood cuttings taken in spring root easily. Divide clumps in spring or fall.

**Phlox paniculata**     Garden phlox. Hardy perennial. Plants do not come true from seed. Sow seeds as soon as ripe in fall to germinate the next spring. They will germinate in three to four weeks at 20°C (68°F). Grow plants one season and transplant in fall. Softwood cuttings taken from young shoots in spring or summer root easily, but they are subject to damping-off if kept too wet. It is best to propagate from root cuttings. Dig clumps in fall; remove all large roots to within 5 cm (2 in.) of crown (which is replanted). Cut roots into 2-in. lengths and place in flats of sandy soil; cover 13 mm (½ in.) deep. Transplant next spring. Divide clumps in fall or spring. Phlox can be propagated in vitro using shoot explants.

**Polemonium spp.**     Jacob's Ladder. Hardy perennial. Seeds germinate in three to four weeks at 20–30°C (68–86°F). Divide clumps or root stem cuttings.

**Polianthes tuberosa**     Tuberose. Tender bulbous perennial. Propagated by removing offsets at planting time. The small bulbs take more than one year to flower. Divide clumps every four years.

**Portulaca grandiflora**     Rose moss. Half-hardy annual. Seeds germinate in two to three weeks at 20–30°C (68–86°F); they respond to light (*4*). Reseeds itself.

**Primula obconica**     Top Primrose. Tender perennial grown as an annual. Seeds germinate well in a cool greenhouse or after three to four weeks at 12–32°C (54°–90°F). The very tiny seeds respond to light and should not be covered. *Primula malacoides.* Fairy Primrose. Seeds will germinate in two to three weeks at 20–30°C (68–86°F). *Primula sinensis.* Chinese Primrose. Seeds will germinate in three to four weeks at 20°C (68°F). In these species double-flowering cultivars do not produce seeds but are propagated by cuttings taken in spring, or by division. Other *Primula* spp. Hardy perennials grown outdoors. Seeds germinate in three to six weeks at 20°C (68°F), but some species may require lower temperatures. It is best to collect and sow seeds as soon as they are ripe in fall. Cuttings taken in spring root easily. Clumps can be divided just after flowering.

**Ranunculus asiaticus**     Turban or Persian Ranunculus. Tender perennial. Seeds germinate in one to four weeks at 20°C (68°F). Grows from tuberous roots which can be divided. *Ranunculus* spp. Buttercup. Hardy perennial. Seeds germinate in one to four weeks at 20°C (68°F). Plant seeds as soon as they are ripe. Divide plants in spring or fall.

**Reseda odorata**     Mignonette. Hardy annual. Seeds should not be covered; they germinate in two to three weeks at 12°C (54°F) and respond to light (*4*).

**Ricinus communis**     Castor Bean. Soak seeds in water for 24 hours or nick with file before planting. Plant each seed in individual pot and after frost danger is past transplant to outdoor location. Seeds are poisonous.

**Rudbeckia spp.**     Coneflower. Hardy annual, biennial, and perennial species. Seeds germinate in two to three weeks at 20–30°C (68–86°F). Perennial kinds are propagated by division. Will reseed naturally.

**Saintpaulia ionantha**     African violet (*31, 34, 66*). Tropical perennial. Propagation is by seed, division, or cuttings. The very fine seeds, which germinate in three to four weeks at 30°C (86°F), should not be covered. Seedlings are subject to damping-off. Vegetative methods are necessary to maintain cultivars. Plants can be divided. Leaf cuttings (blade and petiole) are easily rooted, either in a rooting medium or in water. Rapid, large-scale propagation can be accomplished using leaf petiole sections by in vitro culture techniques (*9, 14, 56*).

**Salpiglossis sinuata**     Painted tongue. Semihardy annual. Seeds are difficult to germinate but some will start in one to two weeks at 20–30°C (68–86°F). Start indoors in peat pots. Can be propagated by in vitro culture techniques (*40*).

**Salvia splendens**     Scarlet sage. Tender perennial grown as an annual. Germinate seeds at 20–30°C (68–86°F), then grow at 13°C (55°F) night temperature. Softwood cuttings taken in fall root readily.

**Salvia spp.**     Sage. Annual, biennial, and perennial species. Seeds germinate in two to three weeks at 20–30°C (68–86°F) and may respond to light (*4*). Soak flats thoroughly, and do not rewater until sprouted; they also grow well under mist (*6*). Softwood cuttings of young shoots 3 to 4 in. long root readily. Plants can be divided, but such divisions are slow to recover.

**Sanvitalia procumbens**     Creeping zinnia. Hardy annual. Seeds germinate in one to two weeks at 20°C (68°F) and may respond to light (*4*). Sow in place in spring, or in fall in mild climates.

**Sansevieria trifasciata and S. ' hahnii' (dwarf form)**     Bowstrip hemp. Snakeplant. Tropical perennial. Plants grow from a rhizome, which can be readily divided. Leaves may be cut into sections, several inches long, and inserted into a rooting medium; a new shoot and roots will develop from base of leaf cutting. The variegated form, *S. t.* 'Laurentii', is a chimera, which can be maintained only by division.

**Saponaria officinalis**     Bouncing Bet. Hardy perennial. Seeds germinate in two to three weeks at 20°C (68°F). The plant spreads rapidly by an underground creeping stem which can be divided. *Saponaria vacaria.* Soapwort. Hardy annual. Seeds germinate in two to three weeks at 20°C (68°F).

**Saxifraga spp.**     Many interesting and unusual species and hybrids. Mostly hardy perennials. Seeds germinate easily; they are sown preferably when ripe. Some hybrids and cultivars are maintained only by vegetative methods. Most plants grow as small rosettes and are easily propagated by making small cuttings involving single rosettes which are taken after bloom. Plants can be divided in spring or fall. *S. stolonifera,* Strawberry geranium, is a tender perennial that reproduces by runners.

**Scabiosa spp.**     Pincushion Flower. Annual and hardy perennial species. Seeds germinate in two to three weeks at 20–30°C (68–86°F). Perennial kinds can be divided.

**Scarlet Sage**     *See* Salvia.

**Schefflera arboricola**     Can be propagated easily by seeds, cuttings, or air layering (*60*).

**Schizanthus spp.**     Butterfly flower. Tender annual. Seeds germinate in one to two weeks at 12°C (54°F) and are sensitive to high temperatures (*4*). Sow seeds in fall for early spring bloom indoors, or sow in early spring to be transplanted outdoors for summer bloom.

**Scilla spp.**     Squill. Includes several kinds of bulbous hardy spring-flowering perennials. Dig plants when leaves die down in summer and remove the bulblets. *S. autumnalis* is planted in spring and blooms in fall.

**Sedum**     *See* Succulents.

**Sempervivum**     *See* Succulents.

**Senecio cineraria**     Dusty Miller. These tender perennials are grown as annuals. Seeds germinate in two to three weeks at 20°C (68°F). Stem-tip cuttings root rapidly if treated with IBA and placed under mist over bottom heat.

**Sinningia speciosa**     Gloxinia. Tropical perennial. Commonly grown from seeds, which are very fine and require light. Sow in uncovered, well-drained, peat moss medium; they germinate in two to three weeks at 20°C (68°F). Vegetative methods are required to reproduce cultivars. Plant grows from a tuberous root on which a rosette of leaves is produced. The root can be divided as described for tuberous begonia. Softwood cuttings or leaf cuttings taken in spring from young shoots starting from the tubers root easily.

**Snapdragon**    *See* Antirrhinum majus.

**Spider Plant**    *See* Chlorophytum comosum.

**Stock**    *See* Matthiola.

**Stokesia laevis**    Stokes Aster. Hardy perennial. Seeds germinate in four to six weeks at 20–30°C (68–86°F) and the plants bloom the first year. Make root cuttings or divide clumps in spring.

**Strelitzia reginae**    Bird-of-Paradise. This tropical perennial grows from a rhizome, which can be divided in the spring, or small offsets can be detached and placed in a potting mixture. Seeds should be sown under warm conditions. Freshly harvested seed should be used to avoid seed coat impermeability.

**Succulents** *(20, 22, 48, 59)*    This loosely defined horticultural group includes many genera, such as: *Agave, Aloe, Crassula, Echevaria, Euphorbia, Hoya, Kalanchoe, Portulacaria, Sedum, Yucca.* These are plants with fleshy stems and leaves that store water, or plants that are highly drought-resistant. Mostly they are half-hardy or tender perennials. Seed propagation is possible, although young plants are often slow to develop and to produce flowers. It is best to germinate seeds indoors at high day temperatures 29 to 35°C (85 to 95°F). Seedlings are susceptible to damping-off.

Cuttings of most species root readily—either stem, leaf-bud, or leaf—in a 1:1 peat-perlite medium. They should be exposed to the open air or inserted into dry sand for a few days to allow callus to develop over the cut end. Some protection from drying is needed during rooting. Some species can be reproduced by removing offsets. Grafting is possible as described for cacti (see Figure 11–5). Some of the succulents, such as *Kalanchoe,* can be propagated by in vitro culture methods (*55*).

**Sweet Alyssum**    *See* Lobularia maritima.

**Sweet Pea**    *See* Lathyrus.

**Tagetes spp.**    Marigold. Tender annuals. Seeds germinate readily in one week at 20–30°C (68–86°F) and sometimes respond to light (*4*). Sow in place in spring after frost.

**Thalictrum spp.**    Meadow rue. Hardy perennial. Seeds germinate in four to six weeks at 20°C (68°F). Sometimes hard seeds are present. Plants can be divided in spring or fall.

**Thermopsis caroliniana**    Tender or hardy perennials. Use fresh seeds, which will germinate in two to three weeks at 20–30°C (68–86°F). Plants can be divided, but it is best to leave them undisturbed.

**Thunbergia spp.**    Clockvine. Tender perennials grown as annuals. Seeds germinate in two to three weeks at 20–30°C (68–86°F), but seedlings grow slowly. Softwood cuttings taken from new shoots root readily.

**Thymus spp.**     Thyme. Hardy perennials. Seeds germinate in one to two weeks at 12–32°C (54–90°F). Sometimes germination is promoted by light. May be increased by division or by softwood cuttings taken in summer.

**Ti**     *See* Cordyline terminalis.

**Tigridia pavonia**     Tiger Flower. Tender bulbous perennials. Plant bulbs in spring and dig in fall when leaves die. Increase by removal of small bulblets just before they are planted. Easily started by seed.

**Tithonia rotundifolia**     Mexican Sunflower. Tender perennial grown as an annual. Seeds germinate in two to three weeks at 20–30°C (68–86°F).

**Tolmiea menziesii**     Piggyback Plant. New plantlets form on upper surface of leaves. Such leaves are removed and petiole is stuck in rooting medium to depth of new plantlet, which then resumes growth.

**Torenia fournieri**     Wishbone Flower. Tender perennial. Seeds germinate in two weeks at 20–30°C (68–86°F).

**Trachymene caerulea**     Laceflower. Tender perennial grown as an annual. Seeds germinate in two to three weeks at 20°C (68°F).

**Trollius spp.**     Globeflower. Hardy perennial. Plant seeds in fall. Increases also by clump division.

**Tropaeolum majus**     Nasturium. Tender perennial grown as an annual. Plant seeds in place; they germinate in one to two weeks at 20°C (68°F), but are difficult to transplant. Double-flowering kinds must be propagated vegetatively, usually by softwood cuttings.

**Tulipa spp. and hybrid cultivars**     Tulip. Hardy bulbous perennials. Plant bulbs in fall for spring bloom. Seeds are used to reproduce species and for breeding new cultivars. They germinate readily if planted as soon as ripe. Vegetative methods include removal of offset bulbs in the fall. Different bulb sizes are planted separately, since the time required to produce flowers varies with size. For details of procedure see chapter 15.

**Valeriana officinalis**     Valerian. Hardy perennial. Seeds germinate in three to four weeks at 12–32°C (54–90°F). New plants can be obtained by clump division.

**Venidium fastuosum**     Venidium. Half-hardy annual. Seeds germinate in four to six weeks at 20–30°C (68–86°F); or sow outdoors at 10 to 13°C (50 to 55°F).

**Verbascum spp.**     Mullein. Hardy perennials and biennials. Seeds are slow to germinate; best temperature is 30°C (86°F). Propagate named cultivars by root cuttings taken in early spring.

**Verbena hybrida**     Verbena. Tender perennial grown as an annual. Seeds germinate in three to four weeks at 20–30°C (68–86°F); sometimes promoted by light (4). May be propagated by softwood cuttings taken in summer. *V. canadensis.* Clump verbena. Hardy perennial that blooms first year from seed. Seeds germinate in two to four weeks at 12–32°C (54–90°F). Plants can be propagated by division or by softwood cuttings.

**Veronica spp.**     Speedwell. Hardy perennials. Seeds germinate in two weeks at 12–32°C (54–90°F). Plants are increased by division in spring or fall or by softwood cuttings taken in the spring or summer.

**Vinca major**     Tender perennial. Propagate by division or by softwood cuttings taken in summer. *V. minor.* Periwinkle. Hardy perennial. Seeds germinate in two to three weeks at 20°C (68°F). Easily propagated by softwood cuttings or by division. *V. rosea.* Madagascar periwinkle. Tender perennial grown as an annual. Propagation is the same as for other species. Started by seed germinated at 20° (68°F).

**Viola cornuta**     Horned Violet, Tufted Pansy. Hardy perennial. Seeds germinate in two to three weeks at 12–32°C (54–90°F). Seeds of some cultivars need light. Vegetative propagation is by cuttings taken from new shoots obtained by heavy cutting back in the fall. Clumps may also be divided. *V. tricolor hortensis.* Pansy. Hardy or semi-hardy, short-lived perennial often grown as an annual. Usually propagated by seeds as described for *V. cornuta* but may also be increased by cuttings or by division. *V. odorata.* Sweet violet. Tender to semi-hardy perennials. Grows by rhizome-like stems, which can be separated from others on the crown and treated as a cutting with some roots present. *Viola* species. Many hardy perennial kinds. These are grown by seeds as above, but germination may be slow and seeds are best exposed to cold before planting. Many species produce seeds in inconspicuous, enclosed (cleistogamous) flowers near the ground, whereas the conspicuous, showy flowers produce few or no seeds. These plants can also be reproduced by cuttings or by division.

**Yucca spp.**     Yucca. Tender to semihardy perennials. Seeds germinate at 20°C (68°F) but rather slowly and require four to five years to flower. Plants are monocots; some are essentially stemless and grow as a rosette, while others have either long or short stems. Offsets growing from the base of the plant can be removed and handled as cuttings; sometimes entire branches or the top of the plant can be detached a few inches below the place where leaves are borne and replanted in sandy soil. Sections of old stems can be laid on sand or other medium in a warm greenhouse, and new side shoots that develop can be detached and rooted.

**Zantedeschia spp.**     Calla. Several species, which have similar propagation requirements. Tropical perennials. Plants grow by thickened rhizomes, which produce offsets or rooted side shoots; these are removed and planted.

**Zebrina Pendula**     Wandering Jew. Easily propagated at any season by stem cuttings.

**Zinnia elegans and other species**     Zinnia. Half-hardy hot weather annual. Seeds germinate outdoors in one week at 20–30°C (68–86°F). Sometimes seeds respond to light (*4*).

# REFERENCES

*1.* Apps, D. A., and C. W. Heuser. 1975. Vegetative propagation of *Hemerocallis*— including tissue culture. *Proc. Inter. Plant Prop. Soc.* 25:362–67.

*2.* Arditti, J. 1967. Factors affecting the germination of orchid seeds. *Bot. Rev.* 33:1–97.

*3.* ———. 1977. *Orchid biology.* Ithaca, N. Y.: Cornell Univ. Press.

4. Assn. Off. Seed Anal. 1970. Rules for seed testing. *Proc. Assn. Off. Seed Anal.* 60(2): 1–115.

5. Bailey, L. H., E. Z. Bailey, and Staff of Bailey Hortorium. 1976. *Hortus third.* New York: Macmillan.

6. Ball, V., ed. 1980. *The Ball red book* (13th ed.). Chicago: G. V. Ball Co.

7. Ball, V. 1977. Seed geraniums. Chapter 17 in *Ball bedding book.* West Chicago, Ill.: Geo. J. Ball, Inc.

8. Besemer, S. 1980. Carnations. In *Introduction to floriculture,* R. A. Larson, ed. New York: Academic Press.

9. Bilkey, P. C., B. H. McCown, and A. C. Hildebrandt. 1978. Micropropagation of African violet from petiole cross-sections. *HortScience* 13(1):37–38.

10. Burr, R. W. 1975. Mass production of Boston fern through tissue culture. *Proc. Inter. Plant Prop. Soc.* 25:122–24.

11. Bush, S. R., E. D. Earle, and R. W. Langhans. 1976. Plantlets from petal segments, petal epidermis, and shoot tips of the periclinal chimera, *Chrysanthemum morifolium* 'Indianapolis'. *Amer. Jour. Bot.* 63:729–37.

12. Carter, F. M. 1973. Grafting cacti. *Horticulture* 51(3):34–35.

13. Conover, C. A., and R. T. Poole. 1978. Production of *Ficus elastica* 'Decora' standards. *HortScience* 13(6):707–8.

14. Cooke, R. C. 1977. Tissue culture propagation of African violets. *HortScience* 12(6):549.

15. Dallon, J., Jr., and D. Durkin. 1972. Culturing geranium from seed. *Proc. Inter. Plant Prop. Soc.* 21:324–30.

16. Darrow, G. M., and F. G. Meyer, eds. 1968. *Day lily handbook.* American Horticultural Society, Vol. 47, No. 2.

17. Deburg, P., and J. DeWall. 1977. Mass propagation of *Ficus lyrata. Acta Hort.* 78:361–64.

18. Dodge, M. 1978. Propagation of named delphinium cultivars. *Proc. Inter. Plant Prop. Soc.* 28:496–98.

19. Earle, E. D., and R. W. Langhans. 1975. Carnation propagation from shoot tips cultured in liquid medium. *HortScience* 10(6):608–10.

20. Edinger, P., ed. 1970. *Succulents and cactus.* Menlo Park, Calif.: Lane.

21. Gilbertson-Ferriss, T. L., and H. F. Wilkins. 1977. Factors influencing seed germination of *Freesia refracta* Klatt cv. Royal Mix. *HortScience* 12(6):572–73.

22. Haage, W. 1963. *Cacti and succulents.* New York: E. P. Dutton.

23. Hammer, P. A. 1976. Stolon formation in *Chlorophytum. HortScience* 11(6):570–72.

24. Hawkes, A. D. 1961. *Orchids, their botany and culture.* New York: Harper & Row, Pub.

25. Henny, R. J. 1978. *In vitro* propagation of *Peperomia* 'Red Ripple' from leaf discs. *HortScience* 13(2):150–51.

26. Higaki, T., and D. P. Watson. 1973. Anthurium culture in Hawaii. *Univ. Hawaii Coop. Ext. Ser. Cir. 420.*

27. Holley, W. D., and R. Baker. 1963. *Carnation production.* Dubuque, Iowa: William C. Brown Co., Publishers.

28. Hosoki, T., and T. Asahira. 1980. *In vitro* propagation of bromeliads in liquid culture. *HortScience* 15(5):603–4.

29. Hull, Helen. 1969. Handbook on ferns. *Plants and Gardens* 25(1):1–77.

30. Hussey, G. 1975. Totipotency in tissue explants of some members of the Liliaceae, Iridaceae, and Amaryllidaceae. *Jour. Exp. Bot.* 26:253–62.

31. Jackson, H. C. 1975. Propagation and culture of African violets. *Proc. Inter. Plant Prop. Soc.* 25:269–71.

32. Johnson, B. B. 1978. *In vitro* propagation of gloxinia from leaf explants. *HortScience* 13(2):149–50.

33. Knauss, J. F. 1976. A partial tissue culture method for pathogen-free propagation of selected ferns from spores. *Proc. Fla. State Hort. Soc.* 89:363–65.

34. Kramer, J. 1971. *How to grow African violets* (4th ed.). Menlo Park, Calif.: Lane.

35. Kunisaki, J. T. 1975. *In vitro* propagation of *Cordyline terminalis* (L.) Kurth. *HortScience* 10(6):601–2.

36. ———. 1980. *In vitro* propagation of *Anthurium andreanum* Lind. *HortScience* 15(4):508–9.

37. Kusey, W. E., Jr., P. A. Hammer, and T. C. Weiler. 1980. *In vitro* propagation of *Gypsophila paniculata* L. 'Bristol Fairy'. *HortScience* 15(5):600–601.

38. Lane, B. C. 1981. A procedure for propagating ferns from spore using a nutrient agar solution. *Proc. Inter. Plant Prop. Soc.* 30:94–97.

39. Lyons, R. E., and R. E. Widmer. 1980. Origin and historical aspects of *Cyclamen persicum* Mill. *HortScience* 15(2):132:35.

40. Lee, C. W., R. M. Skirvin, A. I. Soltero, and J. Janick. 1977. Tissue culture of *Salpiglossis sinuata* L. from leaf discs. *HortScience* 12(6):547–49.

41. Makino, R. K., P. J. Makino, and T. Murashige. 1977. Rapid cloning of *Ficus* cultivars through application of *in vitro* methodology. *In Vitro* 13(3):160.

42. Meyer, M. M. 1976. Propagation of daylilies by tissue culture. *HortScience* 11(5):485–87.

43. Mikkelsen, E. P., and K. C. Sink, Jr. 1978. *In vitro* propagation of Rieger Elatior begonias. *HortScience* 13(3):242–44.

44. McMillan, R. N. 1980. Cyclamen production problems. *Proc. Inter. Plant Prop. Soc.* 29:173–76.

45. McRae, E. A. 1978. Commercial propagation of lilies. *Proc. Inter. Plant Prop. Soc.* 28:166–69.

46. Morel, G. M. 1966. Meristem culture: Clonal propagation of orchids. *Orchid Digest* 30(2):45–49.

47. Murashige, T., M. Serpa, and J. B. Jones. 1974. Clonal multiplication of gerbera through tissue culture. *HortScience* 9(3):175–80.

48. Perl, P. 1978. *Cacti and succulents.* Alexandria, Va.: Time-Life Books.

49. ———. 1977. *Ferns.* Alexandria, Va.: Time-Life Books.

50. Post, K. 1949. *Florist crop production and marketing.* New York: Orange-Judd Publishing.

51. Raward, I. D. 1975. Propagation of *Codiaeum* (croton) by tip cuttings. *Proc. Inter. Plant Prop. Soc.* 25:386.

52. Roberts, D. J. 1965. Modern propagation of ferns. *Proc. Inter. Plant Prop. Soc.* 15:317–21.

53. Rockwell, F. F., G. C. Grayson, and J. de Graff. 1961. *The complete book of lilies.* Garden City, N. Y.: Doubleday.

54. Sheerin, P. 1974. Propagation of various types of begonia. *Proc. Inter. Plant Prop. Soc.* 24:292-93.

55. Smith, R. H., and A. E. Nightingale. 1979. *In vitro* propagation of *Kalanchoe. HortScience* 14(1):20.

56. Start, N. D., and B. G. Cumming. 1976. *In vitro* propagation of *Saintpaulia ionantha* Wendl. *HortScience* 11(3):204-6.

57. Stoltz, L. P. 1977. Growth regulator effects on growth and development of excised mature iris embryo *in vitro. HortScience* 12(5):495-96.

58. Skelsey, A. 1978. *Orchids.* Alexandria, Va.: Time-Life Books.

59. Stefanis, J. P., and R. W. Langhans. 1980. Factors affecting the production and propagation of xerophytic succulent species. *HortScience* 15(4):504-5.

60. Stewart, J., and E. Hennessy. 1981. *Orchids of Africa.* Boston: Houghton Mifflin.

61. Stimart, D. P., and P. D. Ascher. 1978. Tissue culture of bulb scale sections for asexual propagation of *Lilium longiflorum* Thumb. *Jour. Amer. Soc. Hort. Sci.* 103:182-84.

62. Thorn-Horst, A., R. K. Horst, S. H. Smith, and W. A. Ogleree. 1977. A virus-indexing tissue culture system for geraniums. *Flor. Rev.* 160(4148):28-29, 72-74.

63. Tjosvold, S. 1978. Uniform spore dispersal on warm nutrient agar solution. *Univ. Calif. Nursery and Flower Rpt.* Summer issue.

64. Widmer, R. E. 1980. Cyclamens. In *Introduction to floriculture,* R. A. Larson, ed. New York: Academic Press.

65. Wildern, J. A., and R. A. Criley. 1975. Cytokinins increase shoot production from leaves of begonia. *The Plant Propagator* 20(4) and 21(1):7-9.

66. Wilson, H. V. 1980. *Saintpaulia* species. *Amer. Hort.* 58(6):35-39.

## SUPPLEMENTARY READING

BALL, V. 1980. *Ball bedding book* (13th ed.). West Chicago, Ill.: Geo. J. Ball, Inc.

CROCKETT, J. U. 1971. *Annuals.* New York: Time-Life Books.

———. 1972. *Perennials.* New York: Time-Life Books.

———. 1972. *Foliage house plants.* New York: Time-Life Books.

GILES, F. A., R. M. KEITH, and D. C. SAUPE. 1980. *Herbaceous perennials.* Reston, Va.: Reston.

GRAF, A. B. 1974. *Exotica 3: Pictorial cyclopedia of exotic plants from tropical and near-tropical regions* (7th ed.). East Rutherford, N.J.: Roehrs Company.

INTERNATIONAL BEDDING PLANT CONFERENCE. *Annual Proceedings.*

INTERNATIONAL PLANT PROPAGATORS' SOCIETY. *Proceedings of Annual Meetings.*

JOINER, J. N., ed. 1981. *Foliage plant production.* Englewood Cliffs, N.J.: Prentice-Hall.

LARSON, R. A., ed. 1980. *Introduction to floriculture.* New.York: Academic Press.

LAURIE, A., D. C. KIPLINGER, and K. S. NELSON. 1979. *Commercial flower forcing* (8th ed.). New York: McGraw-Hill.

MASTALERZ, J. W., ed. 1976. *Bedding plants* (2nd ed.). University Park, Pa.: Pennsylvania State Univ.

NEEL, P. L. 1979. Macropropagation of tropical plants as practiced in Florida. *Proc. Inter. Plant Prop. Soc.* 29:468–80.

NELSON, K. S. 1978. Flower and plant production in the greenhouse (3rd ed.). Danville, Ill.: Interstate.

REILLY, A. 1978. *Park's success with seeds.* Greenwood, S. C.: Geo. W. Park Seed Co., Inc.

ZILLIS, M., D. ZWAGERMAN, D. LAMBERTS, and L. KURTZ. 1979. Commercial propagation of perennials by tissue culture. *Proc. Inter. Plant Prop. Soc.* 29:404–13.

# Index